ANNALS OF THE NEW YORK ACADEMY OF SCIENCES

Volume 974

EDITORIAL STAFF

Executive Editor
BARBARA M. GOLDMAN

Managing Editor
JUSTINE CULLINAN

Associate Editors
JOHN W. KENNEDY
MARION L. GARRY

The New York Academy of Sciences
2 East 63rd Street
New York, New York 10021

THE NEW YORK ACADEMY OF SCIENCES
(Founded in 1817)

BOARD OF GOVERNORS, September 2002–September 2003

TORSTEN N. WIESEL, *Chairman of the Board*
JOHN T. MORGAN, *Treasurer*

Honorary Life Governors
WILLIAM T. GOLDEN   JOSHUA LEDERBERG

Governors

ELEANOR BAUM	KAREN E. BURKE	LAWRENCE B. BUTTENWIESER
PRAVEEN CHAUDHARI	BRIAN FERGUSON	GERALD FISCHBACH
JOHN H. GIBBONS	MICHAEL GOLDEN	RONALD L. GRAHAM
MARNIE IMHOFF	JACQUELINE LEO	BRUCE McEWEN
PAUL MARKS	RONAY MENSCHEL	JOHN F. NIBLACK
SANDRA PANEM	PETER RINGROSE	JOHN J. ROCHE
LEE G. VANCE		DEBORAH WILEY

HELENE L. KAPLAN, *Counsel* [ex officio]

MICROGRAVITY TRANSPORT PROCESSES IN FLUID, THERMAL, BIOLOGICAL, AND MATERIALS SCIENCES

ANNALS OF THE NEW YORK ACADEMY OF SCIENCES
Volume 974

MICROGRAVITY TRANSPORT PROCESSES IN FLUID, THERMAL, BIOLOGICAL, AND MATERIALS SCIENCES

Edited by S. S. Sadhal

Associate Editors
Vijay K. Dhir, Haruhiko Ohta,
Reginald W. Smith, and Johannes Straub

The New York Academy of Sciences
New York, New York
2002

Copyright © 2002 by the New York Academy of Sciences. All rights reserved. Under the provisions of the United States Copyright Act of 1976, individual readers of the Annals *are permitted to make fair use of the material in them for teaching and research. Permission is granted to quote from the* Annals *provided that the customary acknowledgment is made of the source. Material in the* Annals *may be republished only by permission of the Academy. Address inquiries to the Permissions Department (permissions@nyas.org) at the New York Academy of Sciences.*

Copying fees: *For each copy of an article made beyond the free copying permitted under Section 107 or 108 of the 1976 Copyright Act, a fee should be paid through the Copyright Clearance Center, Inc., 222 Rosewood Drive, Danvers, MA 01923 (www.copyright.com).*

⊚ *The paper used in this publication meets the minimum requirements of American National Standard for Information Sciences—Permanence of Paper for Printed Library Materials. ANSI Z39.48-1984.*

Library of Congress Cataloging-in-Publication Data

Microgravity transport processes in fluid, thermal, biological, and materials sciences / edited by S.S. Sadhal.
    p. cm. — (Annals of the New York Academy of Sciences ; v. 974)
Includes bibliographical references and index.
  ISBN 1-57331-422-6 (hard : alk. paper) — ISBN 1-57331-423-4 (pbk. : alk. paper)
  1. Reduced gravity environments. 2. Liquids—Effect of reduced gravity on. 3. Matter—Effect of reduced gravity on. I. Sadhal, S.S. II. Series.
  Q11 .N5 vol. 974
  [TA357.5.R44]
  500 s—dc21
  [620/.4

2002013019
CIP

K-M Research/PCP
Printed in the United States of America
ISBN 1-57331-422-6 (cloth)
ISBN 1-57331-423-4 (paper)
ISSN 0077-8923

ANNALS OF THE NEW YORK ACADEMY OF SCIENCES

Volume 974
October 2002

MICROGRAVITY TRANSPORT PROCESSES IN FLUID, THERMAL, BIOLOGICAL, AND MATERIALS SCIENCES

Editor
S. S. SADHAL

Associate Editors
VIJAY K. DHIR, HARUHIKO OHTA,
REGINALD W. SMITH, AND JOHANNES STRAUB

This volume is the result of a conference entitled **Microgravity Transport Processes in Fluid, Thermal, Biological, and Materials Sciences**, held September 30–October 5, 2001 in Banff, Alberta, Canada.

CONTENTS

Preface. *By* SATWINDAR SINGH SADHAL, VIJAY K. DHIR, HARUHIKO OHTA, REGINALD W. SMITH, AND JOHANNES STRAUB.................... xi

Part I. Space Systems: Electrostatic, Electromagnetic, Fluid, and Thermal Phenomena

Use of an Electric Field in an Electrostatic Liquid Film Radiator. *By* S.G. BANKOFF, E.M. GRIFFING, AND R.A. SCHLUTER 1

Large-Scale Geophysical Flows on a Table Top. *By* JOHN HEGSETH AND KAMEL AMARA 10

Transport from Higher Order g-Jitter Effects. *By* ROBERT J. NAUMANN 29

The Use of Pulsatile Flow to Separate Species. *By* AARON M. THOMAS AND R. NARAYANAN.. 42

Part II. Materials Processing and Crystal Growth

The Influence of Gravity on the Precise Measurement of Solute Diffusion Coefficients in Dilute Liquid Metals and Metalloids. *By* REGINALD W. SMITH, XIAOHE ZHU, MARK C. TUNNICLIFFE, TIMOTHY J.N. SMITH, LOWELL MISENER, AND JOSEE ADAMSON.......... 57

Structures of Solid and Liquid during Melting and Solidification of Indium.
  By G. COSTANZA, F. GAUZZI, AND R. MONTANARI 68

Unidirectional Solidification of Magnetostrictive Materials Using a
  Magnetic Field in Microgravity. *By* HIDEKI MINAGAWA, KEIJI KAMADA,
  HIDEAKI NAGAI, YOSHINORI NAKATA, AND TAKESHI OKUTANI........... 79

Dielectric *in Situ* Monitoring of Microgravity Polymerizations.
  By ALVIN P. KENNEDY AND SOLOMON TADESSE 87

The Influence of Gravity on Composition Uniformity and Microstructure in
  Immiscible Al–In Alloys. *By* J. BARRY ANDREWS AND LETITIA J. HAYES... 102

The Influence of Marangoni Flows on Crack Growth in Cast Metals.
  By B.J. YANG, R.W. SMITH, M. SAHO, AND M. SADAYAPPAN 110

Thermal Diffusivity Coefficient of Glycerin Determined on an Acoustically
  Levitated Drop. *By* K. OHSAKA, A. REDNIKOV, AND S.S. SADHAL 124

Thermophysical Property Measurements by Electromagnetic Levitation
  Melting Technique under Microgravity. *By* SAYAVUR I. BAKHTIYAROV
  AND RUEL A. OVERFELT... 132

Mass and Thermal Diffusivity Algorithms: Reduced Algorithms for Mass
  and Thermal Diffusivity Determinations. *By* R. MICHAEL BANISH,
  LYLE B.J. ALBERT, TIMOTHEE L. POURPOINT, J. IWAN D. ALEXANDER,
  AND ROBERT F. SEKERKA.. 146

ATEN: A New High Temperature Materials Processing Facility for the
  International Space Station. *By* WAYNE N.O. TURNBULL,
  DONALD L. MISENER, TIMOTHY J.N. SMITH, GUY R.J. ORAM,
  AND REGINALD W. SMITH.. 157

Microgravity Effects on Thermodynamic and Kinetic Properties of
  Colloidal Dispersions. *By* TSUNEO OKUBO AND AKIRA TSUCHIDA......... 164

Gravity-Induced Anomalies in Interphase Spacing Reported for Binary Eutectics.
  By REGINALD W. SMITH 176

Part III. Interfacial Phenomena and Two-Phase Flows

Optical Measurement of Concentration Gradient Near Miscible Interfaces.
  By N. RASHIDNIA AND R. BALASUBRAMANIAM 193

Coupled Mean Flow-Amplitude Equations for Nearly Inviscid Parametrically
  Driven Surface Waves. *By* EDGAR KNOBLOCH, CARLOS MARTEL,
  AND JOSÉ M. VEGA... 201

Thermocapillary Convection Around Gas Bubbles: An Important Natural
  Effect for the Enhancement of Heat Transfer in Liquids Under Microgravity.
  By J. BETZ AND J. STRAUB 220

Experimental Investigation of the Liquid Interface Reorientation upon
  Step Reduction in Gravity. *By* MARK MICHAELIS, MICHAEL E. DREYER,
  AND HANS J. RATH... 246

Microscale Heat Transfer to Subcooled Water: 200–6,000 psia, 0–3,500 W/cm^2.
  By ROBERT H. LEYSE 261

Condensate Removal Mechanisms in a Constrained Vapor Bubble Heat Exchanger. *By* LING ZHENG, YINGXIN WANG, PETER C. WAYNER, JR., AND JOEL L. PLAWSKY 274

Numerical Simulations of Bubble Motion in a Vibrated Cell Under Microgravity Using Level Set and VOF Algorithms. *By* TIMOTHY J. FRIESEN, HIROYUKI TAKAHIRA, LISA ALLEGRO, YOSHITAKA YASUDA, AND MASAHIRO KAWAJI . 288

Annular Flow Film Characteristics in Variable Gravity. *By* RYAN M. MACGILLIVRAY AND KAMIEL S. GABRIEL 306

A Study of Gas–Liquid Two-Phase Flow in a Horizontal Tube Under Microgravity. *By* BUHONG CHOI, TERUSHIGE FUJII, HITOSHI ASANO, AND KATSUMI SUGIMOTO 316

Part IV. Boiling Phenomena

Vapor Bubbles in Flow and Acoustic Fields. *By* ANDREA PROSPERETTI AND YUE HAO. 328

Origin and Effect of Thermocapillary Convection in Subcooled Boiling: Observations and Conclusions from Experiments Performed at Microgravity. *By* JOHANNES STRAUB. 348

High Heat Flux Cooling by Microbubble Emission Boiling. *By* KOICHI SUZUKI, HIROSHI SAITOH, AND KAZUAKI MATSUMOTO 364

Dynamics of Single and Multiple Bubbles and Associated Heat Transfer in Nucleate Boiling Under Low Gravity Conditions. *By* D. QIU, G. SON, V.K. DHIR, D. CHAO, AND K. LOGSDON 378

Rapidly Expanding Viscous Drop from a Submerged Orifice at Finite Reynolds Numbers. *By* NIVEDITA R. GUPTA, R. BALASUBRAMANIAM, AND KATHLEEN J. STEBE 398

Review of Existing Research on Microgravity Boiling and Two-Phase Flow: Future Experiments on the International Space Station. *By* HARUHIKO OHTA, ATSUSHI BABA, AND KAMIEL GABRIEL 410

Pool Film Boiling Experiments on a Wire in Low Gravity: Preliminary Results. *By* P. DI MARCO, W. GRASSI, AND F. TRENTAVIZI 428

Mechanisms of Steady-State Nucleate Pool Boiling in Microgravity. *By* HO SUNG LEE 447

Heat Transfer Mechanisms in Microgravity Flow Boiling. *By* HARUHIKO OHTA. 463

Criteria for Approximating Certain Microgravity Flow Boiling Characteristics in Earth Gravity. *By* HERMAN MERTE, JR., JAESEOK PARK, WILLIAM W. SHULTZ, AND ROBERT B. KELLER 481

Part V. Biotransport Processes and Protein Crystal Growth

Microgravity Studies of Cells and Tissues. *By* GORDANA VUNJAK-NOVAKOVIC, NANCY SEARBY, JAVIER DE LUIS, AND LISA E. FREED 504

Flow Field Measurements in the Cell Culture Unit. *By* STEPHEN WALKER, MIKE WILDER, ARSENIO DIMANLIG, JUSTIN JAGGER, AND NANCY SEARBY . 518

Compact Optical Sensor for Real-Time Monitoring of Bacterial Growth for
  Space Applications. *By* R.C. VAN BENTHEM, D. VAN DEN ASSEM,
  AND J. KROONEMAN   541

Bioactive, Degradable Composite Microspheres: Effect of Filler Material
  on Surface Reactivity. *By* QING-QING QIU, PAUL DUCHEYNE,
  AND PORTONOVO S. AYYASWAMY   556

Vapor Transport Growth of Organic Solids in Microgravity and Unit Gravity:
  Some Comparisons and Results to Date. *By* MARIA ITTU ZUGRAV,
  WILLIAM E. CARSWELL, GLEN B. HAULENBEEK, MOHAN SANGHADASA,
  SUE K. O'BRIEN, BOGDAN C. GHITA, AND WILLIAM E. GATHINGS.   565

Sliding-Cavity Fluid Contactors in Low-Gravity Fluids, Materials, and
  Biotechnology Research. *By* PAUL TODD, JOHN C. VELLINGER,
  SHRAMIK SENGUPTA, MICHAEL G. SPORTIELLO, ALAN R. GREENBERG,
  AND WILLIAM B. KRANTZ   581

Microgravity Protein Crystallization: Are We Reaping the Full Benefit of
  Outer Space? *By* NAOMI E. CHAYEN AND JOHN R. HELLIWELL   591

Space-Grown Protein Crystals Are More Useful for Structure Determination.
  By JOSEPH D. NG   598

Mathematical Model for Diffusion of a Protein and a Precipitant about a
  Growing Protein Crystal in Microgravity. *By* ONOFRIO ANNUNZIATA
  AND JOHN G. ALBRIGHT   610

Index of Contributors   625

Financial assistance was received from:

- NATIONAL AERONAUTICS AND SPACE ADMINISTRATION
- NATIONAL SCIENCE FOUNDATION
- UNITED ENGINEERING FOUNDATION, INC.

> The New York Academy of Sciences believes it has a responsibility to provide an open forum for discussion of scientific questions. The positions taken by the participants in the reported conferences are their own and not necessarily those of the Academy. The Academy has no intent to influence legislation by providing such forums.

Attendees of the Microgravity Transport Processes Conference, Banff, Alberta, Canada, October 2001

Preface

With several Space-Shuttle experiments that have been conducted during the past two decades, significant new insights have been gained into various physical phenomena under microgravity conditions. There is at present a need to provide direction for microgravity research in which fluid flow and other transport processes form a common basis for many investigations. For example, in biomedical engineering and materials processing, fluid, thermal, and mass-transport aspects have gained primacy among various researchers, and the sharing and complementing of expertise have become a necessity for technical progress. In addition, with long-term manned space missions in the near future, technical problems encompassing several of these disciplines have been envisioned. The Banff conference provided a forum for the free flow of ideas under this common theme of *transport phenomena* and focused on interdisciplinary aspects.

One of the main objectives of the conference was to promote the exchange of technical information and ideas among the various scientists and engineers working in microgravity fluid, thermal, biological, and materials sciences. These areas have become of importance to various disciplines within the broader realm of microgravity research. The conference, therefore, addressed the cross-cutting aspects of microgravity science and technology and provided a forum for the synthesis of knowledge that has been acquired during the past several years. It was intended to effectively cover the growing interdisciplinary aspects of microgravity science and technology. As a result, many of the participants were able to gain considerable knowledge and insight into new areas, and it is expected that such sharing of expertise will create new frontiers in the emerging science.

We were indeed fortunate to have scientific diversity among the participants, and the high level of interdisciplinary activity that emanated from interaction among the participants. By crossing traditional boundaries of scientific expertise, it is our hope that new grounds for collaborative research will continue to emerge, and healthy research programs will develop. The scientific community has indeed recognized that *transport phenomena* as a science has reached a point whereupon it has a very significant role in many different areas.

The conference organizing committee expresses its sincere gratitude to Professor Frank Schmidt who, as the United Engineering Foundation's technical liaison, gave enormous support. In addition, we appreciate the valuable input from various members of the scientific committee in putting the conference together, and we are grateful to all the participants for sharing their technical expertise. Thanks are also due to NASA and the U.S. National Science Foundation for financial support. On behalf of all the participants, we thank Barbara Hickernell and the rest of the Foundation's staff in New York who, in spite of the catastrophic tragedy of September 11, 2001, remained resilient in carrying out the final planning for the conference. Finally, we thank the editorial staff of the New York Academy of Sciences for their role in publishing these keynote articles and scientific papers in the *Annals of the New York Academy of Sciences* in archival form.

SATWINDAR SINGH SADHAL (USA), Chair & Scientific Secretary
VIJAY K. DHIR (USA), Co-Chair
HARUHIKO OHTA (Japan), Co-Chair
REGINALD W. SMITH (Canada), Co-Chair
JOHANNES STRAUB (Germany), Co-Chair

Use of an Electric Field in an Electrostatic Liquid Film Radiator

S.G. BANKOFF,[a] E.M. GRIFFING,[a] AND R.A. SCHLUTER[b]

[a]*Chemical Engineering Department, Northwestern University, Evanston Illinois, USA*

[b]*Physics and Astronomy Department, Northwestern University, Evanston Illinois, USA*

ABSTRACT: Experimental and numerical work was performed to further the understanding of an electrostatic liquid film radiator (ELFR) that was originally proposed by Kim et al.[1] The ELFR design utilizes an electric field that exerts a normal force on the interface of a flowing film. The field lowers the pressure under the film in a space radiator and, thereby, prevents leakage through a puncture in the radiator wall. The flowing film is subject to the Taylor cone instability, whereby a cone of fluid forms underneath an electrode and sharpens until a jet of fluid is pulled toward the electrode and disintegrates into droplets. The critical potential for the instability is shown to be as much as an order of magnitude higher than that used in previous designs.[2] Furthermore, leak stoppage experiments indicate that the critical field is adequate to stop leaks in a working radiator.

KEYWORDS: electric field; electrostatic liquid film radiator

INTRODUCTION

Various possible applications have spurred a general interest in electrohydrodynamics. Examples include separation enhancement, electrostatic spraying to produce fine droplets or powders, ink-jet printers, spray painting, electrohydrodynamic pumps, and a variety of MEMS devices. Electrostatic fields produce tensile stresses at fluid interfaces. Typically, these forces are smaller than gravitational forces on Earth. However, in a microgravity environment, electrostatic fields can be quite important. The principal subject of this review is the application of strong electrostatic fields to flowing thin liquid films (usually liquid metal), including the stability of the free surface. This is in connection with a very lightweight enclosed radiator concept by the author, termed the *electrostatic liquid film radiator* (ELFR). An extensive review of free-surface, thin-liquid-film dynamics and long-wave instability has been given by Oron, Davis, and Bankoff.[3] Comprehensive treatments of electrohydrodynamics were provided by Melcher[4] and by Woodson and Melcher.[5] A more recent review of the literature, including details of the leaky dielectric model, is given by Saville.[6]

Lightweight radiator concepts include droplet,[7–10] controlled free-stream,[11,12] and falling film.[13–19] Generally, however, losses over time, in an unenclosed radiator of hot coolant, by evaporation into free space and/or droplet collision have not been

Address for correspondence: S.G. Bankoff, Chemical Engineering Department, Northwestern University, Evanston, IL 60208, USA.
gbankoff@northwestern.edu

quantified. Two experimentally observed modes of droplet breakup are (1) the formation of conical points (Taylor cones) that eject a fine spray of liquid droplets along field lines, and (2) the separation of one drop into two drops that are temporarily connected by a thin thread of fluid. Miksis,[20] Sherwood,[21] and others calculate a critical value of the dielectric constant that separates the two modes of breakup. When the dielectric constant is greater than the critical value, the charge accumulates on the fluid, and tangential stresses induce circulatory flow. Both authors predict the formation of conical interfaces, but the methods break down due to a singularity when the cone tip is too sharp. More recently several authors have used finite element methods to more accurately calculate droplet break-up and Taylor cone dynamics. Notz and Basaran,[22] measured droplet formation from a capillary tube in the presence of an electric field and used a finite element method to calculate droplet breakup.

Far less work has been done involving the deformation of a planar fluid interface. An important difference between these problems is the disparity in length scale that is often encountered in planar interface problems. For instance, typical length scales are 15 cm, whereas the Taylor cones often form over less than 1 cm. This presents an added difficulty in obtaining numerical solutions. Taylor and McEwan[23] observed Taylor cones on flat non-moving fluids. They showed that linear stability theory could accurately determine the threshold potential for cones. Mayberry and Donohoe[24] measured the wavelength of the instability of liquid gallium, and found it to be consistent with linear stability predictions. Melcher and Smith[25] investigated the effect of surface charge on the stability of non-conducting fluids by turning on a dc field, both slowly and quickly. The dc fields had similar effects regardless of how quickly they were administered.

ELECTROSTATIC LIQUID FILM RADIATOR DESIGN

The EFLR design is a pumped-loop, double-membrane radiator in which a thin coolant film runs along the inside of an outer membrane that radiates heat into space. The inner membrane, which is very thin, has an array of metallized rectangular electrodes, any one of which can be charged to a high voltage if a leak is detected by means of a conductive layer on the exterior surface. The electric field pulls the liquid surface away from the puncture and can overcome the combined vapor pressure of the coolant and the hydrostatic head of the thin coolant layer. The voltage is supplied by a small van de Graaf generator in the interior of each module.

Because of thin walls (about 100 μm) and a large view factor (0.85–0.95), the ELFR is extremely light in weight. According to preliminary calculations, the specific weight for a lithium-cooled radiator, in kg/kW thermal power, for a rotating disk radiator with inflow temperature of 800 K, outflow temperature of 725 K, and inlet Reynolds number of 300, is estimated to be 0.094 kg/kW without the circulating coolant, and 0.141 kg/kW including the coolant. Note that all radiator types require a circulating coolant to remove waste heat from the power cycle. This implies that a 1-MW thermal radiator would weigh only about 150 kg, in part because of its excellent view factor. Even this number can be substantially reduced if a thinner niobium lining to protect against corrosion by the lithium can be used (possibly 25, rather than 50 microns). If proven practicable, the ELFR could provide dramatic cost savings for a variety of missions.

Several ultralight systems, including droplet, conveyor belt, and thin film radiators have been proposed and studied in recent years to accomplish this goal. Tagliafico and Fossa suggest liquid sheet or droplet radiators.[18,19] They report design weights as low as 1 kg/kW at 300–400 K.

The international space station generates about 75 kW of power. More ambitious designs might require one megawatt of power or more, necessitating the rejection of two or more megawatts of waste heat. Practical radiators for this amount of radiation might use temperatures of 700 to 1,000 K. We have made two designs to satisfy both power generation and housekeeping operating conditions. A high temperature radiator would use a liquid metal such as lithium and operate at about 700 K, and a low temperature radiator utilizing either oil, NaK, or gallium would operate at about 300 K. Lithium and NaK have both been used in space designs to date. They are very light (Li, about 0.45 g/cm^3 and NaK, about 0.85 g/cm^3) and have good heat transfer properties. A carbon coating on the exterior of the radiator would be used to decrease the wettability (and hence the tendency to leak), and at the same time increase the emissivity of the surface. With oil, a Teflon coating could be used to obtain non-wetting, but in general a liquid metal coolant is preferred because its high surface tension and conductivity. However, the long-term chemical reactions, and hence wettability, of the exterior surface of the radiator after leakage is arrested from a particular puncture are still unknown. Note that with a liquid metal all punctures are initially non-wetting, and probably remain so, because with a continuous meniscus, the required driving pressures are many orders of magnitude greater than that available.

Previous work by Kim, Bankoff, and Miksis[13–15] showed fluid film profiles for flow under finite electrodes (see FIGURE 1). They derived lubrication and Kármán-Pohlhausen models to calculate the height profiles and compared the results to a numerical solution of the Navier-Stokes equations as calculated with the SOLA code.[26] They solved for the height profile and pressure under the film for conical and cylindrical radiator geometries. Designs were then optimized based on minimum specific weight with constraints on fluid depth and maintaining the ability to generate a negative pressure on the wall at the leak location with an electric field.

In the current work, we consider an electrohydrodynamic instability similar to Taylor cones, leak dynamics within the hole, and electrical breakdown. Critical fields for the Taylor cone type instability are obtained experimentally for mercury and oil films. These fields are representative for the liquid metal and hydrocarbon type radiators suggested. A large margin is observed for the liquid metal, which is greatly stabilized by surface tension. However, the critical field for Taylor cones on oil films is comparable to that used in previous designs.

Our leak stoppage experiments were very encouraging. For a non-wetting substrate, the positive pressure required to start or stop a leak was simply the capillary pressure, $2\sigma/r$, for small holes. For holes that were significantly larger than the fluid depth, an even greater pressure was required to start a leak. The capillary pressure was very large compared to the hydrostatic head and internal pressure that force leaks. Therefore, choosing a non-wetting substrate would be a great advantage in an ELFR. For a wetting substrate, the pressure required underneath the film was nearly

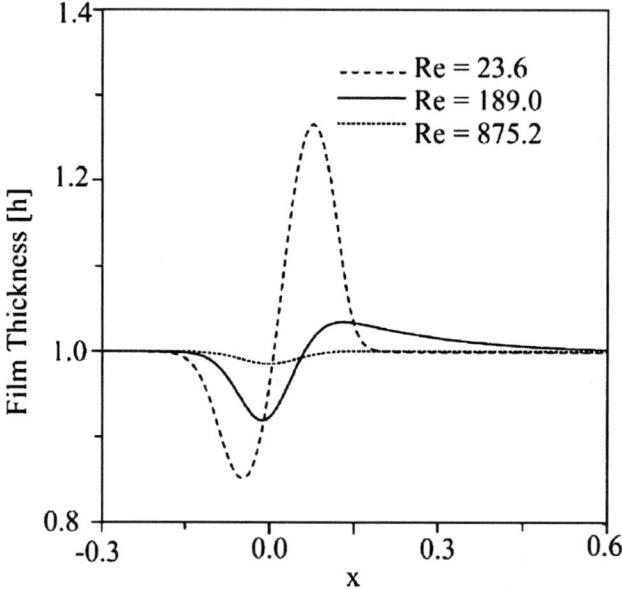

FIGURE 1. Fluid height profiles for liquid lithium (700 K) flowing in the presence of a fixed normal electric field. The field is taken to be Gaussian centered at 0, and the height is calculated using the Reynolds lubrication model. Taken from Kim et al.,[13] who provide further details.

zero, implying that leaks could be stopped in an organic, as well as metal, based radiator with wetting problems.

Gas phase breakdown can be avoided by going to lower pressures. On the low pressure ($P < 10^{-1}$ mTorr) side of the Paschen curve, vacuum breakdown takes over as the pressure is reduced. Fields on the order of (200kV/cm) can be exerted in this region, making the Taylor cone threshold (8 to 70kV/cm) the limiting factor on the electric field. From these experiments, we were able to impose additional constraints on the previous design work, specify new design parameters, and calculate new estimates of weight/power.

PREVIOUS WORK

Kim et al.[13] numerically investigated the effect of a fixed electric field or pressure profile on a fluid flowing down an inclined plane. Using lubrication and Kármán-Pohlhausen models, as well as a Navier-Stokes solver, they showed steady-states with a depression in the fluid interface underneath the leading edge of the electrode and a peak in the interface underneath the trailing edge. The unsteady solutions exhibited the formation of a traveling wave that washed away downstream. This work neglected surface tension, as well as the coupling of the electric field with the fluid interface, thereby precluding the formation of short wave instabilities.

When fully developed, these are similar to the well-known Taylor cones that have been observed on a horizontal plate and droplets. Infinite height values were observed under the trailing edge of a finite electrode as the field was increased. However, the fields required for this behavior were significantly greater than that measured experimentally.

Kim et al.[14,15] and Bankoff et al.[2] also investigated other geometries that are more applicable to an ELFR. A spinning conical design, in which the coolant flowed from the apex of a cone to the wide end, had a fluid height solution that typically thinned as the fluid moved along the radiator wall. The magnitude of the depth change depended on the cone angle and the rotation rate. A stationary cylindrical design, in which the fluid thickened as it moved tangentially along the radiator wall, was also studied. In both cases, the effect of the field on the interface was qualitatively the same as that for flow down an inclined plane.

CURRENT WORK

In the current work, we compared the height and pressure calculations of Kim et al.[13] to coupled solutions of the fluid and electric field equations and experimentally demonstrated the feasibility of the ELFR by using the electric field to stop a leak. We used diffusion pump oil to simulate the low temperature design and mercury to simulate the high temperature design. The oil flowed down an inclined plate, passing in front of a glass electrode, which allowed the quantified fluorescence imaging method[27] to be used for local film thickness measurements. Pressure and stability measurements were made with ethanol and diffusion pump oil, and stability measurements were made with flowing mercury films. The low temperature design could be used for electronics and habitat cooling. The high temperature design for power generation, which incorporates lithium, appears to be quite robust but may present handling problems.

Taylor cones pose a potential problem in a radiator. Taylor and McEwan[23] showed that the stability of a horizontal layer of fluid under an electrode could be calculated with linear stability theory. When the potential is increased beyond a critical value, the fluid forms waves that develop into a cone and emit a jet of fluid. Each cone shorts the electrode requiring it to be recharged. Fluid would possibly leak during the recharge time, which is determined by the power (or weight) of the recharging device. The energy produced during the short could cause irreversible damage to the electrode as material evaporates. This was observed in a tin-oxide coated glass electrode in our laboratory. The liquid film could also ionize, enhancing corrosion chemistry or carbonization.

Taylor cones were investigated on a horizontal non-moving layer and on an inclined flowing film (see FIGURES 2 and 3). In both cases, we used a finite electrode to determine edge effects. In the horizontal case and small electrode heights, the cones preferentially formed under the corner of the electrode. This edge effect was previously noted by Taylor and McEwan.[23] We found that the critical potential for cone formation was not significantly altered and that the stability results can be predicted well using the linear analysis. Non-linear stability checks with the lubrication theory also proved to be effective in predicting cone formation.

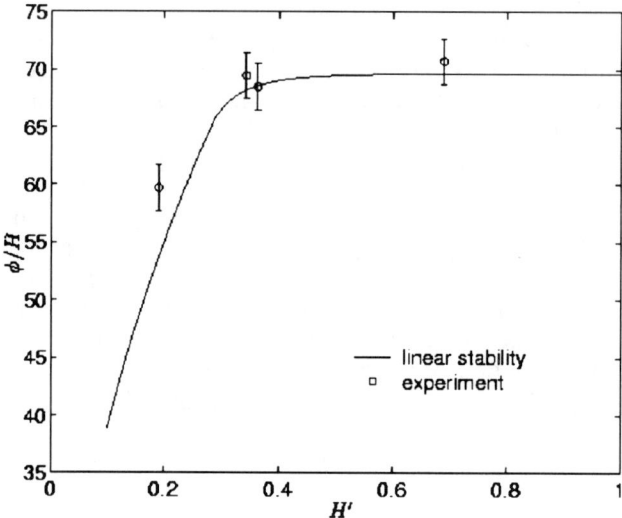

FIGURE 2. Experimental values of the critical field are compared to linear stability predictions for a finite rectangular electrode held above a horizontal mercury film. H' is the height of the electrode above the static film height under the electrode corner.

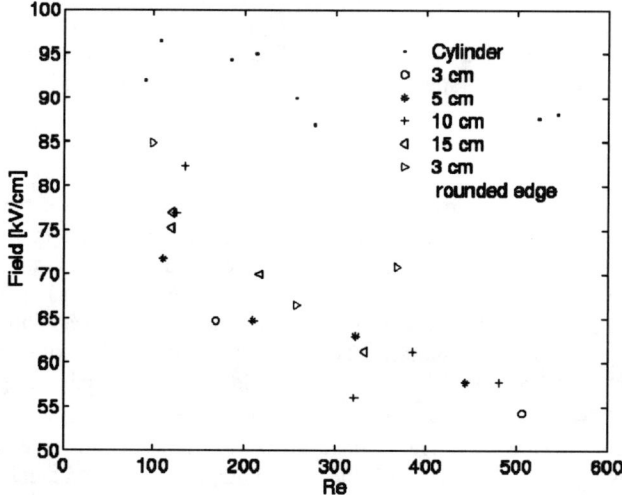

FIGURE 3. Experimental values of the critical field for a flowing mercury film passing in front of a charged electrode. The data series are for various electrode lengths in the direction of flow. The critical field is not strongly dependent on the length above 3 cm. The decrease in critical field with increasing Reynolds number (Re) is thought to be due to surface waves.

Taylor cones were observed on a vertical oil film when the electric field exceeded a critical value, $E \approx 13\,\text{kV/cm}$. Below this value, the height profile showed less than 1% variation along the direction of fluid flow, which was consistent with lubrication theory. Liquid metals have much higher surface tension than oils. Lithium, which was considered in Bankoff et al.,[2] has a surface tension of 395 dynes/cm at 700 K compared to about 30 dynes/cm for oil. This allows for a much higher critical potential and a more robust design. A dimensional balance of the surface tension pressure with the electric field pressure shows $\sigma \approx \varepsilon \varepsilon_0 E^2 / 2\kappa$, where σ is the surface tension, E is the electric field, ε is the dielectric constant, ε_0 is the permittivity of free space, and κ is the curvature of the conical interface. Neither theoretical work nor experimental data giving Taylor cone thresholds on flowing films exist in the literature. Therefore, we measured Taylor cone thresholds for flowing mercury. Stable films were observed with fields of 55 to 90 kV/cm, a factor of two higher than the square root dependence predicted by the dimensional analysis. This is not surprising considering that the Reynolds number (Re) changed by a factor of 10^3, the Taylor cones observed on the oil film were three-dimensional, and the wave number is likely to vary. The relative stability of the liquid metal film to Taylor cones allows for a large margin of error.

The following results support design feasibility.

- Fluid film profiles for flow under an electrode were calculated by Kim et al. for a variety of design types and parameters.

- Film stability has been demonstrated at sufficient electric fields to stop leaks. In our experiments, vertical flowing mercury films were stable to electric fields of $E > 50\,\text{kV/cm}$, compared to 8 kV/cm, used in previous design work. (Values of the field listed in these works are higher due to an error in units.)

- Leaks have been started and stopped, even with the complication of Earth gravity, for wetting and non-wetting substrates.

- Electrical breakdown in the vapor phase has been considered and can be avoided without large increases in weight.

Before such a radiator is considered safe, further work should be done concerning the nature of the leak holes, long-term chemistry of realistic materials, the effect of reduced gravity, and electrical breakdown in a realistic design. However, it is thought that these issues will affect design details rather than present significant problems.

Since liquid lithium is toxic, explosive in contact with water and flammable in contact with air, the recommended experiments should be done in a properly equipped facility with experienced operators.

ACKNOWLEDGMENTS

This work was supported by the Microgravity Division at NASA with Allen Wilkinson as the program manager. Helpful comments were made by Professor M.J. Miksis.

REFERENCES

1. KIM, H., S.G. BANKOFF & M.J. MIKSIS. 1991. Lightweight space radiator with leakage control by internal electrostatic fields. Proc. Eighth Symp. On Space Nuclear Power Systems, University of New Mexico, Albuquerque, NM. 1280–1285.
2. BANKOFF, S.G., M.J. MIKSIS, H. KIM & R. GWINNER. 1994. Design considerations for the rotating electrostatic liquid-film radiator. Nuc. Eng. Des. **149**(1–3): 441–447.
3. ORON, A., S.H. DAVIS & S.G. BANKOFF. 1997. Long-scale evolution of thin liquid films. Rev. Mod. Phys. **69**(3): 931–980.
4. MELCHER, J.R. 1981. Continuum Electrohydromechanics. MIT Press.
5. WOODSON, H.H. & J.R. MELCHER. 1968. Electromechanical Dynamics. Wiley, New York.
6. SAVILLE, D.J. 1997. Electrohydrodynamics: the Taylor-Melcher leaky dielectric model. Annu. Rev. Fluid Mech. **29**: 27–64.
7. MATTICK, A.T. & A. HERTZBERG. 1982. The liquid droplet radiator—an ultra-lightweight heat rejection system for efficient energy-conversion in space. Acta. Astronaut. **9**(3): 165–172.
8. MATTICK, A.T. & A. HERTZBERG. 1985. Liquid droplet radiator performance studies. Acta. Astronaut. **12**(7–8): 591–598.
9. ORME, M. & E.P. MUNTZ. 1987. New technique for producing highly uniform droplet streams over an extended range of disturbance wave-numbers. Rev. Sci. Instrum. **58**(2): 279–284.
10. PERSSON, J. 1990. The liquid-droplet radiator—an advanced future heat-rejection system. ESA J.–EUR Space Agency **14**(3): 271–288.
11. MUNTZ, E.P. & M. DIXON. 1986. Applications to space operations of free-flying controlled streams of liquids. J. Spacecraft Rockets **23**(4): 411–419.
12. MUNTZ, E.P. & M. ORME. 1987. Characteristics, control, and uses of liquid streams in space. AIAA J. **25**(5): 746–756.
13. KIM, H., S.G. BANKOFF & M.J. MIKSIS. 1991. The effect of an electrostatic field on film flow down an inclined plane. Phys. Fluids **4**(10): 2117–2130.
14. KIM, H., S.G. BANKOFF & M.J. MIKSIS. 1993. Interaction of an electrostatic field with a thin film flow within a rotating conical radiator. J. Propul. Power **9**(2): 245–254.
15. KIM, H., S.G. BANKOFF & M.J. MIKSIS. 1994. Cylindrical electrostatic liquid film radiator for heat rejection in space. J. Heat Trans. **116**: 441–447.
16. GONZALEZ, A. & A. CASTELLANOS. 1996. Nonlinear electrohydrodynamic waves on films falling down an inclined plane. Phys. Rev. E **53**(4): 3573–3578.
17. GONZALEZ, A. & A. CASTELLANOS. 1998. Nonlinear electrohydrodynamics of free surfaces. IEEE T. Dielect. El. In. **5**(3): 334–343.
18. TAGLIAFICO, L.A. & M. FOSSA. 1997. Lightweight radiator optimization for heat rejection in space. Heat Mass Transfer **32**(4): 239–244.
19. TAGLIAFICO, L.A. & M. FOSSA. 1999. Liquid sheet radiators for space power systems. P.I. Mech. Eng. G. **213**(G6): 399–406.
20. MIKSIS, M.J. 1981. Shape of a drop in an electric field. Phys. Fluids **24**(11): 1967–1972.
21. SHERWOOD, J.D. 1988. Breakup of fluid droplets in electric and magnetic fields. J. Fluid Mech. **188**: 133–146.
22. NOTZ, P.K. & O.A. BASARAN. 1999. Dynamics of drop formation in an electric field. J. Col. Int. Sci. **313**(1): 218–237.
23. TAYLOR, G.I. & A.D. MCEWAN. 1965. The stability of a horizontal fluid interface in a vertical electric field. J. Fluid Mech. **22**(1): 1–15.
24. MAYBERRY, C.S. & G.W. DONOHOE. 1995. Measurements of the electrohydrodynamic instability in planar geometry using gallium. J. Appl. Phys. **78**(9): 5270–5276.
25. MELCHER, J.R. & C.V. SMITH. 1969. Electrohydrodynamic charge relaxation and interfacial perpendicular-field instability. Phys. Fluids **12**(4): 778–786.
26. HIRT, C.W., B.D. NICHOLS & N.C. ROMERO. 1975. SOLA—a numerical solution algorithm for transient fluid flows. Technical Report. Los Alamos Scientific Laboratory of the University of California. Los Alamos.

27. JOHNSON, M.F.G., R.A. SCHLUTER & S.G. BANKOFF. 1997. Fluorescent imaging system for global measurement of liquid film thickness and dynamic contact angle in free surface flows. Rev. Sci. Instrum. **68**(11): 4097–4102.

Large-Scale Geophysical Flows on a Table Top

JOHN HEGSETH AND KAMEL AMARA

Nonlinear Science Laboratory, Department of Physics,
University of New Orleans, New Orleans, Lousiana, USA

ABSTRACT: Results from two similar experimental systems that attempt to create laboratory geophysical analogs in spherical geometry are presented. In the first system, real time holographic interferometry and shadowgraph visualization are used to study convection in the fluid between two concentric spheres when two distinct buoyancy forces are applied to the fluid. The heated inner sphere has a constant temperature that is greater than the outer sphere's constant temperature by ΔT. In addition to the usual gravitational buoyancy from temperature induced density differences, another radial buoyancy is imposed by applying an ac voltage difference, ΔV, between the inner and outer spheres. The resulting electric field gradient in this spherical capacitor produces a central polarization force. The temperature dependence of the dielectric constant results in the second buoyancy force that is especially large near the inner sphere. The normal buoyancy is always present and, within the parameter range explored in our experiment, always results in a large-scale cell that is axisymmetric about the vertical axis. We have found that this flow becomes unstable to toroidal or spiral rolls that form near the inner sphere and travel vertically upward when ΔT and ΔV are sufficiently large. These rolls start near the center sphere's equator and travel upward toward its top. In the second experimental system, the central force is applied to a highly compressible near-critical fluid in weightlessness (parabolic flight) and normal gravity. Although a geophysically similar density distribution could not be obtained in the limited time of a parabolic flight, clear influences of the central force on the fluid were observed in both weightlessness and terrestrial experiments.

KEYWORDS: large-scale geophysical flow

NOMENCLATURE:
r_i inner sphere radius
r_o outer sphere radius
d gap between spheres, $r_o - r_i$
Γ aspect ratio, d/r_i
ΔT temperature difference across capacitor
T_c critical temperature
ΔT_c temperature difference from T_c, $T_c - T$
β thermal expansion coefficient
ρ_a ambient density
γ dielectric variability
D_{th} thermal diffusivity
H heat flux
Nu Nusselt number
Nu^* modified Nusselt number

Address for correspondence: John Hegseth, Nonlinear Science Laboratory, Department of Physics, University of New Orleans, New Orleans, LA 70148, USA. Voice: 504-280-6706; fax: 504-280-6048.
jhegseth@uno.edu

E	electric field
ΔV	voltage difference across capacitor
ε	permittivity
κ	dielectric constant
κ_a	ambient dielectric constant

INTRODUCTION

In the following, results from our experiments performed on the ground and during parabolic flight are described. These experiments were motivated by the goal of creating laboratory geophysical analogs that could complement and check numerical codes, thus greatly contribute to understanding many important geophysical flow systems, such as the Earth's ocean and the atmosphere. These experimental systems are performed in spherical geometry at a small scale, in a spherical capacitor, both on the ground and during parabolic flights using NASA's KC-135A. Using a central polarization force, the experiments attempted to induce a density gradient in a highly compressible fluid in low gravity and study the buoyancy driven flow of incompressible fluid under $1g$ (g is the acceleration due to terrestrial gravity, 9.8 m/sec). The former fluid is highly compressible because its temperature and pressure are close to the liquid–gas critical point. Near this point the difference between gas and liquid disappears. The resulting presence of density fluctuations makes the fluid highly compressible. Near the critical point these density fluctuations become larger than the wavelength of light and efficiently scatter light so that the fluid appears pearl white in color, the so-called critical opalescence. Inside the near-critical fluid is a spherical capacitor. A large ac potential difference is applied to the capacitor. In this spherical geometry, the electric field between the spheres is non-uniform, so that a polarization force is applied to the fluid and act on a molecular dipole in an electric field gradient. The potential difference results in a central force that is analogous to the gravitational field of a planet. When applied to a compressible fluid a spherical density gradient should be produced, given enough time and an environment free of large perturbations.

Under normal terrestrial gravity, a highly compressible near-critical fluid stratifies under its own weight. By placing a fluid sample in a weightless environment and applying a spherically symmetric body force on this fluid we attempted to induce a radial density gradient in analogy to a geophysical system, such as a planetary atmosphere. The weightless environment was produced during parabolic flight. The entire apparatus was placed in a rack bolted to the floor of a NASA KC-135A aircraft. This aircraft flies up and down on a sinusoidal trajectory producing approximately 20 seconds of weightlessness near the top parabola of each cycle. These cycles are typically repeated 40 times during one flight. We used a simple optical diagnostic to observe the density changes. The experimental variables are then changed to see how voltage changes and temperature changes affect the density gradient.

This same central polarization force is also capable of inducing a buoyancy force in a normal fluid. When the central sphere of the spherical capacitor is heated, the decreased density of the fluid also produces a decreased dielectric constant. The reduced dielectric constant results in a smaller force than at the cooler fluid farther away from the central sphere. The cooler fluid is attracted to the center whereas the hotter fluid is repelled, in analogy to normal buoyancy from relative weight changes. As discussed in the first part of this paper, we have observed and characterized an instability in our system that is caused by heating the inner sphere of our capacitor in the presence of the polarization force.

INCOMPRESSIBLE FLUID

An electric field influences a dielectric fluid in several ways. These influences include the modification of the thermodynamics of a fluid and the mechanical body force exerted on a fluid in a non-uniform field.[1] Because the electric field produces both attractive and repulsive forces on positive or negative charges, it can result in various bulk effects. These effects include the force on any charged fluid element and the polarization of the charge distribution of non-polar molecules to produce a dipole moment. The induced dipole moment is created because the average positions of the positive and negative bound charges in a molecule become spatially separated as the field pulls the nucleus in one direction and the electrons in the opposite direction. If a molecule has a permanent dipole moment then the field produces a torque that tends to align the moment with the electric field. In other words, the positively charged portion of the molecule is forced in the direction of the field and vice versa for the negatively charged portion. In a non-uniform field, or an electric field gradient, such a polarized molecule feels a net force from the field toward the region of higher field intensity (i.e., the force due to charge at the higher field intensity is always greater than that on the other charge). The result, in the continuum limit is a body force in the fluid.[1] In an ac field, the dipole moment remains aligned with the field, preserving the attraction toward the higher field gradient.

In a spherical geometry, such as concentric spheres with dielectric fluid between them (the spherical capacitor), the electric field gradient produces a central body force, as shown in FIGURE 1. When the central sphere is heated the change in the dielectric constant with temperature (i.e., $\gamma = \gamma(T)$ where γ is the dielectric variability) produces a buoyancy in the presence of the polarization force field. This effect has been previously used to obtain interesting results of geophysical significance by additionally rotating the system. The results include banana cells (rapid rotation and heating), soccer-pattern (heating and slow rotation), turbulence and Hadley cells,[2] (low rotation and buoyancy from terrestrial g).[3] Buoyancy from the Earth's gravity was avoided in these previous experiments by performing them in the weightlessness of an orbiting spacecraft. If the system is too large it is difficult to get an appreciable effective gravity compared to g using an electric field below the breakdown voltage. We have found, however, that it is possible to get such a large effective gravity in a small system, creating the possibility for terrestrial experiments where normal buoyancy is comparable to the buoyancy from the polarization force.[4] Unfortunately, such a small system makes it difficult to obtain a small aspect ratio, Γ, similar to that typical of atmospheres or oceans, Γ is defined in FIGURE 1. There are, however, other interesting geophysical systems, such as a planetary core, that have a large aspect ratio. In this paper, we present an experimental study of an instability caused by the radial buoyancy from an imposed electric field gradient and an imposed temperature

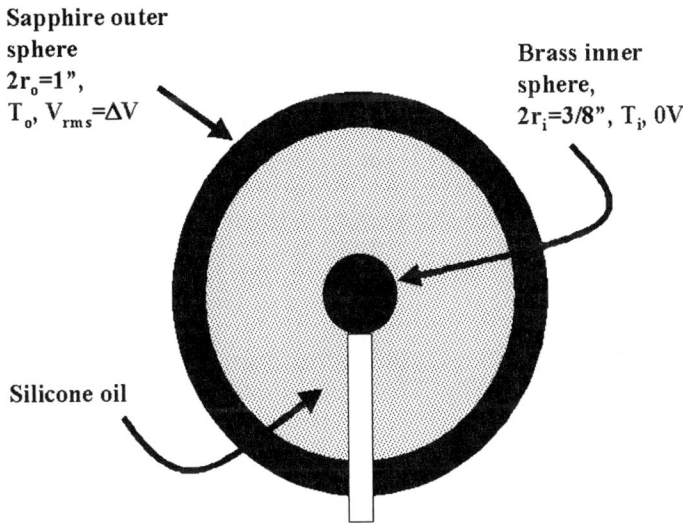

FIGURE 1. The figure shows a cross-section of the experimental apparatus. The inner sphere, made of brass, is electrically grounded, has an imbedded thermocouple to measure temperature, T, and is heated with a nichrome wire glued near its center. A high voltage is applied to the outer sapphire sphere. The volume between the spheres is filled with Dow Corning (DC-200 10 cs) silicone oil. r_i is the outer radius of the inner sphere, r_o is the inner radius of the outer sphere, and the gap between the sphere, d, is $d = r_o - r_i$. The dimensionless gap or the aspect ratio, Γ, is defined using r_i as a length scale, $\Gamma = d/r_i$.

gradient in spherical geometry without rotation, as first step in studying large scale geophysical flows.

If the dielectric properties of the fluid change from point to point then a uniform field may exert a force on the fluid. By contrast, a non-uniform field exerts a force on a uniform dielectric due to the net force that such an electric field gradient exerts on each dipole, as discussed above. The net pondermotive force (per unit volume) is,

$$f = \rho_q E - \frac{\varepsilon_0}{2} E^2 \nabla \left(E^2 \frac{d\kappa}{d\rho} \rho \right) \tag{1}$$

where ρ_q and ρ are the charge and mass density, respectively; E is the electric field; ε_0 and ε are the permittivity of free space and the fluid, respectively; and $\kappa = \varepsilon/\varepsilon_0$ is the dielectric constant. $\rho_q E$ is the force per unit volume on any free charges in the fluid (the electrophoretic force), $(\varepsilon_0/2)E^2\nabla\kappa$ is sometimes called the dielectrophoretic force, and $(\varepsilon_0/2)\nabla\{E^2(d\kappa/d\rho)\rho\}$ is sometimes called the electrostrictive force.[5]

The electrophoretic force is typically large compared to the other forces in Equation (1). All dielectric media have a small amount of charge creating an electrophoretic force. This effect can complicate or dominate the dielectrophoretic and electrostrictive effects. However, it is possible to define a relaxation time, τ, for free charge carriers subjected to an electric field, $\rho_q = \rho_{q_0} e^{-t/\tau}$, where $\tau = \kappa\varepsilon_0/\sigma$, and σ is the conductivity of the fluid. If the relaxation time, τ, is large compared to the period, T_f, of an ac field ($\tau \gg T_f$) then the charge carriers do not have time to respond to the applied field, essentially eliminating the $\rho_q E$ force, any current, and Ohmic heating from such a current. The other two forces, depending on E^2, are not similarly affected.

The inner sphere is positioned at the center of the outer sphere, to within 0.2%, by using several stiff electrically conductive wires for grounding, heating current, and a thermocouple. In an ideal spherical capacitor with a potential difference, ΔV, applied between the inner sphere of radius r_i and the outer sphere of radius r_o, the electric field is in the radial direction with a magnitude

$$E = \frac{\Delta V r_i r_o}{d r^2} \tag{2}$$

with a gap between the spheres of $d = r_o - r_i$. This electric field is distorted from a radial field in the region of the wires. We have minimized this distortion by twisting the wires together when they pass through the fluid to the inner sphere. We have also run our experiment both with the wires going up and with the wires going down. We have not seen any difference in the experimental results between these two orientations.

As discussed above, the dielectrophoretic force in an incompressible fluid generates a buoyancy force associated with the temperature change in the dielectric constant. This change, $\kappa = \kappa_a(1 - \gamma \Delta T)$, is analogous to the buoyancy from density changes $\rho = \rho_a(1 - \beta \Delta T)$ (κ_a is the ambient dielectric constant, γ is the dielectric variability $\gamma = \Delta \kappa / (\kappa_a \Delta T)$, ρ_a is the ambient density, and β is the thermal expansion coefficient). In fact, the buoyancy from dielectric variations reproduces the Bousinesq equations of motion with the force on the fluid in this approximation given by $F = g_a \gamma \Delta T r$.[4,5] This force has the same form as the usual buoyancy force, $F = g\beta \Delta T z$, from density variations under normal gravity, but has a different orientation. In this incompressible case the equivalent gravity, g_e, is given by

$$g_e = \frac{2\kappa \varepsilon_0}{\rho} \left(\frac{\Delta V r_o r_i}{d} \right)^2 \frac{1}{r^5}. \tag{3}$$

Equation (3) gives appreciable equivalent gravity values for small cells at the inner sphere, where the force has a maximum value. Normal buoyancy dominates most of the volume of the sphere (i.e., there is a very small g_e value at the outer sphere). Very close to the inner sphere, however, the polarization force gets comparably large. In fact, at the surface of the inner sphere g_e reaches as large as about $2g$ at 4kV. Because β and γ are nearly equal in Dow Corning 200 10cs silicone oil (DC-200 10 cs, Prandtl number, $P_r = 10.5$) used in this experiment, one can see that our small system has a large possibility for instability near the inner sphere.

Our set-up, shown in FIGURE 1, uses a spherical capacitor built from two concentric spheres. We use the inner sphere radius ($r_i = 4.76$ mm) as the characteristic length of the spherical capacitor so that the aspect ratio is $\Gamma = 1.67$. The characteristic time for heat transport in the inner sphere is $r_i^2/\kappa_\beta = 0.6$ seconds, where κ_β is the thermal diffusivity of brass. Because this time is much smaller than $d^2/\kappa_S = 624$ seconds, where d is the gap between the spheres and κ_S is the thermal diffusivity of DC-200 10cs silicone, we see that T_i is the same throughout the inner sphere to a very good approximation. The outer sphere, of inner diameter $2r_o = 25.4$ mm, is made of two 6.35mm-thick transparent sapphire hemisphere domes to allow optical

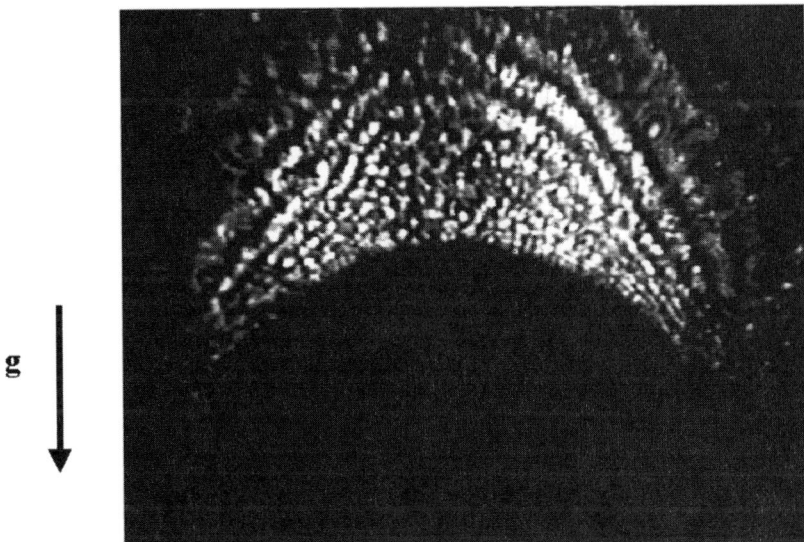

FIGURE 2. Shown are interference fringes that form above the inner sphere from hot fluid that flows upward under the influence of normal buoyancy at $R_g = 4.8 \times 10^4$ ($\Delta T = 4°C$). The thermal force exists without an imposed ΔV and creates a large axisymmetric Hadley cell.

diagnostics and good thermal conduction (i.e., thermal conduction time of 2.8 seconds). A thermocouple is in contact with the outer sphere wall to monitor the temperature of the outer sphere. The entire capacitor is put inside a silicone oil bath of DC-200 10cs, so that this fluid fills the capacitor gap ($d = 7.94$ mm). The bath temperature is controlled by a temperature controlled external water bath. The experiments are conducted by keeping the outer sphere temperature, T_o, constant and slightly elevated from room temperature (usually at 19°C ± 0.1°C). The inner sphere temperature, T_I, is computer controlled in the range 19 ± 0.1°C to 47 ± 0.1°C, so that the temperature difference $\Delta T = T_i - T_o$ can range between 0°C and 28°C. The outer sapphire sphere is coated with transparent and electrically conductive indium–tin–oxide (ITO). The outer sphere is connected to a high voltage ac transformer driven by a variable voltage and variable frequency power supply. The high voltage, ΔV, applied across the capacitor could be set from 0 kV to 5 kV (± 10 V) at a frequency of 300 Hz. The field thus reverses polarity in a period much shorter than the charge relaxation time, $\tau = 221$ seconds for the fluid. This ensures an absence of dc electric currents and associated effects, as discussed above.

Real-time holographic interferometry was used to visualize the flow when the heating and/or the voltage were applied. This optical technique allows *in situ* observation of a continuously changing process, whereas the usual double exposure technique is only able to capture phenomena at an instant in time.[6] This technique gives a quantitative measure of the average change in index of refraction along the beam path. The beam path is perpendicular to the image that shows the fringes. This index change can be directly related to the average temperature field. Shadowgraph visualization was also used at higher ΔT when light refraction became large. At higher ΔT our visualization typically consisted of a shadowgraph image near the inner sphere, where refraction effects where large, together with interference fringes farther away from the inner sphere.

We have observed a large scale cell in the system whenever ΔT is applied between the two spheres by tracing dust particles in the spherical capacitor and observing the interference pattern (see FIGURE 2). Because this cell may exist without an imposed ΔV, we conclude that this is a convective cell driven by normal buoyancy from density changes. We characterize this buoyancy in the usual way, by using the Rayleigh number, R_g, the ratio of the buoyancy force from g to the viscous force. Because our system has a ΔT between two spheres, there is always a horizontal temperature gradient, so that there is always a flow. In our case the Rayleigh number is defined by

$$R_g = \frac{\beta g \Delta T r_i^3}{\nu D_{th}}, \tag{4}$$

where r_i, β, and D_{th} are the characteristic length, the thermal expansion coefficient, the kinematic viscosity, and the thermal diffusivity, respectively. The fluid heated close to the inner sphere moves upward under normal buoyancy, where g points in the downward direction (FIG. 2). The returning flow moves downward near the outer sphere, driven by pressure gradients (reaction forces required by continuity) and the downward buoyancy from cooling near the outer sphere. These hot and cold currents form an axisymmetric toroidal cell that fills the gap between the spheres. We call this cell a "Hadley" cell, by analogy with the Hadley cell in atmospheric dynamics.[7] When a Hadley cell is established at a given ΔT, we apply a large central polarization force by applying a large ΔV between the spheres (as explained above, $g_e \approx 2g$ near the inner sphere at 4kV). We characterize the effects of the dielectric buoyancy force using another Rayleigh number, $R_{E_{max}}$,

$$R_{E_{max}} = \frac{2\gamma\kappa\varepsilon_0}{\nu D_{th}\rho(1-\eta)}\left(\frac{r_o}{d}\right)^2 \Delta V^2 \Delta T.$$

Because $R_{E_{max}} \propto \Delta V^2 \Delta T$, it depends on both external parameters. Additional details concerning $R_{E_{max}}$ are given in Reference 8. Just after applying a large voltage for a given temperature difference ΔT, we have seen rolls form close to the inner sphere. Individual rolls in this pattern are transient, lasting for only several seconds (e.g., 2 seconds for $R_g = 13.2 \times 10^4$ [$\Delta T = 11°C$], and $R_{E_{max}} = 1,045$ [$\Delta V = 6kV$]) and then vanishing. A roll can appear at all places around the inner sphere, including the bottom pole, with no apparent preferred position. This pattern of transient rolls also disappears after several seconds and a steady state convection pattern appears (see FIGURE 3). The steady state convection consists of traveling rolls that only occur in the upper hemisphere. The traveling rolls start near the equator and propagate upward to the top pole. Because the rolls only exist when both ΔV and ΔT are applied, and since they only exist close to the inner sphere where the polarization force is large, we conclude that the dielectric buoyancy causes this flow.

We have systematically studied the onset of the traveling rolls by slowly increasing the central polarization force. At a given ΔT, with a steady Hadley cell present, we slowly increased ΔV between the spheres. At a threshold value for ΔV, we observed perturbations that start near the equator and travel along the inner sphere a

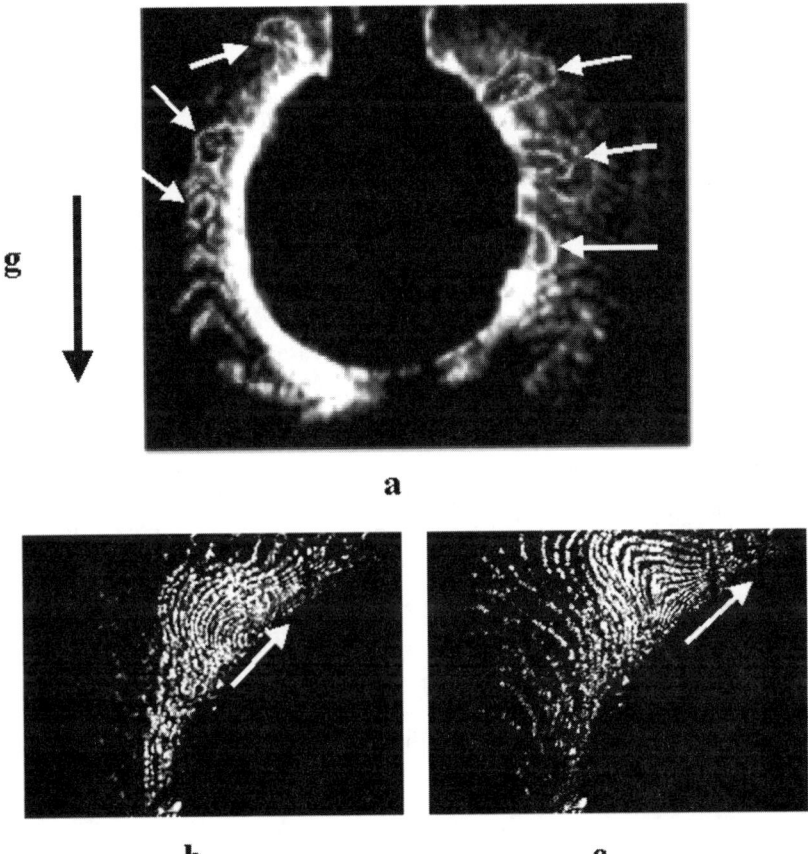

FIGURE 3. The top image (**a**) shows a global view of the traveling rolls, indicated by the *arrows*, at $R_g = 3.6 \times 10^4$ ($\Delta T = 3.0°C$) and $R_{E_{max}} = 198.0$ ($\Delta V = 5{,}000\,V$), where R_g and $R_{E_{max}}$ are defined in the text. These rolls form at all values of $R_g \propto \Delta T$ when $R_E \propto \Delta V^2 \Delta T$ is sufficiently large. They also form near the equator and travel upward along the circumference of the inner sphere. Images (**b**) and (**c**) illustrate the propagation of these rolls by showing two close-up interferograms of a traveling roll (taken about 1 second apart) as it travels toward the top pole, indicated by the *arrows*, at $Rg = 1.3 \times 10^4$ ($\Delta T = 1.1°C$) and $R_{E_{max}} = 72.6$ ($\Delta V = 5{,}000\,V$).

slight distance from it. The interference fringes near the inner sphere show a slight radial displacement so that the oscillatory perturbations appear as transverse waves traveling in the polar direction. The amplitude of these apparently transverse waves also increases as they travel. At higher ΔV, we see localized roll structures that first appear near the equator, propagate toward the top pole where they disappear simultaneously on both sides of the inner sphere. FIGURE 3a shows a global view of the shadowgraph and interference pattern formed by these rolls. FIGURE 3 also shows several close-up images of the interference pattern of the traveling rolls.

Simultaneous shadowgraph observations, perpendicular to the holographic optic axis, have shown the same flow pattern (i.e., rolls appear, propagate upward, and disappear simultaneously in both orientations). We conclude that these rolls extend around the sphere forming toroidal or spiral roll patterns that travel upward. The traveling rolls always appear as ΔV is increased at all ΔT values possible in this experiment. We have also observed, prior to the onset of the propagating rolls, that increasing ΔV slightly decreases the interference fringe spacing.

We have measured the ΔV values for the onset of convection for various ΔT (at given T_o). The results are shown in FIGURE 4. The values plotted are the ΔV values when the first slow oscillations appear close to the inner sphere. The onset ΔV values, shown in FIGURE 4, were found by fixing the temperature difference, ΔT, and increasing ΔV until the oscillations could be seen in the interference pattern. As can be seen in FIGURE 4, at higher ΔT values ΔV_c for onset decreased. Because the rolls disappear when ΔV is decreased below ΔV_c, with apparently no hysteresis, we conclude that we have observed a supercritical Hopf bifurcation to traveling rolls.

As the rolls propagate upward toward the top pole they produce a periodic signal in the interferogram. We measured this oscillation frequency using a photodiode placed at a given point in the interferogram. We found several regimes where this frequency showed interesting behavior when external parameters were changed. The

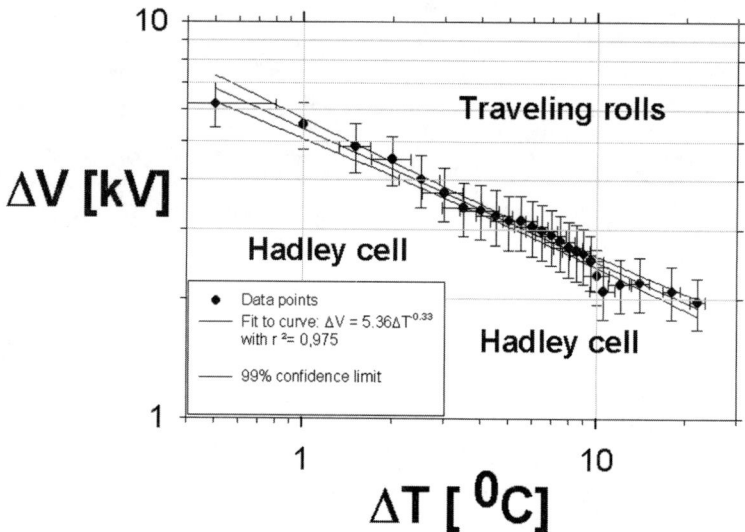

FIGURE 4. Critical voltage amplitudes, ΔV, for the onset of propagating rolls as a function of the temperature difference ΔT (data taken at an ac frequency $f = 300\,\text{Hz}$ and an ambient temperature $T_o = 21.6°\text{C}$). The onset values where found by fixing ΔT and increasing ΔV until oscillating roll structures could be seen in the interference pattern. The error bars show the errors in ΔV and ΔT values. The *middle solid line* is the fit to the curve: $\Delta V \propto \Delta T^{-0.33}$. The lines above and below this line indicate 99% confidence intervals. This interval is where 99% of the data fall in repeated measurements based on the statistics of the given data. The limits are used to estimate the error of the exponent, 0.33 ± 0.03. The correlation coefficient r^2 is used as a goodness of fit parameter.

frequency of the propagating rolls was measured when ΔT was varied, at a fixed value $\Delta V_c = 4\,\text{kV}$ close to the onset of the propagating rolls. The frequency of the traveling rolls was also measured far above the onset of the traveling rolls, when ΔV was varied, at a fixed value for $\Delta T = 22°\text{C}$. In both of these examples the frequency monotonically increases. In the first case the frequency spectra was sharply peaked, indicating a well defined periodic signal, whereas in the latter case, far above the onset of the traveling rolls, the frequency spectra is broad banded and the frequencies measured were the peak frequencies.

We further characterized this flow by measuring the heat transport from the inner sphere. This effect is usually characterized by measuring the Nusselt number (Nu), as a function of ΔT. The Nusselt number for heat transfer from the inner sphere is defined by

$$Nu = \frac{H}{\kappa_s \left[\frac{\partial T}{\partial r}\right]_{r_i}} = \frac{H(1 - r_i/r_o)r_i}{\kappa_s \Delta T}. \tag{5}$$

H is the heat flux transferred between the inner and the outer spheres and κ_s is the heat transfer coefficient for heat conduction, otherwise known as the thermal conductivity. This number is the ratio of the actual heat transfer to the heat transfer that would occur by conduction alone if the fluid remained at rest. Normally, the onset of convection is marked by Nu increasing above unity for a bifurcation from a conductive state. The Hadley cell, however, is present at any $\Delta T > 0$ and prevents a conductive state from occurring in our system. Nevertheless, we measured Nu as a function of ΔT by recording the voltage, V_h, and the current, I_h, to the inner sphere heater. For $\Delta T < 2°\text{C}$ the heat transport stays roughly constant, implying that the Hadley flow is steady so that the combined conduction and convection heat transport remain about the same. For $\Delta T > 2°\text{C}$ heat transport increases, probably mostly due to more appreciable Hadley convection. At higher ΔT, corresponding to the onset of traveling rolls, Nu increases at a greater rate. The change is more evident at higher ΔV; for example, at $\Delta V = 3\,\text{kV}$ there is a small change in the Nu slope at $\Delta T \approx 5°\text{C}$, whereas at $\Delta V = 4\,\text{kV}$ the change in the Nu slope at $\Delta T \approx 3°\text{C}$ is larger. More striking is the behavior of Nu far above onset, where Nu may actually decrease. These decreases appear to be smaller at higher ΔV.

We note that the presence of the Hadley cell complicates the classical method of finding the onset of convection by measuring Nu.[9] Because of this, we also measured a modified Nusselt number Nu^* by calculating, for each ΔT value, the ratio

$$Nu^* = \frac{\text{Heat flux with } \Delta V}{\text{Heat flux without } \Delta V} = \frac{H_{\Delta V}}{H_{\Delta V = 0}}. \tag{6}$$

In this way, we obtain a dimensionless measure of the heat transport in the presence of the high voltage—and the propagating rolls, if they exist—normalized by the heat transport by conduction and Hadley convection when $\Delta V = 0$. The results of Nu^* are shown at $\Delta V = 3\,\text{kV}$ and $\Delta V = 4\,\text{kV}$ in FIGURE 5. Below onset of the propagating rolls Nu^* is unaffected by ΔV and above onset, Nu^* increases from the convective heat transport of the propagating rolls. A more detailed discussion of this experiment can be found in Reference 8.

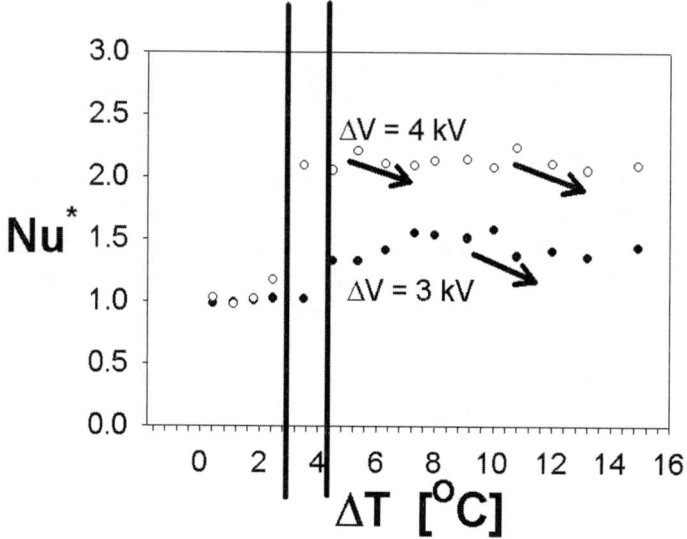

FIGURE 5. Nu^* versus ΔT. The two vertical lines show the onset of traveling rolls and these values are approximately the same as those shown in FIGURE 4. The onsets of the instability clearly corresponds to the values of Nu^* greater than one. The traveling rolls significantly increase the heat transported from the inner sphere (Nu^* increases 30% to 100%) as measured by what it would have been without the traveling rolls. At $\Delta T > 5.5°C$ for $\Delta V = 4kV$, where the rolls first appear disordered is space and time, Nu^* begins to decrease. The appearance of disordered rolls at $\Delta V = 3kV$, $\Delta T > 6.61°C$, correlates with a small increase in Nu^*.

COMPRESSIBLE FLUID

The compressible near-critical fluid in our cell is SF_6 and was chosen for its relatively low critical pressure ($P_c = 37.7$ bars), low critical temperature ($T_c = 45.54°C$), and high breakdown voltage (greater than 100kV/cm). A cylindrical cell was filled with SF_6 to within 10% of the critical density, ρ_c, prior to the parabolic flight and then sealed. More carefully performed ground based experiments used a cell filled to within 0.01% of ρ_c. This precision is obtained by observing the volume fraction as a function of T.[10] The closed cell constrains the fluid to a constant volume and constant mass (a constant average density), so that temperature is the only independent thermodynamic variable.

Previous experiments in near-critical fluids showed a density change when an electric field gradient was applied.[4] The density change was observed using interferometry under a relatively weak electric field gradient. In these experiments the electric field gradient was increased and better characterized.

Inside the critical fluid cell is a spherical capacitor, as described above except for the following differences: the brass inner sphere has an outer diameter 1/4″ and the outer sphere is grounded with ΔV applied to the inner sphere. This capacitor is placed inside the cell, as shown in FIGURE 6, so that it contains critical SF_6 fluid between the electrodes. The outer sphere is positioned inside of the critical fluid cell using two Teflon rings, one that is pressed into the cylindrical cavity of the critical fluid cell and the other is glued to the copper cylinder so that it may be removed later without damaging the sapphire hemispheres. An ac voltage of 5 kV may then be applied to the inner sphere, this being the limit of a high voltage feed through.

The test cell is set up so that white light may shine through the fluid between the inner and outer sphere of the spherical capacitor, as shown in FIGURE 6. An external white light source shines on to a light-diffusing sheet of white paper that has a test pattern of concentric circles and radial lines printed on it. This produces a uniform illumination and well-defined reference positions on the dark grid. The image of the test pattern may then be focused into a CCD camera so that it may serve as the object for the optical system. Other image planes may also be selected by adjusting a 200-mm telephoto lens on the CCD camera. The deflection of the white light by

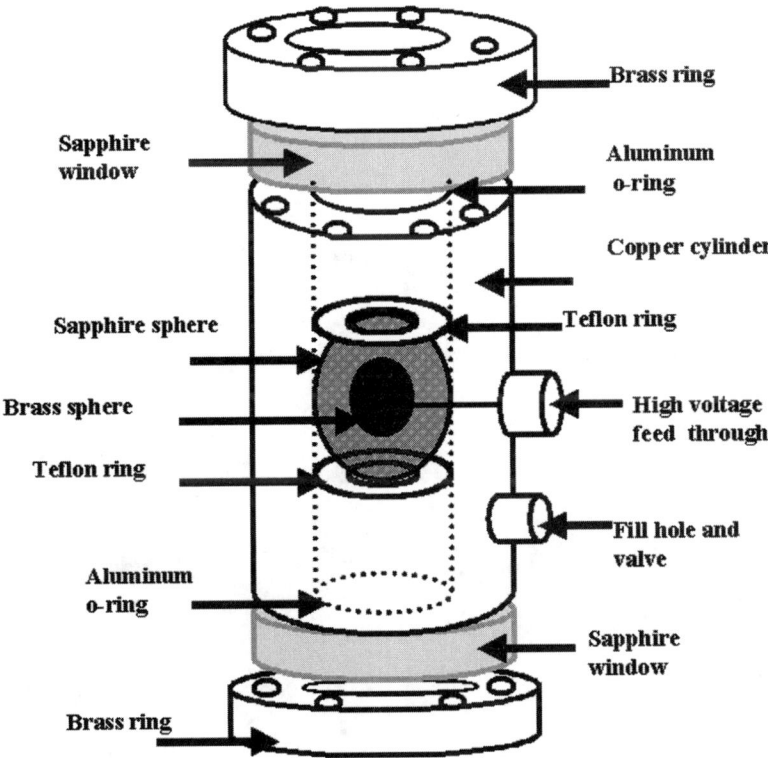

FIGURE 6. Shown is the critical fluid cell that is a sealed cylindrical vessel containing the SF_6. This copper cylinder is 3″ in diameter, 3″ in length, and has a 1.5″ concentric cylindrical hole milled along the cylinder axis. Two 2.0″ diameter sapphire windows at each end of the critical fluid cell are held in place with a brass ring that is bolted to the cylinder ends. Aluminum O-rings, sandwiched between the copper and sapphire window, seal the windows of the critical fluid cell maintaining the SF_6 at the critical density.

refraction distorts this image if there is a density gradient in the fluid. The test pattern is recorded in zero gravity for reference and is then recorded after a large ac voltage is applied to the compressible fluid in the capacitor. The distortion of the test pattern indicates density changes in the fluid.

The critical-fluid cell is placed inside of several thermal shields so that the temperature of the fluid is precisely controlled to be near the critical point. The critical-fluid cell is placed in a temperature-controlled aluminum "can" that is 1/8"–3/16" thick and has multiple foil heaters glued to it to produce a fairly uniform heat flux. Acrylic and Teflon spacers keep the critical fluid cell centered in the can.

The aluminum can is placed in a 6"-cubical Plexiglas box with 1"-thick walls. This box provides thermal isolation and support for the aluminum can. Angle iron pieces support the Plexiglas box in the rack and allow its position to be aligned with optical axis of the CCD camera.[11] The Plexiglas box and the optical axis of the cell

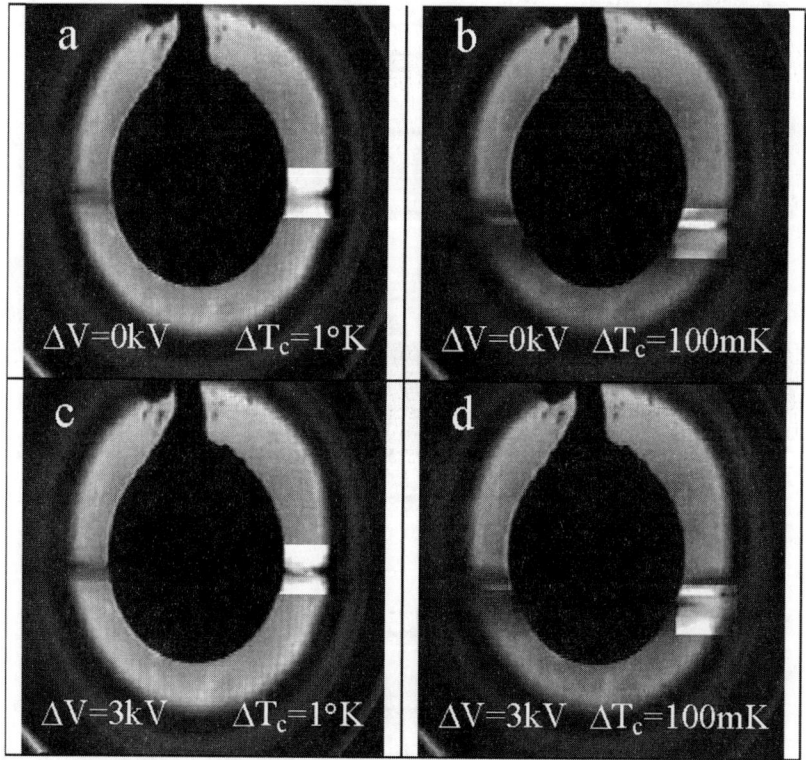

FIGURE 7. The figure shows images of the two-phase near critical fluid at 1 g. The dark horizontal band is the meniscus. The meniscus without an electric field is horizontal (**a** and **b**). The meniscus with an electric field applied (**c** and **d**) deflects upward near the center electrode. The meniscus has been enhanced on right side of the inner sphere in (**c**) and (**d**). The ΔV and $\Delta T_c = T_c - T$ values are given at the bottom of each image. The columns have the same temperatures; the rows have the same voltages.

and CCD camera can be orientated in either the horizontal or vertical directions. This is accomplished by placing the CCD camera on support beams that may be attached to the rack in either orientation while allowing some adjustment of the camera axis. A white light external source outside the Plexiglas box is attached to the rack in either orientation so that it shines onto a test pattern at the entrance to the aluminum can. The test pattern light is detected by a CCD camera placed at the other end of the aluminum can.

At 1 g and far below the critical temperature, a meniscus exists between the liquid and the gas, as is normally the case. The higher density liquid resides below the lower density gas leaving a meniscus that appears as a horizontal line (see FIGURE 7a). As the temperature is increased above T_c, the region near the meniscus becomes a large density gradient. FIGURES 7a, 7b, 8a, and 8b show this process without an electric field applied. Near the critical point a large density gradient forms near the

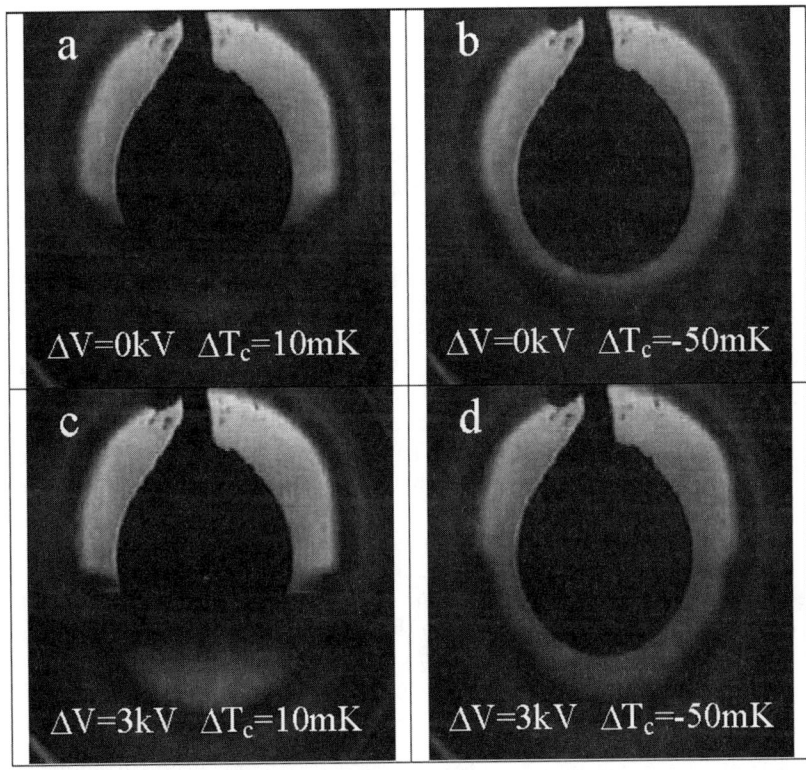

FIGURE 8. Shows images of the fluid just below (**a** and **c**) and just above (**b** and **d**) the critical temperature, T_c. The meniscus height decreases as T is increased so that the dark horizontal band of the meniscus in (**a**) and (**c**) is below the equator. The dark region is larger near T_c from critical opalescence. Above T_c the meniscus develops a density gradient and becomes clearer. Light refraction occurs from the density gradient and the outer sphere. Some differences in the phase distribution and the density distribution can be seen between $\Delta V = 0$ kV (**a** and **b**) and $\Delta V = 3$ kV (**c** and **d**).

meniscus that significantly refracts light. This density gradient forms because of the very high compressibility of the fluid near T_c. The region also appears darker when viewed in the horizontal orientation near T_c (FIGS. 7, 8a, and 8c). The darker region in the image is caused by density fluctuations (critical opalescence) that scatter light out of the image (light extinction or "absorption"). Near T_c, the density gradient, combined with the outer sphere, produces considerable light refraction, as can be seen in FIGURE 8. Slightly above T_c ($-\Delta T_c = T - T_c = 50$ mK in FIGS. 8b and 8d) only the single-phase supercritical fluid with a density gradient is present. In this case, the meniscus has transformed into a density gradient and there is less critical opalescence but still significant light refraction.

Because a near-critical fluid always wets a solid, the contact angle between the center electrode and the meniscus is zero.[12] When the electric field is applied, the interface shape is modified near the center electrode, as shown in FIGURE 7, so that the contact angle appears different. The "apparent" contact angle changes as the cell temperature, T, is increased toward T_c. Because the liquid–gas surface tension, the density difference, and the dielectric constant difference disappear at the critical temperature with universal power laws, the interface shape changes as T_c is approached. Specifically as T_c is approached and the interface softens and the interface away from the center electrode flattens. Very close to the critical point the contact angle could not be seen. The meniscus height also decreases as T is increased because the average fluid density is somewhat below the critical density, ρ_c.

Qualitative results were also observed in the weightlessness of the parabolic flights. FIGURE 9 shows the effect that the electric field has on the phase distribution. As can be seen the electric field attracts the higher density liquid toward the center electrode where the electric flux is higher. Our cell was filled at a density slightly greater than ρ_c because the gas volume fraction decreased as we heated the cell. FIGURE 9a and b show the meniscus somewhat above a center that is approximately at the equator of the inner sphere.

In FIGURE 9c and d the axis is in a vertical orientation so that the horizontal meniscus is not visible at 1 g. This figure also shows the buoyancy force that the center electrode exerts on the system with the lower density gas bubble repelled from the center electrode. The repulsion is evident even in the presence of the large perturbations that the KC-135A produces as it bumps and jerks the cell during the flight.

We also attempted to visualize a density change in the supercritical fluid (above the critical temperature) in the horizontal configuration. In this view, the region that would have been the liquid–gas meniscus becomes a large horizontal density gradient that strongly refracts light. If a spherical density gradient were produced under the action of the central polarization force, we would expect this region to form a circle around the center electrode. In this horizontal configuration we were able to see that the about 20 seconds of weightlessness is clearly not enough time in induce a spherical density gradient. This is because the horizontal density gradient normally present at 1 g is not significantly changed during about 40 cycles of a parabolic flight. The heavy fluid remains at the bottom of the cell. In fact, about 20 seconds of weightlessness at the top of each cycle is followed by about 20 seconds of 2 g weight that reinforces the usual horizontal stratification. FIGURE 10 shows the influence of the electric field on the horizontal density gradient in weightlessness. We see clear evidence that the central force significantly influenced the density distribution. When the fluid was just above the critical temperature, the refracting region thickens during

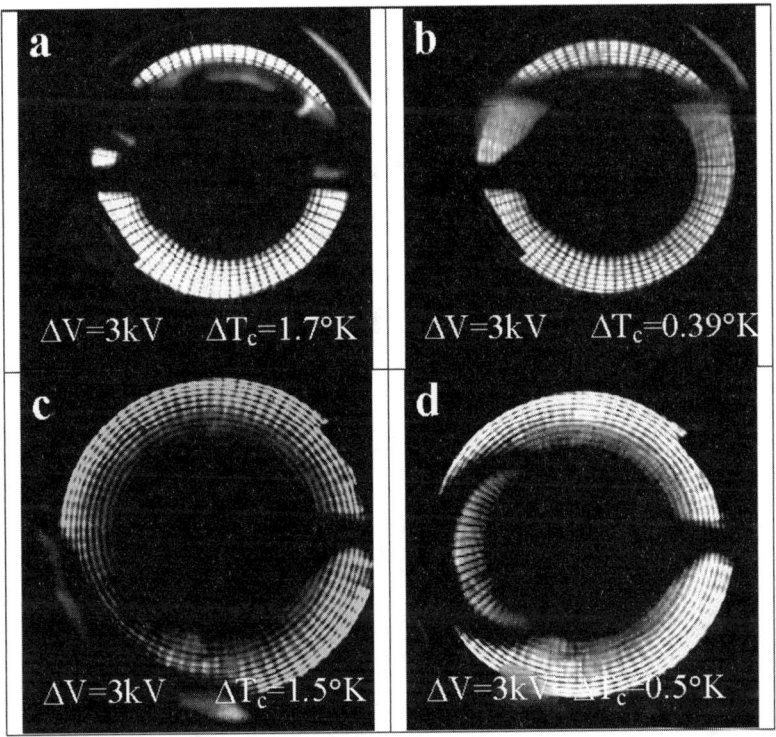

FIGURE 9. Shows the two-phase, near critical fluid when viewed with the optical axis horizontal (**a** and **b**) and vertical (**c** and **d**).

weightlessness, as is shown in FIGURE 10a. Because our cell was filled to an average density that is above the critical density the refracting region that corresponds to the large density gradient was above the equatorial plane of the inner sphere. In order to observe the effect in the very limited time available in weightlessness we switched voltage on and off producing an approximate square-wave voltage with a period of two seconds. When the 3-kV ac potential was applied, this refracting region was distorted toward the center electrode. This is what should happen—that is, the fluid above the center electrode that is below the critical density should become denser. In principle the fluid elsewhere around the sphere should also become denser. The highly compressible near-critical fluid only exists at the critical density near this refracting region and so the density change is restricted to the horizontal region. We also note that even though our cell was not filled exactly at the average critical density, the existence of a horizontal density gradient assures that there will be fluid at the critical density somewhere in the fluid. Fortunately this critical density layer appeared near the center electrode so that we could see the effect. FIGURE 10 shows a roughly synchronous response of the density gradient to the oscillating electric field strength when we applied a two-second square wave (one second on and one second off).

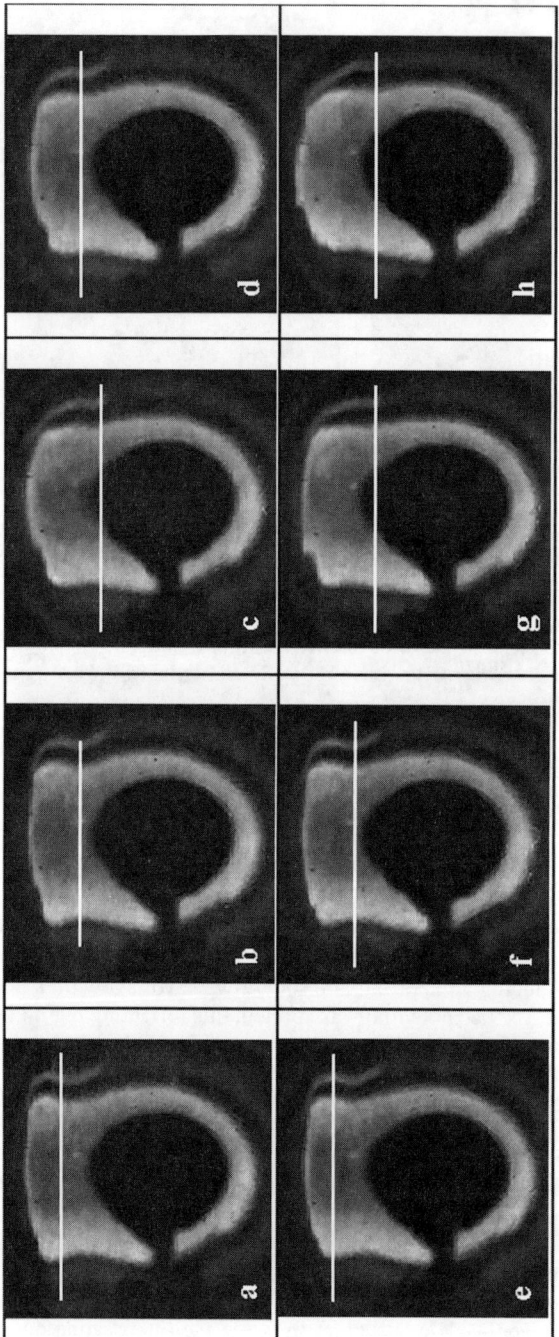

FIGURE 10. Shows the response of the density gradient to the central force when an electric field is applied. In this example $\Delta V = 3\,\text{kV}$ and $\Delta T_c = -100\,\text{mK}$. The images were obtained at 0.5 second intervals with a voltage of 3 kV applied for one second and then released for one second. The region of large vertical density change is also where the image is most distorted by refraction. A *horizontal line* is drawn in each image to mark this region. The line changes height as the field is repeatedly applied and removed.

We also attempted to visualize the density change from the high voltage on supercritical fluid in the vertical configuration. We saw, however, only insignificant density changes in these experiments using the grid image diagnostic. We believe that this is because the density is horizontally stratified with large changes occurring only above the center of the center electrode. Light refraction from density changes could not be seen because the center electrode is blocking the light, or equivalently the strong density changes only occur in the shadow of the center electrode in this configuration.

CONCLUSION

The fluid in the first system had two forces applied to it when heated, one large scale and one small scale. A large-scale axial force from normal buoyancy always drives a Hadley cell in the fluid. A smaller scale radial force from dielectric buoyancy drives the traveling rolls at all ΔT when sufficiently large ΔV is applied. The Hadley cell drives the traveling rolls so they travel upward, whereas the small-scale flow pumps hot fluid into the Hadley cell, producing a large increase in heat transfer. There are similarities between this system and large-scale geophysical flows. This system also has similarities with a planetary liquid outer core subjected to a strong external gravitational field. The radial force in spherical geometry, together with a mixing convective layer near the inner sphere also exhibits similarities to the atmospheric boundary layer. Although this system has significantly different geometry from most atmospheres, a planet may have a thin unstable layer near the planetary surface surrounded by a thicker stable layer, as in the case of the earth. Our system also has an unstable layer close to the surface surrounded by a stable flow in the outer layer. In fact the Earth also has a large scale Hadley cell, driven by the large-scale horizontal temperature gradient from the non-uniform heating of the sun. The large heat transport increase from the inner sphere shown in our experiment shows that this surface convection can be quite important in such a system; that is, a highly conductive solid in contact with less conductive fluid still looses significant heat when convection is present. Another interesting result is the decrease in heat transport at larger ΔT.

By varying the ambient temperature of the fluid and the voltage applied to the capacitor, we observed the effects that these changes had on a near-critical fluid. Although we had hoped to observe and verify that a spherical density gradient is produced in the fluid, the actual conditions in flight did not allow this to happen. We did, however, see clear evidence of the existence of the influence of this force on the fluid. Below the critical temperature the liquid was clearly attracted by the polarization force whereas the gas was repelled. This suggests that geophysical analogies using two-phase flows in microgravity may be possible when this system in heated and rotated. Above the critical temperature the horizontal near-critical fluid layer is clearly modified by the attraction of higher density fluid toward the inner sphere. This suggests that geophysical analogies using supercritical fluid near the liquid–gas critical point in microgravity is possible when this system in heated and rotated. We can now be confident that geophysical flow analogies can be made using near-critical fluid when the spherical capacitor is rotated and heated.

We believe that the convective effect clearly demonstrated here has a much greater applicability than geophysical similarity. The enhanced heat transport could be used in many engineering applications, especially in space vehicles where normal buoyancy is absent; for example, increasing heat transport in forced convection. The convection could also be used to manage fluid in two-phase systems and compressible supercritical fluid chemical reactors. There is, however, a much greater potential for application in the emerging field of microfluidic devices. We showed that the polarization force greatly increases at smaller scales in spherical geometry. This increase is also true in any diverging geometry. Already the contact angle effects shown above have been used to manipulate and transport nanoliter scale water droplets in electrode arrays.[13] It is likely that this dielectric convection could also be used in micro-fluidic devices.

ACKNOWLEDGMENTS

We gratefully acknowledge the support of NASA through NASA-OBPS grants NAG3-1915 and NAG3-2447.

REFERENCES

1. LANDAU, L.D. & E.M. LIFSHITZ. 1982. Statistical Physics, Part 1, 3rd edit. Pergamon Press, New York. (See also L.D. Landau & E. M. Lifshitz. 1960. Electrodynamics of Continuous Media. Addison-Wesley, Reading.)
2. HART, J.E., G.A. GLATZMAIER & J. TOOMRE. 1986. Space-laboratory and numerical simulations of thermal convection in a rotating hemispherical shell with radial gravity. J. Fluid Mech. **173:** 519–529.
3. HART, J.E. 1976. Studies of Earth simulation experiments. NASA-Contractor Report, NASA CR-2753.
4. HEGSETH, J., L. GARCIA & M.K. AMARA. 1999. A compressible geophysical flow experiment. Presented at the Fourth Microgravity Fluid Physics Conference, Cleveland, OH, August 9. Available at: <http://www.ncmr.org>.
5. STRATTON, J.A. 1941. Electromagnetic Theory. McGraw-Hill, New York.
6. VEST, C.M. 1979. Holographic Interferometry. Wiley, NewYork.
7. GUYOT, G. 1998. Physics of the environment and climate. Wiley, NewYork.
8. AMARA, M.K. & J. HEGSETH. 2001. Convection in a spherical capacitor. J. Fluid Mech. **450:** 297–317.
9. AHLERS, G., BERGE, L.I. & D. CANNELL. 1993. Thermal convection in the presence of a first-order phase change. Phys. Rev. Lett. **70:** 2399–2402. (See also G. Ahlers, D. Cannell, L. I. Berge & S. Sakurai. 1994. Phys. Rev. E **49:** 545–555.)
10. MORTEAU, C., M. SALZMAN, Y. GARRABOS & D. BEYSENS. 1997. In Proceedings of the 2nd European Symposium on Fluids in Space. A. Viviani, Ed.: 327–334. Congressi srl, Rome.
11. HEGSETH, J. & M.K. AMARA. 1999. Geophysical flow experiment, test equipment data package for experiments on the NASA KC-135A Aircraft. NASA-Lewis, Cleveland.
12. GARRABOS, Y., C. CHABOT, J. HEGSETH, et al. 2001. Gas spreads on a heated wall wetted by liquid. Phys. Rev. E **64:** 51609–51619.
13. JONES, T.B., M. GUNJI, M. WASHIZU & M.J. FELDAMN. 2001. Dielectric liquid actuation and nanodroplet formation. J. Appl. Phys. **89:** 1441–1448.

Transport from Higher Order *g*-Jitter Effects

ROBERT J. NAUMANN

Center for Microgravity and Materials Science, University of Alabama in Huntsville, Huntsville, Alabama, USA

ABSTRACT: Large complex spacecraft, such as the International Space Station (ISS), will have a rich spectrum of vibrational modes that will be excited by crew activity as well as by on-board mechanical systems. The response of various experiments to this vibratory environment is not completely understood and has been a subject of concern to the users community. Since these vibrations arise for the most part from internal forces, the net acceleration must time-average to zero. Steady state periodic accelerations applied to a fluid in a completely filled container with an imposed density gradient will drive first-order flows that time-average to zero. Likewise, the resulting first-order thermal and solutal fluctuations time-average to zero. Over the range of frequencies in the vibrational environment expected on the ISS, the velocity oscillations tend to be nearly 90° out of phase with the thermal and/or solutal fluctuations; consequently, little net transport occurs from the first-order effects. We must, therefore, examine possible higher-order effects that can lead to significant net transport. Second-order flows with non-zero time averages arise from the non-linear terms in the flow equations and from incomplete cancellation of first-order flows if the density gradient changes with time, or if the periodic acceleration has both axial and transverse components relative to the imposed density gradient. The time-average of such flows can be expressed in terms of the sum of the square of each periodic acceleration, weighted by the appropriate function of the frequency, taken over all frequencies; or as the weighted integral over the power spectral distribution (PSD) of the frequency spectrum. Approximate analytical solutions for these second-order flows, which agree closely with numerical computations, have been found using a perturbation analysis. These analytical models are then used to predict the effects of the anticipated vibratory environment on various classes of experiments planned for the ISS. Even though these second-order flows are much smaller than the first-order flows, it is shown that they can produce significant transport.

KEYWORDS: transport; higher order *g*-jitter effects

NOMENCLATURE:
a	half-width, $W/2$
C	concentration
D	solutal diffusivity
f	applied vibrational frequency
g	acceleration
Gr_y	Grashof number, $g_y(\beta \nabla T + \gamma \nabla C)a^4/\nu^2$
Gs	Gershuni number, $(g_y \beta \Delta T L/\nu \omega)^2 Pr/2$
	(also called the vibrational Rayleigh number)
L	length of the flow chamber
Nu	modified Nusselt number equal to convective thermal transport/conductive thermal transport

Address for correspondence: Robert J. Naumann, Center for Microgravity and Materials Science, University of Alabama in Huntsville, Huntsville, AL 35899, USA.

naumannr@email.uah.edu

Pr	Prandtl number, ν/κ,
Ra_x	thermal Rayleigh number, $g_x \beta \nabla T a^4 / \nu\kappa$
Ra_x^S	solutal Rayleigh number, $g_x \gamma \nabla C a^4 / \nu D$
Sc	Schmidt number, ν/D
Sh	modified Sherwood number equal to convective solutal transport/diffusional solutal transport
T	temperature
u	x-component of core velocity
x	position along L
y	position along W
U	a dimensionless velocity, ua/ν
W	width of the flow chamber
Greek Symbols	
β	coefficient of thermal expansion
γ	coefficient of solutal expansion
η	a dimensionless coordinate, y/a
Θ	a dimensionless temperature perturbation, $\delta T/(a\nabla T)$
ν	kinematic viscosity
ρ	density
χ	a dimensionless compositional perturbation, $\delta C/(a\nabla C)$
ψ	stream function
ω	angular frequency, $2\pi f$
Ω	a dimensionless frequency, $\omega a^2 / \nu$
Sub- and Superscripts	
x, y	components of acceleration
tilde	amplitude of periodic term
net	integrated time average

INTRODUCTION

Experiments flying on orbiting spacecraft are subject to low level, quasi-steady accelerations due to atmospheric drag and gravity gradient effects, as well as the spectrum of periodic accelerations known as g-jitter. If the intent of the experiment is to examine a process in the virtual absence of gravity driven convection, it must be designed to tolerate both the quasi-steady acceleration and the g-jitter spectrum. Since the first-order effects of the periodic accelerations tend to time-average to zero, an experiment that can tolerate the quasi-steady accelerations will generally not be affected by the first-order effects of periodic accelerations, even though the amplitudes of these periodic accelerations may be orders of magnitude larger than the quasi-steady acceleration. We must then look for various second-order effects that can result in non-zero time average flows and examine their effects to determine if they are comparable or larger than the effects of the quasi-steady acceleration.

QUASI-STEADY FLOWS AND TRANSPORT

An analytical solution for the first-order core flow resulting from a steady acceleration with a y-component that is normal to the axial direction was developed for a two-dimensional slot whose length is very much greater than its width.[1] A uniform density gradient is assumed to exist along the axis resulting from either a temperature gradient or a solutal gradient or both. Under these circumstances, the momentum equation is decoupled from the transport equations in the first-order. This allows the first-order flow to be obtained directly from the momentum equation. This solution for the flow can then be plugged into the transport equations to obtain the first-order perturbations to the thermal and solutal fields. These perturbations can then be multiplied by the flow and integrated over the cross section to obtain the net transport.

The first-order core flow is given by

$$U(\eta) = -\frac{Gr_y}{6}(\eta^3 - \eta).$$

The perturbation to the temperature field in the core region is given by

$$\Theta(\eta) = -\frac{Gr_y Pr}{360}(3\eta^5 - 10\eta^3 + 15\eta)$$

for adiabatic walls, and by

$$\Theta(\eta) = -\frac{Gr_y Pr}{360}(3\eta^5 - 10\eta^3 + 7\eta)$$

for conductive walls. The perturbation to the solutal field is the same as the perturbation to the thermal field with adiabatic walls if the Schmidt number Sc is substituted for the Prandtl number Pr. The net transport is expressed in terms of a modified Nusselt (or Sherwood) number defined as the ratio of convective transport to diffusive transport. These ratios are obtained by multiplying the flow by the resulting perturbation and integrating across the chamber width to give

$$Nu = \frac{1}{4725} Gr_y^2 Pr^2$$

for conductive walls,

$$Nu = \frac{2}{2835} Gr_y^2 Pr^2$$

for adiabatic walls, and

$$Sh = \frac{2}{2835} Gr_y^2 Sc^2$$

for solutal transport.

The influence of the axial component of acceleration was omitted from the above for simplicity. The axial component itself does not drive flows (unless the critical Rayleigh number is exceeded), but will either accelerate or retard the flows and transport expressed above, depending on whether the acceleration is along or opposed to the density gradient. A stability analysis shows that the critical Rayleigh number (thermal plus solutal) is 31.284 for the case of impervious walls and is 97.406 for conductive walls.[2,3]

The above analysis is based on the limiting case of an infinite aspect ratio (length to width ratio) A. However, numerical computations confirm that the approximation is remarkably good for the core flow for aspect ratios as low as $A = 2$ and can even be extended to $A = 1$ by dividing the both flow and its resulting perturbation by 2. A similar analysis for the core flow in a high aspect ratio cylinder obtained a maximum flow velocity that is 3/4 the velocity in the rectangular slot, and the maximum perturbation to the temperature/solutal field was 0.47 of the maximum value for the rectangular slot.[4] Therefore, it can safely be assumed that the use of the rectangular slot for a model will always over predict the transport in actual systems, which are generally in the form of cylinders with aspect ratios on the order of 1.

In most microgravity experiments it is desirable to operate under diffusive limited transport conditions, which dictates that the experiment must be designed such that Nu and Sh are on the order of 0.01. This requires that the combined thermal and solutal Rayleigh number for the y-component of the quasi-steady acceleration must be less than 3.76 using the above model.

FIRST-ORDER PERIODIC FLOWS AND TRANSPORT

The analytic solution described above was extended to periodic accelerations.[2] In the high frequency limit, $\Omega \gg \pi^4$ (which corresponds to frequencies of about 1 Hz and greater for most experiment configurations of interest), the first-order flow as well as the thermal and solutal fluctuations reach their maximum near the container walls. The maximum amplitude of the flow velocity is given by $\tilde{U}_{max} = \tilde{G}r_y/\Omega$ and the maximum amplitude of the temperature and solutal fluctuations are given by $\tilde{\Theta}_{max} = \tilde{\chi}_{max} = \tilde{G}r_y/\Omega^2$, where the tilde represents the amplitude of the term. Note that the Pr or Sc does not appear in the equations for the thermal and solutal perturbations in the high frequency limit. Also, the thermal profiles for the case of adiabatic walls is virtually the same as for conductive walls because at high frequencies, the isotherms (or iso-concentrates) simply do not have time to move very far from their equilibrium position.

The net transport from these first-order flows is found by integrating the product of the velocity and temperature (or solutal) fluctuation over the cross section and then taking a time average over a cycle. Both the flow velocity and the fluctuations in the thermal and solutal field time-average to zero. The phase between the velocity fluctuation and the resulting temperature and solutal fluctuations approaches 90° as the frequency gets large; however, at finite frequencies, a small net transport results given by

$$Nu = Sh \approx \frac{\tilde{G}r_y^2}{2^{3/2}\Omega^{7/2}}$$

for Pr or Sc very much greater than 1.

The first-order solutions are valid so long as the perturbations to the density field are small. Since Ω is large, these first-order solutions may be extended to much larger values of $\tilde{G}r_y$ than would be permitted under steady accelerations. Since the first-order terms are all linear, the effects of multiple frequencies are simply additive. This means that the maximum velocity amplitude can be expressed as

$$\tilde{U}_{max} = \sum_i \tilde{G}r_y(g_i)/\Omega(f_i)$$

and

$$\tilde{\Theta}_{max} = \tilde{\chi}_{max} = \sum_i \tilde{G}r_y(g_i)/\Omega(f_i)^2,$$

where the sum is taken over each individual frequency component suitably weighted by the frequency. Since the accelerations are applied at random times and with random phase, it is reasonable to assume that the probability distribution of velocities can be represented by a Gaussian with a root-mean-square given by

$$\tilde{U}_{max} = \tilde{G}r_y(1\mu g)/\Omega(1\,\text{Hz})\left[\int\frac{PSD(f)}{f^2}df\right]^{1/2},$$

where $\tilde{G}r_y(1\mu g)$ is the Grashof number calculated at $1\mu g$, $\Omega(1\,\text{Hz})$ is the dimensionless frequency computed at 1 Hz, and $PSD(f)$ is the power spectral density of the g-jitter spectrum in μg^2/Hz. Similarly, the rms of the distortion of the thermal (or solutal) field is

$$\tilde{\Omega}_{rms} = \tilde{\chi}_{rms} = \tilde{G}r_y(1\mu g)/\Omega(1\,\text{Hz})^2\left[\int\frac{PSD(f)}{f^4}df\right]^{1/2},$$

and the net transport can be expressed as

$$Nu = Sh = \tilde{G}r(1\mu g)^2/\Omega(1\,\text{Hz}_i)^{7/2}\int\frac{PSD(f)}{f^{7/2}}df.$$

It is shown later that these first-order effects are negligible for virtually any experiment that can tolerate the quasi-steady acceleration of the ISS, even under the most pessimistic assumptions for the g-jitter environment.

EFFECTS FROM THE NON-LINEAR TERMS IN THE NAVIER-STOKES EQUATIONS

Second-order flows can arise from the second-order coupling between the first-order flows and the perturbations they cause to the thermal and compositional fields in the transport equations as well as from the non-linear advection terms in the momentum equation. Although these various first-order terms fluctuate periodically with a zero time average, their products have a non-zero time average that gives rise to small, but continuous second-order flows. Even though such flows are much smaller than the first-order oscillating flows, their persistence can cause significant heat and mass transport, just as smaller flows from the quasi-steady acceleration can produce much more transport than much larger first-order periodic flows.

Such second-order flows have been studied extensively in two-dimensional rectangular cavities.[3,5–8] In 1981 Kamotani, Prasad, and Ostrach approached this problem by expanding the solutions to the Navier-Stokes (N-S) equations into first and second-order oscillatory terms.[5] The first-order equations were linearized; that is, the advective momentum terms were omitted and a constant density gradient was assumed. second-order corrections were then obtained by time-averaging the complete NS equations with the first-order terms. In a two-dimensional square with differentially heated ends and adiabatic sides, for $Pr = 0.01$, which emphasizes the second-order momentum terms, they found a time average flow with zero net stream function in the form of eight counter rotating cells distributed with mirror symmetry about the 0, $\pi/4$, and $\pi/2$ axes. The rotation was such that fluid was swept from the hot and cold ends of the cell along the center line toward the middle of the cell and then out along the diagonals, as shown in FIGURE 1A. For $Pr = 1$, the cells near the hot and cold ends had weakened somewhat, but still persisted as shown in FIGURE 1B.

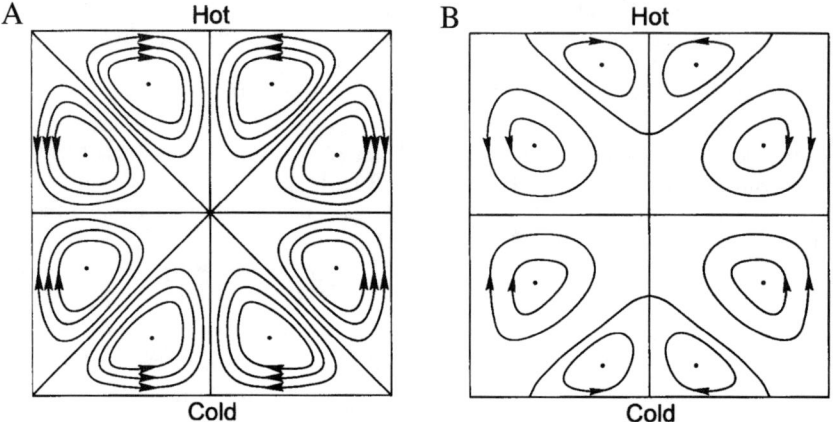

FIGURE 1. Streamlines computed by Kamotani *et al.* for $Pr = 0.01$ (**A**) and for $Pr = 1$ (**B**) for adiabatic walls. Note the direction of flow is away from the hot and cold faces.

Gershuni developed an alternative method for time averaging the Navier-Stokes equations in the high-frequency limit in order to obtain the time-average flows directly.[3] The convective to diffusive transport is characterized by what Gershuni terms a vibrational Rayleigh number that is proportional to $Gr^2 Pr/\Omega^2$. For Prandtl number $Pr = 1$ and low values of Gershuni's vibrational Rayleigh number, the time average flows were found to be in the form of four countercirculating eddies with no

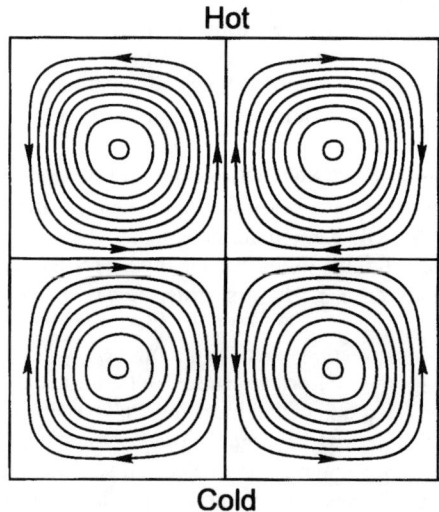

FIGURE 2. Streamline computed from Gershuni's averaging method for $Pr = 1$ and adiabatic walls. Note that the flow is from the center toward the hot and cold faces. Streamlines for $Pr = 0.01$ are the same.

net stream function, located near the corners of the chamber, as shown in FIGURE 2. The rotation of the eddies is such that fluid is swept from the middle of the chamber and projected along the center line toward the hot and cold ends, where it is circulated along the hot and cold faces and along the sides to be returned to the middle of the chamber. The result is to carry the isotherms along the center line closer to both the hot and cold ends. At higher values of Gershuni's number, a bifurcation occurs in which two of the cells on one of the diagonals merge to form an asymmetrical global flow pattern with a non-zero stream function. The result is a significant increase in the Nusselt number signifying mixing of the fluid on a global scale.

Monti has checked Gershuni's time-averaging method by direct integration and finds that the time average method gives the proper value for the stream function for $Pr = 1$ provided $\Omega > 600$.[8] For the case of adiabatic sides, he also finds that the critical value for Gershuni's number for the bifurcation to occur ranges from 8,100 to 8,300 as the Pr is varied from 0.1 to 1,000.

There seems to be some discrepancy between the analysis carried out by Kamotani *et al.* and the results obtained by integrating Gershuni's time averaged equations for the $Pr = 1$ case. In Kamotani's result, there was still an octapole flow at $Pr = 1$ with fluid moving away from the hot and cold walls, whereas Gershuni's result finds a quadrapole flow with fluid moving toward the hot and cold walls. Integrating Gershuni's time averaged equations for $Pr = 0.01$ at small values of Gershuni's number, the flow remains in a quadrapole configuration.

Even though the second-order flows are nonlinear and, therefore, are not directly additive, their time averages are additive. The effective value for Gershuni's number for a multispectral distribution would be $\tilde{\imath} s_{\text{eff}} = \tilde{G} r_y (1 \mu g)^2 \int PSD(f)/\Omega(f)^2 df$

INCOMPLETE CANCELLATION EFFECTS FROM AXIAL ACCELERATIONS

Another form of second-order flow with a non-zero time average was first recognized by Alexander who showed numerically that such a flow could result if the periodic acceleration had both a transverse and a longitudinal component.[9] This comes about because, during half of a cycle, the axial flow encounters a destabilizing gradient that increases the flow, then during the second half of the cycle, it encounters a stabilizing gradient that decreases the flow. The result is an oscillating flow, similar to the flows produced by transverse accelerations, except there will also be a non-zero time average, which in the high-frequency average, is given by

$$U_{\text{net}} = \frac{\tilde{G}r_y \tilde{R}a_x}{12\Omega^2 Pr}(\eta^3 - \eta).$$

This expression holds for thermal and/or solutal convection which either adiabatic or conducting walls.[10] These continuous flows drive a net distortion in the thermal and solutal fields similar to the case of quasi-steady accelerations. For dilute systems, the compositional distortion is given by

$$\chi_{\text{net}} = -\frac{\tilde{G}r_y \tilde{R}a_x Sc}{720\Omega^2 Pr}(3\eta^5 - 10\eta^3 + 15\eta).$$

For non-dilute systems, this becomes

$$\chi_{net} = -\frac{\tilde{Gr}_y \tilde{Ra}_x^S}{720\Omega^2}(3\eta^5 - 10\eta^3 + 15\eta).$$

The net transport for dilute systems is given by

$$Sh = \frac{\tilde{Gr}_y^2 \tilde{Ra}_x^2 Sc^2}{5670\Omega^4 Pr^2},$$

and for non-dilute systems by

$$Sh = \frac{\tilde{Gr}_y^2 \tilde{Ra}_x^{s^2}}{5670\Omega^4}.$$

The thermal transport in a system with adiabatic walls can be obtained by substituting Pr for Sc in the above expressions. For thermal transport in a system with conducting walls, the constant 5,670 becomes 18,900.

Note that the direction of the flow depends on the signs of the x and y components. For example, if the acceleration is applied at 45° relative to the axial direction, both components are positive; however, if the acceleration is applied at 135°, the flow would be reversed. Summing over an isotropic distributions of accelerations cancels the net flow, but any asymmetry in the distribution produces a net flow and transport. Therefore, one must be able to determine the directional, as well as the spectral, distribution of the g-jitter spectrum in order to determine its effect.

APPLICATIONS

The International Space Station (ISS) will be subjected to two types of periodic accelerations, a low-level, low-frequency acceleration with a period of 90 minutes from the diurnal atmospheric bulge, and a spectrum of high-frequency accelerations with frequencies ranging from subHz to tens of Hz from the vehicular vibration modes excited by crew activities, operating machinery, antenna and solar cell motion, and so forth. The low-frequency effects are considered quasi-steady accelerations. The effects of these accelerations are considered for three different classes of experiments: (1) a simple fluid flow heat transfer experiment similar to the JUST-SAP experiment flown on a recent space shuttle flight;[11,12] (2) a directional solidification of a dilute system; and (3) a directional solidification of a nondilute or alloy-type system. Estimated parameters for a typical experiment in these classes are listed in TABLE 1.

The actual vibratory environment of the International Space Station is not yet known, but modal analyses have been carried out[13] to give an estimated rms value for the acceleration in each of 10 intervals per log decade of frequency. From these estimates a preliminary power spectral density (PSD) was inferred and weighted integrals were evaluated to obtain $PSD(f)/f^2 df = 6.64\times10^8\,\mu g^2/Hz^2$, $PSD(f)/f^4 df = 1.66\times10^8\,\mu g^2/Hz^4$, $PSD(f)/f^{7/2} df = 3.57\times10^8\,\mu g^2/Hz^{7/2}$.

The quasi-steady acceleration was assumed to be 1 μg acting perpendicularly to the experiment axis. The full spectrum of periodic accelerations was also assumed to be applied either perpendicularly to the axis or at 45° to the axis, in the case of both axial and transverse accelerations. Obviously this represents a worst case situation since the PSD is the vector sum of the acceleration components along all three

TABLE 1. Experimental conditions

Parameter	Case 1	Case 2	Case 3
a (cm)	2.5	0.8	0.65
L (cm)	10	8	8
$-\beta$ (1/°C)	4.0×10^{-4}	2.5×10^{-4}	2.5×10^{-4}
ΔT (°C)	40	100	200
γ (1/At%)	—	0.002	0.002
ΔC (At%)	—	1	50
ν (cm^2/sec)	0.01	0.003	0.003
Pr	7	0.01	0.01
Sc	—	30	30

axes. Also, the assumption that all of the periodic accelerations are applied at 45° to the axis is also unrealistic, but is done to illustrate the importance that a highly anisotropic g-jitter spectrum can have as well as the necessity to measure both the directionality as well as the frequency distribution of this spectrum.

The experimental cases were configured such that convection from a transverse quasi steady acceleration of 1 μg produced about 1% of the diffusive transport as may be seen in TABLE 2. Even under these circumstances, the thermal distortion (δT_{max}) in Case 1 is easily measurable, which indicates the sensitivity of such a flow cell for measuring quasi-steady accelerations.

The dimensionless frequency Ω (1 Hz) is $O(10^3) \gg \pi^4$ in all cases, which allows the use of the high frequency approximation. It may be seen in TABLE 3 that the first-order effects produce negligible transport in all cases. The rms velocities can be quite large, but they time average to zero. The thermal and compositional fluctuations would be virtually undetectable.

Gershuni's number is much less than the critical value for bifurcations to occur in square container in all cases. The critical values for $L > W$ have not been determined but since the computed values are so much less than critical values for the square container, it is reasonable to assume that no global flows will occur from the second-order effects of the type considered by Gershuni and others.

TABLE 2. Effects from quasi-steady state accelerations

Parameter	Case 1	Case 2	Case 3
Gr_y (1 μg)	−0.612	−0.128	0.121
v_{max} (μm/sec)	1.57	0.308	0.360
δT_{max} (°C)	0.953	2.8×10^{-4}	4.4×10^{-4}
δC_{max} (At%)	—	8.5×10^{-3}	0.329
Nu	0.013	1.2×10^{-9}	1.0×10^{-9}
Sh	—	0.010	0.0094

TABLE 3. Effects from first-order flows

Parameter	Case 1	Case 2	Case 3
Gr_y (1 µg)	−0.612	−0.128	0.121
Ω (1 Hz)	3,930	1,340	885
v_{rms} (µm/sec)	160	92.4	160
δT_{rms} (°C)	5.1×10^{-3}	9.2×10^{-3}	0.032
δC_{rms} (At%)	—	9.2×10^{-5}	8.1×10^{-3}
Nu	1.2×10^{-5}	2.3×10^{-5}	9.0×10^{-5}
Sh	—	2.3×10^{-5}	9.0×10^{-5}

As shown in TABLE 4, the flows and transport from combined axial and transverse vibrations can be quite significant if the is a large anisotrophy in the directional distribution. Such a situation could occur if, for example, there were more modes of vibration along one axis than the others (a likely prospect) and the experiment was mounted diagonally to this axis.

OTHER POSSIBLE EFFECTS

An interesting effect was found in analyzing the effect of the start-up transient of the periodic acceleration.[2] Little additional transport during the start-up transient occurs if the acceleration is applied as a cosine function. The flow velocity acquired during the initial quarter-cycle of positive acceleration is quenched and reversed during the following half-cycle of negative acceleration. As a result, the velocity oscillates about zero as it builds up to its steady-state amplitude. However, starting the acceleration with a sine function makes a dramatic difference. Now the initial flow build-up during the initial half-cycle of acceleration is brought back to zero

TABLE 4. Effects from second-order flows

Parameter	Case 1	Case 2	Case 3
Gr_y (1 µg)	−0.443	−0.091	0.086
Ra_x (1 µg)	−3.03	-9.8×10^{-4}	-8.6×10^{-4}
Ra_x^s (1 µg)	—	2.37	5.54
Ω (1 Hz)	3,930	1,340	885
Gs	113	0.061	375
V_{net} (µm/sec)	10.36	3.97	9.26
δT_{net} (°C)	10.0	3.6×10^{-3}	0.011
δC_{net} (At%)	—	0.11	16.9
Nu	0.564	1.9×10^{-7}	6.9×10^{-7}
Sh	—	1.73	24.8

during the opposite half-cycle, but is not reversed. Consequently, the net velocity is positive until the transient dies out. Thus, there is a significant net displacement of fluid, which carries the isotherms and isoconcentrates with it. The resulting initial disturbance in the concentration field decays very slowly and, even though the transport from the periodic accelerations becomes small after the transient in the periodic flow velocity dies out, the lingering distortion in the thermal and concentration fields will be seen as radial segregation in a solidification experiment. This effect was shown rather dramatically in the numerical simulation of periodic flows by Alexander et al.[9]

How these various start-up transients might effect actual flight experiments is not clear. Since periodic accelerations may occur with random phases, one would expect to see a mixture of the two extremes. Perhaps of greater consequence is the impulse imparted when a crew member pushes off from one wall and soars to another. Flow and a displacement of fluid will result until a compensating impulse is received when the other wall is reached. The net transport depends on the time between compensating impulses.

All of the above analyses assume that the imposed density gradient remained constant in time (except for the perturbations resulting from the flows). An important exception to this model would be experiments involving a diffusion couple. In this case the density gradient would be extremely large when the material is first melted and would diminish as the inverse square root of the time. This time varying density gradient would lead to incomplete cancellation of the first-order periodic flows which would result in increased transport, especially during the initial melting. This type of effect could possibly explain the lower diffusion coefficients measured by Smith when using the vibration isolation on MIR.[14]

FLIGHT RESULTS

There have been virtually no flight experiments to explore the effects of g-jitter. The JUSTSAP experiment flown on STS-95 consisted of an instrumented flow cell designed to measure the integrated effect of the acceleration environment on the fluid in a differentially heated cylinder.[12,13] During the times the Shuttle was held in Earth-fixed orientations, the differential temperatures in the core region were consistent with the flows predicted from pure gravity gradient accelerations. No significant difference could be detected in the core differential temperatures between sleep and active periods that could be ascribed either to start-up transient effects or to periodic accelerations with axial and transverse components. An attempt was made to see if a change in the thermal gradient at the two ends of the cylinder could be seen during periods of high activity. From Gershuni's analysis, the flows should carry the isotherms closer to hot and cold ends. As may be seen in FIGURE 3, a very slight shift in temperatures near the hot and cold region was detected, but the isotherms tended to move away from the hot and cold walls during periods of activity and toward the walls during sleep periods—exactly opposite of what would be expected from Gershuni's analysis. Clearly, more experimental work is need to complete our understanding of these second-order flows.

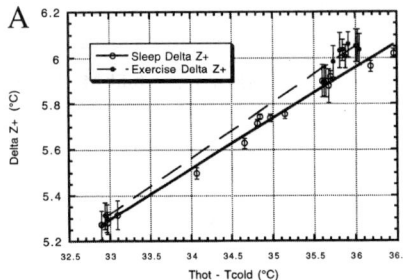 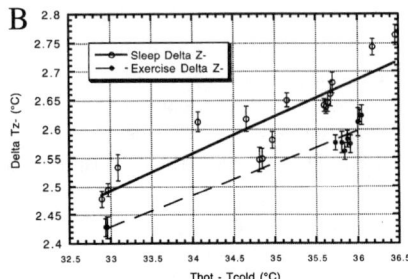

FIGURE 3. Measured δT_{z+} and δT_{z-} versus the total temperature difference for sleep versus exercise on STS-95. Despite the scatter in the data, due primarily to the changing heat sink temperature, the linear least squares fits indicate a statistically significant increase in δT_{z+} and decrease in δT_{z-}, just the opposite of what would be expected from the symmetric Gershuni flows.

CONCLUSIONS

Experiments designed to withstand the quasi-steady acceleration environment should not be affected by the first-order flows even under the worst estimate of the g-jitter spectrum. The second-order flows of the Gershuni-type may produce some local mixing, but the effective Gershuni number for a well-designed experiment should be sufficiently far from the critical value at which point the flow bifurcates and becomes global, that this should not be a problem. However, more study is required to extend the analysis of such flows to different geometries, especially to higher-aspect ratios, and to cylindrical geometries. The additional cross terms in the momentum equation in cylindrical coordinates can produce additional non-zero time average flows. Above all, more experimental data are needed.

The non-zero time average flows produced by the combination of transverse and longitudinal accelerations seem to be the most likely to produce unwanted convection from g-jitter. However, such flows could be avoided if the directional as well as the spectral distribution were known so that the experiment could be oriented to minimize such an effect. Such a mapping should be a high priority for the early phases of the ISS operation.

Open issues remain concerning the effect of start-up transients. No such effects were seen in the JUSTSAP experiment on STS-95, but the g-jitter environment on the Shuttle is much less than the expected environment on the ISS. Similarly, the effects of periodic accelerations with a time-varying density gradient need more study to determine if this might be the mechanism for the g-jitter effects seen in measuring diffusion coefficients.

ACKNOWLEDGMENTS

This work was done under the sponsorship of the Alliance for Microgravity Materials Science and Applications, NCC-8, MSFC, NASA.

REFERENCES

1. CORMACK, D., L. LEAL & J. IMBERGER. 1974. Natural convection in a shallow cavity with differentially heated end walls, part 1. Asymptotic theory. J. Fluid Mech. **65**: 209–229.
2. NAUMANN, R.J. 2000. An analytical model for transport from quasi-steady and periodic accelerations on spacecraft. Int. J. Heat Mass Transfer **43**: 2917–2930.
3. GERSHUNI, G.Z. & D.V. LYUBIMOV. 1998. Thermal Vibration Convection. John Wiley & Sons Ltd., West Sussex, England.
4. BEJAN, A. & C. TIEN. 1978. Fully developed natural convection in a long horizontal pipe with different end temperatures. Int. J. Heat Mass Transfer **21**: 701–708.
5. KAMOTANI, Y., A. PRASAD & S. OSTRACH. 1981. Thermal convection in an enclosure due to vibrations aboard a spacecraft. AIAA J. **19**: 511–516.
6. FAROOQ, A. & G.M. HOMSY. 1994. Streaming flow due to g-jitter–induced natural convection. J. Fluid Mech. **271**: 351–378.
7. FAROOQ, A. & G.M. HOMSY. 1996. Linear and non-linear dynamics of a differentially heated slot under gravity modulation. J. Fluid Mech. **313**: 1–38
8. NAPOLITANO, L.G. 1996. Study on g-jitter for the evaluation of tolerability limits, Universita Degli Studi di Napoli "Federico II", Departmento di Scienza e Ingegneria dello Spazio. Final Report, ESA Contract 11597/95/F/FL, Dec. 1996.
9. ALEXANDER, J.I.D. 1994. Residual gravity jitter effects on fluid process. Microgravity Sci. Technol. **VII/2**: 131–136.
10. NAUMANN, R.J., H. KAWAMURA & K. MATSUNAGA. 2002. First-order perturbation solution for periodic accelerations with axial and transverse components. Int. J. Heat Mass Transfer. Submitted.
11. ISS. 1998. DAC-6 integrated loads and dynamic analytical model properties. Boeing Report A92-J195-STN-M-WLP-980050, April 1998.
12. NAUMANN, R.J., G. HAULENBEEK, H. KAWAMURA & K. MATSUNAGA. 2002. The JUSTSAP experiment on STS-95. Microgravity Sci. Technol. **XIII**(2): 22–32.
13. NAUMANN, R.J., G. HAULENBEEK, H. KAWAMURA & K. MATSUNAGA. 2000. A new concept for measuring quasi-steady microgravity accelerations. Presented to the First International Symposium on Microgravity Research and Applications in Physical Sciences and Biotechnology, Sorrento, Italy, 15 September 2000, ESA Special Report SP-454, Vol. II, 835–845.
14. SMITH, R.W. 1998. The influence of g-jitter on liquid diffusion—the QUELD/MIM/MIR Programme. Microgravity Sci. Technol. **XI/2**: 78.

The Use of Pulsatile Flow to Separate Species

AARON M. THOMAS[a] AND R. NARAYANAN[b]

[a]*Department of Chemical Engineering, University of Idaho, Moscow, Idaho, USA*
[b]*Department of Chemical Engineering, University of Florida, Gainesville, Florida, USA*

ABSTRACT: Pulsatile motion greatly enhances the mass transfer of a dilute species compared to that due to pure molecular diffusion. If two dilute species are present in a carrier, the mass transfer of the faster diffusing species may be higher, lower, or the same as the slower diffusing species. This depends on the time constants associated with the system and the ability of a species to remain in the fast moving portion of the flow field. The difference in the mass transfer of each species can lead to a separation that can be used in a number of processes including the removal of carbon dioxide from the air. This phenomenon is modeled in an open tube geometry and in the annular space between two concentric cylinders. In annular pulsatile flow, the effect of the inner cylinder being off center from the outer cylinder on the mass transfer and separation is also analyzed. Finally, experimental results are presented to prove the validity of the models and the separation that can be achieved using this process.

KEYWORDS: pulsatile flows; separations; Taylor diffusion

NOMENCLATURE:

A	peak-to-peak amplitude, cm
c	species concentration, mol/cm^3
D	molecular diffusion coefficient, cm^2/sec
$i = \sqrt{-1}$	
$k = (i\omega/\nu)^{1/2}$	imaginary inverse of viscous length, cm^{-1}
L	tube length, cm
Q	mass transfer, mol/sec
r	radial coordinate or radial component
R	radius for the open tube, cm
R_{in}	inner radius for the annular configuration, cm
R_{out}	outer radius for the annular configuration, cm
$Sc = \nu/D$	Schmidt number, dimensionless
t	time, sec
V	velocity, cm/sec
$W = R(\omega/\nu)^{1/2}$	Womersley number for an open tube, dimensionless
$W = R_{out}(\omega/\nu)^{1/2}$	Womersley number for an annulus, dimensionless
z	axial coordinate or axial component in circular geometry
ε	eccentric displacement, cm
$\lambda = (i\omega/D)^{1/2}$	imaginary inverse of diffusive length, cm^{-1}
ν	kinematic viscosity, cm^2/sec
θ	angular coordinate or angular component
ρ	density, g/cm^3
ω	frequency of oscillation, rad/sec

Addresses for correspondence:
Aaron M. Thomas, Department of Chemical Engineering, University of Idaho, PO Box 441021, Moscow, ID 83844-1021, USA. Voice: 208-885-7652; fax: 208-885-7462.
amthomas@uidaho.edu
R. Narayanan, Department of Chemical Engineering, University of Florida, PO Box 116005, Gainesville, FL 32611-6005, USA. Voice: 352-392-0881; fax: 352-392-9513.
ranga@che.ufl.edu

INTRODUCTION

The removal of carbon dioxide from air is important in producing a habitable environment for the self-supporting space stations of the space program. Pulsatile flow is a novel way to separate out various species from air by using a purely mechanical method. The advantage of this is that no chemicals are needed. Pulsatile flow also has the advantage that it can handle large volumes. Although it is not expected that this process will replace existing methods of separation, it can surely be used as a means to assist in the overall separation process, possibly as a precursor to conventional methods. This work specifically focuses on the physics of pulsatile flow, its effect on the mass transfer of species, and the separation that can be achieved. From the theoretical model that predicts the mass transfer and separation of species, we provide a physical explanation of the phenomena predicted by the models. Experiments were also conducted to verify the validity of the models and the viability of pulsatile flows as a separations procedure.

PHYSICS OF THE FLUID FLOW–MASS TRANSPORT INTERACTION

Picture a tube with a gas occupying a reservoir at each end. One of the reservoirs can be considered to hold a pure gas, called the carrier, the other is a mixture of the carrier gas with a dilute amount of a single species as shown in FIGURE 1. According to Fick's law for dilute species, the non-interacting particles will move from the mixture to the carrier gas, from high to low concentration, in a process involving pure

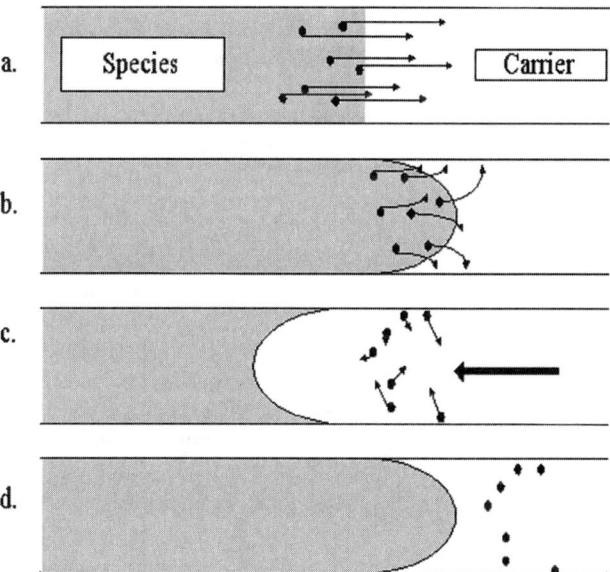

FIGURE 1. Dispersion mechanism for dilute species in pulsatile flow.

molecular diffusion. Now, suppose that the fluid in the tube oscillates with no net flow from one reservoir to the other. This, for example, can be achieved with a piston. In the first half of a cycle of the piston stroke, a nearly parabolic flow profile is produced in the tube, provided that the frequency is not large. This in turn causes radial concentration gradients. The dilute species then diffuses from the core of the tube to the boundary. In the second half of a cycle, the flow profile is reversed and the species moves from the boundary to the core where the concentration of the species is small compared to its value at the boundary. In the first half of the next cycle of the piston stroke, the species that is in the fast moving core of the tube is convected down the tube and again radially diffuses towards the wall of the tube. The species thus proceeds to move in this zigzag fashion down the tube giving it a higher transport than by pure molecular diffusion, yet with no net flow between the two reservoirs. If another non-interacting dilute species were added to the mixture, the time constants of the system become very important due to the differing diffusion coefficients of the species, the frequency of oscillation, and the kinematic viscosity of the fluid. These different time constants can give rise to a separation of species due to the periodic flow. It is the relationship of the different time constants that govern the mass transfer for each species as well as an analysis of the mass transfer and separation of species for different geometries that are the focus of this research.

PULSATILE FLOW IN AN OPEN TUBE

Numerous researchers have examined pulsatile flow in an open tube.[1-7] Although a majority of the work consists of pulsatile motion due to an oscillating pressure drop, this work is concentrated on inducing pulsatile flow by oscillating the boundary. The mass transfer and separation are analyzed, along with a brief explanation of the physics associated with the phenomena observed.

Picture a tube of radius R oscillating with a velocity

$$V_z = \frac{1}{2}A\omega\cos(\omega t). \tag{1}$$

The equation of motion for a boundary driven configuration in cylindrical coordinates, assuming an incompressible Newtonian fluid, is

$$\frac{\partial V_z}{\partial t} = \frac{\nu}{r}\frac{\partial}{\partial r}\left(r\frac{\partial V_z}{\partial r}\right). \tag{2}$$

This can be solved analytically, assuming that the no-slip boundary condition holds and that the velocity is finite at $r = 0$, to find the time and space dependant velocity profiles of the fluid. For the frequencies of oscillation that we consider, the velocity profiles are assumed to be linear in that they are unidirectional in z and vary only with the radial coordinate and time.

The concentration field can be found by solving the species continuity equation, that is

$$\frac{\partial c}{\partial t} + V_z\frac{\partial c}{\partial z} = D\left[\frac{\partial^2 c}{\partial r^2} + \frac{1}{r}\frac{\partial c}{\partial r} + \frac{\partial^2 c}{\partial z^2}\right], \tag{3}$$

where Fick's law is assumed to hold. The boundary condition that implies that the tube wall is impermeable to the transport of the dilute species is

$$\frac{\partial c}{\partial r} = 0 \text{ at } r = R. \tag{4}$$

To complete the problem, it is given that

$$c \text{ is finite at } r = 0. \tag{5}$$

Now, to enable us to get a solution, the concentration field in the fluid is assumed to be the sum of two parts, one due to a mean axial concentration gradient and the other due to a transverse variation that also fluctuates with time. This implicitly assumes that the end effects are neglected and that the calculation can be considered to be the same as examining the middle portion of the tube, thereby giving us a heuristic model. This assumption can be justified by the method of moments; the model is similar to that developed by Harris and Goren,[1] Watson,[2] and Kurzweg and Jaeger[4] among others. This also assumes that the amplitude to spacing ratio is large. The concentration field is then written as follows

$$c = \frac{(c_2 - c_1)z}{L} + c^*(r, t), \tag{6}$$

where $(c_2 - c_1)/L$ is the mean axial concentration gradient. The concentration can then be found analytically with the above assumption and the velocity known.

The time-averaged mass transfer of the system is defined by

$$\bar{Q} = -\frac{D(c_2 - c_1)\pi R^2}{L} + \frac{\omega}{2\pi}\int_0^{2\pi/\omega}\int_0^{2\pi}\int_0^R V_z C r \, dr \, d\theta \, dt. \tag{7}$$

The final expression for the total time averaged mass transfer is then

$$\bar{Q} = \frac{D(c_2 - c_1)\pi R^2}{L}$$
$$+ \frac{\omega^2 A^2 (c_2 - c_1)\pi R}{4DL} \times \Re\left\{\frac{i\lambda J_1(i\lambda R)[kJ_0(\lambda R)J_1(kR) - \lambda J_1(\lambda R)J_0(kR)]}{kJ_0(i\lambda R)J_0(\lambda R)J_1(kR)(k^4 - \lambda^4)}\right\}, \tag{8}$$

where

$$\lambda = \sqrt{\frac{i\omega}{\nu}} \tag{9}$$

and

$$k = \sqrt{\frac{i\omega}{D}}. \tag{10}$$

Similar models for the mass transfer produced by a boundary driven system can also be found in Poplasky[8] and Rader.[9]

In the final expression for the time- and space-averaged mass transfer, a complicated relationship exists between the mass transfer, the gap width, the frequency of motion, and the kinematic viscosity of the carrier fluid. This can be written in terms of dimensionless groups, such as the Womersley number, $r(\omega/\nu)^{1/2}$, the Schmidt number, and the ratio A/R. However, for the purposes of explaining the physics, it is easier to retain the calculations in unscaled rather than scaled form. Furthermore, although we mentioned the use of pulsatile flow to remove carbon dioxide from the air, the physics will be explained using species that best exhibit the unusual nature of the effect of pulsatile flow on the mass transfer and separation.

A depiction of the total time-averaged mass transfer versus frequency of oscillation for a dilute CO_2–He binary system in a nitrogen carrier is given in FIGURE 2. Observe first that the total mass transfer for both species continually increases for increasing ω. There is nothing to limit the increase in $\overline{Q}$ with increasing ω except for the onset of secondary flows that are disallowed because of the assumption that the velocity is unidirectional and only a function of r and t. A more striking observation is that "crossovers" occur in the total mass transfer in these binary systems. Kurzweg and Jaeger[10] and Jaeger et al.[6] have shown the existence of a crossover frequency, where the axial transport is the same. In their 1987 paper,[10] there appears to be an existence of a second crossover in their Figure 2 for large Schmidt numbers and small Womersley numbers. We have demonstrated that up to three crossovers may occur.[11] These crossovers are frequencies at which the total mass transfer for each species is the same and, therefore, reveal that there are frequencies at which the total mass transfer of two different species varies in relation to one another. Depending upon the magnitude of the frequency of oscillation, the fast diffusing species helium can either have a lower or higher total mass transfer than the slower diffusing CO_2 species. The ability for a slower diffusing species to overtake a faster diffusing species in its transport down a tube lies in each species ability to remain more in the faster moving flow in the tube and away from the slower moving boundaries, in order to take advantage of the convection. As each species radially diffuses toward the boundary (FIG. 1), the species that remains closer to the faster moving core gains more from the convective flow. However, as the frequency increases, there is less time for either species to reach the boundary before the cycle repeats itself. Furthermore, changing the frequency also changes the flow profiles that will further alter the mass transport of each species. It is the time constants associated with the system that determine the total mass transfer and the relation between a slower diffusing species and a faster moving one. These time constants arise from the diffusion

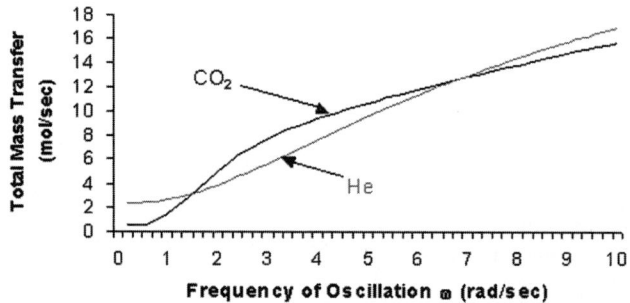

FIGURE 2. Total calculated mass transfer versus frequency for carbon dioxide and helium in an open tube configuration: $A = 10$ cm, $R = 0.995$ cm, $\Delta c/L = 1$ mol/cm^4, $D_{He} = 0.710$ cm^2/sec, $D_{CO_2} = 0.165$ cm^2/sec, $\nu = 0.15$ cm^2/sec. The value $\Delta c/L$ is normalized and used for calculation purposes only. In reality, this concentration difference would violate the assumption that the species are dilute. In this section, it is only the qualitative results that are of interest.

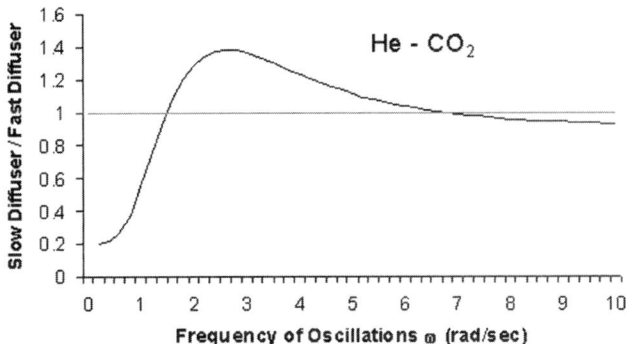

FIGURE 3. Separation ratio versus frequency for helium and carbon dioxide in and open tube configuration.

coefficients that describe the response time for the movement of species in a concentration gradient; the kinematic viscosity, which indicates the response of the fluid to a mechanical perturbation; and the frequency of oscillation. The relation of the time constants associated with these parameters dictate the mass transfer of each species and the number of crossovers that can be achieved between a fast and slow diffusing species. A more complete explanation of the physics and evidence of systems that contain one, two, and even three crossovers can be found in the paper by Thomas and Narayanan.[11]

Since each species transports at a different rate, a separation can then be produced between the helium and carbon dioxide. This can be seen in taking the ratio of the mass transfer of the slow diffuser to the fast diffuser, as illustrated in FIGURE 3. This can be thought of as a separation curve, where a maximum in the separation of the CO_2 can clearly be observed. No separation occurs for values of unity since these points are the crossover frequencies seen in FIGURE 2, where the mass transfer for both species is then the same. Notice a maximum occurs in the separation curve, indicating the region where separation of carbon dioxide is most efficient.

PULSATILE FLOW IN AN ANNULUS

Another geometry studied is the flow in the annular region between two concentric cylinders oscillating in phase with the same velocity as Equation (**1**). The velocity and concentration profiles are found again by using Equations (**2**) and (**3**), respectively. However, the boundary conditions are the no slip condition and the impermeability of the walls on both the outer *and* the inner tubes. The total mass transfer can be found in a similar fashion to that in the open tube configuration, and is defined by

$$\overline{Q} = -\frac{D(c_2 - c_1)}{L}\pi(R_{out}^2 - R_{in}^2) + \frac{\omega}{2\pi}\int_0^{2\pi/\omega}\int_0^{2\pi}\int_0^R V_z Cr\, dr\, d\theta\, dt. \quad (11)$$

This geometry also exhibits crossovers that were again evident in the open tube configuration.

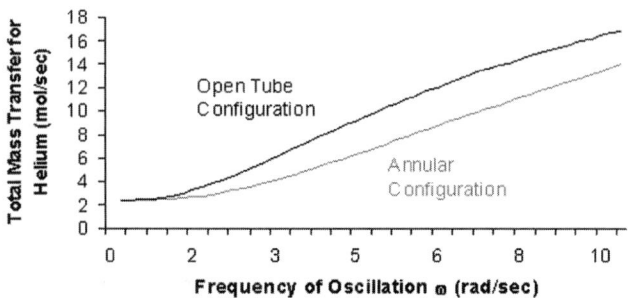

FIGURE 4. Total mass transfer for helium in both the open tube and annular configuration. The parameters are the same as in FIGURE 2. The radius of the open tube is 0.995 cm. The outer tube radius for the annulus is 1.0 cm and the inner tube radius is 0.1 cm to ensure the cross-sectional area of both configurations are the same.

We asked whether there is an advantage to using an annular configuration to the open tube configuration on the mass transfer and separation of species. To make a fair comparison, the cross-sectional flow area for the open tube and the annulus are the same. In terms of the mass transfer, FIGURE 4 shows that the open tube configuration exhibits a higher mass transfer than the annular configuration. However, in analyzing separation in a He–CO_2 system, the maximum in the separation of the slow diffusing species CO_2 is slightly higher for the annulus than the corresponding maximum in the open tube as shown in FIGURE 5. Since the maximum occurs at a higher frequency for the annulus, the mass transfer of both dilute species is higher for the annular configuration at this point as well. This indicates that an annular configuration will give a greater maximum separation with a higher throughput of species than the open tube geometry even though the mass transfer for each species in the open tube geometry is higher for all frequencies. This motivates the use of an annular geometry as a viable method to achieve a reasonable separation at a high throughput.

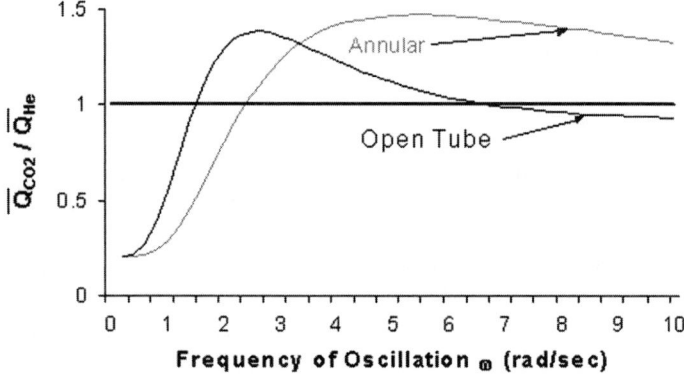

FIGURE 5. Separation ratio for a He–CO_2 system for both circular configurations.

THE EFFECT OF ECCENTRICITY ON PULSATILE ANNULAR FLOW

In conducting experiments on pulsatile annular flow, it is difficult to ensure that the cylinders are concentric. We wanted to explore the effect of the inner tube being slightly off center, or eccentric, from the outer tube and its effect on the mass transfer and separation. Suppose that in the annular configuration the inner tube is slightly displaced from the center of the outer tube by an amount ε as shown in FIGURE 6. The outer radius remains unchanged, but the inner radius is now a function of ε, and from the law of cosines,

$$R_{in,0}^2 = R_{in}^2 + \varepsilon^2 - 2\varepsilon R_{in}\cos\theta, \qquad (12)$$

where $R_{in,0}^2$ is the radius of the inner cylinder in the unperturbed state. Since the boundary has shifted slightly from its centered base state, the radial position variable, r, is redefined as a function of ε with the Taylor series expansion

$$r = r_0 + \varepsilon r_1 + \frac{\varepsilon^2}{2!}r_2 + \dots . \qquad (13)$$

In a sense, the domain region between the two tubes and the inner boundary has been "remapped" because of the shift. The other position variables in cylindrical coordinates and the time variable, t, remain unchanged in the new geometry.

The z-component of the velocity can now be written as a Taylor series expansion about $\varepsilon = 0$ as follows:

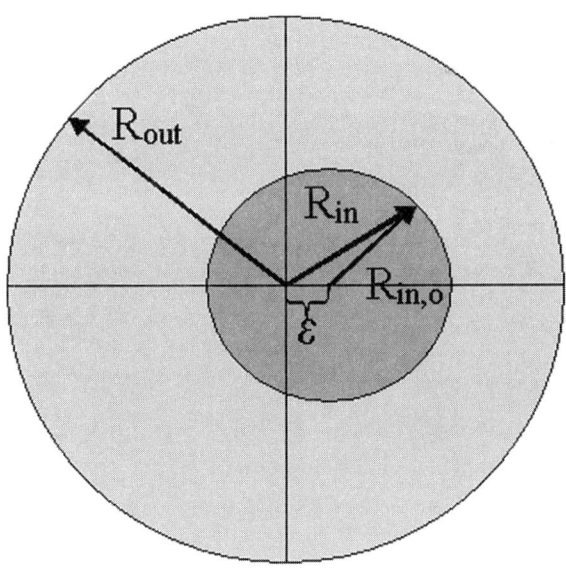

FIGURE 6. Eccentric annular geometry with inner cylinder displaced from the center of the outer cylinder by an amount ε. R_{in} is now the new inner radius of the displaced inner cylinder.

$$V_z(r, \theta, t, \varepsilon) = V_0 + \varepsilon \frac{dV_z}{d\varepsilon}\bigg|_{\varepsilon = 0} + \frac{\varepsilon^2 d^2 V_z}{2! d\varepsilon^2}\bigg|_{\varepsilon = 0} + \ldots . \tag{14}$$

However, not only is the velocity directly a function of ε, as immediately seen in the series expansion, it is also implicitly a function of ε through the position variable r as given by Equation (13). Because total derivatives of the velocity with respect to ε are needed, the "mapping" of the new geometry needs to be taken into account. Using the second essay of Johns and Narayanan[12] as a guide, the velocity equations with the proper boundary condition can now be found. For the zeroth order representing the velocity for a centered inner tube, the domain equations are

$$\frac{\partial V_0}{\partial t_0} = \frac{\nu}{r_0}\frac{\partial}{\partial r_0}\left(r_0\frac{\partial V_0}{\partial r_0}\right), \tag{15}$$

with the following boundary conditions

$$V_0 = \frac{1}{2}A\omega\cos(\omega t) \text{ at } r_0 = R_0 \tag{16}$$

$$V_0 = \frac{1}{2}A\omega\cos(\omega t) \text{ at } r_0 = R_{out}. \tag{17}$$

The first-order domain equation is

$$\frac{\partial V_1}{\partial t_0} = \nu\left(\frac{\partial^2 V_1}{\partial r_0^2} + \frac{1}{r_0}\frac{\partial V_1}{\partial r_0} + \frac{1}{r_0^2}\frac{\partial^2 V_1}{\partial \theta_0^2}\right), \tag{18}$$

with the corresponding boundary conditions

$$V_1 = 0 \text{ at } r_0 = R_{out} \tag{19}$$

$$V_1 + R_1\frac{\partial V_0}{\partial r_0} = 0 \text{ at } r_0 = R_0. \tag{20}$$

Finally, the second-order domain equation is

$$\frac{\partial V_2}{\partial t_0} = \nu\left(\frac{\partial^2 V_2}{\partial r_0^2} + \frac{1}{r_0}\frac{\partial V_2}{\partial r_0} + \frac{1}{r_0^2}\frac{\partial^2 V_2}{\partial \theta_0^2}\right),$$

and on the boundary

$$V_2 = 0 \text{ at } r_0 = R_{out} \tag{22}$$

$$V_2 + \frac{\partial^2 V_2}{\partial r_0^2}R_1^2 + 2\frac{\partial V_1}{\partial r_0}R_1 + \frac{\partial V_0}{\partial r_0}R_2 = 0 \text{ at } r_0 = R_0. \tag{23}$$

Thus, the velocity profiles in the perturbed state can be found analytically. The concentration profiles can also be found analytically in a similar manner.

To find the total mass transfer in the perturbed state, recall that the total mass transfer is defined as

$$Q = \int_{R_{in}}^{R_{out}}\int_0^{2\pi}\left(-D\frac{\partial C}{\partial z} + V_z C\right)r\,dr\,d\theta. \tag{24}$$

If the system is perturbed, like the eccentric system, the mass transfer is also a function of the perturbation in the same way as the velocity and concentration.

$$Q = Q_0 + \varepsilon Q_1 + \frac{1}{2}\varepsilon^2 Q_2 + \dots, \qquad (25)$$

where

$$Q_1 = \left.\frac{dQ}{d\varepsilon}\right|_{\varepsilon=0} = \frac{d}{d\varepsilon}\left[\int_{R_m(\varepsilon)}^{R_{out}}\int_0^{2\pi} V_z(\varepsilon, r', \theta') C(\varepsilon, r', \theta') r' dr' d\theta'\right]_{\varepsilon=0} \qquad (26)$$

and

$$Q_2 = \left.\frac{d^2 Q}{d\varepsilon^2}\right|_{\varepsilon=0} = \frac{d^2}{d\varepsilon^2}\left[\int_{R_m(\varepsilon)}^{R_{out}}\int_0^{2\pi} V_z(\varepsilon, r', \theta') C(\varepsilon, r', \theta') r' dr' d\theta'\right]_{\varepsilon=0}, \qquad (27)$$

taking only the *convective* (second) term of Equation (**24**). A problem arises in taking the total derivative of the integral because one of the limits of integration is a function of ε in $R_{in}(\varepsilon)$. Employing Leibnitz's rule aids in overcoming this problem so that the mass transfer can be calculated. The details of this calculation can be found in Thomas' dissertation[13] and only the results are presented here.

FIGURE 7 is a plot of the second-order perturbation on the mass transfer versus the frequency of oscillation for two species. The second-order term is needed because no difference is seen mass transfer for the eccentric annulus in the first-order term. As can be seen, the second-order perturbation is initially negative and decreasing and then begins to increase towards zero. This demonstrates that the eccentricity initially decreases the total mass transfer. If the inner tube is slightly off center, this modifies the flow in the annular region and causes reduced total mass transfer. Because the eccentricity affects each species differently, this will have an effect of the separation that can be achieved in an eccentric geometry.

FIGURE 8 demonstrates the separation that can be achieved for a CO_2–He system in a nitrogen carrier for a centered annulus and an eccentric annulus displaced by separate amounts. The value of ε is chosen so that it is large enough to give a noticeable difference between the perturbed and unperturbed case yet not too large to discount the Taylor series expansion about small ε that is the basis for the perturbation

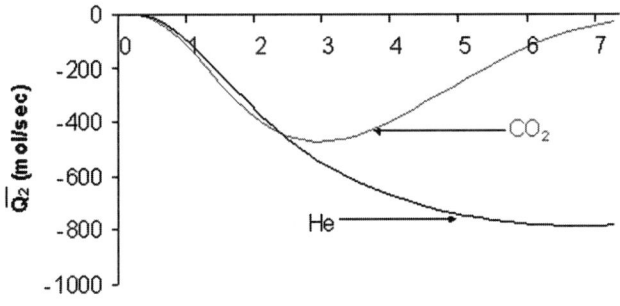

FIGURE 7. Curves for the second-order perturbation of the mass transfer versus frequency for helium and carbon dioxide.

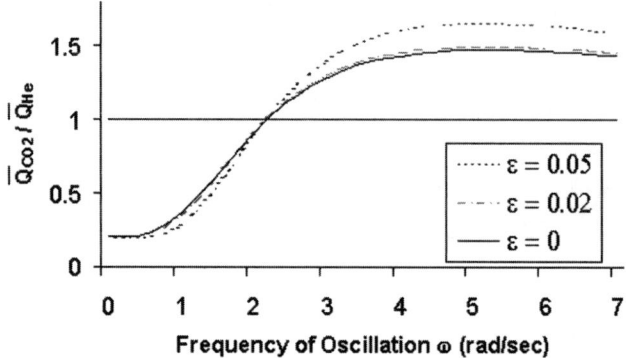

FIGURE 8. Separation ratio for a centered annulus ($\varepsilon = 0$) and configurations where the inner cylinder has been displaced by two different amounts.

analysis. As can be seen in the ratio of the slow diffusing CO_2 to the fast diffusing He, the separation of species increases for an eccentric geometry and continues to increase for greater displacements.

Although an eccentric annulus in periodic flow decreases the mass transfer for each species in comparison to a centered inner cylinder, it does enhance the separation achieved in a binary system. This information will prove useful for those that use this geometry and method in the separation processes.

EXPERIMENTAL RESULTS

The apparatus used to conduct experiments is shown in FIGURE 9. The reservoir on the right was filled with pure nitrogen, and the reservoir on the left had nitrogen as its carrier with dilute amounts of either CO_2, CH_4, or helium present in a binary mixture. The gases continually flowed through the reservoirs with constant flow rates controlled by mass flow controllers. The reservoirs were also well mixed as recirculation fans cycled the gas in the chambers. Pulsatile motion was achieved by the oscillation of the tubes walls. Converting the angular motion of the flywheels turned by the motors into linear oscillatory motion produced the desired effect. Now, oscillating the pressure in the chambers or oscillating the boundary both give rise to periodic motion, and both give the same qualitative results and are physically similar. The crossovers seen theoretically and the separation curve depicted in FIGURE 3 should be reproduced experimentally.

The following results are for separation curves for a CO_2–CH_4 binary system and a He–CH_4 binary system both in a nitrogen carrier. Since the physics has already been explained in unscaled terms, the experimental results are now in scaled quantities with the effective diffusion coefficient plotted versus Womersley number for a fixed Schmidt number and amplitude to radius (A/R) ratio. The effective diffusion coefficient is related to the total mass transfer by

$$D_{\text{eff}} = -\frac{\overline{Q}L}{(c_2 - c_1)\text{Area}}. \tag{28}$$

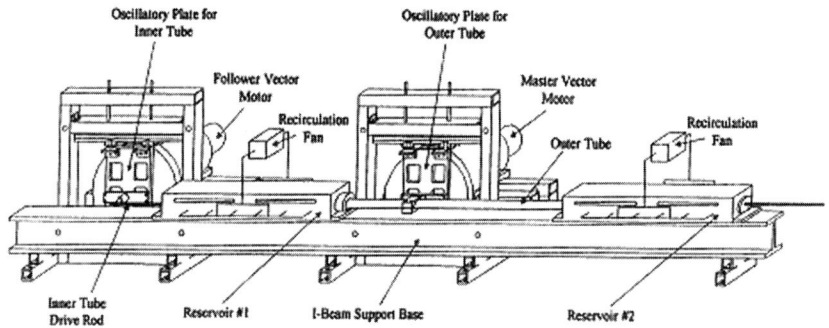

FIGURE 9. Experimental apparatus.

This helps normalize the time- and space-averaged total mass transfer with the concentration difference.

FIGURE 10 shows the experimental results and the theoretical separation curve for CO_2–CH_4 binary system in a nitrogen carrier. As can be seen, the experiments closely agree with the theoretical prediction. The results are slightly higher than expected, and the second crossover occurred experimentally at a Womersley number of about 5.3, whereas the theory predicted a value of about 4.3. The experiments were unable to produce a first crossover because the inertia of the flywheels prevented a steady angular frequency at such low Womersley numbers. However, the general shape of the separation curve, especially the maximum in the separation and the second crossover are clearly evident in experiments, as expected from the theory. FIGURE 11 depicts the results for a He–CH_4 binary system. This also shows that the experiments closely agree with the theoretical model even for a different system. This indicates

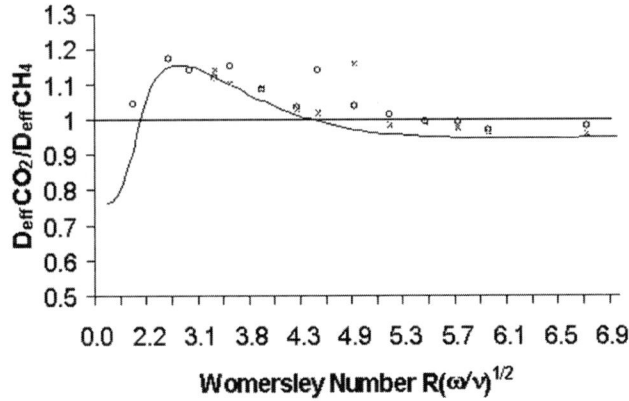

FIGURE 10. Ratio of the effective diffusion coefficients versus Womersley number for CO_2 and CH_4. $Sc_{CO_2} = 0.965$, $Sc_{CH_4} = 0.739$, $A/R = 19$. Schmidt numbers are corrected for the pressure and temperature of the experiments.

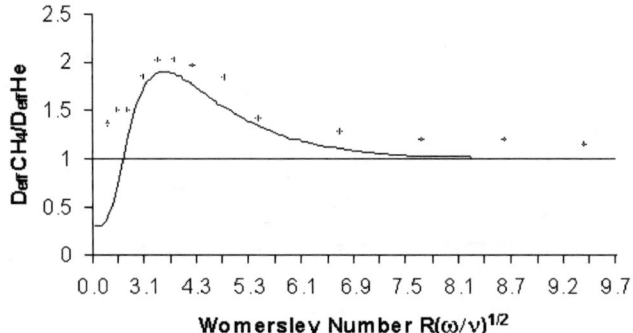

FIGURE 11. Ratio of the effective diffusion coefficients versus Womersley number for He and CH_4. $Sc_{He} = 0.224$, $Sc_{CH_4} = 0.739$, $A/R = 19$. Schmidt numbers are corrected for the pressure and temperature of the experiments.

that the models developed and the physical explanation behind the mass transfer and separation of species correctly predict experimental behavior.

Finally, the separation ratio for the experimental results is presented in FIGURE 12 for an annular experiment. In this case, only the first crossover is achieved for the range of frequencies shown. The reason the first crossover is seen and not the second as in the open tube case is the cross-sectional flow area for these experiments is much less than the flow area for the open tube experiments. It is encouraging that the crossovers are still apparent in the annular configuration as expected.

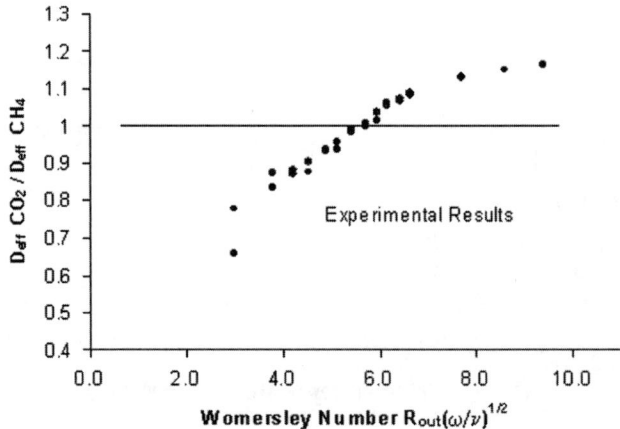

FIGURE 12. Experimental results for the separation ratio of methane and carbon dioxide versus Womersley number. $R_{in}/R_{out} = 0.79$.

CONCLUSIONS

Pulsatile flow does enhance the mass transfer of dilute species present in a carrier, and it proves to be a mechanical means to separate out different species. It has been shown that in a binary mixture, the faster diffusing species can have either a higher, lower, or the same total mass transfer as the slower diffusing species. This depends on the interaction of the diffusing species with the pulsatile motion, and is dictated by the time constants associated with the diffusing coefficients, kinematic viscosity, and frequency of oscillation. In an annular configuration with the same cross-sectional flow area as an open tube, the mass transfer is less, but the separation is slightly higher for the annulus. Although the overall mass transfer is lower than the open tube configuration, the mass transfer at the maximum in the separation curve is higher for both species making the annular configuration a viable option for high throughput and reasonable separation of species. If the inner cylinder in an annulus is slightly eccentric, this causes the mass transfer to decrease and the separation to increase. Finally, experiments have shown that crossover frequencies do indeed exist in both the open tube and annular configuration. In fact, for the open tube, the experiments closely agree with the theoretical models predicting the behavior of the system.

ACKNOWLEDGMENTS

This work is supported by NASA under Grant No. NAG3 2415. Support for Thomas was given through an NSF Fellowship under Grant No. 9616053 during his time on this project.

REFERENCES

1. HARRIS, H. & S. GOREN. 1967. Axial diffusion in a cylinder with pulsed flow. Chem. Eng. Sci. **22**: 1571.
2. WATSON, E.J. 1983. Diffusion in oscillatory pipe flow. J. Fluid Mech. **133**: 233.
3. JOSHI, C.H., R.D. KAMM, J.M. DRAZEN & A.S. SLUTSKY. 1983. An experimental study of gas exchange in laminar oscillatory flow. J. Fluid Mech. **133**: 245.
4. KURZWEG, U.H., G. HOWELL & M.J. JAEGER. 1984. Enhanced dispersion in oscillatory flows. Phys. Fluids **27**: 1046.
5. KURZWEG, U.H. & M.J. JAEGER. 1985. Tuning effect in enhanced gas dispersion under oscillatory conditions. Phys. Fluids **29**: 1324.
6. JAEGER, M.J., T. SOEPARDI, A. MADDAHIAN & U.H. KURZWEG. 1991. Diffusional separation of gases and solutes in oscillatory flow. Sep. Sci. Technol. **26**: 503.
7. RICE, R.G. & L.C. EAGLETON. 1970. Mass transfer produced by laminar flow oscillations. Can. J. Chem. Eng. **48**: 46.
8. POPLASKY, M.S. 1996. Enhanced Diffusion in the Annular Space Between Oscillating Concentric Tubes. M.S. Thesis, University of Florida, Gainesville, Florida.
9. RADER, C.E. 1997. A study of the effect of oscillatory flow on dispersion. Project Report. University of Florida. Available from R. Narayanan on request.
10. KURZWEG, U.H. & M.J. JAEGER. 1987. Diffusional separation of gases by sinusoidal oscillations. Phys. Fluids **30**: 1023.
11. THOMAS, A.M. & R. NARAYANAN. 2001. Physics of oscillatory flow and its effect on the mass transfer and separation of species. Phys. Fluids **13**: 859.

12. JOHNS, L.E. & R. NARAYANAN. 2001. Interfacial Instabilities. Springer Verlag, New York.
13. THOMAS, A.M. 2001. Mass Transfer and Separation of Species in Oscillating Flows. Ph.D. Thesis. University of Florida, Gainesville, Florida.

The Influence of Gravity on the Precise Measurement of Solute Diffusion Coefficients in Dilute Liquid Metals and Metalloids

REGINALD W. SMITH,[a] XIAOHE ZHU,[a] MARK C. TUNNICLIFFE,[a] TIMOTHY J.N. SMITH,[b] LOWELL MISENER,[b] AND JOSEE ADAMSON[b]

[a]*Department of Materials and Metallurgical Engineering, Queen's University, Kingston, Ontario, Canada*

[b]*Millenium Biologix Inc., 785 Midpark Drive, Kingston, Ontario, Canada*

ABSTRACT: It is now well known that the diffusion coefficient (D) measured in a laboratory in low earth orbit (LEO) is less than the corresponding value measured in a terrestrial laboratory. However, all LEO laboratories are subject to transient accelerations (g-jitter) superimposed on the steady reduced gravity environment of the space platform. In measurements of the diffusion coefficients for dilute binary alloys of Pb–(Ag,Au,Sb), Sb–(Ga,In), Bi–(Ag,Au,Sb), Sn–(Au,Sb), Al–(Fe,Ni,Si), and In–Sb in which g-jitter was suppressed, it was found that $D \propto T$ (temperature) if g-jitter was suppressed, rather than $D \propto T^2$ as observed by earlier workers with g-jitter present. Furthermore, when a forced g-jitter was applied to a diffusion couple, the value measured for D increased. The significance of these results is reviewed in the light of recent work in which *ab initio* molecular dynamics simulations predicted a $D \propto T$ relationship.

KEYWORDS: impurity diffusion; liquid metals; liquid metalloids; gravity; microgravity; g-jitter

INTRODUCTION

In order to achieve optimum control of crystal growth and casting processes on earth, it is necessary to develop numerical control models. For this, accurate diffusion data are required. The currently available data, where they exist at all, are often widely inaccurate, the published values for any particular liquid metal diffusion coefficient varying by 50–100%. This is primarily due to the influence of buoyancy-driven convection on the experimental system used to obtain the coefficient.[1,2] Most of these experiments have used the long capillary diffusion couple and, in order to reduce convective transport, the diameters of the capillaries have been progressively decreased. However, it has been reported that if the diffusion capillary is less than about 1 mm in diameter, then the rate of diffusion is reduced—the so-called wall

Address for correspondence: Reginald W. Smith, Department of Materials and Metallurgical Engineering, Queen's University, Kingston, Ontario, Canada, K7L 3N6. Voice: 613-533-2753; fax: 613-533-6610.
smithrw@post.queensu.ca

effect. Thus, it is conventional to write

$$D_{\text{effective}} = D_{\text{intrinsic}} + D_{\text{buoyancy}} + D_{\text{wall effect}} + D_{\text{thermal (Sorêt effect)}} \quad (1)$$

It has been supposed that if a diffusion couple is isothermally processed in microgravity, then D_{buoyancy} and D_{thermal} will be absent and thus only $D_{\text{intrinsic}}$ and $D_{\text{wall effect}}$ should be present. Consequently, by choosing capillaries of different diameters, an estimate of $D_{\text{wall effect}}$ might be made and, hence, an accurate value of $D_{\text{intrinsic}}$ could be obtained. Furthermore, if the diffusion couples are processed at a number of temperatures, then the temperature dependence of the transport process taking place could be determined. This can provide clues to the mechanism(s) by which diffusion occurs in non-ionic liquids.

Unfortunately, this proposition takes no account of the influence on $D_{\text{effective}}$ of the ever-present g-jitter on a space platform, to be explored in this paper, which reports experimental data obtained on the Russian Space Station MIR using the Canadian Space Agency (CSA) materials processing facility QUELD II, designed and built by Queen's University in collaboration with Millenium Biologix Inc, in conjunction with the CSA Microgravity Isolation Mount (MIM). The MIM facility may be used in three modes: (1) *latched* in which the so-called "flotor" (platform to which the experiment is strapped) is rigidly attached to the "stator" containing the magnetic levitation devices and control computer; (2) *isolating* in which g-jitter is attenuated; and (3) *forcing* in which a selected forced oscillation is superimposed upon the isolating condition. The use of the MIM in the isolating mode reduces the effect of g-jitter on the measured value of D. The use of MIM in forcing mode permits the correlation of fluid dynamics predictions with the physical consequences of the actual g-jitter experienced by the sample and hence can be used to refine the theory.

The Principal Investigator (PI) in these experiments (R.W.S.) had flown long capillary diffusion couples earlier as a get-away-special (GAS)—QUESTS—(12 samples) and as a mid-deck experiment—QUELD I—(36 samples) on the NASA shuttle. With the development of MIM, the Canadian Space Agency invited the PI to develop QUELD II, a semi-automated version of QUELD I, the operator still being required to manually load the two furnaces.[3] Subsequently it operated on MIR for more than two years and processed a total of 200 samples of various types. Of these, 121 were long capillary diffusion samples.

Thus, the objectives of the experiments to be described were to: (1) measure the D values for selected molten metals and semiconductors at various temperatures and (2) derive (1) for (a) MIM in latched mode (raw g-jitter), (b) MIM in isolating mode (minimized g-jitter), and (c) MIM in forcing mode (to induce predictable fluid transport).

EXPERIMENTAL

The long-capillary method[3,4] was selected for use with QUELD II samples. In this, a 1.5–3.0 mm diameter × 38 mm length specimen was maintained at a fixed temperature for a given period of time to permit the solute in the 2-mm solute slug attached to one end of the solvent rod to diffuse into the body of the specimen. The diffusion anneal period was selected to ensure that the semi-infinite diffusion condition was not violated; that is, no solute gets to the far-end of solvent charge. The specimen was then quenched, returned to earth, and chemically analyzed using regular atomic absorption spectrophotometry techniques (AAS) in order to determine the solute distribution along the specimen. This solute distribution was then used to calculate the D value for that particular alloy, anneal temperature, and MIM condition used.

To obtain high-quality processed diffusion couples, it was necessary to ensure that (1) the diffusion interface was planar, (2) the quenching process used to capture the solute distribution in the specimen following the isothermal anneal was effective, and (3) the furnace was preheated before sample insertion so that, upon insertion, the sample comes almost instantly to the desired processing temperature.

RESULTS

Lead–Gold System

FIGURE 1 depicts the results previously obtained for Pb–Au with the STS flights 47 and 52, respectively, of QUESTS I and QUELD I and the equivalent 1g values. These were obtained for long capillary (semi-infinite) 1-mm diameter samples in which a 2-mm section of Pb–1wt%Au alloy was attached to a 38-mm pure lead filament. The Pb and Au were of at least 99.999wt% purity. It is seen that the microgravity results lie below those corresponding to terrestrial measurements.[4] For

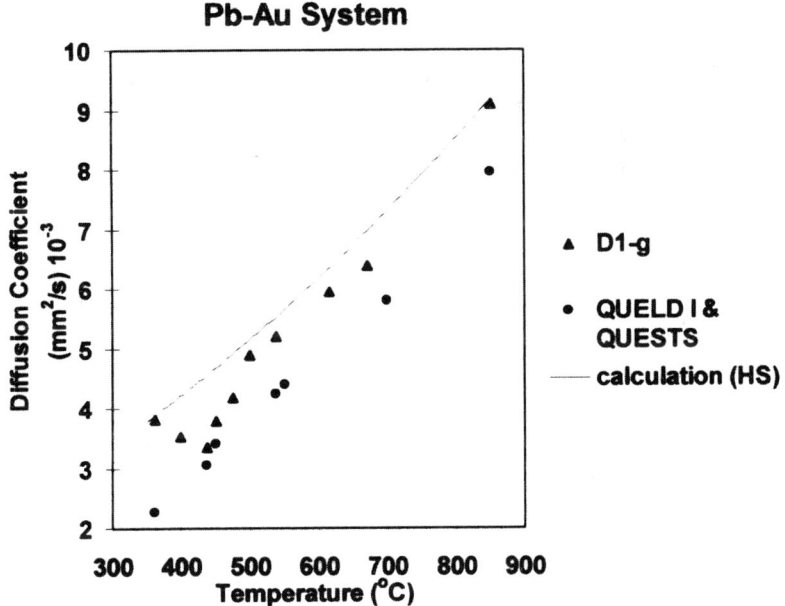

FIGURE 1. Diffusion coefficients of gold in lead space and ground-based experimental results.

comparison, the variation of D with temperature for Pb–Au, calculated using the hard sphere model (HS) of liquid metals,[5] is also given. It can be seen that the latter quite closely represents the $1g$ behavior, but is significantly higher than the microgravity-derived data.

FIGURE 2 shows data obtained from the MIR/MIM/QUELD mission. The previous QUESTS I/QUELD I data of FIGURE 1 are represented by the solid line in FIGURE 2. Significantly below this lie the data points obtained with MIM in isolating mode but using 1.5-mm diameter specimens.

The results from g-jitter isolation experiments are very interesting. The diffusion coefficients obtained are significantly lower than those from latched experiments. Of particular note is the fact that, whereas the curves for QUELD I and QUESTS I data are concave upwards, that for g-jitter isolation is essentially a straight line; that is, $D \propto T$ over the range of temperature studied.

We note that the D values produced with MIM in latched mode correspond closely with the equivalent values derived from earlier QUESTS I/QUELD I experiments on the Shuttle. This suggests that, as far as gathering high-fidelity liquid diffusion data using small diameter specimens is concerned, the reduced gravity environment of the MIR is no worse than that of the Shuttle. It can also be seen that the diffusion coefficient obtained with a larger specimen diameter (3 mm), and with MIM in latched mode, is somewhat larger than that of the equivalent specimen of smaller diameter (1 mm) used previously with QUESTS I/QUELD I. This is to be expected since g-jitter is likely to be able to create more convective transport in larger diameter diffusion couples. Also shown in FIGURE 2 is the effect of using the MIM in forcing mode to superimpose a forced oscillation of 0.1 Hz upon the MIM isolating condition for a 3 mm sample. The forced oscillation resulted in a mean flotor acceleration of $4\mu g$ at 45° to the long axis of the diffusion couple. The resulting D value is twice that of a 1.5-mm diameter MIM-latched sample, a reflection of the increased mass transport induced by the forced oscillation.

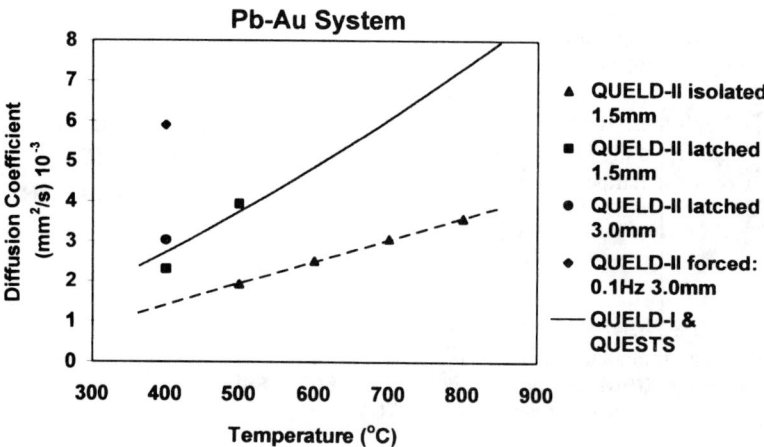

FIGURE 2. Diffusion coefficients of gold in lead (3MIM modes).

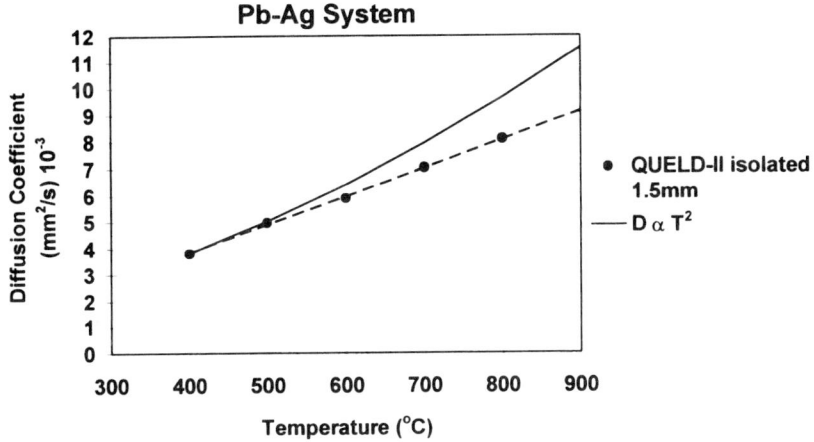

FIGURE 3. Diffusion coefficients of silver in lead.

Lead–Silver System

As with the Pb–Au system described above, here a 2-mm section of Pb–1wt% Ag alloy was attached to a 38-mm pure lead filament. Again, metals of 99.999% purity were used. The diffusion coefficient data obtained with the Pb–Ag diffusion couples are shown in FIGURE 3. The data were obtained with the MIM in the isolating mode. It can be seen that these data points may be fitted closely to a $D \propto T$ (linear) relationship. For comparison, the full curve shown in FIGURE 3 is that for $D \propto T^2$, based on the D value for 400°C.

Other Alloy Systems

Similar experiments to those using Pb–Au and Pb–Ag systems were performed on the following: Sb-1wt%In; In-1wt%Sb; Pb-1wt%Sb; Sn-1wt%Au; Sn-1wt%Sb; Sb-1wt%Ga; Bi-1wt%Sb; Bi-1wt%Ag; Bi-1wt%Au; Al-1wt%Ni; Al-1wt%Fe.

Although there were insufficient data to show clearly the nature of the D versus T relationship in the last four of the above alloy systems, all the others showed that $D \propto T$ over the temperature range examined.

DISCUSSION

There have been a number of models proposed for the structure(s) of non-ionic liquids (e.g., metals and semiconductors) and the manner in which this controls the rate of solute diffusion,[1,2] see TABLE 1.

As is seen, each predicts a particular relationship of D with T—for example, D proportional to $e^{-1/T}$, T, T^2, or $T^{(1.7 \rightarrow 2.3)}$. However, in order to comment on the validity of any one model it is necessary to have high quality data for selected alloy systems. As noted already, stray convection has confounded terrestrial experiments and

TABLE 1. Theories/models of diffusion in liquids

Diffusion Coefficient–Temperature Relationship	Author(s)
$D = D_0 \exp(-Q/RT)$, where Q is the activation energy, R is the gas constant, and D_0 is a system constant	conventional description of data
$D = kT/6\pi R\eta$, where R is the particle radius and η is the viscosity of the medium	Stokes–Einstein Theory
$D = B\sqrt{T}\exp(-bV^*/V_f)$, where V^* is the critical volume associated with diffusing atom, V_f is the free volume of the liquid, and B and b are constants	model of critical volume
$D = AT^2$, where A is a system constant	fluctuation theory—Swalin 1959
$D = -a + bT$, where a and b are system constants	fluctuation theory—Reynik 1969
$D = \alpha\sqrt{T}(9.385 T_m/T)^{-1}$, where T_m is the melting point and α is a system constant	hardsphere models
$D = A^1 T^m$, where $m = 1.7$–2.3 and A^1 is a system constant	molecular dynamics

so degraded the quality of the data obtained. The use of microgravity research facilities has permitted an increase in data quality to be obtained in liquid diffusion experiments.

FIGURE 1 shows that the effects of convection in terrestrial experiments on Pb–Au alloys is significant, D decreasing markedly when similar experiments are performed on spacecraft. However, the experimental examination of the physical effects of the time-dependent variations in the gravity level aboard a spacecraft has only been possible with the relatively recent development of the MIM-type disturbance isolating system. Although the experimental procedures adopted and reported here may not reveal the truly intrinsic value of D for a particular alloy system, they give values that need to be much closer to this ideal value and so should more easily permit a detailed correlation of experiment with theory.

The influence of low frequency (0.1 Hz) oscillation at 45° to the specimen axis was found to produce an observable increase in the measured value of D for the Pb–Au alloy, as had been anticipated. Since a detailed analysis of the effects of forced vibration on liquid diffusion couples of the dimensions used in this series of experiments was not available when planning the experiments, a forced oscillation of 0.1 Hz, but with a square waveform, was requested for processing some of the diffusion samples. This was to provide a fundamental oscillation mode of maximum amplitude 4 mg and at 0.1 Hz, plus a range of harmonics at various frequencies and amplitudes. This was done on the basis that, since there was no clear description of the atomic movements involved in liquid diffusion, a variety of forcing modes might all contribute to the solute redistribution process, in addition to the bulk convective transport induced in the sample by the low-frequency components of the forcing oscillation. As is seen in FIGURE 2, the value of D was approximately doubled for Pb–Au.

One of the most systematic studies of self- and inter-diffusion in liquid metals was Frohberg's work in the 1980s on the Sn system using the long capillary technique with 1g and μg processing. Frohberg concluded that the temperature dependence of the diffusion coefficient appeared to obey the predictions of the fluctuation theory ($D \propto T^2$) more closely than an Arrhenius-type relationship for both self-diffusion in Sn and inter-diffusion in liquid Sn–In alloys. When considering this marked reduction in diffusion coefficient in the absence of significant g-jitter, and the fact that it was found that $D \propto T$ for all the alloy systems examined here, it is interesting to note recent experimental work by the Frohberg group[7] in which it was shown that diffusion coefficients measured at 1g with a recently developed shear cell for the In–Sn system fell nicely on the extrapolated D versus T curve of long capillary measurements made at lower temperatures, indicating that the method of measurement does not significantly influence the results obtained. In the 1g case, with considerable buoyancy-driven solute transport at higher temperatures, the variation was found to be a power law relationship that may be approximated by an exponential, whereas in the case of microgravity experimental results with buoyancy convection much reduced but with g-jitter present, it was best represented by a $D \propto T^2$ relationship. However, the present authors claim that, as the contribution to mass transport by g-jitter is reduced still further by using the MIM in the isolating mode, the diffusion coefficient and temperature relationship becomes linear.

Returning to the $D \propto T$ behavior reported here, it should be noted that, although Frohberg et al.[7] reported a $D \propto T^2$ relationship for experiments conducted in low earth orbit without benefit of a MIM, $D \propto T$ is commonly observed to describe both fluidity (reciprocal viscosity) and diffusivity in non-electrolyte liquids if care is taken to suppress buoyancy-driven liquid flows. This view has been championed by a number of researchers, particularly Hildebrand,[8] see for example FIGURE 4. The measured diffusion rates for various chemical species in CCl_4 and C_6H_6 were found to be closely linear with temperature over the temperature range tested. The corruption of data by buoyancy effects was rendered insignificant by measuring the diffusion coefficient using the experimental conditions governed by Fick's First Law—the flow of the chosen species through a membrane. In these experiments, the membrane between the species source and the large solvent-vessel beneath it consisted of a 9-mm thick slab of 2,962 lengths of fine bore stainless steel hypodermic tubing in a parallel array embedded in solder. The small diameter of the hypodermic tubing effectively damped out any buoyancy contributions in the needles to the observed value of the diffusion coefficient.

In FIGURE 4, the diffusion coefficients are expressed in cm^2/sec, where the symbol ○ represents the self-diffusion coefficients of C_6H_6 and CCl_4, as shown near their corresponding lines. The symbol ● represents the diffusion coefficient of CCl_4 in C_6H_6; + represents the diffusion coefficient of I_2 in C_6H_6.

In further support of the proposition that D varies directly with T in the absence of convective transport, it is noted that recent work by Munejiri et al.[9] shows that the results of *ab initio* molecular-dynamics studies of self-diffusion in liquid tin at

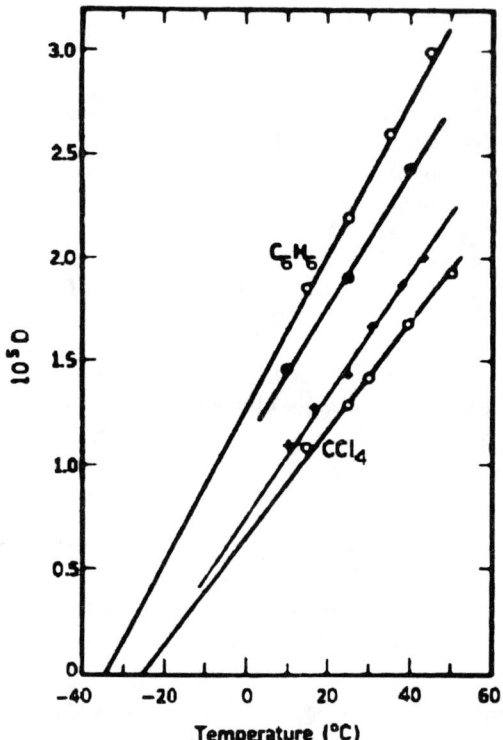

FIGURE 4. Diffusion coefficients of C_6H_6 and CCl_4.[14]

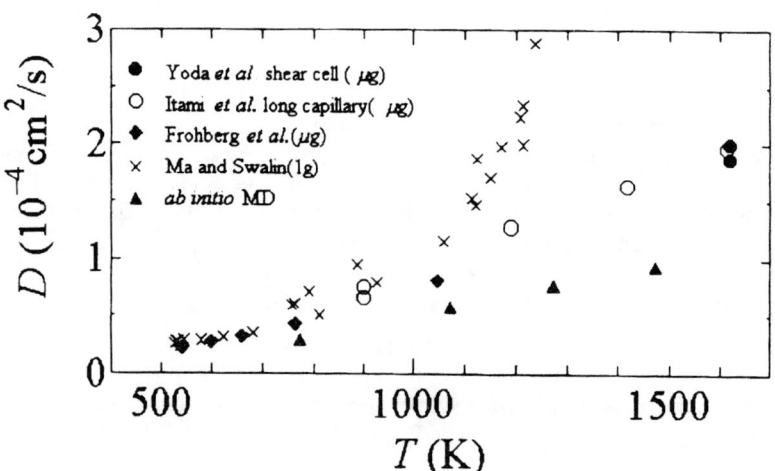

FIGURE 5. The variation of the self-diffusion coefficient of tin with temperature.[9]

various temperatures correlate well with neutron scattering results and suggest that D increases monatonically with T (see FIGURE 5).

FIGURE 5 also shows the results of various experimental studies using long capillary specimens. In preparing to carry out such experiments, it is necessary to join together the two sections constituting the diffusion couple. Usually great care is taken in effecting this to ensure that the common interface between the two sections is complete; that is, free from bubbles or foreign matter that might hinder the free passage of the diffusing species from each section into the other. Recent numerical simulations[10] of the influence of such barriers on the measured diffusion coefficient found that local obstructions of the transport path have to be in excess of half the transport path's cross-section to result in the measured diffusion coefficient being less than 95% of the true/intrinsic value. This relative insensitivity of D to the completeness of the union between the full cross-sections of the two parts of the diffusion couple will render the preparation for such diffusion experiments more easily accomplished.

As noted earlier, Frohberg's research group, including other ESA associates, have been using the shear cell in combination with long capillary specimens. In this, the as-diffused liquid specimen is separated into many segments by the relative rotation of adjoining discs (of the crucible). Thus, the sample is sheared into a number of segments that subsequently freeze and are chemically analyzed. Although the shear cell technique has some advantages over the quenching technique used in this study from the point of view of possible solute segregation during the quenching process, there appears to be concern that the free surface created on the molten specimen at each gap between the discs of the shear cell could provide undesirable Marangoni convection.

In passing, it should be noted that there have been a number of attempts to model the influence of g-jitter on the results obtainable from a liquid metal diffusion couple used on an orbiting earth platform. Some of these were reported at the ESA, First International Symposium on Microgravity Research and Applications in Physical Science and Biotechnology, Sorrento Italy, September, 2000.[11,12] These analyses are of technical interest for the manner of solution of complex fluid-dynamics problems; however, the simplification of the physical situation pertaining to actual g-jitter effects to permit solution render their results somewhat equivocal. Another paper given at the symposium, entitled "Benefits of a uniform magnetic field for measuring impurity diffusion coefficients in liquid metals," sought to report the results of a study of the use of strong magnetic fields (exceeding 2 Tesla) to quench convective liquid movements. Unfortunately, this paper did not appear in the published proceedings of the conference and so is not available for closer inspection.

CONCLUSIONS

The results obtained to date support the following conclusions:

1. The value of D, the impurity diffusion coefficient, is reduced markedly when measured in low earth orbit.
2. The D values with the "raw" g-jitter of MIR (corresponding to MIM latched) are similar to those obtained in like alloy diffusion couples in 1992 on STS 47 and

52. Thus, as far as liquid diffusion in narrow capillaries is concerned, both space vehicles have provided similar reduced gravity environments.
3. The reduction of g-jitter afforded by MIM, reduces the measured value of D significantly.
4. In the absence of significant g-jitter, D varies monotonically with temperature.

ACKNOWLEDGMENTS

The efforts of many people have gone into the development of equipment, the preparation of samples, certification of flight samples, isothermal annealing, chemical analysis, and the theoretical review of the results of these experiments. The names of the principal individuals involved have been captured in the collective authorship of this paper. Another important group has been Dr. W.G.M. Gallearneault and other staff at Alcan International Limited, Research and Development Laboratory, Kingston, Ontario. However, none of this work would have been possible without the direct financial assistance of the Canadian Space Agency, the National Science and Engineering Research Council, and Queen's University and the access to the STS and MIR space platforms made available to the CSA through NASA and the Russian Space Agency. All this assistance is most gratefully acknowledged.

REFERENCES

1. MALMEJAC, Y. & G. FROHBERG. 1986. Mass Transport by Diffusion, Chap. V, Fluid Sciences and Materials Science in Space. H.U. Walter, Ed.: 159. Springer-Verlag, Berlin.
2. FROHBERG, G., K.H. KRAATZ & H. WEVER. 1986. Microgravity experiments on liquid self- and inter-diffusion. Norderney-Symposium, August 1986. 144–151.
3. SMITH, R.W., T.J.N.S. SMITH, L. MISENER, et al. 1998. The QUELD II MIR Materials Processing Facility. Tenth International Symposium on Experimental Methods for Microgravity Materials Science, TMS.
4. ZHU, X. & R.W. SMITH. 1996. Diffusion in liquid Pb–Au binary system. Materials Science Forum. **215–216**: 113.
5. ZHU, X. & R.W. SMITH. 1998. Impurity diffusion of gold in liquid lead. Adv. Space Res. **22**(8): 1253.
6. SMITH, R.W. 2000. Solute diffusion in liquid metals and metalloids—the influence of g and g-jitter. Proceedings of the 12th International Symposium on Experimental Methods for Microgravity Materials Science, TMS.
7. GRIESCHE, A., K.H. KRAATZ & G. FROHBERG. 1998. A modified shear cell for mass transport measurements in melts. Rev. Sci. Instrum. (American Institute of Physics) **69**(1): 315.
8. HILDEBRAND, J.L. 1977. Viscosity and Diffusivity, Chap. 1. John Wiley, New York.
9. MUNEJIRI, S., T. MASAKI, T. ITAMI, et al. 2000. Self-diffusion coefficient and structure of liquid Sn: ab initio molecular dynamics simulation and neutron scattering experiment. Proceedings of Spacebound 2000. D.W. Brooks, Ed.: 1. Canadian Space Agency, Vancouver.
10. JALBERT, L.B., F. ROSENBERGER & R.M. BANISH. 1998. On the insensitivity of liquid diffusivity measurements to deviations from ID transport. J. Phys. Condensed Matter **10**: 7113–7120.

11. SEKERKA, R.F. & V.M. SOFENEA. 2001. Effect of convection on diffusion in liquids in microgravity. Proceedings of First International Symposium on Microgravity Research and Applications in Physical Sciences and Biology, Sorrento, Italy, Sept. 2000, Vol. I. Bridgette Schurmann, Ed.: 31. Publ. ESA, SP – 454.
12. MONTE, R. & R. SAVINO. 2001. The fluidynamic disturbances due to the microgravity environment prevailing on the international space station. Proceedings of First International Symposium on Microgravity Research and Applications in Physical Sciences and Biology, Sorrento, Italy, Sept. 2000, Vol. II. Bridgette Schurmann, Ed.: 813. Publ. ESA, SP – 454.

Structures of Solid and Liquid during Melting and Solidification of Indium

G. COSTANZA, F. GAUZZI, AND R. MONTANARI

Dipartimento di Ingegneria Meccanica, Università di Roma Tor Vergata, Roma, Italy

ABSTRACT: During melting and solidification of polycrystalline indium the structures of the solid and liquid phases were investigated by X-ray diffractometry (XRD) in 1g conditions. The experiments showed that melting is immediately preceded and accompanied by an abnormal increase of vacancy concentration in the solid that enhances the diffusion. At the melting point (T_M) the lattice planes facing the first formed liquid appear to be {0 0 2} and {1 0 1}; that is, those planes allocating first and second neighbors around a given atom, with shell radii very close to the mean distance of nearest neighbors in liquid as obtained from the radial distribution function (RDF). On this basis some types of solid–liquid interface (sharp or diffuse) are discussed; the possibility of a sudden texture change in the solid is also considered. On the other hand, the evolution of RDF curve of liquid indium, cooled down through T_M, shows that there are correlations between the structure of the liquid and of the forming solid during solidification. To avoid convective motions in the liquid and to achieve the best experimental conditions, we discuss the possibility of repeating the measurements in microgravity (μg) using thin oxide films as containers. This technique was already successfully tested by one of the investigators in the experiment ES 311 A-B carried out during the mission SPACELAB-1.

KEYWORDS: solid structure; liquid structure; melting; solidification; indium; X-ray diffraction

INTRODUCTION

The phase transformations in metals from solid to liquid and vice versa have been extensively investigated (see, e.g., the review in Ref. 1). Both melting and solidification involve a discontinuous jump in thermodynamic parameters, such as enthalpy and volume, even if precursor effects due to structural changes occur on either side of the discontinuity. The knowledge of the structural aspects in solid and liquid is of the utmost importance for the explanation of the great variety of behaviors observed in different materials and to validate theoretical models.

This work describes investigations by XRD at high temperature of the structural features of solid and liquid phases during melting and solidification of indium. Since convective motions in the melt are a serious disturbance to X-ray measurements, the same experiments should be repeated in conditions of reduced gravity aboard the International Space Station (ISS).

Address for correspondence: R. Montanari, Università di Roma Tor Vergata, Dipartimento di Ingegneria Meccanica, Via di Tor Vergata 110, Roma 00133, Italy. Voice: +39 (0)6 7259-7182; fax: +39 (0)6 2021-351.
roberto.montanari@uniroma2.it

EXPERIMENTAL

The investigated material is polycrystalline indium with a purity of 99.99 wt%. Melting experiments were performed on 1 mm-thick sheets. Before measurement runs the samples were molten and slowly cooled to room temperature to avoid the presence of extended defective structures in the base material and the effects on XRD patterns due to their recovery during heating. The samples examined exhibited a strong {002} texture. A high-temperature X-ray camera with a sample holder adapted to contain molten metals was employed. Diffraction patterns were collected at increasing (melting) or decreasing (solidification) temperatures with intervals of 1 K around T_M (429 K). During each test a constant temperature was maintained. The changes of state took place under conditions of certainly heterogeneous nucleation, so it was possible to neglect superheating and supercooling effects that were estimated to be comparable with the accuracy of temperature measurements (±0.5 K).

XRD spectra were recorded by step-scanning with 2Θ angular intervals of 0.05° and counting intervals of 5 sec per step. Mo–Kα radiation ($\lambda = 0.7093$ Å) was used. From the diffraction pattern of liquid collected at various temperatures, the RDF curve, which is the average number of atom centres between distances r and $r + dr$ from the centre of a given atom, was determined according to the following procedure.[2]

For a monoatomic liquid, such as indium melt, the RDF, $4\pi r^2 \rho(r)$, can be expressed as

$$4\pi r^2 \rho(r) = 4\pi r^2 \rho_0 + \frac{2r}{\pi}\int_0^\infty Qi(Q)\sin(Qr)dQ, \qquad (1)$$

where $Q = 4\pi\sin\theta/\lambda$ and ρ_0 is the mean density; that is, the average number of atoms per unit volume.

The measured XRD pattern of liquid $I(Q)$ was corrected for the effects due to absorption A and polarization P to obtain the intensity $I^*(Q)$

$$I^*(Q) = \frac{I(Q) - I_B}{PA}, \qquad (2)$$

where I_B is the background intensity, and P and A are expressed by Equations (3) and (4), respectively,

$$P = \frac{1 + \cos^2 2\theta}{2} \qquad (3)$$

$$A = 1 - \frac{1 - e^{-x}}{x}, \qquad (4)$$

with $x = 2\alpha\mu/\sin 2\theta$, where μ is the linear absorption coefficient and α is the beam divergence.

The corrected experimental intensity $I^*(Q)$, expressed in arbitrary units, was then converted to $I^*_{eu}(Q)/N$, the intensity in electron units per atom. To obtain $I^*_{eu}(Q)/N$, the experimental curve $I^*(Q)$ was smoothly interpolated to $Q = 0$ and then multiplied by a suitable factor so that at large Q it oscillated about the curve $I_{tot}(Q)$ representing the total independent scattering per atom, which is the sum of elastic and Compton scattering. The function $i(Q)$ is determined from the relationship

$$i(Q) = \frac{I^*_{eu}(Q)/N - I_e(Q) - I_C(Q)}{I_e(Q)} = \frac{I^*_{eu}(Q)/N - I_{tot}(Q)}{I_e(Q)}, \quad (5)$$

where $I_e(Q)$ and $I_C(Q)$ are the components due to elastic and Compton scattering, respectively.

Finally, the RDF curve can be determined from Equation (1). The integration is made to $Q = 7\,\text{Å}^{-1}$ and ρ_0 is set equal to 0.037 by assuming that the volume expansion due to melting is 2%.

RESULTS

Melting

The XRD spectrum recorded at T_M is the overlapping of diffraction patterns due to liquid and solid phases. As is shown in FIGURE 1, the two contributions can be separated by subtracting the pattern of liquid, obtained when the solid–liquid (S–L) transformation has been completed, from the overall spectrum. Contributions from solid and liquid were then examined separately.

This approach assumes that the shape of liquid contribution remains substantially unchanged after completion of the S–L transformation. Of course, the possibility has been considered that the contribution due to liquid phase may exhibit specific features when it is in presence of the solid phase. Therefore, XRD measurements performed independently on both the phases were carried out employing a different heating system with a thermal gradient along the sample.

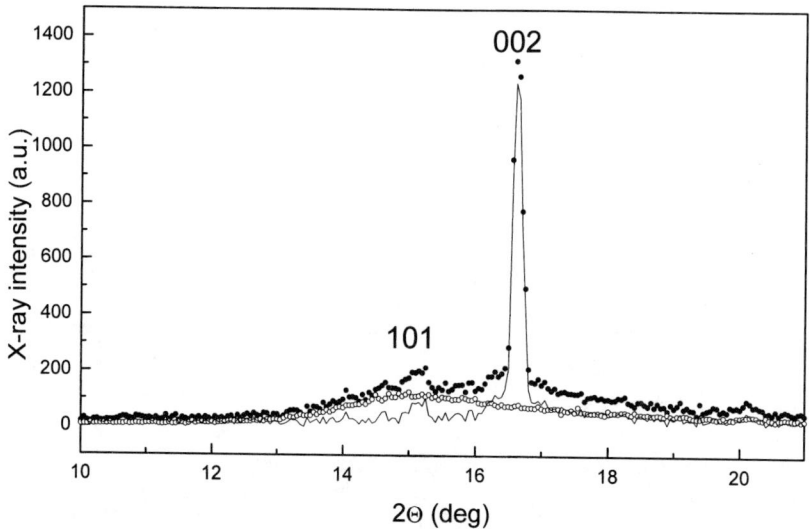

FIGURE 1. The diffraction pattern of indium collected during melting (●) is the sum of liquid (○) and solid (*continuous line*) contributions.

The RDF curve obtained from the liquid contribution of FIGURE 1 is plotted in FIGURE 2. It oscillates above and below the average density curve $4\pi r^2 \rho_0$. Bars indicate the distribution of first and second neighbors in crystalline indium.

The RDF curve in FIGURE 2 exhibits the first peak centered at 3.3 Å. FIGURE 3 shows the lattice positions of first and second neighbors in the solid. indium is a tetragonal body centered ($a = 3.25$ Å, $c = 4.95$ Å), the first neighbors (4 atoms) are at a

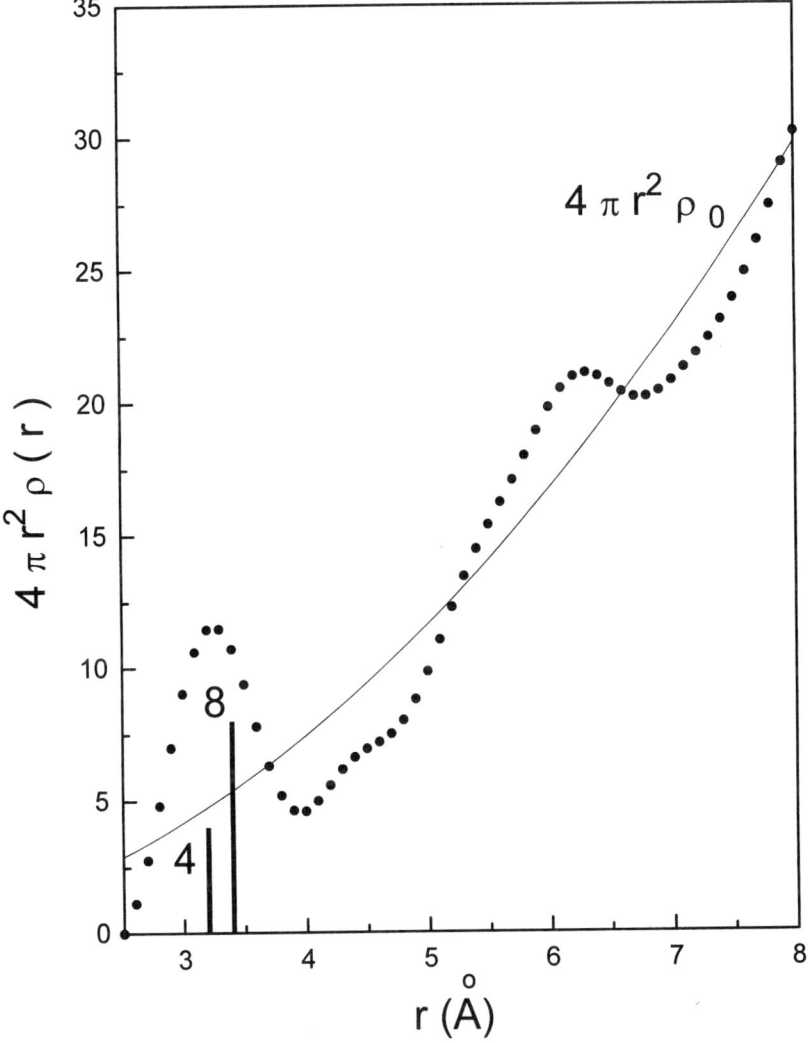

FIGURE 2. The RDF curve, $4\pi r^2 \rho(r)$ versus r, obtained from the liquid diffraction pattern in FIGURE 1. *Bars* indicate the distribution of first and second neighbors in crystalline indium.

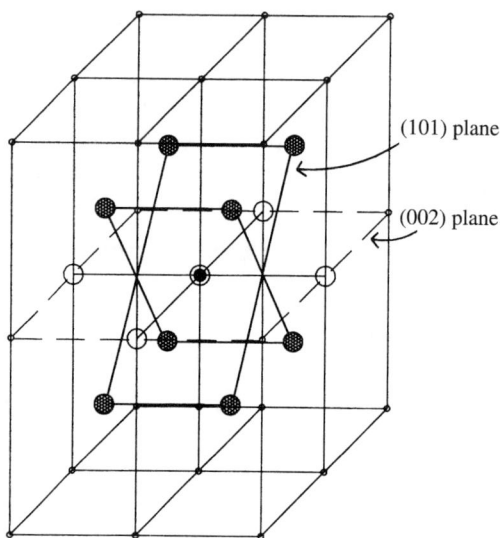

FIGURE 3. First and second neighbors around a given atom in the in lattice: ◉, reference atom; ○, first neighbors (4 atoms), $d = 3.25$ Å; ●, second neighbors (8 atoms), $d = 3.38$ Å.

distance of 3.25 Å and the second neighbors (8 atoms) at 3.38 Å. Therefore, the shell radii of the first and second neighbors in the solid are very close to the mean distance of nearest neighbors in liquid as obtained from the RDF.

When compared to the spectrum recorded at a temperature (428 K) just below T_M, the contribution due to solid phase exhibits some important features (see FIGURE 4).

1. The {101} reflection, absent in the spectrum at 428 K, appears at 429 K. Its intensity is relatively low and a rough estimation indicates that the thickness of diffracting material is about 0.3 μm. The presence of the {101} reflection involves an abrupt reorientation of the diffracting solid, which accompanies or immediately precedes the formation of the first liquid.
2. The {002} line profile is broadened, in particular, in the lower part of the peak.
3. During melting the {002} peak is shifted toward lower angles.

Graham et al.[3] found that thermal expansion of indium is unusual because the c-axis passes through a maximum near room temperature. Their data, in the range 90–408 K, are fitted by the following equations, which give a and c, expressed in Å, as a function of the absolute temperature T

$$a = 3.2182 + 8.47 \times 10^{-5} T + 3.61 \times 10^{-10} T^3 \quad (6)$$

$$c = 4.9201 + 15.15 \times 10^{-5} T - 6.64 \times 10^{-10} T^3 \quad (7)$$

Equation (7) describes the progressive shortening of c-axis on approaching the melting point. The lengths of c-axis measured at various temperatures up to 428 K in the present work are substantially in agreement with that prediction but they turn away just before melting. The observed peak shift corresponds to a remarkable change in

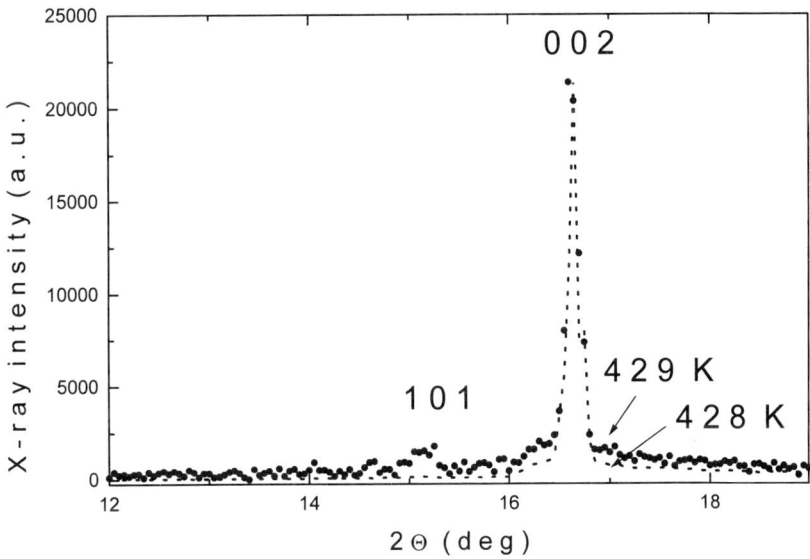

FIGURE 4. The XRD pattern recorded at 428 K, just below T_M, is compared with the contribution from solid to the overall pattern (see FIG. 1) at $T_M = 429$ K. The {002} peak at 429 K is broadened and shifted toward lower angles.

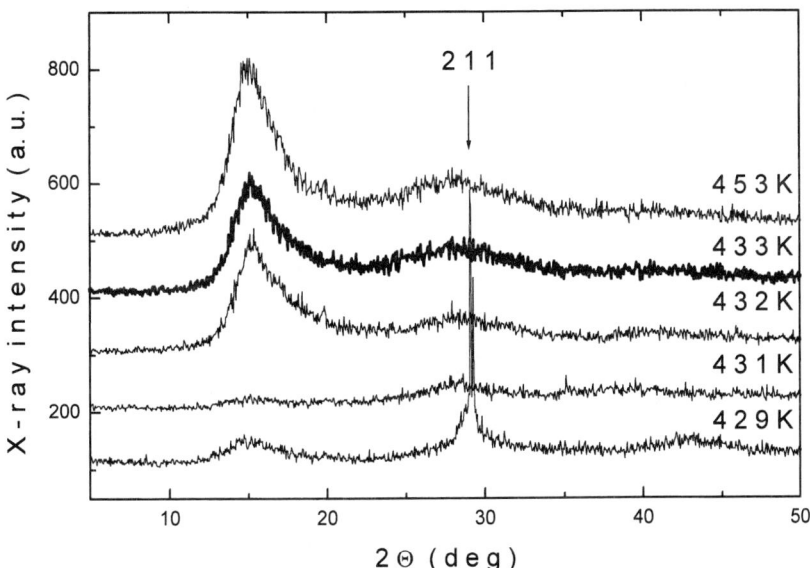

FIGURE 5. XRD spectra recorded at various temperatures during cooling to T_M.

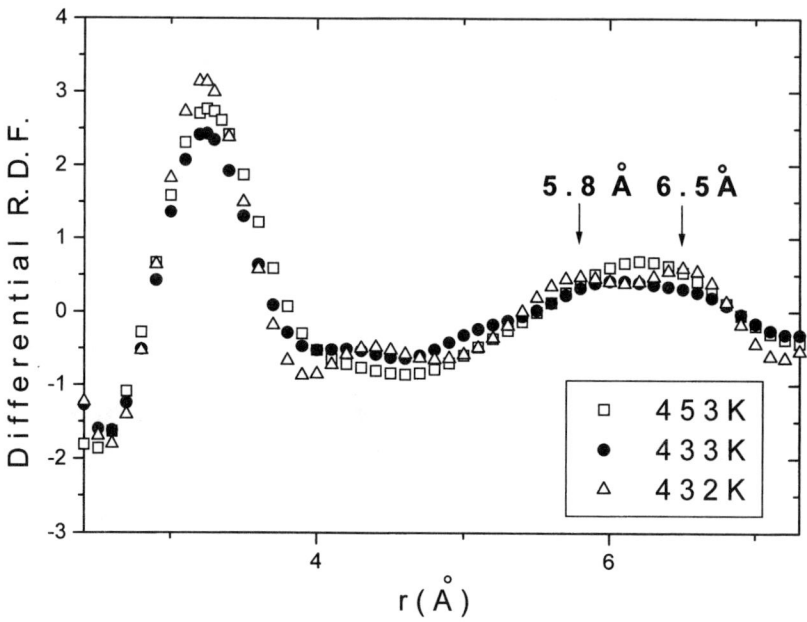

FIGURE 6. Differential RDF curves of liquid indium obtained from patterns in FIGURE 5.

{002} interplanar spacing (d_{002} = 2.4494 Å at 428 K, d_{002} = 2.4568 Å at 429 K) thus it involves an abrupt increase of the c-axis.

Solidification

FIGURE 5 shows the XRD spectra collected during cooling to T_M, where the first solid forms with a strong {211} orientation. The diffraction pattern recorded at 431 K—that is, immediately before solidification—shows an anomalous trend: the second maximum is more intense than the first. The differential RDF curves obtained from diffraction patterns in FIGURE 5 are plotted in FIGURE 6.

DISCUSSION

The peak shift observed at T_M in melting experiments indicates that the c-axis length suddenly turns away from the trend predicted by the equations of Graham et al.[3] This phenomenon can be explained by the occurrence of an abnormal increase in vacancy concentration in the solid.[4] In fact, the main effect of vacancies on crystal lattice consists in its isotropic dilatation with consequent increase of all the interplanar spacings, which is evidenced by a general shift of XRD reflections towards lower angles. The case of indium is of particular interest due to the unusual thermal expansion exhibited by this metal: the c-axis decreases as temperature increases (Eq. 7).

Because {0 0 2} interplanar spacing corresponds to the c-axis half-length, shortening of c-axis involves shift of the {0 0 2} peak to higher angles. When the concentration of vacancies undergoes the abnormal increase at T_M, the lattice expansion is strong enough to push the {0 0 2} peak in the opposite direction, to lower angles ($\Delta 2\theta \cong 0.05°$). As reported in Reference 5, the peak shift has also been observed in samples with an initial {1 0 1} texture.

As well as lattice expansion, an abnormal increase of vacancies at T_M should involve extremely favorable conditions for vacancy clustering. Due to higher concentration and migration velocity, vacancies usually tend to aggregate forming complex defects as temperature increases. For instance, positron annihilation experiments on indium performed by Kishimoto and Tanigawa[6] evidenced a remarkable increase in mean lifetimes at 423 K, near T_M, caused by divacancies. The broadening of XRD peaks, in particular in the lower part, is consistent with vacancy clustering in small dislocation loops. As shown theoretically by Dederichs,[7] this type of defect produces a diffuse scattering that affects the tails of diffraction profiles. Unfortunately, the counting statistics of present experiments is too poor to allow quantitative evaluations.

As shown in FIGURE 4, the {1 0 1} peak appears when the first liquid forms. To explain the change two hypothesis connected to different morphologies (diffuse or sharp) of the S–L interface can be considered. In the case of a diffuse interface, liquid advances in the solid along preferential directions and solid blocks, detached from the bulk, may be dragged away by the liquid. This process produces a reorientation of the blocks that can give rise to the appearance of new diffraction peaks. If the S–L interface is sharp, the presence of the {1 0 1} peak may be explained by a sudden texture change in the solid phase. Because of a preliminary treatment, consisting of slow cooling to room temperature from the liquid state, the examined material has a low dislocation density. Therefore, the driving force for recovery and recrystallization is low and the possible texture change (second hypothesis) connected to the appearance of a {1 0 1} reflection at 429 K is ascribable to grain growth, made easier by the initial strong texture.[8] In this case the presence of {1 0 1} peak is due to the growth of grains with this orientation at the expense of those with a {0 0 2} orientation. Since grain growth is a thermally activated process, the abrupt appearance of the {1 0 1} peak is a further indication of the abnormal increase of vacancy concentration observed immediately before metal melting. The enhancement of boundary mobility by a vacancy flux, or supersaturation, is extensively discussed in several review papers (e.g., see Ref. 9).

Regardless of the interface morphology, the lattice planes facing the first formed liquid are the {1 0 1} and {0 0 2} planes that allocate first and second neighbors (see FIG. 3) with shell radii very close to the mean distance of nearest neighbors in the liquid. On the basis of present results, a correlation between the structures of solid and liquid seems to exist.

On the other hand, the solidification experiments also seem to confirm this correlation. The split of the second RDF maximum indicates that a structural change in the liquid phase precedes solidification. Atom clusters form with mean distances corresponding to those of nearest neighbors in those planes, {2 1 1} (see FIGURE 7), that will result to face the liquid. These are low-atomic-density planes and grow faster

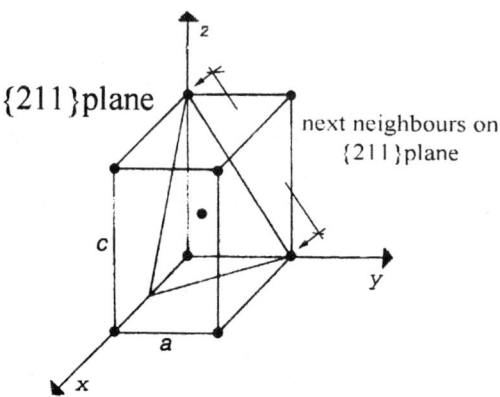

FIGURE 7. The distance between nearest neighbors on the {2 1 1} plane of indium is 5.9 Å.

than the close-packed planes, such as {0 0 2}. As a consequence,[10] slow-growing close-packed faces are the most developed, and they determine the final texture. Other experiments are underway to verify the occurrence of such phenomenon when the solid forms with different preferred orientations.

FURTHER DEVELOPMENTS AND EXPERIMENTS IN MICROGRAVITY

Present results are consistent with interactions between solid and liquid and promise to shed more light on mechanisms governing melting and solidification of metals. However, the counting statistics of present XRD measurements is too poor. In conventional diffractometry the pattern is collected by moving the counter to increasing angular positions. Thus, different parts of the spectrum are recorded at different times. During melting and solidification the relative amounts of solid and liquid progressively change and XRD monitoring of the process should be performed in a sufficiently short time. Under these conditions it is impossible to study the diffuse scattering along peak tails to evaluate average density and size of dislocation loops forming by vacancy aggregation. This drawback can be overcome by employing time analysis diffractometry (TAD). A sketch of the TAD apparatus is provided in FIGURE 8.

The experimental apparatus substantially consists of a side-window position-sensitive counter and a multichannel analyzer (MCA).[11] The TAD technique permits collection, simultaneously, of a large part of the spectrum (some tens of degrees) with a resolution comparable to that obtained in conventional diffractometry. In this way it is possible to get longer counting times for each angular position with improved statistics.

Study of the observed phenomenon will be simplified in µg by the absence of convective motions in the liquid phase. However, Marangoni motions may represent a serious disturbance for X-ray measurements, also in µg. Furthermore, the

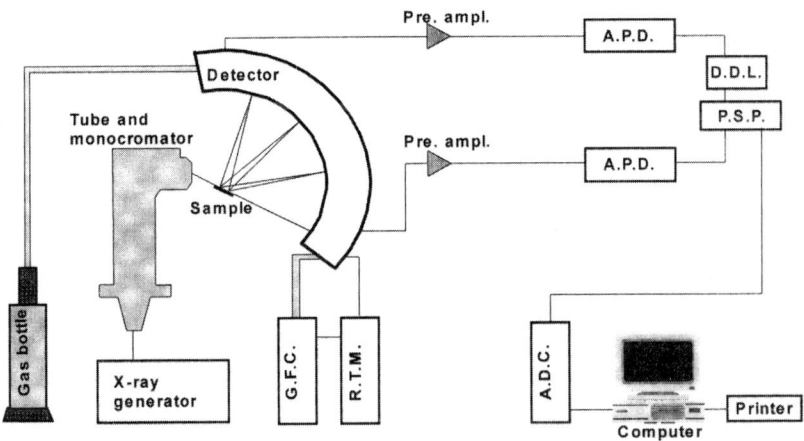

FIGURE 8. TAD apparatus. The analog position discriminator (A.P.D.) enables the determination of pulse centroid, the position sensitive processor (P.S.P.) produces pulses of amplitude proportional to the position on the detector of X-photons, the analog to digital converter (A.D.C.) transforms in a digital format the amplitude of pulses coming from P.S.P., the double delay line (D.D.L.) permits to expand part of the diffractogram. R.T.M. and G.F.C. are the ratemeter and the gas flow controller, respectively.

shape change due to surface tension occurring during melting may also cause liquid movements. To achieve the best experimental conditions in μg, thin oxide films grown on the sample surface will be used as containers during melting. The method was successfully tested in the experiment ES 311 A-B carried out in the mission SPACELAB-1.[12]

CONCLUSIONS

During melting and solidification of pure indium the structures of solid and liquid seem to be correlated. The improvement of counting statistics by employing the technique TAD will allow more details about these processes to be achieved.

To obtain the best experimental conditions it is necessary to eliminate convective motions in the melt, thus, the same experiments should be repeated in μg. For the specific characteristics of XRD measurements, the ISS appears to be the ideal laboratory.

It is suggested to control the shape change due to surface tension, which may cause liquid movements also in μg, by employing thin oxide films grown on the metal surface as containers during melting and solidification.

ACKNOWLEDGMENTS

The present investigation was carried out with funds made available by A.S.I. (Italian Space Agency).

REFERENCES

1. UBBELOHDE, A.R. 1978. The Molten State of Matter. Wiley Interscience.
2. WARREN, B.E. 1969. X-ray diffraction. Addison-Wesley Publishing Company.
3. GRAHAM, J., A. MOORE & G.V. RAYNOR. 1955. The effect of temperature on the lattice spacings of indium. J. Inst. Metals **84:** 86–87.
4. GONDI, P., R. MONTANARI & G. COSTANZA. 2000. X-Ray characterization of indium during melting. Adv. Space Res. **29**(4): 521–525.
5. MONTANARI, R. & G. COSTANZA. 2000. XRD investigation on structural aspect of indium melting. Proceedings of the 1st International Symposium on Microgravity Research and Applications in Physical Sciences and Technologies, Sorrento-Italy: 417–424.
6. KISHIMOTO, Y. & S. TANIGAWA. 1982. *In* Point Defects and Defect Interactions in Metals. J. Takamura, M. Doyama & M. Kiritani, Eds.: 281–283. North-Holland Publishing Co., Amsterdam-Netherlands.
7. DEDERICHS, P.H. 1971. Diffuse scattering from defect clusters near Bragg reflections. Phys. Rev. B **4**(4): 1041–1050.
8. HUMPHREYS, F.J. & M. HATHERLY. 1996. Recrystallization and Related Annealing Phenomena. 321. Pergamon Press, London.
9. HAESSNER, F. & S. HOFMAN. 1978. Recrystallization of Metallic Materials. Haessner, Ed.: 63. DR Riederer Verlag, Stuttgart-Germany.
10. REED-HILL, R.E. & R. ABBASCHIAN. 1994. Physical Metallurgy Principles, 3rd edit. 439. International Thomson Publishing, Boston.
11. GONDI, P., R. MONTANARI, E. EVANGELISTA, *et al.* 1997. X-ray study of structures in liquid metals with controlled convective motions. Micrograv. Quart. **7**(4): 155–160.
12. BARBIERI, F., P. GONDI, R. MONTANARI, *et al.* 1984. Experiment ES 311: bubble reinforced materials. Proceedings of 5th European Symposium on Material Sciences under Microgravity, Schloss Elmau-Germany. 101–107.

Unidirectional Solidification of Magnetostrictive Materials Using a Magnetic Field in Microgravity

HIDEKI MINAGAWA,[a] KEIJI KAMADA,[b] HIDEAKI NAGAI,[a] YOSHINORI NAKATA,[a] AND TAKESHI OKUTANI[a]

[a]*Microgravity Materials Laboratory,*
National Institute of Advanced Industrial Science and Technology,
Tsukisamu-Higashi, Toyohira-ku, Sapporo Japan

[b]*Japan Space Utilization Promotion Center, Nishiwaseda,*
Shinjuku-ku, Tokyo 169-8624, Japan

ABSTRACT: Unidirectional solidification experiments of magnetostrictive materials of $Tb_{0.3}Dy_{0.7}Fe_{2-x}$, (Terfenol-D) alloy in a static magnetic field and microgravity were performed using the Japan Microgravity Center (JAMIC) free-fall facilities. When the magnetic field strength was increased from 0 T to 3.7×10^{-2} T during unidirectional solidification in microgravity, a <1 1 1> crystallographic alignment of the grains dominated, and the maximum magnetostriction constant increased from 500 ppm to 2,000 ppm. For unidirectional solidification in normal gravity, the maximum magnetostriction constant remained at 1,000 ppm for increases in the magnetic field.

KEYWORDS: $Tb_{0.3}Dy_{0.7}Fe_{1.9}$; Terfenol-D; magnetostrictive material; unidirectional solidification; microgravity; magnetic field

INTRODUCTION

$TbFe_2$ and $Tb_{0.3}Dy_{0.7}Fe_{2-x}$ are drawing much attention due to their excellent magnetostrictive properties at room temperature.[1] The alloy composition with $Tb_{0.3}Dy_{0.7}Fe_2$ (Terfenol-D alloy) exhibits an especially large magnetostriction and high sensitivity to an external magnetic field.[1,2] These materials have found many applications, such as linear actuators and sensor transducer systems. To obtain excellent magnetostriction, $TbFe_2$ and the $Tb_{0.3}Dy_{0.7}Fe_{2-x}$ alloys having [1 1 1] orientation and large crystals are needed.[2] However, large, [1 1 1] oriented crystals of $TbFe_2$ or $Tb_{0.3}Dy_{0.7}Fe_{2-x}$ are very difficult to synthesize in normal gravity because of heterogeneous nucleation and subsequent randomly oriented polycrystalline structure produced by thermal convection in the melt. In microgravity, thermal convection and mass transfer are suppressed.[3] In our previous study, we demonstrated unidirectional solidification by rapid cooling in microgravity to produce single crystals.[4] However,

---

Address for correspondence: Hideki Minagawa, Microgravity Materials Laboratory, National Institute of Advanced Industrial Science and Technology, 2-17-2-1 Tsukisamu-Higashi, Toyohira-ku, Sapporo 062-8517, Japan. Voice: +81-11-857-8942; fax: +81-11-857-8968.
  h-minagawa@aist.go.jp

when solidifying magnetic materials, there are many difficulties in forming homogeneous or crystallographically arranged structures, even in microgravity.

A magnetic field applied to a melt is known to suppress convection during solidification, thereby reducing segregation. For example, much research on the effect of a magnetic field during Czochralski and Bridgeman crystal growth has been reported.[5–9] Kim[10] showed that convective mass transfer can be inhibited, but that it is very difficult to totally suppress it in a weak magnetic field. There have also been several reports describing the application of a high magnetic field, up to 8T, and resulting macroscopic segregation following diffusion-controlled growth.[7,8,11] Taking a different point of view, we plan to produce oriented structures and large crystals in microgravity and/or a static magnetic field, because the convection in the melts is suppressed in microgravity and a magnetic field is sufficiently strong to produce oriented structures in an environment without gravity. Thus, we proposed unidirectional solidification as a promising crystal growth method in microgravity and a magnetic field. In this paper, we report an experiment in the crystal growth of $Tb_{0.3}Dy_{0.7}Fe_{2-x}$ by unidirectional solidification in a static magnetic field and microgravity.

EXPERIMENTAL

We performed microgravity experiments in the free-fall facility of the Japan Microgravity Center (JAMIC).[12] The details of the apparatus are described elsewhere.[13] Melting and solidification experiments were performed in an argon atmosphere at a pressure of 0.15 Pa with a flow rate of 20 sccm (3.4×10^{-2} Pam^3/sec). The

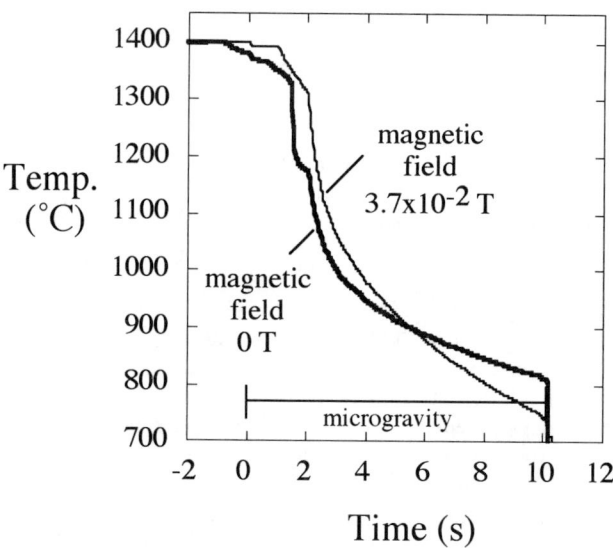

FIGURE 1. Cooling curve for solidification in a magnetic field and in microgravity.

temperature of the sample was monitored by Pt–Pt/Rh (13%) thermocouples (type P) and an infrared sensor. The sample was heated, using two infrared heaters, from room temperature to 1,550°C in 60 sec. Unidirectional solidification was induced by direct contact with an iron chill block around which a magnetic coil is placed for the static magnetic field experiment. A magnetic field was applied by an electromagnet parallel to the solidification direction. The maximum magnetic field was 0.1 T. The samples were produced by alloying the three constituents, Tb, Dy, and Fe, by arc melting in Ar using a nonconsumable electrode and a water-cooled copper hearth and mold. The atomic ratios of the constituents were Tb:Dy:Fe = 3:7:19. For the solidification experiment, we cut the matrix into 10-mm diameter × 2-mm thick disks. FIGURE 1 shows the cooling curve for solidification in a magnetic field and in microgravity. Initially, the sample was heated by two infrared heaters from room temperature to 1,400°C within 40 sec. When the sample reached 1,400°C and was fully molten, the system was dropped. After a 2.0 sec free fall, heating was stopped and solidification was induced by contact with the chill block. In normal gravity, the sample cooled from 1,400°C to 800°C within 8.0 sec.

We examined the samples using a scanning electron microscope (JEOL, JXA-8600) and an optical microscope for microstructure (Nikon, Measurescope 20), an electron probe microanalyzer (JEOL, JXA-8600) for composition data, and X-ray diffractometer (Rigaku, Geigerflex RAD system) for crystalline structure. A magnetostriction measurement instrument (Tamakawa Co., Ltd.) was used to measure the magnetostriction constant, using the *three-terminal capacitance method*.[14,15]

For the optical microscopy observation, we polished the surface of the sample and chemically etched it using nital solution (HNO_3 [5.0 vol.%]/C_2H_5OH).

RESULTS AND DISCUSSION

FIGURE 2 shows the microstructures of the $Tb_{0.3}Dy_{0.7}Fe_{1.9}$ sample solidified in 0 and 3.7×10^{-2} T magnetic fields in normal gravity and microgravity. In normal gravity solidification, we observed randomly aligned columnar structures in the sample solidified at 0 T and 3.7×10^{-2} T. For solidification in microgravity, the samples produced under 0 T and 3.7×10^{-2} T exhibited a mixture of dendrite structure and columnar structure. The grain lengths along the solidified direction of the sample solidified at 3.7×10^{-2} T exceeded 1 mm, whereas those of the sample solidified at 0 T were a few hundred microns with slightly disordered dendrite structures.

FIGURE 3 shows X-ray diffraction patterns of sample surfaces that had contacted the chill block in normal gravity and microgravity. One broad peak at around 20 degrees, caused by diffraction from the support material (resin) of the samples, was measured as a background and superimposed on the spectra of $TbFe_2$, $DyFe_2$ (1 1 1). The structure of the product solidified at 0 T in microgravity was amorphous, but we observed (Tb–Dy)Fe_2 (0 2 2) and (Tb–Dy)Fe_2 (1 1 3) peaks in the sample produced in 9.5×10^{-3} T. Furthermore, the (Tb–Dy)Fe_2 (2 2 2) peak increased considerably in the sample solidified in the 3.7×10^{-2} T magnetic field.

These results indicate that crystallographic alignment along the [1 1 1] direction was achieved for unidirectional solidification in microgravity and a static magnetic field. Shi *et al.*[16] reported crystallographic alignment along the [1 1 0] direction and

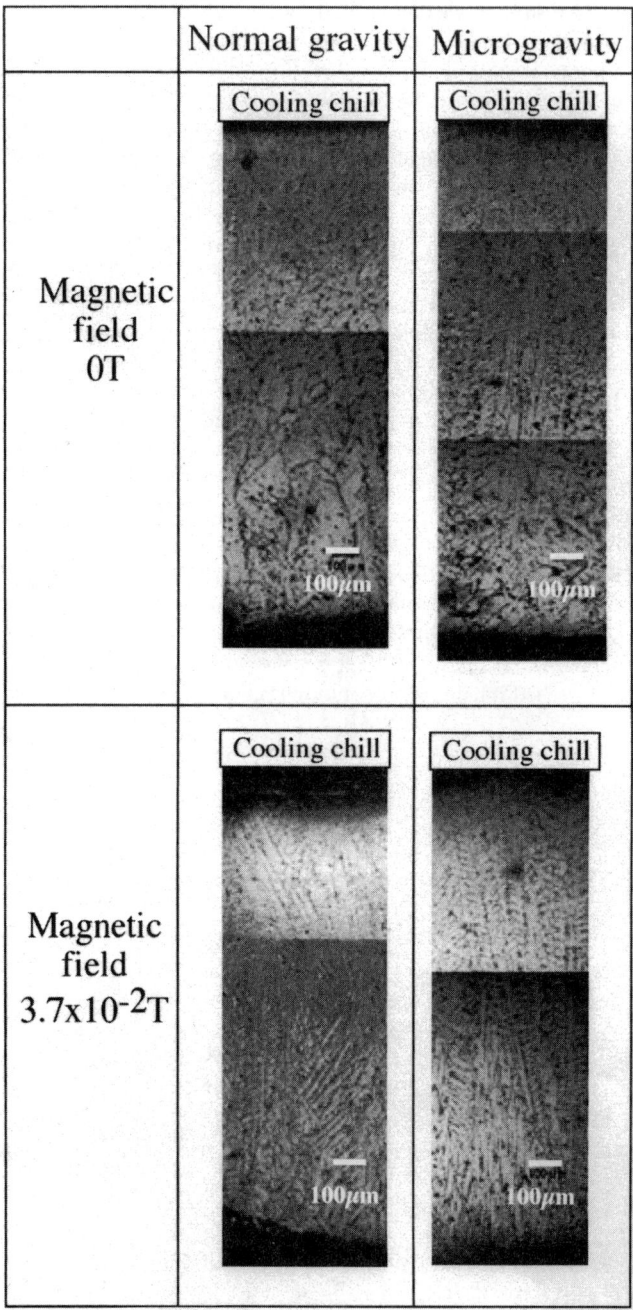

FIGURE 2. Microstructures of the $Tb_{0.3}Dy_{0.7}Fe_{1.9}$ sample solidified in 0 and 3.7×10^{-2} T magnetic fields in normal gravity and microgravity. (Etched by nital solution.)

Verhoven et al.[17–19] reported [1 1 2] texture observed in a directionally solidified (Tb–Dy)Fe$_2$ alloy.

FIGURE 4 shows typical magnetostriction curves of Tb$_{0.3}$Dy$_{0.7}$Fe$_{1.9}$ produced in normal gravity and microgravity. The magnetostriction curves of TbFe$_2$ are superimposed.[13] For unidirectional solidification in normal gravity, the maximum magnetostriction remained below 1,000 ppm despite the increased magnetic field used during solidification. For unidirectional solidification in microgravity, the

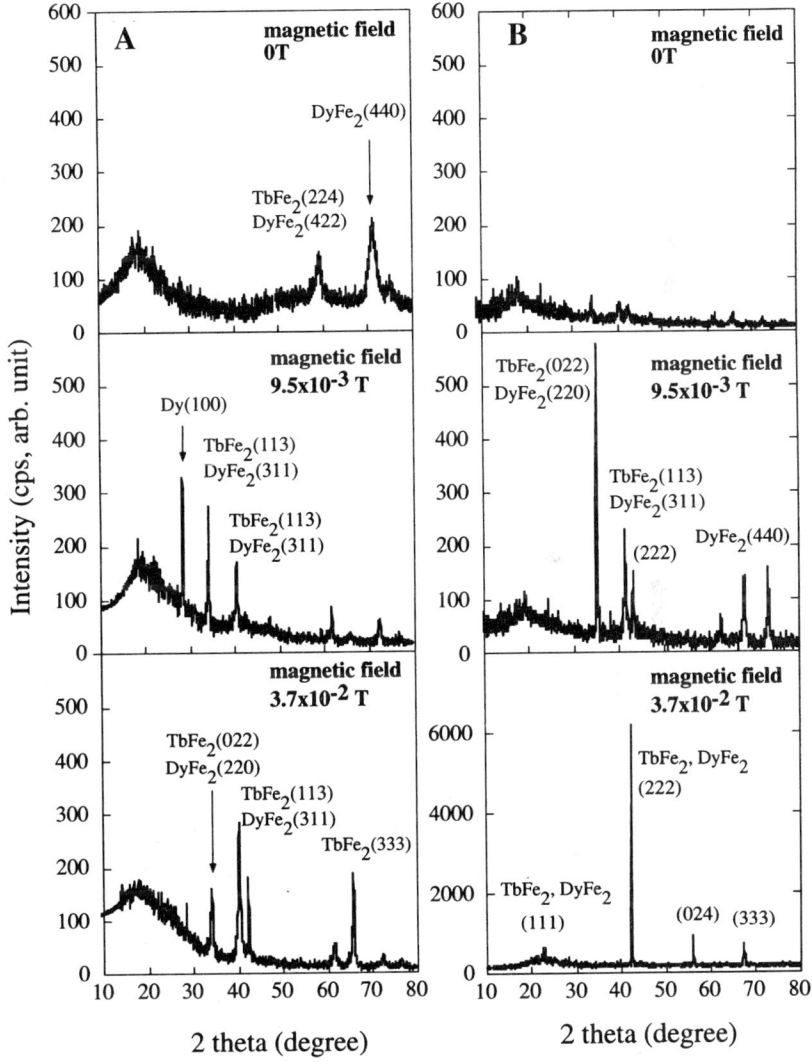

FIGURE 3. X-ray diffraction patterns from the sample surfaces that had contacted the chill block in normal gravity (**A**) and microgravity (**B**).

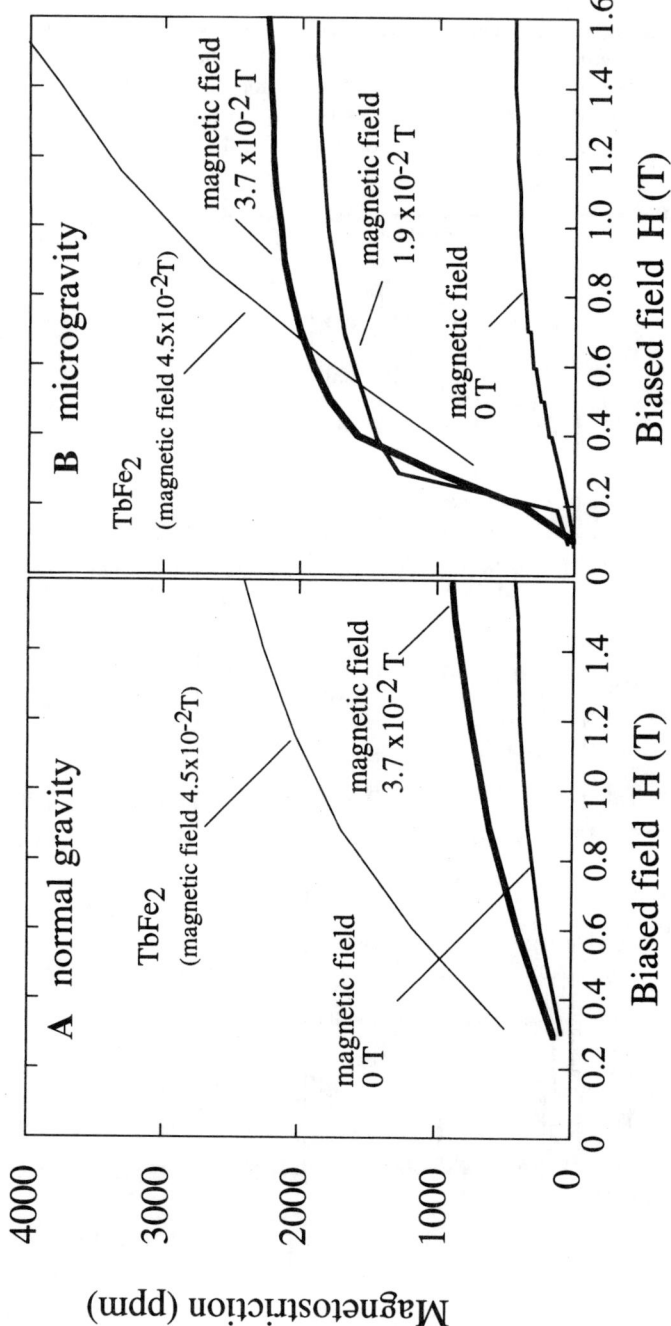

FIGURE 4. Typical magnetostriction curves of $Tb_{0.3}Dy_{0.7}Fe_{1.9}$ produced in normal gravity (**A**) and microgravity (**B**). The magnetostrictive characteristics were measured at room temperature (25°C).

maximum magnetostriction constant increased from 500 ppm to 2,000 ppm as the magnetic field strength during solidification increased from 0T to 3.7×10^{-2} T. The maximum magnetostriction is about 2,000 ppm, which is about half of that of $TbFe_2$,[13] whereas the magnetostriction constants of $Tb_{0.3}Dy_{0.7}Fe_{1.9}$ are saturated at a biased field of 0.4T. The saturated magnetostriction constant at 0.4T of $Tb_{0.3}Dy_{0.7}Fe_{1.9}$ was about 1,700 ppm, which was much larger than that of $TbFe_2$. These results reveal that unidirectional solidification in a static magnetic field and microgravity is of great advantage in synthesizing $Tb_{0.3}Dy_{0.7}Fe_{1.9}$ for magnetostrictive materials.

CONCLUSIONS

Columnar grain-aligned $Tb_{0.3}Dy_{0.7}Fe_{1.9}$ discs, 10 mm diameter × 3 mm depth, were produced by unidirectional solidification using a static magnetic field in microgravity. The structure of $Tb_{0.3}Dy_{0.7}Fe_{1.9}$ alloy unidirectionally solidified in microgravity changed with the applied magnetic field during solidification from amorphous to [1 1 1] crystallographic alignment as the magnetic field during solidification was increased. This increase in magnetic field during solidification also markedly improved the magnetostrictive properties of the materials.

ACKNOWLEDGMENTS

The authors would like to thank Professor R. W. Smith of Queen's University for helpful discussion, remarks, and suggestions. This work was supported by the Japan Space Utilization Promotion Center entrusted by NEDO.

REFERENCES

1. CLARK, A.E. & H. BELSON. 1972. Giant room-temperature magnetostrictions in $TbFe_2$ and $DyFe_2$. Phy. Rev. B **5**: 3642–3644.
2. CLARK, A.E. 1980. Magnetostrictive rare earth–Fe_2 compounds. *In* Ferromagnetic Materials, Vol. 1. E.P. Wohlfarth, Ed.: 531–581. North-Holland Publishing company.
3. SHIBUYA, M., Y. NAKATA, H. NAGAI, *et al.* 2001. Effect of non-equillibrium gravity on convection behavior. J. Jpn. Soc. Micrograv. Appl. **18**(Suppl.): 51.
4. OKUTANI, T., H. MINAGAWA, H. NAGAI, *et al.* 1999. Synthesis of crystalline materials with high quality under short-time microgravity. Ceramic Eng. Sci. Proc. **20**(4): 215–226.
5. MOTAKEF, SH. 1990. Magnetic field elimination of convective interference with segregation during vertical-bridgman growth of doped semiconductors. J. Cryst. Growth **104**: 833–850.
6. ELURLE, D.T.J. & R.W. SELIES. 1994. Handbook of Crystal Growth, Vol. 2, 261. Elsevier Science, Amsterdam.
7. MATTHESEN, D.H., M.J. WARGO, SH. MOTAKEF, *et al.* 1987. Dopant segregation during vertical bridgman-atockbarger growth with melt stabilization by strong axial magnetic fields. J. Cryst. Growth **85**: 557–560.
8. BECLA, P., J.-C. HAN & SH. MOTAKEF. 1992. Application of strong vertical magnetic fields to growth of II-VI pseudo-binary alloys: HgMnTe. J. Cryst. Growth **121**: 394–398.
9. KANG, J., Y. OKANO, K. HOSHIKAWA & T. FUKUDA. 1994. Influence of a high vertical magnetic field on Te dopant segregation in Sb grown by the vertical gradient freeze method. J. Cryst. Growth **140**: 435–438.

10. KIM, K.M. 1982. Suppression of thermal convection by transverse magnetic field. J. Electrochem. Soc. **129:** 427–429.
11. YASUDA, H., I. OHNAKA, Y. FURUKUBO, et al. 1996. Macrosegregation of unidirectionally solidified (Bi, Sb_2)Te_3 in a magnetic field. *In* Solidification Science and Processing. I. Ohnaka & D.M. Stefanescu, Eds.: 311–317. The Minerals, Metals and Materials Society, Warrendale, PA.
12. NAGAI, H., F. ROSSIGNOL, Y. NAKATA, et al. 1998. Wetting of molten indium under short-time microgravity. Mater. Sci. Eng. A **248:** 206–211.
13. MINAGAWA, H., K. KAMADA, T. KONISHI, et al. 2001. Unidirectional solidification of $TbFe_2$ alloy using magnetic field in microgravity. J. Magn. Magn. Mater. **234:** 437–442.
14. O'CONNOR, M. & H.S. BELSON. 1970. Magnetostriction of commercial ferrite memory cores. J. Appl. Phys. **41:** 1028–1029.
15. TSUYA, N., K.I. ARAI, K. OHMORI & Y. SHIRAGA. 1974. Magnetostriction measurement by three terminal capacitance method. Jpn. J. Appl. Phys. **13:** 1808–1810.
16. SHI, ZH., ZH. CHEN, XI. WANG, et al. 1997. Directionally solidified $Tb_xDy_{1-x}(MyFe_{1-y})_{1.9}$ giant magnetostrictive materials and their applications in transducers. J. Alloy Comp. **258:** 30–33.
17. VERHOVEN, J.D., E.D. GIBSON, O.D. MCMASTERS & H.H. BARKER. 1987. The growth of single crystal terfenol-D crystals. Metall. Trans. **18A:** 223–231.
18. VERHOVEN, J.D., E.D. GIBSON, O.D. MCMASTERS & G.E. OSTENSON. 1990. Directional solidification and heat treatment of Terfenol-D magnetostrictive materials. Metall. Trans. **21A:** 2249–2255.
19. MEI, W., T. UMEDA, S.Z. ZHOU & R. WANG. 1995. Magnetostriction of grain-aligned $Tb_{0.3}Dy_{0.7}Fe_{1.95}$. J. Alloy Comp. **224:** 76–80.

Dielectric *in Situ* Monitoring of Microgravity Polymerizations

ALVIN P. KENNEDY AND SOLOMON TADESSE

Department of Chemistry, Morgan State University, Baltimore, Maryland, USA

> ABSTRACT: The objective of the research described here is to develop dielectric spectroscopy to monitor polymerization reactions and processes in microgravity. Ground based measurements have been made on both neat thermoset resins and solution polymerization reactions. Polymerization of diglycidyl ether of bisphenol A (DGEBA) using two curing agents, *m*-phenylenediamine (*m*-PDA), and 4,4'-diamino-diphenyl-methane (DDM) were investigated to determine the relationship between the dielectric spectra and the thermoset polymer network and morphology. The results for the epoxy thermosets indicate that this technique can be used to monitor the extent of polymerization and the qualitative nature of polymer network. The photopolymerization of 6-(2-methyl-4-nitroanilino)-2,4-hexadiyn-1-ol (DAMNA) in solution was also investigated to determine if gravitational effects could be observed using dielectric spectroscopy. The technique is able to distinguish between surface and solution polymerizations during formation of the polydiacetylene thin films. In addition, it is capable of monitoring convective effects during solution polymerization.
>
> KEYWORDS: dielectric monitoring; microgravity polymerization

NOMENCLATURE:

ε'	dielectric constant
ε''	dielectric loss
ε_r	relaxed (uncured) dielectric constant
ε_u	unrelaxed (cured) dielectric constant
ε_0	permitivity of free space
σ	ionic conductivity
ω	angular frequency
τ	characteristic relaxation time
DGEBA	diglycidyl ether bisphenol A
m-PDA	m-phenylenediamine
DDM	4,4'-diamino-diphenyl-methane
DAMNA	6-(2-methyl-4-nitroanilino)-2,4-hexadiyn-1-ol

INTRODUCTION

As human space exploration continues it becomes important to understand how microgravity affects the processing of polymers and polymer composites used in space based devices and structures. Various technologies need to be developed to monitor the fabrication and integrity of polymer based devices and structures produced and used in space. *In situ* processing of polymer structures in space is

---

Address for correspondence: Alvin P. Kennedy, Department of Chemistry, Morgan State University, Baltimore, MD 21251, USA.
akennedy@morgan.edu

necessary to build and repair the space station, to develop optic networks and devices, as well as to develop radiation shielding on the Moon and Mars.[1]

Polymeric systems have not been extensively investigated in microgravity because polymers were believed to be too viscous for any effects such as convection or sedimentation to be important.[2] However, Briskman has recently shown that four mechanisms: sedimentation of polymeric microglobules, thermal and concentration convection, instability of polymerization front, and the effect of mass forces on a deformable polymer can influence polymerization processes and polymer structures.[3] The first two effects are important in solution polymerization. To understand the effects of these gravitational mechanisms, polymerization processes should be monitored throughout the duration of the reaction. At this time most polymerization experiments conducted under microgravity conditions are not monitored during flight. *In situ* monitoring techniques can make these experiments more cost effective by providing real-time data to compare to ground based experiments. Many different techniques have been used to investigate polymerization processes on earth, however dielectric (impedance) spectroscopy provides a simple yet powerful technique for following the complete polymerization process from monomer through fully cured polymer.

Dielectric measurements are usually performed at constant temperature or constant frequency. Under these conditions changes in the dielectric spectra are associated with molecular motions. The high frequency responses are due to alpha and/or glass transitions, whereas other molecular motions, such as side group rotations and ionic diffusion, occur at lower frequencies. The resulting response has been correlated with the well-established Debye theory. This theory is based on a static number of dipoles in a fully polymerized system that responds to variations in the applied frequency. The dielectric constant, ε', is given by

$$\varepsilon' = \varepsilon_u + \frac{\varepsilon_r - \varepsilon_u}{1 + (\omega\tau)^2}, \tag{1}$$

where ε_u and ε_r are the unrelaxed and the relaxed dielectric constants, respectively, ω is the angular frequency given by $\omega = 2\pi f$, and τ is the characteristic relaxation time. The dielectric loss ε'' is given by

$$\varepsilon'' = \frac{\sigma}{\omega\varepsilon_0} + \frac{(\varepsilon_r - \varepsilon_u)\omega\tau}{1 + (\omega\tau)^2}, \tag{2}$$

where σ is the ionic conductivity and ε_0 is the permittivity of free space. The first term represents the ionic conductivity of the material and becomes dominant at low frequencies. It varies with the viscosity of the polymer and the ionic concentration. The second term is the dipolar component of the dielectric loss. It is associated with segmental motion of the polymer chain. A maximum is observed in the dielectric loss spectra when $\omega\tau = 1$ and, experimentally, the relaxation time can be determined from the loss peak.

Dielectric spectroscopy has also been used to study the polymerization of a variety of thermoset resins, most of this research has been conducted on epoxy thermosets. The dielectric spectra of epoxy resins measured during cure, as a function of time, are similar to the isothermal dielectric spectra measured as a function of

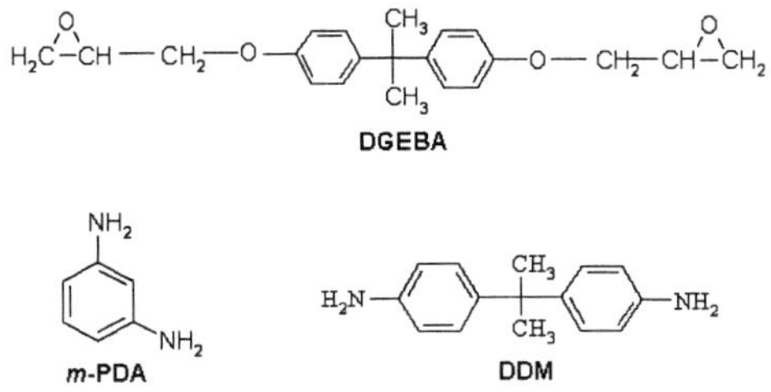

FIGURE 1. *m*-PDA and DDM.

frequency for fully cured polymers. In the Debye formalism ε_r and ε_u correspond to the neat resin and cured polymer, respectively. Although the dipoles in the monomer and curing agent are consumed during the polymerization process, the model has worked well for epoxy resins. The molecular origin of the dipolar component during cure is not fully understood and it is important to determine this in order for the technique to be used successfully to monitor polymerization reactions. The first part of this research investigates how the dielectric properties change with network morphology and structure by using two different curing agents to form an epoxy thermoset of diglycidyl ether of bisphenol A (DGEBA). The curing agents, *m*-phenylenediamine (*m*-PDA), and 4,4′-diamino-diphenyl-methane (DDM) (see FIGURE 1), have different reactivities and produce different network morphologies and structures.

Most gravitational effects do not influence the high viscosity polymers produced by these resins; however, when polymerized in solution, the effects become more prevalent. There have been a few investigations involving the solution polymerization of epoxy resins.[4,5] In the second part of this research the solution photopolymerization of 6-(2-methyl-4-nitroanilino)-2,4-hexadiyn-1-ol (DAMNA) (see FIGURE 2) was investigated to determine if dielectric spectroscopy can monitor the surface and solution polymerization reactions and gravitational effects. Frazier *et al.*[6,7] reported the surface polymerization of polydiacetylene films from the photopolymerization of DAMNA monomer solutions. This unique photopolymerization process produces amorphous thin films of polydiacetylene, which can be used in the fabrication of waveguides and photonic devices. However, films formed on earth are low quality and have defects that appear to be the result of buoyancy-driven convection. Microgravity experiments on the same system produced very high quality optical films. The lack of *in situ* measurements limited the analysis of the effect of this gravitational mechanism on the development of defects in the polydiacetylene films.

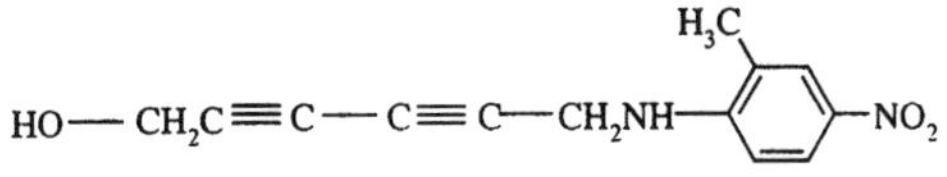

FIGURE 2. DAMNA monomer.

EXPERIMENTAL

Epoxy Resin Polymerization

The diepoxide prepolymer used in this study was EPON Resin 825 supplied by Shell Chemical Company. EPON Resin 825 is high purity low molecular weight liquid DGEBA with an epoxide equivalent weight in the range of about 175–180 (as reported by the vendor). The curing agents, m-phenylenediamine (m-PDA), and 4,4′-diamino-diphenyl-methane (DDM), were purchased from Fisher Scientific Company. Both the resin and the curing agents were used as supplied.

Stoichiometric amounts of DGEBA and each curing agent were mixed mechanically in small glass vials for five minutes at a constant temperature about one degree above the melting temperature of the curing agent. The resulting homogeneous fluid sample was cooled to room temperature by immediately quenching in ice water before it was used for the investigation of dielectric behavior.

Capacitance and dissipation factor values were measured using Hewlett Packard Model HP 4284A LCR meter interfaced to personal computer for automatic data acquisition. Measurements were taken using a small planar interdigitated electrode (a DekDyne gold–ceramic sensor), which was connected to the LCR meter using mini-test clamps. Dielectric measurements were performed at 60°C. Ten different frequencies were measured between 100 Hz and 1 MHz at five-minute intervals. The experimental values were converted to dielectric constant and dielectric loss using a calibration method developed in-house.

Polydiacetylene Polymerization

The impedance measurements were made using a HP4274A LCR meter. Measurements were taken at 400 Hz, and at 1, 2, 4, 10, 20, 40, and 100 kHz. Gold plated quartz interdigitated electrodes were purchased from DekDyne and used to measure the impedance of the solutions. A 100 W UV lamp was used at a distance of 10 cm from the cell. Data was taken at five-minute intervals for at least 24 hours. Solutions containing 2.5 mg/mL of DAMNA in 1,2 dichloroethane were used.

A temperature-controlled cell was designed to permit irradiation from both sides. A 2.54 cm quartz window was placed on one side and the interdigitated electrode on the other. Solution polymerization was monitored by irradiating through the quartz window while the electrode was in contact with the solution on the opposite side (see FIGURES 3A and 4A). It should be noted that when monitoring solution polymerization, only processes occurring near the surface of the electrode could be monitored. Polymerization at the surface was monitored by measuring the impedance

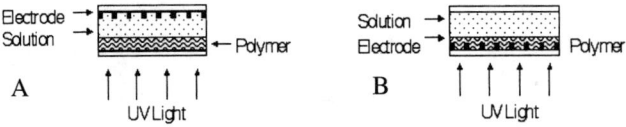

FIGURE 3. Bulk and surface polymerization with convection.

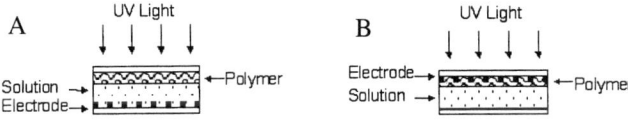

FIGURE 4. Bulk and Surface polymerization without convection.

while irradiating the solution through the back of the electrode (FIGS 3B and 4B). Because of the differences in transmission characteristics of the quartz window and the electrode, quantitative comparisons could not be made. The dielectric properties could not be calculated because of the irregular surface area that was irradiated.

DISCUSSION

Epoxy Resin Polymerization

The general features and overall trends of both dielectric constant (ε') and dielectric loss and (ε'') versus time plots are identical for both curing agents. Only those results for intermediate frequency ranges (1 kHz to 1 MHz) are reported here for the purpose of clarity and uniformity. FIGURES 5 and 6 show plots of ε' and ε'' against curing time (t) for m-PDA and DDM, respectively, over the nine intermediate frequencies at 60°C.

During the early cure time ε' decreased gradually. As the cure progressed, ε' decreased rapidly until it reached a characteristic minimum value, after which it leveled off. The magnitude of the relaxed dielectric constant (ε'_r) and that of the unrelaxed dielectric constant (ε'_u) changed slightly with frequency during cure for all the two curing agents. The change in the dielectric constant ($\Delta\varepsilon$) for both systems is presented in TABLE 1. It increased as the frequency of measurement is increased and the highest values were for the m-PDA system. The time required to reach the onset, as well as the completion, of the monotonic fall in ε' was also shorter for the DGEBA/m-PDA system. For measurements up to 200 kHz the ε'' versus time plots showed two peaks, one during the early cure time and one during the later stage in cure. The first peak results from ionic conductivity effect and the second peak is due to dipolar orientation.[8–10] At the beginning of cure, ε'' initially increased rapidly in magnitude to reach a maximum that began to decrease, passed through a minimum followed by a peak, as cure progressed, and ultimately decreased to a lower value characteristic of a fully cured polymer.[9] For measurements at frequencies of 500 kHz and 1 MHz, neither the initial rise in the ε'' values nor noticeable minima were observed. As a result only one peak was observed in the ε'' versus time plots of the two higher frequency measurements. These phenomena were observed in both systems. The initial steady increase in ε'' value is attributed to the rise in conductivity, which resulted from the decrease in viscosity due to rise in the temperature[10] of the sample as it equilibrated to the cure temperature. The decrease in ε'' values during the early cure time is due to the decrease in the conductivity of the system because

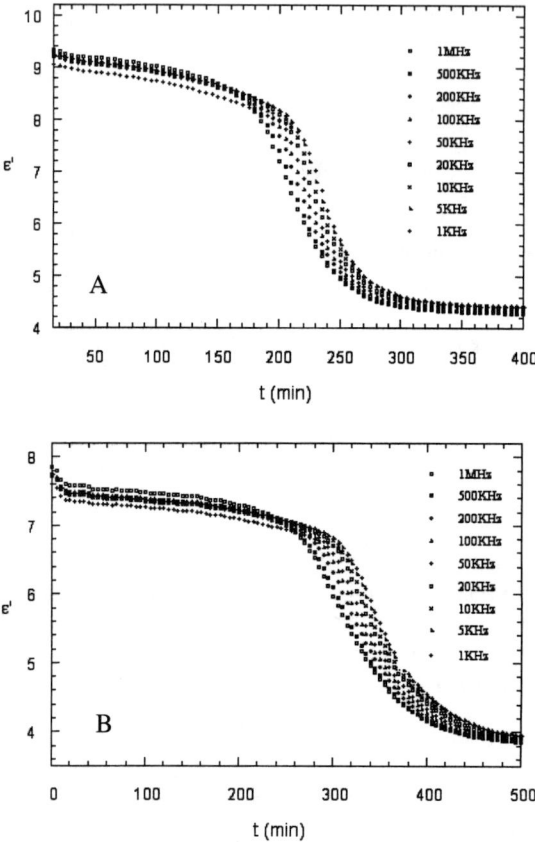

FIGURE 5. Dielectric constant versus cure time for DGEBA/*m*-PDA, and DGEBA/DDM cure at 60°C.

of the increase in viscosity due to polymerization. This begins to dominate the intrinsic temperature dependence of the viscosity[10] and results in the consumption of the crosslinking component by the cure reaction.[11]

Comparison of the individual ε'' versus time plots showed an increase in both the peak area and peak height as the frequency increased. The time to reach the loss peak (t_{max}) is given in TABLE 2. This time decreased as the frequency of measurement was

TABLE 1. Changes in the dielectric constant for DGEBA/*m*-PDA and DGEBA/DDM at 60°C

Material	$\Delta\varepsilon = \varepsilon'_r - \varepsilon'_u$								
	1 MHz	500 kHz	200 kHz	100 kHz	50 kHz	20 kHz	10 kHz	5 kHz	1 kHz
DGEBA/*m*-PDA	5.970	5.894	5.847	5.832	5.824	5.815	5.806	5.819	5.696
DGEBA/*m*-PDA	4.02	3.948	3.906	3.891	3.879	3.868	3.863	3.861	3.798

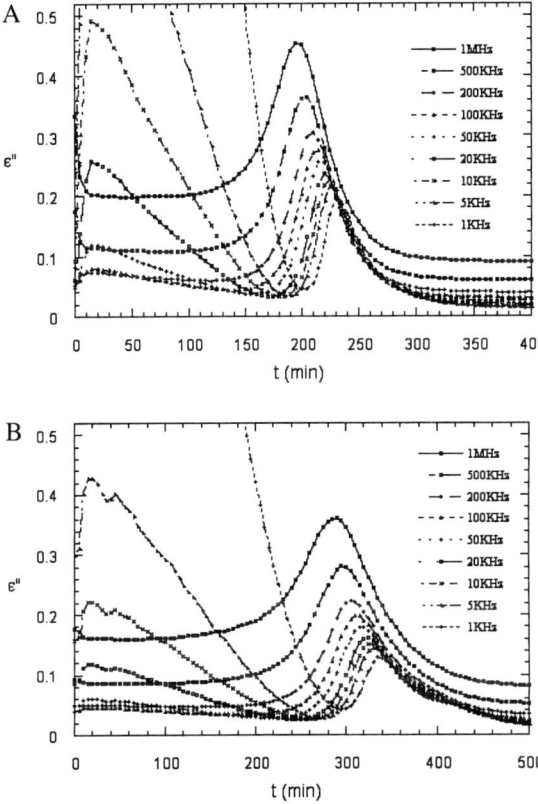

FIGURE 6. Dielectric loss versus cure time for DGEBA/m-PDA, and DGEBA/DDM during cure at 60°C.

increased and changed with the type of the curing agent. Similar behavior with frequency dependence has been reported in a number of studies[10,12] involving epoxy/amine systems. Wu, et al.[14] reported similar observation for systems formulated using hydroxyalkylated polyamine and dodoceneyl-sucinicanhydride, both of which are a different class of compounds from that used in this study.

TABLE 2. The time to the dielectric loss peak maximum during an isothermal curing of DGEBA/m-PDA and DGEBA/DDM at 60°C

Material	t_{max} (min)								
	1 MHz	500 kHz	200 kHz	100 kHz	50 kHz	20 kHz	10 kHz	5 kHz	1 kHz
DGEBA/m-PDA	200	205	210	215	220	220	225	230	230
DGEBA/m-PDA	290	295	305	310	315	325	325	330	335

In the final stage of cure, as the ε'' peak began to fall after reaching the peak maximum, a unique coalescence behavior was seen in the plot of ε'' versus time of DGEBA/m-PDA formulation. For the DGEBA/m-PDA formulation, the ε'' peaks coalesced at all frequencies except at 1 MHz. No such coalescence was seen for DGEBA/DDM samples.

The two curing agents differ in their structure, molecular weight, and in the location of the reactive amine functional groups. The curing of DGEBA depends upon the structure of the diamine.[15] Therefore, it was assumed that the thermosets formed during the cure of DGEBA with the two curing agents would differ in their network morphology and crosslink density. The separation between the reactive amine functional groups can determine the chain length between network junctions resulting in either loose or tight network formation. Based on the C–C separation between the two primary amine groups in the molecule of DDM, the resulting network consists of a structure with longer chain segments between junctions compared to that produced by the resin cured with m-PDA. Although the two systems may have similar network morphologies, the volume of space per crosslink unit is always larger for the DGEBA/DDM system as a result of longer chain segment between network junctions compared with DGEBA/m-PDA system. Therefore, the crosslink density for the DGEBA/DDM system is less than that of stoichiometric formulation of DGEBA/m-PDA system.

FIGURES 7 and 8 compare the ε' and ε'' versus cure time plots for DGEBA/m-PDA, and DGEBA/DDM systems for 1 MHz, 100 kHz, 10 kHz, and 1 kHz frequencies during an isothermal cure at 60°C, respectively. As can be seen from the ε'' versus time plots of the two systems, for all measurement frequencies, DGEBA/m-PDA samples required less time to cure than the DGEBA/DDM samples. The differences in time for the appearance of the dispersion region in the ε' versus time plots and the time for the appearances of the dielectric loss peak in the ε'' versus time plots could be qualitatively correlated to the reactivities of the curing agents. At both cure temperatures m-PDA cures faster than DDM. The lower reactivity of DDM can be explained in terms of its lower basisity or nucleophilicity compared to m-PDA. The nuclophilicity of m-PDA is enhanced by the presence of the two primary amine groups within the same benzene ring compared with the DDM, in which the amine groups are attached to separate benzene rings in the molecule. Therefore, because of its lower nucleophilicity, DDM reacts slower than m-PDA.

Polydiacetylene Polymerization

A comparison of the surface and bulk polymerizations measured at 1 kHz is shown in FIGURES 9 and 10. There is a distinct difference in both the capacitance and the dissipation factor between the surface and bulk polymerizations at the lower frequencies. When the lamp was turned on there was an initial decrease in the capacitance (FIG. 9) and an increase in the dissipation factor for the surface polymerization (FIG. 10). This was due to the heating of the electrode by the UV lamp. A similar decrease was observed when an empty cell was illuminated. There is a rapid increase in the capacitance for the surface polymerization that indicates the responses are due to the deposition of polydiacetylene on the surface by the photopolymerization reaction. There is a peak in the dissipation factor at 120 minutes during the surface

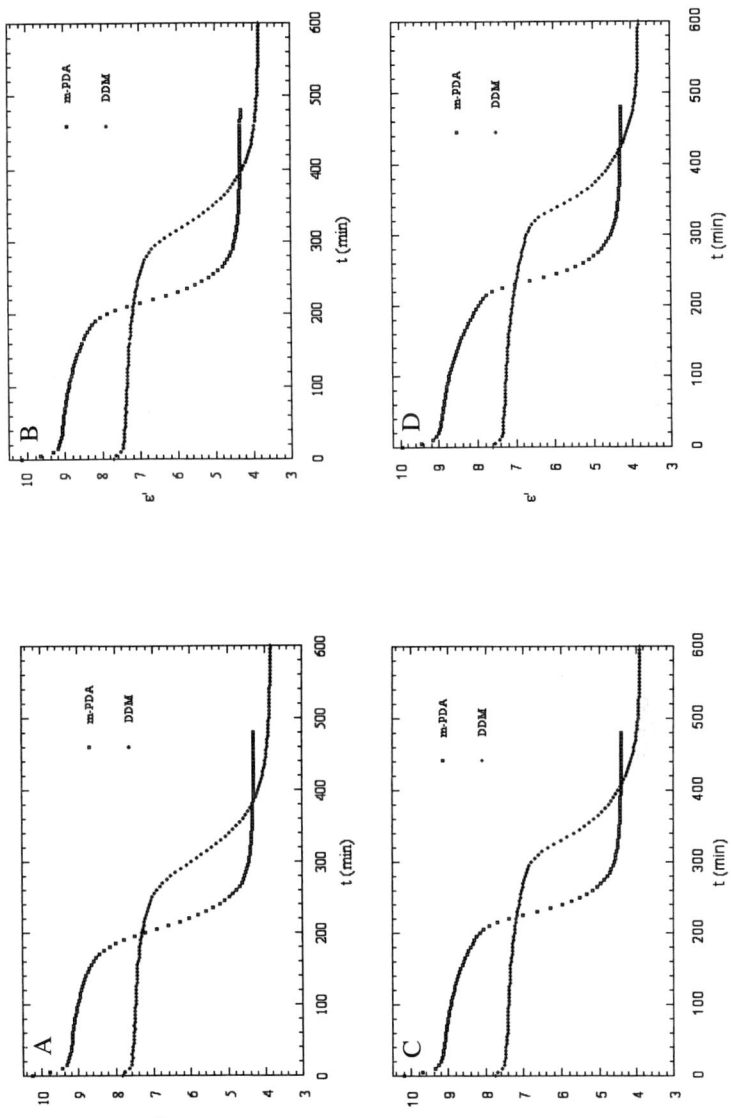

FIGURE 7. Comparison of curing agents (ε' versus time) for cure at 60°C and (**A**) 1 MHz, (**B**) 100 kHz, (**C**) 10 kHz, and (**D**) 1 kHz.

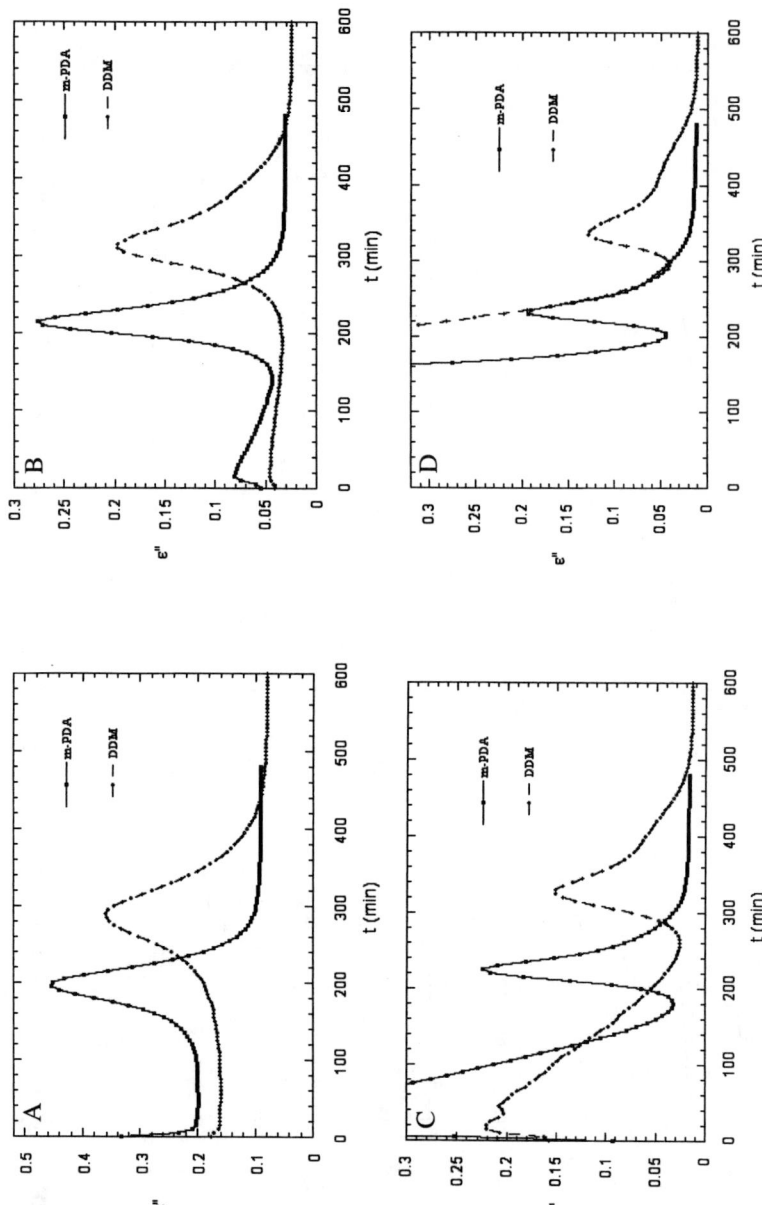

FIGURE 8. Comparison of curing agents (ε'' versus time) for cure at 60°C and (**A**) 1 MHz, (**B**) 100 kHz, (**C**) 10 kHz, and (**D**) 1 kHz.

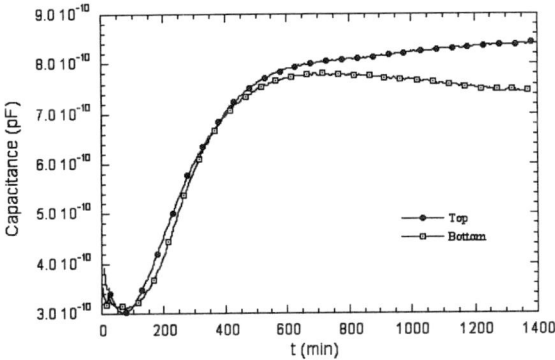

FIGURE 9. Capacitance for surface and bulk solution at 1 kHz.

polymerization. A maximum is observed in the case of the bulk polymerization but it decreases slowly with time.

The capacitance of the surface polymerization is a factor of four larger than the bulk. This clearly indicates that polymer film is growing on the surface of the electrode. The capacitance increases rapidly during the first 500 minutes and then begins to plateau at longer times. Paley et al.[6] studied the kinetics of the thin film growth for polydiacetylene and reported the rate to be first order in intensity and half-order in monomer concentration. The estimated thickness of the film after 500 minutes was 1 micron. It is reasonable to assume that after 500 minutes the film is opaque to the UV radiation and film growth becomes significantly decreased with a resulting reduction in capacitance. A similar inverse trend was observed for the dissipation factor.

To determine if convective effects can be observed using this technique, the sample cell was placed in various orientations. FIGURE 11 shows the comparison in

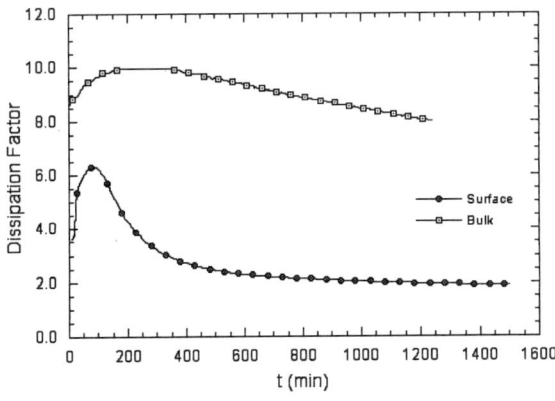

FIGURE 10. Dissipation factor surface and bulk solution polymerization at 1 kHz.

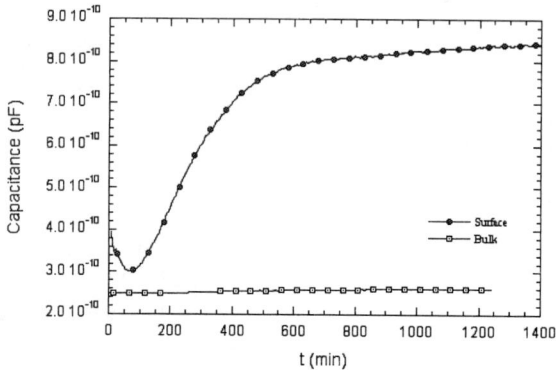

FIGURE 11. Capacitance for the surface polymerization of 2.5 mg/mL of polydiacetylene in horizontal orientation.

surface polymerization irradiated from above and below when the sample is in the horizontal orientation (see FIGS. 3B and 4B). The capacitances for both measurements are identical except for differences at longer times. This suggests that gravitational effects do not influence surface polymerization at early times. However, when convection is dominant (irradiated from bottom) the capacitance does decrease at longer times. Since the reaction has slowed at this point, the decrease in capacitance is probably due to the removal of polymer from the surface by convection.

FIGURES 12 and 13 show the results for the solution polymerization at various orientations. There is a distinct difference between irradiation from above and from below in this case (see FIGS. 3A and 4A). Gravitational effects exert a more pronounced influence on polymerization in solution than at the surface. When irradiated from below convection is dominant and capacitance increased. The increase in

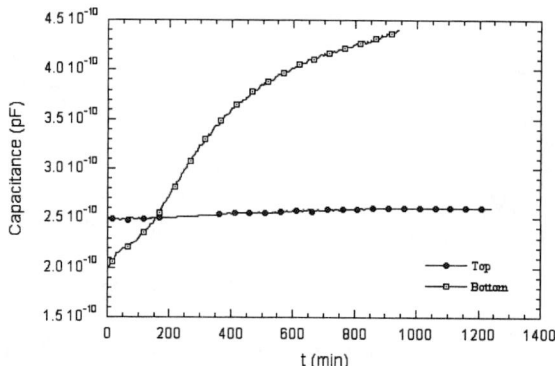

FIGURE 12. Capacitance for the bulk solution polymerization of 2.5 mg/mL of polydiacetylene in various orientations.

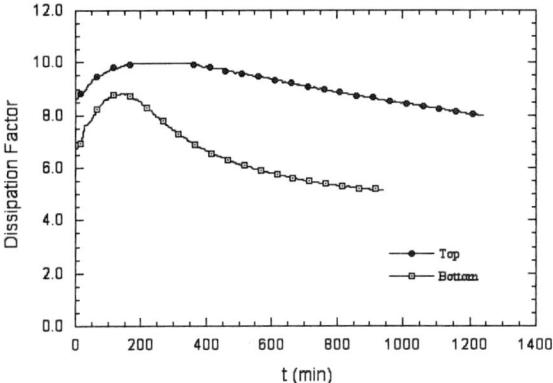

FIGURE 13. Dissipation factor for the bulk solution polymerization of 2.5 mg/mL of polydiacetylene in various orientations.

capacitance is more gradual and changes less than the surface polymerization. When irradiated from above convection is minimized and there is no change in the capacitance. This suggests that the polymer is formed at or near the surface, and no significant amount of polymer is transported to the bottom of the solution when convection is suppressed. This also suggests that sedimentation of insoluble crystallites is not significant in this orientation. However, when convection is dominant it is possible to form a significant concentration of oligomers in solution, which could form defects in the thin film surface.

FIGURE 14 shows the change in capacitance versus frequency. Both the capacitance and the dissipation factor decrease monotonically with increasing frequency. This frequency dependence is indicative of ionic conductivity. The increase in capacitance as a function of time at the lower frequencies appears to be due to the formation of a thin film on the surface of the electrodes.

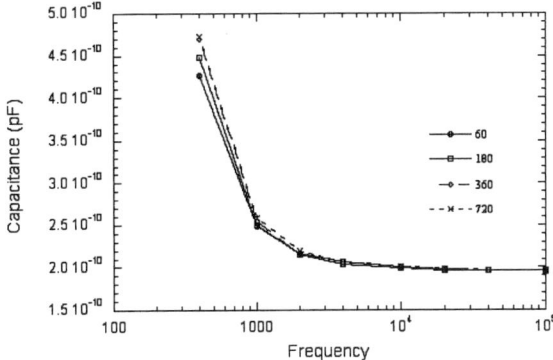

FIGURE 14. Capacitance versus frequency at various times (minutes).

CONCLUSION

In situ dielectric measurements were made using two different polymerization reactions, epoxy resins and solution polymerization of polydiacetylene. The results from epoxy resin system indicate that the changes in the frequency response of the dielectric loss with time are related to the network morphology. A coalescence of the dielectric loss spectra was observed for the more tightly crosslinked network formed by *m*-PDA. For the loose networked formed by DDM, various frequencies of dielectric loss spectra decrease separately in time. The time to reach the dielectric loss maximum depends on the rate of the epoxy–amine reaction. Since the reaction rate is partially determined by basisity *m*-PDA reacts faster and has a shorter time to peak maximum.

The results from the solution polymerization of polydiacetylene showed a distinct difference between the response of the surface and bulk polymerization. Surface polymerization is not affected by convection except at longer times when the polymer film appears to be decreasing in thickness. This suggests that photopolymerization is occurring primarily at the surface and polymerization in solution does not contribute significantly to film growth. Convective effects have a strong influence on bulk polymerization. When convection is minimized the reaction products appear to be localized near the surface.

These results suggest that dielectric spectroscopy is capable of monitoring important parameters (i.e., network morphology and reaction rate) that determine the properties of polymer materials. In addition, it can detect convective effects in solution polymerization reactions. This technique can potentially be used in the fabrication and monitoring of polymer structures and devices developed in space.

ACKNOWLEDGMENTS

A special thanks to Drs. Don Frazier and Steve Paley of NASA Marshall Space Flight Center.

REFERENCES

1. KIM, M.H.Y., S.A. THIBEAULT, J.W. WILSON, *et al.* 2000. High Perfor. Polym. **12**(1): 13–26.
2. FRAZIER, D.O., C.E. MOORE & B.H. CARDELINA. 1993. NASA Conference Publication. 3250.
3. BRISKMAN, V.A. 1999. Gravitational Effects in Materials and Fluid Sciences. **24**(10): 1199–1210.
4. MATEJKA, L. & K. DUSEK. 1995. J. Poly. Sci. Part A **33**: 461.
5. MATEJKA, L., S. PODZIMEK & K. DUSEK. 1995. J. Poly. Sci. Part A **33**: 473.
   SHEPPARD, N.F., M.C.W. COLN & S.D. SENTURIA. 1984. Proc. 29th Natl. SAMPE Symp. 1243–1250.
6. ABDELDAYEM, H., D. FRAZIER, M. PALEY, *et al. 1996.* SPIE 2809, 126.
7. FRAZIER, D.O., R.J. HUNG, M.S. PALEY & Y.T. LONG. 1997. J. Crystal Growth **175**: 172.
8. PRIME, R.B. 1997. *In* Thermal Characterization of Polymeric Materials, 2nd edit. Vol. 2. E.A. Turi, Ed. Academic Press, New York.

9. MANGION, M.B.M., M. WANG & G.P. JOHARI. 1992. J. Polym. Sci. Part B, Polym. Phys. **30:** 433–443.
10. SENTURIA, S.D., N.F. SHEPPARD, H.L. LEE & S.B. MARSHALL. 1983. SAMPE J. July/August 22–26.
11. PARTHUN, M.G. & G.P. JOHARI. 1992. J. Polym. Sci. Part B Polym. Phys. **30:** 655–667.
12. OLYPHANT, M., JR. 1985. Effects of cure and aging on dielectric properties. Proc. of the 6th Elec. Insulation Conf., Supplement, Oct.
13. HARAN, E.N., H. GRINGRAS & D. KATZ. 1965. J. Appl. Polym. Sci. **9:** 3505–3518.
14. WU, S., S. GEDEON & R.J. FOURACRE. 1988. IEE Trans. Elect. Insulation **23**(3): 409–417.
15. GUPTA, N. & I.K. VARMA. 1998. J. Appl. Polym. Sci. **68:** 1789–1766.

The Influence of Gravity on Composition Uniformity and Microstructure in Immiscible Al–In Alloys

J. BARRY ANDREWS AND LETITIA J. HAYES

*Materials Science and Engineering Department,
University of Alabama at Birmingham, Birmingham, Alabama, USA*

ABSTRACT: The results obtained from flight experiments on hypermonotectic Al–In alloys carried out during the Life and Microgravity Spacelab mission are presented and compared to those predicted from models that have been developed for solute redistribution in eutectic alloy systems. Experimental results from microgravity experiments are also compared with results obtained from unit gravity directional solidification experiments. Unit gravity directional solidification of hypermonotectic alloys results in macrosegregation due to convective instability inherent in these alloy systems. Identical samples directionally solidified under microgravity conditions reveal a much more uniform composition profile. Results indicate steady-state coupled growth is possible in hypermonotectic alloys through microgravity processing.

KEYWORDS: monotectics; hypermonotectics; immiscible; microgravity processing; directional solidification; coupled growth; solute redistribution; convective instability; Al–In

NOMENCLATURE:
δ boundary layer thickness
ρ density
C_m monotectic composition
C_0 alloy composition
D diffusivity in the liquid
f_s fraction solidified
g acceleration due to gravity
p V/D
V growth velocity

INTRODUCTION

The establishment of coupled growth conditions in monotectic alloy systems through directional solidification has been investigated for several decades. Coupled growth in these systems leads to the formation of a fibrous composite-like microstructure, where solute-rich fibers form in a matrix composed of nearly pure solvent. The majority of research in this area has been on alloys of monotectic composition.[1–10]

Address for correspondence: J. Barry Andrews, Materials Science and Engineering, University of Alabama at Birmingham, BEC 254, 1530 3rd Avenue South, Birmingham, AL 35294-4461, USA. Voice: 205-934-8452; fax: 205-934-8485.
  bandrews@eng.uab.edu

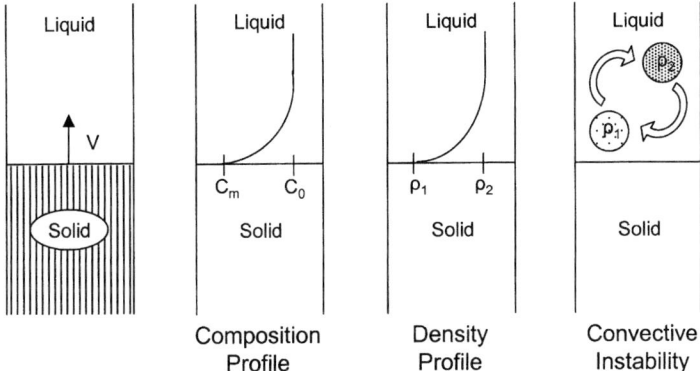

FIGURE 1. Expected composition profile during steady-state directional solidification of a hypermonotectic and the resulting density variation in the liquid that may lead to convective instability.

Although these monotectic composition alloys have potential for use in many applications, higher solute content alloys (i.e., hypermonotectics) should be capable of yielding enhanced performance. Much less information is available on the establishment of steady state coupled growth conditions in hypermonotectic alloys. This lack of information is primarily due to difficulties encountered in processing these alloys. The goal of the coupled growth in hypermonotectics (CGH) flight experiment is to improve our understanding of the physics controlling solidification processes in these intriguing alloy systems. The investigation involves both a modeling segment and an experimental segment where hypermonotectic Al–In alloys are processed under conditions that lead to coupled growth.

Steady-state growth conditions in a hypermonotectic alloy should lead to the development of a macroscopically planar growth interface and a solute depleted boundary layer adjacent to the interface as is shown in FIGURE 1. In almost every known monotectic alloy system, the solute element has a significantly higher density than the solvent. The density difference causes the boundary layer to have a lower density than that of the bulk liquid. Obviously, the density variation, when sufficient, may cause convective instability during normal vertically upward directional solidification. The mixing caused by convective instability could then lead to continuous compositional variations in a solidified sample as a function of fraction solidified. In this study, samples were directionally solidified under normal processing conditions where convection may be expected and under low gravity conditions that minimize buoyancy driven convection.

APPROACH

The aluminum–indium alloy system was selected for this investigation. This system has been used by several investigators to study coupled growth in monotectic composition alloys[3–8,10] and at least a limited amount of thermophysical data is available for the system. The phase diagram for the aluminum–indium system is shown in FIGURE 2. Hypermonotectic alloy compositions 1.2 wt%In and 2.4 wt%In above the monotectic composition (18.5 and 19.7 wt%In, respectively) were used. An approach was desired that was capable of determining whether buoyancy driven convection occurred in hypermonotectic samples. One sample of each alloy composition was directionally solidified vertically upward on the ground. Samples of the same compositions were also processed under microgravity conditions.

Samples were prepared by induction melting the 5-nines pure constituents together in a high-purity alumina tube under vacuum conditions. After melting and holding for 10 minutes, samples were rapidly cooled, removed from their containers, inverted and remelted in order to aid in homogenization.

Both the flight and ground based samples were processed in specially designed aluminum nitride ampoule assemblies. All samples were 10 mm in diameter by approximately 75 mm in length. The aluminum nitride ampoule material was selected to avoid difficulties that occur with most ampoule materials due to perfect wetting by the indium-rich, minor liquid phase. A piston and spring were utilized within the ampoule in an attempt to minimize void formation resulting form thermal contraction of the melt and solidification shrinkage. Details are provided elsewhere.[12,13]

The advanced gradient heating facility (AGHF) was utilized to directionally solidify the samples. All samples were heated to 1,100°C and held for six hours to promote homogenization prior to processing. Samples were solidified using a furnace translation rate of 1 μm/sec. Processing conditions were identical for all samples with the exception that two were processed under one-g conditions and two under microgravity conditions.

After processing and retrieval, samples were removed from the AlN ampoules and longitudinally sectioned using a low-speed diamond wafering saw. One half of

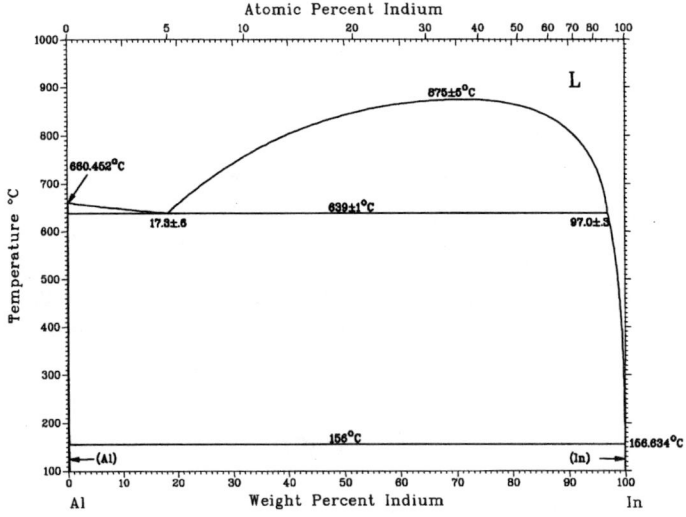

FIGURE 2. Phase diagram for the Al–In alloy system.[11]

the sectioned sample was retained for metallographic analysis. The remaining half was sectioned into 5-mm long segments. These segments were then used for compositional analysis through precision density measurements following the approach of Bowman and Schoonover.[14]

RESULTS

The results obtained from precision density composition analysis on the ground processed 18.5 wt%In sample are shown in FIGURE 3A. It is apparent from this plot that the first portion of the sample to solidify had the highest indium content. This result would be expected for extensive mixing in the melt. As solidification

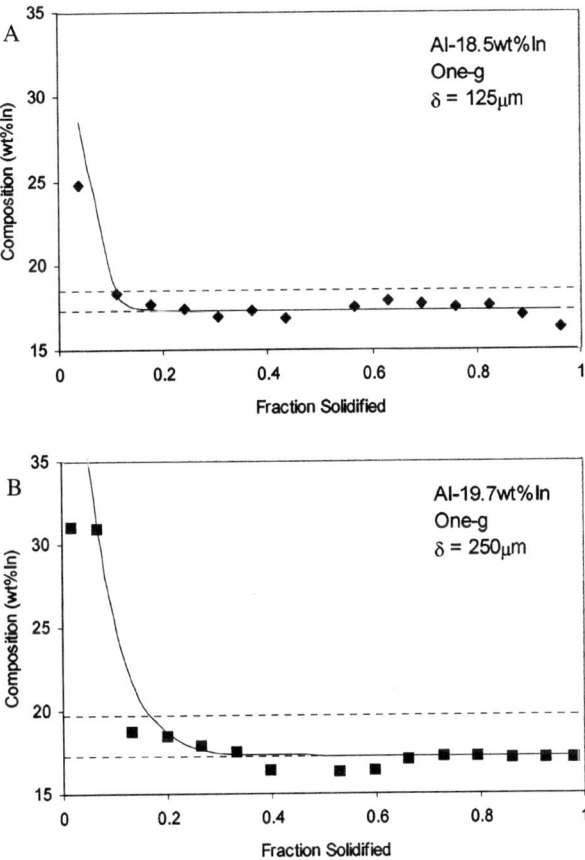

FIGURE 3. (**A**) Composition profile observed in an Al–18.5 wt%In ground processed sample. The solid curve represents results obtained using Equation (**1**) and a δ value of 125 μm. (**B**) Composition profile observed in an Al–19.7 wt%In ground processed sample. The solid curve represents results obtained using Equation (**1**) and a δ value of 250 μm.

continued, the composition of the sample gradually dropped eventually approaching the composition of the monotectic (17.3 wt%In). Although solute redistribution in hypermonotectic systems has not been investigated to any extent, work has been carried out by several investigators on eutectic systems.[15–18] Monotectic systems fit into the eutectic category and are expected to behave in a similar fashion. Verhoeven and Homer's investigation on solute redistribution in the convectively unstable Pb–Sn system revealed similar compositional variations to those seen in this study.[15] In addition, Verhoeven and Homer developed a relationship describing solute redistribution in eutectic systems

$$\overline{C}_s = C_e - \frac{C_E - C_0}{1 - e^{-p\delta}}(1 - f_s)^{\frac{1}{1 - e^{p\delta}}}. \tag{1}$$

This relationship uses one adjustable parameter, δ, which is related to the extent of convective mixing in the system. For our analysis, a small value of δ can be assumed to signify a relatively short solute boundary layer due to extensive convection. An attempt has been made to utilize Verhoeven and Homer's relationship to describe the compositional variations in the convectively unstable Al–In hypermonotectic alloys. The result is included as the solid curve in FIGURE 3. This line represents a best fit using Equation (1) and was obtained using the value 125 µm for δ.

Figure 3B shows the compositional data obtained from precision density measurements on a 19.7 wt%In ground processed sample. As seen previously, the first-to-freeze region contained a higher percentage of indium than the remainder of the sample. In addition, the indium content gradually dropped approaching the monotectic composition as solidification continued. Using Equation (1) to fit this data yields the solid curve shown in FIGURE 3B. This fit was obtained using the value 250 µm for δ.

Both of the above composition profiles indicate that convective mixing occurred during directional solidification of these hypermonotectic Al–In alloys. In order to determine whether steady-state growth conditions and a uniform composition along the sample could be obtained by minimizing the effect of buoyancy driven convection, two identical samples were processed under microgravity conditions. Microgravity processing should dramatically reduce the effects of convective mixing. The same ampoule design, growth rates, homogenization times and furnace facility were used for these tests.

Composition variations along the 18.5 wt%In microgravity processed sample are shown in FIGURE 4A. Analysis of this data reveals that under microgravity conditions, the first-to-freeze portion of the sample had a composition very close to the overall alloy composition. This result implies mixing in the liquid was insignificant and that steady-state growth conditions were present. In addition, this sample revealed an aligned fibrous microstructure from the beginning of solidification. As solidification progressed, the microstructure remained fibrous and the alloy composition began to rise. This unanticipated rise in composition is thought to be due to the influence of a void found in the sample in this general location. Once the solidification front passed the void, the microstructure changed from fibrous to dispersed. However, the sample composition continued to proceed at roughly the overall alloy composition.

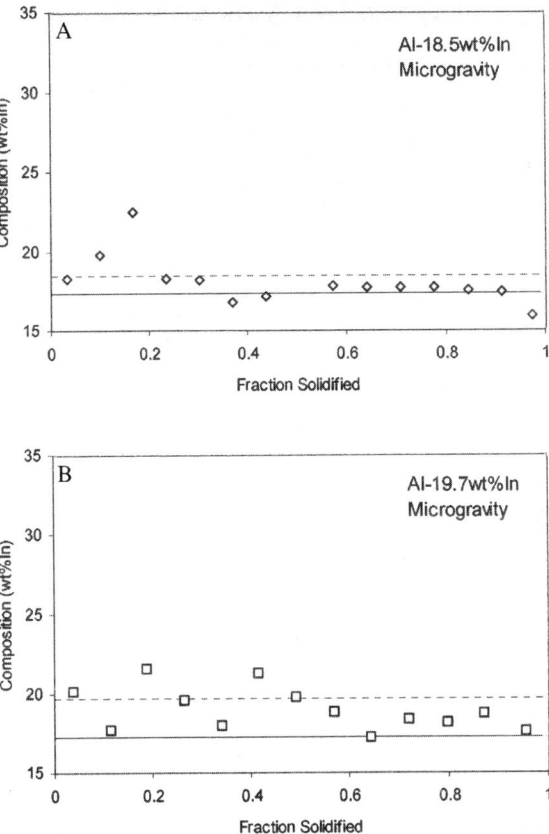

FIGURE 4. (A) Composition profile observed in an Al–18.5 wt%In microgravity processed sample. (B) Composition profile observed in an Al–19.7 wt%In microgravity processed sample.

The composition variation obtained for the 19.7 wt%In microgravity processed sample is shown in FIGURE 4B. It is important to note that this sample also revealed a first-to-freeze composition very close to the overall alloy composition. Relatively small composition variations were seen along the remainder of this sample. Growth conditions for this higher composition sample were not capable of maintaining an aligned fibrous microstructure. Instead, a partially aligned, primarily dispersed structure was obtained.

Results from both of the microgravity processed samples indicate a much more uniform composition profile than those obtained for ground processed samples. In particular, the high initial composition seen in ground processed samples followed by a decrease in composition to the monotectic composition indicative of convective instability was eliminated.

CONCLUSIONS

Several conclusions can be drawn from the results obtained in this study:
1. Convective instability appears to be a factor during ground processing of hypermonotectic aluminum–indium alloys.
2. The resulting mixing in the melt leads to significant composition variations along the length of ground processed samples.
3. Verhoeven and Homer's relationship that was developed to model solute redistribution in eutectic systems can be used to approximate the composition variations observed in ground processed hypermonotectic alloys.
4. Steady-state coupled growth resulting in a uniform sample composition and, in some cases, aligned fibrous growth, appears to be attainable in hypermonotectic alloys processed under microgravity conditions.

ACKNOWLEDGMENTS

The authors wish to acknowledge the National Aeronautics and Space Administration for their support of this research effort through contract NAS8-99059.

REFERENCES

1. LIVINGSTON, J.D. & H.E. CLINE. 1969. Monotectic solidification of Cu–Pb alloys. Trans. TMS-AIME **245:** 351–357.
2. PARR, R.A. & M.H. JOHNSTON. 1978. Growth parameters for aligned microstructures in directionally solidified aluminum-bismuth monotectic. Metall. Trans. A **9:** 1825–1828.
3. GRUGEL, R.N. & A. HELLAWELL. 1981. Alloy solidification in systems containing a liquid miscibility gap. Metall. Trans. A **12:** 669–681.
4. GRUGEL, R.N. & A. HELLAWELL. 1982. The occurrence of aligned microstructures in directionally solidified aluminum-bismuth alloys. Metall. Trans. A **13:** 493–494.
5. TOLOUI, B., A.J. MACLEOD & D.D. DOUBLE. 1982. Microstructural perturbations in directionally solidified Al–In, Al–Bi, and Zn–Bi monotectic alloys. *In* In Situ Composites IV. F.D. Lemkey, H.E. Cline & M. McLean, Eds.: 253–266. Elsevier, Boston.
6. KAMIO, A., H. TEZUKA, S. KUMAI & T. TAKAHASHI. 1984. Unidirectional solidification structure of Al–In monotectic alloys. Trans. Jpn. Inst. Metals **25:** 569–574.
7. GRUGEL, R.N., T.A. LOGRASSO & A. HELLAWELL. 1984. The solidification of monotectic alloys—microstructures and phase spacings. Metall. Trans. A **15:** 1003–1012.
8. GRUGEL, R.N. & A. HELLAWELL. 1984. The breakdown of fibrous structures in directionally grown monotectic alloys. Metall. Trans. A **15:** 1626–1631.
9. DHINDAW, B.K., *et al.* 1988. Directional solidification of Cu–Pb and Bi–Ga monotectic alloys under normal gravity and during parabolic flight. Metall. Trans. A **19:** 2839–2846.
10. KAMIO, A., S. KUMAI & H. TEZUKA. 1991. Solidification structure of monotectic alloys. Mater. Sci. Eng. A **146:** 105–121.
11. MASSALSKI, T.B. 1990. Binary Alloy Phase Diagrams. ASM International. Materials Park, Ohio. 162.
12. ANDREWS, J.B. & S.R. CORIELL. 2000. The coupled growth in hypermonotectics flight experiment. *In* Space Technology and Applications, International Forum-2000. M.S. El-Genk, Ed.: 511–518. American Institute of Physics, College Park, MD.
13. ANDREWS, J.B., *et al.* 1999. Microgravity solidification of Al–In alloys. *In* Solidification and Gravity 2000. A. Roosz, M. Rettenmayr & D. Watring, Eds.: 247–254. Trans Tech Publications, Zurich-Vetikon, Switzerland.

14. BOWMAN, H.A. & R.M. SCHOONOVER. 1967. Procedure for high precision density determinations by hydrostatic weighing. J. Res. NBS **71C:** 179–198.
15. VERHOEVEN, J.D. & R.H. HOMER. 1970. The growth of off-eutectic composites from stirred melts. Metall. Trans. **1:** 3437–3441.
16. BOETTINGER, W.J., F.S. BIANCANIELLO & S.R. CORIELL. 1981. Solutal convection induced macrosegregation and the dendrite to composite transition in off-eutectic alloys. Metall. Trans. A **12:** 321–327.
17. VERHOEVEN, J.D., K. KURTZ KINGERY & R. HOFER. 1975. The effect of solute convection upon macrosegregation in off-eutectic composite growth. Metall. Trans. B **6:** 647–652.
18. VANDENBULCKE, L., R.J. HERBIN & G. VUILLARD. 1976. The growth of off-eutectic Al–Si and Pb–Sn alloys from stirred melts. J. Cryst. Growth. **36:** 53–60.

The Influence of Marangoni Flows on Crack Growth in Cast Metals

B.J. YANG,[a] R.W. SMITH,[a] M. SAHO,[b] AND M. SADAYAPPAN[b]

[a]*Department of Materials and Metallurgical Engineering, Queen's University, Kingston, Canada*

[b]*Materials Technology Lab (CANMET), National Resources Canada, 568 Booth Street, Ottawa, Canada*

> ABSTRACT: In previous work with copper-based alloys, the authors showed that a necessary part of the hot-tearing process is the creation of a void in association with an inclusion and its subsequent growth to form the crack observed in the as-cast state. The work reported here is an examination of hot tearing in aluminum alloy A201 and various similar Al-based aerospace alloys to determine (1) if a similar process to that seen in copper alloys takes place; (2) the extent to which buoyancy forces influence the movement of solute-enriched liquid and so contribute to the development of voids which are subsequently observed as hot tears; and (3) the influence of the local variations in liquid/void surface tension arising due to local composition and temperature changes on interdendritic fluid flow—that is, the effects of Marangoni convection on void size and movement during solidification. Hot cracking of aluminum alloy A201 has been examined under standardized experimental conditions. In addition, experiments were conducted in which fluid flow in test castings was controlled by magnetic fields. The results from these various investigations are presented in this paper. Similar experiments are planned for reduced gravity aircraft parabolic flights. To assist in the planning and interpretation of the results, numerical modeling simulations have been developed for Al–Cu alloys.
>
> KEYWORDS: Marangoni flow; hot tear; modeling

NOMENCLATURE:
C_L solute concentration
T temperature
σ surface tension

INTRODUCTION

Hot tearing is a common casting defect encountered in both ferrous and nonferrous materials. Long freezing range alloys are usually prone to hot tearing as tensile/shear stresses, arising due to contraction restraints, propagate cracks/tears in the mushy zone. Much effort has been made to understand the hot-tearing phenomenon since the work of Pellini and coworkers.[1,2] Recently, Farup and Mo[3] developed a model associated with hot tearing and concluded that none of the hot-tearing criteria suggested in the literature is able to predict the variation in hot-tearing susceptibility

---

Address for correspondence: B.J. Yang, Department of Materials and Metallurgical Engineering, Queen's University, Kingston, Canada K7L 3N6. Voice: 613-533-2753.
smithrw@post.queensu.ca

resulting from variations in experimental parameters such as solidification interval, contraction of solid phase on freezing, freezing rate, and liquid fraction at coherency.

In order to illustrate the problem of analyzing the formation of hot tears, Rappaz et al.[4] recently reviewed the experimental results of microscopy observations of the freezing of a succinonitrile–acetone alloy. They noted that the solidification process leads to the creation of a multitude of discrete liquid pockets, and so it is likely that the frequency of crack formation is determined by the ease of cavity nucleation. This latter point was shown to be important in our own early work on the hot tearing of the long freezing range copper alloy C83600, containing 5%Sn, 5%Zn, and 5%Pb.[5] Our overall conclusion from this work was that for the hot tear to be initiated, it is first necessary to create a cavity in the residual liquid. For this to occur, a nucleation substrate must be present. In practice this is usually an exogenous inclusion such as a folded flake of oxide, or a void. This cavity can then grow into an observable macroscopic crack by the precipitation of gas into it from the supersaturated liquid; the cavity then proceeds to drive out any liquid surrounding the nucleation site and so expand the void. In 1992, Holt et al.[6] proposed a model to describe the influence of the interfacial tension-driven fluid flow on the hot cracking, but it is noted that they only considered the second-order Marangoni force in their model. For most of casting conditions, the second-order Marangoni forced convection resulting from temperature or solute gradients along solid/liquid interface is relatively small, whereas the first-order Marangoni induced flow at any gas/liquid interface is much larger, and so more important.

As noted above, a hot tear tends to appear between those columnar grains that are orientated parallel to the maximum local shear stress.[2] However, since the columnar grains tend to grow up the maximum local temperature gradient, as the major axis of the cavity forms it is also along this maximum temperature gradient. The probable freezing pattern of an alloy prone to hot tearing is shown schematically in FIGURE 1. Here two dendrites are shown growing from the mold wall. There are minor temperature gradients from the tips of the secondary arms to the stem of the dendrite and relatively major temperature gradients between the primary stem tips and the mold wall. As a result, the surface tension of the liquid varies with position around the cavity, resulting in conditions ripe for Marangoni-induced convective flow, toward the

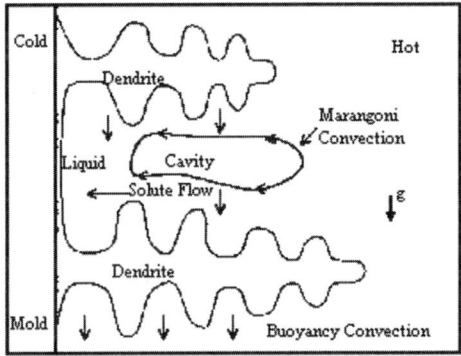

FIGURE 1. Probable freezing pattern of an alloy prone to hot tearing.

base of the interdendritic groove. Clearly, somewhere in the interdendritic network there may be complementary flows away from the mold wall, but normally this will be reduced by the need to supply liquid mass down the temperature gradient to compensate for the natural contraction occurring in the alloy on solidification. Complicating this rather simple picture is the role of solute in the alloy. In principle, particular solutes may increase or decrease the surface tension of the base material. Although the data on these effects are very limited, some data for the variation of the surface tension of liquid aluminum with temperature and solute content are available.[7-9]

As noted earlier, the Marangoni convective flow is also along the maximum temperature gradient and hence parallel to the columnar grain boundaries. This is significant in scavenging the solute rejected by the growing dendrite arms and moving it down into the roots of the columnar boundaries, thereby increasing the local segregation and the freezing range of the alloy, and so its propensity to permit void nucleation and the creation of a hot tear.

A possible competitor to Marangoni forces in shaping the macroscopic solute profile of a cast metal is gravity-induced (buoyancy-driven) solute flow. If the solute is denser than the solvent, the enriched liquid tends to fall as shown in FIGURE 1, and thus certain regions of the casting develop larger than normal concentrations of the alloying element and, hence, extend the freezing range of the alloy even further.

In order to investigate the role of Marangoni convection on the hot-tearing tendency, and to unambiguously separate it from effects arising from buoyancy-driven solute flows, it is necessary to adopt two distinct experimental strategies: the investigation of (1) buoyancy-driven solute flow and (2) Marangoni convective solute flow under both $1g$ and μg conditions. The buoyancy flow can be controlled with a reduction in gravity but, at $1g$, both buoyancy and Marangoni flows can be reduced with magnetic fields due to Lorenz damping. If the gravitational field is much reduced, as in an aircraft parabolic flight, then buoyancy-driven flows will be minimal and the only flows present should be those associated with Marangoni convection, which can be suppressed at will if the casting is made in an apparatus in which a magnetic field can be applied.

As noted in the ABSTRACT, this project involved both an experimental study in which the effects of changes in various casting parameters on the hot-tearing tendencies of a series of aluminum-rich alloys were determined and the development of a numerical model to interpret the experimental data.

EXPERIMENTAL WORK

Melting and Casting

Commercial alloy A201 ingots, supplied by ORMET, USA, were used in this investigation.[10,11] Pure aluminum, copper, magnesium, silver, and Al–Mn master alloys were used to prepare model binary and ternary alloys. In the Materials Technology Laboratory (MTL) of CANMET, melting was carried out in a push-up type, 100 kW-induction furnace using clay-graphite crucibles. Degassing was carried out using chlorine tablets. The gas content was measured using a reduced pressure test. Grain refinement was achieved using Al–5Ti–1B master alloy. Ceramic flow-through filters were used. Some of the melts were carried out at Queen's University to confirm the results obtained at MTL.

Hot-tearing resistance was measured using a pattern, similar to that used at MIT[12] to study hot tearing of aluminum alloys. The pattern was further developed at Queen's to accommodate six bars with lengths of 3, 4, 5, 6, 7, and 8 inches, respectively, and was found suitable for this investigation. The pattern and final casting are shown in FIGURE 2.

The hot-tearing rating was determined as follows: for a fully sound bar, 1; a fully fractured bar, 4. Bars with minor cracks (visible with the aid of a magnifying glass or at 10×) were rated as 2. Bars with relatively large cracks had a rating of 3. Since each mold had six bars, a rating of 6 would indicate good hot-tear resistance whereas a rating of 24 would indicate poor hot-tear resistance.

The molds were prepared in green sand. Four molds were made of which two were provided with filters. Castings were poured in four different conditions; as-cast, degassed, degassed + filtered, and degassed + grain refined + filtered, representing

FIGURE 2. Photographs of (**A**) the hot-tearing pattern and (**B**) casting produced.

different melt quality varying from poor to very clean. The pouring temperature was maintained between 710°C and 720°C. In this investigation, hot-tearing resistance of six alloys, A201, binary Al–Cu, and Al–Mg, and ternary Al–Cu–Mg, Al–Cu–Mn, and Al–Cu–Ag was evaluated. Also, the effect of Mischmetal addition on the hot tearing resistance of A201 was studied.

Dendritic Coherency

The dendritic coherency was determined using thermal analysis technique as described elsewhere. The experiments were conducted at University of Windsor. The cooling curves of three alloys, namely A201, Al–Cu, and Al–Cu–Mg were measured.

The progression of dendrite growth and entrapment of liquid pockets in between them during solidification could be visualized using quench experiments. Efforts were made to develop an experiment to quench aluminum alloy samples at various stages of the solidification. A graphite cup 25 mm diameter and 40 mm height was preheated to about 85°C and filled with the alloy. One thermocouple was placed at the wall of the cup, and water was sprayed on the cup at predetermined temperatures. Sample sections were observed using an optical microscope.

EXPERIMENTAL RESULTS AND DISCUSSION

Hot Tearing

The compositions of the alloys used in this investigation are presented in TABLE 1. The ratings for various alloys from the hot-tearing studies are presented in TABLE 2. The results indicate that only grain refinement appears to improve the hot tearing resistance of A201 when virgin metal is used. This is evident from the rating reduced to 13 from a high value of 16.

The effect of degassing and filtering was not significant except in the case of binary Al–Cu alloy. In this binary alloy, after degassing the hot-tearing resistance increased (rating reduced from 14 to 11). When the risers from this melt were

TABLE 1. Composition of alloys prepared to evaluate hot tearing

Alloy	Cu	Ag	Mn	Mg	Ti
A201	4.8	0.5	0.33	0.30	0.3
		2.0			
Al–Cu	5.0	—	—	—	—
Al–Cu–Mg	4.9	—	—	0.42	—
	5.3			0.05	
				0.3	
Al–Cu–Ag	4.8	0.6	—	—	—
Al–Cu–Mn	4.7	—	0.39	—	—
Al–Mg	—	—	—	0.55	—
				9.0	

TABLE 2. Effect of alloying elements on hot tearing

Alloy	As Melted	Degassed	Degassed and Filtered	Degassed, Filtered, and Grain Refined
A201	16	16	16	13
Al–Cu	14	11	11	10
Al–Cu–Mg	17	19	17	13
Al–Cu–Mn	12	11	10	9
Al–Cu–Ag	12	11	10	10

sectioned and polished, the effect of gas content on the cracking pattern could be clearly seen (see FIGURE 3). The same trend was not observed in other alloy castings.

The above assumption was confirmed by conducting an experiment as follows: returns from one melt was melted and held for 30 minutes; the oxide layer formed on the surface was mixed back into the melt; two test castings were poured one with foam filter and one without any filters. Then the melt was degassed and cleaned and the gas content was confirmed by a reduced pressure test. Again two castings were poured as before. The ratings are reported in TABLE 3.

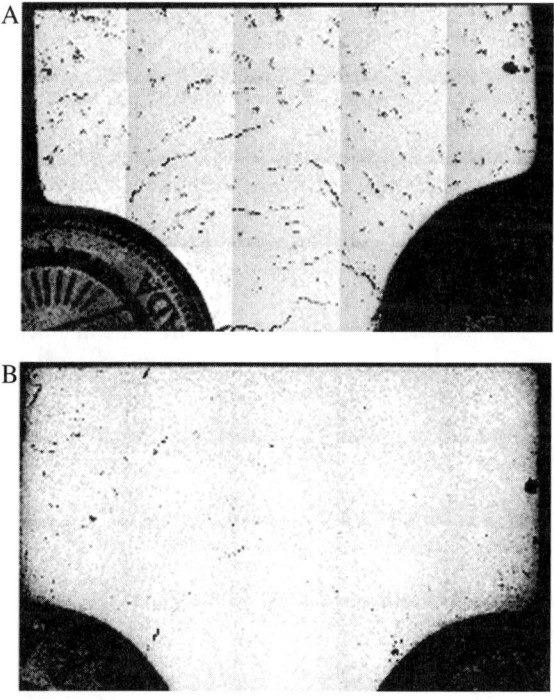

FIGURE 3. The effect of gas content on the hot tearing of Al–Cu alloy: (**A**) as-melted, (**B**) degassed.

TABLE 3. Effect of melt processing

Alloy	Processing	Rating
A201	remelt and holding	14
	filtered	12
	degassed	12
	degassed and filtered	12

It is evident from TABLE 3 that the inclusions cause the major effect on hot tearing as compared with gas content. Filtering the undegassed melt improved the hot-tearing resistance. Degassing, which also removes some of the inclusions present in the melt, did not change the rating. Thus, it can be concluded that gas content may not have much effect in hot tearing in small castings if inclusions are not abundant. This relative insensitivity to gas content also been reported elsewhere.[13]

The second major finding is the effect of magnesium on the hot tearing. The data in TABLE 2 indicates that the ternary Al–Cu–Mg alloy is the least resistant to hot tearing. When 0.5% Mg was added to the Al–Cu alloy the rating increased from 14 for the binary alloy to 17 for the ternary alloy, indicating reduced resistance to hot tearing. The other additions to the binary Al–Cu alloy, namely, Mn and Ag, marginally improved the resistance. However, the effect of magnesium on hot tearing of Al–Cu alloy was not completely removed by either Mn or Ag.

Magnesium is the only element lighter than aluminum in alloy A201 and forms a eutectic with aluminum at 450°C. It extends the solidification range of binary Al–Cu alloy and promotes incipient melting. It also changes the surface tension of the molten aluminum. To understand the effect of magnesium one more melt was conducted in which magnesium was added in steps. The melt was degassed and filtered to remove the effect of any gas and inclusions. The ratings are presented in TABLE 4. The results from earlier trials are also presented for comparison. Even a small amount of 0.06% magnesium could make the alloy prone to hot tearing. Further magnesium addition increased the hot-tearing tendency. This is shown in FIGURE 4 where the cracking tendencies of Al–Cu and Al–Cu–Mg are compared.

The surface tension of the liquid aluminum controls the interdendritic fluid flow. To change the surface tension, Mischmetal was added to A201 and Al–Cu–Mg alloys in this study. The results are presented in TABLE 5. Mischmetal improved the hot-tearing resistance of the alloy significantly. An addition of 2% Mischmetal produced

TABLE 4. Effect of magnesium on hot tearing of Al–Cu alloy

Alloy	Cu (%)	Mg (%)	Rating
Al–Cu	5.1	0	11
Al–Cu–Mg	5.0	0.44	17
Al–Cu	5.3	0	11
	5.3	0.06	14
Al–Cu–Mg	5.3	0.16	16
	5.3	0.33	16

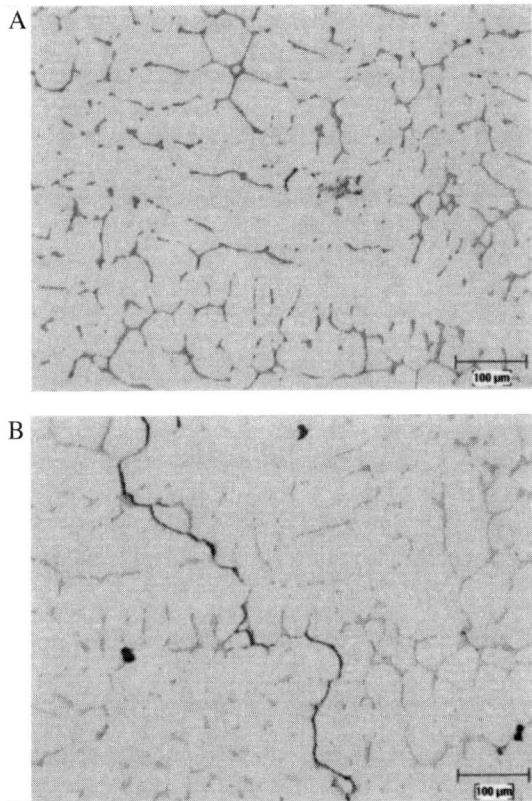

FIGURE 4. Effect of magnesium on intergranular cracking: **(A)** Al–Cu alloy, **(B)** Al–Cu–Mg alloy.

TABLE 5. **Effect of Mischmetal addition on the hot tearing of A201**

Alloy	Melt Condition	Mischmetal Addition (%)	Rating
A201	degassed and filtered	0	14
		0.05	13
		0.1	16
		0.3	16
A201	degassed and filtered	0	15
		1.0	14
		1.5	9
		2.0	6
Al–5%Cu–0.3%Mg	degassed and filtered	0	17
		0.3	16
		0.6	13
		1.0	12

a casting free of hot tearing even in 8" long rod. However, the improvement could be realized only when the addition exceeded 0.3%. The amounts of Ce and La retained in these alloys for the various levels of Mischmetal additions have been determined.

The other two binary alloys, Al–0.5%Mg and Al–9%Mg, did not show any cracking at the bar/ball junction, usually observed in other alloys. However, extensive shrinkage was observed in the retaining disc together with some cracking. Hence, steel chills were used with these alloys and caused them to be free from hot tearing. However, these alloys were not rated since they were cast with chills. On the other hand, chills did not affect the rating of alloy A201.

Dendrite Coherency

The cooling curve experiments conducted at University of Windsor revealed that the dendritic coherency temperatures of unrefined A201, Al–5%Cu and Al–5%Cu–0.4%Mg alloys are 643.2°C, 642.3°C, and 641.7°C, respectively. These temperatures indicate a fraction of solid of approximately 20%. This is quite close to the values reported in the literature for Al–Cu alloys.[14]

Preliminary trials from quenching experiments also revealed the formation of a dendritic network between 640°C and 620°C. Further work is in progress to study the growth of this dendritic network. These alloys were not grain refined.

NUMERICAL MODELING

The Model

In this work, numerical modeling is used to investigate the role of Marangoni convection arising from the variation of interfacial stress around a cavity on the convective flows within the interdendritic region. This is being done to better understand the behavior of a cavity in terrestrial and space conditions—that is, the extent to which it will move and grow to become a hot tear during the solidification of an alloy.

The physical model to be investigated numerically can be abstracted from the interdendritic region shown schematically in FIGURE 1, which is simplified to the form shown in FIGURE 5. The diameter of the gas bubble is assumed to be 100µm, and the computational domain assumed to be 1.0×0.5 mm. In initial studies, the Navier-Stokes equations, including the mass continuity, momentum, energy, and species equations, were used with the following assumptions:

1. columnar dendritic solidification takes place but with no solidification shrinkage;
2. one bubble exists in the interdendritic fluid and its position is fixed relative to the dendrites;
3. the interfacial stress along the interface between solid and liquid, known as the second-order Marangoni force, is ignored;
4. the fluid is incompressible but the thermosolutal buoyancy flow is taken into account through the Boussinesq approximation.

The present model was implemented with a general purpose CFD software CFX4. A fully implicit control volume based finite difference method was used to

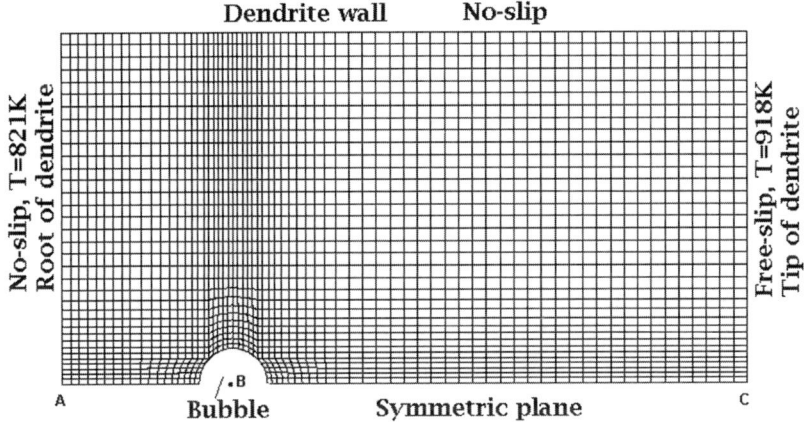

FIGURE 5. Grid system and boundary conditions.

discretize the strongly coupled transport equations. The multiblock technique was used to mesh the selected computational domain. The left side of the computational domain is the root of dendrite, and the right side of the domain is the line through the dendrite tip at the solidification front. The top of the domain is the dendrite wall, and the bottom surface is the symmetric plane of the domain. For the Al–4.5%wtCu binary alloy, the temperature at the left side is set to the eutectic temperature (821 K) and composition (32.7%Cu), and at the right side to be liquidus temperature (918 K) and bulk composition 4.5%Cu. An insulated boundary and no-slip conditions are set on the top and left surfaces. A free-slip condition is set on the right side. The curved surface at the lower side is the interface between the bubble and liquid, and surface tension is present. FIGURE 5 shows the physical model and the calculation conditions. The thermophysical properties used in the simulation are taken from References 3, 9, and 15.

As mentioned in the INTRODUCTION, the Marangoni effect caused by a cavity such as a gas bubble could drastically change the transport processes in the melt. From the available measured data about the surface tension for Al-based alloys, the surface tension is a function of temperature and the concentration gradients along the free surface. According to Poirier and Speiser's work,[9] the surface tension for Al–4.5%wt Cu can be approximated by

$$\sigma = 868 + 0.712 C_L + 1.29 \times 10^{-2} C_L^2 \tag{1a}$$

or

$$\sigma = 1151 - 0.54T + 1.70 \times 10^{-4} T^2, \tag{1b}$$

with $548°C \leq T \leq 645°C$. The units of surface tension in the above equations are dyne·cm^{-1}. This expression was used to set the surface tension values in the interdendritic liquid during the solidification of Al–Cu binary alloys, providing $C_L \leq 32.7$wt%Cu.

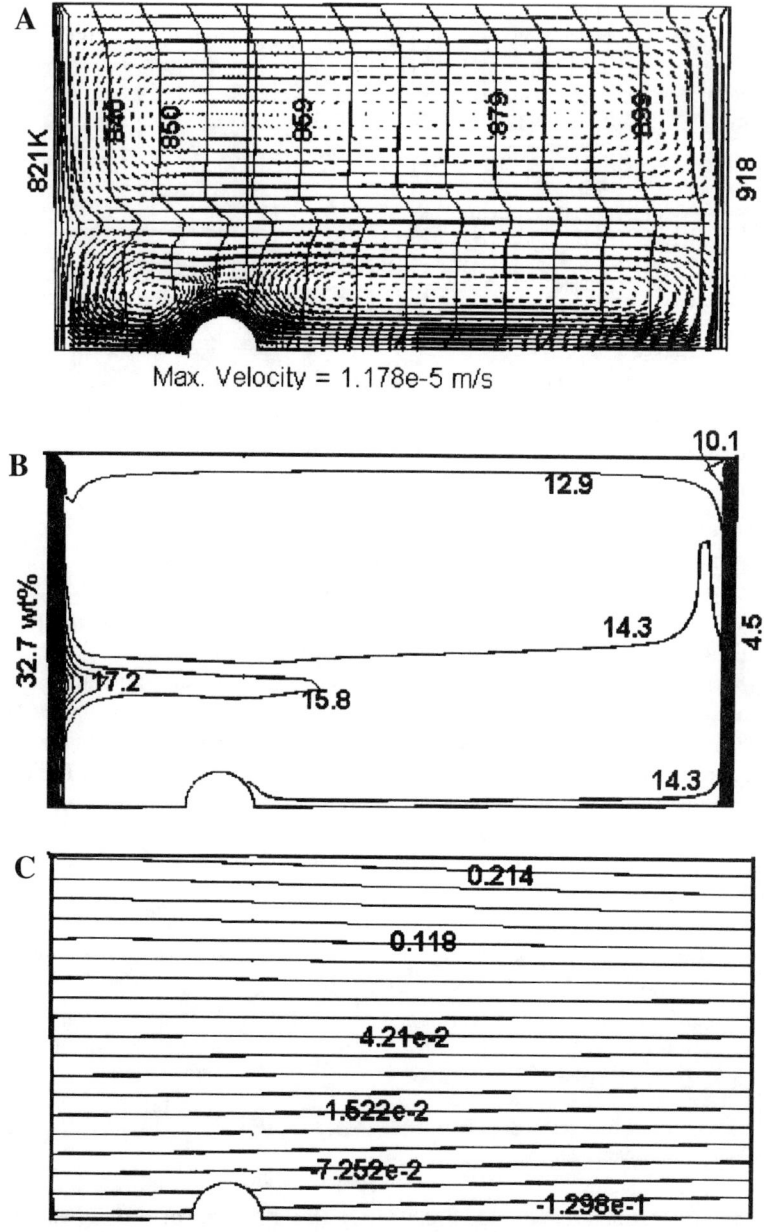

FIGURE 6. The computed results without surface tension: **(A)** calculated flow pattern and isotherms; **(B)** solute distribution, wt%; and **(C)** pressure distribution.

Computed Results and Discussion

Two cases for terrestrial freezing have been implemented to investigate the role of surface tension in determining the hot-tearing tendency in the Al–4.5%wt Cu alloy. The first case assumes that solidification is carried-out without surface tension—that is, only buoyancy-driven flow—and the second case is performed by considering the variation of surface tension with the local temperature gradient—that is, both buoyancy-driven and Marangoni convection. The computed results are shown in FIGURES 6 and 7.

From FIGURE 6A, it can be seen that if the surface tension is ignored along the bubble/liquid interface, there are three circulations in the domain due to the existence of bubble. The maximum velocity appears along the plane of symmetry. However, the velocity is relatively small with only buoyancy effects, and thus the temperature distribution is not very much affected by the convective flow. The influence of convective flow on the solute distribution is shown in FIGURE 6B; the distribution of pressure is linear from the upper to lower side, as shown in FIGURE 6C. The pressures at the left and right of the bubble are almost the same, thus the bubble is not going to move. When the variations in surface tension are taken into account, the results are quite different from the case without surface tension, as illustrated in FIGURE 7. Very high velocities occur around the bubble due to the shear stress. Because of relatively high velocity, the distributions of temperature and solute are more uniform as compared with the case without surface tension. It can be observed from FIGURE 7C that the pressure distribution around the bubble varies. The surface tension variation produces a high pressure at the left of the bubble and a low pressure at its right side. The velocity behavior and the pressure distribution around the bubble are expected to pump liquid towards the roots of the dendrites, thus displacing the bubble up the temperature gradient away from the base of the dendrite and out of the interdendritic liquid, and hence, reducing any hot-tearing tendency. This prediction is supported by the observation that the hot-cracking tendency of A201 increased when freezing took place in a unidirectional magnetic field, which prevented the fluid movement that would have displaced any cavities out of the mushy region into the body of the melt.

SUMMARY AND FUTURE WORK

The experimental work thus far has shown that the major influences on cracking reduction have been grain refinement of the melt and the addition of Mischmetal. Since this "mixture" contains elements that are most likely to influence solid/liquid surface tension, future experimental work at 1g will explore them quantitatively, in preparation for reduced gravity activities. The presence of magnesium increases any hot cracking tendency in an Al–Cu alloy.

Numerical modeling has been used to investigate the role of Marangoni convection arising from the variation of interfacial stress around a cavity on the convective flows within the interdendritic region. From the numerical modeling results it can be seen that the surface tension variations have a very significant influence on the velocity field and the pressure distribution in the interdendritic fluid. The surface tension around a bubble in the model alloy Al–4.5%Cu is shown to be playing a positive role

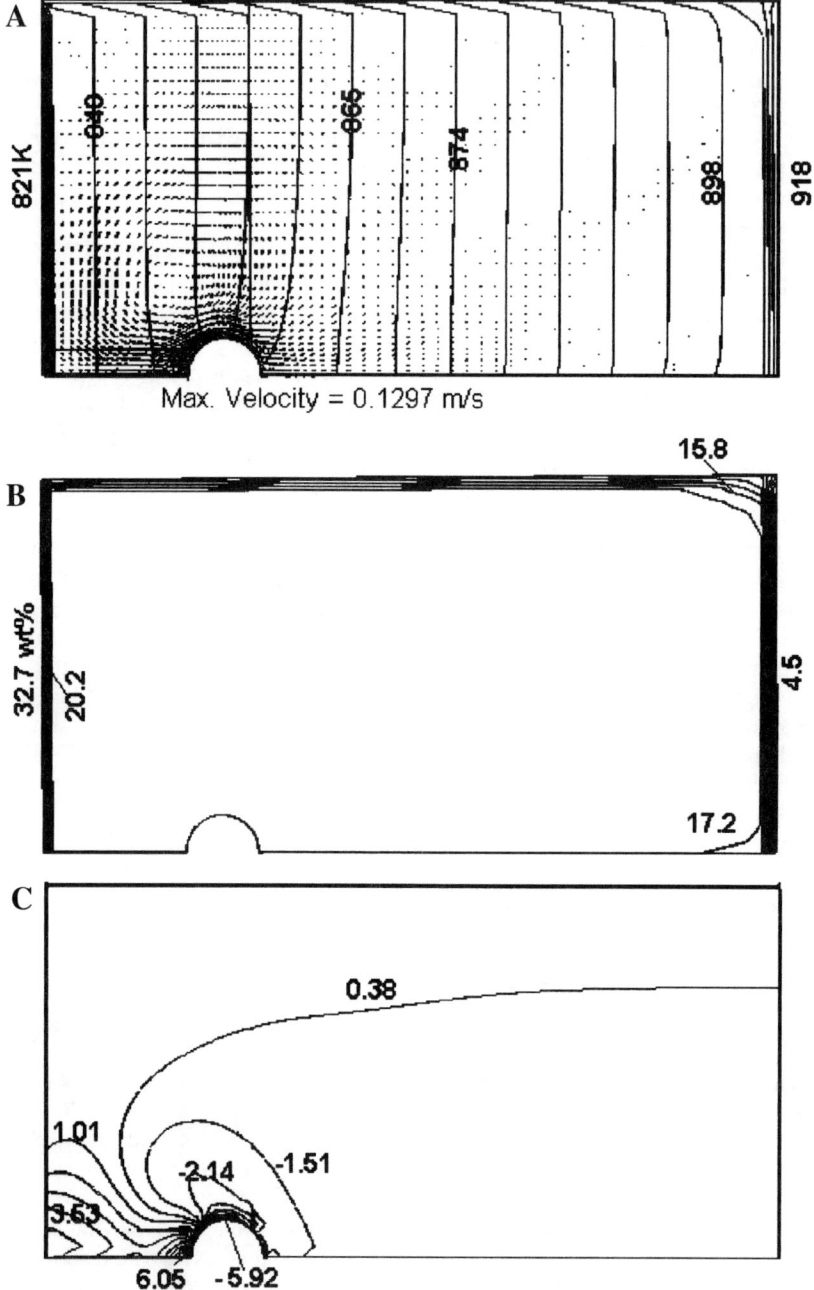

FIGURE 7. The computed results with surface tension: **(A)** calculated flow pattern and isotherms; **(B)** solute distribution, wt%; and **(C)** pressure distribution.

in reducing the hot-tearing tendency. Future modeling activities will consider the influences of bubble movement, the inflow of liquid with the dendritic mesh to compensate for solidification contraction, and the lateral displacement of adjoining dendrite stems due to lateral shear forces.

ACKNOWLEDGMENTS

This work was supported by NASA /AFS Solidification Design Control Consortium (SDCC) through Auburn University.

REFERENCES

1. PELLINI, W.S. 1952. The Foundry **80:** 124–199.
2. BISHOP, H.F., C.G. ACKERLAND & W.S. PELLINI. 1960. AFS Trans. **68:** 818–833.
3. FSRUP, I. & A. MO. 2000. Metall. Mater. Trans. A **31A:** 1461–1472.
4. RAPPAZ, M., J.-M. DREZET & M. GREMAUD. 1999. Metall. Mater. Trans. A **30A:** 449–455.
5. LIU, Q., R.W. SMITH & M. SAHOO. 1993. AFS Trans. **101:** 759–770.
6. HOLT, M., D.L. OLSON & C.E. CROSS. 1992. Scripta Metallurgica Materialia **26:** 1119–1124.
7. LATY, P., J.C. JOUD, P. DESRE & G. LANG. 1977. Surface Sci. **69:** 508–520.
8. GOICOECHEA, J., C. GARCIA-CORDOVILLA, E. LOUIS & A. PAMIES, 1992. J. Mater. Sci. **27:** 5247–5252.
9. POIRIER, D.R. & R. SPEISER. 1987. Metall. Mater. Trans. A **18A:** 1156–1160.
10. SADAYYAPPAN, M., M. SAHOO, G. LIAO, et al. 2001. AFS Trans. **109:** 579–591.
11. SADAYYAPPAN, M., M. SAHOO & R.W. SMITH. 2001. Presented at Light Metals 2001 Symposium, Conference of Metallurgists, Toronto, August 26.
12. ROSENBERG, R.A., M.C. FLEMMING & H.F. TAYLOR. 1960. AFS Trans. **68:** 518–528.
13. SPITTLE, A. & A.A. CUSHWAY. 1983. Metals Technol. **10:** 6–13.
14. ARNBERG, L., L. BACKERUD & G. CHAI. 1996. American Foundrymen's Society Inc., Des Plaines, IL.
15. CHANG, S. & D.M. STEFANESCU. 1996. Metall. Mater. Trans. A **27A:** 2708–2721.

Thermal Diffusivity Coefficient of Glycerin Determined on an Acoustically Levitated Drop

K. OHSAKA, A. REDNIKOV, AND S.S. SADHAL

University of Southern California University Park, Los Angeles, California, USA

ABSTRACT: We present a technique that can be used to determine the thermal diffusivity coefficient of undercooled liquids that exist at temperatures below their freezing points. The technique involves levitation of a small amount of liquid in the shape of a flattened drop using an acoustic levitator and heating it with a CO_2 laser. The heated drop is then allowed to cool naturally by heat loss from the surface. Due to acoustic streaming, heat loss is highly non-uniform and appears to mainly occur at the drop circumference (equatorial region). This fact allows us to relate the heat loss rate with a heat transfer model to determine the thermal diffusion coefficient. We demonstrate the feasibility of the technique using glycerin drops as a model liquid.

KEYWORDS: non-contact thermal diffusivity measurement; glycerin; acoustically levitated drop

INTRODUCTION

There is increasing interest in measuring the thermophysical properties of undercooled liquids that are in metastable liquid states below their freezing points. As an example of fundamental importance, these properties correlate the atomic structures of liquids, which tend to exhibit increasing short-range order, with the clustering of atoms as undercooling increases. Conventional measuring techniques are not applicable to undercooled liquids simply because bulk liquids seldom attain significant undercooling before solidification takes place at one of numerous solid nucleating sites, including the container walls. It has been known that levitation of a small liquid sample is an effective way to attain a large degree of undercooling because the small sample volume reduces the number of potential nucleation sites and the self-contained sample is free from the possibility of nucleation at container walls.[1,2]

Physical forces, such as the electromagnetic, acoustic, electrostatic, and aerodynamic forces, that act against gravity, may be used for levitation of small samples on Earth. Non-contact diagnostic techniques are required for property measurements of undercooled liquids because any contact with the liquids may initiate solidification. The thermophysical properties that have been measured on levitated undercooled liquid drops include density,[3] surface tension,[4] viscosity,[5] and heat capacity.[6] In this paper, we present an exploratory technique that may allow us to determine the thermal diffusivity coefficient of levitated liquids. We employ an acoustic levitator for levitation and an infrared (IR) camera as a diagnostic device. We use glycerin drops

Address for correspondence: K. Ohsaka, University of Southern California University Park, RRB-107 Los Angeles, CA 90089-1191, USA.

whose thermal diffusivity coefficient is well known as a model liquid to demonstrate the feasibility of the technique.

EXPERIMENTAL

Settings

FIGURE 1 schematically shows the present experimental settings. An ultrasonic transducer oscillates at approximately 23 kHz and generates a standing wave between the transducer head and the reflector. A liquid drop can be levitated near a pressure node of the standing wave. Two video cameras, located at the top and side of the levitator, capture the images of the levitated drop. The captured images are then used to determine the aspect ratio and actual dimensions of the drop. A CO_2 laser aimed at the drop is used to heat the drop. An infrared (IR) camera looking down the drop is used to continuously monitor the surface temperature of the drop.

Procedure

A small amount of glycerin (approximately 15 µl) is introduced into the levitation field with a hypodermic syringe. Glycerin is rapidly detached from the syringe and formed into an oblate drop due to the force balance among gravity, surface tension, and acoustic pressure. A typical aspect ratio, a/b, is 1.3, where a and b are the equatorial and polar radii of the drop, respectively. It is possible to additionally flatten the drop by increasing the acoustic pressure. FIGURE 2 shows a flattened glycerin drop

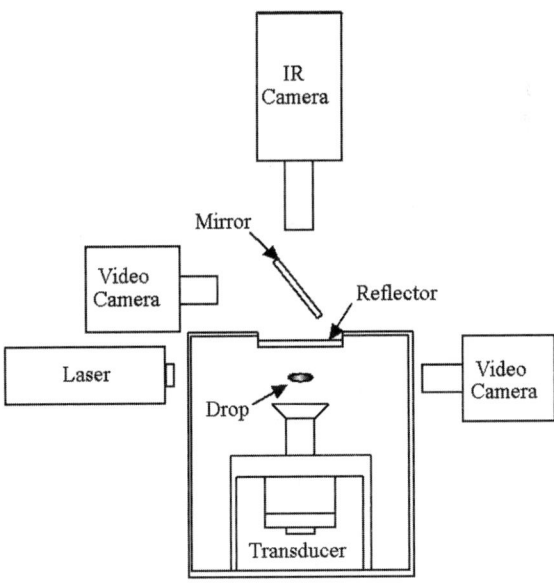

FIGURE 1. Schematic diagram of the experimental apparatus showing the key parts.

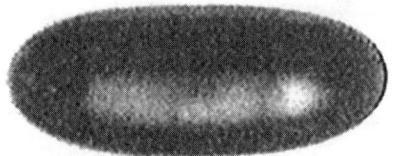

FIGURE 2. Acoustically levitated glycerin drop (side view).

whose aspect ratio is approximately $a/b = 2.6$. Additional increase in the pressure causes the top surface to take on a concave shape. The CO_2 laser that produces a narrow beam is turned on to locally heat the drop at a small area of the circumference. When the drop is heated locally, it usually starts rotating along the polar axis due to a torque created by local heating. As a result, the laser equally heats entire circumference. The absorbed heat is transferred inward by conduction and convection and raises the drop temperature as a whole. Once the drop reaches a certain temperature, the laser is turned off and the drop is allowed to cool naturally. An IR camera is used to record the temperature change of the top surface during cooling.

Observation

The IR camera produces multicolor images that reflect the temperature distribution on the top surface. Each color region represents a certain temperature range that presents on the surface. FIGURE 3 shows a sequence of the IR images of the top surface during the cooling process. The original color images are transformed into gray scale for this paper. The time interval between two consecutive images is 0.5 seconds. The white circles superimposed on the first and last images indicate the circumference positions. The concentric rings that represent specific temperature ranges are slightly distorted in appearance. This deformation is not real but developed during image processing in the IR camera system, thus it is ignored. The images reveal that elevated temperatures exist outside of the actual circumference. This is because the IR camera sees the surrounding air that is heated above room temperature. The arrows pointing to the ring assist in seeing how the ring moves as time elapses. As can be seen in the figure, the ring gradually recedes and eventually disappears at the center. Although the interior of the drop continuously cools, the temperature at the circumference remains approximately constant, at just above room temperature.

It appears that heat loss from an acoustically levitated drop is not uniform on every part of the drop surface, and occurs mainly at the circumference. The receding rings are the result of internal heat flow toward the circumference. Enhanced heat loss at the circumference can be understood as an act of acoustic streaming. It is known that when an object is placed in an acoustic standing wave, a local streaming is generated around the object.[7] This streaming is distinguished from global streaming that exists even when the object is not present. The streaming flow pattern depends on the object shape. FIGURE 4 shows schematically the flow pattern of streaming around a flattened drop. The intensity of streaming is strongest at the circumference where the surface curvature is the largest.

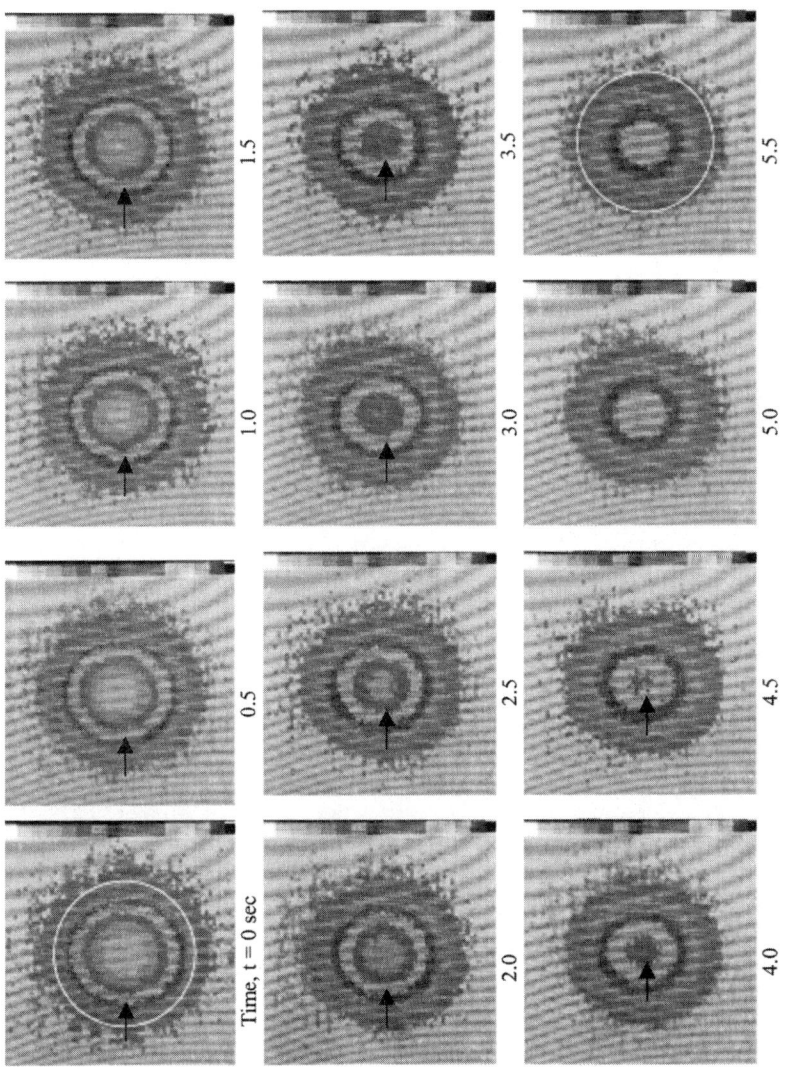

FIGURE 3. A sequence of the IR images showing the receding thermal rings.

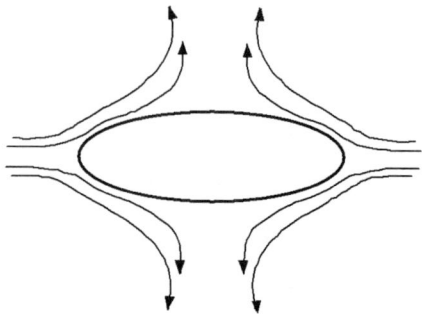

FIGURE 4. Schematic description of acoustic streaming around a flattened drop.

Heat Flow Model

If heat loss at the circumference is predominant and the circumference temperature remains constant at near room temperature, there must be a correlation between the velocity of receding rings and heat transfer in glycerin. We present a heat transfer model that allows us to determine the thermal diffusivity coefficient from the velocity of receding rings. We assume that the flattened drop can be modeled as a short cylinder (disk) with the same radius. We also assume that no heat is lost from the top and bottom surfaces of the cylinder. Under these assumptions, we may write the radial heat conduction equation as

$$\frac{\partial T^*}{\partial t^*} = \frac{\partial^2 T^*}{\partial r^{*2}} + \frac{1}{r^*}\frac{\partial T^*}{\partial r^*}, \tag{1}$$

where T^*, t^*, and r^* are the normalized temperature, time, and radial coordinate originating from the center of the cylinder, respectively, and are defined by

$$t^* = \frac{t\chi}{a^2} \tag{2}$$

$$r^* = \frac{r}{a} \tag{3}$$

$$T^* = \frac{T - T_0}{T_{max} - T_0} \tag{4}$$

where T, t, r, χ, a, T_{max}, and T_0 are the temperature, time, radial coordinate, thermal diffusivity coefficient, cylinder radius, the initial maximum temperature (at the center), and circumference temperature, respectively. We also assume that T_0 remains constant during the cooling process. The initial temperature distribution, determined experimentally from the IR image, is also normalized as follows:

$$T^*_{init}(r^*) = \frac{T_{init}(r) - T_0}{T_{max} - T_0}, \tag{5}$$

where $T_{init}(r)$ is the initial temperature at the coordinate, r. Equation (1) is numerically solved using the following boundary conditions:

$$T^* = 0 \text{ at } r^* = 1 \text{ and } t^* > 0$$
$$T^* = T^*_{init}(r^*) \text{ at } t^* = 0 \text{ and } r^* < 1.$$
(6)

We obtain the time evolution of the temperature profile, $T^*(r^*, t^*)$ that allows us to locate the coordinates of a specific temperature representing a particular color ring in the IR images. We perform the calculation using the known thermal diffusivity coefficient of glycerin and compare the result with the experimental result.

RESULTS

We measured the coordinates of the receding ring (indicated by the arrow) shown in FIGURE 3. The side view of the drop is shown in FIGURE 2. For the numerical calculation, the initial temperature distribution is determined from the first image. The property parameters are $a = 0.26$ cm, $\chi = 0.001$ cm^2/sec. The experimental and numerical results are given in FIGURE 5 that shows the coordinates of the receding ring as a function of the elapsed time in normalized dimensions. The experimental results show that the levitated glycerin drop has an apparent thermal diffusivity coefficient that is larger than the accepted value. The fact, however, is that the present model fairly describes the heat loss of the acoustically levitated glycerin drop. If the thermal diffusivity coefficient of glycerin were not known and the present heat transfer model were used to determine it by matching the calculated receding velocity to the experimental value, the value obtained would be approximately 25% larger than the accepted value.

DISCUSSION

The present model relies on the highly non-uniform heat loss at the drop surface due to forced convection caused by acoustic streaming. The intensity of streaming varies depending on the overall acoustic pressure and the local surface curvature. As

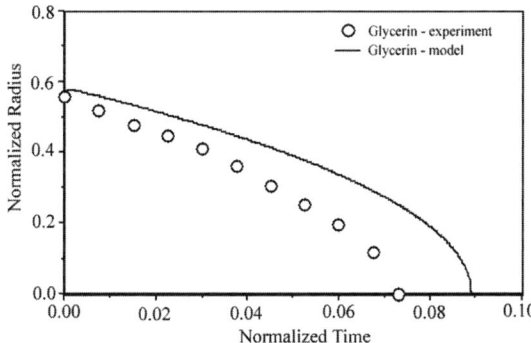

FIGURE 5. The normalized radius of the receding ring representing a particular temperature as a function of the normalized time.

a result, because of the drop shape (flattened drop), major loss takes place at the circumference. The heat loss at the top and bottom surfaces is relatively minor. The model also assumes that the heat loss at the circumference is so strong that its temperature remains constant at near ambient temperature. Under these conditions, thermal conduction from interior to the circumference controls the heat loss rate and, therefore, the model can be justified. It is likely that the above conditions are not met for every levitated drop. For example, the model may not work for high thermal conductive liquids, such as metallic drops, in which the heat transfer at the surface usually controls the heat loss rate. Even in low thermal conductive drops, if the heat loss at the top and bottom surfaces becomes significant due to the combination of drop shape and acoustic pressure, the model may not work.

Another shortcoming of the present technique is its incapability of setting the acoustic pressure and the aspect ratio of the drop independently. A certain level of the acoustic pressure is needed for levitation in Earth-based experiments, and an additional pressure is required for flattening the drop. For a given acoustic pressure, the aspect ratio of the drop varies depending on the surface tension of the drop. In other words, we cannot perform the measurement on two different liquids under the exactly same acoustic pressure and aspect ratio of the drops. This fact prevents us from applying the error analysis performed on the first liquid to the second liquid. For example, the 25% discrepancy found for glycerin drops in the present measurement cannot be applied to water drops.

Fortunately, the intended applications of the present technique are for liquids in the undercooled state; thus, the following procedure largely eliminates the shortcoming involved in the technique: First, we levitate a liquid drop at above its freezing point. We apply an additional pressure and fix the acoustic pressure and the aspect ratio. Then, we determine the thermal diffusivity coefficient using the present technique. Since the thermal diffusivity coefficient of the liquid in the normal state is presumably known or can be measured with conventional techniques, we can use it as a reference to calibrate the value determined by the present technique. Second, we vary the temperature to undercool the drop but keep the acoustic pressure and the aspect ratio at the fixed values. Then, we repeat the measurement and apply the same calibration to the measured values.

There is no doubt that more sophisticated heat transfer models capable of quantitatively determining the intensity of acoustic streaming around the drop and incorporating it in three-dimensional heat conduction inside of the drop are required for the present technique to be generally applicable to every liquid. We have initiated an experiment in which we will investigate the intensity and flow pattern of acoustic streaming around the flattened drops, with various aspect ratios, using solid drops. The results will be compared to a theoretical study that will be performed simultaneously.

CONCLUSIONS

We have demonstrated a technique that can be used to determine the thermal diffusivity coefficient of undercooled liquids. The technique involves levitation of a liquid drop using an acoustic levitator and heating it using a CO_2 laser. The heat loss during the non-uniform cooling caused by acoustic streaming is correlated with the radial heat conduction model to determine the thermal diffusivity coefficient. The measurement using glycerin as a model drop has demonstrated the feasibility of the technique. The technique is well suited when the thermal diffusivity coefficient of the liquid in the normal state (above the freezing point) is known or can be measured by conventional techniques.

ACKNOWLEDGMENTS

The authors would like to thank S.K. Chung for his help in the early stages of this study. This study is supported by the Microgravity Program, Office of Biological and Physical Research, National Aeronautics and Space Administration.

REFERENCES

1. TRINH, E.H. 1985. Compact acoustic levitation device for studies in fluid dynamics and materials science in the laboratory and microgravity. Rev. Sci. Instr. **56:** 2059.
2. OHSAKA, K., E.H. TRINH & M.E. GLICKSMAN. 1990. Undercooling of acoustically levitated molten drops. J. Cryst. Growth **106:** 191.
3. OHSAKA, K., S.K. CHUNG & W.K. RHIM. 1998. Specific volumes and viscosities of the Ni–Zr alloys and their correlation with the glass formability of the alloys. Acta Mater. **46:** 4535.
4. TRINH, E.H., P.L. MARSTON & J.L. ROBEY. 1988. Acoustic measurement of the surface tension of levitated drops. J. Colloid Interface Sci. **124:** 95.
5. RHIM, W.K. & K. OHSAKA. 2000. Thermophysical properties measurement of molten silicon by high-temperature electrostatic levitator: density, volume expansion, specific capacity, emissivity, surface tension and viscosity. J. Cryst. Growth **208:** 313.
6. RHIM, W.K., K. OHSAKA, P.-F. PARADIS & R.E. SPJUT. 1999. Noncontact technique for measuring surface tension and viscosity of molten materials using high temperature electrostatic levitation. Rev. Sci. Instr. **70:** 2796.
7. TRINH, E.H. & J.L. ROBEY. 1994. Experimental study of streaming flows associated with ultrasonic levitators. Phys. Fluids **6:** 3567.

Thermophysical Property Measurements by Electromagnetic Levitation Melting Technique under Microgravity

SAYAVUR I. BAKHTIYAROV AND RUEL A. OVERFELT

Solidification Design Center, Auburn University, Alabama, USA

ABSTRACT: The purpose of this paper is to review the advances in electromagnetic levitation techniques as tools for determination of the thermophysical properties of molten metals under standard gravity and microgravity conditions. We show that a combination of containerless processing and microgravity offers the exclusive possibility to measure critical thermophysical properties of molten metals, both above and below their melting points, over a wide temperature range, and with high accuracy.

KEYWORDS: thermophysical properties; measurement; electromagnetic levitation melting; microgravity

INTRODUCTION

Strong inhomogeneous magnetic fields are necessary to generate sufficient levitation force against $1g$ in ground-based electromagnetic levitation techniques. The asymmetric external magnetic pressure field influences droplet oscillation behavior, drives turbulent internal flows, and induces unstable translational vibrations. Under microgravity conditions the magnetic field strength around a liquid droplet is significantly lower than that required to position the same specimen against Earth's gravity. Therefore, the use of electromagnetic levitation techniques in microgravity conditions minimizes many of the disturbances seen in $1g$ and offers a number of advantages for critical containerless measurement of thermophysical properties of liquid metals. The low-strength and uniformly distributed electromagnetic fields do not perturb the spherical shape of the droplet. In this case, the many theories that assume spherical droplet shape can be directly applied to determine thermophysical properties, such as surface tension, viscosity, heat capacity, and electrical conductivity. A low magnetic field strength also minimizes excessive stirring of the molten specimen and reduces the turbulence of fluid motion. The combination of containerless processing and microgravity offers an exclusive possibility to measure thermophysical properties of molten metals, both above and below their melting points, over a wide temperature range, and with high accuracy.[1]

The purpose of this paper is to review advances in electromagnetic levitation techniques, assess the current status of the technology, and discuss the future exploitation

---

Address for correspondence: Sayavur I. Bakhtiyarov, Solidification Design Center, Auburn University, Auburn, AL 36849-5341, USA. Voice: 334-844-6198; fax: 334-844-3400.
sayavurb@eng.auburn.edu

of the International Space Station for determination of the thermophysical properties of molten metals under microgravity conditions.

THERMOPHYSICAL PROPERTY MEASUREMENTS USING ELECTROMAGNETIC LEVITATION UNDER MICROGRAVITY

Herlach and Feuerbacher[2] considered containerless processing of melts in a microgravity environment as the most effective technique to obtain extended degrees of undercooling. Levitation against the Earth's strong gravity field applies significant power to samples; therefore, the electromagnetic levitation technique under terrestrial conditions is applicable only to metals of high melting point. Since heat transfer takes place in vacuum by radiation only, the temperature of a sample cannot be decreased beyond a minimum of about 1,300 K. Positioning forces under microgravity conditions are at least three orders of magnitude lower than those required under $1g$ conditions. This significantly reduces the heat input, so that samples can be levitated at much lower temperatures. The low positioning forces also prevent undesirable violent agitation of the liquid and translational sample vibrations that can result in disruption of the sample. Space experiments will advance our understanding of heterogeneous and homogeneous nucleation processes over a wide range of temperature. It will allow researchers to develop new alloy compositions with new microstructural properties in stable and metastable phases.

In their review work, Carruthers and Testardi[3] discussed materials processing in the reduced-gravity environment of space. These authors postulated that any use of space facilities must evolve from a comprehensive and competent ground-based research program in materials processing.

El-Kaddah and Szekely[4] extended the theoretical predictions of the electromagnetic force field, the velocity field and the temperature field in a levitated liquid droplet reported previously[5] under zero gravity conditions. The formulation and the computation procedure used were the same as described in their previous report. The computation results were used for the interpretation of experimental measurements made by the space processing applications rocket (SPAR) 1 (rocket-based, see below) flight program. The authors found a substantial difference in the velocity fields and turbulence in levitated droplet under Earthbound and zero-gravity conditions. Zero-gravity simulations revealed a symmetric circulation pattern of four recirculating loops, contrary to an asymmetric circulation pattern and two circulation loops obtained under $1g$ conditions. In addition, under zero-gravity conditions the spatial variation in the turbulent kinetic energy was significantly reduced compared to that under Earthbound conditions.

Herlach et al.[6] conducted undercooling experiments on Ni within (1) an Al_2O_3 crucible and (2) in a containerless state. The experiments revealed the effect of heterogeneous nucleation due to container walls. Containerless processing using electromagnetic levitation techniques eliminated heterogeneous nucleation and enabled a significant increase of undercooling by a factor of about four. Due to the high power absorption necessary to levitate metal samples, the electromagnetic levitation technique was limited to high melting point metals.

Lundgren and Mansour[7] applied a boundary-integral method to study numerically nonlinear oscillations and other motions of large axially symmetric liquid drops in zero gravity conditions. By maintaining first-order viscous terms in the normal stress boundary conditions, the effect of small viscosity was included in the computations. In Earthbound flows of contained liquids with free surfaces the main source of viscous effects comes from boundary layers at the walls, with smaller inputs from the free surface. In containerless zero gravity conditions the free surface is the only damping source. The effect of viscosity is smaller in the fourth-mode computations, and it increases for higher modes.

In microgravity, electromagnetic levitation and undercooling are possible under vacuum conditions.[8] Hence, the heat losses are due to radiation entirely

$$P_{out} = P_{rad} \sim 4\pi R^2 \varepsilon T^4, \tag{1}$$

where P_{out} is the heat output and ε is the total emissivity. If temperature and emissivity are known, this allows calibration of the power input to the specimen with respect to the power input to the heating coil. The latter is the only quantity that can be measured directly. The levitating magnetic fields unavoidably introduce disturbances into a liquid specimen. A microgravity environment can minimize these disturbances.

The use of the electromagnetic levitation technique under terrestrial conditions for measuring the electrical conductivity of liquid metals is limited, because under $1g$ conditions the specimen is aspherical. However, variation of the total impedance with the sample electrical conductivity is known for simple geometric shapes, such as spheres. Therefore, the electromagnetic levitation technique is more suitable for the microgravity environment, where the electrical conductivity of liquid metals can be determined in a wide temperature range.[8]

Most theories for prediction of the surface tension of liquid metals are valid for spherical samples only. Under terrestrial conditions, the levitated specimens are almost always deformed. On the oscillation spectrum, instead of a single peak for each value of n, as in the case of spherical samples, there are often 3–5 peaks. The electromagnetic levitation under microgravity conditions simplifies a quantitative evaluation of the oscillation spectra. The effective thermal conductivity is affected by convection under $1g$ conditions. Therefore, microgravity conditions allow separation of the actual thermal conductivity from convection effects.

Under terrestrial conditions, the electromagnetically induced flow in the liquid metal may be turbulent, which affects the apparent viscosity measurements. Because microgravity conditions allow measurements with small magnetic fields, it is possible to reduce a turbulence of the flow inside the sample. If the density measurements in the undercooled liquid regime are extrapolated by a straight line, there is a hypothetical temperature at which the densities of liquid and solid phase become equal. The phenomenon, when the density of the liquid phase is higher than that of the solid phase, is called *density catastrophe*. Measurements under microgravity conditions can reveal the relationship between this phenomenon and *entropy catastrophe*.[6]

In the microgravity experiments at Tiegelfreies Elektro-Magnetisches Prozessleren unter Schwerelosigkeit (TEMPUS) facilities, metallic specimens were electromagnetically positioned and heated to the temperature above their melting point. The current through the heating coils was then switched off and the sample allowed to cool. The main objectives of the experiments were to allow the specimens to undercool and study the structures obtained by the recalescence and solidification.[9] It is very important to know:[10,11]

- the time scale of the velocity decay,
- the time scale of the heat transfer,
- whether the recalescence and solidification take place in a stagnant or in an agitated liquid,
- whether significant temperature gradients are sustained in the sample.

Zong et al.[12] presented the results of the mathematical modeling of the behavior of electromagnetically levitated metal droplets under the microgravity conditions. The model allowed computation of the actual shape of a specimen that had been deformed due to the action of the electromagnetic field generated by a "squeezing coil". Knowing the force field and the droplet shape permitted calculation of the resultant velocity fields using the FIDAP computational package, on the assumption that the velocity field does not affect the droplet shape. The simulations show that for the operating conditions in the TEMPUS facilities, the positioning coils introduced a circulation with a very small velocity (10^{-2} m·sec^{-1}). However, the heating coils produced a turbulent flow with high velocity (0.3 m·sec^{-1}). In the turbulent range the velocity depends linearly on the applied current. The maximum liquid velocity is independent of the sample size and the sample conductivity, but inversely proportional to the square root of the density.

Zong et al.[12] presented computational results for the transient behavior of levitation-melted and electromagnetically positioned liquid metal droplet, when the supply of the current to the heating coils had been turned off but the current through the positioning coils was sustained at its original level. The numerical solution of the transient Navier–Stokes equations and the associated differential thermal energy balance equations showed the decay of both the velocity and the temperature fields. The decay of the velocity fields was related to the driving force for recirculatory motion produced by the heating coils, which is larger than that of the positioning coils. The radiative heat exchange with the surroundings generates the temperature decay in the specimen.[11] According to the results, the overall rate of the thermal energy loss is limited by thermal radiation at the surface. Hence, the droplet may be considered isothermal, and convection within the droplet does not affect the heat transfer. The velocity field decays more rapidly than the temperature field. When the velocity field approaches the "positioning asymptote", the heat loss from the droplet is quite small. Cooling and undercooling of the droplets take place in a stagnant liquid or in a liquid very gently agitated by the positioning coils.

Zong et al.[12] performed a test of the computation algorithms developed for predicting the electromagnetic force fields and the heating rates that may be achieved in the electromagnetic levitation device TEMPUS. The theoretical predictions of the lifting forces and the power absorption were compared with those determined experimentally on TEMPUS facilities. The simulations and the experiments were conducted for a range of conditions, such as coil configuration, sample position, sample conductivity, and current levels. Excellent agreement was found between the theoretical predictions and the measurements. It also revealed that the analytic solution for the spherical droplets agrees well with the computed data. Egry and Sauerland[13] discussed the advantages of microgravity environment for surface tension and viscosity measurements by electromagnetic levitation technique.

In their review work, Herlach et al.[6] discuss the most frequently used containerless techniques, such as free fall facilities and levitation facilities, including potential applications in the microgravity environment in space. A special emphasis is made on the thermophysical properties measurements, nucleation experiments, crystal growth, and the solidification of non-equilibrium states of undercooled melts. The authors highlight the following advantages of the microgravity environment:

- there is no deformation of the sample and the spherical shape is maintained,
- there is no turbulent flow in the liquid specimen,
- no gas stream is required for cooling the sample, undercooling is possible in UHV,
- wider temperature range becomes accessible and temperature control is facilitated, and
- processing and undercooling of low melting point metals is possible.

Drop Tube Experiments

A drop tube (drop shaft or drop tower) apparatus complements orbital space experiments to study containerless processes in high-temperature metals and alloys. A drop tube is an enclosed space where molten droplets are allowed to cool and solidify while falling freely down a tube within which there is a controlled atmosphere. In a drop tower or drop shaft an entire experimental setup (furnace, sensor electronics, instrumentation, and specimen) is dropped within an enclosure. The technique has some advantages over orbital experiments, since its simple construction and operation allows samples to be tested at low cost and parameters to be changed during the experiment. It is shown that the cooling rate in a falling droplet technique can be increased significantly by back filling the drop tube with an inert gas at a fixed pressure. The existing drop tower facilities give a residual gravitation level of 10^{-5}–$10^{-6}g$ and free fall time up to 10–12 sec. Under microgravity conditions in drop tubes, pure metals solidify as single crystals at high undercooling. Beyond a critical undercooling the grain refinement effect was not observed in most pure metals and alloys. Cech and Turnbull[14] first employed the drop tube technique to study phase formation in Fe–Ni alloys.

Lacy et al.[15] constructed a high vacuum 32 m drop tube apparatus to conduct various containerless low-gravity solidification experiments. Niobium droplets with 2–5 mm diameter have been undercooled by 525 K. A relatively high (320 m·sec^{-1}) dendritic solidification speed was obtained for test specimens. Solidification at large undercooling resulted in single crystalline spheres with the formation of interdendritic shrinkage channels on the sample surface rather than internal shrinkage voids. The authors propose that this may be related to the purity of the sample, its crystallographic structure or the containerless microgravity environment. A containerless low-gravity environment eliminates container induced kinematic forces that result in cavitation and grain refinement under 1 g conditions.

Naumann and Elleman[16] described two free-fall facilities: the Marshall Space Flight Center (MSFC) drop tube and the Jet Propulsion Laboratory (JPL) aerodynamic drop facility. The free-fall techniques eliminate difficulties, such as shape distortions and impact of electromagnetic field on heating and cooling of the sample. However, the limited free-fall distance restricts the sample to a size that can be solidified in the time available.

Fujii et al.[17] investigated the surface tension using the electromagnetic levitation technique under microgravity a using high-speed video camera. To eliminate an eddy current flow in the liquid sample, the authors used pure silicon. Microgravity conditions were obtained using the drop-shaft at the Japan Microgravity Center. The droplet shape was controlled by changing the current ratio in coils. It was shown that the surface oscillation of the sample is much simpler under microgravity ($10^{-5}g$) condition than under terrestrial ($1g$) conditions. The authors established a relationship between the number of oscillation peaks and sphericity of the droplet.

Sounding-Rocket Flight Experiments

Wouch et al.[18] performed a low-gravity experiment on a sounding rocket SPAR 3. A mixture of Be and BeO was melted and resolidified. A sample, 9.22 mm in diameter, was levitated in the NASA electromagnetic containerless processing payload (ECPP) during 230 sec. It was shown that under microgravity conditions a more uniform dispersion can be achieved than at $1g$ conditions. Piller et al.[19] pointed out that the microgravity interval during a sounding-rocket flight is sufficient for one cycle of a complete melting–solidification test.

The SPAR 1 flight experiments used a rocket to generate microgravity (about $10^{-5}g$) conditions for short (about 5–7 min) free-fall time periods.[4] Beryllium spheres, 9 mm in diameter, were levitated using spherical coil made of 2.5 mm copper tubing. The coil consisted of two turns wound on a 16 mm diameter sphere. The azimuth angles formed by the two coil turns were 46° and 74°. A radio frequency generator (1.2 kW and 107 kHz) was used as a power supply. To achieve 1,400°C sample temperature, 46 W power absorption rate at 300 A current) was needed.

To understand containerless processing of molten materials in space, Jacobi et al.[20] conducted an experimental study on a 2.5-cm diameter water drop levitated in a triaxial acoustic resonance chamber during a 240-sec low gravity SPAR flight. Oscillation and rotation of the droplet were induced by modulating and phase shifting the signals to the speakers. The film records were digitized and analyzed. The normalized equatorial area of the rotating drop was plotted as a function of a rotational parameter, and was in excellent agreement with values derived from the theory of equilibrium shapes of rotating liquid drops.

Space Flight Experiments

According to Naumann and Elleman,[16] the use of the microgravity environment of an orbiting vehicle extends the free-fall time indefinitely and allows much larger samples to be processed, as well as close observation and manipulation of the sample during processing. However, the small residual accelerations in an orbiting vehicle require the use of some form of non-contacting positioning force to hold the sample in the confines of its processing chamber for extended periods of time.

Rodot and Bisch[21] described an electromagnetic levitation facility in the fluids physics module during the Spacelab 1 mission. Good agreement was obtained between theoretical predictions and experimental results obtained under the terrestrial conditions.

Piller et al.[19] described a facility for containerless processing of liquid metals using electromagnetic positioning and inductive heating under the TEMPUS program. It was shown that the primary objective of the program was a study of high undercooling in the lower temperature region, which was not accessible under $1g$ conditions. A concept of dividing a desired composite field into a superposition of a dipole field for heating and a quadrupole field for positioning is discussed in detail.

TEMPUS is a multiuser and multipurpose Spacelab facility designed for containerless processing of metallic samples. It offers an opportunity to investigate the characteristics of undercooled metallic melts and their solidification behavior under microgravity conditions. Up to 24 different metallic samples, 6–10 mm diameter, can be containerlessly melted, undercooled, and stimulated, under inert purified gas atmosphere (Ar or He) or UHV environment of 10^{-9} mbar. The facility was designed to provide independent control of heating and positioning by use of two independent coils. Switching off the heating coil reduced both the input power into the sample, allowing cooling of the sample without a cooling gas, and the forces on the droplet surface. An additional heating coil between the positioning coils has been provided in this facility to generate a magnetic dipole field of high strength around the sample. This allows for independent control of the heating power and positioning forces under microgravity. A two-color radiation pyrometer was used for the contactless temperature measurements in the range 300–2,400°C with high measuring frequency (about 1 MHz). A charge coupled device (CCD) camera (50 Hz monochrome) was used to observe the process along the symmetry axis of the coil system.

Electromagnetic levitation techniques require contactless temperature measurement by pyrometry. At high temperatures and under high-vacuum conditions evaporation losses of the specimen occur. The evaporated material condenses on the cold parts of the experimental apparatus, including the front lenses of the pyrometer. This leads to a reduction of the radiation intensity reaching the pyrometer and can be mistakenly interpreted as a decrease in temperature. There are several methods to protect the pyrometer against contamination:

- In the replaceable shielding windows technique, a contaminated window is replaced by a clean one when the measurement errors become too large. The technique can be improved if a rotor with high rotation speed is inserted in the optical path between the sample and the pyrometer. Then the system becomes quite complex, particularly in a UHV environment.

- A diffraction grating technique is an alternative to the shielding mechanism without moving parts. In this technique a transmission diffraction grating is placed into the optical path normal to the incident radiation of the sample. It will diffract the light, but not metal atoms. The radiated intensity losses are possible due to the unused undiffracted light.

- A gas jet in front of the pyrometer offers another method to protect the pyrometer against contamination. However, this method cannot be used in a UHV environment.

- In a periscopic mirror system, contamination can be avoided by a mirror, which blocks the straight line between the sample and the pyrometer. The emissivity and reflectivity properties of the mirror depend on the type of the material, its surface structure and appearance. Under certain vapor pressure and vacuum conditions the evaporated sample material does not significantly deteriorate the optical quality of the mirror surface, consisting of the same material. The absorption and reflection properties of the metallic mirror surface remains unchanged during the condensation of sample material.[22]

- In the double mirror system, two mirrors are placed between the sample and the pyrometer, and the straight line between sample and pyrometer is blocked. The evaporated particles condense onto the mirror facing the sample and do not reach the pyrometer. It is assumed that the contamination of the mirror does not severely influence the optical system quality because the condensate metal vapor has the same reflectivity as the mirror itself.

To protect the pyrometer against contamination by condensation of sample vapor onto its exposed optical components of the TEMPUS facility, various protection mechanisms: replaceable shielding windows, single mirrors, and double mirror system were tested.[22] The TEMPUS laboratory module was equipped with a two-color pyrometer with a sampling rate of 1 MHz, which operates in the infrared region from 300 to 2,400°C. The temperature ratio was calculated from the first and the second temperature signals with wavelength range 1–2.5 µm and 3–4 µm, respectively. The temperature–time profile obtained for aluminum sample using CaF_2 shielding windows revealed that this technique could not solve the problem of reproducible and correct temperature data. Metal mirrors of nickel and aluminum on glassy substrates were prepared in specially designed equipment. Nickel and aluminum mirrors were facing the specimen and the pyrometer, respectively. The nickel–aluminum combination was chosen because nickel was going to be processed and, hence, the evaporated nickel particles condense onto the nickel mirror surface. The aluminum mirror has higher reflectivity and will not be coated with sample vapor. A high-speed video camera with frame-repetition frequencies up to 500 Hz has been used to observe the droplet from the side in the TEMPUS facility. The signals were digitized to obtain sharp contours of the droplet for further image analysis in a computer. The experiments conducted with nickel and aluminum samples for 15 minutes at 1,570 and 1,100°C, respectively, showed that the optical quality of the double mirror does not change significantly during the evaporation process.

Lohöfer et al.[23] discusses the methods and advantages of TEMPUS facility in the measurement of thermophysical properties of liquid metals. The temperature uncertainty for measuring thermal expansion of liquid metals, using a method proposed by Shiraishi and Ward,[24] was about 10–45 K. The optical method proposed by Ruffino[25] for thermal expansion measurements, with only one CCD camera is considered preferable for use in the TEMPUS facility. Precise measurement of the generator input power, as well as of the frequency, make it possible to determine the electrical conductivity of molten metals. During the measurements the voltage drop over the condenser can be controlled and kept constant. Ruffino also proposes another method to determine the electrical conductivity from the resonant frequency, which changes with the inductance of the coil generated by a change of the sample conductivity.

Egry et al.[26] summarize a proposal for electromagnetic levitation experiments in microgravity to measure viscosity and surface tension of undercooled liquid metals as a function of temperature. The authors analyze the applicability of the Arrhenius, the Vogel–Fulcher and the power-low models for the temperature dependence of the viscosity. It is assumed that in alloys, the intermolecular forces may lead to short range order and association of clusters, which has a significant effect on the viscosity of the sample and leads to the concentration dependence of the viscosity. This effect is more pronounced when the freezing temperature is approached from above. The surface tension and the viscosity can be determined by exciting oscillations of a freely floating droplet. The frequency of the oscillations is related to the surface tension, whereas the viscosity is determined from the damping. The authors propose the following advantages of the microgravity environment:

- The sample maintains its spherical shape due to absence of the deformation. This allows direct application of the theory.

- There is no turbulent flow in the droplet. The flows generating centrifugal forces on the surface are significantly reduced.

- Processing in UHV and at lower temperatures becomes possible due to the small amount of heat induced in the specimen.

- The absence of the additional damping due to magnetic field of oscillations.

Diefenbach et al.[27] presented a survey of the TEMPUS user support tasks. TEMPUS provides high resolution, fast three-color pyrometer and a high-speed video system for diagnostic measurements at the metastable state of a melt below its melting point.

Egry et al.[28] presented results of the surface tension measurements for pure gold ($T = 1,225°C$, $m = 5.1 g$), gold–copper ($Au_{56}Cu_{44}$, $T = 970–1,080°C$, $m = 4.21 g$) and zirconium-nickel ($Zr_{64}Ni_{36}$, T=980-11500 C, m=1.94 g) alloys in microgravity using the oscillating drop technique. The experiments were conducted during the Spacelab IML-2 mission in 1994 using the electromagnetic containerless processing facility TEMPUS. It was shown that in microgravity the magnetic pressure of the levitation field on the droplet is negligible, whereas under $1 g$ conditions some corrections (for example, Cummings and Blackburn, 1991) are necessary. The data obtained are in good agreement with those measured by conventional methods. The correction formula accurately accounts for the magnetic field and the gravity effects.

During the MSL-1 mission, the surface tension of Zr, $Co_{80}Pd_{20}$, $Pd_{82}Si_{18}$, $Pd_{78}Cu_6Si_{16}$, $Zr_{11}Ti_{34}Cu_{47}Ni_8$, and Fe–Ni–Cr alloys was measured using TEMPUS facilities.[29,30] Non-contact modulation calorimetry was used to measure the specific heat of a number of glass-forming alloys both in the equilibrium melt and in the undercooled region. The viscosity data on $Co_{80}Pd_{20}$, $Pd_{82}Si_{18}$, $Pd_{78}Cu_6Si_{16}$, and $Zr_{11}Ti_{34}Cu_{47}Ni_8$ were parametrized according to an Arrhenius law. The high-resolution radial video camera equipped with telecentric optics was used to measure density and thermal expansion of glass-forming alloys ($Zr_{65}Cu_{17.5}Al_{7.5}Ni_{10}$, $Zr_{60}Al_{10}Cu_{18}Ni_9Co_3$, $Zr_{11}Ti_{34}Cu_{47}Ni_8$, $Zr_{57}Cu_{15.4}Ni_{12.6}Nb_5Al_{10}$, $Pd_{78}Cu_6Si_{16}$, and $Pd_{82}Si_{18}$) in the undercooled regime. A non-contact, inductive method allowed measurement of the electrical resistivity of $Co_{80}Pd_{20}$ at both solid and liquid states. The electrical resistivity varies with temperature linearly as follows

$$\rho_s(T) = 51.9 + 0.044T, \qquad (2)$$

$$\rho_l(T) = 69.8 + 0.058T. \qquad (3)$$

Egry et al.[31] discussed the experimental results obtained during MSL-1 mission, and the capabilities of the Advanced Electromagnetic Levitation Facility (former TEMPUS) for thermophysical property measurements of liquid metals on the International Space Station (ISS). Due to the permanent nature of the ISS, a modular design is proposed to exchange the consumables without exchanging the entire facility. Non-contact diagnostic tools, such as pyrometry, videography, and inductive measurements, will be used to measure specific heat, surface tension, viscosity, thermal expansion, and electrical conductivity of liquid metals and alloys.

Lohöfer et al.[32] discusses the results obtained during two Spacelab missions, STS 83 and STS 94, in 1997 in the TEMPUS facility. In these experiments the electromagnetically positioned $Co_{80}Pd_{20}$ sample, 8-mm diameter, was inductively heated, melted, and overheated. During the cooling cycle an excitation pulse was applied every 50 K, until the sample solidified. From image analysis, the radius of the sample was determined to be a function of time. The damping constant was obtained from the time signal, and the frequency was obtained from the Fourier transform of the signal. Frequencies and damping constants were used to estimate surface tension and viscosity of the liquid sample as functions of temperature. It was shown that the surface tension changes linearly with temperature as follows:

$$\gamma_{Co_{80}Pd_{20}}(T) = 1.69 - 15 \times 10^{-5}(T - 1613), \qquad (4)$$

where T is the temperature in K. The viscosity can be expressed by the Arrhenius type expression

$$\eta_{Co_{80}Pd_{20}}(T) = 0.15 e^{6790/T}. \qquad (5)$$

Linear temperature dependencies were also found for the electrical resistivity of $Co_{80}Pd_{20}$ in both the solid and liquid states

$$\rho_s(T) = 122 + 0.042(T - 1563), \qquad (6)$$

$$\rho_l(T) = 146 + 0.050(T - 1613). \qquad (7)$$

Parabolic Flight Experiments

TEMPUS was not designed to levitate samples against $1g$.[23] Therefore, the positioning capabilities of TEMPUS were tested during parabolic flights on a KC-135 airplane, which provides up to 20 sec of low gravity ($10^{-2} g$) during each parabola.[33] However, the microgravity time was sufficient to position and melt a 10-mm diameter 75%Fe–25%Ni specimen with a mass of $4.2 g$ in a helium atmosphere, and to excite its surface oscillations. In the first 2 sec of the experiment, the sample was in solid state without any oscillation. Then the sample was heated to about 1,580°C (above the melting point) and oscillations excited at about 3 sec. At $t \approx 6$ sec. when gravity sets in again and the sample hits the sample holder, the oscillations stop. Fourier analysis reveals only one sharp peak in the frequency spectrum, which corresponds to the underformed spherical shape of the specimen under microgravity conditions. Visual inspection of the video recordings identified this peak as the ($l = 2, m = 0$)-mode. Hence, the sample remains essentially spherical in microgravity conditions. Equation with fundamental mode 2 was used to determine the surface tension of the sample near the melting point. The surface tension value obtained, $\gamma = 1.6 \text{N} \cdot \text{m}^{-1}$, is in good agreement with values reported by Keene,[34] but it is about 10% lower than the value obtained in the same research under $1g$ conditions. It needs to be mentioned that the microgravity time possible during parabolic flights is too short to estimate the viscosity of the liquid droplet from the frequency spectrum.

Overfelt et al.[35] reported the results of the containerless surface tension measurements using an electromagnetic levitator (see FIGURE 1) under microgravity conditions on the KC-135 aircraft. The device was powered by a commercial 1 kW Ameritherm power supply. A vacuum level of 5×10^{-6} torr was maintained during measurements by an Alcatel turbopump. Partial pressures of background gases were measured with a Leybold Transpector residual gas analyzer. Up to eight samples can be processed without breaking vacuum. The temperature of each sample was measured by a two-color pyrometer that had been calibrated for a nickel sample with an internal B-type thermocouple. The Rayleigh equation was used in calculations of

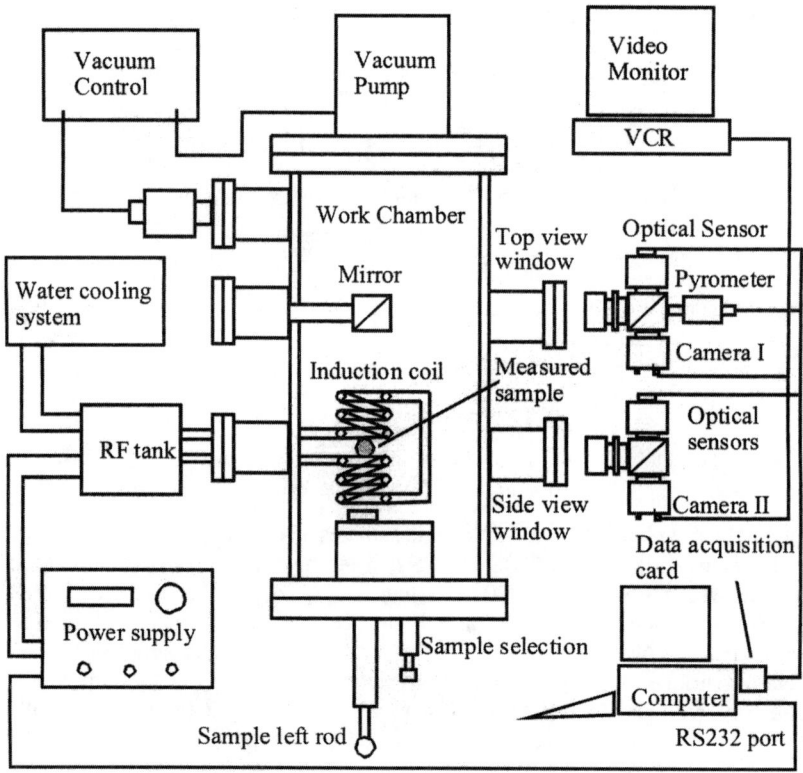

FIGURE 1. Schematics of electromagnetic levitator designed at Auburn University.

nickel droplet oscillation frequency, assuming $\gamma = 1.78\,\text{N}\cdot\text{m}^{-1}$. Good agreement was found between the predicted and experimentally observed oscillation frequencies for nickel samples with masses less than 0.2 g. However, Rayleigh's theory underpredicted the oscillation frequencies for droplets larger than about 0.2 g.

INTERNATIONAL SPACE STATION PROSPECTS

Review of the advances in electromagnetic levitation melting technique since its invention as a containerless processing technique, indicates that the technology offers a unique situation for certain metallurgical systems where many desirable properties of the melt cannot be evaluated by conventional melting methods. Many reliable theories have been developed to predict the lifting forces, the power absorption, the deformation of the droplet, the temperature and the flow fields under both terrestrial and microgravity conditions. Since many thermophysical properties are very sensitive to impurities, electromagnetic levitation melting as a containerless processing tool is definitely the favored method for their measurements. The accuracy of the measurements will be significantly improved when the experiments are conducted in microgravity conditions. Under microgravity conditions: (1) the magnetic field strength around a liquid droplet is significantly lower than that required to position the same specimen against earth gravity, (2) a low magnetic field strength results in a low amount power absorbed by the specimen, (3) a deep undercooling of molten metals in UHV environment is available, (4) there is minimal need to cool samples convectively using a high-purity inert gas, (5) the low-strength and uniformly distributed magnetic force fields do not disturb the spherical shape of the droplet, and (6) a low magnetic field strength minimizes stirring of the molten specimen and reduces the turbulence of fluid motion. The combination of containerless processing and microgravity offers the exclusive possibility to measure critical thermophysical properties of molten metals both above and below their melting points in a wide temperature range and with high accuracy.

ACKNOWLEDGMENTS

The authors gratefully acknowledge the financial support received from the American Foundry Society and NASA's Office of Life and Microgravity Sciences and Applications under Cooperative Agreement NCC8-128. We thank Dr. Deming Wang for providing schematics of electromagnetic levitator designed at Auburn University.

REFERENCES

1. EGRY, I. & J. SZEKELY. 1991. The measurement of thermophysical properties in microgravity using electromagnetic levitation. Adv. Space Rec. **11:** 263–266.
2. HERLACH, D.M. & B. FEUERBACHER. 1986. Nucleation and undercooling. *In* Materials Sciences in Space. B. Feuerbacher, H. Hamacher & R.J. Naumann, Eds.: 168–190. Springer-Verlag, Berlin.

3. CARRUTHERS, J.R. & L.R. TESTARDI. 1983. Materials processing in the reduced-gravity environment of space. Annu. Rev. Materials Sci. **13:** 247–278.
4. EL-KADDAH, N. & J. SZEKELY. 1984. Heat and fluid flow phenomena in a levitation melted sphere under zero gravity conditions. Metall. Trans. B **15B:** 183–186.
5. EL-KADDAH, N. & J. SZEKELY. 1983. The electromagnetic force field, fluid flow field, and temperature profiles in levitated metal droplets. Metall. Trans. B **14B:** 401–410.
6. HERLACH, D.M., R.F. COCHRANE, I. EGRY, et al. 1993. Containerless processing in the study of metallic melts and their solidification. Intl. Mater. Rev. **38:** 273–347.
7. LUNDGREN, T.S. & N.N. MANSOUR. 1988. Oscillations of drops in zero gravity with weak viscous effects. J. Fluid Mech. **194:** 479–510.
8. EGRY, I., G. LOHÖFER & S. SAUERLAND. 1993. Measurements of thermophysical properties of liquid metals by noncontact technique. Intl. J. Thermophys. **14:** 573–584.
9. WILLNECKER, R., D.M. HERLACH & B. FEUERBACHER. 1986. Containerless undercooling of bulk Fe–Ni melts. Appl. Phys. Lett. **49:** 1339–1341.
10. ZONG, J.-H., B. LI & J. SZEKELY. 1992. The electrodynamic and hydrodynamic phenomena in magnetically-levitated molten droplets–I. steady state behavior. Acta Astronautica **26:** 435–449.
11. ZONG, J.-H., B. LI & J. SZEKELY. 1993. The electrodynamic and hydrodynamic phenomena in magnetically-levitated molten droplets—II. transfer behavior and heat transfer considerations. Acta Astronautica **29:** 305–311.
12. ZONG, J.-H., J. SZEKELY & G. LOHÖFER. 1993. Calculations and experiments concerning lifting force and power in tempus. Acta Astronautica **29:** 371–378.
13. EGRY, I. & S. SAUERLAND. 1994. Containerless processing of undercooled melt: measurements of surface tension and viscosity. Mater. Sci. Eng. **A178:** 73–76.
14. CECH, R.E. & D. TURNBULL. 1956. Heterogeneous nucleation of the martensite transformation. J. Metals **8:** 124–132.
15. LACY, L.L., M.B. ROBINSON & T.J. RATHZ. 1981. Containerless undercooling and solidification in drop tubes. J. Crystal Growth **51:** 47–60.
16. NAUMANN, R.J. & D.D. ELLEMAN. 1986. Containerless processing technology. In Materials Sciences in Space. B. Feuerbacher, et al., Eds.: 294–313. Springer-Verlag, Berlin.
17. FUJII, H., T. MATSUMOTO & K. NOGI. 2000. Analysis of surface oscillation of droplet under microgravity for the determination of its surface tension. Acta Materialia **48:** 2933–2939.
18. WOUCH, G., R.T. FROST, N.P. PINTO, et al. 1978. Uniform distribution of BeO particles in Be casting produced in rocket free fall. Nature **274:** 235–237.
19. PILLER, J., R. KNAUF, P. PREU, et al. 1987. Electromagnetic positioning and inductive heating under micro-g. Proceedings of the 6th European Symposium on Material Sciences Under Microgravity Conditions, Bordeaux, France, 2–5 December, 1986. ESA SP-256: 437–443.
20. JACOBI, N., A.P. CROONQUIST, D.D. ELLEMAN & T.G. WANG. 1981. Acoustically induced oscillation and rotation of a large drop in space. Proceedings of the Second International Colloquium on Drop and Bubbles. D.H. LeCroissette, Ed.: 31–38. JPL, Pasadena.
21. RODOT, H. & C. BISCH. 1985. Oscillations de volumes liquides semi-libres en microgravité—Experience ES 236 dans Spacelab 1. Proceedings of the 5th European Symposium on Material Sciences under Microgravity, Schloss Elmau, Germany, November 5–7, 1984. ESA SP-222: 23–29.
22. NEUHAUS, P., I. EGRY & G. LOHÖFER. 1992. Aspects of high-temperature pyrometry for measurements in ultrahigh vacuum. Intl. J. Thermophys. **13:** 199–210.
23. LOHÖFER, G., P. NEUHAUS & I. EGRY. 1991. TEMPUS—a facility for measuring the termophysical properties of indercooled liquid metals. High Temperatures – High Pressures **23:** 333–342.
24. SHIRAISHI, S.Y. & R.G. WARD. 1964. The density of nickel in the superheated and supercooled liquid states. Can. Metall. Q. **3:** 117–122.
25. RUFFINO, G. 1989. Recent advances in optical methods for thermal expansion measurements. Intl. J. Thermophys. **10:** 237–249.

26. EGRY, I., B. FEUERBACHER, G. LOHÖFER & P. NEUHAUS. 1990. Viscosity measurement in undercooled metallic melts. Proceedings of the 7th European Symposium on Materials and Fluid Sciences in Microgravity, Oxford, UK. ESA SP-295: 257–260.
27. DIEFENBACH, A., M. KRATZ, D. UFFELMANN & R. WILLNECKER. 1995. Advanced user support programme—TEMPUS IML-2. Acta Astronaut. **35:** 719–724.
28. EGRY, I., G. JACOBS, E. SCHWARTZ & J. SZEKELY. 1996. Surface tension measurements of metallic melts under microgravity. Intl. J. Thermophys. **17:** 1181–1189.
29. DAMASCHKE, B., K. SAMWER & I. EGRY. 1999. *In* Solidification. W. Hofmeister, J. Rogers, N. Singh, *et al.*, Eds.: 43. TMS, Warrendale.
30. LOHÖFER, G. & I. EGRY. 1999. *In* Solidification. W. Hofmeister, J. Rogers, N. Singh, *et al.*, Eds.: 65. TMS, Warrendale.
31. EGRY, I., A. DIEFENBACH, W. DREIER & J. PILLER. 2000. Containerless processing in space—thermophysical property measurements using electromagnetic levitation. Proceedings of the 14th Symposium on Thermophysical Properties, Boulder, CO.
32. LOHÖFER, G., S. SCHNEIDER & I. EGRY. 2000. Thermophysical properties of liquid undercooled $Co_{80}Pd_{20}$. Proceedings of the 14th Symposium on Thermophysical Properties, Boulder, CO.
33. EGRY, I., G. LOHÖFER, P. NEUHAUS & S. SAUERLAND. 1992. Surface tension measurements of liquid metals using levitation, microgravity, and image processing. Intl. J. Thermophys. **13:** 65–74.
34. KEENE, B. 1988. J. Int. Mater. Rev. **1:** 1.
35. OVERFELT, R.A., R.P. TAYLOR & S.I. BAKHTIYAROV. 2000. Thermophysical properties of A356 aluminum, Class 40 gray iron and CF8M stainless steel. AFS Trans. **00-153:** 369–376.

Mass and Thermal Diffusivity Algorithms

Reduced Algorithms for Mass and Thermal Diffusivity Determinations

R. MICHAEL BANISH,[a] LYLE B.J. ALBERT,[a] TIMOTHEE L. POURPOINT,[a] J. IWAN D. ALEXANDER,[b] AND ROBERT F. SEKERKA[c]

[a]*Center for Microgravity and Materials Research,
University of Alabama in Huntsville, Huntsville, Alabama, USA*

[b]*Department of Mechanical and Aerospace Engineering,
Case Western Reserve University, Cleveland, Ohio, USA*

[c]*Department of Physics, Carnegie Mellon University,
Pittsburgh, Pennsylvania, USA*

ABSTRACT: Mass and thermal diffusivity measurements conducted on Earth are prone to contamination by uncontrollable convective contributions to the overall transport. Previous studies of mass and thermal diffusivities conducted on spacecraft have demonstration the gain in precision, and lower absolute values, resulting from the reduced convective transport possible in a low-gravity environment. We have developed and extensively tested real-time techniques for diffusivity measurements, where several measurements may be obtained on a single sample. This is particularly advantageous for low gravity research were there is limited experiment time. The mass diffusivity methodology uses a cylindrical sample geometry. A radiotracer, initially located at one end of the host is used as the diffusant. The sample is positioned in a concentric isothermal radiation shield with collimation bores located at defined positions along its axis. The intensity of the radiation emitted through the collimators is measured versus time with solid-state detectors and associated energy discrimination electronics. For the mathematical algorithm that we use, only a single pair of collimation bores and detectors are necessary for single temperature measurements. However, by employing a second, offset, pair of collimation holes and radiation detectors, diffusivities can be determined at several temperatures per sample. For thermal diffusivity measurements a disk geometry is used. A heat pulse is applied in the center of the sample and the temperature response of the sample is measured at several locations. Thus, several values of the diffusivity are measured versus time. The exact analytic solution to a heat pulse in the disk geometry leads to a unique heated area and measurement locations. Knowledge of the starting time and duration of he heating pulse is not used in the data evaluation. Thus, this methodology represents an experimentally simpler and more robust scheme.

KEYWORDS: diffusivity; mass diffusion; thermal diffusivity; thermal conductivity

Address for correspondence: R. Michael Banish, Center for Microgravity and Materials Research, University of Alabama in Huntsville, Huntsville, AL 35899, USA. Voice: 256-824-6969; fax: 256-824-6944.
   banishm@email.uah.edu

NOMENCLATURE:

a	diameter of disk
A	constant (cylindrical geometry)
$C(z)$	diffusant concentration distribution, time dependent
$C_0(z)$	diffusant concentration distribution at time 0
C_i	concentration at measurement locations
D	diffusivity (cm^2/sec)
J_0, J_1	Bessel functions
k	thermal conductivity
L	length of sample (cylindrical geometry)
n_i	X-ray photon intensities at measurement locations
$n_i(t)$	X-ray photon intensities versus time
p	ratio of radial location to disk diameter
r	radial location
t	time
T	temperature
T_0	temperature at time 0
$T(r,t)$	temperature at radial location versus time
z_i	measurement location relative to L
α_n	Bessel function root
β_{ij}	constant (disk geometry)
κ	thermal diffusivity (cm^2/sec)
μ_n	n-th positive root of first kind of order Bessel function
ν_n	ratio of Bessel function root to relative radial position
ΔT_{ij}	temperature difference between radial locations

INTRODUCTION

Diffusivities obtained in liquids at normal gravity are prone to contamination by uncontrollable convection. As emphasized for liquid diffusivity measurements by Verhoeven,[1] any horizontal component of a density gradient results in convection without a threshold. Simple scaling arguments illustrate the difficulty of obtaining purely diffusive transport in liquids. In a system of diffusivity 10^{-5} cm^2/sec, representative for mass diffusivities, and a typical diffusion distance of 1 cm, the characteristic diffusion velocity is of order 10^{-5} cm/sec. Hence, if true diffusion is to be observed, convective flow velocities normal to the concentration gradient must be of order 10^{-7} cm/sec or less. Thus, in liquids, the attainment of diffusion-dominated transport over macroscopic distances at normal gravity is obviously not a simple task. Numerical modeling efforts in our group and others has shown that in liquid metals, with their typical viscosities of 10^{-3} poise, temperature nonuniformities of a few hundredths of a degree are sufficient to generate convective contributions equal to the diffusive flux.[2] Thus, even in systems with essentially no *measurable* temperature nonuniformities, convective fluxes may be the dominant mode of transport.[3]

Estimation of the convective contamination of thermal diffusivity determinations follows a similar scaling argument. Although thermal diffusivities may be orders of magnitude higher than mass diffusivities, say 10^{-1} cm^2/sec, temperature gradients of tens of degrees are typically imposed on the systems. Thus, convective contributions are equally severe.

A typical criterion for claiming that convective contributions in species transport are negligible is when the logarithmic plot of concentration versus distance is linear.

(A similar argument is also used for thermal diffusivity or conductivity measurements). In experiments of indium diffusion into liquid gallium[4] the apparent diffusivity of horizontally oriented ampoules was 20% greater than that from the vertical orientations. Despite this, both results yielded perfectly Gaussian $C(x)$ plots. Our numerical modeling efforts have shown that only the most extreme convective contamination (several hundred percent) leads to nonlinearities in these graphs.[2] We have experimentally confirmed these results in horizontally and vertically oriented ampoules.

Several countermeasures have been taken to suppress convective transport in liquid diffusion experiments. The use of narrow capillaries in diffusion experiments has revealed (poorly understood) wall effects.[5–7] Magnetic fields have been widely used to suppress convection in conducting liquids. In general, it has been shown that the convective suppression scales with the fourth power of the Hartmann number. However, it needs to be noted that magnetic fields only dampen the already present fluid motion. Thus, the magnitude of the field strength necessary to suppress flows at the 10^{-6} to 10^{-7} cm/sec level are questionable at this point. We have numerically modeled field strengths up to 3 Tesla, the point at which the diffusive motion may itself be modified. Even at these field strengths only with (unrealistic) temperature non-uniformities of 0.01 K were the convective flows reduced significantly below the diffusive speed. Experiments by Youdelis showed that the field strengths necessary to reduce the convective contribution to a few percent of the diffusive flux (approximately 1 T) modify the diffusive flux itself, even in the solid state.[8] More recent experiments by Lehmann[9] showed that at field strengths of around 1 Tesla the apparent diffusivity values no longer change. However, it is neither clear if the convective motion is totally damped nor if there is any effect on the diffusive motion itself. Numerical modeling performed in our group indicates that even higher field strengths may be necessary.[10] Thus, the use of magnetic fields to suppress convective contributions in liquid diffusion measurements is still questionable. At this point it would be exceedingly helpful to have sufficient low-gravity and terrestrial values for comparison.

Low-gravity conditions currently appear to offer the only means to obtain reliable liquid diffusion data. Previous liquid metal diffusion studies conducted aboard spacecraft have demonstrated the gain in precision afforded by a low-gravity environment.[11–15] The shear cell, essentially a long capillary method,[16–17] has emerged as the dominant technique for obtaining diffusion coefficients in low gravity. One of the drawbacks of this method, given the limited flight opportunities and experiment time, is that a new sample is required for each temperature.

We have developed a technique for the *in situ* measurement of diffusivities in liquid at several temperatures with one sample[18] based on the methodology of Harned and Nuttall[19] and Codastefano, Di Russo, and Zanza.[20] Since this is a real-time technique it circumvents the need for solidification prior to concentration profiling.

Following from the reduced algorithm for mass diffusivity measurements in a long cylinder a similar reduction was derived for a disk geometry. This was applied to thermal diffusivity determinations. Using this methodology, initial conditions, such as the duration and magnitude of the heating pulse, do not need to be known. In addition, only the relative, not the absolute temperature, increase is used. This is very amenable to low-gravity experiments where changes can typically not be made during a mission.

MASS DIFFUSIVITY MEASUREMENT METHODOLOGY

The concept underlying our setup was originally developed for diffusivity measurements with gaseous krypton.[20] As schematically indicated in FIGURE 1 A, we use an initially solid cylindrical diffusion sample that consists mostly of inactive material (solvent) and a short section of activated isotope, as the diffusant, located at one end. After melting the sample and heating to a uniform measurement temperature T, the evolution of the diffusant concentration distribution $C(z)$ is monitored through the intensity of the radiation received through two bores (collimators) in a radiation

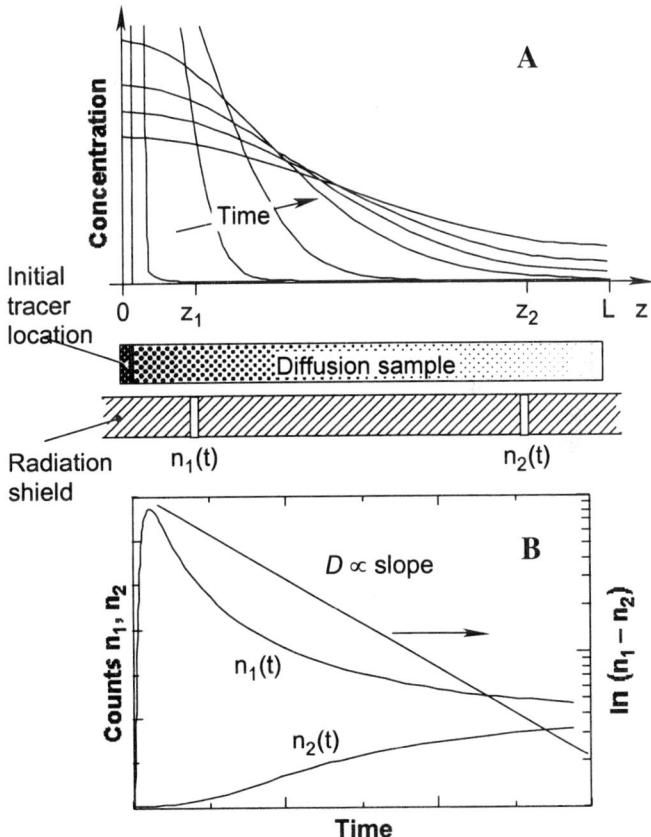

FIGURE 1. (**A**) Schematic presentation of the evolution of the concentration profile and position of the measurement locations. (**B**) Time traces of the signals at the two detector measurement locations, and corresponding presentation of the signal difference according to Equation (2).

shield. The intensities n_1 and n_2 are assumed to be proportional to the concentrations C_1 and C_2, respectively. The characteristic shape of the signal traces $n_1(t)$ and $n_2(t)$ associated with the spreading of the diffusant is illustrated in FIGURE 1 B.

For the case of a (long) cylinder with impermeable boundary conditions at both ends the diffusion equation, solution is simplified by choosing only two measurements locations at the positions

$$z_1 = \frac{L}{6} \text{ and } z_2 = \frac{5L}{6}, \tag{1}$$

where L is the sample length at the measurement temperature. Using symmetric locations causes the even terms in the expansion to cancel. Choosing the measurement locations at either $L/6$ (or $L/3$) cancels the $n = 3$ term in the expansion. Thus, only the $n = 1$ and $n = 5$ and higher terms remain. The $n = 5$ and higher terms are insignificant compared to the first term and are, therefore, neglected. The diffusivity D is then calculated from the difference of the signal traces using the relation

$$\ln[n_1(t) - n_2(t)] = A - \left(\frac{\pi}{T}\right)^2 Dt, \tag{2}$$

where the constant A depends on the concentration profile $C_0(z)$ at the beginning of the measurement. Since $C_0(z)$ does not explicitly enter the D evaluation, diffusivities can be consecutively determined at several temperatures during the spreading of the concentration profile in the same sample. Obviously, before the first measurement, $C_0(z)$ must have spread sufficiently to provide a significant signal at both detector locations. For measurements over a wide temperature range (two) collimation bores straddling the $L/6$ and $5L/6$ positions are necessary.

In addition to its real-time feature, the isotopic labeling technique permits investigations of the perceived "wall effect" mentioned above or convective contamination. By employing an isotope that emits photons at two sufficiently different energies and, thus, different self-absorption behaviors, transport in the bulk of the sample and near the container wall can be distinguished to some extent. Based on the attenuation data for 24 keV and 190 keV photons of ^{114m}In[21] we have calculated the fraction of the total radiation received at the two different emission energies outside a 3-mm thick sample versus the distance of the emitter from the surface of the sample; for details of the one-dimensional slab model see Reference 22. For this model, the 24-keV photons received at the detector are predicted to originate, at most, only from a 300 µm deep surface layer. The 190-keV photons, on the other hand, stem from throughout the whole sample slab.

MASS DIFFUSIVITY EXPERIMENTAL APPROACH

A detailed description of the experiment setup and its optimization by computer simulation is given in.[17] However, rather than only two measurement locations, we employ four total. The use of two separate pairs of radiation detector pairs allows us to input the proper $L/6$ and $5L/6$ counting rate values at any temperature and sample length. A single, low temperature version (185°C), using redundant pairs, was flown during NASA Increment 4 on the Mir Space Station. Three measurements of the self-diffusivity of ^{114m}In/In were completed during this mission.

The diffusion couples consist of 29-mm long 3-mm diameter native abundance section and a 1-mm thick disk of the radiotracer, located at one end of the cylindrical diffusion sample, as the diffusant. High purity (5–6N) starting material was used for each section. High purity boron nitride, which most liquid metals, including indium, does not wet, is used for the ampoule material. A press-fit plug provides for a square edge at the bottom and a spring-loaded close fitting plunger maintains a square edge at the top and keeps the sample touching the bottom plug. To provide a defined atmosphere the ampoule is sealed inside a welded aluminum or silica cartridge. The cartridge is positioned in the concentric isothermal radiation shield with its collimation bores.

In our experiments a 2-mm diameter collimation bore and 16-mm (of Au) radiation shield wall thickness gave the optimum signal to noise (background) ratio. The intensity of the radiation emitted through the collimators was measured versus time with CdZnTe solid state detectors and associated energy discrimination electronics. The counts from the radiation detectors interfaced to counter timers that recorded the counting rate every 10 to 60 seconds depending on the specific activity.

These data are then numerically analyzed and the diffusivity recovered from Equation (**2**). It should be noted that the diffusivity values are only valid after a statistically significant signal is present at the second, far, detector, note that this is well after the "hump" is decreasing at the first detector (i.e., the maximum is reached).

THERMAL DIFFUSIVITY MEASUREMENT METHODOLOGY

The determination of the thermal diffusivity κ (or conductivity k) typically requires precise knowledge of the applied parameters. Such parameters include the time at which a heat pulse was applied, the amount of energy, added or the initial temperature increase. The diffusivity is then obtained by matching the experimentally measured profile to the output of an analytical/numerical model. The resulting diffusivity is sensitive to values of these applied parameters.

Following from the simplified algorithm that was obtained in the (long cylinder) mass diffusion methodology we have derived a corresponding formulation for a disk geometry. In the final formulation the only experimental measurements required are temperature versus time at selected locations. As in the mass diffusivity case, in the (thermal) diffusivity methodology described below, no initial conditions are used directly in the final result—that is, the exact starting time of the temperature pulse is not require for data analysis. The thermal diffusivity is determined using the difference between the temperatures measured at two locations. A heat pulse is applied for an arbitrary time to one region of the sample. Temperature changes are then monitored versus time. The thermal diffusivity is calculated from the temperature differences versus time.

We consider a perfectly insulated circular cylinder of radius a for which the temperature depends only on the distance r from the center of the cylinder. Heat is applied uniformly, heating the cylinder to T_0 out to an arbitrarily chosen fixed fraction, p, of the cylinder radius. Thus, a function giving the initial temperature distribution $f(r) = T(r, 0)$ is defined as follows:

$$f(r) = \begin{cases} T_0 & \text{for } 0 < r/a < p \\ 0 & \text{for } p < r/a < 1 \end{cases} \quad (3)$$

where $0 < p < 1$, a is the cylinder radius, and T_0 is a constant (see FIGURE 2A). For time $t > 0$, radial heat transfer begins. The evolution of the temperature $T(r, t)$ is then given[23] by the expression

$$T(r, t) = \frac{2}{a^2}\left(\int_0^a rf(r)dr\right) + \frac{2}{a^2}\left(\sum_{n=1}^{\infty} e^{-\kappa\alpha_n^2 t}\frac{J_0(r\alpha_n)}{J_0^2(a\alpha_n)}\int_0^a rf(r)J_0(\alpha_n r)dr\right), \quad (4)$$

where $\mu_n = a\alpha_n$ is the n-th positive root of the Bessel function of the first kind of order 1, that is, $J_1(\mu_n) = 0$. Using the initial conditions defined above, the integrals in Equation (4) become

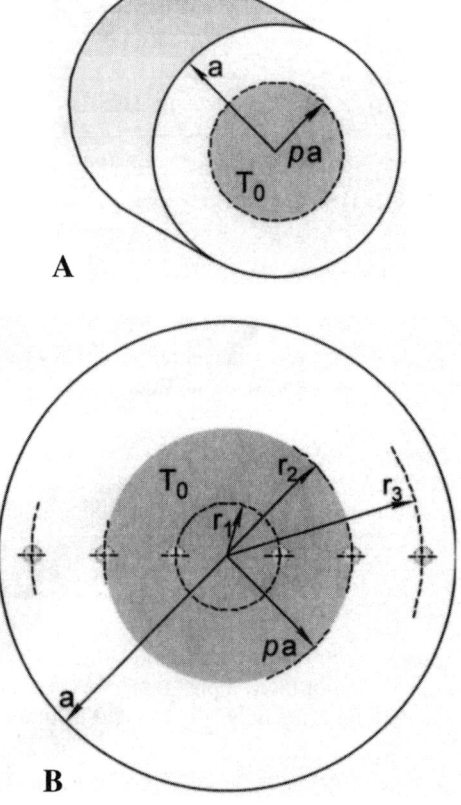

FIGURE 2. (**A**) Schematic of starting analytical model, a is the sample radius and p the heated fraction. (**B**) Plan view of the sample showing the positions of the heated region and temperature measurement locations.

$$\int_0^a r f(r) dr = T_0 \int_0^{pa} r dr = \frac{1}{2} T_0 (pa)^2 \tag{5a}$$

and

$$\int_0^a r f(r) J_0(\alpha_n r) dr = T_0 \int_0^{pa} r J_0(\alpha_n r) dr = T_0 (pa/\alpha_n) J_1(\alpha_n pa), \tag{5b}$$

where

$$\frac{d}{dx}[x J_1(\alpha x)] = \alpha x J_0(\alpha x) \tag{6}$$

has been used. Thus we obtain

$$T(r,t) = T_0 p^2 + T_0 \sum_{n=1}^{\infty} \left(e^{\frac{-\kappa \mu_n^2 t}{a^2}} \right) \frac{2p J_0(\mu_n r/a)}{\mu_n J_0^2(\mu_n)} J_1(\mu_n p). \tag{7}$$

We proceed to simplify the infinite sum in Equation (5) by eliminating the $n = 2$ and $n = 3$ terms. Since the terms for $n > 3$ decay rapidly with time, we are left after a short transient with the $n = 1$ term, *a tractable expression that can be used to analyze experimental data*. Setting p, the ratio of the heated to the sample radius, equal to the special value $\mu_1/\mu_2 = 0.5436$, we can eliminate the $n = 2$ term of the series because

$$J_1(\mu_2 p) = J_1(\mu_1) = 0. \tag{8}$$

The $n = 3$ term can be set to zero by choosing the temperature measurement locations r so that $\mu_3 r/a = \nu_n$ corresponds to the first three zeros of J_0 (that is, $J_0(\nu_n) = 0$) which are 2.4048, 5.5201, and 8.6537, respectively. With $\mu_3 = 10.17347$, this results in three special values of r given by $r_1/a = \nu_1/\mu_3 = 0.2364$, $r_2/a = \nu_2/\mu_3 = 0.5426$, and $r_3/a = \nu_3/\mu_3 = 0.8506$. The next zero of J_0 is at $\nu_4 = 11.792$ and would result in a value $r_4/a > 1$, outside the sample.

The geometry of the heated area and the three special values of r (temperature measurement locations for the experiments) are shown in FIGURE 2B. With the $n = 2$ and $n = 3$ terms set to zero by means of the choices adopted above, the temperature at each r_i is

$$T(r_i, t) = T_0 p^2 + T_0 \frac{2p J_0(\mu_1 r_i/a)}{\mu_1 J_0^2(\mu_1)} J_1(\mu_1 p) e^{\frac{-\kappa \mu_1^2}{a^2}} + hot, \tag{9}$$

where *hot* stands for higher order terms ($n > 3$).

Taking the difference between two locations, for example, between r_1 and r_3,

$$T(r_1, t) - T(r_3, t) = T_0 \frac{2p J_1(\mu_1 p)}{\mu_1 J_0^2(\mu_1)} \times \left(J_0\left(\mu_1 \frac{r_1}{a}\right) - J_0\left(\mu_1 \frac{r_3}{a}\right) \right) e^{\frac{-\kappa \mu_1^2}{a^2}} + \Delta(hot). \tag{10}$$

It can be shown that the difference between the higher order terms in Equation (10) is less than 1% of the $n = 1$ term for $\kappa t/a^2 > 0.015$. This corresponds to a time greater than 3.7 sec for $a = 5$ cm and $\kappa = 0.1$ cm^2/sec. Neglecting such high order terms, and taking the natural logarithms of the differences $\Delta T_{ij} = T(r_i, t) - T(r_j, t)$, we obtain

$$\ln(\Delta T_{ij}) = \ln(\beta_{ij}) - \left(\frac{\kappa \mu_1^2}{a^2}\right) t, \tag{11}$$

where

$$\beta_{ij} = T_0 \frac{2p}{\mu_1} \frac{J_1(\mu_1 p)}{J_0^2(\mu_1)} \left(J_0\left(\mu_1 \frac{r_i}{a}\right) - J_0\left(\mu_1 \frac{r_j}{a}\right) \right). \quad (12)$$

Details of all of the initial conditions, including the initial applied temperature T_0, are contained in β_{ij}. Therefore, the slope of a plot of $\ln(\Delta T_{ij})$ versus t is independent of the details of the initial conditions and is proportional to the value of κ, with a known proportionality constant. The values of thermal diffusivity are obtained from the slopes of the curves of $\ln(\Delta T_{ij})$ as a function of time, as shown in FIGURE 3.

The effect of (1) measurement uncertainties (i.e., noise), (2) a non-ideal heated region and, (3) variation in the measurement locations have been investigated. Surprisingly, in spite of the stringent conditions required for Equation **(10)**, these relaxations had minimal effect on the final results. This is due to the fact that the $n = 1$ term is the predominate one. The $n = 2$ term contributes less than 5% for $\kappa t/a^2 > 0.08$

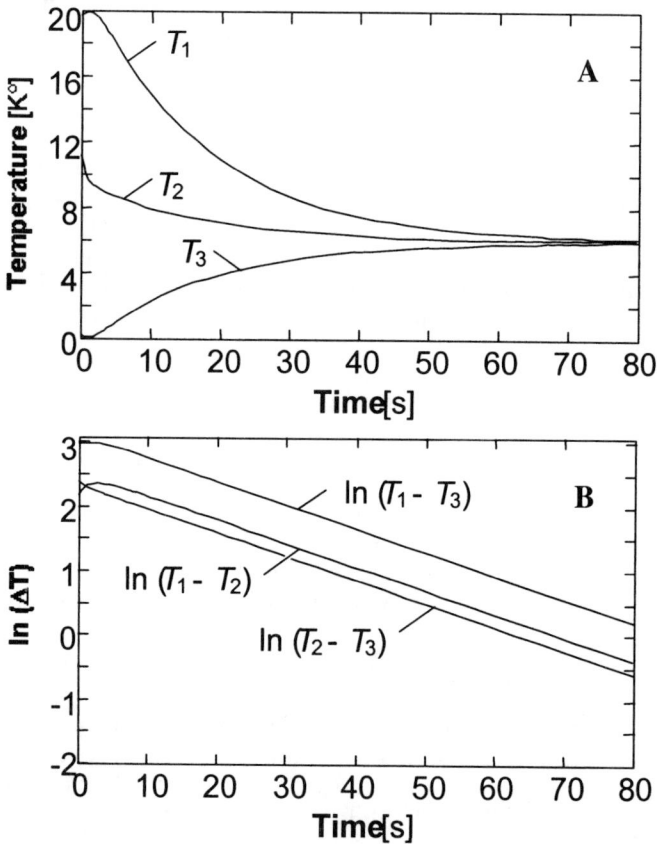

FIGURE 3. Temperature profiles and values of $\ln(\Delta T_{ij})$ versus time, generated numerically by using the special heated region and measurement locations. The thermal diffusivity was $0.1 \, \text{cm}^2/\text{sec}$. T_i is the temperature at special measurement location r_i.

and less than 1% for $kt/a^2 > 0.11$. The effect of the $n = 3$ term is, of course, even less. These results have been confirmed by measurements on boron nitride and graphite.[24]

CONCLUSIONS

We have developed and optimized a technique for the *in situ* measurement of diffusivities in liquids at several temperatures with one sample. This approach circumvents the solidification of the diffusion sample required by other methods. Thus, this method is amenable to low-gravity research where resources are highly limited.

We have also developed and used a novel methodology for measuring thermal diffusivities. This methodology does not require knowing the initial temperature increase or any timing between the applied and measured response. We have shown by numerical simulation that the exact mathematical solution can be considerably relaxed. Multiple, independent, determinations can be made during a single experiment.

ACKNOWLEDGMENTS

This research was supported by the National Aeronautics and Space Administration under grants NCC8-99 and NAG8-1476, and by the State of Alabama through the Center for Microgravity and Materials Research at the University of Alabama in Huntsville. RFS is grateful for support from NASA Grant NAG8-1704. We thank Lynne Carver for expert preparation of the figures. We would also like to thank the reviewers for their helpful comments.

REFERENCES

1. VERHOEVEN, J.D. 1968. Convective effects in the capillary reservoir technique of measuring liquid metal diffusion coefficients. Trans. Met. Soc. AIME **242**: 1937–1941.
2. ALEXANDER J.I.D., J.-F. RAMUS & F. ROSENBERGER. 1996. Numerical simulations of convective contamination of diffusivity measurements in liquids. Microgravity Sci. Technol. **9**: 158–162.
3. BANISH, R.M. & L.B. JALBERT. 2002. Submitted.
4. PERSSON, T., P.E. ERIKSSON & L. LINDSTROM. 1980. Convection and other disturbing effects in diffusion experiments in liquid metals. J. Physique **41**: 374–377.
5. CARERI G., A. PAOLETTI & M. VINCENTINI. 1958. Further experiments on liquid indium and tin self-diffusion. Nuovo Cimento **10**: 4050–4061.
6. FOSTER, J.P. & R.J. REYNIK. 1973. Self-diffusion in liquid tin and indium over extensive temperature ranges. Metall. Trans. **4**: 207–216.
7. NACHTRIEB, N.H. 1967. Self-diffusion in liquid metals *In* Proceedings International Conference on Properties in Liquid Metals. Adams, Davies & Epstein, Eds.: 309–323. Taylor and Francis, London.
8. YOUDELIS W.V., D.R. COLTON & J. CAHOON. 1964. On the theory of alloy solidification in a magnetic field. Can. J. Phys. **42**: 2238–2258. On the theory of diffusion in a magnetic field. Can. J. Phys. **42**: 2217–2237.
9. ALBOUSSIERE, T., J.P. GARANDET, P. LEHMANN & R. MOREAU. 1999. Measurement of solute diffusivity in electrically conducting liquids. *In* Transfer Phenomena in Magnetohydrodynamic and Electroconducting Flows. A. Alemany, *et. al.* Eds.: 359–372.

10. KHINE, Y.Y., R.M. BANISH & J.I.D. ALEXANDER. 2002. Convective effects during diffusivity measurements in liquids with an applied magnetic field. Int. J. Thermophys. **23:** 649–666.
11. UKANWA, A.D. 1979. Radioactive tracer diffusion. *In* Proceedings III. Space Processing Sympos. Skylab Results. 427–457. Marshall Spaceflight Center, Huntsville.
12. REED, R.E., W. UELHOFF & H.L. ADAIR. 1977. Surface-tension-induced convection. NASA SP-412, Apollo-Soyuz Test Project. Summary Science Report **1:** 367–401.
13. FROHBERG, G. 1986. Diffusion and atomic transport & thermophysical properties. *In* Materials Science in Space. B. Feuerbacher, H. Hamacher & R.J. Naumann, Eds.: 93–128 & 425–446. Springer-Verlag, Berlin.
14. FROHBERG, G., K.H. KRAATZ & H. WEVER. 1984. Self diffusion of Sn112 and Sn124 in liquid Sn. *In* Proceeding of the 5th European Symposium on Materials Science under Microgravity. Schloss Elmau, ESA. 201–205.
15. MALMEJAC, Y. & G. FROHBERG. 1987. Mass transport by diffusion. *In* Fluids Sciences and Materials Science in Space. H. Walter, Ed.: 159–190. Springer-Verlag, Berlin.
16. BROOME, E.F. & H.A. WALLS. 1969. Self-diffusion measurements in liquid gallium. Trans. Met. Soc. AIME **245:** 739–741. Liquid metals diffusion: a modified shear cell and mercury diffusion measurements. Trans. Met. Soc. AIME **242:** 2177–2184.
17. KOHL, J.G. & B. PREDEL. 1978. Measurement of chemical and intrinsic diffusion coefficients in liquid Pb–Bi alloys at 621 K by the shear cell technique. Z. Metallk. **69:** 248–251.
18. JALBERT, L.B., R.M. BANISH & F. ROSENBERGER. 1998. Real-time diffusivity measurements in liquids at several temperatures with one sample. Phys. Rev. E **57:** 1727–1736.
19. HARNED, H.S. & R.L. NUTTALL. 1947. The diffusion coefficient of potassium chloride in dilute aqueous solution. J. Am. Chem. Soc. **69:** 736–740.
20. CODASTEFANO, P., A. DI RUSSO & V. ZANZA. 1977. New apparatus for accurate diffusion measurements in fluids. Rev. Sci. Instrum. **48:** 1650–1653.
21. TAIT, W.H. 1980. Radiation Detection. Butterworths, London.
22. TSOULFANIDIS, N. 1983. Measurement and Detection of Radiation. McGraw-Hill, Washington.
23. CARSLAW, H.S. & J.C. JAEGER. 1959. Conduction of Heat in Solids, 2nd edit. Oxford University Press, London.
24. POURPOINT, T.L., R.M. BANISH, F.C. WESSLING & R. SEKERKA. 2000. Real-time determination of thermal diffusivity in a disk-shaped sample applications to graphite and boron nitride. Rev. Sci. Instrum. **71:** 4512–4520.

ATEN

A New High Temperature Materials Processing Facility for the International Space Station

WAYNE N.O. TURNBULL,[a] DONALD L. MISENER,[a] TIMOTHY J.N. SMITH,[a] GUY R.J. ORAM,[a] AND REGINALD W. SMITH[b]

[a]*Microgravity Systems, Millenium Biologix Inc., Kingston, Ontario, Canada*

[b]*Department of Metallurgy and Materials Science, Queen's University, Kingston, Ontario, Canada*

> ABSTRACT: Since the beginning of microgravity materials research, studies of diffusion in liquids have been performed as the typical research that efficiently uses the microgravity environment. Successful experiments in microgravity have demonstrated the ability of the Canadian Microgravity Program (QUEST I, QUELDs I and II) to make significant contributions to this field of international microgravity research. Recently, Millenium Biologix was selected to develop and build the advanced thermal environment facility (ATEN) for the International Space Station. The design of this new processing facility builds on the considerable experience gained in designing and building the QUELD II furnace and developing sealed samples for use on board a manned space platform. The system requirements for ATEN are presented, along with preliminary test data from a prototype furnace.
>
> KEYWORDS: ATEN; QUELD; QUEST; microgravity furnace; liquid diffusion in microgravity; *g*-jitter

INTRODUCTION

To date, three Canadian furnaces have been successfully flown in microgravity. They are: QUEST I, QUELDs I, and II. The first, QUELD I, was used solely for isothermal processing. Astronaut intervention was required to select the operating temperature and to change the temperature setting immediately after inserting the sample into the furnace. Two single heater zone furnaces were used. However, QUELD II contained a three-zone furnace with sample quench facilities. As a result, it was used for isothermal and gradient-freeze sample processing. QUELD II was a semi-automated, two-channel material processing facility based onboard the MIR space station. The astronaut/cosmonaut had to select the sample from a storage magazine, insert it in the sample carrier, enter the sample program number, and press the start button. Following this, the samples were processed automatically, via the automatic insertion of the sample into a preheated furnace, in one of the following three modes:

Address for correspondence: Donald L. Misener, Millenium Biologix Inc., 785 Midpark Drive, Kingston, ON, K7M 7G3 Canada. Voice: 613-389-6565; fax: 613-389-6825.
mbi@millenium-biologix.com

- isothermal,
- temperature-gradient, and
- gradient-freeze.

The furnace temperature was controlled automatically. Following the required thermal exposure, the samples were automatically withdrawn and subsequently quenched by the automatic closure of paired spring-loaded quench blocks. All QUELD II sample processing routines were contained within a removable memory module, with a single module containing the information necessary to process approximately 30 different samples. The specific sample routine was defined using two digit selector switches on the front panel of the QUELD II hardware.

The capacity of the QUELD II system was enhanced by the use of two fully independent sample processing channels. Due to power limitations on MIR, only one channel was in operation at any given time. However, it was possible to load both channels at the same time and make use of a preprogrammed time offset to sequence the processing routines for each sample. This reduced demands on astronaut/cosmonaut time by only requiring their attention for reloading samples after every second sample had been processed. Further details of the design construction of QUELD II can be found in Reference 1; details of the materials science studies conducted with QUELD II appear in Reference 2.

The success of the QUEST/QUELD payloads led to requests from the scientific community to the Canadian Space Agency (CSA) for an even more capable high temperature material processing facility. The response to this perceived need is the advanced thermal environment, or ATEN.

ATEN FEATURES AND SPECIFICATIONS

Following the success of the QUEST/QUELD generation of microgravity furnaces, design efforts were undertaken by Millenium Biologix to develop a new microgravity furnace facility that would build upon the success of the previous generation of furnaces. The principal requirements of this new payload are:

- Higher operating temperatures, to expand the range of materials to be processed at minimal power input.
- Closely defined temperature profiles, to permit specific materials processing regimes.
- Expanded automation, to handle high capacity sample magazines and minimize crew operational requirements.
- Increased sample capacity.

In order to achieve these requirements, a fully integrated and automated payload concept was developed. This concept is presented in FIGURE 1.

The payload, as presented, consists of a high performance furnace, dual sample cartridge assembly, variable mass quench blocks, robotics to move the sample from its storage location to its processing location, and robotics to move the furnace. It also contains all of the associated electronics and the system CPU. Because the sample

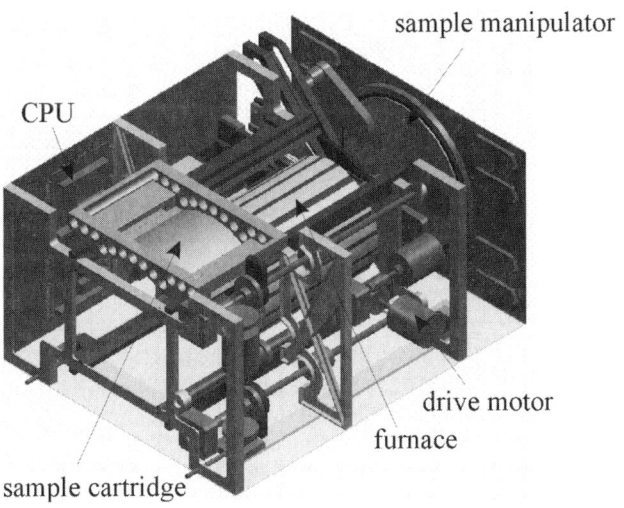

FIGURE 1. Solid model of the ATEN payload.

processing location is fixed, any requirement for relative motion between the sample and the furnace (to achieve a travelling gradient materials processing mode for instance) is achieved by moving the furnace. No astronaut intervention is required for the payload once it has been placed on the microgravity isolation mount and the sample cartridges are in place. Sample processing is automatic, according to instructions contained in the payload software and the sample electronics. If required, ground personnel can change these instructions while the payload is in orbit.

The key features of the ATEN system are:

- Maximum operating temperature of 1,300°C.
- Maximum temperature gradient of 50°C/cm.
- Capable of isothermal, gradient, travelling gradient, and melt zone modes of operation.
- Sample cool down rates in travelling gradient mode adjustable within range of 1°C/cm to 24°C/cm.
- Provide a melt zone length adjustable within the range of 1 and 10mm.
- Maximum of 85 Watts available for the furnace and 120 Watts for the payload.
- Automatic sample manipulation, heating, cooling, and storage.
- Multiple *in situ* temperature measurement points in the sample.
- Complete record of sample thermal history stored in sample onboard memory.
- Each of the two sample cartridges to have a capacity of up to 20 samples (for a total of 40).
- Automatic sample identification.

As was the case with QUELD II, it is intended that ATEN be operated in conjunction with the CSA microgravity isolation mount (MIM) platform. There are several advantages to the combination of the ATEN system with a MIM platform since the MIM may be used in three modes:

Latched. In this condition, the MIM g-jitter of the platform on which MIM is mounted is passed directly to ATEN.

Isolating. Here, the ability of the MIM to improve the quality of the microgravity environment to a quiescent level of $5 \times 10^{-6} g$ permits, for example, the collection of high fidelity thermophysical data.

Forcing. In this mode, the MIM continues to isolate the ATEN unit from local g-jitter but then forces the platform to vibrate in a programmed manner.

This ability of the MIM to generate controlled microgravity fluctuations, not only permits the study of the effects of gravity on physical processes, particularly the triggering of interface instabilities, but also allows the operator to conduct experiments designed to refine theoretical understanding of the effects of gravitationally-induced convection. For example, it may be reasoned that when a given sample is subjected to a particular vibration pattern, then certain convection should result, leading to a particular solute redistribution in a growing crystal. The experiment can be carried-out and the resulting solute distribution determined. This is then compared with that predicted and, if these do not correspond, the theoretical prediction is modified to better reflect the fact. The final advantage of the MIM/ATEN combination is that the available data transfer ports on the MIM permit the recording of experimental parameters important in the postmission analysis of sample properties.

PROTOTYPE FURNACE TESTING

The most critical element, and the most technically challenging area in the design of the ATEN system, is the furnace module. This is due to the limited power available to the ATEN payload for ohmic (resistive) heating. Consequently, a novel approach was required which, when combined with modern insulating materials, would deliver a very efficient furnace. Extensive thermal modeling suggested that Millenium's initial concept of using reflective shields should be capable of producing the required core temperatures within the allowed power budget.

Assembly of the first of the furnace prototypes has been completed and initial testing commenced (see FIGURE 2). Initial test runs demonstrated that a core peak temperature exceeding 1,000°C could be achieved, in a gravity field, and with the furnace open to receive a sample, using about 50 Watts of power. Having the furnace open is equivalent to placing a load on the furnace (because heat radiates from the opening to the cool environment). This load has been estimated to be about 10 Watts. When placed in a vacuum chamber and evacuated to 0.3 Torr the same power input gives a peak furnace temperature of approximately 1,305°C.

Actual on orbit performance of the furnace will fall between the above two temperatures when 50 Watts of power is applied. However, it is reasonable to assume that actual performance will be much closer to that of the vacuum case. This gives confidence that the overall requirement of achieving a flat axial temperature profile of 1,300°C will be achievable using less than the allotted 85 Watts of power.

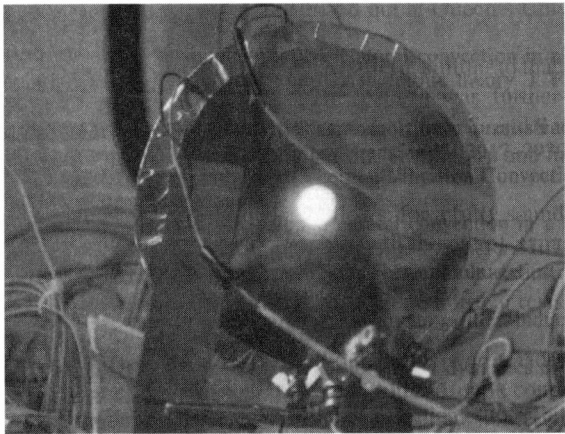

FIGURE 2. Prototype furnace after heating to 1,300°C.

ATEN SAMPLES

The ATEN system has as its focus the accurate thermal processing of a wide series of specimen-types. Each type of specimen must be encased in a crucible and three additional levels of metallic containment to permit processing without hazard to the operator or the equipment. The whole assembly is referred to as a "sample". Its specifications are shown in TABLE 1. A schematic drawing of a typical sample is shown in FIGURE 3.

Chemical Containment

The requirements for containment are that the specimen, when liquid, does not reach beyond its immediate environment and that none of the specimen constituents are permitted to leak into the ATEN payload or elsewhere. This requires that the specimen be contained in a suitably inert crucible, which is closed with a spring-loaded plug to ensure that no free surface exists during microgravity processing to avoid any potential Marangoni convection.

The selection of the crucible and sheathing materials is based on meeting the requirement that the specimen materials, when in the liquid state, do not seriously affect the crucible or integrity of the sheathing materials when exposed in direct contact for a period of at least twice the duration of the planned experiment.

TABLE 1. ATEN sample primary specifications

Parameter	Description
Sample diameter	13 mm
Specimen diameter	1 to 10 mm
Specimen length	45 to 80 mm
Material sheathing	noble metal

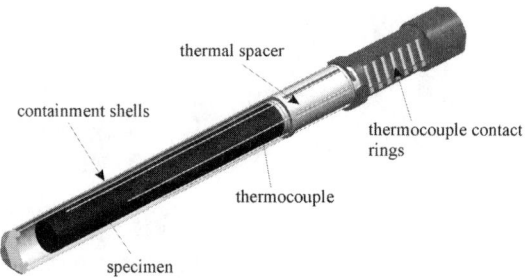

FIGURE 3. Solid model of the ATEN sample.

Structural Containment

The materials and assembly techniques selected for any sample must be able to withstand the internal physical and chemical stresses likely to arise within the capsule in the unlikely event of a furnace run-away condition. For ATEN, with a maximum operating temperature of 1,300°C, this represents a significant challenge. Thus, in order to be accepted for microgravity processing, all specimens must be enclosed in three impervious sealed shells that can withstand exposure to the ISS laboratory atmosphere at the runaway temperature for 12 hours. In practice this will mean that at least the outermost shell be made from an alloy based on the platinum group of metals. Specimens are best placed, first in a graphite or boron nitride crucible, and then sealed by laser welding in three vacuum-tight layers of the noble metal. Springs are included, not only to prevent Marangoni convection within the crucible, but also to gently force the crucible to make good thermal contact at its base with the central thermal plug to which a thermocouple is welded. This plug is required to ensure that the required specimen isothermal temperature/thermal gradient has been achieved.

SUMMARY

Canada has gained considerable experience in the high temperature processing of materials in microgravity through its QUEST I and QUELD I and II programs. In order to build upon this expertise a newly designed high performance furnace facility for the ISS is being developed and built by Millenium Biologix Inc. for the Canadian Space Agency. This new facility will permit the automatic manipulation, heating, cooling, and storage of up to 40 samples per payload operation. Significantly higher temperatures will be achievable at much lower power levels than possible with past furnaces. Astronaut involvement is minimal.

ACKNOWLEDGMENTS

The development of ATEN by Millenium Biologix Inc. is funded by the Canadian Space Agency under Contract Number 9F007-006090/001/SR. Project Management on behalf of the CSA is provided by Ms. Catherine Casgrain with support from the CSA ATEN team.

REFERENCES

1. SMITH, R.W., *et al.* 1998. The QUELD II—MIR Processing Facility. *In* Tenth International Symposium on Experimental Methods for Microgravity Materials Research. R. Schiffman, Ed.: Paper 22. The Metals and Materials Society, Warrendale, PA.
2. HERRING, R.A., *et al.* 1999. Recent measurements of experiment sensitivity to g-jitter and their significance to ISS facility development. J. Jpn. Microgravity Appl. **16**(4): 234–244.

Microgravity Effects on Thermodynamic and Kinetic Properties of Colloidal Dispersions

TSUNEO OKUBO AND AKIRA TSUCHIDA

Department of Applied Chemistry, Gifu University, Yanagido, Gifu, Japan

ABSTRACT: Microgravity experiments on the physicochemical properties of colloidal dispersions reported hitherto have been reviewed. In microgravity, reliability and reproducibility of experimental data have been improved significantly by the elimination of convection in the suspension. Vanishment of the excess downward diffusion in microgravity has also produced a significant effect on properties of colloidal suspensions. For example, colloidal crystallization rates decreased in microgravity, whereas colloidal alloy crystallization rates increased. These results demonstrate the important role of the segregation effect in normal gravity. Colloidal silica formation reactions are retarded in microgravity, a phenomenon that is correlated deeply with disappearance of the downward diffusion of heavy products.

KEYWORDS: colloidal dispersion; microgravity; thermodynamics; kinetic properties

INTRODUCTION

The effect of microgravity on the physicochemical properties of most colloidal dispersion systems has not been discussed in detail hitherto. However, we note that the gravitational effect is significant for the translational diffusion of colloidal particles larger than 0.1 µm and heavier than 0.1 in the relative density of the particles with reference to solvent ($\Delta\rho$). The Peclet number, Pe, the ratio of the time of sedimentation of a sphere in gravity against that of translational Brownian diffusion, is one of the convenient parameters to use in understanding the gravitational effect compared with Brownian diffusion.[1–4]

$$Pe = \frac{rS}{D_t} = \frac{4\pi r^4 \Delta\rho g}{3k_B T}, \tag{1}$$

where r is the sphere radius; S and D_t are, respectively, the sedimentation and translational diffusion coefficients; k_B is the Boltzmann constant; and T is the absolute temperature. When $\Delta\rho$ values are 0.1 and 1, for example, Pe values evaluate to 10^{-8} (r = 10 nm), 10^{-4} (100 nm), 1 (1 µm), 10^4 (10 µm), 10^{-7} (10 nm), 10^{-3} (100 nm), 10 (1 µm), and 10^5 (10 µm).

For colloidal crystals, crystal-like distribution of colloidal particles in suspension, for example, the sedimentation of single crystals is quite fast in normal gravity

---

Address for correspondence: Tsuneo Okubo, Department of Applied Chemistry, Gifu University, Yanagido, Gifu 501-1193, Japan. Voice: +81-(0)58-293-2620; fax: +81-(0)58-293-2628.

okubotsu@apchem.gifu-u.ac.jp

compared with that calculated from the Stokes–Einstein equation for appropriate spheres forming the single crystals. As is discussed later, colloidal crystals are packed densely with the single crystals (just like other crystals, such as metals, ice, and proteins).[5–7] When the lattice structure of single crystals is assumed to be simple-cubic and the nearest-neighbor intersphere distance is 500nm, for example, the number of spheres composing a single crystal, $0.1\text{mm} \times 0.1\text{mm} \times 0.1\text{mm}$ in volume, is estimated to be 8×10^6. Thus, it is easily understood that each single crystal is composed of a huge number of colloidal particles that sediment quite quickly in comparison with the respective spheres. However, we note that the sedimentation rate of single crystals should be different from that of single crystals alone.[8] Furthermore, the grain boundaries should move by dynamic fluctuation of the spheres in equilibrium with the single crystal regions.[9,10] Furthermore, the rotational diffusion coefficient of an anisotropic-shaped colloidal particle is often coupled with translational sedimentation in normal gravity.

In order to compensate for the sedimentation effect in normal gravity, some researchers have studied colloidal dispersions in which the densities of the particles and solvent are matched, using sucrose–water mixtures instead of water as solvent; or a dc electric field applied perpendicularly for anionic charged colloidal particles. However, these ideas do not really mimic the microgravity environment, since *convection* of suspensions is not eliminated and additional interactions of the particle with solvent are added. We note that the influence of convection is quite significant in normal gravity, exerting a substantial effect on suspension properties. Critical sphere concentration of melting and/or melting temperature for colloidal crystals, for example, are influenced substantially by convection in colloidal suspensions.[8] Thus, studies on colloidal suspensions in microgravity are very important for the determination of precise parameters in physicochemistry and also for syntheses of highly monodisperse colloidal spheres.

Microgravity has been achieved by several methods: (1) free-fall experiments of a capsule through an evacuated tube (time duration, 3 to 10 seconds, quality of microgravity, about 10^{-5} to $10^{-4}g$); (2) parabolic flight in an aircraft (20 to 30 seconds, about 0.01–$0.05g$); (3) a rocket flight (10 to 20 minutes, about $10^{-4}g$); (4) Space Shuttle flight (several days, about $10^{-3}g$); and (5) Space Station experiments scheduled in near future (several years, about $10^{-3}g$).

Some researchers have studied thermodynamic and kinetic properties of colloidal dispersions in microgravity hitherto.[11] Vanderhoff *et al.*[12] succeeded in preparing monodisperse polystyrene spheres, five to eighteen micrometers in diameter, with narrower particle size distributions in microgravity than those prepared in normal gravity. A series of polyacrylamide gels has been synthesized in microgravity.[13] Recently Zhu *et al.*[14] reported a microgravity effect on the morphology of colloidal crystals using Space Shuttle Columbia. Our group has made several microgravity experiments on colloidal dispersion systems, such as colloidal crystallization kinetics of single- and two-component spheres in exhaustively deionized aqueous suspensions,[15–18] polymerization kinetics of colloidal silica spheres formed,[19] syntheses of nylon vesicles,[20] and the rotational relaxation times of anisotropic-shaped particles of tungstic acid colloids.[21] In this article, microgravity experiments on the thermodynamic and kinetic properties of colloidal suspensions are reviewed and discussed.

COLLOIDAL CRYSTALLIZATION IN MICROGRAVITY

Colloidal crystallization takes place for monodisperse colloidal particles in suspension. Many researchers have clarified that the colloidal crystals are formed by Brownian movement of colloidal-sized particles, resulting in interparticle *repulsion* by the principle of *minimizing dead space*—space not occupied by particles.[5–10,22–30] In other words, particles form a crystal-like distribution, automatically, with the help of Brownian movement and in order to *maximize the packing density*.

When repulsive interactions, such as electrostatic repulsion, are in effect among colloidal particles in addition to the repulsion forces from Brownian movement, colloidal crystallization takes place easily, even at very low particle concentrations. Most colloidal particles get negative charges on their surfaces in a polar solvent like water. The ionic groups leave their counterions, and these excess charges accumulate near the surface forming an electrical double layer. The counterions in the diffuse region are distributed according to a balance between the thermal diffusion forces and the forces of electrical attraction with colloidal particles. FIGURE 1 demonstrates clearly that the lattice spacing of colloidal crystal changes with the particle concentration exclusively, which strongly supports the importance of the interparticle repulsion induced by Brownian movement of particles themselves and also electrostatic interparticle repulsive forces. Furthermore, the observed nearest neighbor intersphere distance of the colloidal crystal, l_{obs}, is equal to or a little less that the effective diameter of spheres, including the electrical double layers, $d_{eff} = d_0 + 2 \times D_l$. Here, D_l is the thickness of the electrical double layers, given by

$$D_l = \left(\frac{4\pi e^2 n}{\varepsilon k_B T}\right)^{-1/2}, \qquad (2)$$

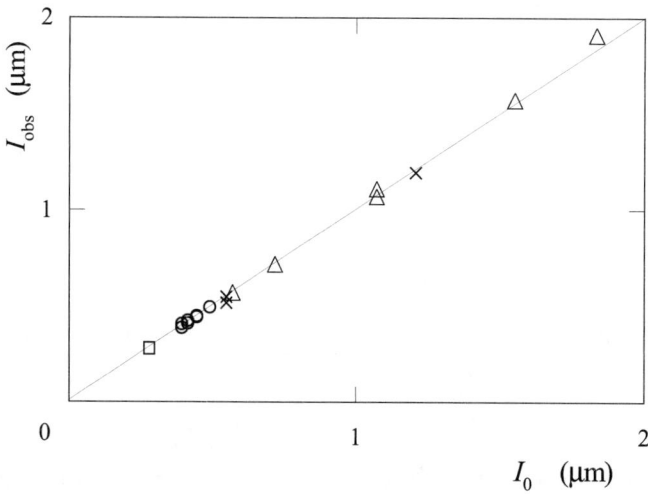

FIGURE 1. Comparison of I_{obs} and I_0 for the colloidal crystals of silica spheres. ○, reflection spectroscopy;[31] ×, light-scattering;[32] △, light-scattering;[33] □, reflection spectroscopy.[34]

where e is the electronic charge and ε is the dielectric constant of the solvent; n is the concentration of "diffusible" or "free-state" cations and anions in suspension.

It should be mentioned that the colloidal crystals are densely packed and separated by grain boundaries. Some researchers, including us, have reported close-up or microscopic photographs of colloidal single crystals.[7,35–40] At first glance, colloidal crystals seemed to locate separately. However, the patterns and their colors change rapidly and beautifully with very slow rotation of the cell and/or on slowly changing the angle of incident light. Some single crystals change their colors and then disappear suddenly; other single crystals appear instead in the previously blank regions. These features are observed because of a change in the Bragg condition of the crystals; that is, single crystals emit colored light only when the Bragg condition fits.

Few researchers have reported existence of the two-state structure or boid structure; that is, coexistence of gas- or liquid-like structures and crystal-like structures in colloidal crystals, from *microscopic* observation.[8] However, from our experience it is highly plausible that these observations are attributed to erroneous artifacts resulting from several experimental problems. First, we note the strong attractive interaction between colloidal spheres and observation cover glass plate. The most dense Bragg plane is always oriented parallel to the glass wall. Second, most observations showing the two-state structure have not been made under a sedimentation equilibrium, but for an agitated unstable state by heat from iridescent light of a microscope. Third, in most microscopic observations, the angle of the incident light is fixed at 180°. When the angle is changed slowly, a packed state of the single crystals is observed.[35] Fourth, coexistence of crystal and liquid regions is observed at the beginning stages of crystallization. However, single crystals move in a sea of *super saturated* liquid by translational and rotational movements, resulting in a blinking phenomenon.[38] Fifth, these void structures are observed quite easily when the sphere concentration is insufficient for a sphere to be surrounded by other colloidal spheres, especially when the cover glass is not rinsed completely. We note that colloidal crystals are formed only in the closed system filled with colloidal particles that hit each other by Brownian motion. Therefore, we remark that the microscopic method has many limitations.

The colloidal crystals are so soft (10^{-3} to 10^3 Pa) and packed so densely that they are compressed homogeneously and easily, even by gravitational force. From the microscopic and reflection spectroscopic measurements of the lattice constants in the crystals of colloidal spheres and rods, static elastic moduli have been evaluated under sedimentation equilibrium in normal gravity.[41–43] In microgravity no compression should take place and a strictly homogeneous crystal structure should be formed, although the wall effect—most dense crystal planes orient parallel to the cell wall—remains. We point out that diffusion equilibrium measurements have been made in normal gravity to determine the elastic moduli of colloidal crystals.[44] Centrifugal compression has been discussed as a means to obtain elastic information and is deeply correlated to gravitational compression.[45]

Zhu *et al.*[14] have reported structure and morphology properties of colloidal single crystals of hard-sphere systems in microgravity using Space Shuttle Columbia. Whereas in normal gravity colloidal crystals grown just above the volume fraction at melting show a mixture of random stacking of hexagonally close-packed planes (rhcp) and face-centered cubic (fcc) packing, those in microgravity exhibit the rhcp structure alone, suggesting that the fcc component may be induced by gravity-induced stresses. They also observed dendritic growth instabilities that are not evident in normal gravity, presumably because they are disrupted by shear-induced stresses as the crystals settle under gravity. (Dendritic crystals have been often observed for small and rather dispersed polystyrene spheres in highly deionized and diluted aqueous suspensions in normal gravity.[7]) Furthermore, glassy samples at high volume fraction that fail to crystallize after more than a year on Earth, crystallize fully in less than two weeks in microgravity. Zhu et al. explained the results in terms of the fact that gravity masks or alters some of the intrinsic aspects of colloidal crystallization.

Ishikawa et al.[15,16] made aircraft and rocket experiments in order to study the effect of microgravity on the colloidal crystallization rates of soft-sphere systems; that is, polystyrene spheres in deionized and diluted aqueous suspensions. Microgravity effects on several kinetic parameters of colloidal crystallization have been discussed in terms of the important role of the gravitational diffusion of colloidal spheres in the nucleation and crystallization processes in classical crystallization theory.

In 1996, Okubo et al.[17] have made microgravity experiments, produced by parabolic flights of an aircraft, on colloidal crystallization kinetics of colloidal silica spheres in highly deionized and diluted aqueous suspensions. Sphere concentrations ranged from 0.0016 to 0.0045 in volume fraction. Nucleation rates decreased in microgravity. Crystal growth rates of the fcc lattices decreased in microgravity (μg) by about 25% compared with those in normal gravity ($1\,g$). One of the main causes for the retardation in microgravity was suggested to be elimination of the downward diffusion of spheres, which may enhance intersphere collisions. Absence of convection of the suspensions in microgravity was also (or much more) important than the elimination of the downward diffusion effect.

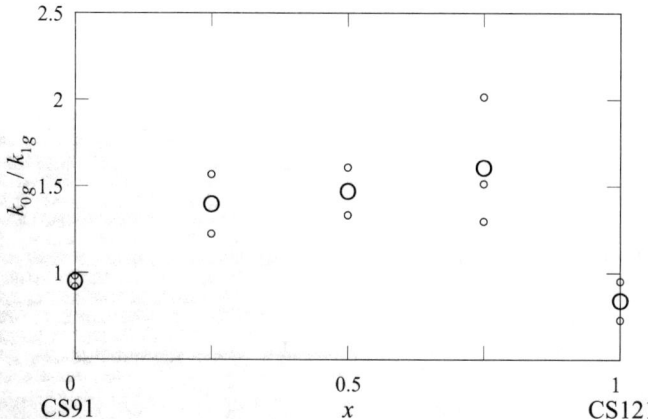

FIGURE 2. Growth rate ratio in microgravity and normal gravity as a function of sphere volume fraction in CS91 (110 nm in diameter)–CS121 (136 nm) mixtures at 25°C. [CS91] = [CS121] = 0.005. *Large circles* show the mean of each run shown by *small circles*.

Crystal growth rates of the colloidal alloys of binary mixtures of monodisperse polystyrene and/or silica spheres have been studied in microgravity using aircraft.[18] The rates increased substantially up to about 1.7-fold in microgravity, compared with those in normal gravity, as is shown in FIGURE 2. We note that the segregation effect is familiar for binary mixtures of colloid and powder science; that is, large spheres segregate upward and small spheres downward in normal gravity.[46–48] In microgravity such segregation effects should disappear and homogeneous mixing take place favorably, leading to fast alloy crystallization. The substantial acceleration effect of microgravity in the alloy crystallization again strongly supports the idea that colloidal crystallization takes place by the packing model described above (alloy structure is determined by the condition that maximum packing density is achieved or not for a given ratio of the sphere sizes, including surrounding electrical double layers). Quite recently, we made aircraft experiments to determine the microgravity effect on the kinetics of colloidal crystallization in the presence of sodium chloride. Crystal growth rates decreased in microgravity, which is similar to those observed for the deionized systems.

SYNTHESES OF COLLOIDAL SPHERES IN A MICROGRAVITY

Recently, keen attention has been paid to chemical reactions, especially to polymerization reactions in microgravity, as a new technology in space.[11] In microgravity, using the Space Shuttle, Vanderhoff et al.[12] succeeded in preparing monodisperse polystyrene spheres, 5 to 18 micrometers in diameter, with narrower particle size distributions than those prepared in normal gravity. A series of polyacrylamide gels have been synthesized in microgravity.[13] All these microgravity experiments are, however, restricted to the studies on morphologies, such as size and shape of the polymers and gels formed, and their reaction yields. Kinetic analyses of polymerization reactions, such as rate constants and the reaction mechanisms, have not been reported to date.

Recently, polymerization rates of colloidal silica spheres from tetraethylorthosilicate (TEOS), ammonia and water in ethanol were studied in microgravity using parabolic flight of an aircraft.[19] The specific gravity of colloidal silica spheres is about 2.2, high enough compared with that of the solvent, ethanol (0.82). Thus, the spheres formed are sedimented in normal gravity and in most cases the reaction mixtures must be agitated throughout the polymerization process. In microgravity, on the other hand, the products do not sediment and the reaction proceeds homogeneously without stirring. Another advantage of microgravity experiments is the substantial decrease in convection of reaction mixtures. For example, it is clearly observed using a CCD camera, that the movement of the very small air bubbles formed by chance in the reaction mixtures stop completely in microgravity within $\pm 0.01\,g$ produced by the parabolic flight of an aircraft. FIGURE 3 shows the apparent reaction rates, v, of the formation reaction of colloidal silica spheres in microgravity and in normal gravity.[19] Surprisingly, the polymerization rates were lowered greatly when the reaction mixtures experienced the microgravity in the beginning stages in the reaction processes. It is highly plausible that a decrease in the formation rate of the primary particles and also a decrease in their number in microgravity influenced the

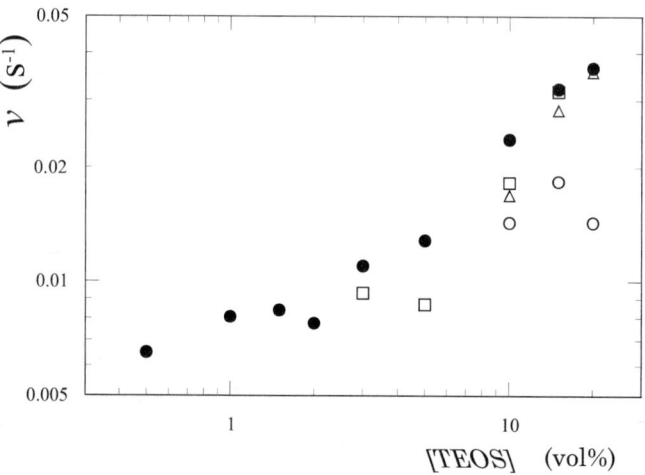

FIGURE 3. Polymerization rates in the course of colloidal silca formation at 25°C. [NH3] = 8.75 wt%; ○, μg starts at $t = 0$ sec; △, 15 sec; □, 30 sec; ●, at 1 g; TEOS denotes tetraethylorthosilicate.

polymerization (*coalescence*) rates. Increase in the induction times, where the primary particles are formed, were also observed in microgravity. One of the main reasons for these observations is the fact that the translational Brownian movement of product spheres is free from downward translational movement in microgravity. No convection in the reaction suspensions in microgravity is another important factor. The microgravity effect on the formation reactions of colloidal silica spheres was much more substantial than our expectation.

PHYSICOCHEMICAL PROPERTIES OF COLLOIDAL PARTICLES IN MICROGRAVITY

As described in the INTRODUCTION, experimental clarification of the microgravity effect on the translational diffusion coefficient of colloidal particles is important. Krutzer *et al.*[49] reported that the colloidal particles larger than 2 μm coagulate faster in pseudomicrogravity, which is created by the density-matching method. Static and dynamic light-scattering measurements in microgravity on colloidal gelation have been made on the US Space Shuttle and Russian orbiter Mir.[50] However, the results are not yet published except for details of the instrument used.[51]

Recently, we made free-fall experiments to determine the rotational diffusion coefficients of anisotropic-shaped colloids; that is, tungstic acid colloids in microgravity.[21] The Peclet number of tungstic acid colloids is roughly estimated to be 230, which shows that the contribution of sedimentation is much larger than Brownian translational diffusion. Clearly, sedimentation of non-spherical particles is complicated by the coupling of the translational and rotational motions. The particles sediment in a direction that depends on their orientation. Brownian motion changes the

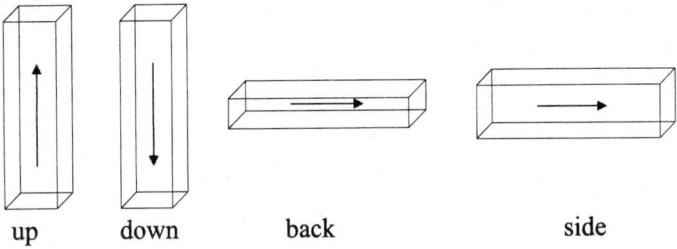

FIGURE 4. Directions of the sample flow in the cell.

orientation of the particles and, hence, their sedimentation speeds and directions. However, on average, the particles move in the direction of gravity only. The rotational relaxation times (τ) were evaluated from the relaxation traces of the optical transmittance of the suspension using the stopped-flow technique. For four flow directions of the sample suspension in the observation cell, shown in FIGURE 4 (abbreviated *up*, *down*, *back*, and *side*), τ values were observed in microgravity (μg) and in normal gravity (1 g). The probing light always passed through the cell perpendicular to the largest planes of the cell. FIGURE 5 shows the τ values thus obtained at μg and 1 g. All the solid symbols show the τ values observed at 1 g. A decreasing tendency of τ was observed as the suspension temperature increased. Clearly, the observed τ values at 1 g scattered rather significantly, which is explained by the convection of the suspension. It should also be noted that τ values at 1 g were about 6%

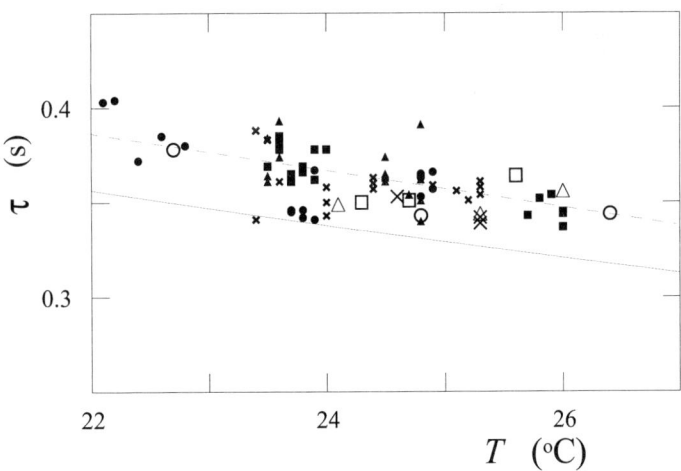

FIGURE 5. Rotational relaxation time (τ) of W29 tungstic acid colloids, 0.20 wt%. ○, up, μg; ×, down, μg; △, back, μg; □, side, μg; ●, up, 1 g; ×, down, 1 g; ▲, back, 1 g; ■, side, 1 g. *Solid curve*, calculated from Perrin's equation ($D_l = 0$ nm); *broken curve*, calculated ($D_l = 21$ nm).

larger than those evaluated theoretically using Perrin's equations. This difference demonstrates the importance of the electrical double layers formed around the colloidal particles. The broken curve shows the τ values when the Debye-length was taken to be 21 nm. Conductivity measurements of the colloidal suspension gave 36.7 μS/cm (23.3°C) at $\phi = 0.20$ wt%, which corresponds to 2.5×10^{-4} mol/L KCl and a Debye length of 27 nm. Thus, agreement between the observed and calculated values of the double layers is satisfactory. Clearly, the τ values in microgravity were not scattered as much as those in normal gravity, and there was no difference between the τ values observed at $1 g$ and μg. Again, the observed τ values were larger than the theoretical values due to the significant contribution from the electrical double layers. The experimental errors in τ values at μg were smaller than those at $1 g$. Furthermore, the τ values in microgravity were quite insensitive to the flow directions in the cell. Very reliable values of τ were observed in microgravity, which supports the fact that the lack of sedimentation of the colloidal particles and absence of convection in the suspension, raise reliability of the experimental data significantly.

CONCLUDING REMARKS

A few microgravity experiments have been made for colloidal dispersion systems to date. Microgravity duration longer than 20 sec was highly desired for the comprehensive understanding of thermodynamic and kinetic properties of colloidal dispersion systems. However, several important findings can be noted. First, the very important role of *no convection* of suspension in microgravity has been clarified for a significant improvement of the reliability and reproducibility of experimental data. Second, *release of the excess downward diffusion* of colloidal particles in microgravity exerted a significant influence on the orientation properties, such as colloidal crystallization, especially alloy crystallization.

In the near future, many kinds of microgravity experiments on colloidal systems are open: (1) translational and rotational diffusion coefficients, (2) colloidal aggregation rates, (3) chemical reaction rates of reactive colloidal particles with various kinds of reactants, (4) reaction rates of colloid–colloid associations as models of antigen–antibody, duplication of nucleic acids, and/or enzymatic reactions. Furthermore, studies on the effects of microgravity on other kinds of thermodynamic properties of colloidal systems, such as activity, surface tension, and adsorption of colloidal suspensions, are exciting, since the following fundamental relation holds.[52]

$$dU = TdS - PdV + \sigma dA + g(h - h_0)dM - \frac{r^2\omega^2}{2}dM + \mu_i dn_i, \quad (3)$$

where U is the internal energy. T, S, P, V, σ, A, g, h, h_0, M, r, ω, μ_i, and n_i are the temperature, entropy, pressure, volume, surface tension, area, acceleration of gravity, height, reference height, molecular weight, radius, angular velocity, chemical potential of component i, and number of moles of component i, respectively. We further note that experimental studies between μg and $1 g$, and above $1 g$ are also important to understanding microgravity effects.

ACKNOWLEDGMENTS

Financial support from the Ministry of Education and Culture, Japan is deeply appreciated. The author thanks Professor Emeritus Sei Hachisu of Tsukuba University for his interest, encouragement, and for valuable comments. National Space Agency of Japan (NASDA), Japan Space Utilization Promotion Center, Tokyo (JSUP), Japan Space Forum, Tokyo (JSF), Diamond Air Service Co., Toyoyama, Aichi-pref. (DAS) and the Microgravity Laboratory of Japan Co., Toki, Gifu (MGLAB) are acknowledged for their financial and/or technical support of our microgravity experiments.

REFERENCES

1. HUNTER, R.J. 1981. Zeta Potential in Colloid Science. Principles and Applications: 117–118. Academic Press, London.
2. RUSSEL, W.B., D.A. SAVILLE & W.R. SCHO-WALTER. 1989. Colloidal Dispersions. 394–428. Cambridge University Press, Cambridge.
3. VAN DE VEN, T.G.M. 1989. Colloidal Hydrodynamics. 78–96. Academic Press, London.
4. HIEMENZ, P.C. & R. RAJAGOPALAN. 1997. Principles of Colloid and Surface Chemistry, 3rd edit. 176–181. Marcel Dekker, New York.
5. VANDERHOFF, W., H.J. VAN DE HUL, R.J.M. TAUSK & J.TH.G. OVERBEEK. 1970. The Preparation of monodisperse latexes with well characterized surfaces. In Clean Surfaces: Their Preparation and Characterization for Interfacial Studies. G. Goldfinger, Ed.: 15–44. Dekker, New York.
6. PIERANSKI, P. 1983. Colloidal crystals. Contemp. Phys. **24:** 25–73.
7. OKUBO, T. 1994. Giant colloidal single crystals of polystyrene and silica spheres in deionized suspension. Langmuir **10:** 1695–1702.
8. OKUBO, T. 1988. Extraordinary behavior in the structural properties of colloidal macroions in deionized suspension and the importance of the Debye screening length. Acc. Chem. Res. **21:** 281–286.
9. KOSE, A., M. OZAKI, K. TAKANO, et al. 1973. Direct observation of ordered latex suspension by metallurgical microscope. J. Colloid Interface Sci. **44:** 330–338.
10. KOSE, A. & S. HACHISU. 1976. Ordered structure in weakly flocculated monodisperse latex. J. Colloid Interface Sci. **55:** 487–498.
11. SCHIFFMAN, R.A., Ed. 1998. Experimental Methods for Microgravity. Materials Science Research. Second to 8th Intern. Symp. **2–8** (1988–1996). TMS (Technology, Minerals, & Materials Soc.) CD-ROM Library. TMS. Warrendale.
12. VANDERHOFF, J.W., M.S. EL-AASSER & F.J. MICALE. 1986. Preparation of Large-particle-size monodisperse latexes in space. Polymer. Kinet. Process Devel. **54:** 584. PMSE Proc. Am. Chem. Soc., Div. Polym. Mat. Sci. Eng.
13. BRISKMAN, V., K. KOSTAREV, V. LEONTYEV, et al. 1997. Polymerization in microgravity as a new process in space technology. Proc. 48th 1–11. Intern. Astron. Cong. Turin, Italy. Intern. Astron. Fed.
14. ZHU, J., M. LI, R. ROGER, et al. 1997. Crystallization of hard-sphere colloids in microgravity. Nature (London) **387:** 883–885.
15. ISHIKAWA, M., H. NAKAMURA, S. KAMEI, et al. 1996. Growth of colloidal crystals under low-gravity (Japanese). J. Jpn. Soc. Microgravity Appl. **13:** 149–157.
16. ISHIKAWA, M., H. NAKAMURA, T. OKUBO, et al. 1998. Growth of colloidal crystals under microgravity. J. Jpn. Soc. Microgravity Appl. **15**(Suppl. II): 168–172.
17. OKUBO, T., A. TSUCHIDA, T. OKUDA, et al. 1999. Kinetic analyses of colloidal crystallization in microgravity. Aircraft experiments. Coll. Surf. **153:** 515–524. **160:** 311–320.

18. OKUBO, T., A. TSUCHIDA, S. TAKAHASHI, *et al.* 2000. Kinetic analyses of colloidal alloy crystallization of binary mixtures of monodispersed polystyrene and/or colloidal silica spheres having different sizes and densities in microgravity using aircraft. Coll. Polym. Sci. **278:** 202–210.
19. OKUBO, T., A. TSUCHIDA, K. KOBAYASHI, *et al.* 1999. Kinetic study of the formation reaction of colloidal silica spheres in microgravity using aircraft. Coll. Polym. Sci. **277:** 474–478.
20. OKUBO, T., A. TSUCHIDA, H. YOSHIMI & K. OIWA. 2002. Polymerization of nylon in microgravity. Coll. Polym. Sci. Submitted.
21. OKUBO, T., A. TSUCHIDA, H. YOSHIMI & H. MAEDA. 1999. Rotational diffusion of anisotropic-shaped colloidal particles in microgravity as studied by free-fall experiments. Coll. Polym. Sci. **277:** 601–606.
22. WILLIAMS, R., R.S. CRANDALL & P.J. WOJTOWICZ. 1976. Melting of crystalline suspensions of polystyrene spheres. Phys. Rev. Lett. **37:** 348–351.
23. LINDSAY, H.M. & P.M. CHAIKIN. 1982. Elastic properties of colloidal crystals and glasses. J. Chem. Phys. **76:** 3774–3781.
24. AASTUEN, D.J.W., N.A. CLARK, L.K. COTTER & B.J. ACKERSON. 1986. Nucleation and growth of colloidal crystals. Phys. Rev. Lett. **57:** 1733–1736.
25. SOOD, A.K. 1991. Structural ordering in colloidal suspensions. Solid State Phys. **45:** 1–73.
26. OTTEWILL, R.H. 1989. Colloidal stability and instability. "Order disorder". Langmuir **5:** 4–9.
27. MONOVOUKAS, Y. & A.P. GAST. 1991. A study of colloidal crystal morphology and orientation via polarizing microscopy. Langmuir **7:** 460–468.
28. PALBERG, TH., R. SIMON, M. WURTH & P. LEIDERER. 1994. Colloidal suspensions as model liquids and solids. Prog. Coll. Polym. Sci. **96:** 62–71.
29. HENDERSON, S.I., T.C. MORTENSEN, S.M. UNDERWOOD & W. VAN MEGEN. 1996. Effect of particle size distribution on crystallization and the glass transition of hard sphere colloids. Physica A **233:** 102–116.
30. REUS, V., L. BELLONI, T. ZEMB, *et al.* 1997. Equation of state and structure of electrostatic colloidal crystals. Osmotic pressure and scattering study. J. Phys. (France) **II:** 603–626.
31. OKUBO, T. 1992. Suspension structures of deionized colloidal silica spheres as studied by the reflection and transmission spectroscopy. Ber. Bunsenges. Phys. Chem. **96:** 61–68.
32. OKUBO, T., K. KIRIYAMA, N. NEMOTO & H. HASHIMOTO. 1996. Static and dynamic light scattering of colloidal gasses, liquids and crystals. Coll. Polym. Sci. **274:** 93–104.
33. OKUBO, T. & K. KIRIYAMA. 1997. Structural and dynamic properties in the complex fluids of colloidal crystals, liquids and gases. J. Mol. Liq. **72:** 347–364.
34. OKUBO, T., A. TSUCHIDA, S. OKADA & S. KOBATA. 1998. Nonlinear electro-optics of colloidal crystals as studied by reflection spectroscopy. J. Coll. Interface Sci. **199:** 83–91.
35. OKUBO, T. 1987. Microscopic observation of crystallites in the deionized suspensions of monodispersed polystyrene spheres. Ber. Bunsenges. Phys. Chem. **91:** 516–519.
36. OKUBO, T. 1990. Phase transition between liquid-like and crystal-like structures of deionized colloidal suspensions. J. Chem. Soc. Faraday Trans. **86:** 2871–2876.
37. OKUBO, T. 1992. Giant single crystals of colloidal spheres in deionized and diluted suspension. Naturwissenscaften **79:** 317–320.
38. OKUBO, T. 1992. Blinking of colloidal single crystals in aqueous suspensions of monodisperse colloidal silica spheres. J. Coll. Interface Sci. **153:** 587–591.
39. OKUBO, T. 1993. Polymer colloidal crystals. Prog. Polym. Sci. **18:** 481–517.
40. OKUBO, T. 1994. Phase diagram of ionic colloidal crystals. *In* Macro-ion Characterization. From Dilute Solutions to Complex Fluids, Chap. 28. ACS Symp. Ser. 548. ACS. Washington, D.C.
41. CRANDALL, R.S. & R. WILLIAMS. 1977. Gravitational compression of crystallized suspensions of polystyrene spheres. Science **198:** 293–295.

42. OKUBO, T. 1987. Microscopic observation of ordered colloids in sedimentation equilibrium and important role of Debye-screening length. I. J. Chem. Phys. **86:** 2394–2399.
43. OKUBO, T. 1989. Elastic modulus of crystal-like structures of deionized colloidal spheres in sedimentation equilibrium as studied by the reflection-spectrum method. J. Chem. Soc. Faraday Trans. 1 **85:** 455–466.
44. OKUBO, T. 1995. Rigidity of colloidal crystals as studied by the diffusion equilibrium method. J. Chem. Phys. **102:** 7721–7727.
45. OKUBO, T. 1990. Centrifugal compression of crystal-like structures of deionized colloidal spheres. J. Am. Chem. Soc. **112:** 5420–5424.
46. HACHISU, S., A. KOSE & Y. KOBAYASHI. 1976. Segregation phenomena in monodisperse colloids. J. Coll. Interface Sci. **55:** 499–509.
47. ROSATO, A., K.J. STRANDBURG, F. PRINZ & R.H. SWENDSEN. 1987. Why the Brazil nuts are on top: size segregation of particulate matter by shaking. Phys. Rev. Lett. **58:** 1038–1040.
48. VANEL, L., A.D. ROSATO & R.N. DAVE. 1997. Rise-time regimes of a large sphere in vibrated bulk solids. Phys. Rev. Lett. **78:** 1255–1258.
49. KRUTZER, L.L.M., R. FOLKERSMA, A.J.G. VAN DIEMEN & H.N. STEIN. 1993. Influence of density differences between disperse and continuous phases on coagulation. Adv. Coll. Interface Sci. **46:** 59–71.
50. ANSARI, R.R. & K.I. SUH. 1998. New generation laser light scattering probes: particle size measurements in transparent and turbid media and medical applications. J. Jpn. Soc. Microgravity Appl. Suppl. II **15:** 186–193.
51. LANT, C.T., A.E. SMART, D.S. CANNELL, *et al.* 1997. Physics of hard spheres experiment: a general-purpose light-scattering instrument. Appl. Opt. **36:** 7501–7507.
52. LEWIS, G.N. & M. RANDALL. 1961. Thermodynamics. 513–514. McGraw-Hill, New York.

Gravity-Induced Anomalies in Interphase Spacing Reported for Binary Eutectics

REGINALD W. SMITH

*Department of Materials and Metallurgical Engineering,
Queen's University at Kingston, Ontario, Canada*

ABSTRACT: It has been reasoned that desirable microstructural refinement in binary eutectics could result from freezing in reduced-gravity. It is recognized that the interphase spacing in a binary eutectic is controlled by solute transport and that, on Earth, buoyancy-driven convection may enhance this. Hence, it has been presumed that the interphase spacing ought to decrease when a eutectic alloy is frozen under conditions of much-reduced gravity, where such buoyancy effects would be largely absent. The result of such speculation has been that many workers have frozen various eutectics under reduced gravity and have reported that, although some eutectics became finer, others showed no change, and some even became coarser. This reported varied behavior will be reviewed in the light of long term studies by the author at Queen's University, including recent microgravity studies in which samples of two eutectic alloy systems, MnBi–Bi and MnSb–Sb, were frozen under very stable conditions and showed no change in interphase spacing.

KEYWORDS: eutectic growth; microgravity; spacing anomalies

NOMENCLATURE:

λ	eutectic phase spacing
V	growth velocity
T	temperature
α	eutectic phase
β	eutectic phase
$C_{\alpha M}$	composition of α
$C_{\beta M}$	composition of β
D_L	diffusion coefficients
$\sigma_{\alpha L}$	interfacial tension α/liquid
T_E	eutectic temperature
ΔH_M	latent heat of crystallization
ρ_s	density of solid

INTRODUCTION

Eutectics find use in many technological applications; for example, structural, magnetic, and optical properties, see FIGURE 1. Since their physical properties are microstructure sensitive, considerable effort has been expended to learn how to control microstructure and so optimize desirable properties. In an ideal world, scientists

Address for correspondence: Reginald W. Smith, Department of Materials and Metallurgical Engineering, Queen's University at Kingston, Ontario, Canada K7L 3N6. Voice: 613-533-2753; fax: 613-533-6610.
smithrw@post.queensu.ca

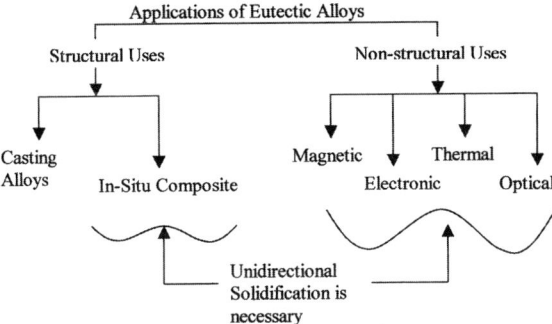

FIGURE 1. A schematic summary of eutectic applications including the more traditional casting alloys.

would formulate the set of rules governing the formation of microstructure during eutectic freezing, hopefully from first principles, and so provide engineers with predictions of what to do in order to optimize a particular desired property. However, this is a less than ideal world, and thus if engineers hope to exploit the properties of eutectics, then they usually have to do so from their own empirically-derived relationships. By way of example, it has been known since 1921[1] that additions of sodium and other such surface active elements, such as Li and Sr,[2] "modify" the manner of freezing of the silicon phase, and hence its habit, in the microstructure of eutectic Al–Si alloys. This modification provides a material with a much finer scale microstructure, much improved ductility, and significantly increased tensile strength. However, the precise manner in which modifying elements directly influence the eutectic silicon morphology is still a matter of speculation.

Some formalism has been applied to the growth of binary eutectics leading to the promulgation of a number of "laws". Brandt[3,4] and Zener,[5] from an analytic description of the solid-state transformation of single-phase austenite in the two-phase

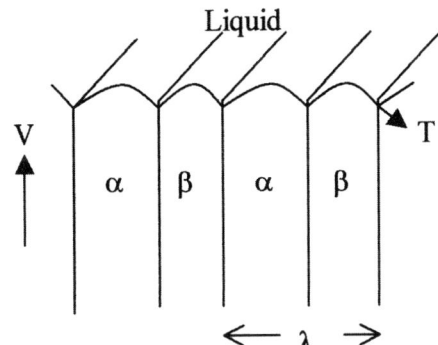

FIGURE 2. A binary eutectic–liquid cross-section (schematic).

lamellar eutectoid pearlite, proposed that the product of the growth velocity, v, and the square of the interlamellar spacing, λ, should be constant, that is

$$\lambda^2 v = \text{constant.} \tag{1}$$

These analyses provided a basic two-phase growth model for Tiller[6] to extend to the eutectic transformation. In parallel, and subsequently, many experimenters (see Ref. 7) tested the extent to which Equation (1) would describe eutectic growth. They reported that both lamellar and rod-like eutectics obeyed the $\lambda^2 v$ = constant relationship fairly well (see GENERAL DISCUSSION).

In view of the limitations of the various previous eutectic growth analyses, Jackson and Hunt[7] developed a steady state solution of the diffusion equation for a lamellar eutectic growing with a macroplanar solid/liquid interface, FIGURE 2. In such a eutectic, for the α and β phases to grow side by side the growth velocities of the individual phases must be *coupled*. This coupling arises because, although the phase diagram tells us that a liquid of composition C_E will solidify to form two solid phases α and β, the reality is that the α phase rejects B solute into the liquid as it forms and, correspondingly, β rejects A. Thus, there is a composition boundary layer in the liquid at the solid–liquid interface. Lateral diffusion gradients arise and so constrain the free growth of each phase—that is, the extension of one phase into the liquid is *coupled* to that of the other. Such solute gradients imply an undercooling (ΔT_D) of the solid–liquid interface, which varies from point to point. At the triple point T, FIGURE 2, the necessary energy balance requires that the liquid–α and liquid–β surfaces are curved. This curvature (Gibbs–Thomson effect) gives rises to a second component of undercooling, ΔT_C below the eutectic temperature T_e. Thus the local undercooling is often given as

$$\Delta T_{(x)} = \Delta T_D + \Delta T_C. \tag{2}$$

For steady state growth, Jackson and Hunt showed that two relationships would be expected: $\lambda^2 V$ = constant, that is, Equation (1), and

$$\frac{\Delta T^2}{V} = \text{constant.} \tag{3}$$

Equation (3) has been tested, but with more limited success.

MICROGRAVITY EXPERIMENTS

With access to reduced gravity research facilities (spacecraft, sounding rockets, and drop tower/tubes) becoming available in the 1970s, materials scientists have speculated on the likely effect of a reduced-gravity environment on the interphase spacing of eutectics. How might reduced g be expected to influence eutectic spacing? To review this we note that for regular eutectics, Jackson and Hunt[7] showed that

$$\lambda^2 = K \frac{\pi^2 D_L (C_{\alpha m} - C_{\beta m}) \sigma_{\alpha L} T_E}{m_L \Delta H_m \rho_s} V^{-1}, \tag{4}$$

that is,

$$\lambda^2 = K' D_L V^{-1} \tag{5}$$

for a given alloy system, where K and K' are constants, $C_{\alpha m}$ and $C_{\beta m}$ are the compositions of phases α and β with the highest solid solubility under the eutectic temperature, respectively. D_L, $\sigma_{\alpha L}$, T_E, m_L, ΔH_m, and ρ_s denote the diffusion coefficient, the interfacial tension between solid α and the liquid, the eutectic temperature, the slope of the liquidus, the latent heat of crystallization, and the density of the solid, respectively.

For a given value of V, since all other influencing parameters are only alloy system-dependent, no influence of g would be expected. However, for a eutectic to form, solute must partition between the solid phases. In Equation (4), the rate at which this occurs is reflected in the value of D. We might ask: Could g influence the value of D? This seems unlikely, but any convection might be expected to increase the rate of mass transport and so we might write

$$\lambda_{\text{observed}} = \lambda_{\text{true}} + \lambda_{\text{convection}}. \qquad (6)$$

In Equation (6), $\lambda_{\text{convection}}$ embraces all convective influences; for example, buoyancy, local fluid flows to meet the volume changes occurring as the liquid changes to the eutectic solid, and forced flow due to stirring. Thus, we might anticipate that $\lambda_{\text{observed}}$ would decrease as the g level was reduced. This is illustrated schematically in FIGURE 3.

However, in normal Bridgman-type crystal growth, with a relatively quiescent liquid where B_c is approximately equal to λ and B_λ is many times the value of B_c, we might expect little influence of g on λ unless we assist in the penetration of the hydrodynamic boundary layer by deliberately reducing B_λ by externally induced convection, such as crystal oscillation about its growth axis (often turned *spin-up* and *spin-down*) or by electromagnetic induction. Penetration would be assisted by the existence of any solid–liquid interface structures of short period (i.e., surface raggedness) or longer period convolutions, such as cellular on dendritic growth projections, FIGURE 4.

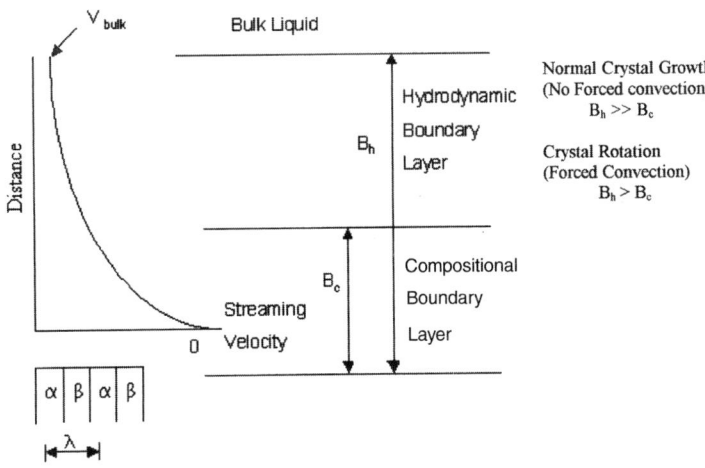

FIGURE 3. Boundary layer ahead of an eutectic–liquid interface.

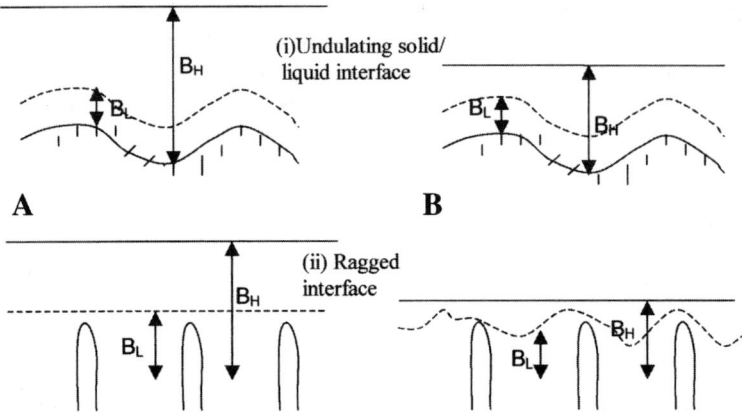

FIGURE 4. Influence of solid–liquid interface shape and phase morphology on penetration of hydrodynamic boundary layer by convective flows: **(A)** no convective stirring, **(B)** convective stirring.

The question then is to what extent does convection cause changes in λ in normal crystal growth and, hence, should any structural refinement result from eutectic processing in microgravity, particularly in those alloys of industrial interest.

To investigate the influence of g, a number of eutectic alloys have to be processed in a low-g environment. The results are shown in TABLES 1 and 2. On inspection of these results, a confusing picture emerges: whereas some systems were reported to have displayed the anticipated spacing reduction, others showed no change, and some even became coarser. The long-term eutectic growth studies of the author led him to expect a more consistent pattern of behavior. In view of this he decided to carry out very careful eutectic growth experiments at 1 g and in space and support these with numerical modeling.

TABLE 1. Previous space work on eutectics

System	%	Reference	Type	Class	V_F%
Al–Al$_2$Cu	0	Favier and DeGoer 1984 Space Lab[8]	NF/NF	1	48
Al–Al$_2$Cu	0	DeGoer and Maguet 1983 Texus VI[9]	NF/NF	1	48
Al–Al$_3$Ni	−15±5[a]	Favier and DeGoer 1984 Space Lab[8]	NF/F	4	10
Al–Al$_3$Ni	−17±5[a]	DeGoer and Maguet 1983 Texus VI[9]	NF/F	4	10
NiSb–InSb	30	Muller and Kyr 1984 Texus X	F/F(NF)	3	2
NiSb–IbSb	20	Muller and Kyr 1984 Space Lab IESA[11]	F/F(NF)	3	2

[a] A negative sign indicates an increase in λ in microgravity.

TABLE 2. Previous space work on the Bi–MnBi system (Type-F/F(NF), Class 3, $V_F = 2\%$

Experiment	Growth Rate	Spacing Change	Reference
ASTP	3 cm/h	50%[a]	Pirich et al.[12]
SPAR VI	30 cm/h	66%	Pirich et al.[12]
SPAR X	50 cm/h	50%	J. Bethin et al.[13]
Shuttle 51-G	(Sample did not translate)		Larson et al.[14]

[a] Spacing at 1 g was twice that in microgravity.

QUEEN'S UNIVERSITY STUDIES

Eutectic Growth

In attempting to rationalize the data of TABLES 1 and 2, note must be taken of the fact that the eutectics reported to undergo spacing changes in reduced gravity were all from the anomalous (or irregular) group of eutectics, FIGURE 5B. In such eutectics, one or both of the phases normally facets when growing freely as a dendrite. This property results in coupling difficulties during eutectic growth since the growth velocity/ undercooling relationship for each phase may differ significantly. The net result is that, rather than having a broad coupled zone that is near symmetrical about the eutectic composition (FIG. 5A), the coupled zone is invariably narrow and skewed towards the faceting phase (FIG. 5B). In addition, it rarely embraces the eutectic composition. For example, to obtain coupled growth with Al–Si eutectic alloys in the absence of a growth modifier, the alloy must be frozen rapidly. An examination of the microstructure produced suggests aluminum dendrites nucleate from the undercooled liquid, grow quickly, and force the composition of the liquid away from the eutectic point and into the coupled zone. Hence, in the liquid between

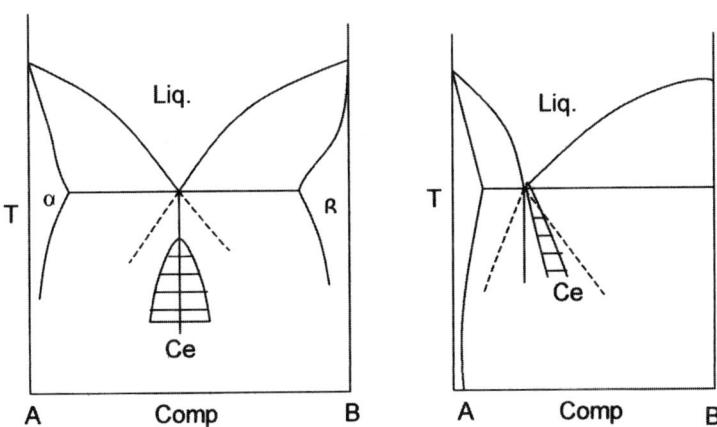

FIGURE 5. Binary eutectic phase-diagram types.

the dendrites, eutectic growth takes place in a coupled fashion more closely akin to that represented by FIGURE 2. In contrast, if the eutectic alloy is frozen slowly, a very ragged interface results with the faceting phase leading locally.

The reason for this markedly different behavior is that the growth velocity of a faceting phase (F) at small undercooling is significantly smaller than that of a non-faceting phase (NF). A comparable growth velocity is only achieved at much larger undercooling where the facets roughen and growth is more akin to that of a non-faceting phase.

Returning to the description of the interface undercooling, it is clear, if we are concerned with a NF/F or F/F eutectic, a kinetic term (ΔT_k) must be introduced. Equation (2) then becomes

$$\Delta T_{(x)} = \Delta T_d + \Delta T_c + \Delta T_k. \tag{7}$$

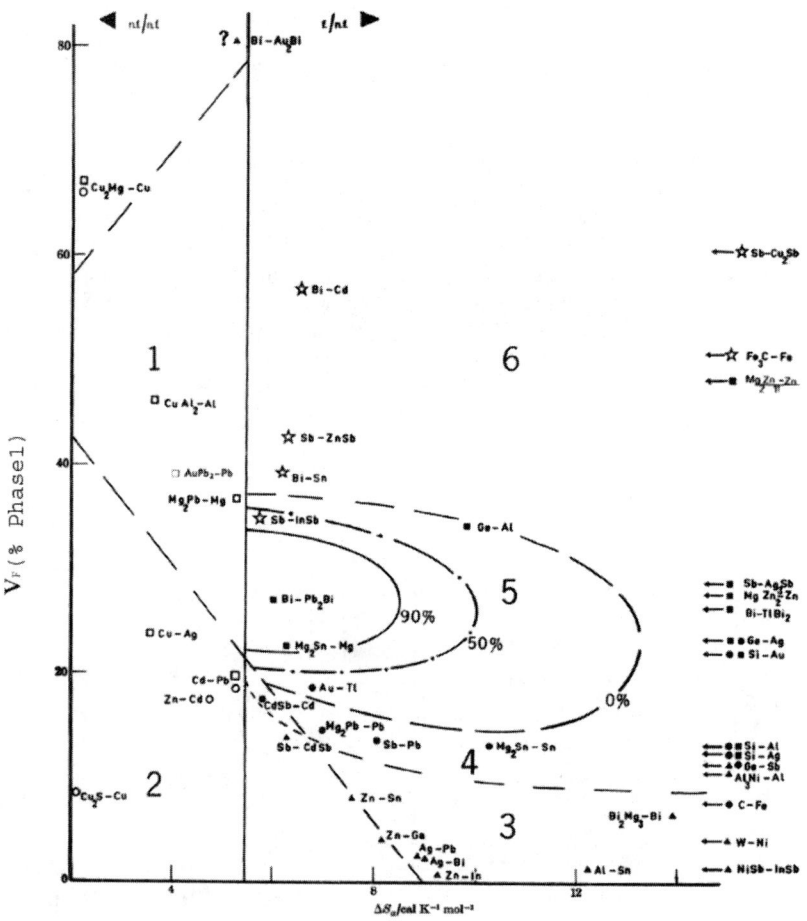

FIGURE 6. Binary eutectic classes.

The inclusion of the kinetic term considerably complicates the analysis of growth in NF/F eutectics to the extent that Liu and Elliott[15] question whether a steady state growth model of the type developed by Jackson and Hunt[7] is even applicable to the growth of irregular eutectics.

A broader approach to the entire problem of predicting the microstructure of a binary eutectic at various growth rates has been developed by the author.[16] In this, the faceting tendency of eutectic phase is characterized by its entropy of solution value, ΔS_α. This parameter, together with consideration of the volume fraction of the faceting phase and the rate of unidirectional growth, permits the development of FIGURE 6. This is a section through a three dimensional plot for a growth rate of 5×10^{-4} cm/sec, a typical growth rate for the study of eutectics by directional solidification. The various microstructural classes are shown in FIGURE 7. The boundaries of the classes move as a function of growth velocity.

In developing FIGURE 6, it was noted that, although the local temperature gradient in the liquid is very important in describing the morphology of the solid–liquid interface in single phase alloys, the microstructural scale of eutectics is much finer and so the direct influence of the temperature gradient is less.

The usefulness of FIGURE 6 lies in the fact that closely similar microstructural responses as a function of growth velocity are expected from all the members of each group. Thus, for example, all the broken-lamellar eutectics would be expected

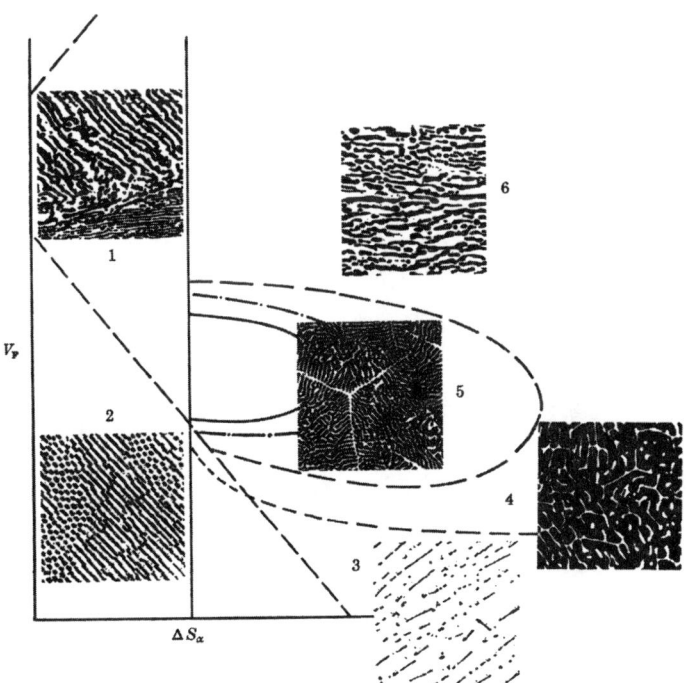

FIGURE 7. Binary eutectic microstructural types.

to behave similarly at 1 g and presumably in reduced gravity; viz, Bi–MnBi and NiSb–InSb.

The data given in TABLE 1 is illustrative rather than inclusive. Experiments have been done in space with Al–Si, Fe–C, and other eutectics of the Class 4 in eutectic alloys. These eutectics were expected to behave in the manner reported for Al–Al$_3$Ni but did not. This suggested to the writer that what was being measured was not an *intrinsic* property of eutectic growth in reduced gravity, but was more likely to be an experimental artifact; that is, had the microgravity and 1 g experiments been done in some other way the results would have been different.

In passing, it should be noted that many of the broken-lamellar eutectics have Bi as the major phase. However, as such, it would not be expected to facet and suffer growth restriction, even though it does so when freezing as a primary phase. Since it forms the matrix in these alloys, microfacets form much less easily[17] and so the matrix phase suffers little or no kinetic restraint (and the associated larger local undercooling).

EXPERIMENTAL RESULTS

In view of the large spacing changes reported for the Bi–MnBi in reduced gravity (TABLE 2) the author decided to use this alloy for gradient-freeze experiments using a get-away special. The three-zone furnaces were designed and built at Queen's University and packaged in the GAS-can by Bristol Aerospace Ltd. of Winnipeg, Manitoba. The samples were of 4 mm diameter and two growth rates were selected, namely, 2.5 cm/h and 4.2 cm/h, in order to span an earlier growth rate of 3 cm/h.[12]

The QUESTS equipment flew on STS 47 (Space lab J) in September 1992. During the flight, the furnaces performed flawlessly, providing very steady growth conditions, FIGURE 8.

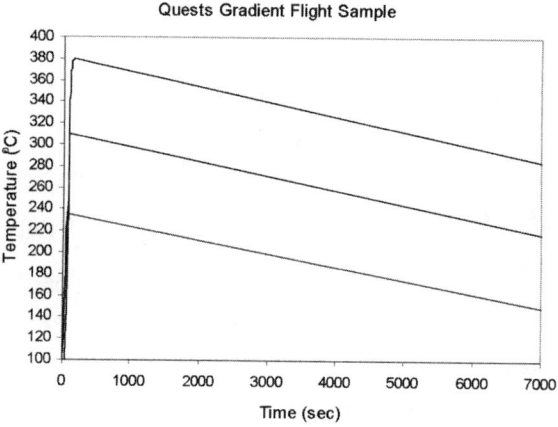

FIGURE 8. Time–temperature profiles in the three zones of the QUESTS gradient-freeze surfaces.

After the flight, the specimens were recovered and cross-sections were prepared for metallographic examination. This was done using a LECO image analysis system. The results are shown in FIGURE 9 for the flight samples and others produced in the flight equipment immediately prior to and after the flight.

The principal conclusion to be drawn from FIGURE 9 is that, for the growth velocities used, there is no change in λ with reduction in gravitational field. In combination with this experimental study, this eutectic system was numerically modelled to simulate the conditions monitored during the flight and ground-based experiments. It was found that for unidirectional growth in a typical Bridgman context with the solid down (i.e., solidification proceeding upward) the hydrodynamic boundary layer extended into the liquid much further than the diffusion boundary layer controlling the eutectic spacing. Thus, growth in reduced gravity where the hydrodynamic

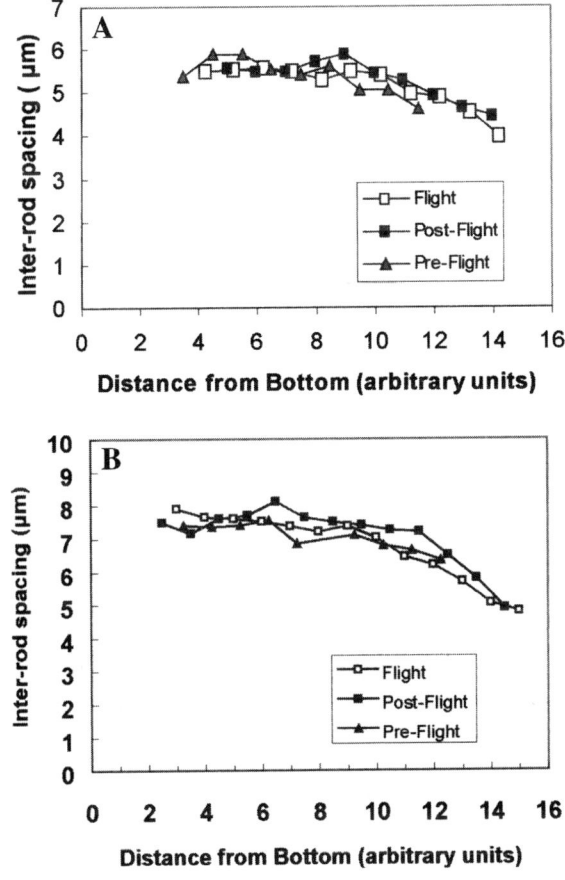

FIGURE 9. Interphase spacing versus distance along the length of the Bi–MnBi sample: **(A)** growth rate of 42 mm/h, **(B)** growth rate of 25 mm/h.

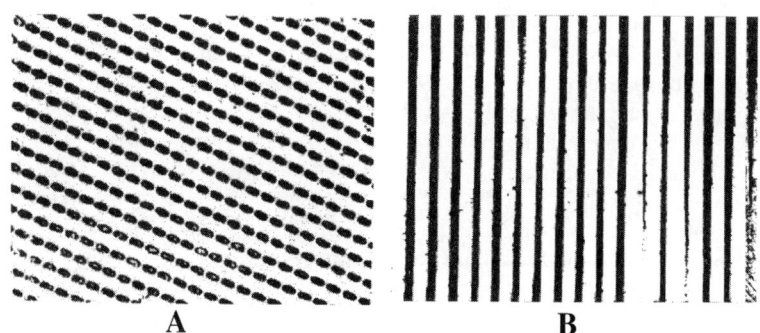

FIGURE 10. Microstructure of the Sb–MnSb eutectic, grown at 25 mm/h: **(A)** transverse section (400×), **(B)** longitudinal section (400×).

boundary layer becomes very large did not cause any change in spacing. Quite clearly, if crystal growth practice at 1 g is such as to reduce the depth of the hydrodynamic boundary layer (e.g., by inducing liquid shear at the interface by oscillatory rotation) then changes in spacing should be expected and have been observed.[19] In support of the experimental data, the numerical studies predicted that λ would not change as a function of g. (The detailed experimental and numerical data appear elsewhere.[18])

Shortly after the STS 47 Flight, the opportunity arose for a reflight of the QUESTS get-away special. Taking note that the volume fraction of the minor phase in Bi–MnBi is only about 2%, the author selected a F/F eutectic in which this was approximately 40%, namely Sb–MnSb. As with Bi–MnBi, the matrix phase would normally facet during growth as a primary dendrite but suffers little or no growth restriction during growth of the eutectic except when forming faceted cells at the solid–liquid interface. This eutectic falls on the borderline of eutectic Regions 5 and 6, close to the other Sb–matrix eutectics shown there. Not surprisingly, the eutectic is rod-like at very small growth rates (see FIGURE 10) and becomes complex regular at large growth rates (see FIGURE 11).

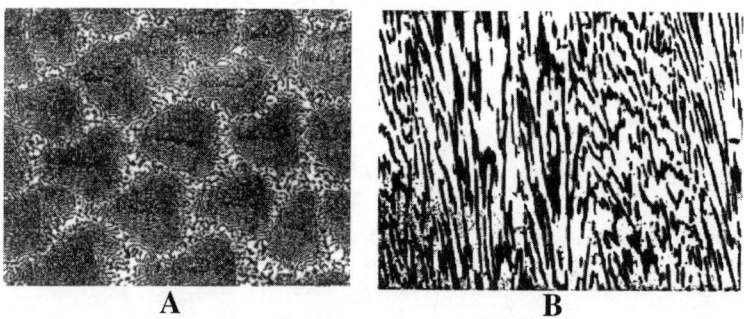

FIGURE 11. Microstructure of the Sb–MnSb eutectic, grown at 40 mm/h: **(A)** transverse section, showing cell structure (100×); **(B)** longitudinal section, showing branching (400×).

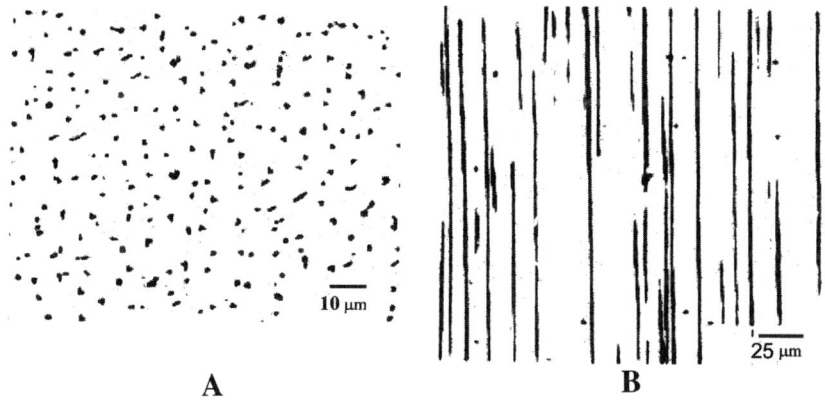

FIGURE 12. Microstructure of the Sb–MnSb eutectic, postflight: (**A**) transverse section, (**B**) longitudinal section.

For comparison, FIGURE 12 shows the results for Bi–MnBi. The much smaller volume fraction MnBi phase at small growth rates is well-faceted in cross-section, whereas the MnSb phase is well rounded, reflecting the much smaller entropy of solution value (i.e., faceting tendency) of this phase.

The Sb–MnSb eutectic was prepared for flight in much the same way as was Bi–MnBi. However, an additional feature was to be the inclusion of a thermocouple in each sample in an attempt to determine the undercooling at the solid/liquid interface. Unfortunately, although the thermocouple system worked well in the ground-based testing, the original microgravity flight date was brought forward and so did not permit time enough for the modified specimen crucible to be used for the flight samples.

The results of preflight/flight/postflight tests are shown in FIGURE 13. Again, as with Bi–MnBi samples, the interrod spacing appears to be independent of the value of the gravitational force.

GENERAL DISCUSSION

There are two questions that might now be asked:
1. How regular are "regular" eutectic microstructures? The answer appears to be not very. Trivedi et al.[20] used a number of lead-based eutectics and found that the spacing variation could be large at small velocities (100%) becoming less at higher velocities. Elliot and coworkers[21] report similar results for the Al–Cu eutectic.
2. Why did Pirich and Larson report finer Bi–MnBi from space? The Clarkson University Group of Wilcox and Regel have addressed this topic.[22] They summarize the predictions of the various eutectic models with respect to the influence of buoyancy convection in the melt ahead of the eutectic. They conclude that (1) decreased convection is always expected to decrease eutectic spacing; (2) gentle convection alters the interfiber spacing of a fibrous eutectic only when it causes a fluctuating freezing

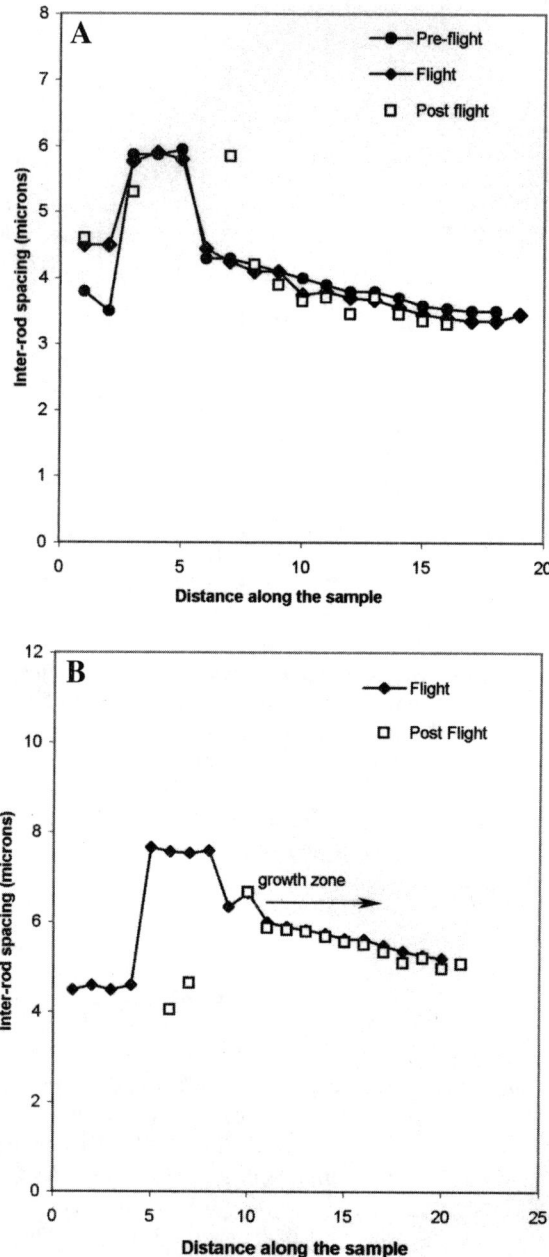

FIGURE 13. Interphase spacing variation along Sb–MnSb eutectic sample: (**A**) sample QG1, (**B**) sample QG2.

rate with a system for which the kinetics of fiber branching differ from that for fiber termination. That such branching and termination should form an integral part of our understanding of the effect(s) of g on eutectic microstructures can be judged from FIGURE 14, which shows how the faceting Ag-rich phase in the Bi–Ag eutectic responds to growth perturbations during Bridgman-type processing at $1\,g$.[23]

In the Queen's University experiments with the Bi–MnBi eutectic, the author observed the following: for Bridgman growth of Bi–MnBi at approximately 3 cm/h; $1\,g$, *solid down*, very stable to convection since the solid is below the liquid giving the usual near-regular broken-lamellar eutectic microstructure; *solid up*, unstable, the solid is above the liquid, leading to banding; μg, very stable growth. Hence, no change of spacing with g-level was observed for stable growth.

Returning to the work of Pirich and Larson,[12] the Clarkson Group pointed out[22] that NASA's advanced directional solidification system has a 14-cm heated center section, with end boosters to give gradients over 3 cm at each end. With this equipment, vigorous convective stirring would be expected at $1\,g$.

Pirich and Larson observed that, at $1\,g$, the solid-up grown specimen gave a 67% larger spacing at a growth rate of 3 cm/h over the solid down value. In μg, they used Peltier demarcation to delineate the solid–liquid interface during growth and so provide an accurate measurement of growth velocity. This was done by passing an electric current pulse at regular intervals. However, when Wilcox and Regel passed pulses of electricity through a Bi–MnBi specimen to simulate the Peltier-effect at the solid/liquid interface, they observed that the pulses produced local temperature changes, irregular growth, and an increased spacing which they ascribe to the presumption that it is easier for MnBi to be terminated than to cause it to branch. The net result is a larger than true spacing for the averaged growth rate. By the same token, branching should occur more easily than termination for Al–Al$_3$Ni, which is reported to coarsen in μg.

More recently, the Clarkson Group have conducted further theoretical analyses and experimental studies of the manner in which certain eutectics respond microstructurally to oscillatory changes in growth rate. Reference 24 describes their attempt to apply the concept of minimum entropy production to the growth of lamellar eutectics with an oscillating freezing rate; Reference 25 reports a phase-field approach to the analysis of this same topic. They showed that when the volume fractions of the two phases in a binary eutectic are very different, freezing rate oscillations would be expected to reduce λ, although in some circumstances, regions where λ is both smaller and larger than the value for steady state growth at the averaged growth rate may be expected.

In Reference 26, the influence of electric current pulses on the microstructure of the MnBi/Bi eutectic was examined. As noted earlier, the volume fraction of the MnBi is about 2%. The most significant experimental finding was that the introduction of oscillations in the freezing rate decreased the average MnBi rod spacing, in keeping with the theoretic predictions of References 24 and 25.

Intrinsic Microgravity Effects

By and large, the microgravity experimental environment has been characterized by activities to explore the absence of Earth-bound effects—e.g., buoyancy-induced convection in various gas–liquid–solid systems. However, it has been proposed that

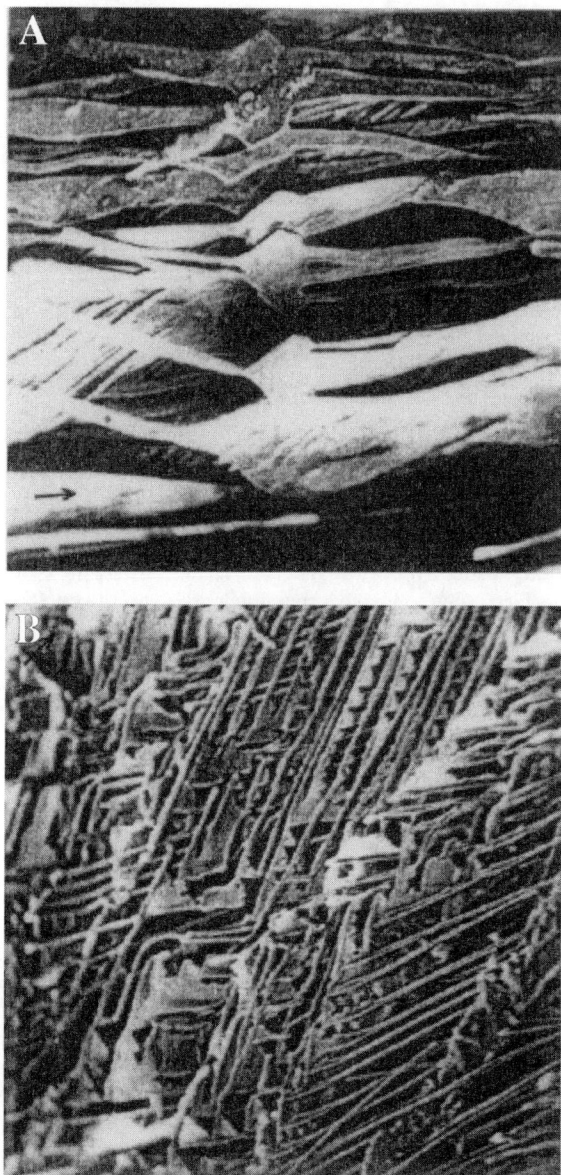

FIGURE 14. Deeply etched longitudinal sections of the BiAg eutectics. **(A)** Growth rate of 6.88×10^{-5} cm/sec (625×). **(B)** Growth rate of 7.04×10^{-4} cm/sec (625×). (Note how the width of the Ag-rich phase broken lamellæ increases and decreases over small distances as a result of the thermal fluctuations in the melt.)

the microgravity state also possess some intrinsic properties. Patuelli and Tognato[27] hypothesize that the effect of a reduction in gravitational field may be akin to the raising of the temperature with respect to the surface free energy of the solid–liquid interface. If this analysis can be substantiated by careful observations, then it is possible that many of the *strange* effects reported from materials processing in space might warrant further study and interpretation. The microstructural forms observed in eutectics are very strongly influenced by surface energy considerations. In this connection, these authors report that the percentage of the rod-like morphology in the normally lamellar Cu–Ag eutectic alloy increased from about 2% at $1g$ to 30% in microgravity.[28] More recently[29] they used a novel method of preparing Cu–Ag alloys as eutectic but with different volume fractions of each phase in order to test their surface energy hypothesis. This was done by sintering various proportions of copper and silver powders 20°C below the eutectic temperature to give volume fractions of 5, 15, 25, and 32% of the Cu-rich phase. These alloys were then partially melted by heating to 850°C and unidirectionally frozen at $1g$. Rod-like morphologies dominated the 5 and 15 volume percent Cu. Lamellar structures were found at 25% Cu and above.

It is unclear why the rod-like eutectic microstructures were formed. Such microstructures are often seen as the result of cellular growth but this was not the case here. From FIGURE 6, it is seen that the Cu–Ag eutectic normally has a volume fraction of the copper-rich phase of about 25%, which should cause it to adopt a rod-like morphology. However, it normally displays strong lamellar features, presumably due to surface energy anisotropy. That a rod-like microstructure was observed in the 5 and 15 $V_F\%$ alloys suggests that some highly mobile solute (such as a gas) was present, which reduced surface energies and surface anisotropy, without causing cellular growth, permitting rod-like growth as predicted by Jackson and Hunt.[7] It will be most interesting to see what happens in reduced gravity.

CONCLUSIONS

It appears likely that the changes in eutectic spacing reported so far are probably due to growth irregularities and that if steady state conditions had been established and maintained, then little microstructural change would have occurred.

ACKNOWLEDGMENTS

It is a pleasure to acknowledge the contributions of various Research Associates of the Microgravity Materials Group at Queen's University, in particular, Drs. Mustafa Kaya, Mark Gallerneault, Stephen Goodman, and Mark Tunnicliffe. In addition, I thankfully acknowledge the financial assistance of the National Science and Engineering Research Council of Canada, the Canadian Space Agency and Queen's University in the support of these extended eutectic experiments.

REFERENCES

1. PACZ, D. 1921. U.S. Patent 1,387,900, 16 August.
2. SMITH, R.W. 1969. The Solidification of Metals. 224. ISI Publication 110, London.
3. BRANDT, W.H. 1945. J. Appl. Phys. **16:** 139.
4. BRANDT, W.H. 1946. AIME Trans. **167:** 405.
5. ZENER, C. 1946. AIME Trans. **167:** 550.
6. TILLER, W.A. 1958. Liquid Metals and Solidification. 276. ASM, Cleveland.
7. JACKSON, K.A. & J.D. HUNT. 1966. AIME Trans. **236:** 1129.
8. FAVIER, J.J. & J. DEGOER. 1984. Fifth European Symposium, Materials Sciences Under Microgravity, Results of Spacelab. I, ESA SP-222 (8–10 Mario Nikis, 75738 Cedex 15, France) 127.
9. DEGOER, J. & R. MAGNET. 1983. Internal Report (Centre d'Etudes Nuclearnes de Grenoble) Solidification dirigee des Systemes Eutectics Al2Cu-Al et Al3Ni-Al en fusee sonde (exploitation de l'experience Texas VI).
10. MULLER, G. & P. KYR. 1984. Internal Report—(DFVLR) Gerichtete Frstarrung des InSb-NiSb Eutectikiums; Statusbericht fur des Texas 10 Experiment.
11. MULLER, G. & P. KYR. 1984. Fifth European Symposium, Materials Sciences Under Microgravity, Results of Spacelab. I, ESA SP-222 (8–10 Mario Nikis, 75738 Cedex 15, France) 141.
12. PIRICH, R.G., D.J. LARSON & G. BUSCH. 1980. SPAR and ASTP studies of plane front solidification and magnetic properties of Bi/Mn-Bi. AIAA 18th Aerospace Sciences Meeting, Pasdena 14–16 January, AIAA-80-10119.
13. BETHIN, J. 1986. SPAR X technical report for experiment 76-22 directional solidification of magnetic composites. NASA-CRr-171271, RE-69, Grumman Aerospace Corporation (1984); also NASA, JM-86548.
14. LARSON, D.J., J. BETHIN & B.S. DRESSLER. 1988. Shuttle mission 51G experiment MPS 772FO75—flight sample characterisation report RE753. Grumman Corporation.
15. LIU, J. & R. ELLIOTT, 1993. Mat. Sci. Eng. **A173:** 129.
16. CROKER, M.N., R.S. FIDLER & R.W. SMITH. 1973. Proc. Roy Soc, (Lond.) **A335:** 15.
17. HUNT, J.D. & D.T.J. HURLE. 1968. The Solidification of Metals. Publ. 110, 162. ISI London.
18. KAYA, M. & R.W. SMITH. 1996. Report to Canadian Space Agency, QUESTS II.
19. APAYDIN, N.A. 1990. An Examination of Microstructure Development in Al–Si and Al–Cu Alloys. Ph.D. Thesis, Queen's University, Kingston, Canada.
20. TRIVEDI, R., J.T. MASON, J.D. VERHOEVEN & W. KURZ. 1991. Met. Trans. A. 2523.
21. OURDJINI, A., J. LIU & R. ELLIOTT. 1994. Mat. Sci. Tech. **10:** 312.
22. WILCOX, W.R. & L.L. REGEL, 1994. Micrograv. Q. **4**(3): 147.
23. FIDLER, R.S. 1969. Ph.D. Thesis, University of Birmingham.
24. POPOV, D.I., L.L. REGEL & W.R. WILCOX. 2000. J. Cryst. Growth **209:** 181–197.
25. POPOV, D.I., L.L. REGEL & W.R. WILCOX. 2001. Cryst. Growth Design **1**(4): 313–320.
26. LI, F., L.L. REGEL & W.R. WILCOX. 2001. J. Cryst. Growth. **223:** 251–264.
27. PATUELLI, C. & R. TOGNATO. 1994. Advanced materials 1993 III/A. Trans. Mat. Res. Soc. (Jpn.) **16A:** 607.
28. BARBERI, F. & C. PATUELLI. 1988. Met. Trans. A **19A:** 2659.
29. CHEN, Q., C. PATUELLI & R. TOGNATO. 1996. Seventh International Symposium on Experimental Methods for Micro-gravity Materials Science, Las Vegas, Feb.

Optical Measurement of Concentration Gradient Near Miscible Interfaces

N. RASHIDNIA AND R. BALASUBRAMANIAM

National Center for Microgravity Research on Fluids and Combustion, NASA Glenn Research Center, Cleveland, Ohio, USA

ABSTRACT: A common path interferometer (CPI) system was developed to measure the diffusivity of transparent liquid pairs. The CPI is an optical technique that can be used to measure changes in the gradient of refractive index of transparent materials. The CPI is a shearing interferometer that shares the same optical path from a laser light source to the final imaging plane. Molecular diffusivity of liquids can be determined by using physical relations between changes in optical path length and liquid phase properties. In this paper, we present results obtained when the diffusivity depends on the local concentration of the miscible fluids.

KEYWORDS: diffusivity; interferometry

INTRODUCTION

Optical measurement techniques have become an integral part of many diagnostic development industries and research laboratories. The point diffraction interferometer (PDI) and its phase shifted version has been used as an instrument for measuring optical wave fronts for lens testing and combustion and fluid flow diagnostics.[1–3] The PDI is considered to be robust (see for example Ref. 4) because it has a common path design. The schlieren technique has been used for many years for flow visualization and gives a qualitative picture of the index of refraction variations within a transparent test section. The common path interferometer (CPI) basically uses a Wollaston prism in combination with a polarizer and an analyzer. The advantage of using a shearing interferometer over the PDI and schlieren photographic techniques is that it is capable of quantitative measurements in liquids with large index of refraction variations, as is often the case in interface dynamics studies. It also has the added benefit of extremely easy alignment of the focussed light on the interferometer.

The dynamics of miscible interfaces is an active area of research that has been identified to benefit from experimentation in reduced gravity. The study of the flow patterns and the shape of the interface when one liquid is slowly displaced by another such that diffusion plays an important role in the dynamics is recognized to be a very important transport phenomenon. It has been suggested that non-traditional stresses in the fluids, caused by the steep variation of the concentration of the miscible fluids in the mixing zone, might be important in the dynamics. The diffusion coefficient

---

Address for correspondence: N. Rashidnia, National Center for Microgravity Research on Fluids and Combustion, NASA Glenn Research Center, Mail Stop 110-3, Cleveland, OH 44135, USA.

(diffusivity) of the miscible fluids is a property that is important in these experiments. Petitjeans and Maxworthy[5] used a variation of Wiener's method, described by Sommerfeld,[6] to measure the diffusion coefficient. Here, the deflection angle of an incident laser light beam is related to the vertical gradient of index of refraction, and hence to the concentration gradient within the test cell. The average diffusion coefficient was obtained by analysis of the measured concentration gradient profile. The method used by Petitjeans and Maxworthy to measure the concentration gradient is a point measurement that requires the light beam and the detector to be traversed through the region of interest. Therefore, instantaneous measurement of concentration gradient cannot be obtained everywhere near the interface. To overcome this limitation, the original method of Wiener, in which the deflection of an inclined light sheet is measured, was used by Rashidnia et al.[7] and Kuang et al.[8]

We have developed a CPI method that uses a Wollaston prism as an alternative technique for obtaining the distribution of index of refraction in the mixing region.[9] Specifically, the CPI, which incorporates a Wollaston prism, operating in its finite fringe mode,[10] is used to visualize the concentration gradient in the mixing zone in real-time. We have shown that finite fringes oriented at a small angle to the interface boundary yields a simple relationship between fringe location and concentration gradient. This novel approach, in which the small angle magnifies the gradient, yielding a nice real-time visualization tool, requires setting the prism in a particular fashion with respect to the orientation of the mixing interface. The set-up of the interferometer is discussed in detail elsewhere.[9]

In this paper we show how the measurement of the concentration gradient in the mixing zone can be used in conjunction with an analysis of the one-dimensional diffusion equation to infer the dependence of the diffusivity on the local concentration of the miscible fluids. For simplicity, we assume that a linear variation of the diffusion coefficient with concentration holds. The technique can be extended to a more general dependence of the diffusion coefficient on the concentration.

TEST APPARATUS AND PROCEDURE

The CPI measures the gradient of index of refraction within a test cell. We have used a Wollaston prism to image the light beam sheared by the concentration gradient near a miscible interface. A collimated, polarized beam of light from a laser is passed through a transparent test cell. The light is then focused on the Wollaston prism, which splits the light into two beams that are slightly displaced from each other. When the beams recombine, they produce interference fringes that indicate gradients of index of refraction in the test medium. The Wollaston prism has an apex angle of 10'; in conjunction with the focal length of the lens used, the beam separation is 0.509 mm. The interferometer is arranged in its finite fringe mode, in which equidistant, parallel interference fringes appear when the index of refraction field in the test cell is uniform. This is the case when only one of the fluids is present in the test cell. When the second fluid is introduced and diffusion occurs, the deviation or shift of a fringe from its undisturbed location is a measure of the index of refraction gradient within the test cell. We assume that the index of refraction within the test cell is a function of only the vertical coordinate. The light beam is passed through

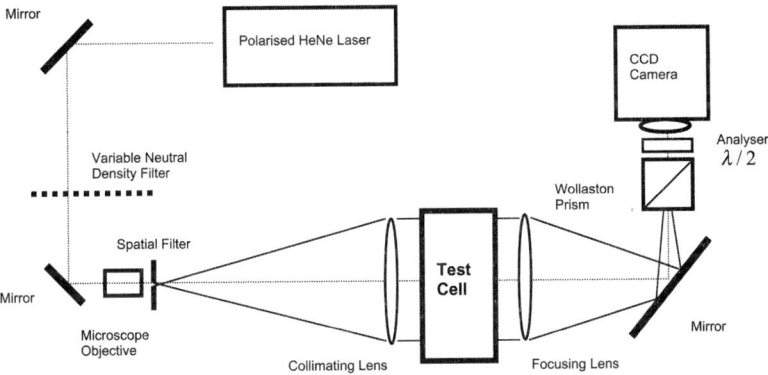

FIGURE 1. Schematic of a common path interferometer (CPI).

the test cell horizontally, and we further assume that each ray of light traverses a path of constant refractive index, without refraction. FIGURE 1 shows a schematic of the CPI. The test cell used to hold the pair of miscible liquids is a quartz container of 10 mm width, 45 mm height, and variable depth. The depth was selected so that severe refraction of the laser light is avoided. An 8-bit, 640×640 pixel CCD camera captures the interferograms of the CPI and a time-lapse super video recorder (S-VHS) stores them on a tape for later analysis.

ANALYSIS OF DIFFUSION WITH A VARIABLE DIFFUSIVITY

We assume that the two fluids are pure and, therefore, only binary diffusion takes place. The total volume of the fluids is assumed to be unchanged by mixing. Let $\phi(x,t)$ denote the volume fraction of one of the fluids (say fluid 1) at a vertical location x and a given time t. At $t = 0$, we assume that the interface between the two fluids is flat and is located at $x = 0$. Far away from the interface, on either side, we assume that the fluids are pure (i.e., $\phi = 1$ and $\phi = 0$). We assume that the density stratification of the fluids is stable, and there is no bulk motion. Joseph and Renardy[11] and Joseph et al.[12] show that by assuming Fick's law of diffusion for each constituent fluid, the one-dimensional diffusion equation for ϕ can be written as follows:

$$\frac{\partial \phi}{\partial t} = \frac{\partial}{\partial x}\left(D(\phi)\frac{\partial \phi}{\partial x}\right). \tag{1}$$

Here

$$D(\phi) = \left(\frac{\rho_1}{\rho_2}\phi + 1 - \phi\right)D_1(\phi)$$

is an overall diffusion coefficient, $D_1(\phi)$ is the diffusion coefficient for fluid 1 in the expression for Fick's law that relates the mass flux of fluid 1 to the gradient of its mass concentration, and ρ denotes the density. The boundary conditions are

$$\phi \to 1 \text{ as } x \to \infty;\ \phi \to 0 \text{ as } x \to -\infty. \tag{2}$$

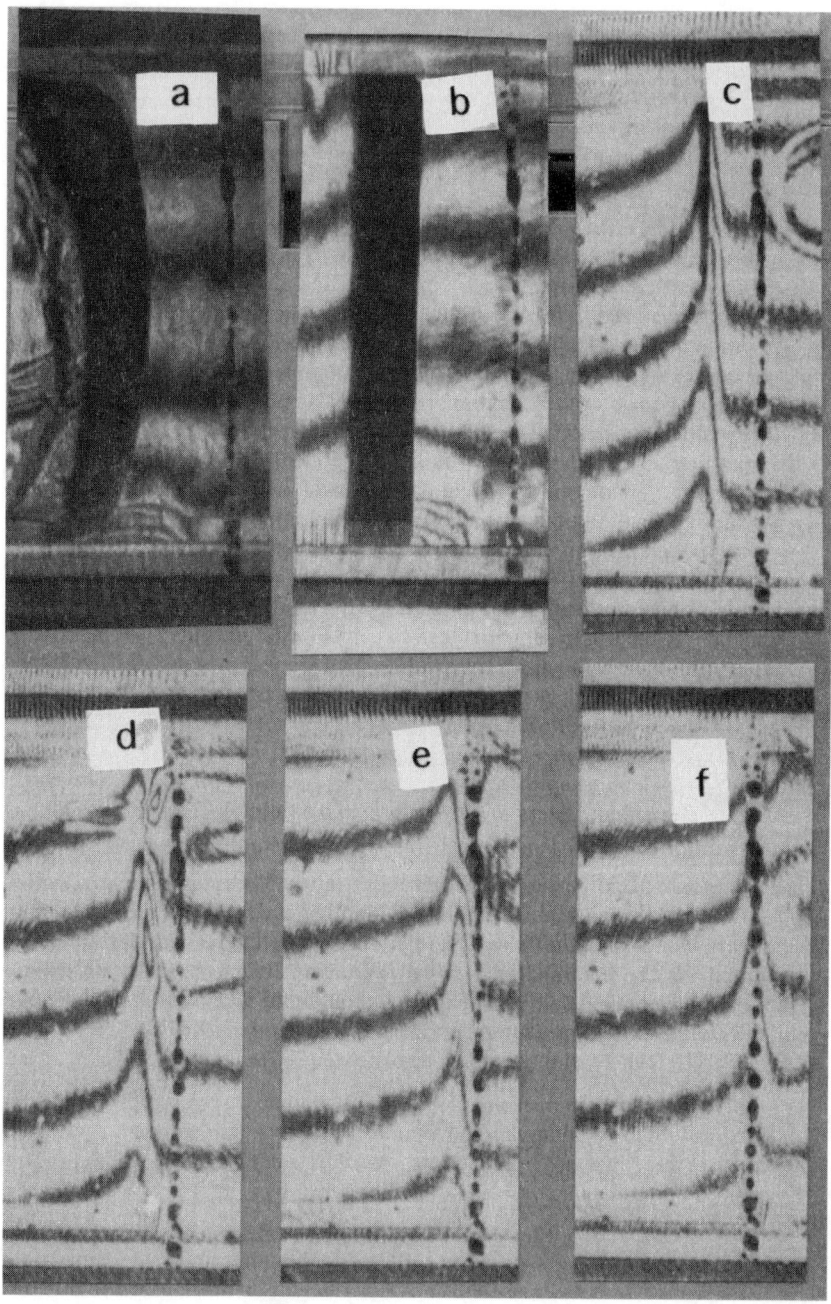

FIGURE 2. Continued on opposite page

The initial condition is

$$\phi = 1 \text{ for } x > 0; \phi = 0 \text{ for } x < 0. \quad (3)$$

Even when the diffusion coefficient is a function of the volume fraction, a similarity transformation exists and reduces the governing equation to an ordinary differential equation. Let D_0 be a reference value for $D(\phi)$. The similarity transformations

$$\eta = \frac{x}{2\sqrt{D_0 t}}, \phi = F(\eta), \quad (4)$$

reduce the governing equation and boundary conditions to

$$\frac{d}{d\eta}\left(\frac{D(F)}{D_0}\frac{dF}{d\eta}\right) + 2\eta\frac{dF}{d\eta} = 0, \quad (5)$$

$$F \to 1 \text{ as } \eta \to \infty; F \to 0 \text{ as } \eta \to -\infty. \quad (6)$$

We assume that $D(\phi) = D_0(1 + a\phi)$. Chen and Meiburg[13] found such a linear relation to be reasonably accurate in describing the experimental results of Petitjeans and Maxworthy[5] for the diffusion of water and glycerine. The governing equation specializes to

$$[(1 + aF)F']' + 2\eta F' = 0. \quad (7)$$

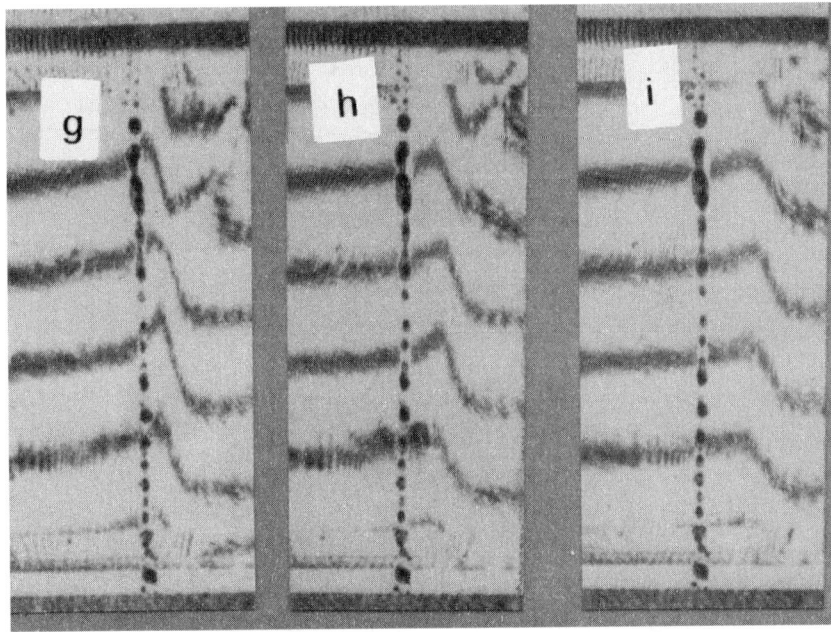

FIGURE 2. Interferograms representing the evolution of the concentration gradient in a water–glycerine system. (a) At time $t = 0$, gylcerine is present at the bottom of the quartz cuvette (*right side*), and water is added to the top (*left side*). The *dashed line* in all the panels is a fixed reference mark on the cuvette. The remaining panels show the evolution of the fringes for (b) $t = 1$ min, (c) $t = 54$ min, (d) $t = 86.5$ min, (e) $t = 118$ min, (f) $t = 145.5$ min, (g) $t = 205.5$ min, (h) $t = 265.5$ min, and (i) $t = 325.5$ min.

For a given value of a, this non-linear equation, together with the boundary conditions in Equation (6) is solved numerically. We used the numerical differential equation solver in Mathematica to obtain the solution. The boundary conditions were applied at $|\eta| = 10$.

RESULTS AND DISCUSSION

FIGURE 2 shows the interferograms obtained in a water–glycerine system. A miscible interface was established by placing water on top of glycerine in a quartz cuvette of depth 1 mm. The fringes depict the evolution of the concentration gradient caused by diffusion. It is clear that the location of the maximum concentration gradient moves toward the glycerine side of the interface. The concentration gradient is sharp in the region where the concentration of glycerine is high, and not as sharp in the region where the concentration of water is high. The diffusion is asymmetric, and more mixing is evident in the water side of the interface.

FIGURE 3 shows the traces of the fringes from panels (c) to (i) of FIGURE 2. The second fringe from the bottom has been traced and superimposed so that the reference mark on the cuvette is in a fixed location. This figure is in good qualitative agreement with the results from Petitjeans and Maxworthy.[5]

FIGURE 4 shows the computed concentration gradient profile for a case with $D_0 = 2.5 \times 10^{-6}$ cm^2/sec and $a = -0.95$. The figure shows the extent to which the concentration gradient profile is broadened, and the peak gradient reduced, by diffusion. The calculations show that the location of the peak gradient moves in the direction of decreasing values of the diffusion coefficient. In the experiments, this corresponds to the direction of increasing viscosity, and is consistent with the results in FIGURE 3.

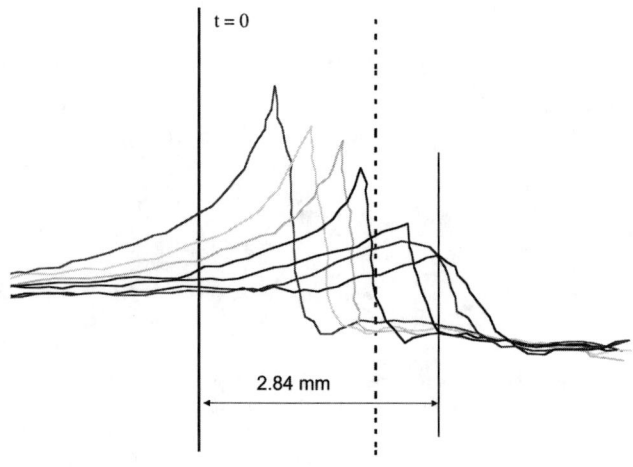

FIGURE 3. Traces of fringe two from the bottom of panels (c) to (i) of FIGURE 2.

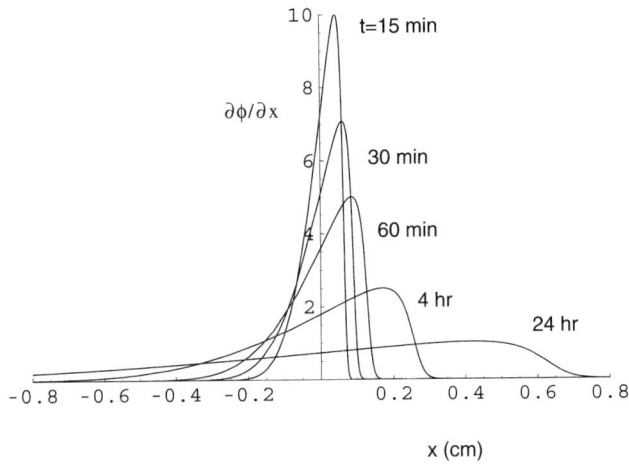

FIGURE 4. Diffusion for the case $D(\phi) = D_0(1 + a\phi)$, with $D_0 = 2.5 \times 10^{-6}$ cm²/sec and $a = -0.95$. At time $t = 0$, the miscible interface is located at $x = 0$.

FIGURE 5 shows the computed concentration gradient profiles for a case with $D_0 = 2.5 \times 10^{-6}$ cm²/sec and $a = -0.75$. In this case, the profiles are more symmetric compared to the case with $a = -0.95$.

Detailed comparison of the experimental and theoretical results to extract the best fit values for D_0 and a are in progress. Our preliminary estimate is that $D_0 = 6.6 \times 10^{-6}$ cm²/sec, $a = -0.93$ for the experimental curves shown in FIGURE 3.

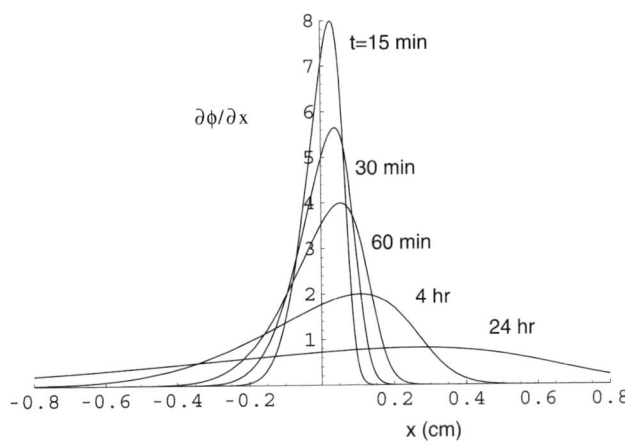

FIGURE 5. Diffusion for the case $D(\phi) = D_0(1 + a\phi)$, with $D_0 = 2.5 \times 10^{-6}$ cm²/sec and $a = -0.75$. At time $t = 0$, the miscible interface is located at $x = 0$.

REFERENCES

1. LINNIK, W. 1933. Simple interferometer for the investigation of optical systems. Comptes Radnus de lAcademic des Sciences d lU.R.SS, **1:** 208–210. Abstract in Z. Instrumentenkd. **54:** 463, 1934.
2. SMARTT, R.N. & J. STRONG. 1972. Point-diffraction interferometer. J. Opt. Soc. Am. **62:** 737.
3. MERCER, C.R., N. RASHIDNIA & K. CREATH. 1996. High data density temperature measurement for quasi steady-state flows. Exp. Fluids **21:** 11–16.
4. MERCER, C.R. & N. RASHIDNIA. 1998. Common-path phase-stepped interferometer for fluid measurements. Proc. Eight International Symposium of Flow Visualization, Flow Visualization VIII. G.M. Carlomango & I. Grant, Eds.: 256.1–256.9.
5. PETITJEANS, P. & T. MAXWORTHY. 1996. Miscible displacements in capillary tubes. Part 1 Experiments. J. Fluid Mech. **326:** 37–56.
6. SOMMERFELD, A. 1964. Optics, Lectures in Physics IV. 347. Academic Press, New York.
7. RASHIDNIA, N., R. BALASUBRAMANIAM, J. KUANG, *et al.* 2001. Measurement of the diffusion coefficient of miscible fluids using both interferometry and Weiner's method. Intl. J. Thermophys. **22**(2): 547–555.
8. KUANG, J., T. MAXWORTHY & P. PETITJEANS. 2002. Measurements of the diffusion coefficient using a modified Weiner's method. Personal communication.
9. RASHIDNIA, N. & R. BALASUBRAMANIAM. 2002. Development of an interferometer for measurement of diffusion coefficient of miscible liquids. Appl. Optics **41**(7): 1337–1342.
10. MERZKIRCH, W. 1987. Flow Visualization. 180–188. Academic Press.
11. JOSEPH, D.D. & Y.Y. RENARDY. 1993. Fundamentals of Two-Fluid Dynamics. Springer-Verlag.
12. JOSEPH, D.D., A. HUANG & H. HU. 1996. Non-solenoidal velocity effects and Korteweg stresses in simple mixtures of incompressible liquids. Physica D **97:** 104–125.
13. CHEN, C.Y. & E. MEIBURG. 1996. Miscible displacements in capillary tubes. Part 2. Numerical simulations. J. Fluid Mech. **326:** 57–90.

Coupled Mean Flow-Amplitude Equations for Nearly Inviscid Parametrically Driven Surface Waves

EDGAR KNOBLOCH,[a] CARLOS MARTEL,[b] AND JOSÉ M. VEGA[b]

[a]*Department of Physics, University of California, Berkeley, California, USA*

[b]*E.T.S.I. Aeronáuticos, Universidad Politécnica de Madrid, Madrid, Spain*

ABSTRACT: Nearly inviscid parametrically excited surface gravity-capillary waves in two-dimensional periodic domains of finite depth and both small and large aspect ratio are considered. Coupled equations describing the evolution of the amplitudes of resonant left- and right-traveling waves and their interaction with a mean flow in the bulk are derived, and the conditions for their validity established. In general the mean flow consists of an inviscid part together with a viscous streaming flow driven by a tangential stress due to an oscillating viscous boundary layer near the free surface and a tangential velocity due to a bottom boundary layer. These forcing mechanisms are important even in the limit of vanishing viscosity, and provide boundary conditions for the Navier–Stokes equation satisfied by the mean flow in the bulk. The streaming flow is responsible for several instabilities leading to pattern drift.

KEYWORDS: Faraday waves; streaming flow; gravity-capillary waves; parametric resonance

INTRODUCTION

Parametrically driven surface gravity-capillary waves in low viscosity fluids are traditionally discussed using a velocity potential formulation. However, this formulation has a serious shortcoming in that it ignores large-scale streaming flows that may be driven by the time-averaged Reynolds stress in the oscillatory boundary layers at the container walls, or at the free surface. We describe a systematic asymptotic technique that includes such flows and show that these flows can interact nontrivially with the waves that are in turn responsible for them. We show that this interaction can lead to new types of instability of parametrically driven waves at finite amplitude, and discuss some of their consequences. We treat only systems with periodic boundary conditions, either in two dimensions or in a cylindrical container, and discuss two cases. In the first the length of the periodic domain is large relative to the wavelength of the instability, and the mean flow contains an important inviscid contribution. In the second the wavelength and container length are comparable and the dynamics of the system is more sensitive to the container shape. In particular, if the shape is perturbed away from circular, the interaction between the streaming flow

Address for correspondence: Edgar Knobloch, Department of Physics, University of California, Berkeley, CA 94720, USA. Voice: 510-642-3395; fax: 510-643-8497.
knobloch@physics.berkeley.edu

and the wave amplitude is enhanced, and complex dynamics may result. These conclusions generalize readily to containers of other shapes suggesting a new class of experiments.

THE FARADAY SYSTEM

Surface gravity-capillary waves or Faraday waves can be excited parametrically by the vertical oscillation of a container.[1-3] We consider a container in the form of a right cylinder with horizontal cross-section Σ, filled level with the brim at $z = 0$. In this geometry the contact line is pinned at the lateral boundary and complications associated with contact line dynamics are reduced. We nondimensionalize the governing equations using the unperturbed depth h as unit of length and the gravity-capillary time $[g/h + T/(\rho h^3)]^{-1/2}$ as unit of time (here g is the gravitational acceleration, T is the coefficient of surface tension, and ρ is the density), and obtain

$$\frac{\partial \mathbf{v}}{\partial t} - \mathbf{v} \times (\nabla \times \mathbf{v}) = -\nabla \Pi + C_g \Delta \mathbf{v}, \quad \nabla \cdot \mathbf{v} = 0, \text{ if } (x, y) \in \Sigma, -1 < z < f,$$

$$\mathbf{v} = 0 \text{ if } z = -1 \text{ or } (x, y) \in \partial\Sigma, \quad f = 0 \text{ if } (x, y) \in \partial\Sigma,$$

$$\mathbf{v} \cdot \mathbf{n} = \frac{\partial f}{\partial t}(\mathbf{e}_z \cdot \mathbf{n}), \quad [(\nabla \mathbf{v} + \nabla \mathbf{v}^T) \cdot \mathbf{n}] \times \mathbf{n} = 0 \text{ at } z = f,$$

$$\Pi - \frac{|\mathbf{v}|^2}{2} - (1 - S)f + S\nabla \cdot \left[\frac{\nabla f}{(1 + |\nabla f|^2)^{1/2}}\right]$$

$$= C_g[(\nabla \mathbf{v} + \nabla \mathbf{v}^T) \cdot \mathbf{n}] \cdot \mathbf{n} - 4\mu\omega^2 f \cos 2\omega t \text{ at } z = f, \quad (1)$$

where $\mathbf{v}$ is the velocity, f is the associated vertical deflection of the free surface (constrained by volume conservation), $\Pi = p + |\mathbf{v}|^2/2 + (1 - S)z - 4\mu\omega^2 z \cos 2\omega t$ is the hydrostatic stagnation pressure, $\mathbf{n}$ is the outward unit normal to the free surface, $\mathbf{e}_z$ is the upward unit vector, and $\partial\Sigma$ denotes the boundary of the cross-section Σ (i.e., the lateral walls). The real parameters $\mu > 0$ and 2ω denote the amplitude and frequency of the forcing. The quantity $C_g \equiv \nu/(gh^3 + Th/\rho)^{1/2}$, where ν is the kinematic viscosity, is a gravity-capillary number and $S \equiv T/(T + \rho g h^2)$ is a gravity-capillary balance parameter; these are related to the usual capillary number $C \equiv \nu\sqrt{\rho/Th}$ and Bond number $B \equiv \rho g h^2/T$ by

$$C_g = \frac{C}{(1 + B)^{1/2}}, \quad S = \frac{1}{1 + B}. \quad (2)$$

The parameter S is such that $0 \leq S \leq 1$ with $S = 0$ and $S = 1$ corresponding to the purely gravitational limit ($T = 0$) and the purely capillary limit ($g = 0$), respectively.

In this paper we consider the (nearly inviscid, nearly resonant, weakly nonlinear) limit

$$C_g \ll 1, \quad |\omega - \Omega| \ll 1, \quad \mu \ll 1, \quad (3)$$

where Ω is an inviscid eigenfrequency of the linearized problem around the flat state. In this case the vorticity contamination of the bulk from the boundary layers at the walls and the free surface remains negligible for times that are not too long, and the flow in the bulk is correctly described by an inviscid formulation but with boundary conditions determined by a boundary layer analysis. In general, this flow consists of an *inviscid* part and a *viscous* part, hereafter called the *streaming flow*. As discussed elsewhere[4,5] the streaming flow enters into the problem because the linearized problem admits *hydrodynamic* (or *viscous*) modes,[6] in addition to the usual *surface modes*. In the nearly inviscid limit the former decay *more slowly* than the surface modes, and so are easily excited, forming the streaming flow. For small C_g, these modes take the form $(\mathbf{v},\Pi,f) = (\mathbf{U}, C_g P, C_g F)\exp(C_g \lambda t) + \ldots$, with the (real) eigenvalue $\lambda < 0$ given by

$$\nabla \cdot \mathbf{U} = 0, \quad \lambda \mathbf{U} = -\nabla P + \Delta \mathbf{U} \text{ if } (x,y) \in \Sigma, -1 < z < 0,$$

$$\mathbf{U} = 0 \text{ if } z = -1 \text{ or } (x,y) \in \partial\Sigma,$$

$$\mathbf{e}_z \cdot \mathbf{U} = 0, \quad [\mathbf{e}_z \cdot (\nabla \mathbf{U} + \nabla \mathbf{U}^T)] \times \mathbf{e}_z = 0 \text{ at } z = 0. \tag{4}$$

The associated (scaled) free surface deflection F is calculated *a posteriori* from the normal stress balance across $z = 0$,

$$S\Delta F - (1-S)F = (-P + [(\nabla \mathbf{U} + \nabla \mathbf{U}^T) \cdot \mathbf{e}_z] \cdot \mathbf{e}_z)_{z=0}, \tag{5}$$

subject to $F = 0$ on $\partial\Sigma$, and $\int_\Sigma F dx dy = 0$. Thus, in contrast to the surface modes, the hydrodynamic modes are nonoscillatory and exhibit $O(C_g)$ free surface deflection. Moreover, these modes decay on an $O(C_g^{-1})$ time scale, in contrast to the $O(C_g^{-1/2})$ time scale of the surface modes, and hence cannot be ignored *a priori* in a weakly nonlinear theory.

MODERATELY LARGE DOMAINS: $L \gg 1$

In this regime it is possible to perform a multiscale analysis of the governing equations using C_g, L^{-1}, and μ as unrelated small parameters. The problem is simplest in two dimensions, where we can use a stream function formulation—that is, we write $\mathbf{v} = (-\psi_z, 0, \psi_x)$. We focus on two well-separated scales in both space ($x \sim 1$ and $x \gg 1$) and time ($t \sim 1$ and $t \gg 1$), and derive equations for small, slowly varying amplitudes A and B of left- and right-propagating waves defined by

$$f = e^{i\omega t}(Ae^{ikx} + Be^{-ikx}) + \gamma_1 A\bar{B}e^{2ikx} + \gamma_2 e^{2i\omega t}(A^2 e^{2ikx} + B^2 e^{-2ikx})$$

$$+ f^+ e^{i\omega t + ikx} + f^- e^{i\omega t - ikx} + \text{c.c.} + f^m + NRT, \tag{6}$$

with similar expressions for the remaining fields. The quantities $f^\pm$ and f^m represent resonant second order terms, whereas NRT denotes nonresonant terms. The superscript m denotes terms associated with the mean flow; f^m depends weakly on time but may depend strongly on x. A systematic expansion procedure[5] then leads to the equations

$$A_t - v_g A_x = i\alpha A_{xx} - (\delta + id)A + i(\alpha_3|A|^2 - \alpha_4|B|^2)A + i\alpha_5\mu\bar{B}$$

$$+ i\alpha_6 \int_{-1}^0 g(z)\langle\psi_z^m\rangle^x dz A + i\alpha_7\langle f^m\rangle^x A, \tag{7}$$

$$B_t + v_g B_x = i\alpha B_{xx} - (\delta + id)B + i(\alpha_3|B|^2 - \alpha_4|A|^2)B + i\alpha_5\mu\bar{A}$$

$$- i\alpha_6 \int_{-1}^0 g(z)\langle\psi_z^m\rangle^x dz B + i\alpha_7\langle f^m\rangle^x B, \tag{8}$$

$$A(x+L, t) \equiv A(x, t), \quad B(x+L, t) \equiv B(x, t). \tag{9}$$

The first seven terms in these equations, accounting for inertia, propagation at the group velocity v_g, dispersion, damping, detuning, cubic nonlinearity and parametric forcing, are familiar from weakly nonlinear, nearly inviscid theories.[7] These theories lead to the expressions

$$v_g = \omega'(k), \quad \alpha = -\omega''(k)/2, \quad \delta = \alpha_1 C_g^{1/2} + \alpha_2 C_g,$$

$$\alpha_1 = \frac{k(\omega/2)^{1/2}}{\sinh 2k}, \quad \alpha_2 = k^2\left[2 + \frac{1+\sigma^2}{4\sinh^2 k}\right],$$

$$\alpha_3 = \frac{\omega k^2[(1-S)(9-\sigma^2)(1-\sigma^2) + Sk^2(7-\sigma^2)(3-\sigma^2)]}{4\sigma^2[(1-S)\sigma^2 - Sk^2(3-\sigma^2)]}$$

$$+ \frac{\omega k^2[8(1-S) + 5Sk^2]}{4(1-S+Sk^2)},$$

$$\alpha_4 = \frac{\omega k^2}{2}\left[\frac{(1-S+Sk^2)(1+\sigma^2)^2}{(1-S+4Sk^2)\sigma^2} + \frac{4(1-S) + 7Sk^2}{1-S+Sk^2}\right],$$

$$\alpha_5 = \omega k\sigma,$$

where $\omega(k) = [(1 - S + Sk^2)k\sigma]^{1/2}$ is the dispersion relation and $\sigma \equiv \tanh k$, that are recovered in the present formulation. In particular, the cubic coefficients coincide with those obtained in *strictly* inviscid formulations.[8–10] The coefficient α_3 diverges at (excluded) resonant wave numbers that satisfy $\omega(2k) = 2\omega(k)$. The detuning d is given by

$$d = \alpha_1 C_g^{1/2} - (2\pi N L^{-1} - k)v_g, \quad N = \text{integer},$$

where the last term represents the mismatch between the wavelength $2\pi/k$ selected by the forcing frequency and the domain length L. The last two terms in Equations **(7)** and **(8)** describe the coupling to the mean flow in the bulk (be it viscous or inviscid in origin) in terms of (a local average $\langle \cdot \rangle^x$ of) the stream function ψ^m for this flow and the associated free surface elevation f^m. The coefficients of these terms and the function g are given by

$$\alpha_6 = k\sigma/2\omega, \quad \alpha_7 = \omega k(1-\sigma^2)/2\sigma,$$

$$g(z) = 2\omega k \cosh[2k(z+1)]/\sinh^2 k, \tag{10}$$

and are real. The new terms are, therefore, conservative, implying that at leading order the mean flow does not extract energy from the system. This result is consistent with the small steepness of the associated surface displacement and its small speed compared with the speed $|\nabla\psi|$ due to the surface waves. The mean flow variables in the bulk depend weakly on time but strongly on both x and z, and evolve according to the equations

$$\Omega_t^m - [\psi_z^m + (|A|^2 - |B|^2)g(z)]\Omega_x^m + \psi_x^m \Omega_z^m = C_g(\Omega_{xx}^m + \Omega_{zz}^m),$$

$$\Omega^m = \psi_{xx}^m + \psi_{zz}^m, \tag{11}$$

with

$$\psi_x^m - f_t^m = \beta_1(|B|^2 - |A|^2)_x, \quad \psi_{zz}^m = \beta_2(|A|^2 - |B|^2),$$

$$(1-S)f_x^m - Sf_{xxx}^m - \psi_{zt}^m + C_g(\psi_{zzz}^m + 3\psi_{xxz}^m) = -\beta_3(|A|^2 + |B|^2)_x \text{ at } z = 0 \quad (12)$$

and

$$\psi_z^m = -\beta_4[iA\bar{B}e^{2ikx} + \text{c.c.} + |B|^2 - |A|^2],$$

$$\int_0^L \Omega_z^m dx = \psi^m = 0 \text{ at } z = -1. \quad (13)$$

Also, $\psi^m(x+L,z,t) \equiv \psi^m(x,z,t)$, $f^m(x+L,t) \equiv f^m(x,t)$, subject to $\int_0^L f^m(x,t)dx = 0$. Here, $\beta_1 = 2\omega/\sigma$, $\beta_2 = 8\omega k^2/\sigma$, $\beta_3 = (1-\sigma^2)\omega^2/\sigma^2$, $\beta_4 = 3(1-\sigma^2)\bar{\omega}k/\sigma^2$. Thus, the mean flow is forced by the surface waves in two ways. The right sides of the boundary conditions **(12a)** and **(12c)** provide a *normal forcing mechanism*; this mechanism is the only one present in strictly inviscid theory[9,11] and does not appear unless the aspect ratio is large. The right sides of the boundary conditions **(12b)** and **(13a)** describe two *shear forcing mechanisms*, a tangential stress at the free surface[12] and a tangential velocity at the bottom wall.[13] Note that neither of these forcing terms vanishes in the limit of small viscosity (i.e., as $C_g \to 0$). The shear nature of these forcing terms leads us to retain the viscous term in **(11a)** even when C_g is quite small. In fact, when C_g is very small, the effective Reynolds number of the mean flow is quite large. Thus, the mean flow itself generates additional boundary layers near the top and bottom of the container, and these must be thicker than the original boundary layers for the validity of the analysis. This puts an additional restriction on the validity of the equations.[5] There is a third, less effective but inviscid, volumetric forcing mechanism associated with the second term in the vorticity equation **(11a)** that looks like a horizontal force $(|A|^2 - |B|^2)g(z)\Omega^m$ and is sometimes called the *vortex force*. Although this term vanishes in the absence of mean flow, it can change the stability properties of the flow and enhance or limit the effect of the remaining forcing terms.

In the following we refer to Equations **(7)–(9)** and **(11)–(13)** as the general coupled amplitude–mean-flow (GCAMF) equations. These equations differ from the exact equations forming the starting point for the analysis in three essential simplifications: the fast oscillations associated with the surface waves have been filtered out, the effect of the thin primary viscous boundary layers is replaced by effective boundary conditions on the flow in the bulk—viz. **(12b)** and **(13a)**—and the surface boundary conditions are applied at the unperturbed location of the free surface—viz. $z = 0$. Thus, only the much broader (secondary) boundary layers associated with the (slowly varying) streaming flow need to be resolved in any numerical simulation.

GRAVITY-CAPILLARY WAVES IN MODERATELY LARGE ASPECT RATIO CONTAINERS

The GCAMF equations describe small amplitude slowly varying wavetrains whenever the parameters C_g, L^{-1}, and μ are small, but otherwise unrelated to one another. Any relation between them will therefore lead to further simplification. To derive such simplified equations we consider the distinguished limit (see the shaded region in FIGURE 1A).

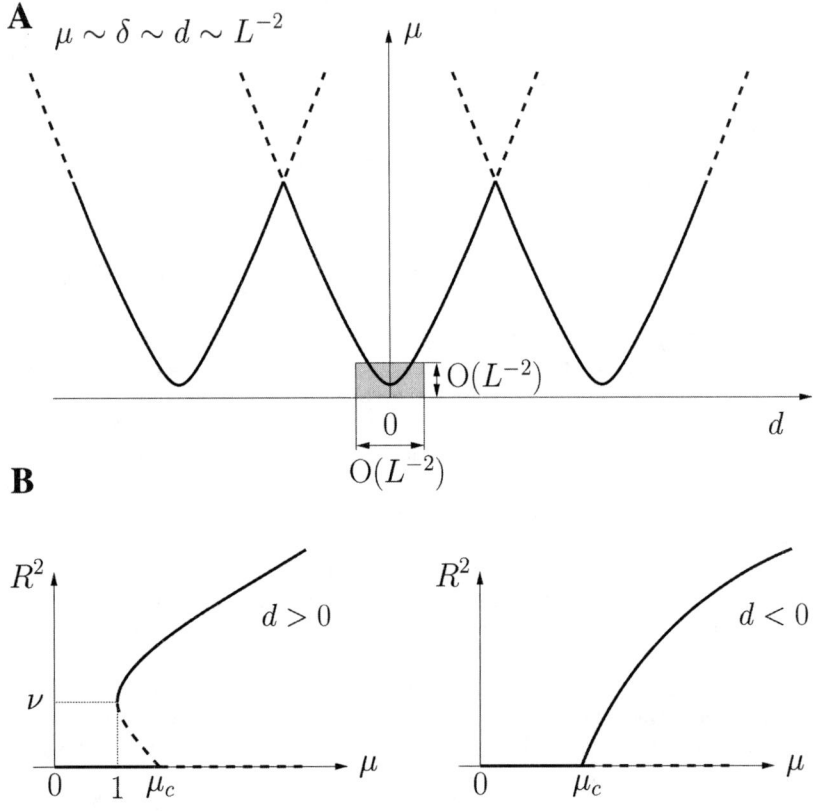

FIGURE 1. (A) Sketch of the primary resonance tongue in the scaling regime of interest. (B) Bifurcation diagrams for the resulting standing waves near onset.

$$\frac{\delta L^2}{\alpha} = \Delta \sim 1, \quad \frac{dL^2}{\alpha} = D \sim 1, \quad \frac{\mu L^2}{\alpha} \equiv M \sim 1, \tag{14}$$

with $k = O(1)$ and $|\ln C_g| = O(1)$. The simplified equations will then be formally valid for $1 \ll L \ll C_g^{-1/2}$ if $k \sim 1$. These are derived under the assumption $1 - S \sim 1$ using a multiple scale method with x and t as *fast* variables and

$$\zeta \equiv \frac{x}{L}, \quad \tau \equiv \frac{t}{L}, \quad T \equiv \frac{t}{L^2} \tag{15}$$

as *slow* variables. In terms of these variables the local horizontal average $\langle \cdot \rangle^x$ becomes an average over the fast variable x. Note that assumption (14) imposes an implicit relation between L and C_g. When $1 - S \sim 1$ the nearly inviscid and viscous mean flows can be clearly distinguished from one another and the viscous mean flow can be identified by taking appropriate averages of the entire mean flow over the intermediate time scale τ, that is, the mean flow variables ψ^m, Ω^m, and f^m take the form

$$\psi^m(x, z, \zeta, \tau, T) = \psi^v(x, z, \zeta, T) + \psi^i(x, z, \zeta, \tau, T),$$
$$\Omega^m(x, z, \zeta, \tau, T) = \Omega^v(x, z, \zeta, T) + \Omega^i(x, z, \zeta, \tau, T), \qquad (16)$$
$$f^m(x, \zeta, \tau, T) = f^v(x, \zeta, T) + f^i(x, \zeta, \tau, T),$$

with integrals over τ of ψ_x^i, ψ_ζ^i, ψ_z^i, Ω^i, and f^i required to be bounded as $\tau \to \infty$. Thus, the nearly inviscid mean flow is *purely oscillatory* on the time scale τ. Since its frequency is of the order of L^{-1} (see Eq. (15)), which is large compared with C_g, the inertial term for this flow is large in comparison with the viscous terms (see Eq. (11)), except in two *secondary* boundary layers, of thickness of the order of $(C_g L)^{1/2}$ ($\ll 1$), attached to the bottom plate and the free surface. Note that, as required for the consistency of the analysis, these boundary layers are much thicker than the *primary* boundary layers associated with the surface waves (see FIGURE 2), which provide the boundary conditions (12) and (13) for the mean flow. Moreover, the width of these secondary boundary layers remains small as $\tau \to \infty$ and (to leading order) the vorticity of this nearly inviscid mean flow remains confined to these boundary layers. Note that without the requirement that the inviscid flow be purely oscillatory, the vorticity would diffuse out of these boundary layers and affect the structure of the whole *nearly inviscid* solution even at leading order. In fact, vorticity does diffuse (and is convected) away from the boundary layers, but this vorticity transport is included in the viscous mean flow. The vorticity associated with the nearly inviscid mean flow is at most of the order of

$$||A|^2 - |B|^2| \text{ and } (|A|^2 + |B|^2)(C_g L)^{-1/2} \qquad (17)$$

in the upper and lower secondary boundary layers, respectively; the jump in the associated stream function ψ^i across each boundary layer is $O(C_g L)$ times smaller, and affects only higher order terms; as a consequence the secondary boundary layers

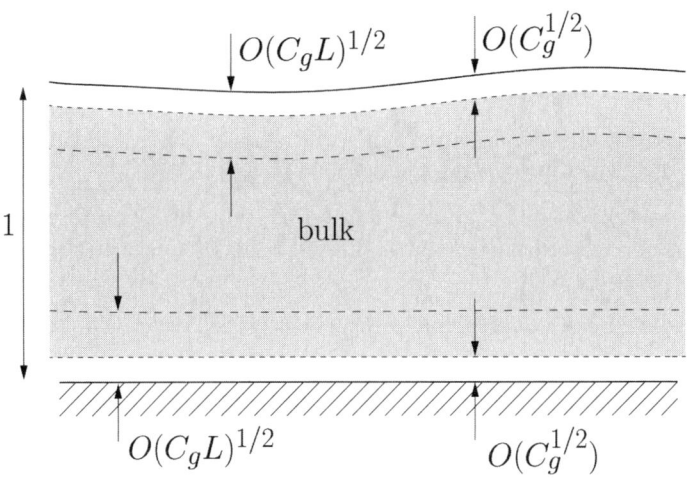

FIGURE 2. Sketch of the primary and secondary boundary layers, indicating their widths in comparison with the layer depth.

can be completely ignored and no additional contributions to the boundary conditions on the nearly inviscid flow need be included in (12) and (13). Outside of these boundary layers, the complex amplitudes and the flow variables associated with the nearly inviscid mean flow are expanded as

$$(A, B) = L^{-1}(A_0, B_0) + L^{-2}(A_1, B_1) + \ldots ,$$
$$(\psi^v, \Omega^v) = L^{-2}(\phi_0^v, W_0^v) + \ldots , \quad f^v = L^{-3}F_0^v + \ldots ,$$
$$(\psi^i, f^i) = L^{-2}(\phi_0^i, F_0^i) + L^{-3}(\phi_1^i, F_1^i) + \ldots ,$$
$$\Omega^i = L^{-3}W_0^i + \ldots .$$
(18)

Substitution of (14)–(16) and (18) into (7)–(13) leads to the following:

(1) From (11)–(13), at leading order,

$$\phi^i_{0xx} + \phi^i_{0zz} = 0 \text{ in } -1 < z < 0,$$

$$\phi^i_0 = 0 \text{ at } z = -1, \quad \phi^i_{0x} = 0 \text{ at } z = 0,$$

together with $F^i_{0x} = 0$. Thus,

$$\phi^i_0 = (z+1)\Phi^i_0(\zeta, \tau, T), \quad F^i_0 = F^i_0(\zeta, \tau, T).$$
(19)

At second order, the boundary conditions (12a) and (12c) yield

$$\phi^i_{1x}(x, 0, \zeta, \tau, T) = F^i_{0\tau} - \Phi^i_{0\zeta} + \beta_1(|B_0|^2 - |A_0|^2)_\zeta$$

$$(1-S)F^i_{1x} - SF^i_{1xxx} = \Phi^i_{0\tau} - (1-S)F^i_{0\zeta} - \beta_3(|A_0|^2 + |B_0|^2)_\zeta$$

at $z = 0$. Since the right hand sides of these two equations are independent of the fast variable x and both ϕ^i_1 and F^i_1 must be bounded in x, it follows that

$$\Phi^i_{0\zeta} - F^i_{0\tau} = \beta_1(|B_0|^2 - |A_0|^2)_\zeta,$$
$$\Phi^i_{0\tau} - v_p^2 F^i_{0\zeta} = \beta_3(|A_0|^2 + |B_0|^2)_\zeta,$$
(20)

where

$$v_p = (1-S)^{1/2}$$
(21)

is the phase velocity of long wavelength surface gravity waves. Equations (20) must be integrated with the following additional conditions

$$\Phi^i_0(\zeta + 1, \tau, T) \equiv \Phi^i_0(\zeta, \tau, T), \quad F^i_0(\zeta + 1, \tau, T) \equiv F^i_0(\zeta, \tau, T),$$
(22)

and the requirements that integrals over τ of $\Phi^i_{0\zeta}$ and F^i_0 remain bounded as $\tau \to \infty$, with $\int_0^1 F^i_0 d\zeta = 0$.

(2) The leading order contributions to Equations (7) and (8) yield

$$A_{0\tau} - v_g A_{0\zeta} = B_{0\tau} + v_g B_{0\zeta} = 0.$$

Thus,

$$A_0 = A_0(\xi, T), \quad B_0 = B_0(\eta, T),$$
(23)

where ξ and η are the characteristic variables

$$\xi = \zeta + v_g \tau, \quad \eta = \zeta - v_g \tau.$$
(24)

Moreover, according to (9),

$$A_0(\xi + 1, T) \equiv A_0(\xi, T), \quad B_0(\eta + 1, T) \equiv B_0(\eta, T). \tag{25}$$

Substitution of these expressions into **(20)** followed by integration of the resulting equations yields

$$\Phi_0^i = \frac{\beta_1 v_p^2 + \beta_3 v_g}{v_g^2 - v_p^2}[|A_0|^2 - |B_0|^2 - \langle|A_0|^2 - |B_0|^2\rangle^\zeta] \tag{26}$$
$$+ v_p[F^+(\zeta + v_p\tau, T) - F^-(\zeta - v_p\tau, T)],$$

$$F_0^i = \frac{\beta_1 v_g + \beta_3}{v_g^2 - v_p^2}[|A_0|^2 + |B_0|^2 - \langle|A_0|^2 + |B_0|^2\rangle^\zeta] \tag{27}$$
$$+ F^+(\zeta + v_p\tau, T) + F^-(\zeta - v_p\tau, T),$$

where $\langle\cdot\rangle^\zeta$ denotes the mean value in the slow spatial variable ζ, that is,

$$\langle G\rangle^\zeta = \int_0^1 G\,d\zeta, \tag{28}$$

and the functions $F^\pm$ are such that

$$F^\pm(\zeta + 1 \pm v_p\tau, T) \equiv F^\pm(\zeta \pm v_p\tau, T), \quad \langle F^\pm\rangle^\zeta = 0. \tag{29}$$

The particular solution of **(26)** and **(27)** yields the usual inviscid mean flow included in nearly inviscid theories;[9] the averaged terms are a consequence of volume conservation[9] and the requirement that the nearly inviscid mean flow has a zero mean on the time scale τ; the latter condition is never imposed in strictly inviscid theories but is essential in the limit we are considering, as explained above. To avoid the breakdown of the solution **(26)** and **(27)** at $v_p = v_g$ we assume that

$$|v_p - v_g| \sim 1. \tag{30}$$

The functions $F^\pm$ remain undetermined at this stage. In fact, they are not needed below because the evolution of both the viscous mean flow and of the complex amplitudes is decoupled from these functions. However, at next order one finds that $F^\pm$ remain constant on the time scale T, but decay exponentially due to viscous effects (resulting from viscous dissipation in the secondary boundary layer attached to the bottom plate) on the time scale $t \sim (L/C_g)^{1/2}$.

(3) The evolution equations for A_0 and B_0 on the time scale T are readily obtained from Equations **(7)–(9)**, invoking **(14)**, **(16)**, **(26)**, **(27)**, **(29)**, and eliminating secular terms (i.e., requiring $|A_1|$ and $|B_1|$ to be bounded on the time scale τ):

$$A_{0T} = i\alpha A_{0\xi\xi} - (\Delta + iD)A_0 + i[(\alpha_3 + \alpha_8)|A_0|^2 - \alpha_8\langle|A_0|^2\rangle^\xi - \alpha_4\langle|B_0|^2\rangle^\eta]A_0$$
$$+ i\alpha_5 M\langle\overline{B_0}\rangle^\eta + i\alpha_6\int_{-1}^0 g(z)\langle\langle\phi_{0z}^\nu\rangle^x\rangle^\zeta dz A_0,$$

$$B_{0T} = i\alpha B_{0\eta\eta} - (\Delta + iD)B_0 + i[(\alpha_3 + \alpha_8)|B_0|^2 - \alpha_8\langle|B_0|^2\rangle^\eta - \alpha_4\langle|A_0|^2\rangle^\xi]B_0$$
$$+ i\alpha_5 M\langle\overline{A_0}\rangle^\xi - i\alpha_6\int_{-1}^0 g(z)\langle\langle\phi_{0z}^\nu\rangle^x\rangle^\zeta dz B_0, \tag{31}$$

subject to **(25)**. Here ξ and η are the comoving variables defined in **(24)**, and $\langle\cdot\rangle^x$, $\langle\cdot\rangle^\zeta$, $\langle\cdot\rangle^\xi$, and $\langle\cdot\rangle^\eta$ denote mean values over the variables x, ζ, ξ, and η, respectively. Note that ζ averages over functions of A_0 are equivalent to ξ averages, whereas those over functions of B_0 are equivalent to η averages.

The real coefficient α_8 is given by

$$\alpha_8 = \frac{\alpha_6(2\omega/\sigma)(\beta_1 v_p^2 + \beta_3 v_g) + \alpha_7(\beta_1 v_g + \beta_3)}{v_g^2 - v_p^2}. \tag{32}$$

Equations **(31)** are independent of $F^{\pm}$ because of the second condition in **(29)**.

When $\Delta > 0$ Equations **(31)** can be used to show that

$$\langle |A_0|^2 - |B_0|^2 \rangle^\tau = \langle |A_0|^2 \rangle^\xi - \langle |B_0|^2 \rangle^\eta \to 0 \text{ as } T \to \infty$$

and the result used to simplify the equations for the viscous mean flow in the long time limit:

$$W_{0T}^v - \phi_{0z}^v W_{0x}^v + \phi_{0x}^v W_{0z}^v = Re^{-1}(W_{0xx}^v + W_{0zz}^v),$$

$$\phi_{0xx}^v + \phi_{0zz}^v = W_0^v \text{ in } -1 < z < 0, \tag{33}$$

$$\phi_{0x}^v = \phi_{0zz}^v = 0 \text{ at } z = 0, \tag{34}$$

$$\langle\langle W_{0z}^v \rangle^x \rangle^\zeta = \phi_0^v = 0, \; \phi_{0z}^v = -\beta_4[i\langle A_0 \bar{B}_0 \rangle^\tau e^{2ikx} + \text{c.c.}] \text{ at } z = -1, \tag{35}$$

$$\phi_0^v(x + L, \zeta + 1, z, T) \equiv \phi_0^v(x, \zeta, z, T), \tag{36}$$

where

$$Re = (C_g L^2)^{-1} \tag{37}$$

is the *effective Reynolds number* associated with this flow.

Several remarks about these equations and boundary conditions are now in order.
1. The viscous mean flow is driven by the short gravity-capillary waves through the inhomogeneous term in the boundary condition **(35c)**. Since $\langle A_0 \bar{B}_0 \rangle^\tau$ depends on both ζ and T (unless either A_0 or B_0 is spatially uniform) the boundary condition implies that ϕ_0^v (and hence W_0^v) depends on both the fast and slow horizontal spatial variables x and ζ. This dependence cannot be obtained in closed form, and one must, therefore, resort to numerical computations for realistically large values of L.
2. Higher order oscillatory terms omitted from the boundary condition **(35c)** oscillate on the intermediate time scale τ, and hence generate secondary boundary layers. However, the contributions from these boundary layers are all subdominant and have no effect on the streaming flow at leading order. Moreover, the free-surface deflection accompanying the viscous mean flow is also small, $f^v \sim L^{-3}$ (see Eq. **(18)**), and so plays no role in the evolution of this flow, as expected of a flow involving the excitation of viscous modes.
3. The dominant forcing of the viscous mean flow comes from the lower boundary. This forcing vanishes exponentially when $k \gg 1$ leaving only a narrow range of wave numbers within which such a mean flow is forced when $\delta = O(C_g)$.[5] Thus, in most cases in which a viscous mean flow is present, one may assume that $\delta = O(C_g^{1/2})$. Note, however, that in fully three-dimensional situations[14] in which lateral walls are included a viscous mean flow will be present even when $k \gg 1$ because the forcing of the mean flow in the oscillatory boundary layers along these walls remains.
4. According to the scaling **(14)** and the definition of δ the effective Reynolds number Re is large, and ranges from logarithmically large values if $k \sim |\ln C_g|$ to $O(C_g^{-1/2})$ if $k \sim 1$. However, even in the latter limit we must retain the viscous terms

in (**33a**) in order to account for the boundary conditions (**34b**) and (**35c**). Of course, if $Re \gg 1$ vorticity diffusion is likely to be confined to thin layers, but the structure and location of all these layers cannot be anticipated in any obvious way, and one must again rely on numerical computations.

5. Note that the change of variables

$$A_0 = \tilde{A}_0 e^{-ik\theta}, \quad B_0 = \tilde{B}_0 e^{ik\theta}, \tag{38}$$

where

$$\theta'(T) = -\alpha_6 k^{-1} \int_{-1}^{0} g(z) \langle \langle \phi_{0z}^y \rangle^x \rangle \zeta dz, \tag{39}$$

reduces Equations (**31**) to the much simpler form

$$\tilde{A}_{0T} = i\alpha \tilde{A}_{0\xi\xi} - (\Delta + iD)\tilde{A}_0$$
$$+ i[(\alpha_3 + \alpha_8)|\tilde{A}_0|^2 - (\alpha_4 + \alpha_8)\langle|\tilde{B}_0|^2\rangle^\eta]\tilde{A}_0 + i\alpha_5 M \langle \overline{\tilde{B}_0} \rangle^\eta,$$

$$\tilde{B}_{0T} = i\alpha \tilde{B}_{0\eta\eta} - (\Delta + iD)\tilde{B}_0$$
$$+ i[(\alpha_3 + \alpha_8)|\tilde{B}_0|^2 - (\alpha_4 + \alpha_8)\langle|\tilde{A}_0|^2\rangle^\xi]\tilde{B}_0 + i\alpha_5 M \langle \overline{\tilde{A}_0} \rangle^\xi,$$

$$\tilde{A}_0(\xi + 1, T) \equiv \tilde{A}_0(\xi, T), \quad \tilde{B}_0(\eta + 1, T) \equiv \tilde{B}_0(\eta, T), \tag{40}$$

from which the mean flow is absent. This *decoupling* is a special property of the regime defined by Equation (**14**), but is not unique to it, as discussed further below. The resulting equations provide perhaps the simplest description of the Faraday system at large aspect ratio, and it is for this reason that they have been extensively studied.[15] FIGURE 3A shows a typical bifurcation diagram obtained from (**40**) when $D = 0$; FIGURE 3B shows the corresponding results obtained for *standing waves*, that is, imposing the requirement that $\tilde{A}_0 = \tilde{B}_0$ ($= \tilde{C}_0$, say) identically. FIGURE 1B sketches the consequence of including a small but nonzero detuning. These diagrams show the L^2 norms of $\tilde{A}_0$, $\tilde{B}_0$, or $\tilde{C}_0$ (after transients have died out) at successive intersections of a trajectory with the hypersurfaces

$$\|\tilde{A}_0\|_{L^2}^2 + \|\tilde{B}_0\|_{L^2}^2 = \mu(\langle \overline{\tilde{A}_0} \rangle \langle \tilde{B}_0 \rangle + \text{c.c.}), \text{ and } \|\tilde{C}_0\|_{L^2}^2 = \frac{1}{2}\mu(\langle \overline{\tilde{C}_0} \rangle^2 + \text{c.c.}),$$

respectively. A finite number of points therefore indicates a steady or a periodic solution, whereas scattered points indicate a chaotic trajectory. The origin of such *amplitude chaos* is discussed elsewhere.[16] In both cases the primary instability is to spatially uniform standing waves (shown by a dashed line when unstable), followed by a supercritical secondary bifurcation to a pattern of spatially nonuniform standing waves, and then a Hopf bifurcation. However, subsequent bifurcations differ as instabilities that break the equality $\tilde{A}_0 = \tilde{B}_0$ set in, producing stable waves that are *not* standing. As discussed elsewhere[15] instabilities that break the equality $\tilde{A}_0 = \tilde{B}_0$ are also responsible for introducing a type of drift into the dynamics, but the origin of this drift is entirely different from that discussed below, being a drift in the temporal phase of the oscillations in contrast to a drift in the spatial phase. The latter drift is described by the decoupled phase-mean flow equations, (**33**)–(**36**) and (**39**) that can exhibit complex dynamics in their own right (see below); this drift can be present regardless of whether $\tilde{A}_0 = \tilde{B}_0$ or not.

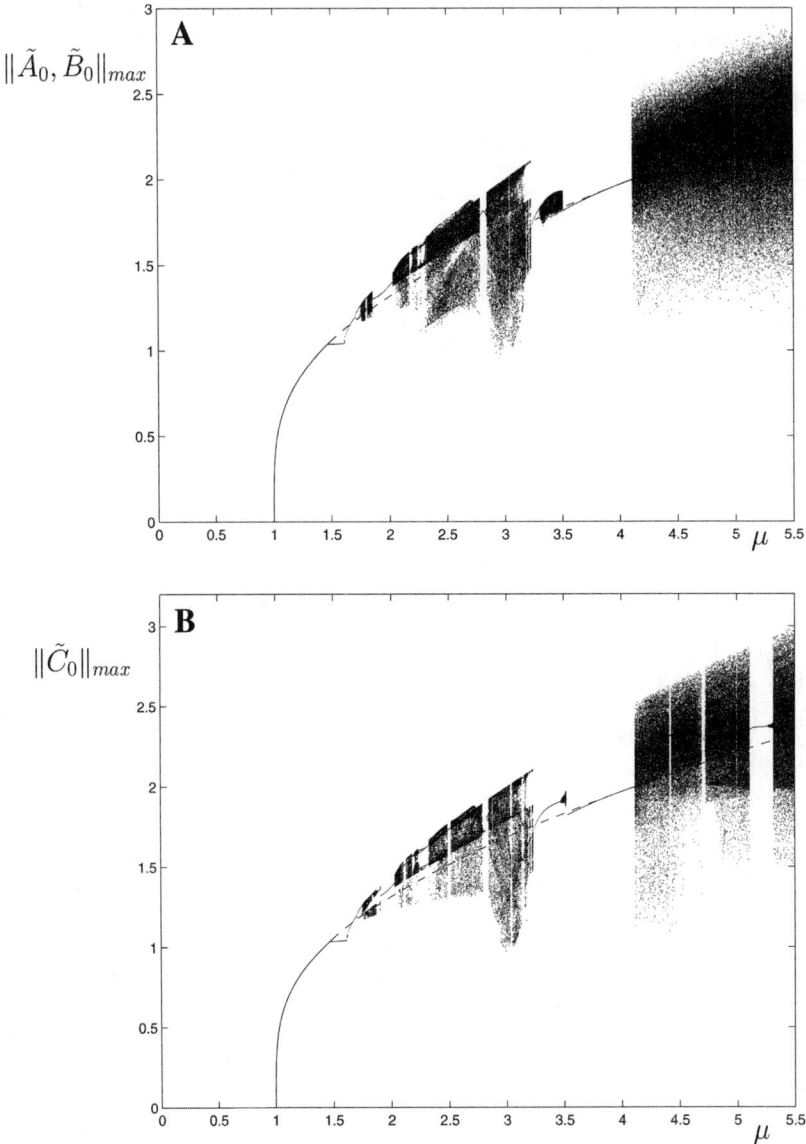

FIGURE 3. Bifurcation diagrams for Equations **(40)** showing successive maxima of the L^2 norm as a function of $\mu \equiv \alpha_5 M$ when $\Delta = 1$, $D = 0$, $\alpha = 0.1$, $\alpha_3 + \alpha_8 = 1.5$, and $\alpha_4 + \alpha_8 = 0.5$. After Martel et al.[15]

SMALL DOMAINS: $L \sim 1$

The theory described above simplifies substantially in small domains, because of the absence of the slow spatial scale ζ and the advection time scale τ. Although the basic set up of the calculation is now quite different, the results are (almost) identical to those just described. This time[17]

$$A_t = -(\delta + id)A + i(\alpha_3|A|^2 - \alpha_4|B|^2)A + i\alpha_5\mu\bar{B} - i\alpha_6 L^{-1}\int_{-1}^{0}\int_0^L g(z)u^v dxdz A,$$

$$B_t = -(\delta + id)B + i(\alpha_3|B|^2 - \alpha_4|A|^2)B + i\alpha_5\mu\bar{A} + i\alpha_6 L^{-1}\int_{-1}^{0}\int_0^L g(z)u^v dxdz B, \quad (41)$$

with A and B spatially constant and the coefficients given by expressions that are identical to those in **(7)** and **(8)**. In these equations t denotes a slow time whose magnitude is determined by the damping $\delta > 0$ and the detuning d, both assumed to be of the same order as the forcing amplitude μ; in the long time limit $|A|^2 = |B|^2 = R^2$. It follows that the mean flow $(u^v(x,z,t), w^v(x,z,t))$ is now entirely viscous in origin, and obeys a two-dimensional Navier–Stokes equation of the form **(33)**. If we absorb the standing wave amplitude R (and some other constants) in the definition of the Reynolds number

$$Re \equiv 2\beta_4 R^2/C_g, \quad (42)$$

this equation is to be solved subject to the boundary conditions

$$u^v = -\sin[2k(x-\theta)], \quad w^v = 0 \text{ at } z = -1,$$
$$u^v_z = 0, \quad w^v = 0 \text{ at } z = 0. \quad (43)$$

Using the dimensional values of the amplitude of the waves A_d, frequency ω_d, wave number k_d, kinematic viscosity ν, and container depth h, the Reynolds number **(42)** of the streaming flow can also be written as

$$Re = \frac{\omega_d A_d^2}{\nu}\frac{6k_d h}{\sinh^2(k_d h)}.$$

Because the structure of Equations **(41)** is identical to that of Equations **(31)** the change of variables **(38)** leads to a decoupling of the amplitudes from the spatial phase θ, which now satisfies the equation

$$\theta_t = \frac{\alpha_6}{kL}\int_{-1}^{0}\int_0^L g(z)u^v(x,z,t)dxdz. \quad (44)$$

FIGURE 4 summarizes the solutions of this coupled phase-streaming flow problem as a function of the wave amplitude R, showing the maxima and minima in the drift speed θ_t as a function of the Reynolds number **(42)** for $k = 2.37$ and $L \equiv 2\pi/k = 2.65$. The primary solution that sets in at $\mu = \mu_c$ consists of stable standing waves with $|A|^2 = |B|^2 = R^2$, defined up to a constant spatial phase θ. The streamlines of the associated streaming flow are shown in FIGURE 5A for $Re = 260$, and take the form of an array of reflection-symmetric counter-rotating eddies with spatial period $L/2$. Such reflection-symmetric states do not drift: $\theta_t \equiv 0$. At $Re = Re^1 \approx 270$ this reflection-symmetric state loses stability at a (secondary) symmetry-breaking Hopf bifurcation that produces *direction-reversing* waves.[18] In this state the standing waves drift periodically, first to the left and then to the right, with no net displacement, but the

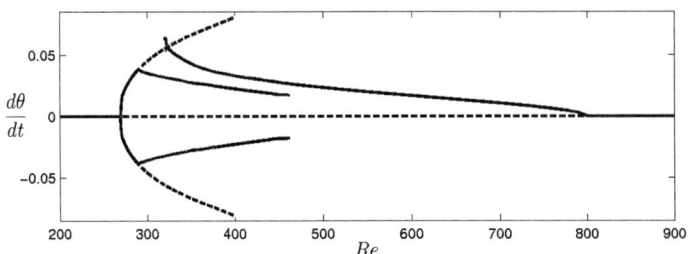

FIGURE 4. Bifurcation diagram for the coupled phase–streaming flow equations showing the drift speed $d\theta/dt$ as a function of the Reynolds number (**42**) for $k = 2.37$ ($L = 2.65$). The secondary Hopf and parity-breaking bifurcations occur at $Re \approx 270$ and $Re \approx 800$, respectively.

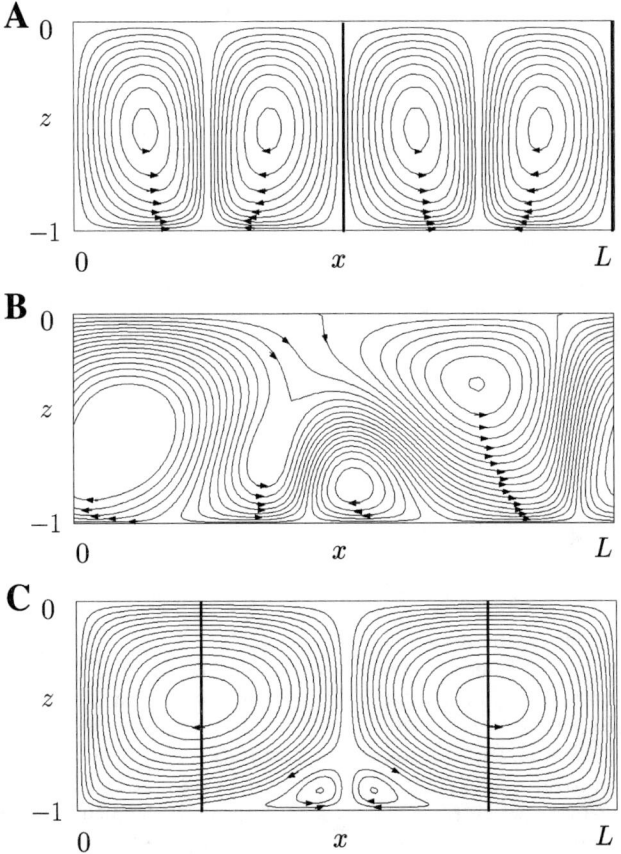

FIGURE 5. Streamlines of the streaming flow when (**A**) $Re = 260$, (**B**) $Re = 325$, and (**C**) $Re = 850$. Vertical lines indicate the location of the nodes of the standing wave.

spatial period $L/2$ of the standing waves is preserved. This bifurcation is supercritical and the direction-reversing waves are, therefore, stable.

The onset of this instability is insensitive to the domain length L once this is large enough. These oscillations lose stability at $Re \approx 291.5$ to a new family of stable direction-reversing waves, this time with spatial period L. These new solutions in turn become unstable at $Re \approx 466$ where a subcritical bifurcation takes the system to a new branch of steadily drifting solutions. These solutions drift *either* to the left *or* to the right, and are stable in the interval $323 \approx Re^3 < Re < Re^2 \approx 800$. In FIGURE 4 only the waves drifting to the right ($\theta_t > 0$) are shown; the corresponding streaming flow streamlines are shown (in the comoving frame) in FIGURE 5B. The bifurcation at $Re = Re^2$ is a parity-breaking steady state bifurcation; at this bifurcation the drift ceases, and a reflection-symmetric streaming flow is restored, but now with spatial period L (FIG. 5C). This flow is stable, but in FIGURE 4 is projected on top of the unstable original symmetric state with period $L/2$. For larger L (e.g., $L = 4\pi/k$) several different stable states coexist, some of which lack reflection symmetry, and exhibit complex time-dependence.[17] Note that neither type of drift is associated with the loss of reflection-symmetry of the waves themselves, at least at leading order; instead they are due to a symmetry-breaking instability of the associated streaming flow.

EXCITATION OF STREAMING FLOWS IN NEARLY CIRCULAR DOMAINS

The results just described are characteristic of domains with periodic boundary conditions, such as circular domains. In such domains the amplitudes decouple from the spatial phase and the streaming flow only causes drift motion in an otherwise standing wave pattern. However, the instabilities found set in only at sufficiently large Reynolds number, necessitating a numerical solution of the coupled phase-streaming flow problem. There is a simple way, however, of pushing these instabilities to much smaller amplitudes and, hence, into a regime where analytic progress is possible. This occurs as soon as the translation (rotation) invariance of the container is broken (eg., by deforming it slightly), thereby forcing the wave amplitudes to couple to the spatial phase. In systems undergoing a symmetry-breaking primary Hopf bifurcation (but with no parametric forcing) this type of forced symmetry-breaking is known to lead to rich dynamics,[19–21] and the present system is no exception. To this end we note that the dominant conservative term that breaks the symmetry $(A, B) \to (e^{-i\phi}A, e^{i\phi}B)$ arising from rotation invariance of the system takes the form $i\Lambda(B,A)$, where Λ is a real coefficient; the addition of this term to the equations breaks rotational invariance, but preserves a plane of reflection symmetry. In the following we assume that $\Lambda \sim \delta \sim d \sim \mu$, and project the streaming flow onto the first hydrodynamic mode. This assumption requires that the effective Reynolds number (42) of the streaming flow be sufficiently small that the nonlinear terms in (33) may be neglected. In the following we denote the (real) amplitude of this mode by v_1. We obtain[22]

$$v_1'(\tau) = \varepsilon(-v_1 + |B|^2 - |A|^2), \tag{45}$$

$$A_t = [-\delta - id + i(\alpha_3|A|^2 - \alpha_4|B|^2)]A + i\alpha_5\mu\bar{B} + i\Lambda B - i\gamma v_1 A, \qquad (46)$$

$$B_t = [-\delta - id + i(\alpha_3|B|^2 - \alpha_4|A|^2)]B + i\alpha_5\mu\bar{A} + i\Lambda A + i\gamma v_1 B, \qquad (47)$$

where $\varepsilon = -\lambda_1 Re^{-1} > 0$ and $\lambda_1 < 0$ is the first hydrodynamic eigenvalue. These equations illustrate well the interplay between breaking of rotation invariance ($\Lambda \neq 0$) and the excitation of streaming flow ($\gamma \neq 0$). When $\Lambda = \gamma = 0$ they reduce to the exact, but decoupled, amplitude equations for A and B that follow from Equations (41) on using (38).

Equations (45)–(47) have solutions in the form of reflection-symmetric standing waves $(A,B) \equiv (C,C)$, $v_1 = 0$, where

$$\delta^2 + [d - \Lambda - (\alpha_3 - \alpha_4)|C|^2]^2 = \alpha_5^2\mu^2, \quad \alpha_3 - \alpha_4 \neq 0; \qquad (48)$$

this state sets in at $\mu = \mu_c$, where

$$\mu_c \equiv \frac{[\delta^2 + (d - \Lambda)^2]^{1/2}}{\alpha_5}. \qquad (49)$$

The amplitude $R \equiv |C|$ increases monotonically for $\mu > \mu_c$ provided $(d - \Lambda)/(\alpha_3 - \alpha_4) \leq 0$; if $(d - \Lambda)/(\alpha_3 - \alpha_4) > 0$ the branch bifurcates subcritically at $\mu = \mu_c$ before turning around towards larger μ at a secondary saddle-node bifurcation.

To determine the linear stability of these states we replace A, B, and v_1 by $C + X_\pm e^{\lambda t} + \bar{Y}_\pm e^{\bar{\lambda}t}$ and $Ze^{\lambda t}$ + c.c., respectively, and linearize. The resulting equations have *even* eigenmodes ($X_+ = X_-$, $Y_+ = Y_-$, $Z = 0$) and *odd* eigenmodes ($X_+ = -X_-$, $Y_+ = -Y_-$, $Z \neq 0$). The former preserve the reflection symmetry of C and are characterized by the dispersion relation

$$(\lambda + \delta)^2 + [d - \Lambda - 2(\alpha_3 - \alpha_4)|C|^2]^2 = \delta^2 + (d - \Lambda)^2. \qquad (50)$$

Such instabilities are, therefore, always nonoscillatory and correspond either to the primary bifurcation at $\mu = \mu_c$ or to the secondary saddle-node bifurcation at

$$|C|^2 = \frac{d - \Lambda}{\alpha_3 - \alpha_4}. \qquad (51)$$

In contrast, the symmetry-breaking perturbations satisfy a cubic dispersion relation,

$$\lambda^2 + 2\delta\lambda + 4d\Lambda - 8\Lambda\left[\frac{\gamma\varepsilon}{\lambda + \varepsilon} + \alpha_3\right]|C|^2 = 0, \qquad (52)$$

with a steady state bifurcation when

$$|C|^2 = \frac{d}{2(\alpha_3 + \gamma)}, \text{ if } d\Lambda \neq 0, \qquad (53)$$

and a Hopf bifurcation when

$$|C|^2 = \frac{4\Lambda d + 2\delta\varepsilon + \varepsilon^2}{4\Lambda(2\alpha_3 - \delta^{-1}\varepsilon\gamma)} > 0. \qquad (54)$$

The oscillations that result are invariant under reflection followed by evolution in time by half the oscillation period $2\pi/|\lambda_I|$, where

$$\lambda_I^2 = -\varepsilon^2 - 4\delta^{-1}\varepsilon\gamma\Lambda|C|^2 > 0, \qquad (55)$$

and are the analogues of the direction-reversing waves discussed in the preceding section. Such *symmetric* oscillations can, therefore, be present only if

$$\gamma\Lambda < 0, \tag{56}$$

and so cannot occur near $\mu = \mu_c$ without *both* the streaming flow and the breaking of the rotation invariance of the system. This is a consequence of the low Reynolds number assumption used to model the streaming flow by a single mode. Recall that FIGURE 4 shows that if Re is *sufficiently* large a symmetry-breaking Hopf bifurcation can destabilize the standing waves even if $\Lambda = 0$.

From Equations (51), (53), and (54) we also find conditions for codimension-two degeneracies:

1. A Takens–Bogdanov bifurcation, corresponding to the coalescence of the symmetry-breaking Hopf and steady bifurcations, occurs if (56) holds and

$$(\gamma + \alpha_3)\varepsilon + 2\delta^{-1}d\gamma\Lambda = 0; \tag{57}$$

2. An interaction between the saddle-node and symmetry-breaking steady bifurcation occurs when

$$d(\alpha_3 + \alpha_4 + 2\gamma) = 2\Lambda(\alpha_3 + \gamma); \tag{58}$$

3. An interaction between the saddle-node and symmetry-breaking Hopf bifurcation, with one zero plus two nonzero imaginary eigenvalues, occurs when (55) holds and

$$\frac{4\delta\Lambda + 2\delta\varepsilon + \varepsilon^2}{4\Lambda(2\alpha_3 - \delta^{-1}\varepsilon\gamma)} = \frac{d - \Lambda}{\alpha_3 - \alpha_4}. \tag{59}$$

The first of these bifurcations contains within its unfolding periodic oscillations that are symmetric in the sense defined above. The second also contains periodic oscillations but this time the oscillations are *asymmetric*. Finally, the third bifurcation contains quasiperiodic solutions[23] that are *symmetric on average*; these correspond to three-frequency states in the Faraday system. Chaotic dynamics are present near global bifurcations with which the quasiperiodic solutions terminate.[24,25] Closely related results apply to containers subjected to *horizontal* vibration as well.[26]

CONCLUDING REMARKS

In this paper we have reexamined the theory of parametrically excited surface gravity-capillary waves in nearly inviscid liquids; the corresponding microgravity results are obtained on setting the Bond number B to zero ($C_g = C, S = 1$). We focused on brimful containers with a pinned contact line and restricted attention to two-dimensional systems with periodic boundary conditions. In systems of this type the primary instability is always to a pattern of standing waves—this is a very general result.[27] In the present case, we have found that these standing waves may lose stability at finite amplitude to two different types of instabilities that break the reflection symmetry of these waves: a Hopf bifurcation producing standing waves that drift back and forth, and a parity-breaking bifurcation that produces waves that drift with a constant speed in one direction or other. We have shown that these instabilities involve the excitation of a viscous mean flow we called *streaming flow*. This flow is driven in the boundary layers at the container walls or at the free surface by time-averaged Reynolds stresses produced by the waves, and in turn couples to the amplitudes of the waves responsible for it. This coupling arises already at third order indicating that it is in general inconsistent to ignore the streaming flow while retaining other cubic terms. We have shown that for both extended ($L \gg 1$) and small ($L \sim 1$) domains the general coupled amplitude-mean flow (GCAMF) equations decouple into equations for the wave amplitudes, and a set of equations for the spatial phase of the waves coupled to the Navier–Stokes equation for the streaming flow. In an extended domain the dynamics in the decoupled amplitude equations can be complex, and could break, at large enough forcing, the instantaneous equality $|A_0| = |B_0|$ characteristic of standing waves (FIG. 3); in a small domain this is not possible and standing waves persist for all time. However, in both cases the coupling to the streaming flow could destabilize the spatial phase of the standing waves resulting in complex drift motion. Our calculations suggest that the uniformly drifting standing waves observed in some experiments[14] could form as a result of a parity-breaking bifurcation of the type described here. We have presented a concrete example of the two types of drift instability that can occur, and described briefly the interaction between the excitation of the streaming flow and forced breaking of the translation invariance of the (annular) container that suggests a variety of new experiments on the Faraday system. These conclusions readily generalize to containers of other shapes, such as square containers perturbed to rectangular ones.[22]

ACKNOWLEDGMENTS

We are grateful to María Higuera and Elena Martín for assistance in obtaining the results reported here, and to the NASA Microgravity Program for support under grant NAG3-2152.

REFERENCES

1. MILES, J. & D. HENDERSON. 1990. Parametrically forced surface waves. Annu. Rev. Fluid Mech. **22:** 143–165.
2. FAUVE, S. 1995. Parametric instabilities. *In* Dynamics of Nonlinear and Disordered Systems. G. Martínez Mekler & T.H. Seligman, Eds.: 67–115. World Scientific.
3. KUDROLLI, A. & J.P. GOLLUB. 1997. Patterns and spatio-temporal chaos in parametrically forced surface waves: a systematic survey at large aspect ratio. Physica D **97:** 133–154.
4. MARTEL, C. & E. KNOBLOCH. 1997. Damping of nearly inviscid water waves. Phys. Rev. E **56:** 5544–5548.
5. VEGA, J.M., E. KNOBLOCH & C. MARTEL. 2001. Nearly inviscid Faraday waves in annular containers of moderately large aspect ratio. Physica D **154:** 313–336.
6. LAMB, H. 1932. Hydrodynamics. Cambridge University Press.
7. EZERSKII, A.B., M.I. RABINOVICH, V.P. REUTOV & I.M. STAROBINETS. 1986. Spatiotemporal chaos in the parametric excitation of a capillary ripple. Sov. Phys. JETP **64:** 1228–1236.
8. MILES, J. 1993. On Faraday waves. J. Fluid Mech. **248:** 671–683.
9. PIERCE, R.D. & E. KNOBLOCH. 1994. On the modulational stability of traveling and standing water waves. Phys. Fluids **6:** 1177–1190.
10. HANSEN, P.L. & P. ALSTROM. 1997. Perturbation theory of parametrically driven capillary waves at low viscosity. J. Fluid Mech. **351:** 301–344.
11. DAVEY, A. & K. STEWARTSON. 1974. On three-dimensional packets of surface waves. Proc. R. Soc. London, Ser. A **338:** 101–110.

12. LONGUET-HIGGINS, M.S. 1953. Mass transport in water waves. Phil. Trans. R. Soc. Ser. A **245**: 535–581.
13. SCHLICHTING, H. 1932. Berechnung ebener periodischer Grenzschichtströmungen. Phys. Z. **33**: 327–335.
14. DOUADY, S., S. FAUVE & O. THUAL. 1989. Oscillatory phase modulation of parametrically forced surface waves. Europhys. Lett. **10**: 309–315.
15. MARTEL, C., E. KNOBLOCH & J.M. VEGA. 2000. Dynamics of counterpropagating waves in parametrically forced systems. Physica D **137**: 94–123.
16. HIGUERA, M., J. PORTER & E. KNOBLOCH. 2002. Heteroclinic dynamics in the nonlocal parametrically driven Schrödinger equation. Physica D. **162**: 155–187.
17. MARTÍN, E., C. MARTEL & J.M. VEGA. 2002. Drift instability of standing Faraday waves. J. Fluid Mech. In press.
18. LANDSBERG, A.S. & E. KNOBLOCH. 1991. Direction-reversing traveling waves. Phys. Lett. A **159**: 17–20.
19. DANGELMAYR, G. & E. KNOBLOCH. 1991. Hopf bifurcation with broken circular symmetry. Nonlinearity **4**: 399–427.
20. DANGELMAYR, G., E. KNOBLOCH & M. WEGELIN. 1991. Dynamics of travelling waves in finite containers. Europhys. Lett. **16**: 723–729.
21. HIRSCHBERG, P. & E. KNOBLOCH. 1996. Complex dynamics in the Hopf bifurcation with broken translation symmetry. Physica D **90**: 56–78.
22. HIGUERA, M., J.M. VEGA & E. KNOBLOCH. 2002. Coupled amplitude-streaming flow equations for nearly inviscid Faraday waves in small aspect ratio containers. J. Nonlinear Sci. In press.
23. GUCKENHEIMER, J. & P. HOLMES. 1983. Nonlinear Oscillations, Dynamical Systems and Bifurcations of Vector Fields. Springer-Verlag.
24. LANGFORD, W.F. 1983. A review of interactions of Hopf and steady-state bifurcations. *In* Nonlinear Dynamics and Turbulence. G.I. Barenblatt, G. Iooss & D.D. Joseph, Eds.: 215–237. Pitman.
25. KIRK, V. 1993. Merging of resonance tongues. Physica D **66**: 267–281.
26. HIGUERA, M., J.A. NICOLÁS & J.M. VEGA. 2002. Weakly nonlinear oscillations of nonaxisymmetric capillary bridges at small viscosity. Phys. Fluids. In press.
27. RIECKE, H., J.D. CRAWFORD & E. KNOBLOCH. 1988. Time-modulated oscillatory convection. Phys. Rev. Lett. **61**: 1942–1945.

Thermocapillary Convection Around Gas Bubbles

An Important Natural Effect for the Enhancement of Heat Transfer in Liquids Under Microgravity

J. BETZ[a] AND J. STRAUB[b]

[a]*Heppstrasse 5, Ingolstadt, Germany*

[b]*Lehrstuhl für Thermodynamik, Technical University of Munich, Boltzmannstrasse 15, Garching, Germany*

ABSTRACT: In the presence of a temperature gradient at a liquid–gas or liquid–liquid interface, thermocapillary or Marangoni convection develops. This convection is a special type of natural convection that was not paid much attention in heat transfer for a long time, although it is strong enough to drive liquids against the direction of buoyancy on Earth. In a microgravity environment, however, it is the remaining mode of natural convection and supports heat and mass transfer. During boiling in microgravity it was observed at subcooled liquid conditions. Therefore, the question arises about its contribution to heat transfer without phase change. Thermocapillary convection was quantitatively studied at single gas bubbles in various liquids, both experimentally and numerically. A two-dimensional mathematical model described in this article was developed. The coupled mechanism of heat transfer and fluid flow in pure liquids around a single gas bubble was simulated with a control-volume FE-method. The simulation was accompanied and compared with experiments on Earth. The numerical results are in good accordance with the experiments performed on Earth at various Marangoni numbers using various alcohols of varying chain length and Prandtl numbers. As well as calculations on Earth, the numerical method also allows simulations at stationary spherical gas bubbles in a microgravity environment. The results demonstrate that thermocapillary convection is a natural heat transfer mechanism that can partially replace the buoyancy in a microgravity environment, if extreme precautions are taken concerning the purity of the liquids, because impurities accumulate predominantly at the interface. Under Earth conditions, an enhancement of the heat transfer in a liquid volume is even found in the case where thermocapillary flow is counteracted by buoyancy. In particular, the obstructing influence of surface active substances could be observed during the experiments on Earth in water and also in some cases with alcohols.

KEYWORDS: thermocapillary convection; Marangoni convection; heat transfer, numerical simulation, subcooled boiling, microgravity, reduced gravity

Address for correspondence: J. Betz, Heppstrasse 5, D-85057 Ingolstadt, Germany.
mail@johannbetz.de
jstraub@thermo-a.mw.tu-muenchen.de

Nomenclature:

B	height of the bubble (diameter at $g = 0$)
Bo	Bond number
c	concentration
Fo	Fourier number, $\alpha t/H^2$
g	gravitational acceleration (against y-direction)
g_0	gravitational acceleration on Earth
H	height of the liquid volume
L	width of the liquid volume
Ma	Marangoni number
n	coordinate perpendicular to the bubble surface
Nu	Nusselt number, Q_{CB}/Q_{HCB}
Pe	Peclet number, wB/α
Pr	Prandtl number, ν/α
Q	heat flux
r	radial coordinate
R	radius
Ra	Rayleigh number
T	temperature
s	circumferential coordinate at the bubble surface
u	velocity component in x-direction
v	velocity component in y-direction
V	volume
w	velocity component in s-direction
x, y, z	coordinates

Greek Symbols

α	thermal diffusivity
β_p	thermal expansion coefficient, isobaric
η	dynamic viscosity
ε	synonym for u, v, and T
γ	contact angle
σ	surface tension
ρ	density of the liquid
φ	angle in circumferential direction
τ	shear stress
ν	kinematic viscosity
$d\sigma/dT$	temperature gradient of surface tension
Θ	dimensionless temperature, $(T - T_B)/(T_T - T_B)$
ΔT	temperature difference, $T_T - T_B$

Subscripts

$1Pro$	1-propanol
$1Pen$	1-pentanol
$1Oct$	1-octanol
$2D$	two-dimensional
$3D$	three-dimensional
Eth	ethanol
Al	liquid aluminium
B	bottom wall
Bl	bubble
c	due to surface active substances
cg	curve g
cl	clean liquid (water)
$crit$	critical for the onset of thermocapillary convection
CB	heat transfer due to thermocapillary convection
g	gas side

HC	heat transfer due to heat conduction in the liquid without bubble
HCB	heat transfer due to heat conduction around an adiabatic bubble
l	liquid side
m	mean value
max	maximum value
M	Marangoni convection
Me	methanol
n	in direction perpendicular to the surface
$r\varphi$	at the radius r in tangential direction
s	in tangential direction of the surface
T	top wall
T_{min}	minimum temperature at the free surface
Abbreviations	
CV-FEM	control volume finite element method
LDV	laser doppler velocimetry
M	pure Marangoni convection
M–B	Marangoni convection acting against buoyancy
M+B	Marangoni convection supported by buoyancy
PIV	particle image velocimetry

INTRODUCTION

Buoyancy, the most important mode of natural convection is driven by unstable density stratifications. In cases, where a gradient of surface tension at a liquid surface or an interface between two liquids is present, a second type of natural convection, called Marangoni convection, occurs. In FIGURE 1 a part of a bubble surface is sketched, where the surface tension σ increases in tangential direction. The local instability at the phase boundary leads to a flow in the direction of increasing surface tension. Due to shear stresses, gas and liquid molecules near the surface are accelerated.

In the steady state the shear stresses in the boundary layers on the liquid $\tau_{r\varphi,l}$ and gas side $\tau_{r\varphi,g}$ and the surface tension gradient $d\sigma/ds$ are balanced, Equations (1) and (2). The surface tension σ of a liquid depends on the temperature T and the concentration c,[1–4] as follows:

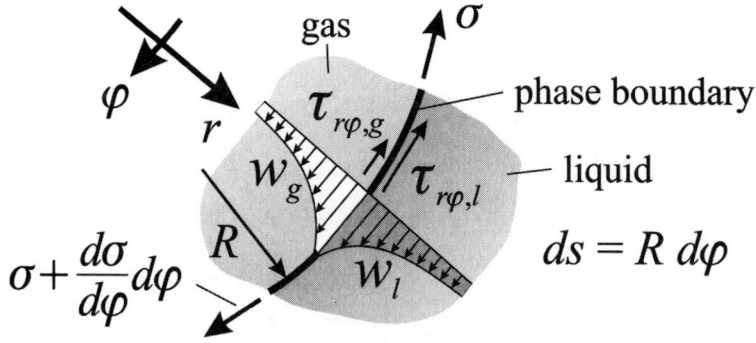

FIGURE 1. Force balance in tangential direction of a bubble surface.

$$\sigma = \sigma(T, c), \quad \frac{d\sigma}{ds} = \frac{\partial \sigma}{\partial T}\frac{\partial T}{\partial s} + \frac{\partial \sigma}{\partial c}\frac{\partial c}{\partial s} \tag{1}$$

$$\frac{d\sigma}{ds} = \frac{1}{R}\frac{d\sigma}{d\varphi} = \tau_{r\varphi,l} + \tau_{r\varphi,g}. \tag{2}$$

If a temperature gradient $\partial T/\partial s \neq 0$ induces interfacial convection a subspecies of Marangoni convection occurs, called "thermocapillary convection". It is predominantly studied at gas bubbles in this article. For most liquids the temperature gradient of surface tension $\partial\sigma/\partial T$ is negative, thus a liquid flow from the warm to the cold side is induced. In cases of local impurities at the bubble surface, concentration gradients $\partial c/\partial s \neq 0$ are present and another subspecies, called diffusocapillary, chemocapillary, concentration-capillary or solution-Marangoni convection[5] occurs. In many cases, the dynamic viscosity of the gas inside the bubble is much smaller than of the surrounding liquid. In consequence, the shear forces on the gas side can be neglected $\tau_{r\varphi,g} \approx 0$. Therefore, in the numerical calculations only those forces on the liquid side are considered, in accord with.[4,6,7]

NUMERICAL SIMULATION

A numerical simulation was made to quantify the flow and the temperature field by thermocapillary convection in the liquid around a single bubble. The aim of the study was to quantify the flow and heat transfer phenomena in pure liquids and to describe the substantial differences between heat transfer and fluid flow on Earth and in microgravity. From this viewpoint, an adapted method was applied that should deliver sufficiently accurate results in an adequate time without making the request to calculate all three-dimensional phenomena precisely.

The geometry under consideration is schematically sketched in FIGURE 2, an axi-symmetric gas bubble curved by the hydrostatic liquid pressure in an angular liquid volume of width L and height H. In principle the gas bubble can be located at various places, either hovering free or touching the upper or lower wall.

Inside the liquid volume, buoyancy convection, and at the liquid–gas interface, thermocapillary convection, can occur. Discretization of the complex 3D geometry would be very complicated and long calculation times are expected. From experimental observations it was known, that the flow is approximately axially symmetric around the vertical bubble axis. The highest velocities and velocity gradients appear near the bubble surface at $x \approx R_{Bl}$, where the horizontal extension of the bubble surface is maximum (FIG. 2). Considering this main flow region, a 2D model of an infinite thin and plain liquid layer with adiabatic and non-viscous front and back walls (2D gas bubble, FIG. 2) seems to be applicable for the following reasons:

$$\frac{\varepsilon}{x} \ll \frac{\partial \varepsilon}{\partial x}, \quad \frac{1}{x}\frac{\partial \varepsilon}{\partial x} \ll \frac{\partial}{\partial x}\left(\frac{\partial \varepsilon}{\partial x}\right), \quad \frac{\varepsilon}{x^2} \ll \frac{\partial}{\partial x}\left(\frac{\partial \varepsilon}{\partial x}\right), \tag{3}$$

where $\varepsilon = u, v, T$; $0.8 < x/R_{Bl} < 1.2$; and $H - 1.2\cdot B < y < H$.

If R_{Bl} is large, the 2D and 3D velocity profiles at the bubble flank resemble each other and the error due to the 2D model approximation is small. The 2D model saves

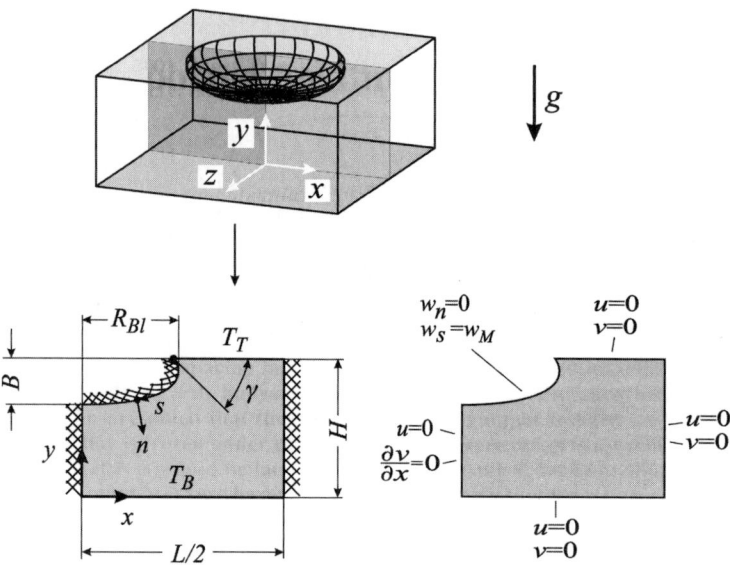

FIGURE 2. Application of a 2D model (2D bubble) of an infinitesimally thin and plain liquid layer with adiabatic and sliding (non-viscous) front and back walls with the shape of the liquid in a cut through the centre of the real 3D bubble. The gas bubble is deformed by the hydrostatic pressure on Earth.[1,8] ▨▨▨, adiabatic boundaries.

computing time and facilitates the possibility of a wide parameter variation. Due to the symmetry only a half plane needed to be considered.

The upper and lower walls are assumed to be isothermal and the side walls adiabatic. The bubble surface is treated as adiabatic. Energy transport through the gas-liquid-interface by heat conduction, convection and thermal radiation is negligible in comparison with the heat transport through the liquid, even evaporation and condensation within the bubble are not considered. At the solid boundaries non-slip condition of the liquid occurs. The flow components perpendicular to the free surface and the symmetry plane vanish due to the non deformable surface and the symmetry condition. At the bubble surface a force balance between surface tension forces and shear forces in the liquid is present in circumferential direction. At this point it is noted, that the shear forces in the s-direction (FIG. 1) of the liquid, due to surface tension forces at the free surface, are similar in the 2D and rotationally symmetric 3D cases.

The unsteady heat and mass transfer problem inside the liquid is described by the 2D conservation laws for mass, energy, and momentum.[8] The numerical method used for integration is described elsewhere.[8] It allows a precise discretization of the curved free surface of the bubble, as well as the rectangular geometry of the liquid cell. Flexibility gains favored triangular elements and the control volume FEM (CV-FEM) method of Baliga and Patankar,[9–11] Baliga,[12] Prakash,[13] and Prakash and Patankar.[14] To guarantee high numerical stability and to avoid possible oscillatory numerical solutions due to an unstable time step, which is discussed in References

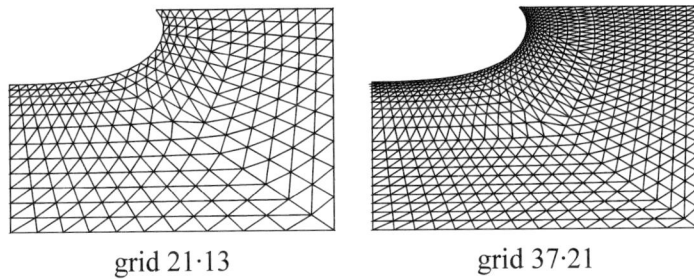

FIGURE 3. Grid used, with 21·13 and 37·21 nodes.

15 and 16, a fully implicit time step was implemented. The pressure–velocity iteration was performed by the SIMPLE algorithm.[17] The numerical grid used is refined at the free surface (see FIGURE 3).[8] Since the precision of the numerical results[8] does not increase significantly for grids finer than 37·21, but the calculation time does increase rapidly, the grid 37·21 was used for all calculations described in this article.

The geometry of the bubble surface is derived from the force balance at a 3D differential element of the bubble surface and a numerical integration with the Runge–Kutta method of fourth order, which is described in detail in Reference 18. The calculated bubble geometry corresponds well with a real 3D bubble geometry, known from experiments.[1,8]

The thermophysical properties of the test liquids were chosen at the mean temperature in the gradient chamber. In accord with the Boussinesq approximation,[19–21] the thermal expansion was only considered in the buoyancy term of the momentum equations.

DIMENSIONLESS NUMBERS

In this article two cases, M and M–B, are discussed. They are explained in TABLE 1. The dimensionless numbers used in this work are arranged in TABLE 2.

The *Ma* number is derived from the force balance between the shear forces on the liquid side and the surface tension forces (2).[8] It describes the strength of thermocapillary convection. The *Ma* number (TABLE 2) is a function of the total temperature

TABLE 1. Definition of the cases M and M–B

Case M	Pure Marangoni convection (M) around a gas bubble. $Ra \approx 0$ and $Bo \approx 0$. This case could only be studied numerically
Case M–B	Marangoni flow acting against buoyancy (M–B). $Ra < 0$ and $Bo < 0$. The temperature stratification is stable. In contrast to case M this configuration can be investigated experimentally on Earth

TABLE 2. Definition of *Ma*, *Ra*, and *Bo* numbers for the cases M and M–B

	Case M	Case M–B
Ma	$-\dfrac{\partial \sigma}{\partial T}\dfrac{B}{\alpha \eta}\|T_T - T_B\|\dfrac{B}{H}$	$-\dfrac{\partial \sigma}{\partial T}\dfrac{B}{\alpha \eta}\|T_T - T_B\|\dfrac{B}{H}$
Ra	0	$\dfrac{g\beta_p(T_B - T_T)B^3}{\alpha \nu}\dfrac{B}{H}$
$T_B - T_T$	all values	< 0
Bo	0	$-\dfrac{\rho g \beta_p B^2}{-(\partial \sigma/\partial T)}$

NOTE: For B, H, T_T, and T_B see FIGURE 2.

gradient in the system, the temperature gradient of surface tension, the characteristic length B of the system, and thermophysical properties. The term $|T_T - T_B|\cdot B/H$ results from the fact that, in a first approximation, only the temperature difference at the free surface of the bubble drives thermocapillary convection. The fact, that thermocapillary convection reduces its own driving temperature gradients at the surface cannot be considered, because the amount of reduction is never known in advance, and is a function of *Ma* and *Ra* itself. To determine the dimensionless numbers in the experiments for the case M-B, the thermophysical properties of the liquid were calculated at a mean temperature value $T_m = T_T - (T_T - T_B)\cdot B/2H$ at the bubble surface.

RESULTS

For the case M-B, the flow field was investigated numerically in the extended parameter range $7 < Pr < 120$. As test liquids, pure alcohols of various chain lengths were used. The main reason for the choice of alcohols was that their thermophysical properties are very well known over a wide temperature and pressure range.[22,23] In the case M-B, the numerical study was confined to the region $11 < Ma < 27{,}800$ because, for $Ma > 30{,}000$, oscillatory numerical flow fields appeared. In reality, oscillatory flow is mostly three-dimensional and cannot be described with a 2D method. In case M, the numerical solution was stable for higher *Ma* numbers, thus the investigation includes the region $11 < Ma < 100{,}000$.

To obtain comparability between the simulations, all bubble contours in the numerical simulations, in the M-B case, had the shape of a real 3D bubble $V_{Bl} = 0.4\,\text{cm}^3$ in methanol.[1] The height of the liquid matrix was $H = 10\,\text{mm}$, the bubble height was $B \approx 3.33\,\text{mm}$, the bubble radius was $R_{Bl} \approx 7.15\,\text{mm}$.

In technical systems (e.g., in the boiling process) a bubble often grows into an existing temperature gradient field. Thus, all 2D simulations started with a bubble in a motionless liquid. In vertical direction, a linear temperature profile and a stable density stratification is present at $Fo = 0$ (see FIGURE 4). Without thermocapillary convection, the theoretical temperature profile due to pure heat conduction around an adiabatic bubble appears for $Fo \to \infty$.

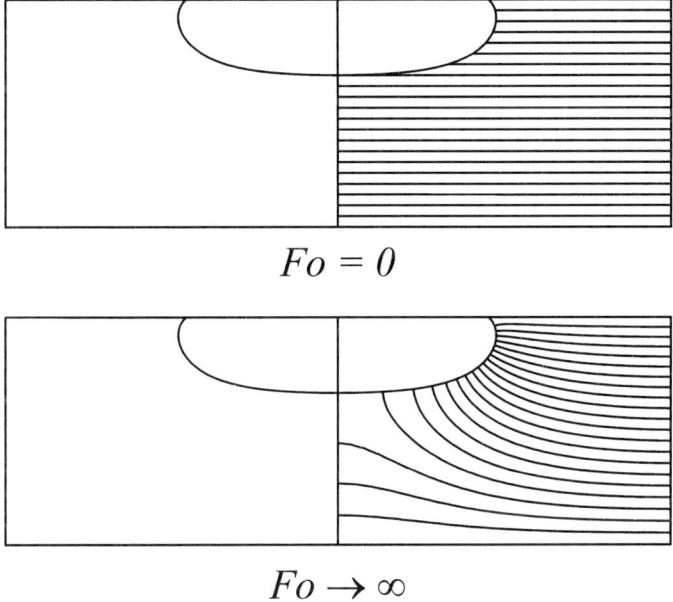

FIGURE 4. Case M-B: isotherms (*right half* of diagrams) of the initial temperature profile around the bubble at $Fo = 0$ and comparison with the theoretical temperature profile at pure heat conduction around an adiabatic bubble for $Fo \to \infty$.

Flow Fields in the Case M-B

In Reference 8 the transient development of the flow and temperature field in methanol, 1-propanol, and 1-octanol is discussed for $Ma = 556$, $Ma = 5{,}560$, and $Ma = 16{,}680$ at small Fourier numbers. Due to lack of space, only the case $Ma = 5{,}560$ is illustrated in FIGURE 5 with the examples of methanol and 1-octanol.

The transient development of the liquid flow and the temperature field is different in methanol and 1-octanol although the *Ma* number is the same. Initially, a clockwise primary recirculation zone is formed at the free surface. Warm liquid is driven from the bubble base in a direction toward the bubble head by surface tension forces. The high local density instability at the bubble head turns the liquid flow up again in the direction of the warm plate. Subsequently, this primary recirculation zone degenerates in the direction perpendicular to the bubble surface. The degeneration occurs at smaller *Fo* numbers in long chained alcohols—for example, in 1-octanol (FIG. 5, $Fo = 8.43 \times 10^{-4}$)—than in short chained alcohols—for example, in methanol. Simultaneously, a secondary recirculation zone develops in the bottom part of the liquid in the opposite direction due to viscous shear stresses. In methanol, complex flow structures are formed temporarily (e.g., FIG. 5, $Fo = 8.16 \times 10^{-3}$) caused by high inertia forces due to the small *Pr* number of methanol.

The velocities in the boundary layer of the thermocapillary flow in circumferential direction at the bubble surface are of most interest. The Peclet number $Pe = wB/\alpha$ serves as dimensionless velocity. In principle, the maximum dimensionless velocity

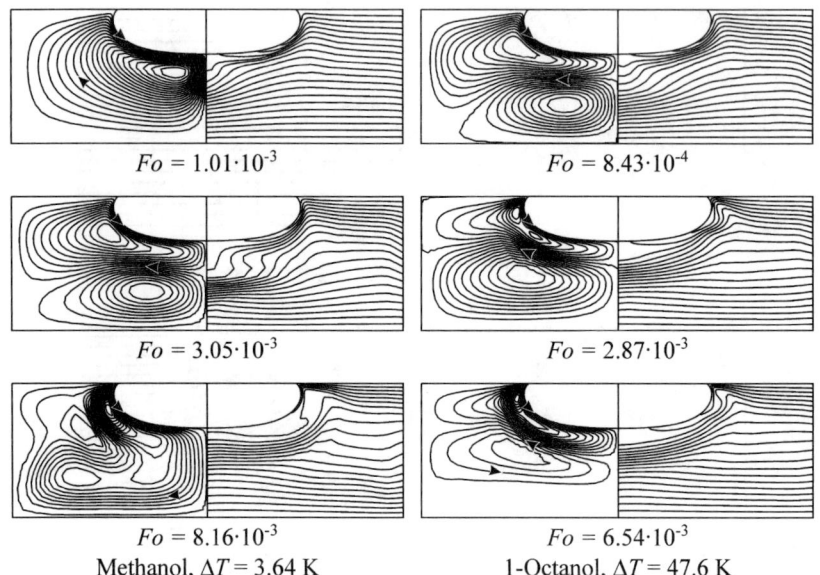

FIGURE 5. Case M–B, $Ma = 5,560$: transient development of the velocity (*left*) and temperature field (*right*) at the 2D bubble in methanol (*left column*) and 1-octanol (*right column*). *Left half* of each diagram, streamlines; *right half*, isotherms. $\Delta T = T_T - T_B$, $Pr_{Me} = 7.3$, $Pr_{1Oct} = 123$, $Bo_{Me} = -1.19$, $Bo_{1Oct} = -1.00$.

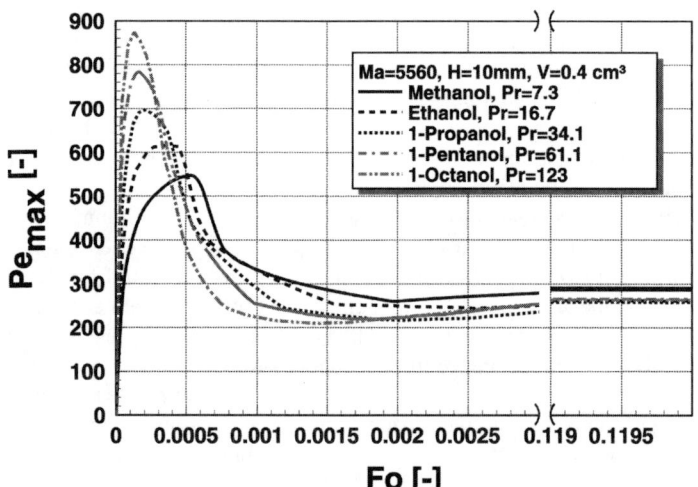

FIGURE 6. Case M–B, $Ma = 5,560$: transient development of the maximum dimensionless velocity at the free surface of a 2D bubble in various liquids. $Bo_{Me} = -1.19$, $Bo_{Eth} = -1.24$, $Bo_{1Pro} = -1.23$, $Bo_{1Pen} = -0.98$, $Bo_{1Oct} = -1.00$.

Pe_{max} always occurs at the bubble surface. The transient development of Pe_{max} is illustrated in FIGURE 6 for various alcohols. For all substances Pe_{max} increases rapidly for $Fo > 0$ due to the initial conditions in the numerical treatment. The maximum value of Pe_{max} in FIGURE 6 increases with increasing Pr number of the alcohol used. For $Fo > 0.12$, Pe_{max} converges to a constant value, nearly independent of Pr. In parallel with the variation of Pe_{max} over Fo, an alternative variation with the Reynolds number as a function of the Strouhal number was investigated, but this did not result in a Pr-independent variation of the transient development of the velocity at the bubble surface.

The steady state flow and temperature field in methanol and 1-octanol is attained for $0.13 < Fo < 0.15$ and is shown for various Ma numbers in FIGURE 7. The high density of the stream function at the free surface indicates high local velocity gradients. The flow in the secondary recirculation zone is rather slow.

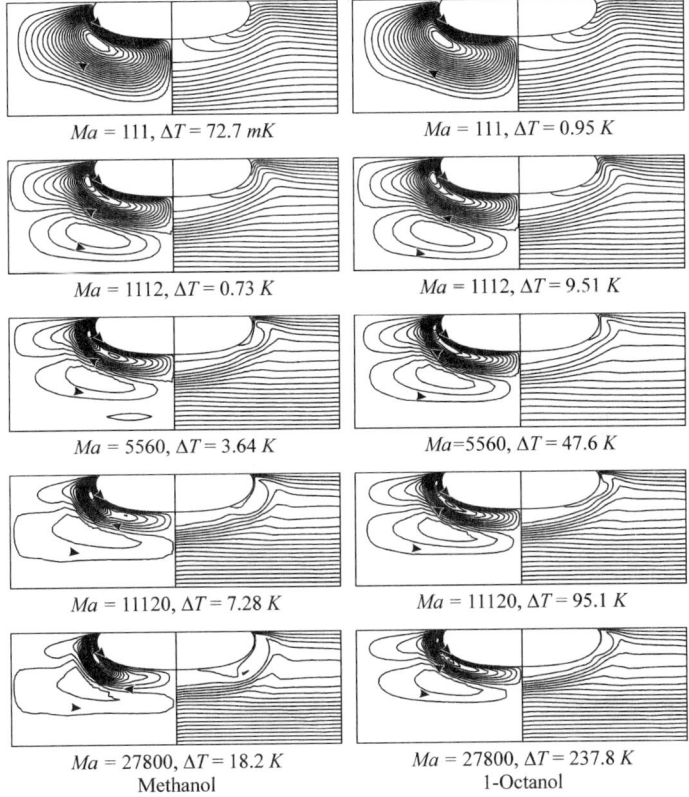

FIGURE 7. Case M–B: steady state flow and temperature field around a 2D bubble in methanol (*left column*, $Fo \approx 0.15$) and 1-octanol (*right column*, $Fo \approx 0.13$), $\Delta T = T_T - T_B$, $Ma = 27{,}800$. $\Delta T = 237.8$ K in 1-octanol is a supplementary theoretical case for comparison with methanol at the same Ma number, which cannot be performed in experiments. Pr and Bo are given in FIGURE 5.

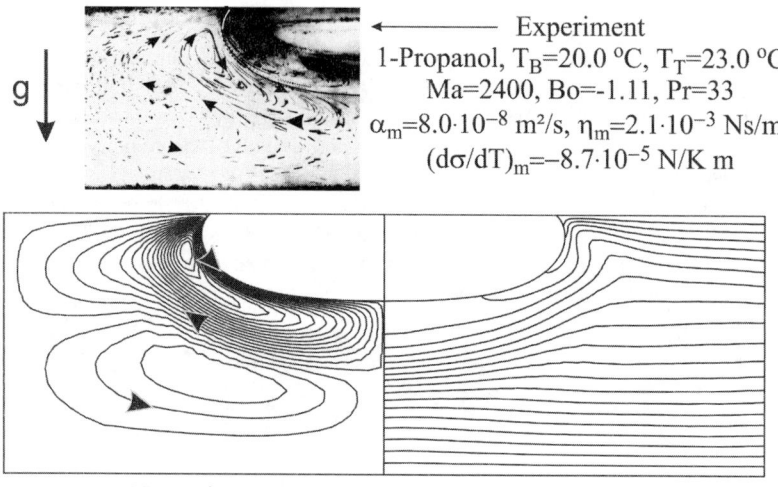

Experiment
1-Propanol, T_B=20.0 °C, T_T=23.0 °C
Ma=2400, Bo=−1.11, Pr=33
α_m=8.0·10^{-8} m²/s, η_m=2.1·10^{-3} Ns/m²
$(d\sigma/dT)_m$=−8.7·10^{-5} N/K m

Numerics, 1-Propanol, T_T-T_B=4.1 K, Ma=2400

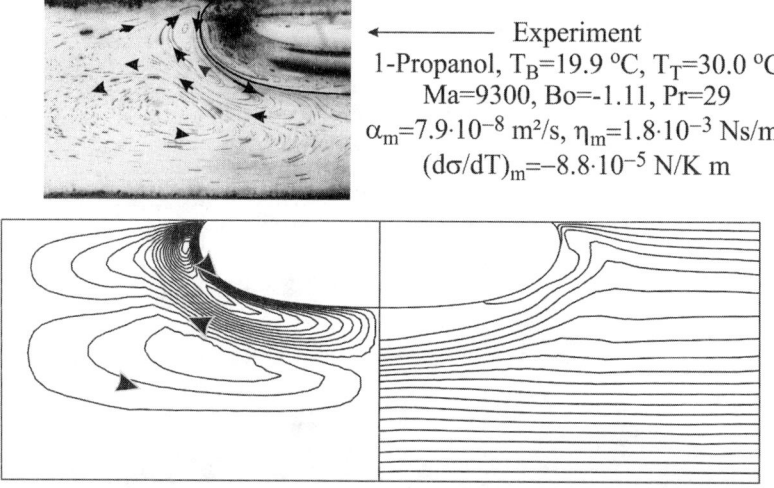

Experiment
1-Propanol, T_B=19.9 °C, T_T=30.0 °C
Ma=9300, Bo=−1.11, Pr=29
α_m=7.9·10^{-8} m²/s, η_m=1.8·10^{-3} Ns/m²
$(d\sigma/dT)_m$=−8.8·10^{-5} N/K m

Numerics, 1-Propanol, T_T-T_B=13.9 K, Ma=9300

FIGURE 8. Case M–B: steady state flow field around a bubble $B \approx 3.33$ mm, experimental ($H = 7.4$ mm) and 2D numerical results ($H = 10$ mm). The thermophysical properties of the liquid in the numerical simulation were chosen to be identical to the experimental values.
Figure continues on opposite page.

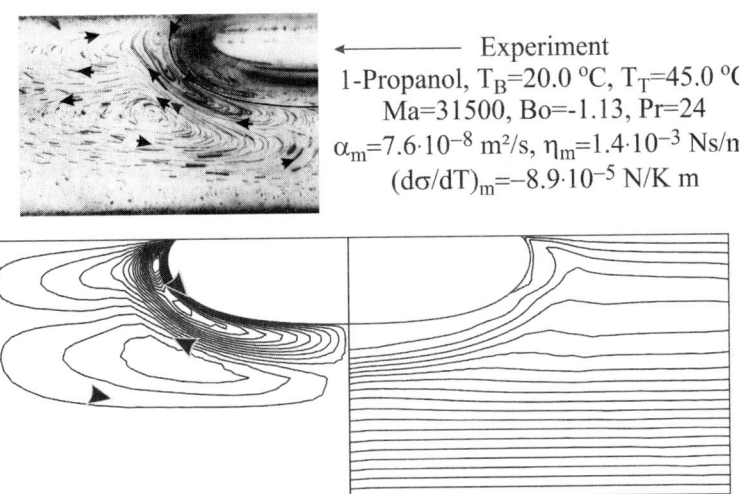

Experiment
1-Propanol, T_B=20.0 °C, T_T=45.0 °C
Ma=31500, Bo=−1.13, Pr=24
α_m=7.6·10^{-8} m²/s, η_m=1.4·10^{-3} Ns/m²
$(d\sigma/dT)_m$=−8.9·10^{-5} N/K m

Numerics, 1-Propanol, T_T−T_B=34.2 K, Ma=31500

FIGURE 8/continued.

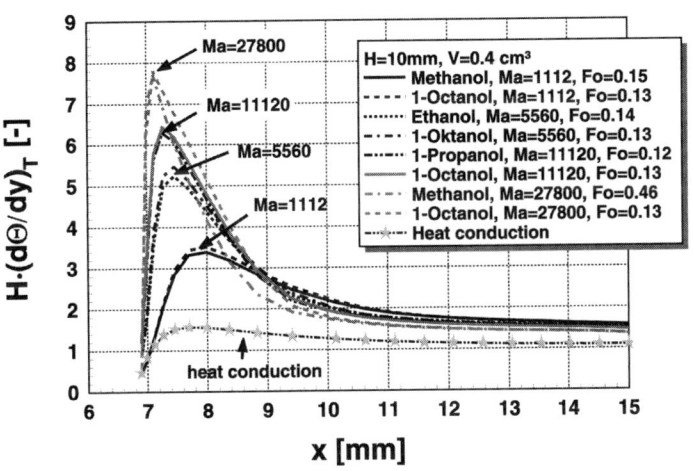

FIGURE 9. Case M–B: the quasi steady state dimensionless temperature gradient perpendicular to the heating surface during 2D thermocapillary convection in different alcohols at various *Ma* numbers.

In the steady state, the primary recirculation zone moves in a direction toward the warm plate and the secondary recirculation zone moves vertically with increasing Ma number (FIG. 7). Furthermore, the primary recirculation zone splits into multiple separated recirculation zones at high Ma numbers.

The strong thermocapillary convection at the free surface reduces the temperature gradients at the free surface from 28% at $Ma = 1,112$ to 65% at $Ma = 27,800$ in 1-octanol, referring to steady state heat conduction without bubbles (FIGS. 4 and 7).

For verification purposes, the numerical flow field in 1-propanol is compared with PIV measurements of the flow field for $Ma = 2,400$, $Ma = 9,300$, and $Ma = 31,500$ (see FIGURE 8). The temperature gradient chamber used in the experiment and the PIV method are discussed in detail in References 1 and 8. The flow paths in the experiment and in the numerical approach resemble each other, although different heights of the liquid matrix were chosen.

A quantitative comparison study of the steady state velocity profile along the bubble surface and perpendicular to the bubble surface between the experimental and the numerical results has already been published.[1,8] The results point out, that the 2D numerical simulation is able to describe the flow at the bubble surface quantitatively.

Temperature Field and Heat Transfer in the Case M-B

In FIGURE 7 the numerical results show a concentration of the isotherms at the warm plate in the region near the 2D bubble base, but also at the cooling plate, with increasing Ma number. This behavior is quantitatively illustrated in FIGURES 9 and 10 in terms of the quasisteady state dimensionless temperature gradient perpendicular to the heating and cooling surface for high Fo numbers. Various alcohols and Ma numbers were investigated. The bubble base touches the heating plate at $x = 6.9$ mm. Thus, all curves in FIGURE 9 originate at $x = 6.9$ mm. $H \cdot (d\Theta/dy)_T$ is smaller than 1 at the bubble base because of the isolating effect of the adiabatic bubble surface and the contact angle of the bubble surface of 40° to the heating plate. The primary recirculation zone transports cold liquid to the heating plate. Thus, higher dimensionless temperature gradients perpendicular to the heating plate occur than in the case of pure heat conduction around the bubble. The maximum dimensionless temperature gradient is located less than 1.2 mm away from the bubble base. With increasing Ma number, the maximum value of $H \cdot (d\Theta/dy)_T$ increases, and the distance of the maximum to the bubble bases decreases as a result of the changing shape of the primary recirculation zone.

The steady state local heat transfer at the cooling plate is strongly affected by the shape and prolongation of the primary and secondary recirculation zones in the liquid volume. At small Ma numbers, for example, for $Ma = 111$, the primary recirculation zone transports warm liquid to the cold plate near the symmetry axis. Thus, the local heat transfer is maximum at $x = 0$ mm (FIG. 10). At $Ma = 1,112$, the secondary recirculation zone transports warm liquid to the cooling plate far away from the symmetry axis, thus the maximum occurs at $x \approx 12$ mm. An additional increase in Ma number leads to a further vertical movement of the secondary recirculation and to a nearly uniform local temperature gradient over the entire free surface.

In general the dependence of the steady state local heat transfer at the heating and cooling plate on the Pr number is small in the investigated region $7 < Pr < 120$ at $0.12 < Fo < 0.15$.

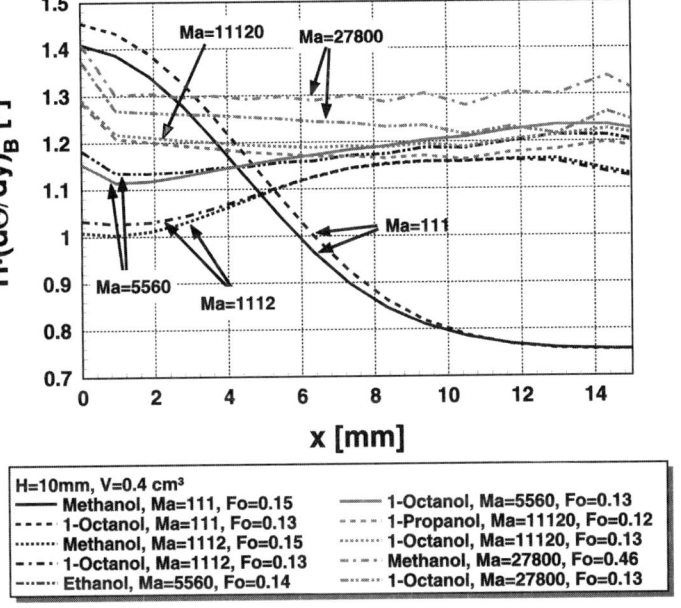

FIGURE 10. Case M–B, 2D bubble: the quasi steady state dimensionless temperature gradient perpendicular to the cooling surface during thermocapillary convection in different alcohols at various Ma numbers.

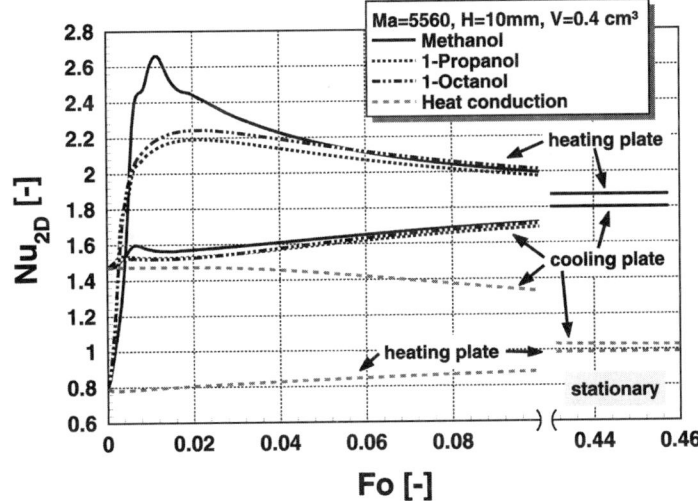

FIGURE 11. Case M–B, 2D bubble, $Ma = 5,560$: transient development of the integral Nusselt number at the heating and cooling plate for various alcohols.

The enhancement of heat transfer by thermocapillary convection is described quantitatively with the Nusselt number, which is defined as integral heat transfer with thermocapillary convection around the bubble (Q_{CB}) in relation to the integral heat transfer due to pure heat conduction in the liquid around an adiabatic bubble (Q_{HCB})

$$Nu = \frac{Q_{CB}}{Q_{HCB}}. \qquad (4)$$

In FIGURE 11 the transient development of the integral Nusselt number at the heating and the cooling plate is shown for methanol, 1-propanol, and 1-octanol at $Ma = 5{,}560$ in the case M-B. The curves at the heating plate start at a value of 0.8 and at the cooling plate at a value of $1.47 = 1/0.68$. This is due to the initial setting of a linear temperature profile in vertical direction at $Fo = 0$. In the case of pure heat conduction with a bubble, the curves Nu over Fo in FIGURE 11 converge to $Nu = 1$ for $Fo \to \infty$ (see corresponding temperature field in FIG. 4). In this case Q_{HCB} is 68% of the value of pure heat conduction without a bubble. For $Fo > 0$ the curves Nu over Fo at the cooling plate and especially at the heating plate, increase more strongly than the corresponding curves for pure heat conduction. The dimensionless heat transfer in 1-propanol and 1-octanol is nearly superimposed. Due to the temporary formation of complex flow structures in methanol (FIG. 5, $Fo = 8.16 \times 10^{-3}$), the integral heat transfer is temporarily higher. In the steady state ($Fo \to 0.5$) the integral Nu number for the three alcohols is higher by a factor of 1.85 compared with pure heat conduction around the bubble.

To avoid long calculation times until a fully converged numerical solution is obtained, the intersection point of the extrapolated curves Nu over Fo (FIG. 11) was evaluated and is depicted as the steady state integral Nu number in FIGURE 12. The

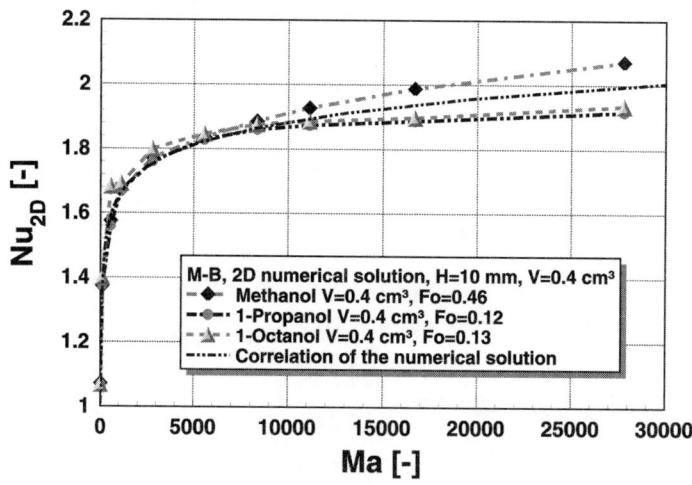

FIGURE 12. Case M–B, 2D bubble: the integral Nu number during steady state thermocapillary convection in different alcohols and various Ma numbers. $H = 10$ mm, $B = 3.33$ mm.

extrapolation was performed at the heating and cooling plate at $Fo \approx 0.12$ in the case of 1-propanol and 1-octanol and at $Fo \approx 0.46$ in the case of methanol.

The stationary Nu number increases with increasing Ma number. The behavior is degressive, because the thermocapillary convection reduces its own driving temperature gradients in increasing manner with increasing Ma number. The dependence of the stationary integral Nu number on the Pr number of the liquids is small in the examined region $7 < Pr < 120$.

In an experimental approach, inside of a temperature gradient chamber, the enhancement of the heat transfer by thermocapillary convection around bubbles of varying volume was measured with a calorimetric arrangement in the case M-B.[1,8] Inside the liquid chamber a constant temperature gradient was established and at the top of it a single air bubble was injected. Due to lack of space only a short summary can be given here. The 2D numerical model shows slightly higher Nu numbers than the experiment. This can be explained with three facts: first, the numerical local heat transfer at the isothermal plates far away from the bubble surface is too high, because the vertical velocity components are smaller in the real 3D geometry due to the mass conservation law. Second, the isolation effect of the 2D bubble is higher than that of the 3D bubble. Third, the concentration of the isotherms at the bubble surface is higher during 2D heat conduction with bubble. This is an additional potential for thermocapillary convection. Nevertheless, the experiments confirm the numerical results very well.

For completeness it should be mentioned that heat transfer by thermocapillary convection was already investigated some years ago in various liquids at a heating wire while a bubble was touched.[7,27] In these publications, the thermocapillary convection was supported by buoyancy (M+B). The enhancement of the heat transfer by the thermocapillary convection was of the same magnitude as in the case M–B here.

INFLUENCE OF GRAVITY

In general, the shape of the bubble surface depends on the hydrostatic pressure difference in the liquid and the surface tension. The aim of this study was to determine the direct influence of Bo and Pr numbers on the steady state flow field and the integral heat transfer. Thus, neglecting the dependence on gravity and the thermophysical properties, the parameters are investigated at a constant bubble shape, as in the case M–B, with $V_{Bl} = 0.4 \text{cm}^3$ in methanol on Earth.

At a 2D bubble in methanol, the overall heat transfer increases with increasing Ma number and decreasing gravity (see FIGURE 13). Illustrations c, d, and e show that heat transfer is supported by an elongation of the thermocapillary convection from the heating plate down to the cooling plate.

A simulation with water on Earth ($Pr = 1.93$, $T_m = 365 \text{K}$) delivers a relatively higher overall heat transfer compared with methanol at the same Bo number $Bo = -0.40$, and a chimney flow down to the cooling plate (FIG. 13a and d) at $Ma = 8,340$ is observed. This result indicates a dependence of the flow field on Pr for small Pr numbers. Nevertheless it should be noted that this strong thermocapillary flow in water is very seldom observed in experiments because of the existence of surface active substances. This fact is discussed in more detail in the following section. At reduced gravity level

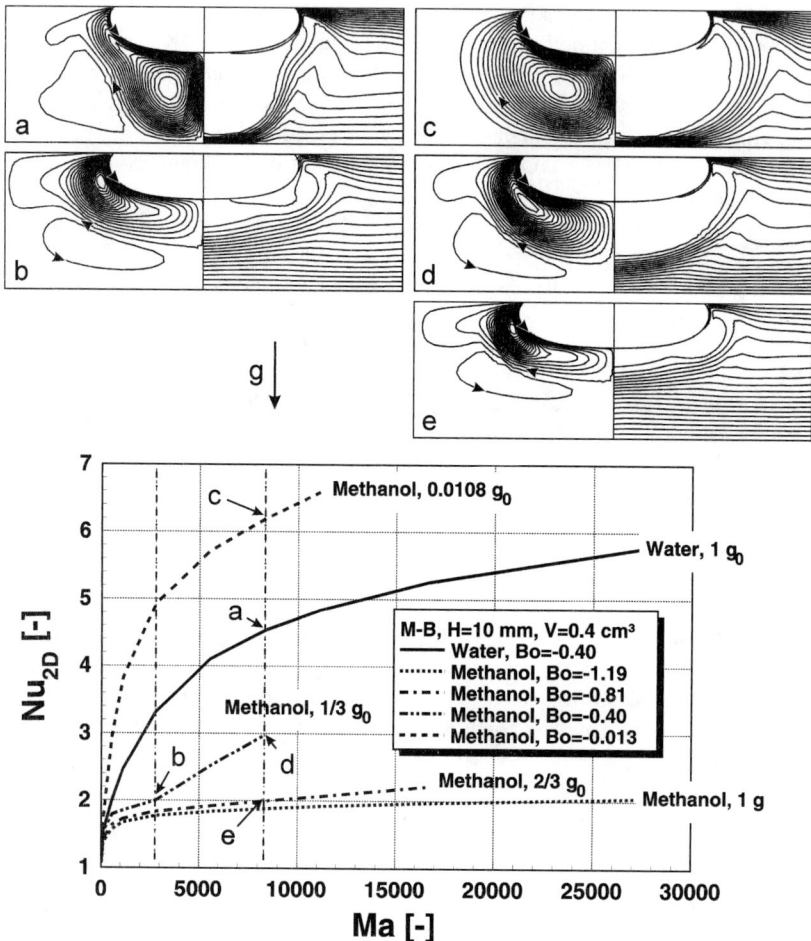

FIGURE 13. Case M–B, 2D bubble, $H = 10$ mm: gravity dependence of the steady state overall heat transfer and examples of the flow and temperature field. Comparison of the results in methanol with those in pure water at a mean temperature of 365 K on Earth (g_0, gravity level on Earth).

$g = 1/3 g_0$, a decrease in the Ma number from $Ma = 8{,}340$ to $Ma = 2{,}780$ in methanol (FIG. 13d and b) causes a considerable back-formation of the primary recirculation zone by the buoyancy. At the same time the overall heat transfer degenerates.

The influence of the Pr number on thermocapillary convection was studied numerically at a spherical stationary 2D gas bubble with a diameter of 20 mm fixed in the centre of a quadratic container with a side length $L = 40$ mm ($B/H = 0.5$) in the case M, $g = 0$, microgravity (see FIGURE 14). In all cases M, the flow fills the entire liquid volume between the upper warm and the lower cold plate. In the case of a gas bubble in melted aluminium the stationary fluid flow concentrates in the region near

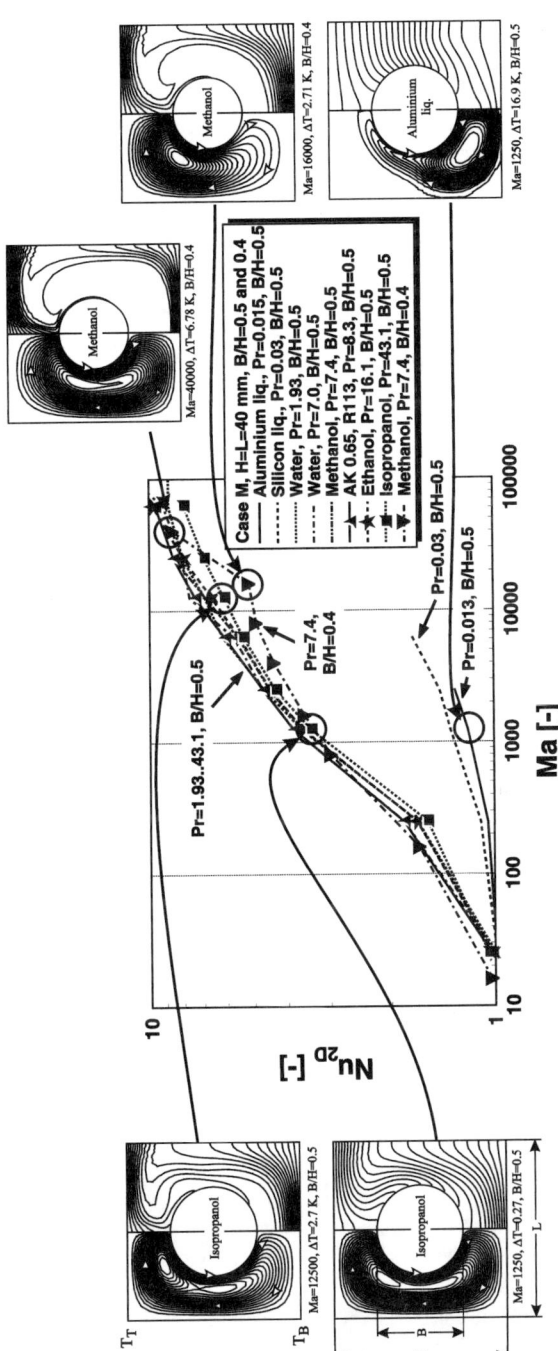

FIGURE 14. Case M ($g = 0$, microgravity), 2D numerical results: the steady state thermocapillary flow and temperature field around a 2D bubble (an infinitesimally thin and plain liquid layer with adiabatic and non-viscous front and back walls and an air hole in the centre, representing the bubble) in various liquids, $H = 40$ mm, $\Delta T = T_T - T_B$.

the cold plate. With increasing Pr number (e.g., for isopropanol, $Pr = 43.1$) it moves in direction toward the heating plate at constant Ma number $Ma = 1,250$. The considerably smaller Pr number ($Pr_{Al} = 0.015$) of liquid aluminium leads to an inertia-dominated flow with a thin thermocapillary boundary layer at the bubble surface. Analogous to the case M–B, the isotherms in the temperature field concentrate near the heating and cooling plate and are reduced at the free surface with increasing Ma number. Although the temperature field in liquid aluminium is strongly formed by the high heat conductivity, a distinct convective heat transfer in isopropanol already occurs at $Ma = 1,250$. Thus, the number of steady state isotherms at the free surface reduces from 15 in liquid aluminium to 8 in isopropanol at $Ma = 1,250$ (FIG. 14).

The stationary overall Nu number increases to values of 7 at $Pr = 43$ and 8 at $Pr = 1.93$ for $Ma = 30,000$ (FIG. 14). At high Ma numbers the relative reduction of the temperature gradients at the 2D bubble surface causes a degressive behavior of the curves Nu over Ma. The stationary overall Nu number is smaller in liquids with very small Pr number. The simulation of the thermocapillary convection at a 2D gas bubble with a diameter $B/H = 0.4$ in methanol (see FIG. 14) shows once more the prerequisite character of the extension of the thermocapillary flow over the entire liquid volume between the warm and the cold plate for a high overall heat transfer.

In summary, it is emphasized that the overall heat transfer in microgravity (case M, FIG. 14) has the same magnitude as if thermocapillary convection acts in opposite direction to gravity (case M–B) in an environment with considerable reduced Bond number or reduced gravity level (FIG. 13, $Bo = -0.013$).

MARANGONI CONVECTION IN POLLUTED TEST LIQUIDS AND IN WATER

In the previous sections, thermocapillary convection in very pure liquids was discussed. In the experiments, the temperature gradient chamber was cleaned meticulously. The cleaning procedure consisted of multiple flushes with chromatic sulfuric acid and finally multiple flushes with clean test liquid.

Nevertheless, for long residence times of the test liquids inside the chamber, a significant decrease in heat transfer by thermocapillary convection was observed. In particular, this effect was observed at long chained alcohols and distinctly in water.

In FIGURE 15 this observation is shown using the example of 1-pentanol in the case M–B ($V_{Bl} = 0.8\,\text{cm}^3$, $H = 2.95\,\text{mm}$, the bubble surface touches both the heating and the cooling plate). Thermocapillary convection in fresh 1-pentanol leads to a maximum in the integral heat transfer and approaches the correlation Nu over Ma for all clean alcohols investigated.[1] However, after a long residence time inside the chamber a significant decrease in the integral heat transfer was recognized. This decrease was observed both when injecting the bubble into the liquid before each single measurement and after leaving the bubble where it is between two measurements. In FIGURE 15 the curves Nu over Ma were investigated for residence times of 5 hours, 14 hours, 1 day, and 2 days.

Each measurement of Nu in FIGURE 15 was independent of the history prior to the measurement. The steady state heat transfer at a certain Ma number is not influenced by a premeasurement at a higher or smaller Ma number, for instance.

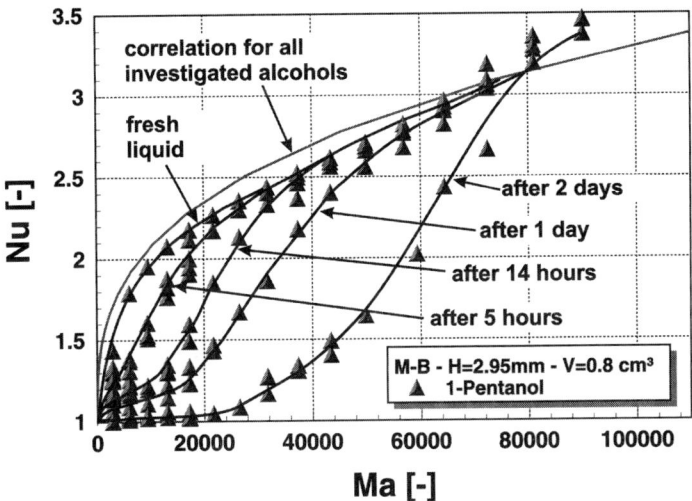

FIGURE 15. Case M–B: the stationary heat transfer at a bubble $V_{Bl} = 0.8\,\text{cm}^3$. Investigation of the influence on the time of residence of the liquid 1-pentanol inside the temperature gradient chamber ($H = 2.95\,\text{mm}$).

For high Ma numbers $Ma > 80{,}000$, the heat transfer around the bubble inside 1-pentanol reached the heat transfer by thermocapillary convection in the fresh liquid, even for a residence time of two days. The heat transfer of the pure 1-pentanol at $Ma < 80{,}000$ could not be reached after a reduction of the Ma number subsequent to a premeasurement at $Ma > 80{,}000$ for instance.

For bubbles in water on Earth, thermocapillary convection could only be observed during a small time period after the cleaning procedure. Multiple flushes with clean and pure water (LiChrosolv, Merck) finished the cleaning process. Subsequently, thermocapillary convection in the LiChrosolv water proceeded continuously for a time period of one to two minutes (see FIGURE 16).

At high Ma numbers oscillatory thermocapillary convection developed in LiChrosolv water (see FIGURE 17, $Ma \approx 65{,}000$).

After five to ten minutes this oscillating thermocapillary convection extinguished and could only be reactivated by a further increase of the temperature gradient. In cases where water from the tap was used instead of LiChrosolv water thermocapillary convection could only be observed at high global temperature gradients $\Delta T/H$ or $Ma > 50{,}000$ as a small oscillating recirculation zone at the bubble base, which extinguished after 10 to 20 seconds. One-step distilled water and sterile water (Ampuwa, Fresenius) was also not suitable for enabling permanent thermocapillary convection.

In water the steady state overall heat transfer was equal to the value of pure heat conduction around the bubble, because the thermocapillary convection discontinued asymptotically before reaching the steady state heat transfer.

Kao and Kenning,[6] Platten and Villers,[24] Platten et al.,[25] and Ito et al.[26] also observed thermocapillary convection only in very pure liquids. Two explanations are available from the literature. The first theory explains that a film exists on the liquid surface, which provides non-slip-conditions for the liquid, and results from air pollution, absorption processes at the liquid surface, or pollution of the water itself. The second theory, which is discussed intensively in Reference 8, is based on the fact that the surface tension is a function of the temperature and the concentration. Due to lack of space only a short summary can be given here. If surface active substances exist, the surface tension of liquids is reduced locally. For example the temperature dependence of the surface tension of pure water $\sigma_0(T)$ is illustrated in FIGURE 18. Huplik and Raithby[28] showed that the surface tension of water with solute surface active substances also decreases with increasing temperature. Thus,

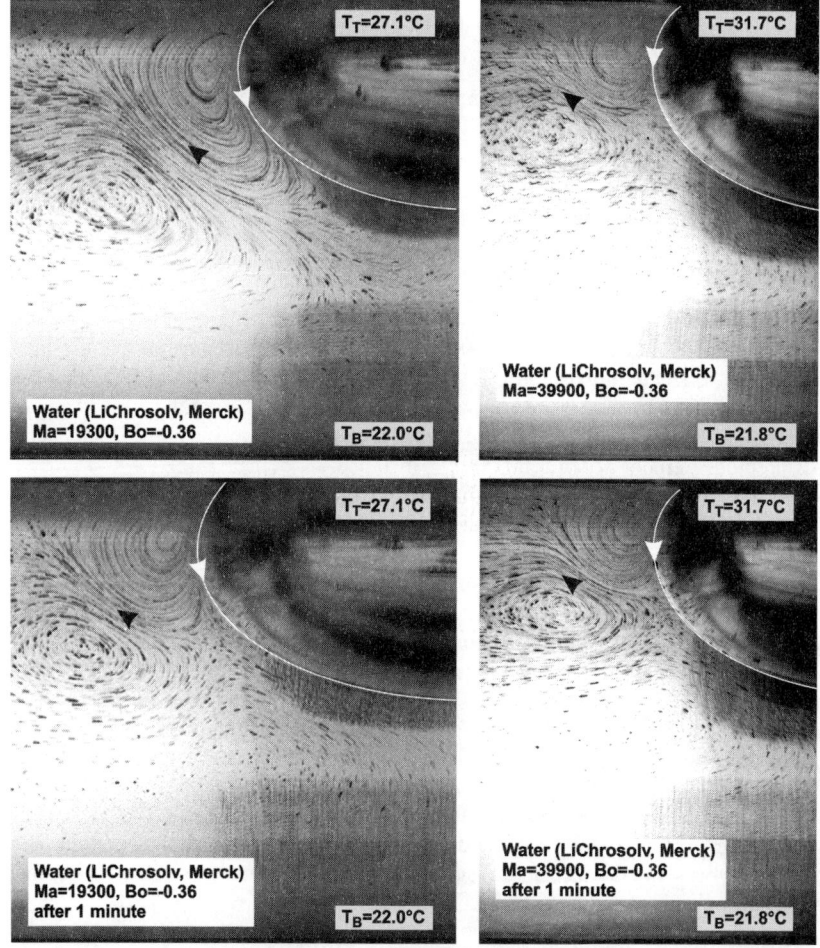

FIGURE 16. Case M–B: thermocapillary convection in water at $Ma = 19,300$ and $Ma = 39,900$, $Bo = -0.36$, $H = 11.2$ mm, $B \approx 5.3$ mm, $V_{Bl} = 0.4$ cm^3. The pictures in the top row were made at $t = 30$ sec, in the bottom row at $t = 60$ sec after injecting the bubble.

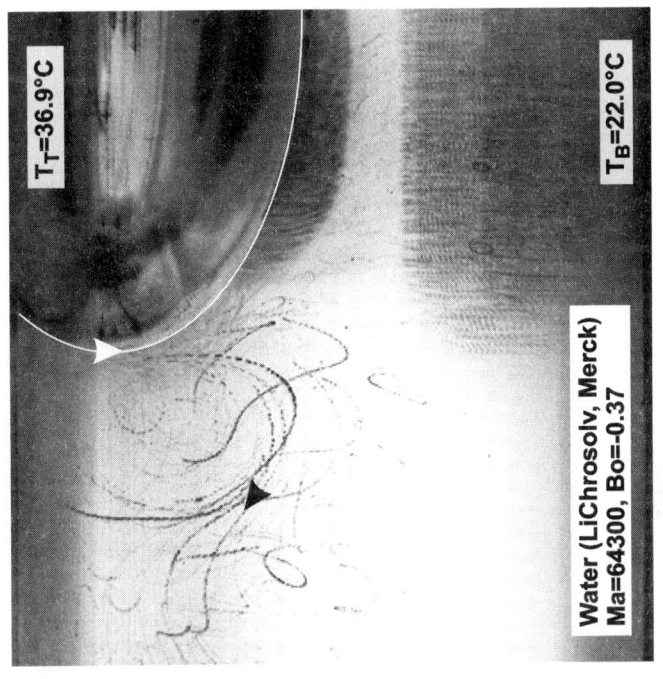
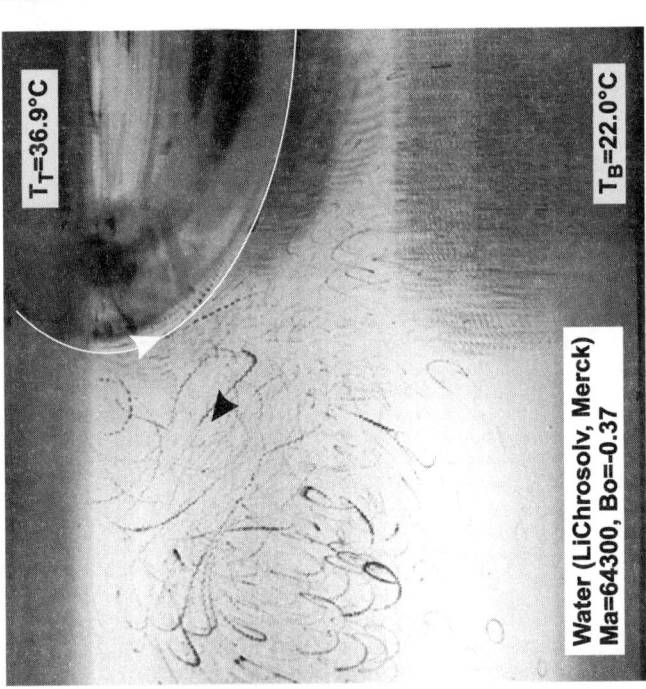

FIGURE 17. Oscillatory thermocapillary convection in water in the case M–B at $t = 30$ sec after injecting the bubble in an existing temperature gradient field. $H = 11.2$ mm, $V_{Bl} = 0.4 \text{cm}^3$, $B \approx 5.3$ mm, $Ma = 64{,}300$, $Bo = -0.36$.

temperature dependencies of the surface tension of water, with varying degrees of solution $\sigma(T) = 0.99\sigma_0(T)$, $\sigma(T) = 0.95\sigma_0(T)$, and $\sigma(T) = 0.9\sigma_0(T)$ are also included in FIGURE 18. On the one hand, thermocapillary convection should result at any degree of solution due to the decreasing surface tension when a temperature gradient is subjected to the surface. On the other hand, the surface active substances are transported in direction of increasing surface tension by the thermocapillary convection itself. If the desorption process at the cold side of the liquid is equal to the convective transport of the surface active substances and if the absorption process for the surface active substances at the warm side of the free surface is slower than the convective transport by the thermocapillary convection, σ would increase in warmer surface areas (FIG. 18a). Other surface tension characteristics (FIG. 18b–d) are also theoretically conceivable for various relations between diffusive and convective processes. In FIGURE 18c, the surface tension gradient extinguishes and the surface behaves like a non slip boundary. In FIGURE 18d, two vortices, like those shown in the bottom row of FIGURE 16, can appear theoretically and temporarily.

Furthermore, if the desorption process at the bubble head is slower than the convective process, surface tension distributions like those shown in FIGURE 18f and g are possible.

In Betz,[8] a critical temperature difference Equation (5) is derived for two special cases. In the first case it is assumed, that the surface active substances decline at the warm side and do not concentrate on the cold side (FIG. 18c). In the second case, the surface active substances uniformly decline on the warm side and concentrate on the cold side of the surface, where thermocapillary convection acts (σ decreases at the bubble head and increases at the bubble base by a value $\Delta\sigma_{cg}$ in curve FIG. 18g).

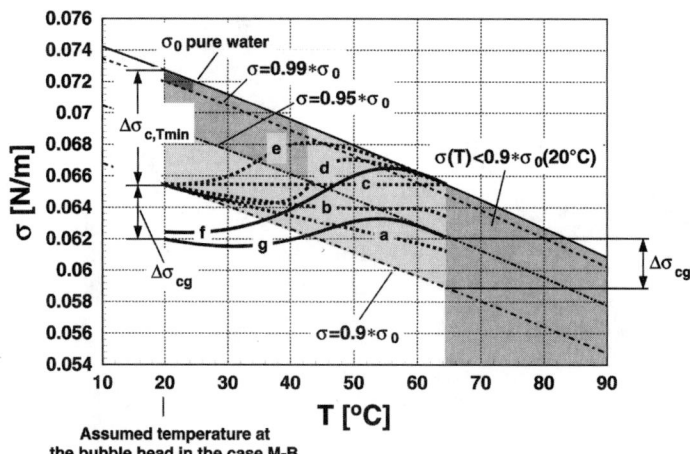

FIGURE 18. Temperature dependence of surface tension of pure and polluted water. Assumption of local distributions of the surface tension at a bubble surface in the case M–B, when the temperature is 20°C at the bubble head, for example, and the temperature at the bubble base is $T > 20°C$.

$$\Delta T_{crit} = \frac{\Delta \sigma_{c, T_{min}}}{\left|\frac{d\sigma}{dT}\right|_{cl, T_{min}}}. \tag{5}$$

In other cases, the onset of thermocapillary convection is a function of $\Delta\sigma_{c, T_{min}}$, the dependence of $d\sigma/dT = (d\sigma/dT)(T,c)$ and the absorption and desorption coefficients of the surface active substances at the surface.

SUMMARY

The combined numerical and experimental investigation showed that the thermocapillary convection is a natural effect in two-phase systems and plays an essential role as a heat transfer mechanism. This is valid both in a microgravity environment and under Earth gravity in liquids whose interfaces are not contaminated by impurities. It is strong enough to drive liquids against the direction of buoyancy on earth. In relation to the situation on Earth the overall heat transfer is higher by a factor of three to four under microgravity if the finite geometric configuration is comparable in size to the extension of the bubbles.

The dimensionless numbers Ma, Bo, Pe, and Pr are suitable for the description of the stationary flow and heat transfer. The dependence of the steady state flow and heat transfer on the Prandtl number is negligible in the region $7 \leq Pr \leq 120$ when the thermophysical properties of the liquids are inserted at an average temperature in the region of the main thermocapillary flow during the evaluation of experiments. This criterion is fulfilled in all simulations due to the Boussinesq-approximation. For small Prandtl numbers $Pr < 7$ a dependence of the flow and heat transfer on Pr must be considered.

Experimental investigations with long chained alcohols and water showed a big influence of the purity of the liquids on the onset and the steady state development of thermocapillary convection. Strong thermocapillary convection and high heat transfer rates could only be reached, in practice, if enormous precautions were taken concerning the purity of the liquids. This behavior was described by a physical model. Nevertheless, the influence of the relevant parameters (e.g., absorption and desorption coefficients of the surface active substances) has not been discussed to date. An extension of the numerical model with a diffusion, adsorption, and desorption model is highly desirable in order to clarify the influence of the surface active substances and the interaction of thermocapillary and diffusocapillary convection. As a result, the real critical Marangoni number for the onset of thermocapillary convection and the reason for the extinguishing thermocapillary flow for old liquids could then be examined more detailed.

ACKNOWLEDGMENTS

This work was supported by the DFG (Deutsche Forschungsgemeinschaft) through grant Str 117/34 - 1-3, which is gratefully acknowledged.

REFERENCES

1. BETZ, J. & J. STRAUB. 2001. Numerical and experimental study of the heat transfer and fluid flow by thermocapillary convection around gas bubbles. Heat Mass Transfer **37**: 215–227.
2. KAHLWEIT, M. 1981. Grenzflächenerscheinungen. Dr. Dietrich Steinkopff Verlag, Darmstadt.
3. LEVICH, V.G. 1962. Physicochemical Hydrodynamics. Prentice-Hall, Englewood Cliffs.
4. RAAKE, D. & J. SIEKMANN & C.-H. CHUN. 1989. Temperature and velocity fields due to surface tension driven flow. Exper. Fluids **7**: 164–172.
5. MAREK, R. 1996. Einfluss thermokapillarer Konvektion und inerter Gase beim Blasensieden in unterkühlter Flüssigkeit. Ph.D. Thesis, Lehrstuhl A für Thermodynamik, Technische Universität München.
6. KAO, Y.S. & D.B.R. KENNING. 1972. Thermocapillary flow near a hemispherical bubble on a heated wall. J. Fluid Mechan. **53**: 715–735.
7. STRAUB, J. & A. WEINZIERL & M. ZELL. 1990. Thermokapillare Grenzflächenkonvektion an Gasblasen in einem Temperaturgradientenfeld. Wärme- und Stoffübertragung **25**: 281–288.
8. BETZ, J. 1997. Strömung und Wärmeübergang bei thermokapillarer Konvektion an Gasblasen. Ph.D. Thesis, Lehrstuhl A für Thermodynamik, Technische Universität München. Herbert Utz Verlag Wissenschaft, München.
9. BALIGA, B.R & S.V. PATANKAR. 1980. A new finite-element formulation for convection-diffusion problems. Num. Heat Transfer **3**: 393–409.
10. BALIGA, B.R & S.V. PATANKAR. 1983. A control volume finite-element method for two-dimensional fluid flow and heat transfer. Num. Heat Transfer **6**: 245–261.
11. BALIGA, B.R & S.V. PATANKAR. 1988. Elliptic systems: finite-element method II. *In* Handbook of Numerical Heat Transfer. John Wiley & Sons, Inc. New York.
12. BALIGA, B.R. 1978. A Control-Volume Based Finite Element Method for Convective Heat and Mass Transfer. Ph.D. Thesis, University of Minnesota, Minneapolis.
13. PRAKASH, C. 1986. An improved control volume finite-element method for heat and mass transfer, and for fluid flow using equal-order velocity–pressure interpolation. Num. Heat Transfer **9**: 253–276.
14. PRAKASH, C. & S.V. PATANKAR. 1985. A control volume-based finite-element method for solving the Navier–Stokes equations using equal-order velocity–pressure interpolation. Num. Heat Transfer **8**: 259–280.
15. OHNISHI, M. & H. AZUMA & T. DOI. 1992. Computer simulation of oscillatory marangoni flow. Acta Astronaut. **26**: 685–696.
16. ZHOU, H. & A. ZEBIB. 1992. Oscillatory convection in soldifying pure metals. Num. Heat Transfer, Part A **22**: 435–468.
17. PATANKAR, S.V. 1980. Numerical Heat Transfer and Fluid Flow. Hemisphere Publishing Corporation, Washington, London. (Mc Graw Hill Book Company, New York, St. Louis, San Francisco.)
18. HARTLAND, S. & R.W. HARTLEY. 1976. Axisymmetric Fluid–Liquid Interfaces. Elsevier Scientific Publishing Company, Amsterdam, Oxford, New York.
19. BIRD R.B. & W.E. STEWART & E.N. LIGHTFOOT. 1960. Transport Phenomena. John Wiley & Sons Inc., New York.
20. JOSEPH, D.D. 1976. Stability of Fluid Motions. Springer Verlag, Berlin, Heidelberg, New York.
21. OVERBECK, A. 1879. Über die Wärmeleitung der Flüssigkeiten bei Berücksichtigung der Strömungen infolge von Temperaturdifferenzen. Annal. Physik Chemie, Neue Folge **7**: 271–292.
22. ALAMGIR, M. & J.H. LIENHARD. 1978. The temperature dependence of surface tension of pure fluids. Trans. ASME **100**: 324–329.
23. DAUBERT, T.E. & R.P. DANNER. 1989. Physical and Thermodynamic Properties of Pure Chemicals—Data Compilation. Design Institute for Physical Property Data, American Institute of Chemical Engineers. Hemisphere Publishing Corporation, New York.

24. PLATTEN, J.K. & D. VILLERS. 1988. On thermocapillary flows in containers with differentially heated side walls. *In* Physicochemical Hydrodynamics—Interfacial Phenomena. M.G. Velarde, Ed.: NATO ASI, Series B, Physics. Vol. **174:** 311–336. Plenum Press, New York.
25. PLATTEN, J.K., D. VILLERS & O. LHOST. 1988. LDV study of some free convection problems at extremely slow velocities: Soret driven convection and Marangoni convection. *In* Laser Anemometry in Fluid Mechanics. R.J. Adrian, T. Asanuma, *et al.*, Eds.: 245–260. Instituto Superior Technico. Ladoan, Lisboa.
26. ITO, A., S.K. CHOUDHURY & T. FUKANO. 1990. Heated liquid film flow and its breakdown caused by Marangoni convection (the characteristic flow of pure water). JSME Intl. J. **33:** 128–133.
27. WEINZIERL, A. 1984. Untersuchung des Wärmeübergangs und seiner Transportmechanismen bei Siedevorgängen unter Mikrogravitation. Ph.D. Thesis, Lehrstuhl A für Thermodynamik, Technische Universität München.
28. HUPLIK, V. & G.D. RAITHBY. 1972. Surface-tension effects in boiling from a downward-facing surface. J. Heat Transfer **94:** 403–409.

Experimental Investigation of the Liquid Interface Reorientation upon Step Reduction in Gravity

MARK MICHAELIS, MICHAEL E. DREYER, AND HANS J. RATH

Center of Applied Space Technology and Microgravity, University of Bremen, Am Fallturm, Bremen, Germany

ABSTRACT: Experiments have been carried out to investigate the settling behavior of a free liquid/gas interface in a partly filled right circular cylinder upon step reduction in gravity. Microgravity conditions were obtained within milliseconds after the release of a drop capsule in the drop tower facility in Bremen. In the experiments the cylinder radius was varied from 10 mm to 20 mm with static contact angles of 2° to 60°. In addition to a series of silicone fluids, two different test liquids with a refractive index matched with the value of the cylinder material were used for experiments. The use of a digital high-speed recording system with a recording frequency of up to 500 fps allowed both an observation of the entire free liquid interface and a detail view on the moving contact line. Digital image processing techniques were applied to detect the contour of the free surface. For the initial condition the system is dominated by hydrostatic forces. In this case the equilibrium of the free liquid surface is characterized by a high Bond number yielding a flat surface and a small liquid ascent at the cylinder wall depending on the static contact angle. After transition to reduced gravity with a very low Bond number, capillary forces govern the flow and a capillary driven reorientation of the liquid to the new equilibrium position is established in a damped oscillation. The particular interest of this study is the investigation of the initial behavior of the free surface reorientation. It was found that the initial rise velocity of the contact point is determined by the Morton number and the static contact angle. Experimental results are presented that show an increasing rise velocity for decreasing Morton numbers. Further results characterize the initial behavior of the free surface at the center point as a function of characteristic time scales.

KEYWORDS: free surface reorientation; step reduction in gravity; capillary rise; circular cylinder; Morton number

NOMENCLATURE:
h_0 fill height
k_z acceleration in z-direction
k_{zi} initial acceleration in z-direction due to gravity
L_c capillary length
p_c capillary pressure
r coordinate
R cylinder radius
n refractive index
t time

Address for correspondence: Mark Michaelis, Center of Applied Space Technology and Microgravity, University of Bremen, Am Fallturm, D-28359 Bremen, Germany.
mm@zarm.uni-bremen.de

t_{pu}	pressure unsteady time scale
t_{puLc}	pressure unsteady time scale with capillary length
u_{pu}	characteristic velocity
u_{puLc}	characteristic velocity for initial flow
z	coordinate
z_c	coordinate of the center point
z_w	coordinate of the contact point
γ_s	static contact angle
γ_{s0}	initial static contact angle
σ	surface tension
ρ	density
ν	kinematic viscosity
Bo	Bond number
Mo	Morton number
Oh	Ohnesorge number

INTRODUCTION

In this study the free liquid/gas interface in a right circular cylinder with radius R, which is partly filled with a liquid (surface tension σ, viscosity ν, and density ρ) to a height h_0 is considered. In the initial situation a gravitational acceleration k_{zi}, aligned with the cylinder axis, is acting on the system. In this case the free liquid interface $h(r, t=0)$ is dominated by hydrostatic forces, and capillary forces can be neglected except at the cylinder wall, where the liquid meets the solid with the initial static contact angle γ_{s0} (see FIGURE 1). The general solution for the equilibrium shape of the free liquid interface in a right circular cylinder is described by the Bond number

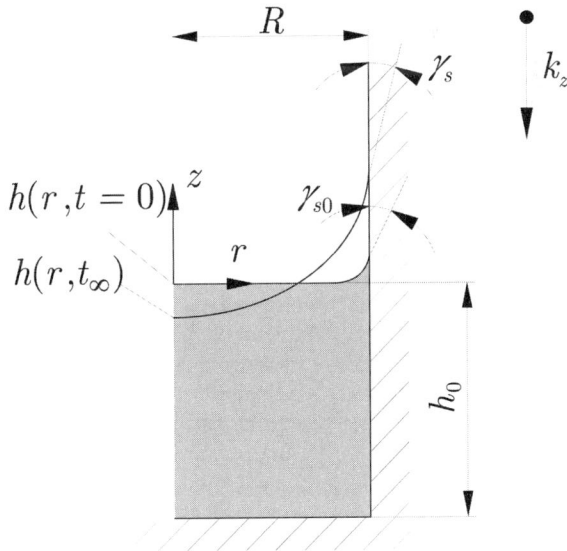

FIGURE 1. Geometry and nomenclature.

$$Bo = \frac{\rho k_{zi} R^2}{\sigma} \qquad (1)$$

and the static contact angle γ_{s0}.[1] The Bond number gives a relation between hydrostatic and capillary forces. Thus, a very high Bond number leads to a flat shape of the interface, whereas a Bond number of zero calls for a perfectly spherical shape with constant curvature. At the wall, where hydrostatic and capillary forces are balanced, the wall coordinate $h(r = R, t = 0) = z_{w0}$ is on the order of $z_{w0} \approx L_c \sqrt{2(1 - \sin\gamma_{s0})}$ (the solution for the rise height at a flat wall), with the capillary length

$$L_c = \sqrt{\frac{\sigma}{\rho k_{zi}}}. \qquad (2)$$

Consequently the local Bond number near the wall is $Bo = \rho k_{zi} L_c^2 / \sigma = 1$ for $L_c < R$.

If a step change in gravity from k_{zi} to k_z (with $k_z/k_{zi} \ll 1$) occurs, capillary forces dominate and a reorientation of the free liquid interface to the new equilibrium position $h(r, t \to \infty)$ of the surface is induced. The free surface reorientation may be separated into two main periods. The first period begins immediately after transition to k_z for the initial capillary rise at the cylinder wall, when the flow is independent of the cylinder radius R (see FIGURE 2). The second main period begins, when the whole system is affected by the disturbance (see FIGURE 3). Thus, the appropriate length

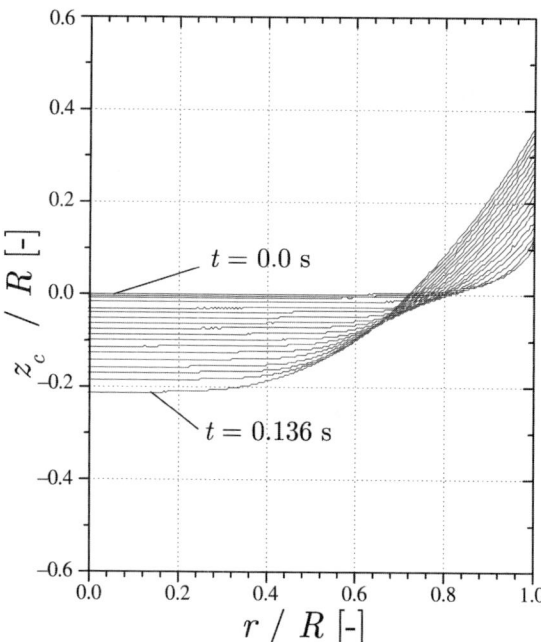

FIGURE 2. Initial development of the free liquid interface upon transition to reduced gravity. Capillary rise at the cylinder wall is basically independent of the cylinder radius. Test liquid: butylbenzene, $R = 20$ mm; $\Delta t = 0.008$ sec. The coordinates are scaled with R.

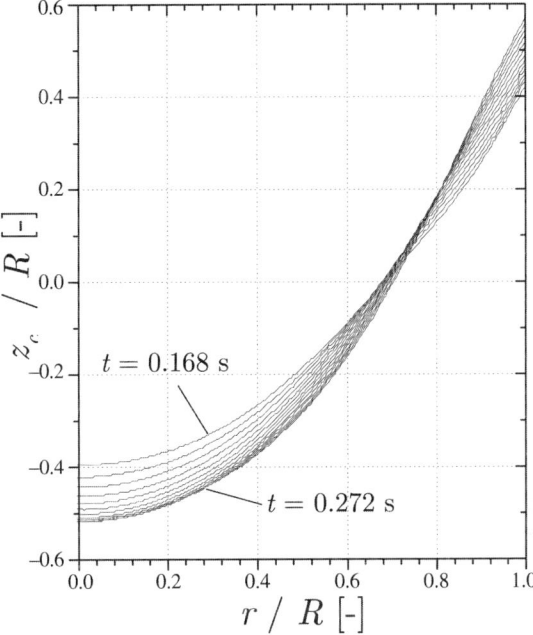

FIGURE 3. Later development of the free liquid interface. Due to the limitation of the bulk liquid the free surface reaches its maximum deflection with a following oscillation around the equilibrium position. In this period the flow depends on the cylinder radius. Test liquid: butylbenzene, $R = 20$ mm; $\Delta t = 0.008$ sec.

scales are L_c for the first period and R for the second period. Scaling the continuity and momentum equations with the pressure unsteady time scale $t_{pu} = \sqrt{\rho R^2/\sigma}$, the characteristic velocity $u_{pu} = R/t_{pu}$, and the capillary pressure $p_c = \sigma/R$ leads to the dimensionless numbers: Bond number $Bo = \rho k_z R^2/\sigma$ and Ohnesorge number

$$Oh = \frac{\mu}{\sqrt{\rho \sigma R}}, \qquad (3)$$

that determine the second period of the reorientation.[2–4] The Ohnesorge number gives a relation for the damping behavior of the system. Low Ohnesorge numbers exhibit an oscillatory reorientation of the free surface with low damping; high Ohnesorge numbers lead to creeping flow without overshot of the equilibrium position. Another important parameter for the reorientation process is the static contact angle γ_s, which affects both the damping and the frequency of oscillation.

The first experimental study for this configuration was carried out by Siegert et al.[5] with containers of varying geometry, partly filled with a liquid with $\gamma_s = 0$. As a result of their study they presented a correlation for the time t_{cr} until the center point $h(r = 0, t)$ passes the first time the equilibrium position:

$$t_{cr} = 0.413 t_{pu}. \qquad (4)$$

In another study from Kaukler,[6] the oscillation behavior of a liquid–liquid interface when exposed to a step reduction in gravity was investigated for various contact angles by varying the ambient temperature. They reported an increasing frequency with rising static contact angles. In an extensive series of drop tower experiments Weislogel and Ross[2] investigated the behavior of the free liquid–gas interface for this configuration by varying the cylinder radius, the test liquids, and the static contact angle. Evaluation of the history of the interface center line also shows an increasing frequency if the static contact angle is increased. As a further result of their study they presented a correlation for the settling time of the liquid interface, when oscillations are no longer detectable, which depends on the Ohnesorge number and γ_s. Since the Ohnesorge number determines the damping behavior, the settling time rises with lower Ohnesorge numbers. A clear trend for the influence of the static contact angle on the settling time is not given.

In a theoretical investigation of Bauer and Eidel[7] the oscillation of a free liquid/gas interface in a circular cylinder for frictionless flow was examined for microgravity conditions. It was found that the natural frequency, scaled with the time t_{pu}, increases with rising static contact angles. When γ_s exceeds the value of 90° the frequency decreases again. In a numerical study, Woelk et al.[3] modelled the entire process of the free surface reorientation upon step reduction in gravity with a finite element code. By introducing a boundary condition for the moving contact line, they were able to validate experimental data from Weislogel and Ross. In a recent numerical work from Gerstmann et al.,[8] the influence of the Ohnesorge number and γ_s on the settling time and the natural frequency of the free surface in the center line were investigated more in detail. Using also a finite element code, they confirmed the behavior of a rising frequency for higher static contact angles in good agreement with experimental data from Weislogel and Ross. However, a strong influence of γ_s on the settling time was not observed in the investigated range of contact angles.

Although there have been many investigations concerning this configuration, the evaluation of experimental data is mainly restricted to the frequency and the settling time of the surface oscillation. Therefore, validation of numerical calculations is rather limited. Moreover, little attention was paid to the initial capillary rise at the cylinder wall (herein already called the first period), which is characterized by a moving three phase contact line with varying dynamic contact angle. Since the flow in the first period determines the first amplitude of oscillation, which in the subsequent reorientation is damped to the equilibrium position, knowledge of the initial flow characteristics is quite important.

Scaling of Initial Capillary Rise

As described above, for the initial situation, the capillary length L_c defines a length scale near the cylinder wall where capillary and hydrostatic forces are balanced. Therefore, for the initial capillary rise L_c should be the appropriate length scale, since other longitudes can not determine the flow in this region. Assuming that capillary and inertia forces dominate the process leads to the pressure unsteady time t_{pu} to be the relevant time scale. With the capillary length L_c as the appropriate length scale, the time scale becomes

$$t_{puLc} = \left(\frac{\sigma}{\rho k_{zi}^3}\right)^{1/4}. \tag{5}$$

Analogous to the scaling for the characteristic velocity u_{pu} in the second period, the velocity scale $u_{puLc} = L_c/t_{puLc}$ for the first period is

$$u_{puLc} = \left(\frac{k_{zi}\sigma}{\rho}\right)^{1/4}, \tag{6}$$

and the capillary pressure becomes $p_c = \sigma/L_c = \sqrt{\sigma\rho k_{zi}}$. Scaling the continuity equation and the Navier–Stokes equation with L_c, t_{puLc}, u_{puLc}, and p_c leads to

$$\frac{\partial}{\partial r}(u_r, r) + r\frac{\partial}{\partial z}(u_z) = 0, \tag{7}$$

and

$$\frac{\partial u_r}{\partial t} + u_r\frac{\partial u_r}{\partial r} + u_z\frac{\partial u_r}{\partial z} = -\frac{\partial p}{\partial r} + Mo\left\{\frac{\partial^2 u_r}{\partial r^2} + \frac{1}{r}\frac{\partial u_r}{\partial r} + \frac{\partial^2 u_r}{\partial z^2} - \frac{u_r}{r^2}\right\}, \tag{8}$$

$$\frac{\partial u_z}{\partial t} + u_r\frac{\partial u_z}{\partial r} + u_z\frac{\partial u_z}{\partial z} = -\frac{\partial p}{\partial r} + Mo\left\{\frac{\partial^2 u_z}{\partial r^2} + \frac{1}{r}\frac{\partial u_z}{\partial r} + \frac{\partial^2 u_z}{\partial z^2}\right\} + \frac{k_z}{k_{zi}}, \tag{9}$$

where all variables are given in dimensionless form. According to this scaling, two dimensionless groups, the acceleration ratio kz/k_{zi} and the Morton number

$$Mo = \left(\frac{k_{zi}v^4\rho^3}{\sigma^3}\right)^{1/4}, \tag{10}$$

determine the capillary rise in the first period of the reorientation. The Morton number (in a notation without the fourth root) is generally applied in the multiphase flow community and gives a relation for the motion of drops and bubbles. Here the Morton number formally corresponds to an Ohnesorge number with the capillary length used as appropriate length scale. It can be considered to be a measure for the velocity of the capillary rise at the cylinder wall. Choosing the above scale parameters leads, for the first period, to a Bond number $Bo = 1$. An additional parameter that has to be considered is the static contact angle, since this determines the initial and final position of the interface.

EXPERIMENTS

Apparatus

The experimental setup was integrated in a drop capsule that was released in the drop tower Bremen to yield reduced gravity conditions for an experiment time of 4.74 sec.

The main components of the setup are a circular cylinder (radius $R = 10$–20 mm), a high-speed digital recording system and a background illumination device. The test setup is shown schematically in FIGURE 4. To minimize optical disturbances resulting from general light reflection and refraction, caused by different refractive indices between vessel material and the test liquid, the cylinder is manufactured from a solid PMMA cube with surface roughness better than $1.0\,\mu m$. The cylinder axis is aligned with the gravity vector by using a levelling system on which the vessel is mounted to keep the flow axial symmetric and, therefore, two-dimensional. In order to prevent impurity of the test fluid and the cylinder wall, the vessel is sealed with a cover plate and a breather hose after preparation of the experiment. Light diffusers between the LED panel and the vessel enable a homogenous light distribution from behind. Black cardboard, mounted at the lateral faces of the vessel, prevents disturbances from lateral light reflection. Filling the vessel with test liquid is achieved by interconnecting the inlet at the bottom of the vessel with a large glass syringe (acting as a reservoir) located on a lower platform. A stepping motor ensures a uniform and slow filling of the vessel just prior to the drop of the capsule to guarantee a uniform initial condition at the contact line of a non-wetted cylinder wall above the contact line. The fill height of the liquid is always $h_0 > 2R$ to prevent interaction of the flow with the bottom boundary of the vessel.

A high speed digital recording system (Kodak Motion Corder Analyzer, Series SR) with two CCD cameras (with Nikon MICRO-NICCOR lenses: $f = 105\,mm/1:2.8$ and $f = 55\,mm/1:2.8$) with 512×480 pixel array was used to observe the reorientation process with a recording frequency between 250 and 500 fps. The application of a beam splitter cube enabled the observation of two different areas of the interface

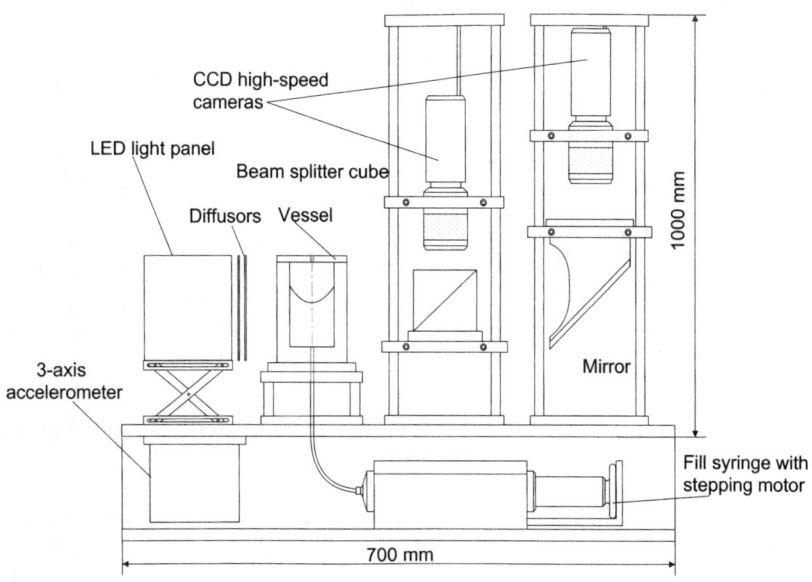

FIGURE 4. Experimental setup.

TABLE 1. Development of the free liquid interface for butylbenzene: $R = 20\,\text{mm}$, $Oh = 1.5 \times 10^{-3}$, $\gamma_s = 1°$, $\Delta t = 0.024\,\text{sec}$

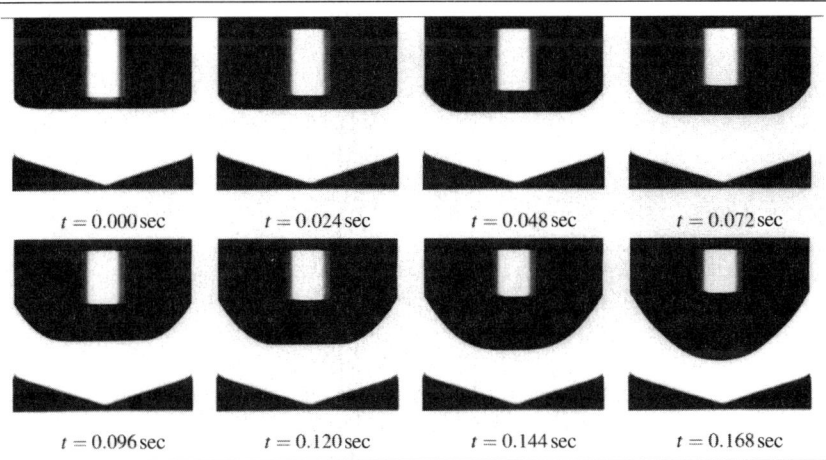

contour. A detail view on the moving contact line and a general view on the meniscus were adjusted, with a pixel resolution of 19–85 µm, depending on the chosen vessel diameter. The optics of the camera system was focussed on the cylinder axis by accurately aligning a high precision glass measure into the cylinder. A series of video images, recorded with the high-speed digital video system, shows the development of the sharp fluid interface after step reduction in gravity in TABLE 1. The constant time step between consecutive images is $\Delta t = 0.024\,\text{sec}$. The test liquid appears bright, the gas above the meniscus appears dark due to total reflection. The dark triangles at the bottom are caused by a screen to enhance the contrast of the liquid interface. Temperature sensors were placed on the bottom of the experiment platform, at the glass syringe and at the vessel to monitor and record the temperature during the experiment. Measurements of the residual accelerations and the angular velocities acting on the drop capsule were carried out with an inertial sensor package (three-axis accelerometer and cyberoptical gyroscope) that was mounted beneath the experiment platform. Accurate cleaning of the test equipment was performed by rinsing the vessel and the glass syringe several times with toluene and ethyl alcohol.

TABLE 2. Properties of the test liquids for $T = 25°$

Test liquid	σ (mN/m)	ρ (kg/m^3)	v (mm^2/sec)	γ_s	n
SF 0.65	15.4	757	0.53	38	1.377
SF 1	16.6	814	1.00	48	1.384
SF 3	18.1	879	3.0	54	1.394
SF 5	18.8	907	4.9	57	1.398
SF 10	19.5	931	9.0	60	1.401
Butylbenzene	25.3	857	1.15	2	1.488
Propylbenzene	28.5	858	0.78	3	1.492

Subsequently, the vessel was exposed to vacuum for a minimum of three hours, to reduce the capacity of solved water in the PMMA vessel.

Test Liquids

A number of different silicone fluids (abbreviated with SF) with a nominal viscosity between 0.65 cSt and 10.0 cSt were used in this investigation, as shown in TABLE 2. In order to vary the contact angle from $\gamma_s = 0$, a thin film of a surface modifier (FC-732 by 3M Co.) with a low surface energy of $11-12 \times 10^{-3}$ N/m was applied to the cylinder wall just prior to each drop, after the cleaning procedure, by filling the cylinder with FC-732 as a liquid and spilling it rapidly to achieve a homogenous film. The thin film of a thickness on the order of 1 μm was dried at ambient air temperature. This surface modification and the assembly of the vessel was carried out in a glove box to prevent impurities from dust. For the silicone fluids static contact angles between 38° and 60° were obtained.

Since for silicone fluids the contact line in the vicinity of the cylinder wall could not be observed accurately because of a different refractive index between the liquid and the vessel material ($n = 1.491$), two additional test liquids (butylbenzene and propylbenzene) were used, with refractive index roughly matching the value of PMMA. For all experiments presented with butylbenzene and propylbenzene no surface modifier was applied to the cylinder wall to achieve contact angles of $\gamma_s = 2°-3°$. All contact angles were geometrically determined from the equilibrium surface shape assuming a spherical geometry. The static contact angles given in TABLE 2 are mean values for experiments with the same liquid. The accuracy for the contact angles is on the order of ±2°, which corresponds to a total error in determination of the interface contour of ±8 pixels for butylbenzene and propylbenzene (low contact angles) and ±4 pixels for silicone fluid (high contact angles). The initial static contact angle γ_{s0} was found to be about 5°±3° higher, than the final static contact angle γ_s, which is caused by the filling procedure of the vessel. The liquid properties given in TABLE 2 were measured with standard laboratory equipment.

DATA EVALUATION

Digital Image Processing

The recorded 8-bit intensity images were stored in Windows Bitmap (BMP) format either in a 512×480 pixel array with a recording frequency of 250 fps or in a 512×240 pixel array with 500 fps. Digital image processing techniques were applied with the commercial software packet Matlab (Version 5.3.0.10183 [R11]) for the detection of the liquid interface by applying a 3×3 sobel filter on the video images.

The strategy of the edge detection routine is to search for the maximum (from a dark region to a bright region) vertical gradient of each column per image, starting at the cylinder axis where the interface is definite and easy to find. For edge coordinates found in adjacent columns, a limited derivation from the latter edge coordinate is accepted to ensure that a coherent interface contour is found. In this way, the interface contour can be detected in less than a second per image. For experiments with index matched liquids the accuracy of the routine for detection of the interface contour is on the order of ±1 pixel. However, due to the low static contact angle with a final thin film, the contact point is difficult to determine. Thus, for the equilibrium configuration, the error of ±8 pixels mentioned above has to be considered for calculation of the static contact angle.

For experiments with silicone fluid the contrast of the interface is less high than for experiments with index matched liquids. Additionally the region near the contact line is afflicted with disturbances due to refraction. However, detection of the contact point on the cylinder axis is definite, since total reflection, caused by a high static contact angle, exhibits an edge of high contrast. Thus, an error of ±4 pixels is considered for the evaluation of the static contact angle.

The calibration of the horizontal and vertical pixel size is carried out by recording a high precision glass measure, which was dipped into the test liquid for identical experiment conditions. In order to prevent impurity of the test fluid, the procedure was done prior to the final cleaning procedure and preparation for drop of the capsule. The first image of the experiment recording series was taken to define the origin of the coordinate system and the angle of rotation of the images. After calibration of the image pixels, the detected interface contour was transformed and, if necessary, rotated from pixel coordinates into real coordinates. The resolution of the detected interface contour is 38–85 μm for the total view and 19–43 μm for the detail view, depending on the chosen configuration. The experimental time is defined to start one image before a significant alteration in the video image (image vibration or change of the free liquid interface) could be observed. The inaccuracies resulting from this definition of the time are on the order of ±2 images, which corresponds to ±0.004–0.008 seconds, depending on the recording frequency. As a result of the digital image processing, contour history plots with high resolution, as shown in FIGURES 2 and 3, were obtained.

Reorientation Characteristics

As described above, the main interest of this study is the investigation of the flow characteristics in the first time period, where the reorientation is mainly independent of the cylinder radius. The reorientation behavior of the liquid interface can be analyzed, if the histories of specific points on the liquid interface, like the center point $h(0,t)$ and the contact point $h(R,t)$ are observed. The histories can be extracted from the contour plots and show significant characteristics. Immediately after step reduction in the acceleration level from k_{zi} to k_z the liquid interface starts to rise along the cylinder wall and induces a disturbance with a wavelength on the order of L_c to move towards the cylinder axis. The velocity of this disturbance is on the order of $L_c/t_{puLc} = u_{puLc}$. Therefore, the disturbance reaches the cylinder axis at a time scale

$$t_{RLc} = \frac{R}{u_{puLc}} = \left(\frac{\rho R^4}{k_{zi}\sigma}\right)^{1/4}. \tag{11}$$

The first characteristic in the history of the center point is a weak kink (change of slope) after a time, denoted by t_{cd} in FIGURE 5, that might give a relation to the arrival of the disturbance. However, the moving disturbance itself cannot be resolved since, due to the camera view, a point $h_1(r_1,t) > h_2(r_2,t)$ with $r_1 < r_2$ is difficult to observe because of the total reflection. Before reaching the first peak of oscillation, the

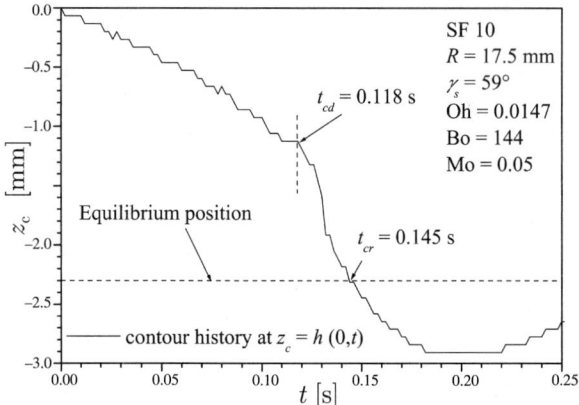

FIGURE 5. Contour history in the cylinder axis at $h(0,t)$ versus time. The time t_{cd} defines the change of the slope in the curve, whereas the time t_{cr} describes the first transit through the equilibrium position.

center point passes the final equilibrium position for the first time at t_{cr}. Since the equilibrium position depends on the cylinder radius R and the static contact angle γ_s, these parameters must influence the time t_{cr}.

In the history of the contact point $h(R,t)$, where the origin of the coordinate system is located at the initial position z_{w0}, two main characteristics can be observed. For experiments with silicone fluids the contact point shows an overshoot beyond the final equilibrium position which decreases with increasing static contact angle and Ohnesorge number (compare FIGURE 6). Thus, in experiments with SF 10 and a small

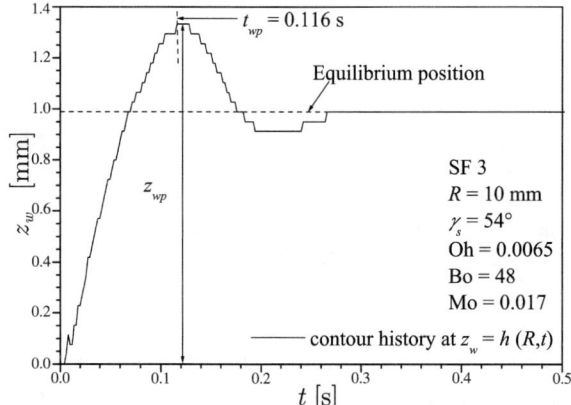

FIGURE 6. Contour history at the contact point at $h(r,t)$ versus time. The time t_{wp} defines the arrival at the first peak of oscillation, whereas z_{wp} defines the deflection of the contact point at the first peak.

cylinder diameter, an immediate fixing of the contact point can be observed, when the equilibrium position is reached. On the other hand for butylbenzene and propylbenzene with a very low contact angle no overshoot of the contact point occurs. In fact, the contact point for test liquids with a low static contact angle rises to a certain point, here also denoted by z_{wp}, with subsequent resting and a new further rise. The mean initial rise velocity of the contact point is defined to be the ratio of the deflection z_{wp} and the time t_{wp} of the contact point at the first peak, or point of resting for experiments with butylbenzene and propylbenzene:

$$\bar{u}_{wr} = \frac{z_{wp}}{t_{wp}}. \tag{12}$$

The velocity $\bar{u}_{wr}$ gives a good relation for the first period, where the flow is almost independent of the cylinder radius. However, the peak in the history of the contact point, or a respective resting point, depends on the cylinder radius, thus with the above definition of $\bar{u}_{wr}$, little influence of the cylinder radius can be expected.

RESULTS AND DISCUSSION

The characteristic time t_{cd} evaluated from the center point history, when a significant kink (change in slope) can be observed, is shown in FIGURE 7 as a function of the characteristic time t_{RLc}. It can be seen that there is a linear dependence between t_{cd} and t_{RLc}, with no obvious influence of the static contact angle, since experiments with silicone fluid and index matched liquids line up very well. However, a linear regression through the data points shows an offset that could be caused by a delay between the intrinsic disturbance arrival and the occurrence in the observed contour history. Obviously, the arrival of the disturbance should lead to a ridge on the free surface in the cylinder axis, which, for the reasons mentioned above, is difficult to observe. In fact, the contour history at $h(0,t)$ shows the minimum of the travelling disturbance until it arrives in the cylinder axis. On arrival the center point is accelerated

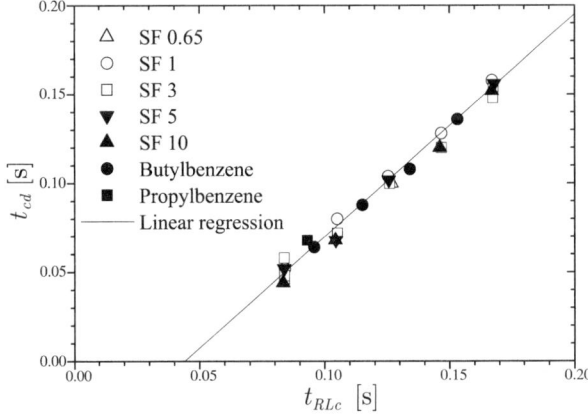

FIGURE 7. Plot of the time t_{cd} versus the time scale t_{RLc}.

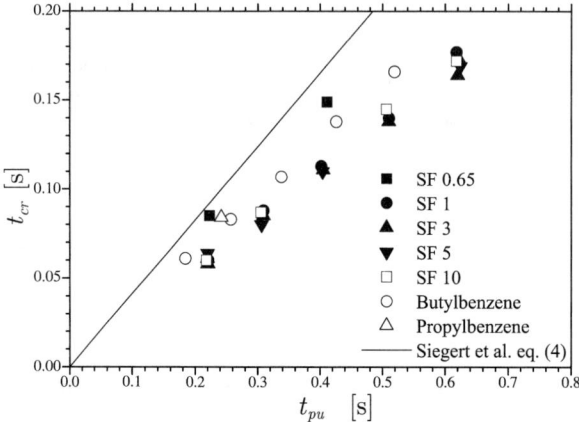

FIGURE 8. Center rise time t_{cr} plotted versus pressure unsteady time scale t_{pu}. The line denotes the correlation from Siegert *et al.* for experiments with a static contact angle of $\gamma_s = 0$.

downward for continuity reasons, which might be the explanation for the observed kink in the history and the offset in FIGURE 7.

The first transit of the free liquid interface through the equilibrium position for reduced gravity conditions (compare FIG. 5), denoted by time t_{cr}, is plotted in FIGURE 8 versus the time scale t_{pu}. As described above, an influence from both the cylinder radius and the static contact angle can be observed. A general trend shows that the time t_{cr} rises with decreasing static contact angle γ_s, which is evident, since the coordinate of the equilibrium position rises as well. The influence of the Ohnesorge number cannot be defined clearly from the present data. It can be seen that experiments with silicone fluid line up very well, with exception of SF 0.65, where

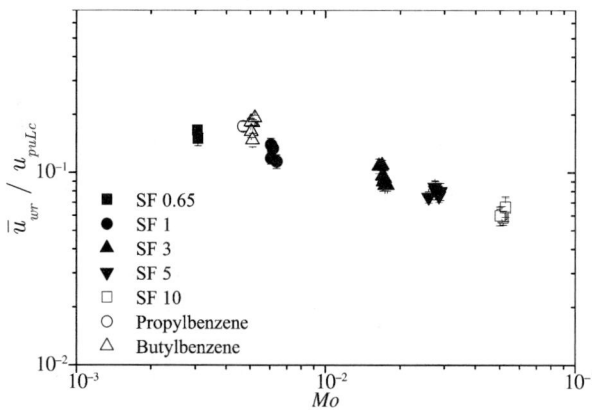

FIGURE 9. Initial mean rise velocity scaled with u_{puLc} versus the Morton number.

high values of t_{cr} are obtained for very low Ohnesorge numbers. These experiments are characterized by a prewetted cylinder wall due to condensation of SF 0.65 in comparison to experiments with other liquids without prewetted walls. The correlation[5] that is valid for experimental data with $\gamma_s = 0$ shows good agreement with the present data, if the trend to higher t_{cd} with decreasing γ_s is considered.

The mean initial rise velocity of the contact point to the first peak t_{wp}, or resting point respectively, is scaled with u_{puLc} and plotted versus the Morton number in FIGURE 9. The graph shows that a decreasing Morton number leads to an increasing dimensionless mean rise velocity of the contact point. This behavior is plausible, since the determining properties in the Morton number (compare Eq. (**10**)) are the viscosity and the density as the hindering factors in the numerator and the surface tension as the driving factor in the denominator. The initial acceleration plays a minor role. Another determining parameter on the rise velocity is the static contact angle. Higher static contact angles lead to lower values for the dimensionless mean rise velocity. Therefore, dependence of $\bar{u}_{wr}/u_{puLc}$ on the Morton number in FIGURE 9 is less strong than could have been supposed, since experiments with silicone fluids show an increasing static contact angle for higher Morton numbers.

CONCLUSIONS

The experiments described in this work were carried in a parameter range of $Mo = 3 \times 10^{-3} - 53 \times 10^{-3}$, $Oh = 0.9 \times 10^{-3} - 20 \times 10^{-3}$, and $Bo = 33-187$ with static contact angles of 2° to 60° of seven different test liquids with PMMA. The aim was to investigate the initial behavior of the free surface reorientation, various characteristics of the history of the center point and the contact point of the liquid interface have been analyzed. Although the initial behavior of the center point cannot be observed, the arrival of the main disturbance in the cylinder axis could be described with the characteristic time scale t_{RLc}. For the first transit of the center point through the final equilibrium position it was observed that the experimental data tend to confirm a correlation from Siegert et al.,[5] which is valid for $\gamma_s = 0$. Variation of the static contact angle to higher values leads to shorter times for first transit through equilibrium. For the initial rise of the contact point, which must be independent of the geometry, it was found by means of scaling the governing equations, that the Morton number, the acceleration ratio, and the static contact angle govern the flow. Plotting of the dimensionless mean initial rise velocity versus the Morton number shows a rising velocity with a decreasing Morton number, whereas variation of the static contact angle to lower values leads to a decreasing velocity. To be able to describe a general relationship between the determining parameters it is desirable to expand the experimental database.

ACKNOWLEDGMENT

This work was funded by the Federal Ministry of Education and Research (BUBF) and the German Aerospace Center DLR contracted under 50 WM 9901. This support is gratefully appreciated.

REFERENCES

1. CONCUS, P. 1968. Static menisci in a vertical right circular cylinder. J. Fluid Mech. **34:** 481–495.
2. WEISLOGEL, M. & H.D. ROSS. 1990. Surface settling in partially filled containers upon step reduction in gravity. NASA TM 103641.
3. WOELK, G., et al. 1997. Damped oscillations of a liquid/gas surface upon step reduction in gravity. J. Spacecraft Rockets **34:** 110–117.
4. GERSTMANN, J., et al. 2000. Surface reorientation upon step reduction in gravity, space technology and applications. International Forum. AIP-Conference 504. 847–853. Albuquerque, NM.
5. SIEGERT, C.E., D.A. PETRASH & E.W. OTTO. 1964. Time response of liquid–vapor interface after entering weightlessness. NASA TN D-2458.
6. KAUKLER, W.F. 1988. Fluid oscillation in the drop tower, Metall. Trans. A. **19A:** 2625–2630.
7. BAUER, H.F. & W. EIDEL. 1990. Linear liquid oscillations in cylindrical container under zero-gravity. Appl. Microgravity Tech. II. **4:** 212–220.
8. GERSTMANN, J., M.E. DREYER & H.J. RATH. 2000. Numerical investigation of the damped oscillation of a liquid/gas surface upon step reduction in gravity. Z. Angw. Math. Mech. **80:** 717–718.

Microscale Heat Transfer to Subcooled Water 200–6,000 psia, 0–3,500 W/cm²

ROBERT H. LEYSE

Inz, Inc., Sun Valley, Idaho, USA

ABSTRACT: Exciting heat transfer phenomena have been discovered with a micron-sized heat transfer element operating in subcooled (20°C) degassed, demineralized water over a wide pressure range (200–6,000 psia) at heat fluxes up to 3,500 W/cm². The platinum heat transfer element (diameter 7.5 microns, length 1.14 mm) is installed within a one-cm³ stainless steel chamber. Sealed electrical terminals penetrate the chamber to effect direct current heating of the platinum element. Pressure is applied pneumatically. The adiabatic heating rate of the element is 6°C per microsecond at 3,700 W/cm²; response is essentially instantaneous for the procedure described herein. The direct current voltage and current are measured from which the power and the resistance (temperature) are determined. The following procedure applies: (1) Pressurize the water-filled stainless steel chamber to 6,000 psia. (2) Apply power at 3,000 W/cm². (3) Maintain constant heat flux as pressure is smoothly reduced from 6,000 psia to 200 psia over a period of 20 seconds. Record voltage, amperage, and pressure at 0.1 second intervals. Heat transfer phenomena thus discovered: (1) Element starting temperature of 370°C at 6,000 psia smoothly increased to 380° as pressure was reduced to 3,970 psia. (2) At 3,970 psia the temperature abruptly stepped upward to 590°C. (3) Temperature smoothly increased to 730°C as pressure was reduced to 3,230 psia. (4) In the vicinity of the critical pressure, the temperature turned around and began smoothly decreasing. (5) At 2,350 psia, the temperature stepped down from 520 to 350°C. (6) Temperature smoothly decreased to 230°C at 190 psia and power was then turned off. Bulk water temperature increased less than 4°C. Controlled gravity (KC-135) tests are planned.

KEYWORDS: microscale heat transfer; ultra-high heat flux; supercritical heat transfer; phase change heat transfer; subcritical boiling heat transfer; intense turbulence

NOMENCLATURE:
C temperature, centigrade
g acceleration due to gravity
psia   absolute pressure, pounds per square inch
W electric power, Watts

INTRODUCTION

The range of applications of phase change heat transfer to industrial separation and chemical reaction processes has recently been substantially broadened. Significant

---

Address for correspondence: Robert H. Leyse, Inz, Inc., P.O. Box 2850, Sun Valley, ID 83353, USA. Voice/fax: 208-622-7740.
Bobleyse@aol.com

phenomena have been discovered in the investigation of microscale heat transfer to subcooled water over a wide pressure range (200–6,000 psia) at ultra high heat fluxes (up to 3,500 W/cm^2). These surprising discoveries open new paths for fundamental research, utilization of *in situ* resources on Mars, and a wide range of industrial applications. The potential for extraterrestrial applications highlights the need to understand the effect of reduced gravity on these phenomena.

The motivation for studies of boiling heat transfer at reduced gravity has largely been directed toward the very well recognized applications of cryogenic liquids for the propulsion, power, and life support systems. The basic studies proposed here will go a long way toward opening up new avenues in applying phase change heat transfer to space process technology.

BACKGROUND

The author has explored several aspects of microscale heat transfer to subcooled water over a wide range of system pressures and heat fluxes. His initial experiments covered the range of heat fluxes from a few thousand W/cm^2 to greater than 4,000 W/cm^2 over a range of pressures up to 6,000 psia. These experiments followed a procedure first reported by Nukiyama,[1] in which a platinum wire served as a heat transfer element as well as a resistance thermometer. That is, the product of the applied direct current and voltage yielded the power to the element, and the quotient of voltage over current yielded the resistance from which the temperature was determined.

Nukiyama varied the power to the platinum heat transfer element and noted a linear increase in the temperature of the element until a certain power was reached, at which point the temperature of the element increased relatively less as the applied power was increased substantially. The steep portion of the curve has subsequently been termed nucleate boiling. Thousands of boiling investigations have followed the pioneering work by Nukiyama. As reported by Dhir[2,3] and others, the work has covered a wide range of fluids, heat fluxes, pressures, geometries, and other factors. However, the work reported herein covers new territory in this field.

Surprised by the ultrahigh heat fluxes that characterized the microscale heat transfer to subcooled water over the wide pressure range, the author turned to an alternate approach to check these phenomena. This led to an alternate procedure; power (heat flux) was held constant as the fluid pressure in the apparatus was varied over a wide range. The wide range of fluid pressure covered subcritical to supercritical pressures in continuous runs. Surprising transitions in heat transfer regimes were thus found.

DISCOVERING THE HEAT TRANSFER PHENOMENA

The Apparatus

These phase change heat transfer phenomena have been discovered with a platinum heat transfer element that is illustrated in FIGURE 1. The platinum element (diameter 7.5 microns, length 1.14 mm) is installed within a one-cm^3 stainless steel chamber that is filled with degassed demineralized water at about 20°C. Sealed

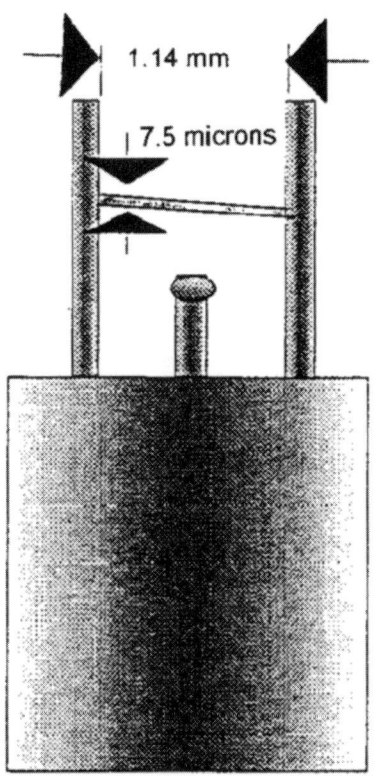

FIGURE 1. Arrangement of 7.5 micron platinum heat transfer element. Note platinum terminals and sheathed thermocouple.

platinum terminals penetrate the chamber to effect direct current heating of the platinum element. The direct current voltage and current are measured from which the power to and the resistance (temperature) of the platinum element are determined. The pressure of the water-filled system is established and controlled with a pneumatic pump and a bleed-off valve. Pressure is measured with a precision pressure transducer. Bulk water temperature is measured with a thermocouple located 0.5 mm below the horizontal platinum heat transfer element.

The physical properties and related characteristics of the platinum element are listed in TABLE 1. Note the adiabatic heating rate of 6°C per microsecond at one watt (heat flux 3,700 W/cm^2) and, therefore, response is essentially instantaneous for the procedures disclosed herein. Note also the extremely high power density at one watt, 20 megawatts per cm^3 or about one megawatt per gram.

The Test Procedure

The surprising transitions in phase change heat transfer were discovered in a series of runs that proceeded as follows: (1) The stainless steel chamber with

TABLE 1. Platinum micro heat transfer element

Property	Dimension
Diameter	7.5 μm
Length	1.14 mm
Area	2.7×10^{-4} cm^2
Volume	5.0×10^{-8} cm^3
Weight	1.07 μg
Density	21.45 g/cm^3
Heat capacity	0.036 cal/g°C
Heat flux at one Watt	3,700 W/cm^2
Power density at one Watt	20 MW/cm^3
Adiabatic heating rate at one Watt	6°C/μsec
Single calibration point	2.61 Ω at 24°C
Calculated resistance	2.75 Ω at 24°C

degassed demineralized water at 20°C was pressurized to 6,000 psia. (2) Power was applied at high heat flux. (3) Power was maintained substantially constant as pressure was smoothly reduced from 6,000 psia to 200 psia over a period of about 20 seconds. (4) Data at heat fluxes ranging from 2,630 to 3,360 W/cm^2 were recorded at 0.1 second intervals.

The Test Results

See FIGURE 2 for a plot of one set of data that serves as an introduction to the transitions in phase change heat transfer. Note the six distinct regions, including the transitions A, B, C, and D in phase change heat transfer at the ultrahigh heat flux, 2,930 W/cm^2. For reference purposes, FIGURE 2 includes key thermodynamic properties of saturated water: Note the plot of saturation temperature of water as a function of pressure (the very lowest plot). Note also the vertical dashed line at the critical pressure of about 3,200 psia. The critical point of water is at the intersection of these plots.

Although high fluid temperatures are reached within the heat transfer films, the bulk water temperature within the one cm^3 increases by less than 5°C during the course of a run. For example, at 3,360 W/cm^2 the power is 0.9 W or about 0.22 calories/sec. Then, in the 20-second duration of the run, the bulk water temperature increase for 1 gram of water is 20 × 0.22 = 4.4°C). Of course, the temperature of the water in the adjacent volumes increases substantially, and the mixing patterns determine an effective local bulk water temperature. These factors are considered later in this paper.

In FIGURE 3, note the plots of four runs. In this discussion, we trace the curves from high pressure to low pressure as the pressure is reduced. In Run 1, the lowest heat flux, 2,630 W/cm^2, the curve is smooth from the starting point of 6,350 psia and 348°C until power is turned off at 193 psia and 230°C. Initially, the temperature of

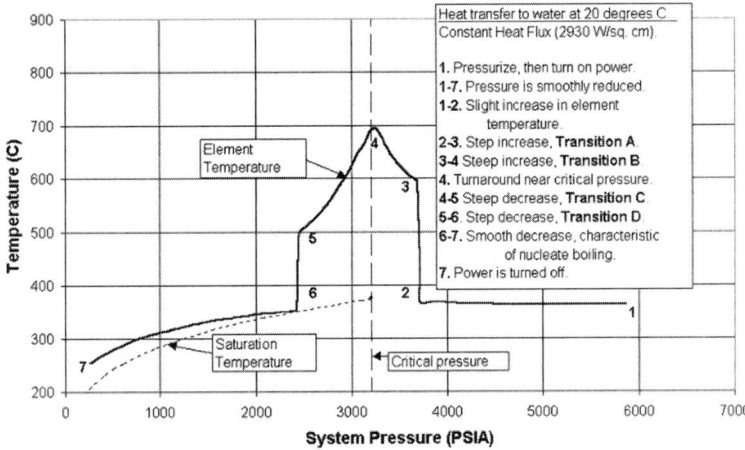

FIGURE 2. Transitions in phase change heat transfer to subcooled water.

the platinum heat transfer element increases gradually as pressure is reduced and at the critical pressure, the temperature peaks at 358°C. At this point the temperature is less than the saturation temperature and begins to smoothly decrease. At about 2,430 psia the temperature is equal to the saturation temperature and then increasingly exceeds the saturation temperature as pressure is reduced until power is turned off at 193 psia and an element temperature of 230°C.

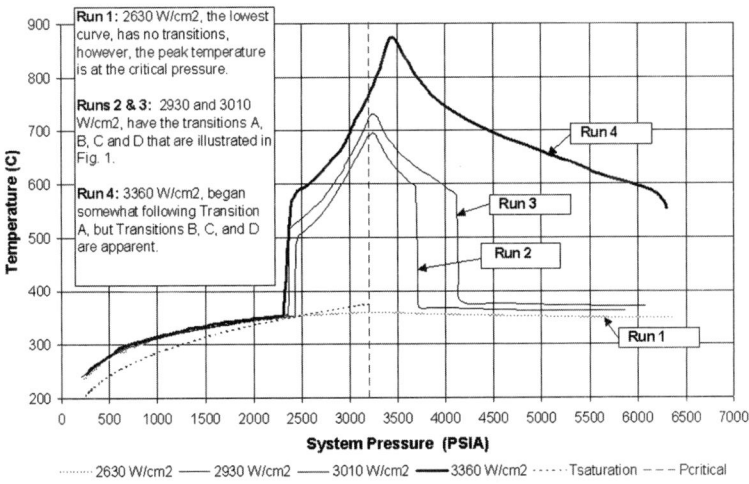

FIGURE 3. Four runs at substantially constant heat fluxes.

In Run 2, 2,930 W/cm^2, note the gradual increase in the element temperature as pressure is reduced. However, at 3,700 psia there is a surprising upward step from 368°C to 594°C (arbitrarily termed *Transition A*). The temperature then smoothly and relatively steeply increases to 695°C at 3,240 psia (*Transition B*). At this point, the temperature "turns around" and smoothly decreases to 482°C at 2,440 psia. (*Transition C*). Next there is a downward step to 352°C at 2,425 psia (*Transition D*). The temperature then smoothly decreases along a path that is narrowly above the path for the run at 2,630 W/cm^2. This run ends at 272 psia and 253°C.

At the next highest power, Run 3, 3,010 W/cm^2, note the positions of the transitions that are defined above. In comparison with 2,930 W/cm^2, note: (1) the higher starting temperature, (2) the higher pressure for the start of Transition A, (3) the lower temperature for the termination of Transition A, (4) the higher temperature of the turnaround at the end of Transition B, (5) the turnaround in the vicinity of the critical pressure, (6) the termination of Transition C at a somewhat higher temperature and lower pressure, (7) the termination of Transition D at a somewhat lower pressure, and (8) the smooth decrease in temperature is narrowly above the first two tracks (for most of the track) and terminates at 195 psia and 239°C.

At the highest power, Run 4, 3,360 W/cm^2, Transition A begins somewhat greater than the starting pressure. The turnaround at 3,435 psia is well above the critical pressure. Transition D begins at a substantially higher temperature, but the pressure is in the range of the prior run. Transition D terminates at a slightly lower pressure than the prior runs, but the temperature is in the same range. The smooth decrease in temperature is narrowly above the first two tracks (for most of the track) and terminates at 261 psia and 245°C.

DISCUSSION

Experts in the field will recognize that this collection of data discloses new territory in the field of phase change heat transfer. The ultrahigh heat fluxes that have been sustained over the wide pressure range are amazing in themselves. Transitions A and B are truly new discoveries. Transitions C and D are likely new discoveries also, although arguments could perhaps be made that other work on the "return to nucleate boiling" also characterizes these discoveries. In any case, the ultrahigh heat fluxes of this work are unprecedented over the range of the four transitions A, B, C, and D. Based on this work and related explorations, the author has documented six primary discoveries and several secondary discoveries, however, there is no need to list these herein. The author claims no fundamental explanations of the phenomena.

The Need to Model and Classify the Unprecedented Turbulence

There is a need to model the turbulence and eddy intensity gradients that characterize the energy transfer to the subcooled water. These analyses should then be extended to other fluids.

The following straightforward heat balance analyses of control volumes surrounding the platinum element clearly demonstrate the presence of intense turbulence. Consider the situation in the vicinity of the element itself. Note from TABLE 1 that at 1 Watt the power density within the platinum heat transfer element is

20 megawatts per cubic centimeter (mW/cm^3). Note also that at 1 watt the heat flux is 3,700 W/cm^2. For Run 4 at 3,360 W/cm^2 the power density within the element is 3,360/3,700 × 20 = 18 mW/cm^3.

Apply the inverse square law to estimate the power density within the fluid surrounding the heat transfer element and for simplicity include the volume of the element itself in the estimate. Thus, doubling the diameter yields a power density of 4.5 mW/cm^3. Moving further outward to a diameter of 25.4 microns (0.001 inch) the power density becomes 1.6 mW/cm^3. Now, consider the adiabatic heating rate of water at 1.6 mW/cm^3. The rate is the order of 400,000°C/sec.

Only very intense turbulence will disperse the energy, because local heating rates are intense. In order to have an adiabatic heating rate as low as 40°C/sec, the volume would have to be uniformly heated at a diameter of about 2,500 microns (2.5 mm) and that is not possible. Very intense turbulence must disperse the energy regardless of the microscopic heat transfer mechanism at the interface between the platinum element and the surrounding water.

Heat Transfer Mechanisms at the Interface Between the Platinum Wire and the Subcooled Water

The impact of this research on theory for phase change heat transfer in the microscale will be astounding. For example, will Marangoni effects be of importance at the ultrahigh heat fluxes and intense eddy intensity gradients, or will such considerations fade into relative insignificance as the modelers suggest mechanisms of far greater impact? In a realistic editorial Tien[4] observed, "The subject matter of experimental heat transfer has shown significant variations ... from macro/mesoscale devices to micro/nano-scale problems, and from continuum behaviors to molecular characteristics." The laws of radiant heat transfer suggest that the radiant heat transfer from the platinum element is negligible, perhaps no greater than 10 W/cm^2 at the highest temperature reached in these runs, about 870°C.

The studies in the field of intermolecular energy transfer by Ohara and Suzuki[5] may not yet be sufficiently inclusive to cover the wide range of the data disclosed herein. Transitions A, B, C, and D and may yield insights that will improve the theory. Their reference to temperature jumps at high heat fluxes at solid–liquid interfaces may be of significance. However, the application to the present work is elusive.

There are no theoretically derived tools that predict the existence of Transitions A, B, C, and D. Furthermore, the microscale heat transfer theoretical analyses that have been reported to date appear to dwell in an arena of low pressure and relatively low power density. Zhang et al.[6] report an analysis of phase-change mechanisms in microchannels using cluster nucleation theory. Zhang also discuses earlier concepts including "fictitious boiling" and departures from classical nucleation theory.

Other investigators have asserted that nucleate boiling does not take place on fine wires, based on data collected with fine wire heat transfer elements that were several times the diameter of the platinum heat transfer element used herein. It is not the function of this proposal to refute those assertions. However, the characteristics of all four runs in the pressure range from about 2,300 psia down to about 200 psia strongly suggest that nucleate boiling is involved. In FIGURE 3, note that the narrow differences in the temperature of the platinum element over the range from 2,300 psia down to 200 psia although the heat fluxes range from about 2,600 W/cm^2 up to about

3,300 W/cm^2. This is characteristic of nucleate boiling and is in striking contrast to the behavior at the pressures greater than 2,400 psia.

In the supercritical pressure region, the onset of Transition A is highly dependent on heat flux over the pressure range, whereas in the subcritical pressure range, the location of Transition D is much less dependent on heat flux. The temperature turnaround in the vicinity of the critical pressure and the strong dependence on heat flux of the peak temperature are further clues in the search for the heat transfer mechanisms at the interface.

Potential Applications

This field is new. It is difficult to anticipate the many avenues of research and applications that will flow from these discoveries during the next decade. However, the impact will be substantial and it is likely that millions of dollars will fund several R&D paths. This is a system in which pressures and heat fluxes may be varied essentially instantaneously in the region of surrounding subcooled fluid. The ultrahigh heat transfer rates and the associated severe turbulence in subcooled fluids suggest several paths to commercialization.

Mixtures of chemical reactants may, thus, be handled in a process of power and pressure pulsing in which the products of chemical reaction will be "frozen" in the subcooled surroundings before reverse reactions can reduce the product yield. Likewise, isotope separation may be achieved via a process of power, pressure and flow pulsing in a fluid such as pressurized uranium hexafluoride. To achieve isotope separation, microscale "skimmers" would separate the internal flow streams and route the products to a following series of cascaded chambers. As other investigators produce microscale devices, these speculations begin to approach reality.

Recently Schubert et al.[7] have reported, "Applications in thermal and chemical process engineering may benefit from microstructure devices such as micro heat exchangers, micromixers, and microstructure reactors when high heat and/or mass transfer rates are mandatory." They describe the steps in producing the microchannel devices. The first step is the micromachining of foils. The broad spectrum of materials includes stainless steel, hastelloy, copper, aluminum, titanium, brass, silver, and palladium. Next the foils are stacked and bonded to yield an assortment of process assemblies. The devices are robust over a wide range of operating pressures and temperatures. The microheat exchanges have heat transfer coefficients up to 54,000 W/m^{2}K.

The basic designs and manufacturing techniques are available for application to chemical production or isotope separation processes. The separation chamber would include internal profiling to direct the severe turbulence to the sampling structures. Thermal hydraulics analysts will be challenged to specify this profiling and the location of the microscale platinum heat transfer element relative to an assembly of sampling grooves. Performance would then be determined in bench tests and the results would lead to altered designs.

PROPOSED INVESTIGATIONS AT CONTROLLED GRAVITY

The proposed work consists of experimental measurements as well as exploratory modeling of the phase change heat transfer phenomena at low gravity over the above ranges of heat flux and pressure in subcooled water and other fluids. The approach is to determine the impact of reduced gravity on Transitions A, B, C, and D in phase change heat transfer over the range of gravitational fields from $1\,g_e$ down to $10^{-2}g_e$.

The complexity of the experiments will be greater than in work conducted to date. The platinum resistance element is an extremely effective tool for determining the overall characteristics of microscale phase change heat transfer as well as the newly discovered transitions. However, the applications to space process technology require that the turbulence and eddy intensity gradients associated with the ultrahigh heat fluxes also be characterized over the range of gravitational fields. Several fine wire platinum resistance thermometers will be built into the test cell. Fiberoptic techniques that will illuminate and monitor turbulent diffraction patterns will be evaluated. At pressures less than 200 psia, the element will be installed within thick-walled glass tubing and turbulence will be measured with the Laboratory's holographic facility and related optics.

Although the author will participate in the design and operation of the new test cells at UCLA by Professor Dhir and his staff, UCLA will have the primary role in the design and operation of the apparatus. UCLA will perform reduced gravity experiments in NASA's KC-135. UCLA will have exclusive responsibility for exploratory modeling and related theoretical work and the associated reporting.

R&D Required for Applications to Space Process Technology

1. Verification of the work thus far performed. This means independent determination of the heat transfer data over the range of heat fluxes and pressures. This includes Transitions A, B, C, and D as well as the approach to Transition A from the high pressure side and the retreat from Transition D (likely nucleate boiling region) on the lower pressure side.

2. Computation of the complex and intense turbulence field surrounding the heat transfer element.

3. Analysis to determine impact of power pulsing on the turbulence, the impact of pressure pulsing on the turbulence, and the impact of simultaneous power and pressure pulsing on the turbulence.

4. Analysis to determine the impact of imposed flow on the turbulence, the impact of pulsing of induced flow on the turbulence, and the impact of simultaneous pulsing of power, pressure, and imposed flow on the turbulence.

Scope of Experiments at $1\,g_e$

The objective of the experiments is to provide, in a very clean manner, the information that is needed to develop empirical, as well as theoretical, models for the heat transfer and intense convection characteristics over the wide range of pressure and power density. Two types of experiments are proposed in the present work. These include experiments at normal gravity and in the KC-135.

The first set of experiments are intended to verify the discoveries that have been claimed including Transitions A, B, C, and D as well as the values of the ultra high heat fluxes. The fluid will be degassed, demineralized water at about 20°C. The test hardware and will be activated at the Phase Change Heat Transfer Laboratory at UCLA.

The second set of experiments will be over a comparable range of heat fluxes and pressures (FIG. 3), however, the test assembly will be far more complex. Dhir and his staff will be commissioned to design and procure high pressure test cells that will include extensive means to monitor the intensive convection. These "extensive means" are themselves the basis for an extensive R&D effort. Bulk convection patterns may be approximately studied via a set of fine wire platinum resistance thermometers dispersed in the anticipated convection patterns. Analysis of the initial set of data and the modeling based thereon will yield a design for a second test cell with an "adjusted" set of locations of the temperature detectors. The detectors will be located at least seven diameters remote from the heat transfer element in order to not excessively influence the flow patterns.

The third set of experiments will utilize the final designs that are developed in the above second set. Provisions for imposed flow will be added and the runs will focus on pulsing of power, pressure and flow. Initially these will be pulsed independently. Following analysis of test results, simultaneous pulsing will be deployed.

Determining the heat transfer mechanisms and the flow patterns in close proximity to the heat transfer element over the range of pressures will require optical techniques that must be developed. A system of illumination and viewing via optical fibers may be feasible. In the vicinity of the critical pressure and at heat fluxes greater than 2,900 W/cm^2 the element itself may provide sufficient illumination for viewing the heat transfer processes. In this limited case, a single fiber or alternate viewing system may be workable. A proof-tested set of thick wall glass pressure vessels may be used for the viewing of heat transfer processes and turbulence at pressures less than 200 psia.

Experiments at Reduced Gravity

The apparatus that is perfected at normal gravity will be employed over a range of gravitational fields in the KC-135. Safety of the high pressure equipment will be assured via shielding. The very low volume of the subcooled non-toxic water has very little stored energy even at 6,000 psia. The tests will be run during the last two years of the four year program. The initial test series will not include induced flow. Tests with pulsing of power, pressure and flow will be conducted following analysis of the conventional tests. The data will be useful in developing empirical and theoretical models for applying the new space process technology on Mars and elsewhere. The details of the test runs will be determined based on the test results as the explorations proceed.

Numerical Simulations/Models

Along with the experimental work, Dhir and his staff will focus on the numerical simulations/modeling of the thermal and hydrodynamic processes. These include:
1. Characteristics of the element-fluid interface over the full range of pressures.

2. Factors that yield the differences in the relationship between heat flux and element temperature at the high end (supercritical) of the pressure range and the lower end (possibly nucleate boiling) of the pressure range.

3. Molecular dynamic simulation of the hydrodynamic and thermal non-equilibrium processes.

4. Factors that lead to the location and shape of Interfaces A, B, C, and D.

5. The velocity field and the temperature field across the intense zones of turbulence and the range of the zones.

6. The impact of power, pressure, and flow pulsing on the dynamics of the interfaces and the flow fields.

7. Generalization of the results.

In carrying out the numerical simulations several modeling regimes will be deployed. The challenge is to cover the vast array of changing mechanisms over the wide pressure range at the ultrahigh heat fluxes. The challenge is amplified by the need to couple all of the regimes in developing power, pressure, and flow pulsing maneuvers for the new process technology. (For example, maneuvers to effect separation of isotopes will employ these predictions.) Clearly, within the scope of this proposal, this will be exploratory modeling. However, with the collaboration of the Mathematics Department at UCLA, this effort will likely impact the shape of most fluid physics modeling of thermally non-equilibrium processes.

Tasks

The proposed exploration has five tasks; several are performed simultaneously.

Task 1. Design and Fabricate a Series of High Pressure Test Cells: Experiments at normal and reduced gravity will employ instrumented test cells. The cells will be designed in sequence as operating data at normal gravity reveals needed modification of the extent and internal location of fine wire platinum resistance thermometers relative to the 7.5 micron diameter platinum heat transfer element. The feasibility of fiberoptics for illuminating and classifying the intense turbulence will be investigated. The stainless steel test cells will have a volume of a few cm^3 and will have a pressure rating of 10,000 psia. This task includes setup of the pressurizing and pressure bleed-off equipment, power supplies and controls, and setup of data acquisition systems.

Task 2. Experiments at Earth Normal Gravity: The experiments will initially employ test cells that have been employed in work to date. The intent of this initial work is to confirm the discoveries that have been reported. In addition, the proof-tested thick-walled glass test cells for runs at less than 200 psia, will use the Laboratory's holographic facility and related optics. Next, the bulk of the experimental work will be in the advanced test cells that are produced in Task 1. The results of the initial test work will be evaluated and later test cells fabricated under Task 1 will be appropriately modified. The initial tests will focus on at least five values of ultrahigh heat flux that are held constant over the pressure range from about 6,000 psia down to 200 psia. Further runs will investigate the impact of power, pressure, and flow pulsing on the intense turbulence.

Task 3. Numerical Simulations/Models: In this task the steady state distribution of turbulence will be modeled for the spectrum of heat transfer regimes that are investigated under Task 2. This includes the data over the full pressure range for the series of at least five ultrahigh heat flux runs in which heat flux is held constant over the range of pressure.

Task 4. Experiments in the KC-135 Aircraft: Reduced gravity will likely have a substantial impact on the intense turbulence that is associated with the ultrahigh heat fluxes over the pressure range. The Transitions A, B, C, and D will likely shift and it will be revealing to compare the data with the predictions from the modeling tasks. Several runs will proceed over a range of gravitational fields that will include Mars gravitation. Power, pressure, and flow pulsing runs will terminate the test series. The modeling will be adjusted as results are analyzed and specific runs will be incorporated per recommendations of the analysts.

Task 5. Merging of the Models: The several hydrodynamic and thermal models will be merged to yield workable applications to space process technology. The merging is required to yield codes for design of processes that employ power, pressure, and flow pulsing. This task will be limited, however, the insights should be applicable to terrestrial projects as well. This is a formidable challenge and is presented as exploratory modeling.

CONCLUSIONS

Surprising discoveries are reported in the field of microscale phase change heat transfer to subcooled water over a wide pressure range and a set of ultrahigh heat fluxes. These discoveries will likely shape significant R&D efforts during the forthcoming decade. Initial reviews of the phenomena suggest two paths for extensive study. One path is to model the intense turbulence that is in the vicinity of the microscale heat transfer element. The other path is to postulate the heat transfer mechanisms at the interface between the heat transfer element and the subcooled water. Several heat transfer regimes were identified over the pressure range. Distinct patterns of turbulence and distinct solid–fluid heat transfer mechanisms likely characterize each regime.

Explorations are proposed over a range of gravitational fields. The KC-135 would be a suitable platform for the 20-second duration that is needed for each of a set of reduced gravity runs. These explorations would enhance the modeling of turbulence, and likely would also yield further insights into the mechanisms at the solid–fluid interface.

The platinum resistance element is an extremely effective tool for determining the characteristics of microscale phase change heat transfer. The applications to space process technology require measurement of the turbulence and eddy intensity gradients. Several fine wire platinum resistance thermometers that are positioned within the test cell are proposed for monitoring the turbulence. Fiberoptic techniques may be applied to monitoring turbulent diffraction patterns.

Finally, the microscale heat transfer technology that is disclosed herein may be integrated with microstructure devices that are being produced by others for thermal and chemical processing.

ACKNOWLEDGMENT

I am very grateful to Professor V.K. Dhir for suggestions and encouragement.

REFERENCES

1. NUKIYAMA, S. 1934. The maximum and minimum values of the heat Q transmitted from metal to boiling water under atmospheric pressure. J. Jpn. Soc. Mech. Eng. **37:** 367–374.
2. DHIR, V.K. 1998. Boiling heat transfer. Annu. Rev. Fluid Mechanics **30:** 365–401.
3. DHIR, V.K. 1989. Framework for a unified model for nucleate and transition pool boiling. J. Heat Transfer **111**(3):
4. TIEN, C.L. 2000. Editorial announcement: EHT—the past and future. Exp. Heat Transfer **13:** 1–2.
5. OHARA, T. & D. SUZUKI. 2000. Intermolecular energy transfer at a solid–fluid interface. Microscale Thermophys. Eng. **4:** 189–196.
6. ZHANG, J.T., X.F. PENG & G.P. PETERSON. 2000. Analysis of phase-change mechanisms in microchannels using cluster theory. Microscale Thermophys. Eng. **4:** 177–187.
7. SCHUBERT, K., J. BRANDNER, M. FICHTNER, *et al.* 2001. Microstructure devices for applications in thermal and chemical process. Microscale Thermophys. Eng. **5:** 17–39.

Condensate Removal Mechanisms in a Constrained Vapor Bubble Heat Exchanger

LING ZHENG, YINGXIN WANG, PETER C. WAYNER, JR., AND JOEL L. PLAWSKY

The Isermann Department of Chemical Engineering, Rensselaer Polytechnic Institute, Troy, New York, USA

ABSTRACT: Microgravity experiments on the constrained vapor bubble heat exchanger (CVB) are being developed for the space station. Herein, ground-based experimental studies on condensate removal in the condenser region of the vertical CVB were conducted and the mechanism of condensate removal in microgravity was found to be the capillary force. The effects of curvature and contact angle on the driving forces for condensate removal is studied. The Nusselt correlations are derived for the film condensation and the flow from the drop to the meniscus at the moment of merging. These new correlations scale as forced convection with $h \propto L^{1/2}$ or $h \propto L_{cd}^{1/2}$. For the partially wetting ethanol system studied, the heat transfer coefficient for film condensation was found to be $4.25 \times 10^4 \, W/m^2 K$; for dropwise condensation at moment of merging it was found to be $9.64 \times 10^4 \, W/m^2 K$; and for single drops it was found to be $1.33 \times 10^5 \, W/m^2 K$.

KEYWORDS: dropwise condensation; capillary force; interferometry

NOMENCLATURE:
- d width of liquid drop
- F driving force
- g acceleration due to gravity
- h heat transfer coefficient
- h_{fg} latent heat of vaporization
- K curvature
- k thermal conductivity
- L distance
- m mass flow rate per unit width of surface
- Nu Nusselt number
- P pressure
- q condensation heat flux
- r radius of curvature
- T temperature
- u velocity
- x axial direction
- y in the direction normal to axial direction
- z in the direction of film thickness

Greek Characters
- θ apparent contact angle
- μ viscosity

Address for correspondence: Joel L. Plawsky, The Isermann Department of Chemical Engineering, Rensselaer Polytechnic Institute, Troy, NY 12180-3590, USA. Voice: 518-276-6049; fax: 518-276-4030.
plawsky@rpi.edu

Π	disjoining pressure
ρ	density
σ	surface tension
τ	shear stress
Subscripts	
0	flat film between the drop and meniscus in the corner
c	liquid film in the corner
cd	between meniscus in the corner and drop
d	liquid drop
l	liquid
lv	liquid-vapor interface
s	solid
v	vapor

INTRODUCTION

Since Schmidt et al.[1] reported that the heat transfer coefficient of dropwise condensation is much higher than that of film condensation, dropwise condensation has been extensively studied. The dropwise condensation process includes initial droplet formation, growth, removal, and renucleation on reexposed sites. Previous studies on the microscopic mechanism of initial droplet formation presented experimental results to support one of two different hypothesis. One hypothesis assumes that droplets are generated and grow at nucleation sites, whereas portions of the surface between the growing droplets remain dry. McCormick and Baer[2] suggested that the vapor could be primarily condensing on submicroscopic drops in the bare areas. Heat is transferred through active areas on the condenser surface that are continually produced by numerous drop coalescences. These areas remain active for a short constant cycle time. During this time numerous submicroscopic drops are growing from randomly distributed sites. This view was strongly supported by Umur and Griffith,[3] who concluded, both by theoretical argument and optical measurement, that condensate films greater than monomolecular thickness do not form on the surface. The second formation scenario is the film fracture hypothesis, first proposed by Jakob.[4] The vapor condenses initially into an extremely thin liquid film on the solid surface, and droplets are formed by the rupture of the liquid film that eventually reaches a critical thickness, estimated to be about 1 μm. Sugawara and Katsuta[5] used the excellent durability of polydimethylsiloxane as a promoter to the copper cooling surface to observe condensation microscopically. They found that a thin liquid film is formed in the bare area between drops. The thin film plays an important part in the coalescence or attraction between drops, and the film becomes unstable and fractures into tiny droplets on reaching a critical thickness of 0.63 μm. In our partially wetting ethanol/quartz system, we observed a very thin film of condensate in the area between the drops and at the spots from which the droplets departed. When a droplet grows to a size that cannot be sustained by surface tension forces,[6] it is removed from the surface by gravity, external forces or drag forces resulting from the surrounding surface conditions. Tanasawa et al.[7] measured the dependence of the heat transfer coefficient on the departing drop diameter, using the gravitational, centrifugal, and steam shear forces to change the departing drop diameter. Yamali and Merte[8] designed a centrifuge to study the effect of body forces on dropwise condensation heat flux under controlled subcooling. Wang et al.[9] gave an initial report on the departure of a single condensed ethanol drop into a concave liquid film due to the intermolecular gradient. Ethanol is chosen as the working fluid due to its partially wetting properties on the heated quartz. It can form both a drop on the flat quartz surface and a concave film in the corner of the CVB. In a partially wetting ethanol–quartz CVB system, the local dropwise condensation process occurs slowly at low heat removal, due to the low thermal conductivity of quartz. In this paper, a dimensionless force balance for viscous shear stress is given to demonstrate the effect of changes in the contact angle and curvature on the driving forces for condensate removal. The Nusselt correlation is derived for film condensation and dropwise condensation at the moment of merging of the drop with meniscus. The driving force for fluid motion in both cases is the interfacial force.

A constrained vapor bubble (CVB) heat exchanger was built to study the dropwise condensation process, see FIGURE 1. A detailed description is given by Wang et al.[10] The quartz cell (square cross section; inside dimensions, 3mm×3mm; outside dimensions, 5.5mm×5.5mm; length, 40mm) was set with its long axis in the vertical direction. The thermoelectric heater was attached on the top of the cell using high thermal conductivity epoxy. The four thermoelectric coolers were pasted

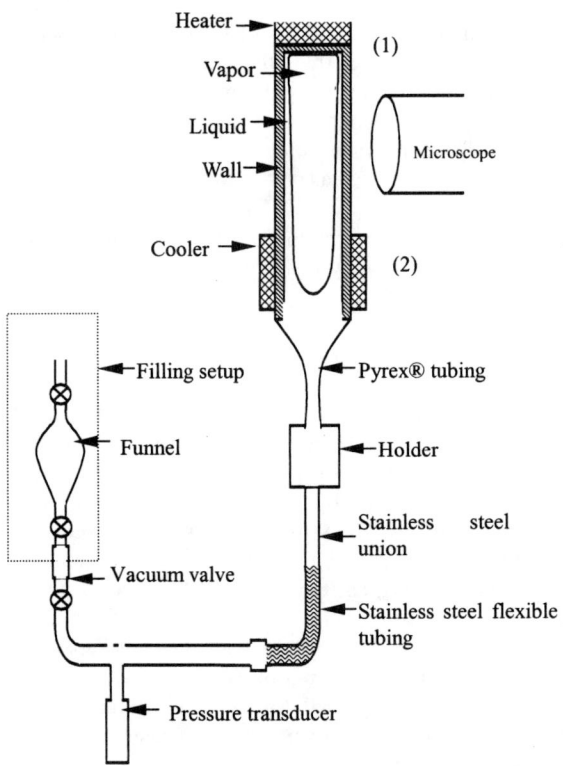

FIGURE 1. The vertical constrained vapor bubble system.

onto each side of the square cell to ensure a symmetric system. Aluminum heat sinks were bonded to the coolers to insure good cooling. A high power microscope was rotated and situated on an adjustable stage, so that the interference fringes in the side corner of the vertical CVB could be viewed and evaluated using image analyzing interferometry.[11,12]

Heat applied at end (1) in FIGURE 1 vaporizes the working liquid. Then energy flows from end (1) to end (2) by means of conduction in the walls and by a combined evaporation, vapor flow, and condensation mechanism. The curvature of the liquid–vapor interface changes continually along the axial length of the cell because of viscous losses, and the evaporation/condensation processes. The resulting capillary pressure gradient between two ends drives the condensate at end (2) back to end (1) along the right-angled corner regions, against the action of gravity. The interfacial force field is a function of the film profile, which is a function of the thermal conditions on the surface. The pressure transducer was connected to the system through a three-way vacuum valve to continuously monitor the vapor–liquid system pressure. A PC-based data acquisition system using LabView® was built to measure the temperature profile. Eleven microscale K-type (Chromel-Alumel) thermocouple beads were pasted onto a side of the cell, the distance between the thermocouples was 2 mm.

FIGURE 2 shows a schematic drawing of three different regions of a vertical CVB in the Earth's gravitational field. These three regions are Region I (the dry region), Region II (the evaporator region), and Region III (the condenser region). Herein, we discuss the optically observed details of film and droplet condensation process occurring in Region III, with a partially wetting fluid, ethanol. FIGURE 3 shows an example of drop condensation process. The straight interference fringes on the right hand side represent the liquid film in the corner.

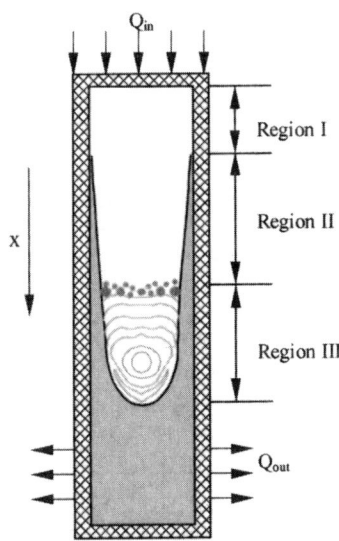

FIGURE 2. A schematic drawing of three different regions of the vertical CVB.

FIGURE 3. Dropwise condensation of ethanol.

On the flat surface of the cell, the vapor condensed into small discrete liquid drops that then grew and coalesced into larger drops. These larger drops were then absorbed into the bulk liquid in the corner, driven by a passive capillary force perpendicular to the direction of the gravitational force. Thus, the removal force is the same in $1g$ and μg. The passive capillary force is critical in dropwise condensation. The dropwise condensation region would be flooded and replaced by film condensation were the drops of FIGURE 3 not removed effectively. Closer to the cooler, the condensate formed a continuous, thicker liquid film on the flat surface of the glass. The condensation process was observable due to the presence of naturally occurring interference fringes resulting from the reflection of monochromatic light at the both the liquid–vapor and liquid–solid interfaces. Monochromatic light ($\lambda = 543.5\,\text{nm}$) from a Hg-arc was used to illuminate the cell. With a 50× objective, each of the 640×480 pixels measures the average reflectivity (thickness) of a region with a diameter of $0.1777\,\mu\text{m}$. The recorded images were analyzed using image processing. The resulting liquid thickness profile yields the curvature and apparent contact angle. The effects of drop curvature and drop contact angle on the driving forces for condensate removal is discussed below.

PRELIMINARY FORCE BALANCE ANALYSIS

Using the following augmented Young–Laplace equation, the shape of the vapor liquid interface gives the pressure field for microscale flow in the liquid:

$$P_v - P_l = \sigma K + \Pi, \tag{1}$$

where K is the curvature, and Π is the disjoining pressure. Changes in the shape dependent intermolecular force field cause a pressure difference between the liquid drop and the liquid in the corner. A macroscopic force balance that relates viscous losses to interfacial forces and apparent contact angles is used to give a general description of the dropwise condensation process occurring in Region III. The detailed derivation was given by Wang, et al.[9] Consider a sessile drop close to the liquid meniscus in the corner. As shown in FIGURE 4, the condensed drop is assumed to be a spherical cap in the region $\delta > 0.1\,\mu m$, whereas the meniscus in the corner is a portion of a cylinder in the region $\delta > 0.1\,\mu m$. The apparent contact angles and curvatures can be obtained from the film thickness profiles based on an analysis of the interference fringe pattern (FIG. 3).

The macroscopic force balance between the liquid drop and the liquid in the corner (control volume shown in FIG. 4) is

$$\sigma_{lv,d}\cos\theta_d + \sigma_{ls,d} + \tau L + P_{l,c}\delta = \sigma_{lv,c}\cos\theta_c + \sigma_{ls,c} + P_{l,d}\delta. \quad (2)$$

In terms of measured quantities, Equation (2) becomes

$$\tau L = \sigma_l(-\cos\theta_d + \cos\theta_c) + \delta(-\sigma_l K_d + \sigma_l K_c), \quad (3)$$

where the length, L, is the distance between the first observable fringes spanning the droplet and meniscus. The film thickness boundaries (δ) in the liquid drop and in the liquid corner are taken to be the thickness of the first destructive interference fringe, $0.1\,\mu m$. This location is also where the apparent contact angles were measured. The right hand side of Equation (3) is the driving force for liquid flow per unit width (F/d_s) between the drop and the meniscus. It is the decrease in chemical potential of the liquid that causes this spontaneous flow. For the purpose of discussion, we assume that d_s is the width of the liquid drop. Equation (3) can then be rewritten in dimensionless form by dividing through σ_l to give the dimensionless difference in free energy for spontaneous condensate removal,

$$\frac{F}{d_s\sigma_l} = (\cos\theta_c - \cos\theta_d) + \delta(K_c - K_d). \quad (4)$$

The left hand side of Equation (4) represents the ratio of the shear resistance force to the surface free energy of a flat bulk liquid. It consists of two terms: the contact angle difference and the capillary pressure difference. Only the term $\cos\theta_d$ opposes the flow of the drop into the corner. As discussed below, the first term can play a major role in the driving force when the contact angle difference becomes large.

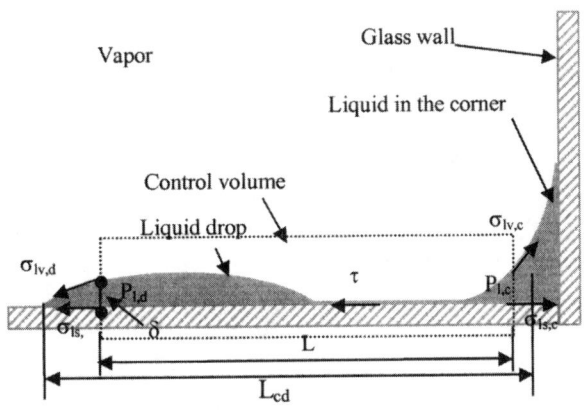

FIGURE 4. A schematic of the force balance.

Near the corner, the spontaneous flow of the drop into the corner occurs rapidly when the growing drop "contacts or gets very close" to the meniscus.

NUSSELT CORRELATION

Film Condensation

In Region III of the CVB, film condensation occurs when the condensation rate is larger than the dropwise removal rate can support. FIGURE 5 shows the spread film of ethanol in the corner of the CVB. The condensate was pumped toward the corner by capillary force.

FIGURE 6 is a cross-sectional view of the meniscus showing the important model parameters. The pressure difference between the convex front and meniscus in the corner is

$$\Delta P = \frac{\sigma_l}{r_{c1}} + \frac{\sigma_l}{r_{c2}}. \tag{5}$$

The distance between the convex front and the corner is

$$L = r_{c1}\sin\theta_{c1} + L_f + \frac{r_{c2}}{2}(\cos\theta_{c2} - \sin\theta_{c1}), \tag{6}$$

where L_f is the distance between the center of convex front and the center of meniscus in the corner. We refer to this as the flat or foot portion of the meniscus. The overall pressure driving force for the flow is

$$-\frac{dP}{dy} \approx \frac{\Delta P}{L} = \frac{\sigma_l\left(\frac{1}{r_{c1}} + \frac{1}{r_{c2}}\right)}{r_{c1}\sin\theta_{c1} + L_f + \frac{r_{c2}}{2}(\cos\theta_{c2} - \sin\theta_{c1})}. \tag{7}$$

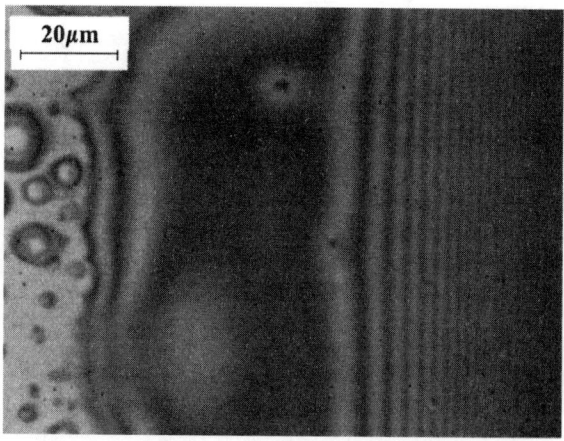

FIGURE 5. Spreading liquid film in the corner.

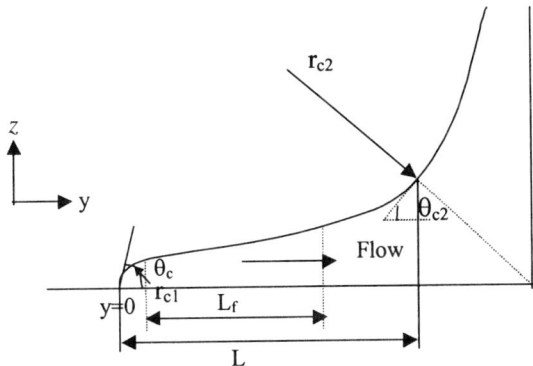

FIGURE 6. A schematic drawing of the film condensation.

A Nusselt-type correlation for the heat transfer coefficient can be derived for film condensation in the CVB. The film condensate profile is shown in FIGURE 6.

The momentum equation for velocity in the y-direction is

$$\frac{d^2u}{dz^2} = \frac{1}{\mu}\frac{dP}{dy}. \tag{8}$$

Here we have assumed no influence of gravity and a one-dimensional velocity profile. The boundary conditions on Equation (8) are

$$z = 0, \ u = 0$$

$$z = \delta, \ \frac{du}{dz} = 0. \tag{9}$$

Integrating Equation (8) from $z = 0$ to $z = \delta$, we obtain the classic parabolic velocity profile

$$u = \frac{1}{\mu}\frac{dP}{dy}\left(\frac{z^2}{2} - \delta z\right). \tag{10}$$

Integrating the velocity over the film thickness yields the mass flow rate per unit width of surface

$$m = -\frac{\rho_l}{\mu}\frac{dP}{dy}\frac{\delta^3}{3}. \tag{11}$$

Differentiating m with respect to δ yields

$$\frac{dm}{d\delta} = -\frac{\rho_l}{\mu}\frac{dP}{dy}\delta^2. \tag{12}$$

The heat flux is given in terms of the condensation rate

$$dq = h_{fg}dm. \tag{13}$$

Assuming that heat transfer across the film is mainly due to conduction, convection effects are neglected, and thus the heat flow dq across the differential element is given by

$$dq = \frac{k_l}{\delta}\Delta T dy, \qquad (14)$$

where $\Delta T = T_{lv} - T_s$. Combining Equations (13) and (14) and rearranging, we obtain a differential equation for the film thickness profile

$$\frac{d\delta}{dy} = \frac{k_l \Delta T \mu}{h_{fg}\rho_l\left(-\dfrac{dP}{dy}\right)}\delta^{-3}. \qquad (15)$$

Integrating the above equation using the condition that $\delta = 0$ at $y = 0$ yields

$$\delta = \left[\frac{4 k_l \Delta T \mu}{h_{fg}\rho_l\left(-\dfrac{dP}{dy}\right)} y\right]^{1/4} \qquad (16)$$

Since heat transfer across the film is by conduction alone, the local heat transfer coefficient is given by $h_l = k_l/\delta$. The local Nusselt number is defined by

$$Nu = \frac{h_l y}{k_l} = \frac{y}{\delta} = \left[\frac{h_{fg}\rho_l\left(-\dfrac{dP}{dy}\right)}{4 k_l \Delta T \mu} y^3\right]^{1/4}. \qquad (17)$$

The mean Nusselt correlation can be obtained by averaging over the entire condensate film:

$$Nu = \frac{1}{L}\int_0^L Nu(y)dy = 0.943\left[\frac{h_{fg}\rho_l L^3\left(-\dfrac{dP}{dy}\right)}{k_l \mu \Delta T}\right]^{1/4}. \qquad (18)$$

Substituting Equation (7) and rearranging

$$h = 0.943\frac{k_l}{L^{1/2}}\left[\frac{h_{fg}\rho_l \sigma_l\left(\dfrac{1}{r_{c1}} + \dfrac{1}{r_{c2}}\right)}{k_l \mu \Delta T}\right]^{1/4}, \qquad (19)$$

Where L can be obtained from Equation (6). Note that the heat transfer coefficient, $h \propto L^{-1/2}$, and thus condensate removal driven by interfacial forces leads to a forced convection-type correlation for h.

Dropwise Condensation

A Nusselt number correlation can also be obtained for dropwise condensation at the moment of drop merging with the meniscus. Neglecting the disjoining pressure, the pressure difference between the drop and meniscus in the corner is

$$\Delta P = \frac{\sigma_l}{r_c} + \frac{\sigma_l}{r_d}. \qquad (20)$$

The distance between the center of drop and the meniscus in the corner is

$$L_{cd} = \frac{r_c}{2}(\cos\theta_c - \sin\theta_c) + 2r_d\sin\theta_d. \quad (21)$$

As shown in FIGURE 4, the pressure drop over the distance L_{cd} is

$$-\frac{dP}{dy} \approx \frac{\Delta P}{L_{cd}} = \frac{\sigma_l\left(\frac{1}{r_c} + \frac{1}{r_d}\right)}{\frac{r_c}{2}(\cos\theta_c - \sin\theta_c) + 2r_d\sin\theta_d}. \quad (22)$$

Substituting Equation (22) into Equation (19) and rearranging

$$h = 0.943 \frac{k_l}{L_{cd}^{1/2}} \left[\frac{h_{fg}\rho_l\sigma_l\left(\frac{1}{r_c} + \frac{1}{r_d}\right)}{k_l\mu\Delta T}\right]^{1/4}. \quad (23)$$

We note again that the heat transfer coefficient follows a forced convection-like correlation with $h \propto L_{cd}^{-1/2}$.

The above correlation is for the heat transfer coefficient between the drop and meniscus in the corner at the moment of merging. Assume condensation resistance at the liquid–vapor interface is negligible. We can also obtain the heat transfer coefficient for the single drop before merging into the corner and the meniscus after merging using $h_l = k_l/\bar{\delta}$. $\bar{\delta}$ is the characteristic length, which can be obtained from volume of the drop or meniscus per unit width divided by the vapor–liquid interfacial area. Based on geometric relations, the characteristic length for the drop is

$$\bar{\delta}_d = \frac{1}{6}(1 - \cos\theta_d)(2r_d + r_d\cos\theta_d). \quad (24)$$

Thus, the heat transfer coefficient for the single drop before merging is

$$h_d = \frac{6k_l}{(1 - \cos\theta_d)(2r_d + r_d\cos\theta_d)}. \quad (25)$$

The characteristic length for the meniscus in the corner is

$$\bar{\delta}_c = \frac{r_c(1 - \cos\phi - \phi + \sin\phi)}{2\phi}, \quad (26)$$

where $\phi = 90° - 2\theta_c$. The heat transfer coefficient for the meniscus after merging is

$$h_c = \frac{2k_l\phi}{r_c(1 - \cos\phi - \phi + \sin\phi)}. \quad (27)$$

RESULTS AND DISCUSSIONS

In the dropwise condensation region of the CVB, the vapor condensed into small discrete liquid drops that then grew and coalesced into larger drops. These drops were then absorbed into the bulk liquid in the corner by the capillary driving force. The liquid was then driven up along the sharp corner to the evaporator against gravitational force and frictional force by the capillary pressure gradient. This paper concerns the condensate removal mechanism in two regimes: film condensation and dropwise condensation.

The departure of the condensed drops into the concave liquid film in the corner of the container is driven by a natural driving force (F) that can be determined using the force balance, Equation (4). The magnitude of F depends on the difference between the shapes of the drop and the meniscus in the corner represented by the contact angles and the curvatures. In FIGURES 7 and 8, pictures of drops at different axial locations along the CVB are presented. The shape of the sessile drops (indicated by the arrows) were analyzed. In FIGURE 7, a picture of drops at $x = 3$ mm from the top is presented. The film thickness profiles of the drop and liquid film in the corner are analyzed using an image analyzing technique developed by Wang, et al.[12] The apparent contact angle at $\delta = 0.1$ μm and the curvature of the drop are $9.6°$ and -2.65×10^4 m^{-1}, respectively. The concave meniscus in the corner has an apparent contact angle of $9.5°$ and a curvature of 2.38×10^3 m^{-1}. Using Equation (4), the dimensionless variable $F/d_s\sigma_l$ is found to be 3.21×10^{-3}. In FIGURE 8, a picture of drops at $x = 3.73$ mm from the top is presented. Obviously, the drop and the liquid film in the corner are more spread out. The magnitude of the apparent contact angle ($\theta_d = 7.3°$) and the curvature ($K_d = -1.42 \times 10^4$ m^{-1}) of the drop become smaller compared to the drop in FIGURE 7. In the corner, the apparent contact angle ($\theta_c = 2.6°$) and curvature ($K_c = 2.34 \times 10^3$ m^{-1}) of the meniscus also decrease. The difference in contact angles between the drop and the meniscus at $x = 3.73$ mm is $4.7°$, which is much larger than that ($0.1°$) at $x = 3$ mm. This causes the contact angle difference term to be 24 times larger. Although the curvature difference decreases slightly at $x = 3.73$ mm, we find that the contact angle difference plays the dominant role in the change in the driving force. Hence, the dimensionless variable $F/d_s\sigma_l$ is 8.75×10^{-3}, which is 2.7 times larger than that at $x = 3$ mm, is presented. For the drop close to the corner, we find that gravity is insignificant. The ratio of the driving force between the drop and the meniscus to the gravitational force on the drop is 8.88×10^2 for the

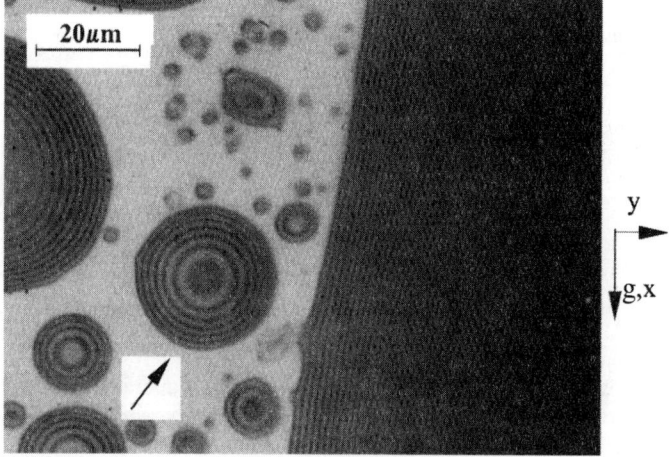

FIGURE 7. Dropwise condensation at $x = 3$ mm.

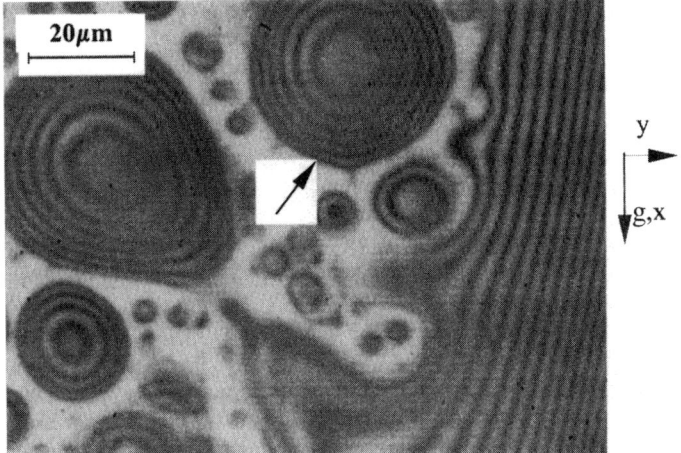

FIGURE 8. Dropwise condensation at $x = 3.73$ mm.

drop in FIGURE 7, and 1.54×10^3 in FIGURE 8. In the direction normal to the vertical direction, the gravitational force has negligible effects on the droplet removal to the meniscus in the corner. However, in the vertical direction, the gravitational force exerts great effects on the pump-up of the axial liquid flow, since the magnitude of gravitational force is equivalent to that of frictional force.[10] In the Earth environment, the liquid is driven up by the capillary pressure gradient against both the gravitational force and the frictional force. Under microgravity, the axial flow rate would increase, the droplets which were absorbed into the meniscus in the corner would be efficiently driven up along the vertical direction, and the surface would not be flooded.

When the condensation rate is larger than the dropwise removal rate can support, film condensation occurs, and the ethanol film spreads out, as shown in FIGURE 5. The first two dark fringes represent the convex front. We notice that the wide dark fringe is uniform, it represents the flat film. The closely spaced fringes on the right represent the meniscus in the corner. The sign of the curvature of the convex front is negative, whereas the curvature of the concave meniscus has positive sign. Thus, the film is absorbed to the meniscus into the corner by capillary force. The overall pressure driving force for the film removal to the corner is $\Delta P/L$ is 5.48×10^5 Pa/m. The heat transfer coefficient is 4.25×10^4 W/m²K based on Equation **(19)**.

The Nusselt numbers for the flow from the drop to the meniscus in the corner at the moment of merging in FIGURES 7 and 8 can be obtained from Equation **(23)**. We found that Nusselt number for the flow at $x = 3$ mm is 112.8, the corresponding heat transfer coefficient is 9.64×10^4 W/m²K, which is larger than that for the film condensation. It is because the driving force for the droplet removal (1.79×10^6 Pa/m) is larger than that for the film removal. Compared with the Nusselt number at $x = 3$ mm, the Nusselt number at $x = 3.73$ mm ($Nu = 72.3$) for dropwise condensation at the moment of merging is much smaller. There are two reasons for this: (1) The temperature difference at 3.73 mm is larger, since it is closer to the cooler. The

Nusselt number is inversely proportional to the temperature difference between the vapor and solid. (2) The film thickness at 3.73 mm is larger. As observed, the film of the drop is very thin at the beginning of the dropwise condensation region, it becomes thicker as it gets closer to the cooler. The heat transfer coefficient at $x = 3.73$ mm is 5.10×10^4 W/m²/K, which is smaller than that at $x = 3$ mm. This is understandable, since the dropwise condensation at 3.73 mm is closer to the film condensation region where the heat transfer coefficient is smaller than that of dropwise condensation.

As can be seen from Equations **(5)** and **(20)**, the pressure difference for film condensation consists of convex front term and meniscus term, for dropwise condensation, the pressure difference consists of meniscus and drop terms. However, the film is more spread out for the film condensation, causing the value of L to increase, thus the driving force ($-dP/dy$) for the film condensation is smaller. Using Equation **(11)**, we found that the mass flow rate is proportional to the driving force, the mass flow rate for the dropwise condensation at 3 mm is 3.3 times larger than that for film condensation at same film thickness. The driving force for the dropwise condensation at 3.73 mm is 8.94×10^5 Pa/m, and the mass flow rate is 1.6 times larger than that for the film condensation.

The heat transfer coefficients for the single drop before merging can be obtained from Equation **(25)**. At 3 mm, the heat transfer coefficient for the single drop before merging is 1.33×10^5 W/m²K, which is larger than that at 3.73 mm (1.20×10^5 W/m²K). It is because the film of the drop at 3 mm ($\overline{\delta}_d = 1.27$ μm) is thinner than that at 3.73 mm ($\overline{\delta}_d = 1.41$ μm). Both the curvature and apparent contact angle of the single drop at 3 mm are larger than those at 3.73 mm. For drops, larger curvature results in thinner film, whereas larger apparent contact angle results in thicker film. The effect of curvature dominates the film thickness of the drop in FIGURE 7, the drop is less spread at 3 mm, and thus the average film thickness is smaller.

The characteristic length of the meniscus after merging at 3 mm ($\overline{\delta}_c = 64.6$ μm) is larger than that at 3.73 mm ($\overline{\delta}_c = 61.4$ μm). For a meniscus, a larger curvature results in a smaller characteristic length; a larger apparent contact angle results in a larger characteristic length. The apparent contact angle of the meniscus at 3 mm is much larger than that at 3.73 mm, whereas the curvature varies little, thus the effect of apparent contact angle dominates and the characteristic length is larger at 3 mm. Using Equation **(27)**, the heat transfer coefficients for the meniscus after merging at 3 mm and 3.73 mm are 2,618 W/m²K and 2,753 W/m²K, respectively. The heat transfer coefficient for the flow at the moment of merging lies in between the heat transfer coefficients for the drop before merging and the meniscus after merging. These values occur because the characteristic length for flow from the drop to the meniscus (h at the moment of merging) is larger than the characteristic length of the drop before merging ($\overline{\delta}_d = 1.27$ μm) but smaller than the characteristic length of the meniscus after merging ($\overline{\delta}_c = 64.6$ μm).

CONCLUSIONS

1. Condensate removal in the CVB driven by interfacial forces was studied for the cases of film condensation and dropwise condensation.

2. Nusselt number correlations were derived for both condensation scenarios. These new correlations scale as forced convection with $h \propto L^{1/2}$ or $h \propto L_{cd}^{1/2}$.

3. For the partially wetting ethanol system studied, the heat transfer coefficient for film condensation was found to be 4.25×10^4 W/m²K, for dropwise condensation at moment of merging was found to be 9.64×10^4 W/m²K and for single drops was found to be 1.33×10^5 W/m²K.

4. As expected, the heat transfer coefficient for the single drop before merging is larger than that at the moment of merging, and the heat transfer coefficient for the meniscus after merging is smaller than that at the moment of merging.

ACKNOWLEDGMENTS

This material is based on work supported by the National Aeronautics and Space Administration under Grant NAG3-1834. Any opinions, findings, and conclusions or recommendations expressed in this publication are those of the authors and do not necessarily reflect the view of NASA.

REFERENCES

1. SCHMIDT, E., W. SCHURIG & W. SELLSCHOP. 1930. Versuche uber die Kondensation von Wasserdampf in Film- und Tropfenform. Tech. Mech. Thermodyn. **1:** 53–63.
2. MCCORMICK, J.L. & E. BAER. 1963. On the mechanism of heat transfer in dropwise condensation. J. Colloid Sci. **18:** 208–216.
3. UMUR, A. & P. GRIFFITH. 1965. Mechanism of dropwise condensation. J. Heat Transfer **87:** 275–282.
4. JAKOB, M. 1936. Heat transfer in evaporation and condensation—II. Mech. Eng. **58:** 729–739.
5. SUGAWARA, S. & K. KATUSUTA. 1966. Fundamental study on dropwise condensation. Proc. 3rd Int. Heat Transfer Conf. **2:** 354–359.
6. MERTE, H., JR. & C. YAMALI. 1983. Profiile and departure size of condensation drops on vertical surfaces. Warme- und Stoffubertragung **17:** 171–180.
7. TANASAWA, I., J. OCHIAI, Y. UTAKA & S. ENYA. 1976. Experimental study on dropwise condensation (effect of departing drop size on heat-transfer coefficients). Trans. JSME **42:** 2846–2851.
8. YAMALI, C. & H. MERTE, JR. 1999. Influence of sweeping on dropwise condensation with varying body force and surface subcooling. Int. J. Heat Mass Transfer **42:** 2943–2953.
9. WANG, Y.-X., J.L. PLAWSKY & P.C. WAYNER, JR. 2001. Optical measurement of microscale transport processes in dropwise condensation. Microscale Thermophys. Eng. **5:** 55–69.
10. WANG, Y.-X., J.L. PLAWSKY & P.C. WAYNER, JR. 2000. Heat and mass transfer in a vertical constrained vapor bubble heat exchanger using ethanol. 34th National Heat Transfer conference. Paper number: NHTC2000-12201.
11. DASGUPTA, S, J.L. PLAWSKY & P.C. WAYNER, JR. 1995. Interfacial force field characterization in a constrained vapor bubble thermosyphon. AIChE J. **41:** 2140–2149.
12. WANG, Y.-X., L. ZHENG, J.L. PLAWSKY & P.C. WAYNER, JR. 2002. Optical evaluation of the effect of curvature and apparent contact angle in droplet condensate removal. J. Heat Transfer **124:** 729–738.

Numerical Simulations of Bubble Motion in a Vibrated Cell Under Microgravity Using Level Set and VOF Algorithms

TIMOTHY J. FRIESEN,[a] HIROYUKI TAKAHIRA,[b] LISA ALLEGRO,[a] YOSHITAKA YASUDA,[c] AND MASAHIRO KAWAJI[a]

[a]*Department of Chemical Engineering and Applied Chemistry, University of Toronto, Canada*

[b]*Department of Energy Systems Engineering, Osaka Prefecture University, Sakai, Osaka, Japan*

[c]*Graduate School of Engineering, Osaka Prefecture University, Sakai, Osaka, Japan*

ABSTRACT: Understanding the stability of fluid interfaces subjected to small vibrations under microgravity conditions is important for designing future materials science experiments to be conducted aboard orbiting spacecraft. During the STS-85 mission, experiments investigating the motion of a large bubble resulting from small, controlled vibrations were performed aboard the Space Shuttle Discovery. To better understand the experimental results, two- and three-dimensional simulations of the experiment were performed using level set and volume-of-fluid interface tracking algorithms. The simulations proved capable of predicting accurately the experimentally determined bubble translation behavior. Linear dependence of the bubble translation amplitude on the container translation amplitude was confirmed. In addition, the simulation model was used to confirm predictions of a theoretical inviscid model of bubble motion developed in a previous study.

KEYWORDS: numerical simulation; bubble motion; microgravity; level set algorithm; VOF algorithm

NOMENCLATURE:

g	acceleration
F	volume-of-fluid function
$f(\varphi_0)$	$(\partial H/\partial \varphi_0)\lvert\nabla\varphi_0\rvert$
H	Heaviside function
$L(\varphi_0, \varphi)$	$S(\varphi_0)(1-\lvert\nabla\varphi\rvert)$
n	normal vector
p	pressure
R_b	initial bubble radius
Re	gTR_b/v_l
S	sign function
t	time
T	period
u	velocity
x, y, z	Cartesian coordinates

Address for correspondence: Masahiro Kawaji, Department of Chemical Engineering and Applied Chemistry, 200 College Street, University of Toronto, Canada M5S 3E5. Voice: 416-978-3064; fax: 416-978-8605.

kawaji@ecf.utoronto.ca

Greek Symbols
δ	delta function
ε	thickness of interface
η	μ_g/μ_l
κ	curvature
λ	ρ_g/ρ_l
ρ	density
μ	dynamic viscosity
ν	kinematic viscosity
σ	surface tension
τ	shear stress
φ	level set function
Ω	grid cell

Subscripts
g	gas
l	liquid

INTRODUCTION

In order to study physical processes that are free from buoyancy-induced convection, experiments can be performed under microgravity aboard orbiting spacecrafts. However, two effects have been identified that may produce fluid motion in systems involving a free surface: Marangoni convection, due to a temperature and/or concentration gradient at the fluid interface, and small accelerations referred to as g-jitter that result from spacecraft vibration.

There has been a great deal of interest in studying and controlling g-jitter levels aboard spacecraft.[1,2] A unique Canadian innovation designed to isolate experiments from g-jitter is the motion isolation mount (MIM) developed by the Canadian Space Agency and used aboard the Space Station Mir and the Space Shuttle Discovery.[3] This device minimizes the transfer of vibration between a stator, which is fixed to the spacecraft, and a flotor, which magnetically levitates relative to the stator.

During the STS-85 mission in August 1997, gas–liquid fluid experiments, referred to as interfacial stability and capillary wave (ISCAP) experiments, were conducted with MIM.[4] A cylindrical fluid cell containing a large air bubble in mineral oil ($\sigma = 0.0207$ N/m, $\mu = 0.058$ kg/msec, and $\rho = 866$ kg/m^3) or water–surfactant solution ($\sigma = 0.022$ N/m, $\mu = 1.0 \times 10^{-3}$ kg/msec, and $\rho = 996$ kg/m^3) was vibrated at controlled frequency and amplitude and the bubble shape and interface position were recorded by a video camera. A photograph of the large air bubble in a water-filled cell is shown in FIGURE 1. The fluid cell dimensions were approximately 48.7 mm in radius and 17 mm in height and the void fraction in the cell was approximately 30%. For the air–water system, a surfactant was added to the water at a concentration of 200 ppm to reduce the surface tension and contact angle. Both air–water and air–mineral oil systems are investigated numerically in the work described here.

The g-jitter recorded during the ISCAP experiments was characterized by acceleration levels of 2–3 mg in all of x, y, and z directions (1 mg corresponds to 1/1000th of normal gravity on Earth). MIM was used to simultaneously isolate the fluid cell from g-jitter and provide controlled vibrations of a frequency between 0.1 and 1.5 Hz, and acceleration levels of 0.1–16 mg. For vibration of the fluid cell perpendicular to the cell axis, the bubble translation amplitude was found to be linearly

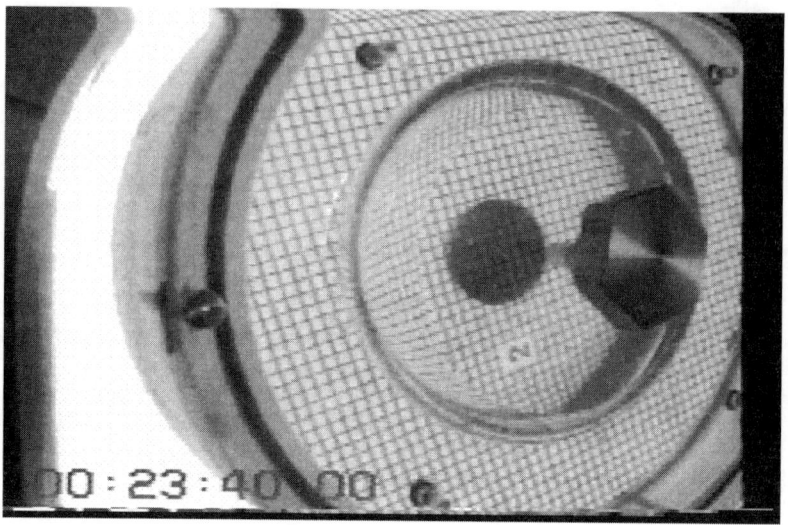

FIGURE 1. A large air bubble in a fluid cell recorded under microgravity during ISCAP experiments aboard the Space Shuttle.

proportional to the cell translation amplitude for both the air–water and air–mineral oil systems.

NUMERICAL SIMULATIONS

Numerical simulations of large bubble translation due to forced vibration at zero gravity were performed using level set and volume-of-fluid algorithms. The main objectives of this investigation were (1) to predict the experimental results for the bubble translation data and (2) to verify the theoretical prediction of the effect of gas–liquid density ratio on the bubble motion.

Level Set Method

The Navier–Stokes equations for a two-fluid system with a deformable interface were first solved by using the level set method. The level set method is based on an Eulerian formulation that describes the interface by the zero level of a Lipschitz-continuous function.[5–9] The interface is captured implicitly on the Eulerian grid by the zero level set. The governing equations in this work are

$$\frac{\partial u_j}{\partial x_j} = 0, \qquad (1)$$

$$\rho(\varphi)\frac{\partial u_i}{\partial t} + \rho(\varphi)\frac{\partial}{\partial x_j}(u_j u_i) = -\frac{\partial p}{\partial x_i} + \frac{\partial \tau_{i,j}}{\partial x_j} - \sigma\kappa(\varphi)\delta(\varphi)\frac{\partial \varphi}{\partial x_i} + \rho(\varphi)g_i, \qquad (2)$$

where

$$\tau_{i,j} = \mu\left(\frac{\partial u_i}{\partial x_j} + \frac{\partial u_j}{\partial x_i}\right), \qquad (3)$$

$$\kappa(\varphi) = \nabla \cdot \vec{n}, \quad \vec{n} = \nabla\varphi/|\nabla\varphi| \qquad (4)$$

σ is the surface tension, δ the delta function, and φ is the level set function obtained by solving the following equation:

$$\frac{\partial \varphi}{\partial t} + u_j \frac{\partial \varphi}{\partial x_j} = 0. \qquad (5)$$

Density and viscosity are functions of φ. They are, respectively,

$$\rho(\varphi) = \rho_l\{\lambda + (1-\lambda)H(\varphi)\}, \qquad (6)$$

$$\mu(\varphi) = \mu_l\{\eta + (1-\eta)H(\varphi)\}, \qquad (7)$$

where $\lambda = \rho_g/\rho_l$ and $\eta = \mu_g/\mu_l$. H is a smoothed Heaviside function, written as

$$H(\varphi) = \begin{cases} 0 & \text{if } \varphi < -\varepsilon \\ \frac{1}{2}\left\{1 + \frac{\varphi}{\varepsilon} + \frac{1}{\pi}\sin\left(\frac{\pi\varphi}{\varepsilon}\right)\right\} & \text{if } -\varepsilon < \varphi < \varepsilon \\ 1 & \text{if } \varphi > \varepsilon \end{cases} \qquad (8)$$

where ε is the thickness of the interface taken to be $\alpha\Delta x$, and α is taken to be 2 in the present calculation.

When we solve the level set equation numerically, the value of φ is distorted by the flow field and diffused by numerical viscosity. To maintain φ as a true distance function, that is $|\nabla\varphi| = 1$ near $\varphi = 0$, we solve the reinitialization equation proposed by Sussman et al.[8]

$$\frac{\partial \varphi}{\partial t} = S(\varphi_0)(1 - |\nabla\varphi|) + \Lambda_{i,j,k} f(\varphi_0) \equiv L(\varphi_0, \varphi) + \Lambda_{i,j,k} f(\varphi_0), \qquad (9)$$

where φ_0 is an initial distribution of φ and $S(\varphi_0)$ is a sign function. $\Lambda_{i,j,k}$ is given by

$$\Lambda_{i,j,k} = -\frac{\int_{\Omega_{i,j,k}} H'(\varphi_0)L(\varphi_0, \varphi)d\Omega_{i,j,k}}{\int_{\Omega_{i,j,k}} H'(\varphi_0)f(\varphi_0)d\Omega_{i,j,k}}, \qquad (10)$$

$$H'(\varphi_0) = \partial H/\partial \varphi_0, \qquad (11)$$

$$f(\varphi_0) = \frac{H'(\varphi_0)}{|\nabla\varphi_0|}, \qquad (12)$$

where $\Omega_{i,j,k}$ is the grid cell. The last term in Equation (9) is used to keep the bubble volume constant during the reinitialization procedure. This procedure is necessary because the density, viscosity, and surface tension are incorporated near the interface.

The solver for these governing equations is basically the same as that used in Takahira and Banerjee.[10] However, we modified the solver for the pressure Poisson equation to consider the high-density ratio. We used CGS[11] and Bi-CGSTAB[12] for the two-dimensional and three-dimensional analyses, respectively.

Volume-of-Fluid Method

Two-Dimensional Simulation

The two-dimensional simulation was performed using the volume-of-fluid algorithm developed by Hirt and Nichols[13] and incorporated into the computational fluid dynamics (CFD) software Fluent™. It is a finite difference code widely used for various types of CFD calculations. In the VOF algorithm, a volume fraction variable, F, is used instead of the level set function, φ, in Equation (5), to track the gas-liquid interface in the computational domain. The value of F in a given control volume is equal to unity when the control volume is filled with a gas, and zero when the control volume is full of liquid. An intermediate value between zero and unity indicates a gas–liquid mixture occupying the control volume.

In the VOF algorithm, the interface orientation and position in each control volume are constructed at each time step based on the F values in the neighboring control volumes in order to calculate the advection of the volume fraction variable while preserving the interface discontinuity. The piecewise linear interface construction (PLIC) method was used to reconstruct the interface and calculate the advection of the volume fraction variable.[14] In the momentum equation, the surface tension force was incorporated using a continuum surface force model of Brackbill et al.,[15] and a source term representing the acceleration of the fluid cell due to the forced vibration was also included. The momentum equations were discretised using the QUICK method, and the semi-implicit method for pressure linked equations (SIMPLE) was used to solve the convection-diffusion equations. The initial velocity and pressure fields were specified to simulate steady bubble oscillation.

An unstructured grid was set up for a circular cell boundary and refined in the region where the interface passed during the simulation in order to minimize the numerical diffusion of the volume fraction variable. This grid was shown to produce grid independent results, such that when higher grid densities were used in the simulation no observable difference in bubble motion was obtained.[16]

Three-Dimensional Model

Three-dimensional VOF calculations were performed using the PLIC interface reconstruction algorithm in a three-dimensional VOF code called the *BUBBLE code*, which is a modified version of the code previously developed by Bussmann et al.[17] to simulate droplet behavior upon impact with a solid surface. The computational domain was one-quarter of the cylindrical cell, full length in the direction of vibration (y) and half lengths in the other two directions (x and z). The contact angle between the gas–liquid interface and the flat cell walls was set to 45 degrees for all calculations. Using a rectangular mesh in Cartesian coordinates, attempts were made to simulate the bubble motion in both air–water and air–mineral oil systems using the actual fluid properties. However, the less viscous air–water system yielded severe deformation of the air bubble unseen in the ISCAP experiments. Thus, only the results of air–mineral oil simulation are presented in this paper.

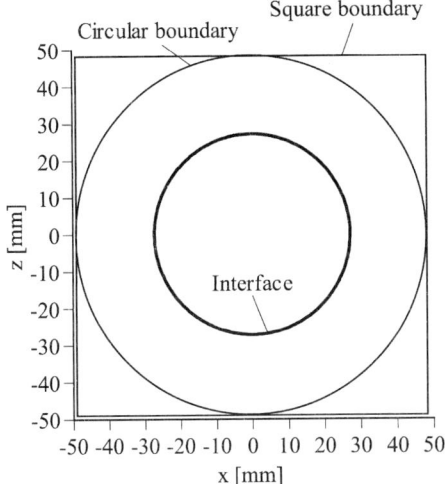

FIGURE 2. Schematic of computational boundaries for the level set simulations.

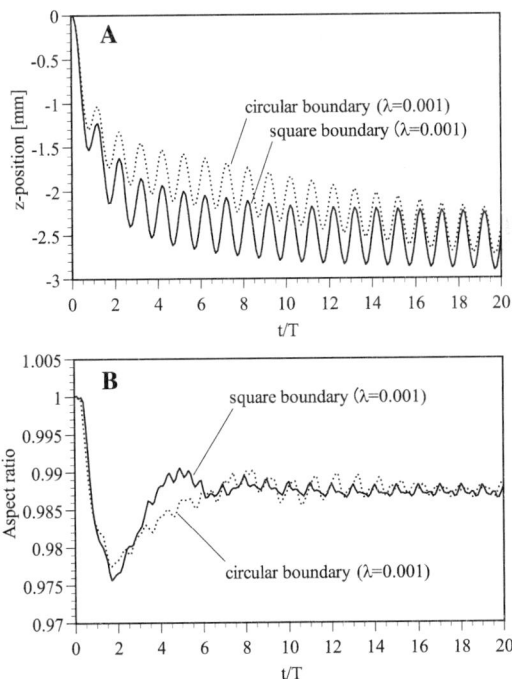

FIGURE 3. Effects of fluid cell shape and gas volume fraction on the two-dimensional bubble motion in mineral oil vibrated at 0.8 Hz, 5 mg: **(A)** bubble centroid, **(B)** aspect ratio.

RESULTS AND DISCUSSION

Level Set Simulation

Two-Dimensional Analysis

The bubble motion was calculated in two types of a two-dimensional fluid cell. FIGURE 2 shows the computational domain for the present analysis. One had a square boundary and the other a circular boundary. The calculations were performed for a mixture of air–mineral oil using a 128×128 rectangular mesh. The bubble volumes were the same in each case. The viscosity of mineral oil used was about 58 times larger than that of water. The bubble was accelerated in the z direction.

The effects of fluid cell shape and gas volume fraction on the position of bubble centroid and the aspect ratio of its shape (vertical length/horizontal length) are shown in FIGURE 3. The amplitude and frequency of the acceleration applied were 5 mg and 0.8 Hz, respectively. The initial bubble radius, R_b, was taken to be 26.2 mm. The vibration Reynolds number, $Re = gTR_b/v_l$, is 24.0 under these conditions. FIGURE 4 shows the two dimensional bubble shape at $t/T = 20$ when the density ratio, λ, is 0.001.

As evident in FIGURE 3A, after about 20 cycles of sinusoidal vibrations, the position of the bubble centroid almost reached a steady periodic state. The magnitude of bubble translation for the square boundary was larger than that for the circular boundary. This is because the bubble in the square cell has a wider space in which to move, since the gas volume fraction in the rectangular cell is smaller than that in the circular cell. On the other hand, little difference was found in the aspect ratio between the rectangular and circular cells in the steady periodic state. The bubble

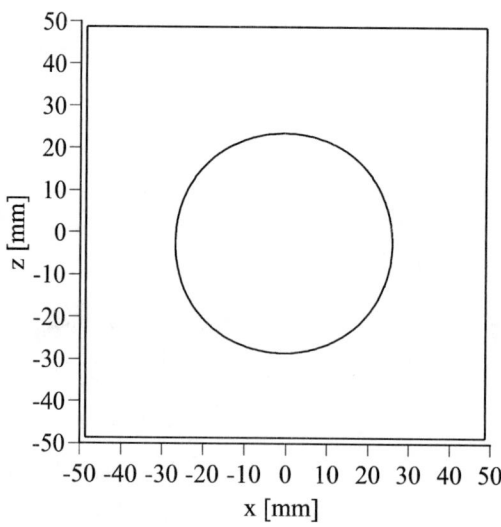

FIGURE 4. Two-dimensional bubble shape in a square cell ($\lambda = 0.001$, $t/T = 20$, 0.8 Hz, 5 mg).

became slightly depressed in the vertical direction (FIGURE 4). The harmonic components were found in the time histories of the aspect ratio shown in FIGURE 3B.

The effects of density ratio are shown in FIGURE 5. The solid and dotted lines indicate the results for density ratios of 0.001 and 0.01, respectively. When the density ratio was high the bubble translation amplitude became larger in the transient state, as expected. However, in the nearly steady periodic state the positions of the bubble centroid were almost the same for the two cases. The amplitude of the centroid for the high-density ratio was about 0.1 mm larger than that for the low-density ratio. The degree of bubble shape deformation was larger when the density ratio was greater, at 0.01.

Three-Dimensional Analysis

One of the difficulties in the three-dimensional analysis may be the treatment of the boundary conditions for the flat walls of the fluid cell. In the first attempt, the contact angle was kept constant so as to determine the bubble motion. This condition

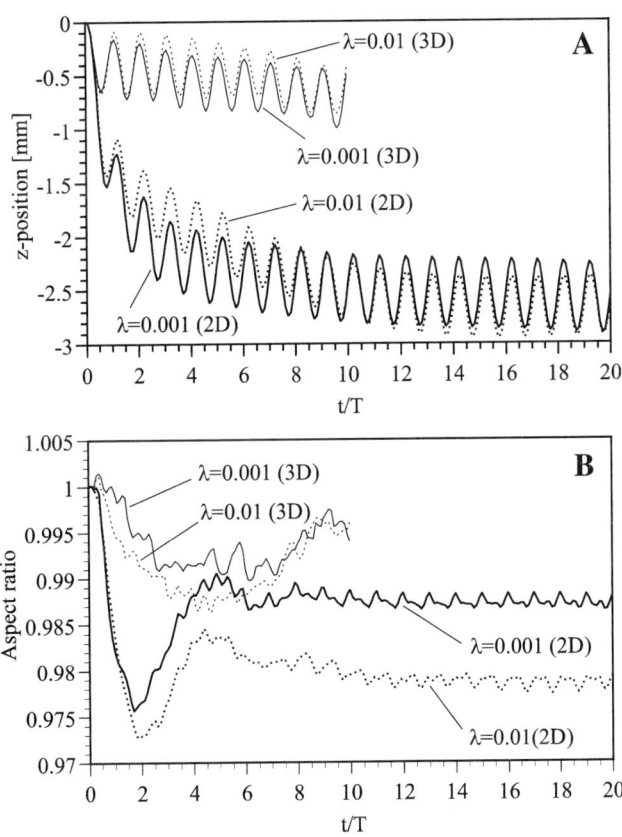

FIGURE 5. Effects of density ratio on the bubble motion in mineral oil vibrated at 0.8 Hz, 5 mg: (**A**) bubble centroid, (**B**) aspect ratio.

is not always correct because the contact angle for a stationary bubble may be different from that for a moving bubble and can change depending on the direction of motion. However, because the detailed information on the contact surface was not available from the experiment, a constant contact angle was assumed. Since the normal derivative on the flat wall is expressed in Equation (4) by using the level set function, the contact angle is easily obtained from the level set function.

Time histories of the bubble centroid position for a shallow rectangular fluid cell containing an air–mineral oil mixture are shown in FIGURE 5 by solid and dotted lines. The cross-section of the three-dimensional cell at constant y was the same as that used in the two-dimensional analysis. The gas volume fraction was identical to that of the two-dimensional analysis. The computational domain was partitioned into a $128 \times 22 \times 128$ mesh. We maintained the contact angle at 50 degrees for the overall motion. Therefore, the initial bubble radius, R_b, in the $y = 0$ plane in the three-dimensional analysis was slightly larger than the initial radius used in the two dimensional analysis. The vibration Reynolds number was 24.9 in the three-dimensional analysis.

The results show that the bubble translation amplitude for the three-dimensional simulation was smaller than that for the two dimensional simulation. However, the amplitude of the z-position in each cycle was almost the same as that in the two-dimensional analysis. The aspect ratio variation for the three-dimensional bubble was also smaller. The bubble shape at $t/T = 6$ is shown in FIGURE 6. The cross-section at $y = 0$ is similar to that of the two dimensional bubble shown in FIGURE 4. Since we kept the contact angle constant, the bubble cross-section at $x = 0$ is similar to that at $z = 0$. When the density ratio was reduced, the bubble translation amplitude increased. This is the same tendency as found for the two-dimensional bubble analysis.

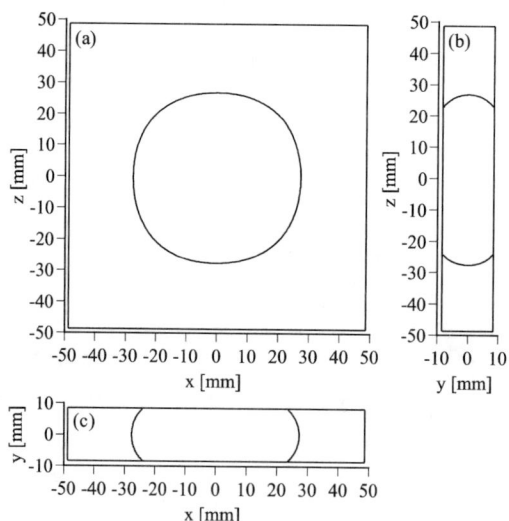

FIGURE 6. Three-dimensional bubble shape in a rectangular cell ($\lambda = 0.001$, $t/T = 6$, 0.8 Hz, 5 mg). (**A**) x–z cross-section at $y = 0$, (**B**) y–z cross-section at $x = 0$, (**C**) x–y cross-section at $z = 0$.

Since the viscosity of water is much smaller than that of mineral oil, it was difficult to obtain steady periodic solutions for the bubble motion in water. FIGURE 7 shows the time histories of the bubble centroid position in a rectangular fluid cell containing an air–water mixture when the acceleration level was 8 mg and frequency 1 Hz. The initial bubble radius at $y = 0$ plane was 29.2 mm and the vibration Reynolds number was 2,283. The density ratio was taken to be 0.01. As can be seen from this figure, the bubble centroid moved rapidly and the bubble translation amplitude became much larger than that for the air–mineral oil system. Since the bubble translation amplitude was too large, the bubble shape was strongly affected by the cell translation. The bubble became elongated in the horizontal direction and surface oscillations of low frequency were observed, as shown in FIGURE 7B. When the viscous damping effect is small, the duration of the transient oscillation state increases. Therefore, the bubble behavior is strongly affected by the initial conditions. This fact makes it difficult to evaluate the characteristics of the bubble behavior numerically.

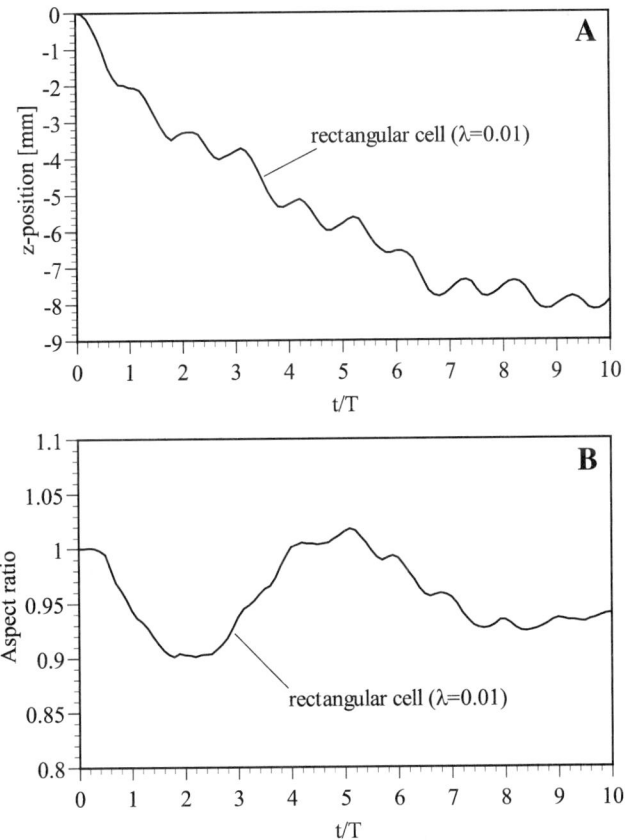

FIGURE 7. Three-dimensional bubble motion in water for a fixed contact angle and 1.0 Hz, 8 mg vibration: **(A)** bubble centroid, **(B)** aspect ratio.

We considered another set of boundary conditions for the present three-dimensional analysis, since a variable contact angle may yield different bubble behavior. In FIGURE 8, the results for the fixed contact line boundary condition are shown. This boundary condition was achieved by keeping the level set function constant at the contact line. This condition may be correct in the very early stage of the bubble motion. That is, if the shear force due to the vibration-induced flow is smaller than the static friction force, the bubble does not move and only the oscillation of the bubble surface may occur. The results obtained under this boundary condition for the air–water system were very different from those shown in FIGURE 7. As shown in FIGURE 8, the position of the bubble centroid oscillated periodically around $z = 0$, and the variation of the bubble aspect ratio was reduced.

The snapshot of the bubble shape at $t/T = 8.6$ obtained under the fixed contact line condition is shown in FIGURE 9. The x–z cross-section of the bubble shape is nearly circular. The y-z cross-section is different from that of FIGURE 6. That is, the lower

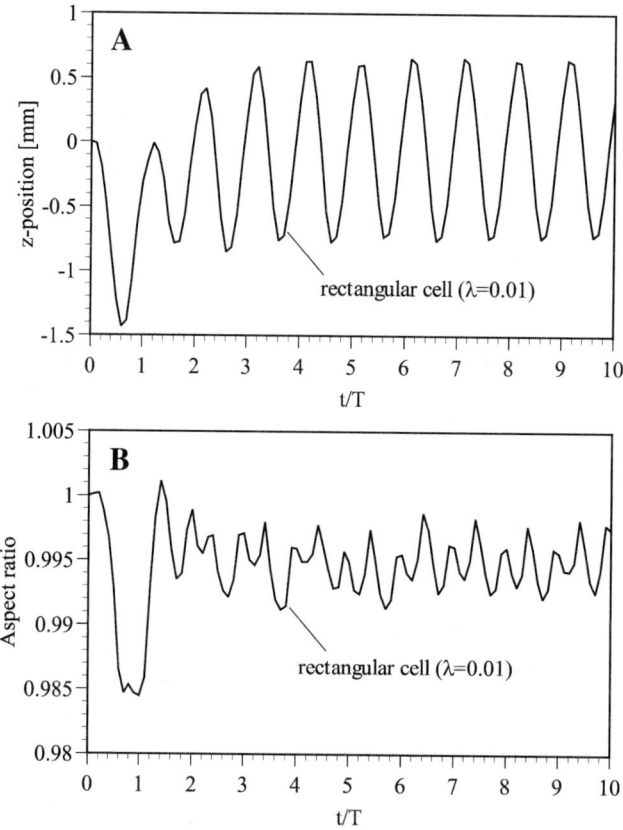

FIGURE 8. Three-dimensional bubble motion in water for a fixed contact line and 1.0 Hz, 8 mg vibration: **(A)** bubble centroid, **(B)** aspect ratio.

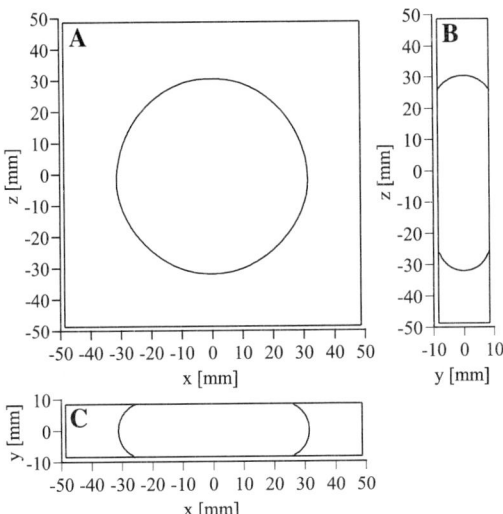

FIGURE 9. Three-dimensional bubble shape in a rectangular cell for a fixed contact line ($\lambda = 0.01$, $t/T = 8.6$, 1.0 Hz, 8 mg), **(A)** x–z cross-section at $y = 0$, **(B)** y–z cross-section at $x = 0$, **(C)** x–y cross-section at $z = 0$.

side of the bubble surface is elongated in the $-z$ direction. The contact angle at this side became less than 50 degrees, whereas it became greater than 50 degrees at the upper side of the bubble surface. These results show that the three-dimensional bubble behavior is affected strongly by the boundary conditions at the contact line.

VOF Simulation

Two-Dimensional Analysis

Simulations were performed for an air–water system subjected to different cell vibration frequencies and acceleration levels using an unstructured grid as shown in FIGURE 10. FIGURE 11 shows the predicted bubble centroid position for 0.6 Hz and 2 mg acceleration (the vibration Reynolds number, $Re = 924$). The bubble always responded to the vibrations at the same frequency as the excitation frequency as experimentally observed. Also, some low frequency oscillation, or migration, of the bubble was initially observed, as in the level set simulation results. This is due to a transient effect as the fluid adjusts to the excitation imposed at time zero.

Other cases of excitation parameters were simulated in order to observe the effect of the vibration characteristics on the bubble motion. As shown in FIGURE 12, an increase in the cell vibration amplitude produced an increase in the bubble translation amplitude. A best-fit line through the numerical results produced a slope of 0.44. Experimentally the relationship was also linear with a slope of 0.45. Thus, the VOF simulation was successful in reasonably predicting the experimentally observed bubble motion. However, it can also be noted that the simulation results shown in

FIGURE 10. Unstructured grid used in a two-dimensional VOF simulation.

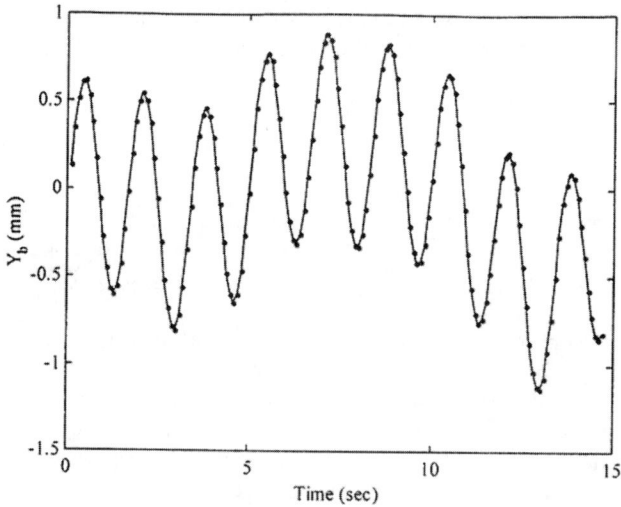

FIGURE 11. Bubble centroid trajectory predicted in a two-dimensional VOF simulation for 0.6 Hz, 2 mg acceleration.

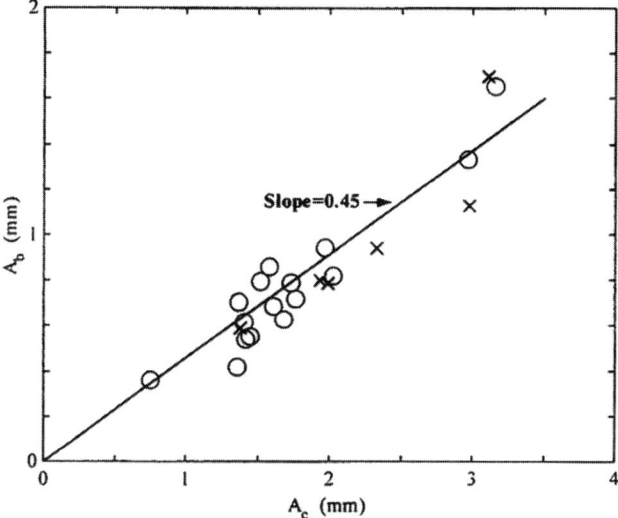

FIGURE 12. Bubble translation amplitude predicted by a two-dimensional VOF model for an air bubble in water: ○, experimental data; —, experimental best-fit line; ×, numerical data.

FIGURE 12 exhibit some non-linearity. It was observed both experimentally and numerically that there is an effect of the excitation frequency on the bubble translation amplitude at very low excitation frequencies. A much greater bubble translation amplitude was predicted for the cell vibration amplitude of 3.2 mm and frequency of 0.2 Hz. However, for vibrations exceeding 0.4 Hz the bubble translation amplitude was predicted to be linearly proportional to the cell vibration amplitude.

Subsequent to the ISCAP experiments, a theoretical model of bubble translation was also developed by solving two-dimensional Navier–Stokes equations for a circular gas bubble in a liquid filled circular container similar to that used in the ISCAP experiments.[4,18] Inviscid and incompressible liquid properties were assumed. The bubble-to-cell translation amplitude ratio, A_b/A_c, was theoretically predicted to be a function of the gas–liquid density ratio, ρ_g/ρ_l, and the ratio of the fluid cell radius to the bubble radius. For given density and radii ratios, the theory predicted a linear relationship between the cell translation amplitude and the bubble translation amplitude.

The effect of the gas–liquid density ratio on the bubble motion was investigated, by considering the case of 0.8 Hz/6 mg vibration for the same bubble-to-cell radius ratio. Simulations were performed for gas–liquid density ratios ranging from 0.001 to 0.5. FIGURE 13 shows the ratio of the bubble translation amplitude to the cell translation amplitude for each simulation. It is evident that the simulation provides evidence that the bubble motion is dependent on the gas–liquid density ratio in accordance with the theoretical model prediction shown by a solid curve. The discrepancy between the theoretical prediction and the simulated bubble translation amplitude was less than 10%.

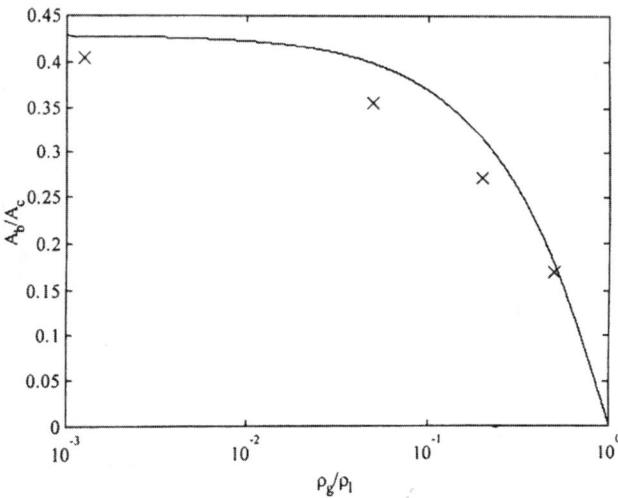

FIGURE 13. Effect of density ratio on the bubble-to-cell translation amplitude ratio for an air bubble in water predicted by a two dimensional VOF model: —, theoretical prediction; ×, numerical data.

Three-Dimensional Analysis

A series of three-dimensional simulations were performed using $42 \times 82 \times 6$ mesh in a computational domain covering one-quarter of the fluid cell. The predicted trajectory of the bubble centroid is shown in FIGURE 14 for an acceleration level of 3.0 mg, vibration frequency of 0.8 Hz, and the vibration Reynolds number of 14.4.

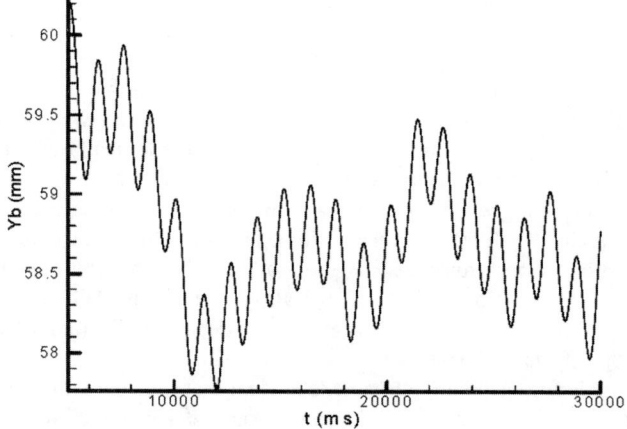

FIGURE 14. Bubble centroid trajectory predicted by a three-dimensional VOF model for an air bubble in mineral oil.

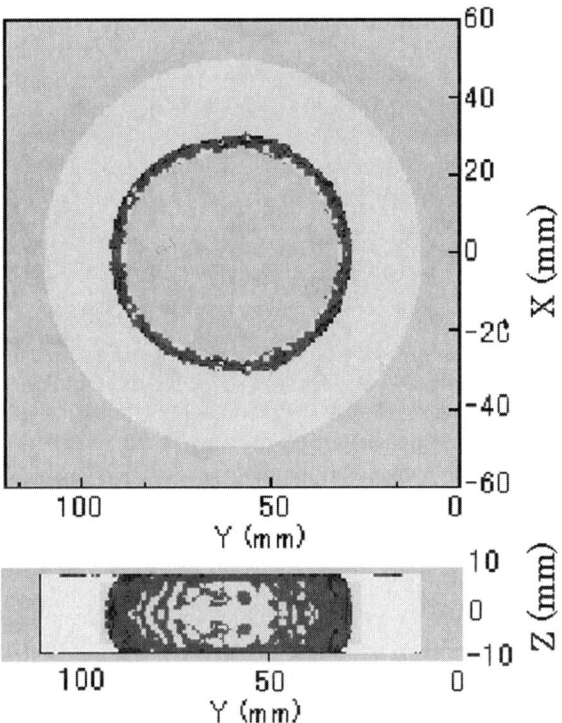

FIGURE 15. Shape of an air bubble in mineral oil predicted by a three-dimensional VOF model.

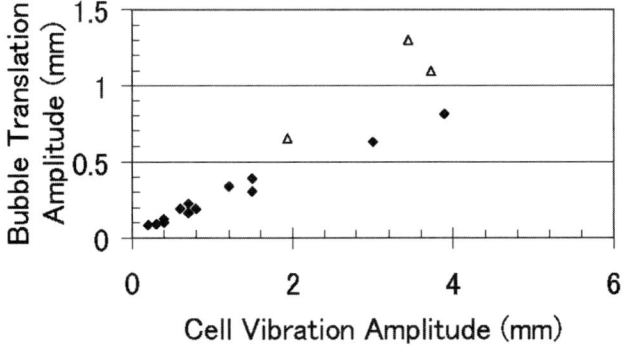

FIGURE 16. Variation of bubble translation amplitude predicted by a three-dimensional VOF model: ◆, predicted amplitude; △, ISCAP data.

Although the bubble is seen to translate at the same frequency as the vibration frequency, low frequency drifting of the bubble was again observed. The bubble shape for this vibration is shown in FIGURE 15. The cell was vibrated in the y direction and the bubble shape was slightly compressed in the same direction. For this viscous system, the bubble translation amplitude also increased linearly with the cell translation amplitude as shown in FIGURE 16. These results were in qualitative agreement with a limited amount of experimental data collected during the ISCAP experiments, also shown in FIGURE 16.

CONCLUSIONS

Two-dimensional and three-dimensional level set and volume-of-fluid simulations were performed to predict the vibration-induced motion of an air bubble in a liquid-filled cell under microgravity. The predicted results were compared with the measurements performed under microgravity aboard the Space Shuttle. Both two-dimensional and three-dimensional simulations using level set and VOF techniques showed satisfactory agreement with the data, reproducing the linear relationship between the bubble translation amplitude and the cell vibration amplitude. Three-dimensional simulations of an air bubble–water system using a level set technique elucidated the importance of boundary conditions related to the contact line at the flat cell walls. In addition, the simulations corroborated the theoretical prediction of how the gas–liquid density ratio affects the bubble motion. This investigation has not only proven the reliability of numerical simulations in predicting free surface motion due to small vibration of the fluid cell under microgravity but also corroborated the theoretical model and experimental data that currently exist for quantifying the interface motion induced by small-amplitude vibrations.

ACKNOWLEDGMENTS

The authors would like to thank the Canadian Space Agency for financially supporting the VOF calculations and Dr. M. Pasandideh-Fard of Simulent Inc. and Prof. J. Mostaghimi of the University of Toronto for providing the BUBBLE code.

REFERENCES

1. MONTI, R. 1990. Gravity jitters: effects on typical fluid science experiments. *In* Low Gravity Fluid Dynamics and Transport Phenomena, Progress in Astronautics and Aeronautics, 130, AIAA.
2. ALEXANDER, I. 1996. Impulse, vibrations and random disturbances: consequences for convective diffusive transport and fluid interfaces in low gravity experiments. Invited paper, AIAA 96-2073, presented at 27th Fluid Dynamics Conference, New Orleans, LA, June 17–20.
3. TRYGGVASON, B.V., *et al.* 1999. Acceleration levels and operation of the microgravity vibration isolation mount (MIM) on the shuttle and the Mir space station. Paper AIAA-99-0578, 37th AIAA Aerospace Sciences Meeting and Exhibition, Jan. 11–14, 1999, Reno, Nevada.

4. KAWAJI, M., et al. 2002. The effects of forced vibration on the motion of a large bubble under microgravity. J. Fluid Mech. Submitted.
5. OSHER, S. & J.A. SETHIAN. 1988. Fronts propagating with curvature-dependent speed: algorithms based on Hamilton–Jacobi formulations. J. Comp. Phys. **79:** 12–49.
6. SUSSMAN, M. & P. SMEREKA. 1997. Axisymmetric free boundary problems. J. Fluid Mech. **341:** 269–294.
7. SON, G. & V.K. DHIR. 1998. Numerical simulation of film boiling near critical pressures with a level set method. ASME J. Heat Transfer **120:** 183–192.
8. SUSSMAN, M., et al. 1998. An improved level set method for incompressible two-phase flows. Computers & Fluids **27:** 663–680.
9. SUSSMAN, M., et al. 1999. An adaptive level set approach for incompressible two-phase flows. J. Comp. Phys. **148:** 81–124.
10. TAKAHIRA, H. & S. BANERJEE. 2000. Numerical simulation of three dimensional bubble growth and detachment in a microgravity shear flow. *In* Microgravity Fluid Physics and Heat Transfer. V. Dhir, Ed.: 80–87. Begell House, New York.
11. SONNEVELD, P. 1989. CGS, A fast Lanczos-type solver for nonsymmetric linear systems. SIAM J. Sci. Stat. Comput. **10:** 36–52.
12. VAN DER VORST, H.A. 1992. Bi-CGSTAB: A fast and smoothly converging variant of Bi-CG for the solution of nonsymmetric linear systems. SIAM J. Sci. Stat. Comput. **13:** 631–644.
13. HIRT, C.W. & J.B. NICHOLS. 1981. Volume of fluid (VOF) method for dynamics of free boundaries. J. Comp. Phys. **39:** 201–225.
14. YOUNGS, D.L. 1982. Time-dependent multi-material flow with large fluid distortion. *In* Numerical Methods for Fluid Dynamics. K.W. Morton & M.J. Baines, Eds.: 273–285. Academic Press, New York.
15. BRACKBILL, J.U., D.B. KOTHE & C. ZEMACH. 1992. A continuum method for modeling surface tension. J. Comp. Phys. **100:** 335–354.
16. FRIESEN, T. 2000. A Numerical Investigation of Large Bubble Translation Behaviour Due to Forced Vibration Under Microgravity. M.A.Sc. Thesis, Department of Chemical Engineering and Applied Chemistry, University of Toronto.
17. BUSSMANN, M., et al. 1999. On a three-dimensional volume tracking model of droplet impact. Phys. Fluids **11**(6): 1406–1417.
18. LYUBIMOV, D.V., et al. 2000. The effects of g-jitter and vibrations on the motion and shape of a large bubble in a cylindrical cell. ISCAP experiments aboard the Space Shuttle (Part II: theoretical predictions). Spacebound 2000 Conference, Vancouver, Canada, May.

Annular Flow Film Characteristics in Variable Gravity

RYAN M. MacGILLIVRAY AND KAMIEL S. GABRIEL

Microgravity Research Group, College of Engineering, University of Saskatchewan, Saskatoon, Saskatchewan, Canada

ABSTRACT: Annular flow is a frequently occurring flow regime in many industrial applications. The need for a better understanding of this flow regime is driven by the desire to improve the design of many terrestrial and space systems. Annular two-phase flow occurs in the mining and transportation of oil and natural gas, petrochemical processes, and boilers and condensers in heating and refrigeration systems. The flow regime is also anticipated during the refueling of space vehicles, and thermal management systems for space use. Annular flow is mainly inertia driven with little effect of buoyancy. However, the study of this flow regime is still desirable in a microgravity environment. The influence of gravity can create an unstable, chaotic film. The absence of gravity, therefore, allows for a more stable and axisymmetric film. Such conditions allow for the film characteristics to be easily studied at low gas flow rates. Previous studies conducted by the Microgravity Research Group dealt with varying the gas or liquid mass fluxes at a reduced gravitational acceleration.[1,2] The study described here continues this work by examining the effect of changing the gravitational acceleration (hypergravity) on the film characteristics. In particular, the film thickness and the associated pressure drops are examined. The film thickness was measured using a pair of two-wire conductance probes. Experimental data was collected over a range of annular flow set points by changing the liquid and gas mass flow rates, the liquid-to-gas density ratio and the gravitational acceleration. The liquid-to-gas density ratio was varied by collecting data with helium–water and air–water at the same flow rates. The gravitational effect was examined by collecting data during the microgravity and pull-up (hypergravity) portions of the parabolic flights.

KEYWORDS: annular flow; frictional pressure drop; film thickness; microgravity; hypergravity; film characteristics; two-phase flow; thermal systems; petrochemical processes; air–water; helium–water; density effect; parabolic flight; terrestrial applications; space applications; conductance probes

NOMENCLATURE:

d	tube diameter (m)
Eu	$\Delta P / \Delta z_f (d/(\rho_l V_{sl}^2))$
g	Earth gravitational acceleration, 9.81 m/sec^2
m	mass (kg)
Re	Reynolds number, $\rho V d / \mu$
V	velocity (m/sec)
We	Weber number, $\rho V^2 d / \sigma$

Address for correspondence: Kamiel S. Gabriel, Ph.D, M.B.A., P.Eng., Professor & Director, Microgravity Research Group, Associate Dean of Graduate Studies, Research and Extension, College of Engineering, University of Saskatchewan, 57 Campus Drive, Saskatoon, SK S7N-5A9, Canada. Voice: 306-966-5280; fax: 306-966-5205.
kamiel@engr.usask.ca   <www.engr.usask.ca/cmore>

x	gas quality, m_g/m_{tot}
Greek Symbols	
α'	pseudo-void fraction, V_{sg}/V_{sl}
δ	film thickness (m)
$\Delta P/\Delta z_f$	frictional pressure drop (Pa/m)
μ	viscosity (kg/m^{2}sec)
ρ	density (kg/m^3)
σ	surface tension (N/m)
Subscripts	
g	gas phase
l	liquid phase
sl	superficial liquid, the property of the liquid flowing alone in the tube
tot	total (gas + liquid)

INTRODUCTION

Two-phase flows occur in many industrial applications, including power generation plants, chemical plants, pipelines, geothermal energy systems, steam generators, air conditioning and refrigeration systems, heat exchangers, and cooling channels of nuclear reactors. Interaction between the phases allows for the characterization of two-phase flows into different flow regimes; shown schematically in FIGURE 1.[3] Among the various flow regimes, annular flow is of particular interest in the present study. Annular flow consists of a liquid film flowing along the tube wall, with a gas and entrained liquid droplets flowing in the core.

Much research[4–8] has been performed over the past forty years to gain a better understanding of the fundamental physics of annular flows for the purpose of developing a model for the design of new two-phase flow systems. Despite extensive study, detailed data on the liquid–gas interface and the interface dynamics are still lacking. These interface dynamics are essential to developing reliable closure laws for flow modeling.

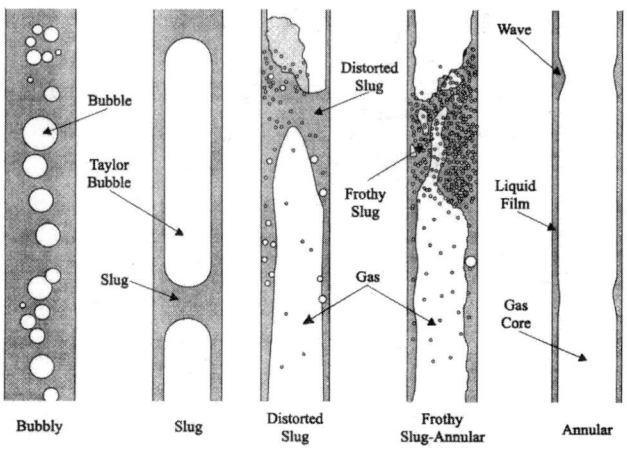

FIGURE 1. Typical microgravity two-phase flow regimes.[3]

de Jong[1] performed a dimensional analysis on annular flow and identified five pertinent dimensionless groups that influence film thickness and pressure drop. These dimensionless groups are:

- the ratio of the gas-to-liquid mass flow rates, $x/(1 - x)$,
- the gas-to-liquid viscosity ratio, μ_g/μ_l,
- the gas-to-liquid density ratio, ρ_g/ρ_l,
- the superficial liquid Reynolds number, Re_{sl}, and
- the superficial liquid Weber number, We_{sl}.

Changing the gas or liquid mass flow rates varies the first and fourth dimensionless groups. However, the remaining groups are dependent on the phase properties. It is also believed that gravitational acceleration influences the annular flow regime. The present research attempts to determine the effect of the gravitational acceleration while also varying the third dimensionless group; the gas-to-liquid density ratio. Annular flow data points were collected using air–water and helium–water combinations. The ratio of the helium-to-air density is approximately 1:7, resulting in a similar reduction in the gas-to-liquid density ratio.

Although annular flow is an inertia-driven flow with little effect of buoyancy, it is still desirable to study it in a microgravity environment. The influence of gravity on upward cocurrent annular flow is to cause an increase in the disturbance waves due to acceleration of the liquid phase in the opposite direction to the fluid motion. By reducing the gravitational acceleration, annular flow is achieved at lower gas velocities and the interface between the two phases becomes more stable. A stable interface is essential to determine the annular flow characteristics; for example, the film thickness and the frictional pressure drop. Knowledge of these two parameters could improve the design of two-phase systems by expanding the knowledge base on annular flow and developing the necessary closure laws for modeling. A microgravity environment also creates more uniform and repeatable wave patterns, further assisting with the development of reliable models.

The effect of both increasing and decreasing the gravitational acceleration on the flow is examined. Microgravity ($\pm 0.05\,g$) and hypergravity ($1.5\,g$ or $1.8\,g$) data were collected for the film characteristics and pressure drop.

EXPERIMENTAL APPARATUS

Previously microgravity data had been collected on the same experimental apparatus.[3,9–11] Recent modifications by the principle author include repositioning of equipment in the test package, reduction of overall mass, and the addition of new hardware and instrumentation. A schematic of the apparatus is shown in FIGURE 2. The flow loop includes a circulating liquid loop, an open gas loop, and a computer based data acquisition and control system. The apparatus is used for collecting data at both microgravity and hypergravity during a parabolic airplane flight, as well as at normal gravity (pre- and postflight data).

The liquid loop consists of a collapsible bladder fuel tank, a gear pump, a by-pass line with a check-valve, four flow venturies for stabilizing the flow, an in-line filter,

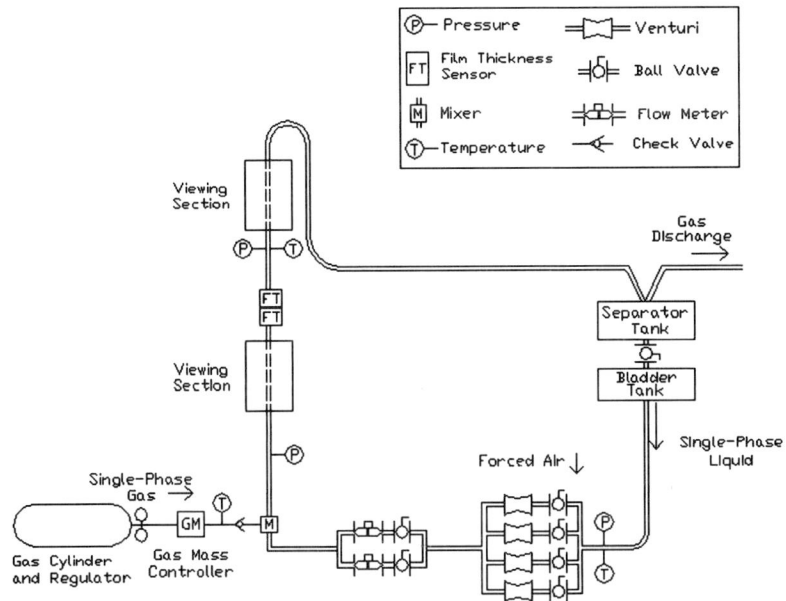

FIGURE 2. Schematic of the experimental apparatus.

two turbine flow meters, a gas-liquid mixer, a vertical developing length, an acrylic viewing section, a pair of two-wire conductance probes, a second viewing section, and a separator tank. The first pressure tap is located 0.72 m (76 D) downstream of the mixer, with the second pressure tap separated by 0.88 m (92 D). The two film thickness probes are separated by 0.019 m (2 D), with the first film thickness probe located 1.44 m (151 D) downstream of the mixer. The separator tank uses the surface tension of the liquid to separate the two phases during the microgravity periods, and allows gravity drainage into the bladder tank during the hypergravity period.

The gas loop consists of a cylinder of compressed gas (either air or helium), a pressure regulator, a gas flow controller, and a check valve. The check valve is located immediately before the gas-liquid mixer to prevent liquid from flowing back into the gas flow controller. The gas is injected radially into the liquid, and is allowed to flow with the liquid until the two-phase flow reaches the separator tank. The gas is then vented overboard.

The data acquisition and control system is a computer based LabVIEW system. The power to the pump is controlled manually by adjusting a variable AC source. However, all other controls are done through the LabVIEW based program (developed by the principal author). Measurements include reference temperature, gas flow rate, pump speed, all-water pressure and temperature, film thickness from the two probes, liquid flow rates, pressure drops, two-phase pressure and temperature, and acceleration levels. Digital images of the flow are recorded using a 30 frame per second digital camera located 0.13 m (14 D) downstream of the second film thickness probe. A 12-bit analog output board is used for control, and a 12-bit input board is used for the data acquisition. Data were collected at a sampling rate of 1,024 Hz.

DATA COLLECTION PROCEDURE

The microgravity and hypergravity data were collected onboard the ESA parabolic aircraft (Airbus A300; operated by Novespace), during the 29th European Space Agency Microgravity Science Mission. The parabolic flight campaign was held during the week of November 20–24, 2000, at Novespace, Bordeaux, France. A typical parabolic maneuver of the A300 is shown in FIGURE 3.[12] Each parabola is preceded by approximately one minute of level flying. The parabola begins with the hypergravity portion, where acceleration levels are approximately $1.8g$ for 15 seconds, followed by 10 seconds of approximately $1.5g$. A transition period of approximately five seconds occurs as the microgravity period is approached. The microgravity period lasts about 22 seconds, during which time the three pilots maintain the acceleration level between $\pm 0.05g$. The recovery portion of the parabola mirrors the entry portion with five seconds of transition, 10 seconds of $1.5g$, followed by 15 seconds of $1.8g$. The parabola ends with approximately 30 seconds of level flight before the next parabola sequence is started. Thirty parabolas were flown each day, with a normal campaign schedule of three flight days (a total of 90 parabolas).

The level flying periods between the parabolas were used to set the liquid and gas flow rates. The data collection was started approximately three seconds prior to the "pull-up" phase and continued until the "pull-out" portion of the parabola. This allowed hypergravity and microgravity data to be collected in the same file without changing any parameters. A typical acceleration trace during the data collection period of a typical parabola is shown in FIGURE 4.

PRELIMINARY RESULTS

All results presented in this section are based on preliminary analysis since additional analyses are currently under way. The preliminary frictional pressure-drop results for microgravity and hypergravity air–water flows are shown in FIGURE 5. The microgravity and hypergravity helium–water frictional pressure drop is shown in

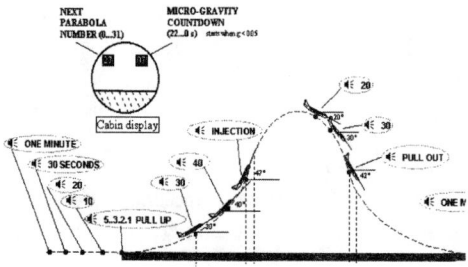

FIGURE 3. Typical Airbus A300 parabolic flight maneuver.[12]

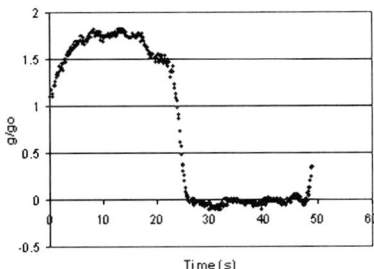

FIGURE 4. Gravity signal during data collection period of a typical parabola.

FIGURE 6. Both plots show the fictional pressure drop plotted against the pseudo-void fraction, α'. The lines represent the best-fit for the data, separated by constant values of gas velocity. The solid symbols represent the microgravity data, the hollow symbols represent the hypergravity data.

Both the air–water and helium–water data follow the same trend for the frictional pressure drop. The frictional pressure drop is increased by maintaining the gas flow rate constant and increasing the liquid flow rate, or by maintaining the liquid flow rate constant while increasing the gas flow rate. This is in agreement with previous data reported by de Jong and Rezkallah,[1] as shown in FIGURE 7. In this case, the data is separated into constant liquid velocities. The helium–water points are taken at comparable pseudo-void fraction values, and the pressure drop values are comparable. However, the helium–water points are at an increased gas velocity. This hints at a significant effect of the gas density on the flow.

The hypergravity data is consistently higher than the microgravity data for both data sets (air–water and helium–water). The air–water annular pressure drop experiences an increase between 16 and 44%, with an average increase of 23%. The helium–

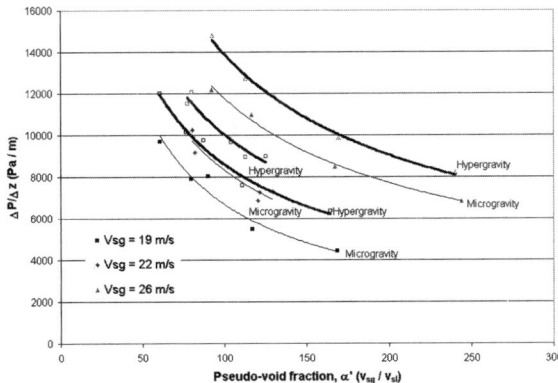

FIGURE 5. Air–water frictional pressure drop.

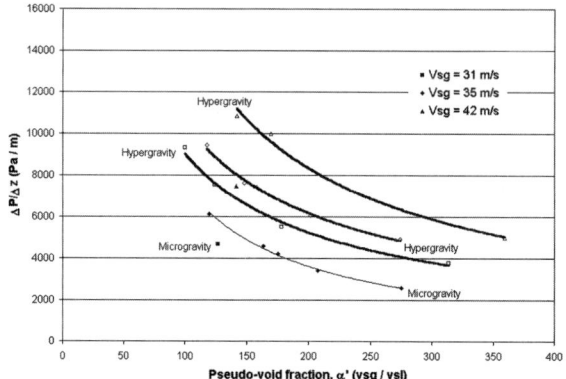

FIGURE 6. Helium–water frictional pressure drop.

water frictional pressure drop increases by 45–90% with the increase in the gravitational acceleration, exhibiting an average increase of 67%.

The average film thickness values are plotted against the pseudovoid fraction in FIGURES 8 and 9. FIGURE 8 shows the air–water micro- and hypergravity values. FIGURE 9 shows the helium–water values. As with the pressure drop figures, the lines represent the best-fit for constant gas velocities. The solid symbols represent the microgravity data and the open symbols correspond to the hypergravity data.

The trends in the average film thickness behavior with increasing gas and/or liquid flow rate are the same for air–water and the helium–water data. They are in agreement with previous studies on air–water annular flow. If the liquid flow rate is kept constant and the gas flow rate is increased, the average film thickness decreases. If the gas flow rate, on the other hand, is kept constant and the liquid flow rate is increased, the average film thickness increases. This trend is less pronounced with the helium–water data due to the higher gas velocities.

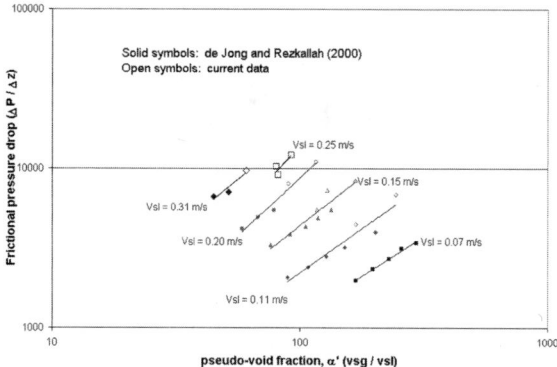

FIGURE 7. Comparison of current and past frictional pressure drop data.

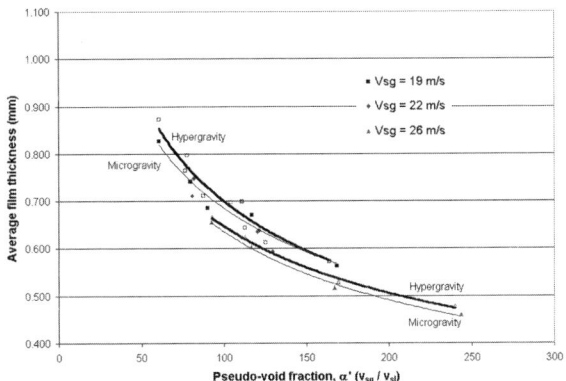

FIGURE 8. Air–water average film thickness values.

Based on the pseudo-void fraction, the hypergravity average film thickness values are only slightly larger than the microgravity data. The air–water film thickness values increase by 1–9%, with an average increase of 4% due to the increase in gravity. The helium–water average film thickness values experience a change of a small 1% decrease to a 15% increase, with an average increase of 7%. The helium–water film thickness values are also larger than the air–water values for the same range of pseudo-void fractions.

As with the average film thickness values, the standard deviation in the film thickness values does not change significantly. The standard deviation in the film thickness gives an indication of the size of the waves in the film; a larger standard deviation represents larger waves. On average, the air–water hypergravity standard deviation of the film thickness gives a 4% increase over the microgravity data. The standard deviation of the helium–water hypergravity data is 3% larger than the microgravity data.

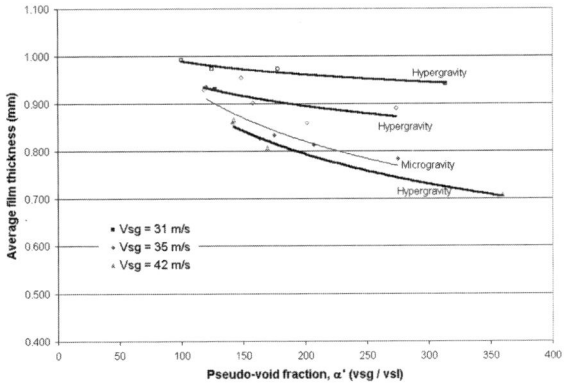

FIGURE 9. Helium–water average film thickness values.

For both the microgravity and hypergravity data, the ratio of the standard deviation to the average film thickness is approximately 33%. This indicates that the relative size of the waves is unaffected by the change in gravitational acceleration.

CONCLUSIONS

Although only preliminary analysis has been performed, some general trends in the behavior of the film in annular flow at both microgravity and hypergravity are reported. The trends previously reported for microgravity data are also present in the hypergravity data. Some differences between air–water and helium–water annular flow are observed; however, there are insufficient data to determine the effect of the gas density. For both flow conditions, the frictional pressure drop increases with increasing gravitational acceleration. The increase for the helium–water set of data is larger than for the air–water data. The average film thickness and the standard deviation of the film thickness signal increase slightly due to the increase in gravity. The relative size of the waves in annular flow appears to be unaffected by the change in acceleration.

ACKNOWLEDGMENTS

The authors gratefully acknowledge the financial support from the Canadian Space Agency, the flight opportunity offered by the European Space Agency, and the superb technical assistance by personnel from Novespace. We also acknowledge the assistance of Dr. Catherine Colin at the Institut de Mécanique des Fluides de Toulouse.

REFERENCES

1. DE JONG, P. & K.S. REZKALLAH. 2000. A dimensional analysis of microgravity annular flow: pressure drop and film characteristics. *In* Proc. of CSA Spacebound 2000. May 2000. Vancouver, BC.
2. BOUSMAN, W.S. 1995. Studies of two-phase gas-liquid flow in microgravity. NASA Contractor Report 195434.
3. LOWE, D. & K.S. REZKALLAH. 1999. Flow regime identification in microgravity two-phase flows using void-fraction signals. Int. J. Multiphase Flow **25**: 433–457.
4. AZZOPARDI, B.J. 1986. Disturbance wave frequencies, velocities and spacing in vertical annular two-phase flow. Nuclear Eng. Design **92**: 121–133.
5. ASALI, J.C., T.J. HANRATTY & P. ANDREUSSI. 1985. Interfacial drag and film height for vertical annular flow. AICE J. **31**(6): 895–902.
6. SHINKAWA, T. 1983. Mean thickness of the liquid film in the annular film-flow region of gas–liquid two-phase flow in a vertical tube. Int. Chem. Eng. **23**(2): 315–322.
7. HEWITT, G.F. & N.S. HALL TAYLOR. 1970. Annular Two-Phase Flow. Pergamon Press, Oxford.
8. HUGHMARK, G.A. 1959. Hold-up and Pressure Drop with Gas–Liquid Flow in a Vertical Pipe. Ph.D. Thesis, Louisiana State University, Baton Rouge.
9. RITE, R. & K.S. REZKALLAH. 1997. Local and mean heat-transfer coefficients in bubbly and slug flows under microgravity conditions. Int. J. Multiphase Flow **23**: 37–54.

10. ELKOW, K.J. & K.S. REZKALLAH. 1997. Void-fraction measurements in gas–liquid flows under 1-g and μ-g conditions using capacitance sensors: part I–measurements. Int. J. Multiphase Flow **23:** 815–829.
11. HUCKERBY, C.S. & K.S. REZKALLAH. 1992. Flow-pattern observations in two-phase, liquid–gas flow in a straight tube under microgravity conditions. *In* Proc. of Spacebound 1992 Conf. Ottawa, ON.
12. NOVESPACE. 2000. ESA Campaign November 2000—Practical and Technical Information. URL: <http://www.novespace.fr/>.

A Study of Gas–Liquid Two-Phase Flow in a Horizontal Tube Under Microgravity

BUHONG CHOI,[a] TERUSHIGE FUJII,[b] HITOSHI ASANO,[b] AND KATSUMI SUGIMOTO[a]

[a]*Department of Mechanical Engineering, Kobe University, Japan*

[b]*Faculty of Mechanical Engineering, Kobe University, Japan*

ABSTRACT: A better understanding of the effect of gravity level on two-phase flow characteristics is crucial for designing two-phase heat transport systems and thermal control systems for space. Hence, an experiment was conducted aboard MU-300 aircraft, as well as at normal gravity using the same horizontal tube, 10 mm internal diameter and flow conditions. The experimental results, obtained under Earth, hypergravity ($2g$), and microgravity environments were compared with previous models. Flow pattern data were compared with data from models for predicting microgravity flow pattern transitions. The mean void fraction under μg and $1g$ conditions was also compared with the Inoue–Aoki model for vertical upward annular flow at normal gravity. Frictional pressure drop fitted well with the Lockhart–Martinelli model, slightly influenced by the change in gravity levels (μg, $1g$, and $2g$); however, the effect of changing gravity on the pressure drop was insignificant for turbulent flow.

KEYWORDS: two-phase flow; microgravity; flow pattern

NOMENCLATURE:

C	Chisholm's parameter
C_o	distribution parameter
f_i	interfacial friction factor
g	Earth gravity, m/sec^2
j	superficial velocity, m/sec
ΔP_f	frictional pressure drop, Pa/m
S	slip ratio
t	temperature, °C
V_{G_j}	drift velocity, m/sec
x	mass flow rate quality
X	Lockhart–Martinelli parameter

Greek Symbols

α	void fraction
β	volumetric fraction
Φ_L	two-phase multiplier, ratio of two-phase to liquid single-phase pressure drop
ρ	density, kg/m^3
σ	surface tension, N/m

Subscripts

G	gas-phase
L	liquid-phase
T	total (gas + liquid)
TP	two-phase

Address for correspondence: Buhong Choi, Department of Mechanical Engineering, Kobe University, Rokkodai Nada, Kobe 657-8501, Japan.

INTRODUCTION

Conventional single-phase heat transport systems based on the sensible heat capacity of the working fluid have been well understood, and are relatively inexpensive and low risk. However, for proper thermal control with small temperature drops, they require large diameter lines, heavy in weight, and high-power pumps. A two-phase system is considered to be an attractive alternative to conventional single-phase systems for handling large heat loads and its ability to provide them at uniform temperatures regardless of the changes in heat load. It is, therefore, recommended that thermal management systems for future large spacecraft have to transport large amounts of dissipated power over large distances. Two-phase flows are strongly influenced by the magnitude and orientation of gravity because there is usually a large difference between the density of the two phases. Even on Earth, the behavior of two-phase flows in vertical tubes is very different from the behavior of the same flows in horizontal or inclined tubes.[1] When gravitational forces are reduced, due to the reduction of buoyancy between two phases, the behavior of two-phase flows is different in that the phase distribution is more symmetric, and the slip ratio is reduced.[2-4] Thus, in order to better understand microgravity two-phase flows, the effect of the change in gravity level on two-phase flow characteristics must be studied.

The purpose of this paper is to clarify the characteristics of microgravity two-phase flows. It is also hoped that these results will help to provide knowledge about the differences and similarities between three gravity levels (μg, $1g$, and $2g$).

EXPERIMENTAL APPARATUS

The flow loop system, shown schematically in FIGURE 1, is used for conducting microgravity and hypergravity ($2g$) experiments on a MU-300 aircraft, as well as under normal gravity conditions. Water and air are used as the working fluids. Water is pumped by the gear pump from the separator tank with stainless steel mesh for separating gas–liquid flows to the mixer through the liquid flow meter and back to the separator tank. Water flow rate is controlled by adjusting the rotational speed of the gear pump. Air is supplied from the compressed air cylinder, passes through the air flow meter, sonic orifices, and is mixed with liquid in the mixer. Air flow rate is controlled by passing the air through sonic orifices depending on the flow rate desired. The two-phase air–water mixture exits the mixer into the test section through the flow development section 500 mm in length. The test section with water-filled viewing box for flow pattern observation is a transparent acrylic horizontal adiabatic tube, 10 mm ID, 600 mm length. The flow patterns are recorded by the high speed video camera system. The differential pressure drop is measured by the differential pressure transducers with pressure tap attached to both ends of the test section.

TABLE 1. Experimental ranges

Fluid	Superficial velocity (m/sec)	Reynolds number
air	$j_G = 0.03–21$	$21–1.4 \times 10^4$
water	$j_L = 0.1–2.6$	$1 \times 10^3–2.8 \times 10^4$

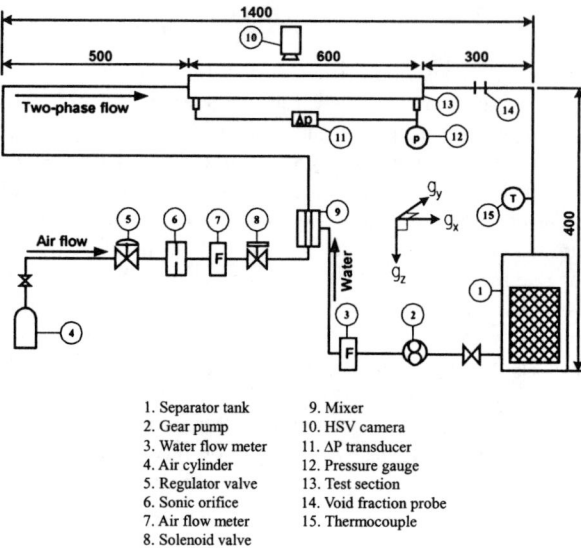

1. Separator tank
2. Gear pump
3. Water flow meter
4. Air cylinder
5. Regulator valve
6. Sonic orifice
7. Air flow meter
8. Solenoid valve
9. Mixer
10. HSV camera
11. ΔP transducer
12. Pressure gauge
13. Test section
14. Void fraction probe
15. Thermocouple

FIGURE 1. Schematic of the experimental apparatus.

The void fraction measurement is made by measuring the electrical resistance between the ring type copper probes.

The complete test loop system including all the elements, is installed in an experimental rack, and then the rack is placed on board the MU-300 aircraft. The parabolic flight trajectories for producing μg and $2g$ environments are performed 57 times during six days in the experimental ranges ($t = 20°C$) of TABLE 1.

Data taken for analysis were those when the flow was stable and fully developed, and where the maximum gravity levels in the x (nose to tail in flight, flow direction), y (wing-tip to wing-tip, transverse to flow direction), and z (floor to ceiling, perpendicular to flow direction) directions were within $g_x = \pm 0.01g$, $g_y = \pm 0.006g$, $g_z = 0.01 \pm 0.01g$ for microgravity conditions; $g_x = 0.05 \pm 0.05g$, $g_y = \pm 0.02g$, $g_z = 2 \pm 0.1g$ for $2g$ conditions. Terrestrial tests were also carried out to compare with the results obtained under microgravity and $2g$ conditions using the same horizontal tube and flow conditions.

EXPERIMENTAL RESULTS

Flow Patterns

Flow patterns under all three gravity levels are shown in FIGURE 2. Microgravity flow patterns for bubble, Taylor bubble, slug, semi-annular, and annular flow are shown in FIGURE 2a. Taylor bubble flow is differentiated from slug flow when the nose of an elongated bubble persists in a bullet shape and is separated from liquid slugs, in which discrete bubbles are not included. Stratified, bubble, plug, semi-annular, and annular flow were observed under $1g$ and $2g$, as shown in FIGURE 2b.

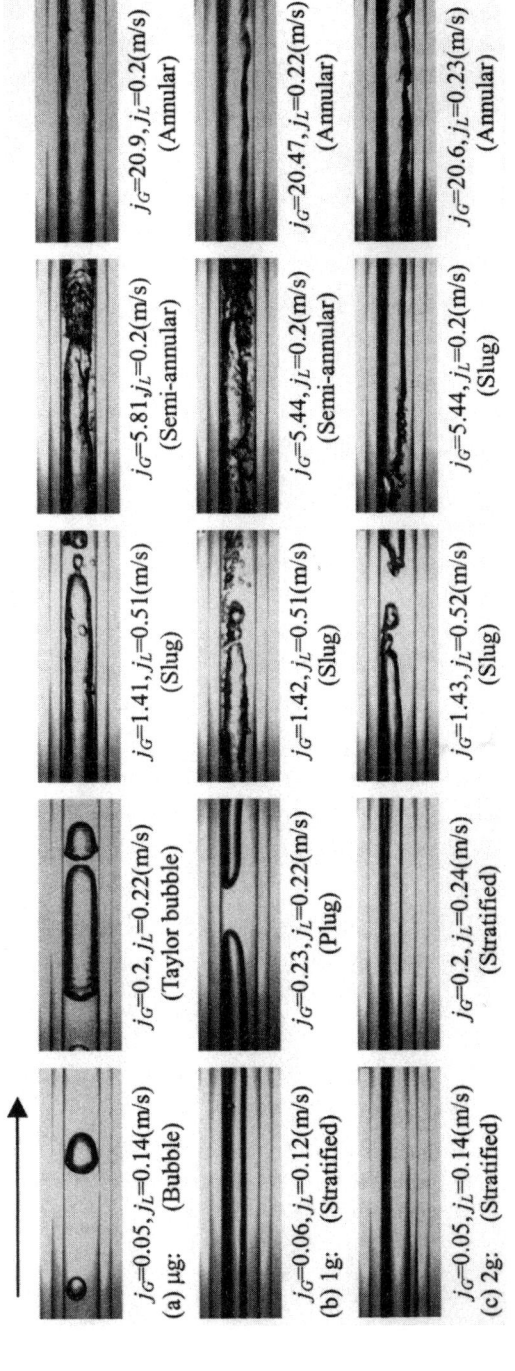

FIGURE 2. Flow patterns under μg, $1g$, and $2g$ conditions.

FIGURE 2 clearly shows the gravity level dependence of flow patterns, in particular, the gas and liquid superficial velocities at approximately $j_G = 0.2$, $j_L = 0.22$ (m/sec).

In FIGURE 3, the flow pattern results obtained under μg conditions were plotted on the flow pattern map based on the gas and liquid superficial velocities, but show only experimental boundaries under $1g$ and $2g$ conditions on the same map. All the transition boundary lines from bubbly to Taylor bubble flow (TB or plug flow) have constant void fraction values with a positive slope 1. The following results were obtained: $j_L = 1.5 j_G$ for μg, $j_L = 5 j_G$ for $1g$, and $j_L = 7.5 j_G$ m/sec for $2g$. This was found that the bubbly to Taylor bubble transition for μg occurred at the highest gas superficial velocities. This was because the slip ratio between the two phases is decreased by the relative reduction of interactions between bubbles resulting from reduction of buoyancy force on two-phase flows under μg (see FIGURE 4), as opposed to $2g$. For the transition from Taylor bubble (TB or plug flow) to slug flow, the transition boundaries of all the three g levels occurred at approximately $j_G = 1.0$ m/sec. For the transition from slug to semi-annular flow, the transition boundary for μg is the line $j_L = 0.12 j_G$, but occurred at $j_G = 3.7$ m/sec for $1g$, $j_G = 7.47$ m/sec for $2g$. It was found that the transition line for $2g$ was located at higher j_G than that for $1g$ because it became difficult to form the liquid bridge as the magnitude of gravity increased. In the transition from semi-annular to annular flow, for all the three g levels, the transition boundaries were located on the same line, $j_L = 0.12 j_G$. The region was dominated by inertia forces. The experimental results obtained at all three g

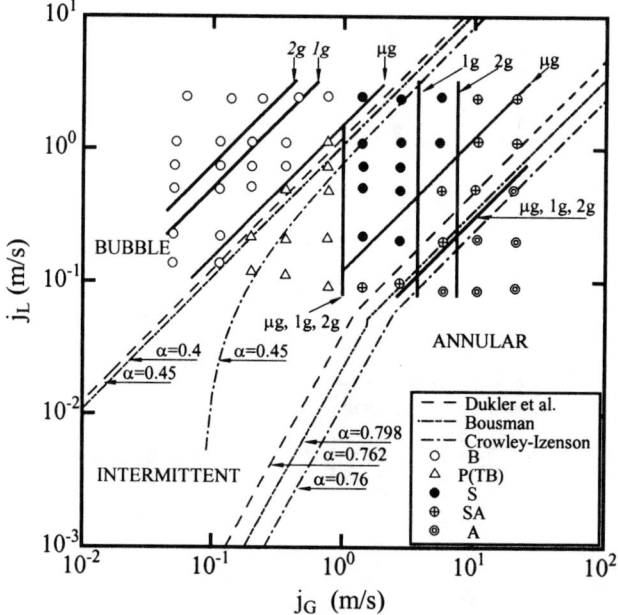

FIGURE 3. Flow pattern map. Comparison of data under μg, $1g$, and $2g$ conditions with the flow pattern transition models.

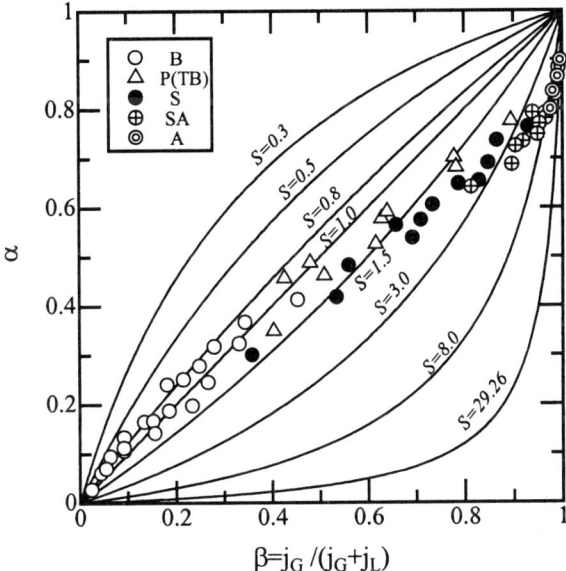

FIGURE 4. Void fraction versus volumetric fraction under μg conditions.

levels were also compared with the models for predicting the microgravity flow pattern transitions proposed by Dukler et al.,[1] Bousman,[5] and Crowely and Izenson,[6] respectively. The detailed equations used in this work are:

(1) Dukler et al.: For the transition from bubbly to slug,

$$j_L = \frac{1-\alpha}{\alpha} j_G. \tag{1}$$

The resulting equation with the recommended void fraction $\alpha = 0.45$ is then

$$j_L = 1.22 j_G, \tag{2}$$

which was also used in this work to map the transition on flow pattern map in (j_G, j_L) coordinates.

For the transition from slug to annular flow, the resulting equation was obtained by using Equations **(5)**, **(6)**, and **(7)** in Reference 5 with the value of $C_o = 1.25$ recommended by Dukler et al.

(2) Bousman: For the transition from bubbly to slug,

$$j_L = \frac{1 - C_o \alpha}{C_o \alpha} j_G. \tag{3}$$

The resulting equation with the recommended void fraction $\alpha = 0.45$ and distribution coefficient $C_o = 1.21$ is then

$$j_L = 1.07 j_G. \tag{4}$$

For the transition from slug to annular flow,

$$\frac{\alpha^{5/2}}{(1-\alpha)^2} = \phi(\alpha)\frac{f_g \rho_g}{f_w \rho_k}\left(\frac{j_G}{j_L}\right)^2 \tag{5}$$

$$\alpha = \frac{1}{C_o} - \alpha\frac{(1-\alpha)^2\phi(\alpha)}{\alpha^{2.4}(v_L/v_G)^{0.2}(\rho_L/\rho_G)}, \tag{6}$$

where the parameter $\phi(\alpha)$ is the Wallis mode:[7]

$$\phi(\alpha) = 1 + 150(1-\alpha^{1/2}) \tag{7}$$

Solving Equation (6) numerically with the recommended value $C_o = 1.21$, yielded the solution $\alpha = 0.798$, and this value of void fraction was used in Equation (5). To map the transition on flow pattern map in (j_G, j_L) coordinates, Equation (5) was converted into dimensional form and solved for the superficial liquid velocity, j_L.

(3) *Crowely and Izenson:* For the transition from bubbly to slug,

$$\frac{j_G}{\alpha} = C_o(j_G + j_L) + 1.41\left(\frac{a\sigma(\rho_L - \rho_G)}{\rho_L^2}\right)^{1/4} \tag{8}$$

was used in this work with the recommended void fraction $\alpha = 0.45$ and $C_o = 1.2$ for a given acceleration magnitude, $a = 0.01\,g$.

For the transition from slug to annular flow,

$$X^2 = \frac{(f_i/f_G)(1-\alpha)^2}{3\alpha^{3.5}}[1 - 1.5(1-\alpha)], \tag{9}$$

where the parameter f_i/f_G is the Wallis model:[7]

$$f_i/f_G = [1 + 75(1-\alpha)]. \tag{10}$$

For this work, Equation (9) was converted into dimensional form to solve for the superficial liquid velocity, j_L.

As can be seen from FIGURE 3, Dukler *et al.*, Bousman, and Crowely and Izenson models all gave equally good predictions. The first gave much better results for the transition from bubbly to Taylor bubble flow, and the second for the transition from semi-annular to annular flow.

Void Fraction

The mean void fraction under μg condition is shown in FIGURE 4 using the volumetric fraction $\beta\,[= j_G/(j_G + j_L)]$. At a constant j_L, the void fraction increases gradually, and the slip approximates to $S = 29.26\,(= [\rho_L/\rho_G]^{0.5}$, Smith)[8] as j_G increases. This also shows that when the range of β is less than 0.15, S is less than 1.0, and when the range of β is at least 0.15 or less than 0.5, S becomes approximately 1.0. On the other hand, for $1\,g$ conditions, when the range of β is more than 0.15, S is greater than 1.0. From the results, we conclude that the slip ratio under μg is lower than that under $1\,g$ for the range $\beta < 0.5$.

The void fraction for the bubbly, Taylor bubble (TB or plug), and slug regions obtained in these experiments were arranged in terms of the drift-flux model, expressed by

$$j_G/\alpha = (C_o(j_G + j_L) + V_{Gj}), \tag{11}$$

where C_o is the distribution coefficient and V_{G_j} is the drift velocity. For bubbly region, the distribution coefficient, C_o, was determined by using the void fraction obtained at all three g levels, as shown in FIGURE 5. The values of C_o for these experiments were determined as follows: $C_o = 1.03$ for μg, $C_o = 1.22$ for $1 g$, and $C_o = 1.35$ for $2 g$. The results explain that the void fraction is greater under μg at the same mixture velocity j_T $(j_G + j_L)$. For the Taylor bubble (TB or plug) and slug region, the values of C_o were determined to be approximately the same, $C_o = 1.29$ for μg, $C_o = 1.30$ for $1 g$ and $C_o = 1.33$ for $2 g$.

A comparison of the void fraction under μg and $1 g$ with the model proposed by Inoue–Aoki[8] is shown in FIGURE 6 as a function of quality x. The root mean square deviation between the void fraction for $1 g$ conditions and the model is 10%. It was also found that the void fraction under μg is greater than that under $1 g$ by about 30%, when the range is $10^{-4} < x < 10^{-3}$.

Friction Pressure Drop

The total pressure drop in two-phase flow can be expressed as the sum of frictional pressure drop, gravitational pressure drop, and accelerational pressure drop. For fully developed, adiabatic, horizontal orientation and steady-state flow in a tube with a uniform cross section, the gravitational and accelerational pressure drop component can be ignored. Therefore, the total pressure drop measured in this experiment can be estimated from the frictional pressure drop alone.

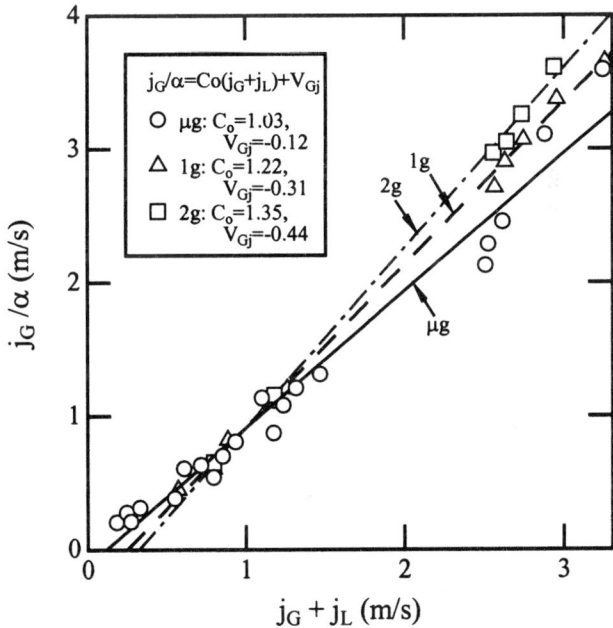

FIGURE 5. Correlation on drift-flux mode with void fraction in the bubble flow under μg, $1 g$, and $2 g$ conditions.

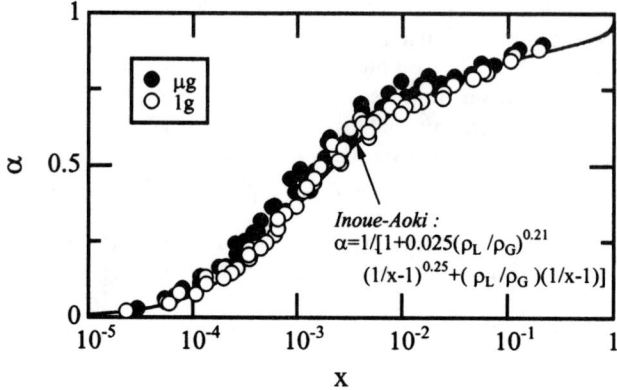

FIGURE 6. Comparison between void fraction under μg and that under $1 g$.

The experimental frictional pressure drop under μg is plotted in FIGURE 7 using the mixture velocity j_T as the independent variable. Corresponding flow patterns are also indicated on each data point. As can be seen from the figure, at a constant j_L, the friction pressure drop increases gradually, as the mixture velocity increases except for $j_L = 0.1, 0.2$ (m/sec). At constant j_G, increasing j_L results in a significant increase in the frictional pressure drops in all cases, for all flow patterns. In the comparison between the frictional pressure drops under μg and $1 g$ conditions, it appears that the

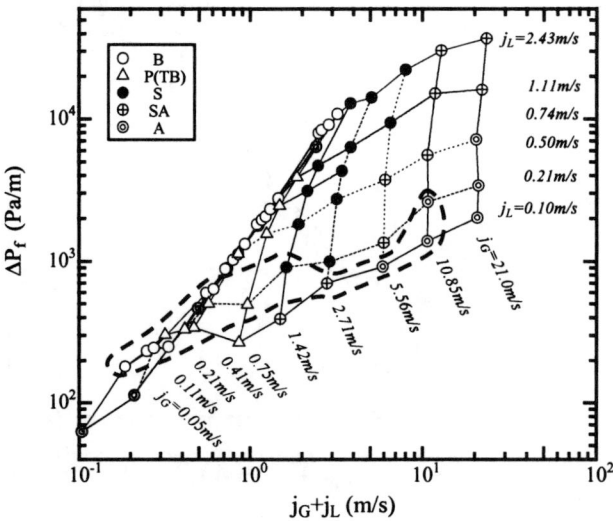

FIGURE 7. Frictional pressure drop under μg conditions.

differences are significant in the region identified by the thick broken line. In this region, since the slip ratio for μg is lower than under $1g$ and $2g$, as mentioned above with respect to the void fraction. The interfacial friction factor should be also reduced and, hence, the pressure drop should be also reduced at microgravity. However, the result explains the reason that the differences in the flow patterns caused by the change in g level more strongly influence the frictional pressure drop rather than the effect of slip ratio.

A comparison between the frictional pressure drops under $1g$ and μg is shown in FIGURE 8 by using the Lockhart–Martinelli parameter X defined in Equation **(13)**. The maximum ratio of the frictional pressure drop under $1g$ to that under μg is significant (approximately $1:1.3$) for the region within the thick broken line, the laminar flow region $j_L \leq 0.2$ (m/sec) (see FIG. 7), whereas it is approximately $1:1$ for the turbulent flow region $j_L \geq 0.5$ (m/sec).

In the comparison between the frictional pressure drops under $1g$ and $2g$, it was also found that there was a great difference in the ratio of the frictional pressure drop between $1g$ and $2g$, than that between $1g$ and μg. For stratified flow under $2g$ versus stratified and plug flows under $1g$, the level of liquid phase is decreased by enhanced gravitational forces acting on the liquid phase as the g level is increased, and then the reduction of liquid level results in great increase of liquid phase velocity. Consequently, the frictional pressure drop under $2g$ becomes greater than that under $1g$ because of the accelerated liquid phase and the increase in the interface roughness under $2g$. Comparing the same plug flow under $1g$ and $2g$ conditions, each point is only significantly different due to the differences in the roughness of two-phase interface. The maximum ratio of the frictional pressure drop under $1g$ to that under $2g$ is approximately $1:1.45$. However, regardless of the change in gravity, the differences among the frictional pressure drops in all three g levels, except for some region at low liquid superficial velocities, are within $\pm 11\%$.

The experimental multiplier Φ_L for μg conditions is plotted in FIGURE 9 using the Lockhart–Martinelli parameter X. The liquid two-phase multiplier Φ_L and X are defined as follows:

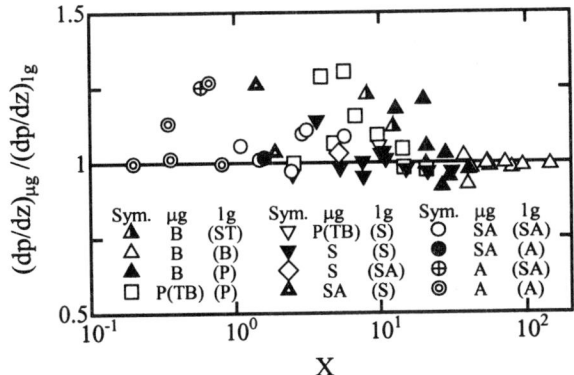

FIGURE 8. Ratio of frictional pressure drop under μg to that under $1g$.

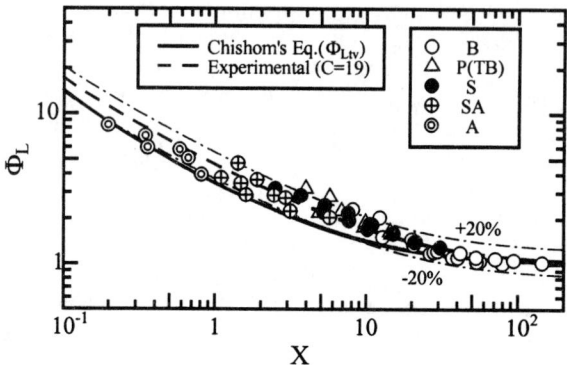

FIGURE 9. Comparison of frictional pressure drop under μg condition with Chisholm's correlation.

$$\Phi_L = \left[\frac{\left(\frac{dp}{dz}\right)_{TP}}{\left(\frac{dp}{dz}\right)_L}\right]^{0.5}, \quad (12)$$

$$X = \left[\frac{\left(\frac{dp}{dz}\right)_L}{\left(\frac{dp}{dz}\right)_G}\right]^{0.5}, \quad (13)$$

where $(dp/dz)_L$ and $(dp/dz)_G$ are, respectively, the single-phase liquid and gas frictional pressure gradients calculated using the liquid and gas flow rates alone. Approximating these relationships, Chisholm's equation is expressed by

$$\Phi_L = \left(1 + \frac{C}{X} + \frac{1}{X^2}\right)^{0.5}, \quad (14)$$

where C is a parameter first introduced by Chisholm. The thick solid line in FIGURE 9 is calculated by using the value of $C = 10$ in Equation **(14)**. The values of C for each g level were obtained by using the least squares approximation method: $C = 19$ for μg, $C = 17$ for $1g$, and $C = 19$ for $2g$.

CONCLUSIONS

Gas–liquid two-phase experiments were conducted aboard a MU-300 and under normal gravity. The following is a summary of the main findings in this study:

1. The effect of the gravity change on the flow patterns and the frictional pressure drop was relatively great at low gas and liquid superficial velocities.
2. The void fraction in microgravity bubble and slug flows was effectively estimated from the drift-flux model.

3. The effect of gravity change on the frictional pressure drops was insignificant for turbulent flows in all three g levels.
4. In the Chisholm correlation, the value of C was estimated as 19 for the microgravity frictional pressure drop.

ACKNOWLEDGMENTS

The facility, support, and guidance of Diamond Air Service are gratefully acknowledged.

REFERENCES

1. DULKER, A.E., J.A. FABRE, J.B. MCQUILLEN & R. VERSION. 1988. Gas liquid flow at microgravity conditions: flow patterns and their transitions. Int. J. Multiphase Flow **14**: 389–400.
2. ZHAO, L. & K.S. REZKALLAH. 1995. Pressure drop in gas–liquid flow at microgravity conditions. Int. J. Multiphase Flow **21**: 837–849.
3. FUJII, T., H. ASANO, et al. 2000. A study of flow characteristics in gas–liquid two-phase flows under microgravity (effect of gravity and liquid surface tension on liquid-film structure). Trans. Jpn. Soc. Mech. Eng. **66**(652-B): 3085–3091.
4. CHOI, B.H., T. FUJII, H. ASANO & K. SUGIMOTO. 2001. A study of flow characteristics in gas–liquid two-phase flows under microgravity. 38th National Heat Transfer Symposium of Japan. Heat Transfer Society of Japan. 775–776.
5. BOUSMAN, W.S. 1995. studies of two-phase gas–liquid flow in microgravity. NASA Contractor Report 195434, 85–228.
6. CROWELY, C.J. & M.G. IZENSON. 1989. Design manual for microgravity two-phase flow and heat transfer. Final report for Small Business Innovative Research Contract AL-TR-89-027, 1–49.
7. WALLIS, G.B. 1969. One dimensional two-phase flow. McGraw-Hill, New York. 318–322.
8. AKAGAWA, K. 1974. Gas–liquid two-phase flow. Corona Publishing Co. Ltd., New York. 36–72.

Vapor Bubbles in Flow and Acoustic Fields

ANDREA PROSPERETTI[a,b] AND YUE HAO[a]

[a]*Department of Mechanical Engineering, The Johns Hopkins University, Baltimore, Maryland, USA*

[b]*Faculty of Applied Physics, Twente Institute of Mechanics, Burgerscentrum, University of Twente, Enschede, the Netherlands*

ABSTRACT: A review of several aspects of the interaction of bubbles with acoustic and flow fields is presented. The focus of the paper is on bubbles in *hot* liquids, in which the bubble contains mostly vapor, with little or no permanent gas. The topics covered include the effect of translation on condensation in a subcooled liquid, the action of pressure gradients on bubbles convected by a liquid flow, rectified heat diffusion on stationary and moving bubbles, the effect of permanent gas on this phenomenon, and the action of pressure-radiation (or Bjerknes) forces.

KEYWORDS: boiling; acoustic cavitation; phase change; rectified heat transfer; vapor bubbles

INTRODUCTION

Ordinarily, a vapor bubble is a labile object: buoyancy forces easily dislodge it from the heated surface where it forms, and natural convection can bring it into contact with slightly colder liquid masses that cause its abrupt condensation. Both effects are drastically reduced in microgravity, and the persistence of bubbles under such conditions can cause major problems: premature transition to film boiling, accumulation of persistent vapor pockets in critical components of the flow loop, two-phase flows in liquid intakes, and others. Although these considerations might suggest minimizing the occurrence of bubbles in microgravity, others argue in the opposite direction. For example, in view of its high efficiency, which permits reducing the size of heat transfer components for a given power rating, boiling heat transfer is highly desirable in space applications, such as thermal energy and power generation, propulsion, and cryogenic storage.

It is, therefore, of considerable interest to investigate whether other forces can be introduced to replace buoyancy and to try to understand their effects. Several possibilities readily come to mind: Marangoni effects, electric or magnetic fields, acoustics, and flow. In this paper, we focus on the last two.

Address for correspondence: Andrea Prosperetti, Department of Mechanical Engineering, 3400 North Charles Street, The Johns Hopkins University, Baltimore, MD 21218, USA. Voice: 410-516 8534; fax 410-516 7254.
   prosperetti@jhu.edu

FORCES

Since the objective of the study is to move bubbles in a liquid, it is necessary to understand the nature of flow and acoustic forces on such entities. We approach this problem by appealing to intuition; a more formal, if partial, justification can be found in Reference 1 and a complete one can be built along similar lines.

Consider a bubble in a liquid flow. In many practical cases, the bubble radius R is much smaller than the characteristic length scale L of the surrounding flow. Similarly, even at several tens of kHz, the wavelength of sound in water is of the order of several centimeters and, therefore, much greater than typical bubble sizes. These considerations motivate a picture in which the bubble is immersed in a locally incompressible and uniform flow in which the *ambient* pressure field is given by the sum of the macroscopic flow pressure P_L and acoustic pressure P_{ac}. With this picture in mind, the Bernoulli equation becomes

$$\frac{\partial \phi}{\partial t} + \frac{1}{2} \mathbf{u} \cdot \mathbf{u} + \frac{P}{\rho_L} = \frac{P_L + P_{ac}}{\rho_L}, \tag{1}$$

where $\mathbf{u} = \nabla \phi$ is the velocity field, ϕ the velocity potential, ρ_L the liquid density (assumed constant), and P the local liquid pressure. Since the mass of the bubble contents is negligible, the total force on it must vanish at every instant and, therefore,

$$\int_S dS \mathbf{n} P + \mathbf{F}_D = 0, \tag{2}$$

where $\mathbf{n}$ is the outward unit normal on the bubble surface S and $\mathbf{F}_D$ the drag force. Exploiting the slow spatial variation of P_L and P_{ac}, this relation may also be written as follows:

$$\int_S dS \mathbf{n}(P - P_L - P_{ac}) + \mathcal{V} \nabla P_{ac} + \mathcal{V} \nabla P_L + \mathbf{F}_D = 0, \tag{3}$$

where $\mathcal{V}$ is the bubble volume. The advantage of this alternative form is that, as is evident from (1), up to an unessential constant $P - (P_L + P_{ac})$ is the *incompressible* pressure field in the neighborhood of a body—the bubble—subjected to a constant ambient pressure. As is well known from fluid mechanics (see e.g., Refs. 2–4) the integral of the negative of this quantity over the bubble surface equals the time derivative of the impulse $\mathbf{I}$ of the bubble, defined by[1]

$$\mathbf{I} = -\int_S dS \phi \mathbf{n}, \tag{4}$$

so that (3) becomes

$$\frac{d\mathbf{I}}{dt} = -\mathcal{V} \nabla P_L - \mathcal{V} \nabla P_{ac} - \mathbf{F}_D. \tag{5}$$

Since, as noted previously, the flow in the neighborhood of the bubble can be approximated as uniform, the situation resembles that of a translating sphere of radius R for which the impulse has the well-known expression

$$\mathbf{I} = \frac{2}{3} \pi C_A R^3 (\mathbf{U}_B - \mathbf{U}_L), \tag{6}$$

where C_A is a normalized added mass coefficient (equal to 1 in the ideal case) and $\mathbf{U}_L$ is the velocity of the imposed macroscopic flow.[2]

By following a parallel line of reasoning, we may find the equation of motion for the bubble radius: on the assumption that deviation from sphericity are small, we may impose the dynamic boundary condition across the bubble wall in the form

$$p = \frac{1}{S}\int_S dS P + \frac{2\sigma}{R} + 4\mu\frac{\dot{R}}{R}, \qquad (7)$$

where p is the bubble internal pressure, σ the surface tension coefficient, μ the liquid viscosity, and dots denote time differentiation following the bubble. We now use (1) to express P and obtain

$$-\frac{1}{S}\int_S dS \left(\frac{\partial \phi}{\partial t} + \frac{1}{2}\mathbf{u}\cdot\mathbf{u}\right) = \frac{1}{\rho_L}\left(p - P_L - P_{ac} - \frac{2\sigma}{R} - 4\mu\frac{\dot{R}}{R}\right), \qquad (8)$$

where P_L and P_{ac} are to be evaluated at the position of the bubble center. Again in view of the scale separation, we may evaluate the term on the left-hand side by using the potential for the flow past a sphere of variable radius translating with a relative velocity $\mathbf{U}_B - \mathbf{U}_L$ to find a form of the Rayleigh–Plesset equation:

$$R\ddot{R} + \frac{3}{2}\dot{R}^2 = \frac{1}{\rho_L}\left(p - P_L - P_{ac} - \frac{2\sigma}{R} - 4\mu\frac{\dot{R}}{R}\right). \qquad (9)$$

It is easy to correct this equation to include weak compressibility effects, but in the applications that follow the bubble motion is relatively mild and these effects are negligible.

Throughout this paper we assume spherical bubbles. For a stationary bubble, the validity of this assumption depends on the stability of the spherical shape. Although a rather complete stability theory for gas bubbles is available,[5] the matter has never been investigated for vapor bubbles. Broadly speaking, stability will prevail provided the velocity and acceleration of the interface are not too large. A translating bubble will tend to be deformed into an oblate ellipsoid with the minor axis in the direction of translation, an effect countered by the action of surface tension. The competition between the two mechanisms is governed by the Weber number, defined by

$$We = \frac{2R\rho_L(\mathbf{U}_B - \mathbf{U}_L)^2}{\sigma}. \qquad (10)$$

THE BUBBLE INTERIOR

In the case of a mixture of gas (index G) and vapor (index V), the mathematical model simplifies considerably if one assumes that the adiabatic indices γ of the two components are equal, and that the ratio of the specific heats c_p is the inverse of that of the molecular masses M, $c_{pG}/c_{pV} \approx M_V/M_G$. These relations are not precisely satisfied in the CO_2–water vapor system that we use as an example since $\mu_G \approx 1.3$, $\mu_V \approx 1.3$ (at 100°C), $M_V/M_G \approx 0.41$, $c_{pG}/c_{pV} \approx 0.45$; nevertheless we are interested here more in general features of the behavior of a gas/vapor mixture than in detailed quantitative predictions and, therefore, in view of the considerable simplification attending this approximation, we feel justified in adopting it.

We further assume that the bubble internal pressure is spatially uniform and only a function of time.[6-8] With the neglect of viscous dissipation, the enthalpy equation for a gas (or a mixture of gases) is

$$\frac{\partial}{\partial t}(\rho h - p) + \nabla \cdot (\rho h \mathbf{v}) = -\nabla \cdot \mathbf{q}, \tag{11}$$

in which ρ the gas density, h the enthalpy, $\mathbf{v}$ the velocity, and $\mathbf{q}$ the heat flux given by

$$\mathbf{q} = -k\nabla T - (c_{pV} - c_{pG}) T \rho \mathcal{D} \nabla c, \tag{12}$$

here k is the thermal conductivity, T the absolute temperature, ρ the total density of the gas-vapor mixture, $\mathcal{D}$ the binary diffusion coefficient, and c the vapor mass fraction. For a mixture of perfect gases

$$\rho h - p = \frac{1}{\gamma_G - 1}\left(p + \frac{\gamma_G - \gamma_V}{\gamma_V - 1} p_V\right), \tag{13}$$

where p_V is the vapor partial pressure. The simplification arising from the assumption $\gamma_G \approx \gamma_V$ consists in the fact that, since p has been taken spatially uniform, $\rho h - p$ also becomes spatially uniform so that (11) is

$$\nabla \cdot (\rho h \mathbf{v} + \mathbf{q}) = -\frac{1}{\gamma_G - 1} \dot{p}. \tag{14}$$

With the further assumption of spherical symmetry, this equation can be integrated to obtain

$$v = -\frac{1}{\gamma_G p}\left[(\gamma_G - 1) q + \frac{1}{3} r \dot{p}\right], \tag{15}$$

in which r is the radial coordinate and v and q the radial components of $\mathbf{v}$ and $\mathbf{q}$. On evaluating this expression at the bubble wall $r = R$, we obtain an equation for the pressure:

$$\dot{p} = -\frac{3}{R}[\gamma_G p v + (\gamma_G - 1) q]_{r=R}. \tag{16}$$

The mathematical model must be closed by adding the energy equation which, with the previously mentioned assumption on the ratio of the specific heats, becomes

$$\frac{\partial T}{\partial t} + v\frac{\partial T}{\partial r} = \frac{\gamma_G - 1}{\gamma_G}\frac{T}{p}\left[\dot{p} + \frac{1}{r^2}\frac{\partial}{\partial r}\left(r^2 k \frac{\partial T}{\partial t}\right)\right] + \frac{c_{pV} - c_{pG}}{c_{pG}}\frac{\mathcal{R} T \rho}{M_{Gp}} \mathcal{D} \frac{\partial T}{\partial r}\frac{\partial c}{\partial r}, \tag{17}$$

in which $\mathcal{R}$ is the universal gas constant, and the diffusion equation

$$\frac{\partial c}{\partial t} + v\frac{\partial c}{\partial r} = \frac{1}{r^2 \rho}\frac{\partial}{\partial r}\left(r^2 \rho \mathcal{D}\frac{\partial c}{\partial r}\right); \tag{18}$$

pressure and temperature diffusion effects are important only in rather extreme cases[9] and have been disregarded.

At the bubble wall, we impose the conservation of energy in the form

$$k_L \frac{\partial T_L}{\partial r}\bigg|_{r=R} + q(R, t) = L\dot{m}, \tag{19}$$

where T_L is the liquid temperature, L denotes the latent heat, and $\dot{m}$ is the vapor mass flux (positive for evaporation)

$$\dot{m} = \rho c[\dot{R} - v(R, t)] + \mathcal{D}\rho \frac{\partial c}{\partial r}\bigg|_{r=R}, \qquad (20)$$

and the conservation of the gas mass

$$\rho(1-c)[\dot{R} - v(R, t)] - \rho\mathcal{D}\frac{\partial c}{\partial r}\bigg|_{r=R} = \mathcal{D}_L \frac{\partial \rho_g}{\partial r}\bigg|_{r=R}, \qquad (21)$$

in which ρ_g is the density of gas dissolved in the liquid. In addition, we impose continuity of temperature, saturation, and Henry's law:

$$\rho_g(R, t) = K(T)p_G, \qquad (22)$$

in which p_G is the gas partial pressure on the gas side of the interface.

THE LIQUID

The Rayleigh–Plesset Equation (9) already incorporates the statements of conservation of mass and momentum for the liquid. What remains to describe is the transport of heat

$$\frac{\partial T_L}{\partial t} + \mathbf{u} \cdot \nabla T_L = D_L \nabla^2 T_L, \qquad (23)$$

where D_L is the liquid thermal diffusivity, and of dissolved gas

$$\frac{\partial \rho_g}{\partial t} + \mathbf{u} \cdot \nabla \rho_g = \mathcal{D}_L \nabla^2 \rho_g, \qquad (24)$$

Since the liquid density is essentially constant irrespective of the amount of dissolved gas, it proves more convenient to use ρ_g directly rather than the mass fraction c as inside the bubble.

As already mentioned, we describe the liquid velocity field in the rest frame of the bubble in terms of a velocity potential. For a single spherical bubble far from other boundaries it is sufficient to assume

$$\mathbf{u} = \nabla\left[-\frac{R^2}{r}\dot{R} + (\mathbf{U}_B - \mathbf{U}_L) \cdot \mathbf{r}\left(1 + \frac{R^3}{2r^3}\right)\right], \qquad (25)$$

where $\mathbf{r}$ is measured from the center of the bubble. With more than one spherical bubble, or in the presence of boundaries the effect of which can be treated by means of image bubbles, the expression of the velocity is much more complex and will not be given here; the reader is referred to Reference 1.

Several forms for the drag coefficient of bubbles exist. Here we use the expression of Levich as augmented by Moore[10]

$$C_D = \frac{48}{Re}\left(1 - \frac{2.211}{\sqrt{Re}}\right), \quad Re = \frac{2\rho_L R|\mathbf{U}_B - \mathbf{U}_L|}{\mu_L}, \qquad (26)$$

in which Re is the Reynolds number.

VAPOR BUBBLES IN FLOW FIELDS

In this section we show several examples of the behavior of pure vapor bubbles in flow fields in the absence of sound; additional information and examples are available elsewhere.[11]

The previous model can be specialized to this case simply by setting $c = 1$ in the vapor equations and dropping the diffusion equation in both phases. A further simplification is applicable when the bubble radius is smaller than the thermal diffusion length in the vapor. In this case, the energy equation in the vapor can be solved[12] approximately to give

$$4\pi R^2 k_L \frac{\partial T_L}{\partial r}\bigg|_{r=R} = L\frac{d}{dt}\left(\frac{4}{3}\pi R^3 \rho_V\right) + \frac{4}{3}\pi R^3 \rho_V c_s \frac{\partial T_S}{\partial t}, \quad (27)$$

where T_S is the bubble surface temperature and $c_s = c_{pV} - L/T_S$ is the specific heat along the saturation line; for water at 100° $c_s/c_p \approx -1.45$.

A vapor bubble placed in a subcooled liquid collapses: an example is shown by the dotted line in FIGURE 1, in which the bubble–liquid relative velocity vanishes and time is normalized by t_c, the characteristic time scale for bubble collapse[13]

$$t_c = \frac{\pi}{4Ja^2}\frac{R^2(0)}{D_L}, \quad (28)$$

in which Ja is the Jacob number defined by

$$Ja = \frac{\rho_L c_{pL} \Delta T}{\rho_{V,sat} L}. \quad (29)$$

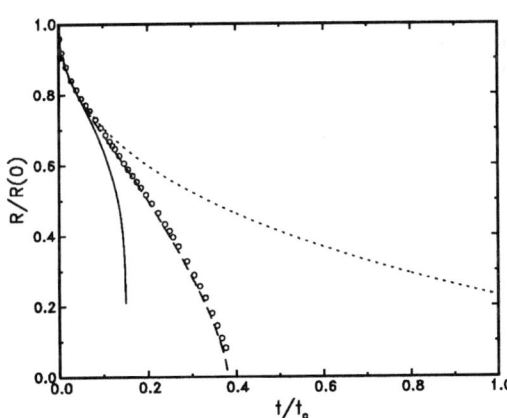

FIGURE 1. Bubble radius versus time in the collapse of a 1 mm-radius bubble in water at a constant pressure of 101.3 kPa for a subcooling of 3.35 K ($Ja = 10$). The dotted line is for a stationary bubble, the *dashed line* for a constant translational velocity of 0.127 m/sec ($Pe = 1495$, $We = 0.293$), and the *solid line* for an initial translational velocity of 0.127 m/sec adjusted at later times; the circles are the numerical results of Wittek & Chao[14] for the constant velocity case. The time scale t_c, defined in (28), equals 46.2 msec.

Here and in the following, the subscript *sat* denotes saturation conditions at the undisturbed liquid pressure P_∞, T_∞ is the liquid temperature far from the bubble, and $\Delta T = T_{sat} - T_\infty$ the liquid subcooling.

In order to understand the significance of the time scale tc introduced in **(28)** we may imagine that subcooled conditions are created instantaneously; that is, by starting with the bubble in a saturated liquid and quickly increasing the pressure. If the first few instants are ignored, one may assume that the surface temperature of the bubble stabilizes at the saturation level T_{sat}, after which the collapse is thermally controlled. In these hypotheses, Equation **(27)** gives

$$L\rho_V \dot{R} = k_L \frac{\partial T_L}{\partial r}\bigg|_{r=R}. \tag{30}$$

Initially, heat diffusion in the liquid occurs in a thermal boundary layer of thickness about $\sqrt{\pi D_L t}$ so that

$$\frac{\partial T_L}{\partial r} \approx \frac{-\Delta T}{\sqrt{\pi D_L t}};$$

the equation can then be integrated to obtain

$$R(t) \approx R(0) - 2Ja\sqrt{\frac{D_L t}{\pi}}, \tag{31}$$

from which, on setting $R(t_c) = 0$, one recovers **(28)**.

The collapse is accelerated by a relative motion between the bubble and the liquid. The dashed line in FIGURE 1 is for a bubble translating with a constant velocity, $U_B = 0.127$ m/sec (the circles are the results of the calculation of Ref. 14). At large Péclet numbers $Pe = 2R(\mathbf{U}_B - \mathbf{U}_L)/D_L$ an approximate expression for the Nusselt number for flow past a sphere of constant diameter is[15,16] $Nu = 2\sqrt{Pe/\pi}$. If convection is assumed to dominate and this expression is used to calculate a heat transfer coefficient in terms of which to express the right-hand side of **(30)**, one finds a collapse time given by

$$t'_c = \sqrt{\frac{\pi}{Pe} \frac{R^2(0)}{2 D_L Ja}}, \tag{32}$$

which will be shorter than **(28)** provided $Ja/\sqrt{\pi Pe} < 1$, that is, at sufficiently large Péclet numbers. For the case of FIGURE 1, the collapse time given by **(32)** is about $0.29 t_c$, in close agreement with the numerical result.

According to Equations **(5)** and **(6)**, when the bubble radius decreases the relative velocity must increase to preserve the impulse. Thus, a collapsing and translating sphere must keep accelerating. When this effect is taken into account, one obtains the radius–time curve shown by the solid line in FIGURE 1. This prediction may not be quite accurate because of the velocity-induced bubble distortion that will tend to slow it down[3] but, with a smaller radius, the effect of surface tension increases tending to restore the spherical shape: an answer as to which effect dominates must await experiment or a numerical simulation in which the bubble is allowed to deform.

The bubble surface temperature for the cases shown in FIGURE 1 is shown in FIGURE 2 (note the much expanded time scale with respect to that of FIG. 1). At the beginning of the process, the vapor pressure equals $p_{sat}(T_\infty)$, which, here, is 89.9 kPa,

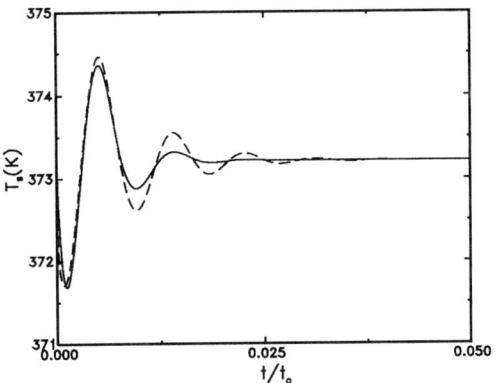

FIGURE 2. Bubble surface temperature for the three cases of FIGURE 1. At $t = 0$ the bubble surface is at the saturation temperature, whereas $T = T_\infty$ elsewhere.

and is therefore significantly lower than the ambient pressure (101.3 kPa): the initial collapse is relatively violent and continues by inertia beyond the point where the bubble surface temperature T_S equals T_{sat} and $p = P_\infty$. As a consequence, more condensation takes place and the surface temperature rises above T_{sat} until the motion is, for a very brief time, reversed (this is barely noticeable in FIG. 1). As heat diffuses away from the interface into the liquid, the surface temperature falls and the cycle repeats a few times until the temperature stabilizes at T_{sat}.

Let us now consider the behavior of a bubble carried by the liquid flow in a region of increasing pressure. For definiteness, we will think of the flow in a duct of variable cross-section

$$\frac{A(x)}{A_\infty} = \frac{1 + \alpha \tanh(x/\ell)}{1 - \alpha}, \tag{33}$$

with A_∞, α, and ℓ suitable constants. We do not intend this to be a model for the actual behavior of a vapor bubble in a variable-area duct, for which one should address the possibility of flow separation when $dA/dx > 0$, the effects of viscosity, turbulence, and so forth; rather, with the assumption of quasi-one-dimensional steady flow, this gives a convenient two-parameter pressure field that enables us to illustrate the general behavior of bubble–liquid heat exchange in flows with pressure gradients. Assuming $U_L(x)A(x) = U_0 A_\infty = \text{const}$, from the liquid momentum equation we have

$$\frac{dP_L}{dx} = -\rho_L U_L \frac{dU_L}{dx}, \tag{34}$$

from which

$$P_L = P_0 - \frac{1}{2}\rho_L(U_L^2 - U_0^2), \tag{35}$$

where P_0 and U_0 are the liquid pressure and velocity far upstream of the region where the pressure gradient is appreciable.

As an example, FIGURE 3 shows the radius versus time for a bubble with an initial radius $R(0) = 1$ mm for an upstream liquid velocity $U_0 = 1$ m/sec, pressure $P_0 = 101.33$ kPa, and $T_\infty = 372.15$ K; here $\alpha = 0.15$, $\ell = 100$ mm (solid line), and $\ell = 10$ mm (dashed line). The motion of this bubble relative to the liquid is computed from (5) (with $P_{ac} = 0$); the dotted line is for a bubble passively convected by the liquid in the same pressure field. The bubble is released at $x(0) = -\ell$ with the same velocity as the liquid; at this position, the ambient pressure is 101.37 kPa and, therefore, the liquid slightly subcooled. When combined with the surface tension overpressure of 0.118 kPa, the effect is to promote the bubble collapse. The dotted line shows that under these conditions, in which there is no liquid–bubble relative velocity, the collapse would be very slow. The situation for a bubble convected into the increasing pressure region is quite different: the bubble acquires a negative (upstream) velocity relative to the liquid, which is amplified due to conservation of impulse: the heat transfer at the bubble surface markedly increases and the collapse is far more rapid. For the case with $\ell = 10$ mm, this negative velocity becomes so large that the bubble actually ends up translating upstream in the laboratory frame. One may say that the bubble tends to be *repelled* by the increasing pressure: the consequent increase of the collapse rate may be expected to be a robust prediction of the model in spite of its idealizations.

In the opposite case of an accelerating flow, the rate of pressure decrease can be so large that the liquid becomes superheated and the bubble ends up growing, rather than collapsing. An example is shown in FIGURE 4 where conditions are as before except that $U_0 = 2$ m/sec, $\alpha = -0.3$. The passively transported bubble (dotted line) initially collapses, but then the radius shows a tendency to stabilize once the pressure has fallen sufficiently. When the pressure gradient is mild ($\ell = 100$ mm, solid line),

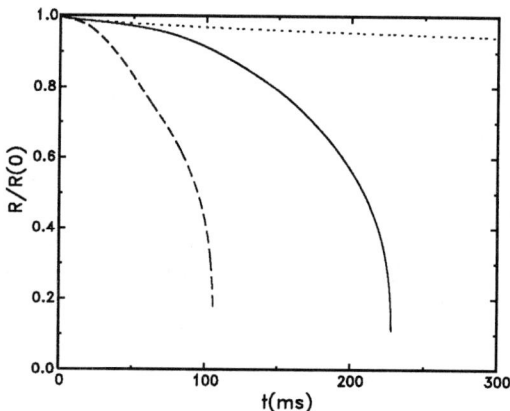

FIGURE 3. Radius versus time for a 1 mm-radius bubble in water convected into a region of increasing pressure. The liquid temperature is 373.15 K and the upstream liquid velocity and pressure 1 m/sec and 1 atm, respectively; the parameter α has the value 0.15. The *dashed lines* are for $\ell = 10$ mm, the *solid lines* for $\ell = 100$ mm, and the *dotted line* for a bubble passively convected by the liquid. The bubble is released at $x(0) = -\ell$ with the same velocity as the liquid.

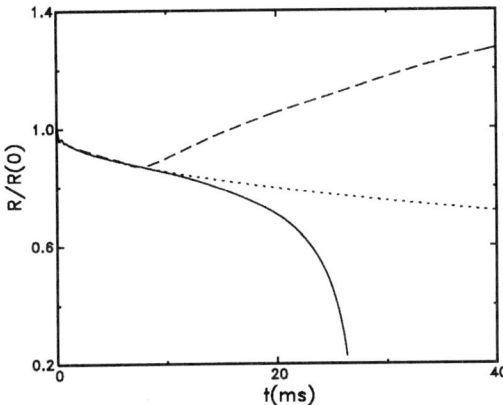

FIGURE 4. Radius versus time for a 1 mm-radius bubble in water convected into a region of decreasing pressure. The liquid temperature is 372.15 K and the upstream liquid velocity and pressure 2 m/sec and 1 atm, respectively; the parameter α has the value -0.3. The *dashed lines* are for $\ell = 10$ mm, the *solid lines* for $l = 100$ mm, and the *dotted line* for a bubble passively convected by the liquid. The bubble is released at $x(0) = -1$ with the same velocity as the liquid.

the bubble picks up a positive velocity with respect to the liquid during the initial condensation period, which is sufficient to cause its total collapse before the benefit of the falling pressure can be felt. In the stronger pressure gradient ($\ell = 10$ mm, dashed line), on the other hand the initial collapse is followed by growth.

VAPOR BUBBLES IN SOUND FIELDS

A gas bubble in a sound field owes its elasticity to the fact that, with a (nearly) constant gas content, its internal pressure is a function of its volume. For a vapor bubble, the situation is quite different: the total vapor content is not constant, even in an approximate sense, and it might therefore appear surprising that a vapor bubble exhibits any elasticity at all. The key to the phenomenon is to be found in the temperature dependence of the saturation pressure: compression of the bubble causes some vapor to condense, which liberates latent heat, heats up the bubble surface, and increases the vapor pressure; the converse happens when the bubble expands. On the basis of this picture it is possible to develop a simple argument that leads to the following estimate of the natural frequency of a vapor bubble of radius R_0:[12]

$$\omega_0^2 R_0^2 \approx \Theta_1 \frac{L^4 \rho_V^4}{\rho_L^3 c_L T_\infty^2 k_L}, \tag{36}$$

where a numerical constant, Θ_1, has been introduced to account for the approximate nature of the derivation. With the value $\Theta_1 = 0.56$, the previous argument gives the upper dashed straight line in FIGURE 5, which is seen to approximate the exact result

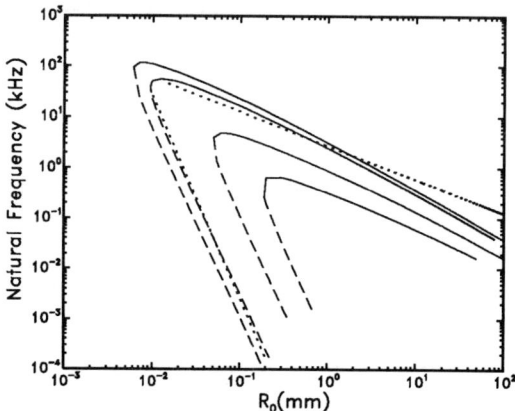

FIGURE 5. Natural frequency of vapor bubbles at temperatures of 50°C, 70°C, 100°C, and 110°C in ascending order. The *lower, dashed* branch of the curves is the so-called second resonance (see text). The *dotted lines* are approximations.

shown by the outermost solid line. For comparison, it may be recalled that the natural frequency of a gas bubble in adiabatic oscillation in an ambient pressure P_∞, with the neglect of surface tension, is given by $\omega_0^2 R_0^2 = 3\gamma P_\infty / \rho_L$. This expression—and in particular its dependence on the bubble size—is very different from **(36)**.

Much has been written about a "second resonance" of vapor bubbles,[17–20] which corresponds to the lower, dashed branch of the lines in FIGURE 5. Actually, as pointed out,[12] this is not a resonance at all but simply corresponds to a situation in which the

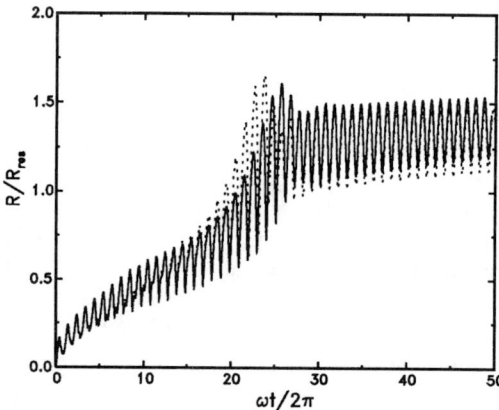

FIGURE 6. Bubble radius (normalized by the linear resonant radius $R_{res} = 2.71$ mm) versus time for saturated water at 101.3 kPa. The sound amplitude is 30.39 kPa and the frequency 1 kHz. The *solid line* is obtained from the complete model; the *dotted line* is the result of an approximation.

destabilizing effect of surface tension is balanced by the stiffness of the bubble. A simple argument gives

$$\omega_0 R_u^4 = \Theta_2 D_L \left(\frac{2\sigma c_L \rho_L T_\infty}{(L\rho_V)^2} \right)^2, \qquad (37)$$

where Θ_2 is another adjustable constant of order 1. This result is shown by the lower straight dashed line in FIGURE 5 for $\Theta_2 = 0.94$. This "pseudo-resonance" is related to bubble stability: if the bubble radius is smaller than this resonant value, the surface tension effects is so large as to give rise to a negative stiffness and the bubble collapses unstably; conversely, it grows for a larger value of the radius.

In analogy with the case of gas bubbles, the previous considerations are based on the idealization of a bubble of constant radius. Actually, this picture is of rather limited value since it ignores the fundamental process of *rectified heat diffusion,* which exerts a very strong effect on the amount of vapor in the bubble. The process of *rectified diffusion of mass* is well known in the dynamics of gas bubbles:[21] rectified heat diffusion is similar, but it is much more intense since the thermal diffusivity of liquids typically exceeds the mass diffusivity by two orders of magnitude.

Two examples are shown by the solid lines in FIGURES 6 and 7. Here the acoustic pressure field is given by

$$P_{ac} = -P_A \sin \omega t, \qquad (38)$$

with $P_A = 30.39$ kPa ($P_A/P_\infty = 0.3$), the initial bubble radius $R(0) = 35$ mm, the ambient pressure $P_\infty = 101.3$ kPa, and $T_\infty = 373.15$ K; the frequencies are $\omega/2\pi = 1$ and 5 kHz, respectively; the radius is normalized by the linear resonant radius (i.e., essentially, by the value given by the exact form of $(36)^{12}$), $R_{res} = 2.71$ mm and 0.42 mm, respectively. The mean radius exhibits a fast growth up to the resonant value, followed by a much slower growth that, as far as can be judged by the numerical

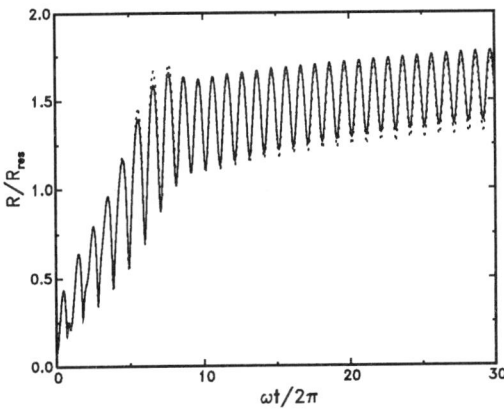

FIGURE 7. Bubble radius (normalized by the linear resonant radius $R_{res} = 0.42$ mm) versus time for saturated water at 101.3 kPa. The sound amplitude is 30.39 kPa and the frequency 5 kHz. The *solid line* is obtained from the complete model, while the *dotted line* is the result of approximation.

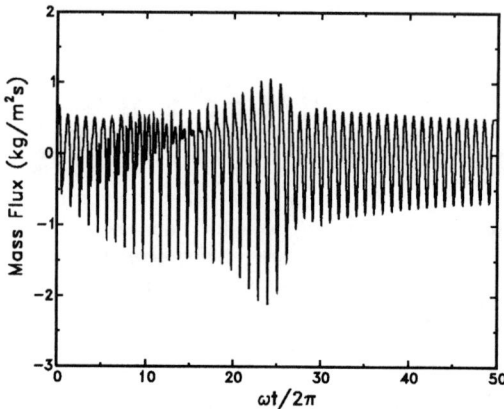

FIGURE 8. Mass flux at the vapor–liquid interface for the case of FIGURE 6.

results (continued much beyond the part shown in this figure) continues indefinitely. From an elaborate perturbation analysis, Gumerov[22] concludes that the bubble growth does not continue indefinitely, but reaches a limit value after a time of the order of hundreds of thousands of cycles. Computing time and error accumulation prevent us from testing this result which, it would seem, can only be examined experimentally. It should be noted that previous approximate calculations,[23–25] in which the perturbation series was arrested earlier than done by Gumerov, predicted a much smaller limit value of the radius, which is clearly disproved by the present results.

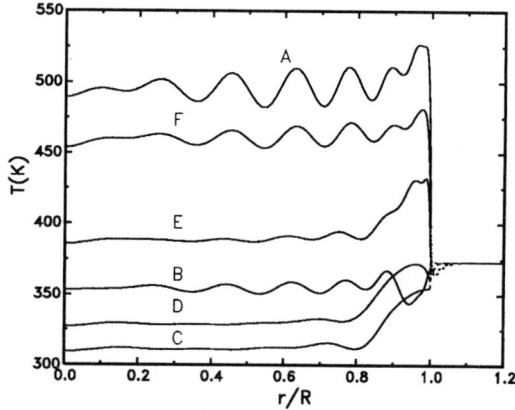

FIGURE 9. Temperature distribution in the bubble (*solid lines*) and in the liquid (*dotted lines*) for the case of FIGURE 6 near the region of maximum amplitude of oscillation; the profiles are taken at times $\omega t/2\pi = 25.1, 25.3, 25.6, 25.8, 26.0$, and 26.2 marked by a *dot* in FIGURE 10.

One of the remarkable features of the deceptively simple process portrayed in FIGURES 6 and 7 is the tremendous mass fluxes that develop at the vapor–liquid interface. An example is shown in FIGURE 8, which corresponds to FIGURE 6. The mass fluxes are of the order of $1\,\text{kg/m}^2\text{sec}$, which translates to a latent heat flux in excess of $2\,\text{MW/m}^2$. In considering this figure it should be remembered that the actual mass flow rate into the bubble is obtained by multiplying the mass flux by the instantaneous bubble area; it is this circumstance that produces a net positive result. It is also interesting to note—especially in FIGURE 6 and, less prominently, also in FIGURE 7—the signature of the first and second harmonic resonances as the bubble grows past the value of the radius where it resonates with 3ω and 2ω, respectively.

FIGURES 6 and 7 serve another purpose, namely to illustrate the effect of using the approximate relation (27), the results of which are shown by the dotted lines, rather than the exact theory (solid lines). Formally, the approximation is justifiable when the parameter $\sqrt{\omega R^2/D_V}$ (where D_V is the thermal diffusivity in the vapor) is small, because in this limit the temperature distribution in the bubble will be nearly uniform. For the two cases of FIGURES 6 and 7, we have $\sqrt{\omega R_{res}^2/D_V} \approx 48$ and 17 and, indeed, the temperature distribution in the bubble is far from uniform, as shown for example in FIGURE 9 for the case of FIGURE 6. Nevertheless the approximation is seen to perform reasonably well, a behavior that we have also found in many other cases.

Rectified heat diffusion also occurs in a subcooled liquid and can confer stability to a vapor bubble that would otherwise condense. Some simple considerations are useful to gain some insight into the pressure amplitudes and frequencies necessary for this purpose. In the first place, the pressure amplitude must be such that, during the expansion phase, the liquid becomes temporarily superheated which, with the aid of the Clausius–Clapeyron relation, gives

$$P_A \geq \frac{L \rho_V}{T_{sat}}(T_{sat} - T_\infty). \tag{39}$$

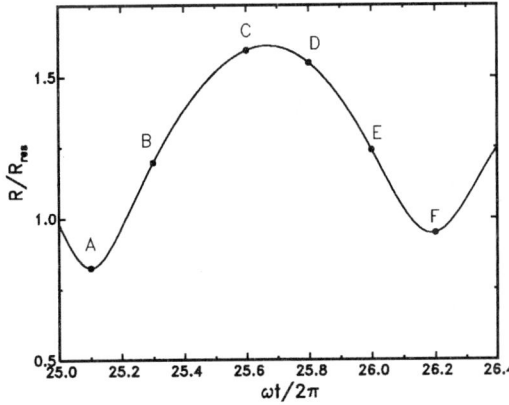

FIGURE 10. Enlargement of one cycle of FIGURE 6 with *dots* marking the times at which the temperature profiles of FIGURE 9 are taken.

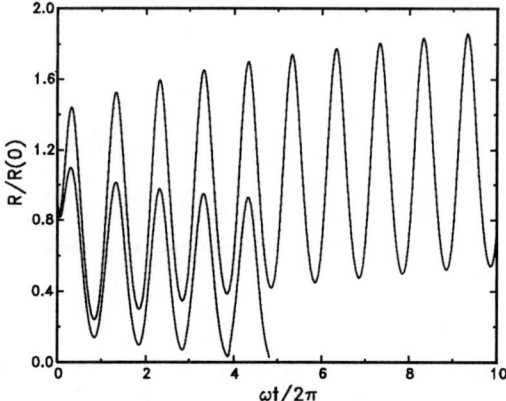

FIGURE 11. Bubble behavior in water subcooled by 3.35 K at 20 Hz; the initial bubble radius is 1 mm and the pressure amplitudes 20.26 (*upper curve*) and 15.16 kPa.

FIGURE 11 shows two examples in which the liquid is subcooled by 3.35 K (the same case as in FIGURE 1), the initial bubble radius is 1 mm, and the frequency 20 Hz. For $P_A = 20.26$ kPa the bubble is seen to slowly grow whereas, for $P_A = 15.16$, it only lasts a few cycles, in fair agreement with (**39**), which predicts $P_A \approx 12$ kPa in this case.

For bubble stability it is also necessary that the sound period not be so large that the bubble collapse completely during the compression half cycle, which, with the previous estimate (**28**), leads to $t_c \geq \pi/\omega$ or

$$\frac{\omega}{2\pi} \geq \frac{2Ja^2 D_L}{\pi R^2(0)}. \tag{40}$$

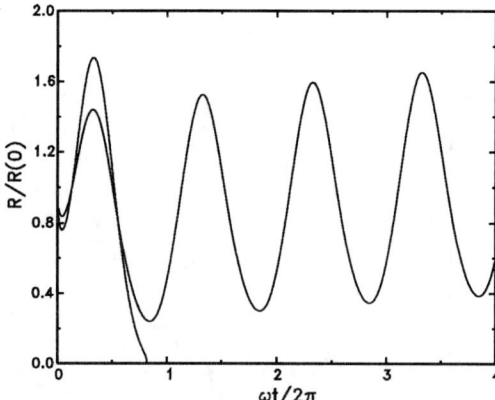

FIGURE 12. Bubble behavior in water subcooled by 3.35 K with $P_A = 20.3$ kPa; the initial bubble radius is 1 mm and the frequencies 20 (*upper curve*) and 8 Hz.

The frequency limit dictated by this inequality is usually low; for example, for the case of FIGURE 1 for which $Ja = 10$ and $R(0) = 1$ mm, one finds $\omega/2\pi \geq 10.7$ Hz. This point is illustrated by the numerical results of FIGURE 12 in which $T_{sat} - T_\infty = 3.35$ K, $R(0) = 1$ mm, $P_A = 20.3$ kPa, and $\omega/2\pi = 8$ and 20 Hz: in the former case the bubble collapses, in the latter it grows.

VAPOR BUBBLES IN FLOW AND SOUND FIELDS

When the action of the sound field is coupled to the effects of a relative motion between the bubble and the liquid the phenomena illustrated above amplify. For example, a bubble that would grow in a subcooled liquid while translating at low velocity, might instead collapse if the velocity is increased. An example is shown in FIGURE 18. Similarly, in a saturated liquid, the process of rectified heat diffusion is stronger and growth is faster with translation, as shown in FIGURE 13.

In the examples shown so far in this and in the previous section, the pressure radiation (or Bjerknes) force, described by the term $V \nabla P_{ac}$ in Equation (5), was disregarded. The effects of this force were studied in a recent paper,[1] from which we take a few examples.

Vapor bubbles are usually encountered in the neighborhood of solid surfaces, such as the heaters that often generate them. Thus, in addition to the driving sound field, one has to account for the pressure field that the oscillations of the bubble itself sets up due to the presence of the boundary. This sound field may be thought of as due to an *image* bubble, and the resulting pressure-radiation force as a bubble–bubble interaction (or secondary Bjerknes force). Due to the proximity to the boundary, this effect can be quite large and overcome the primary force produced by the driving sound even in conditions when, taken by itself, this would drive the bubble

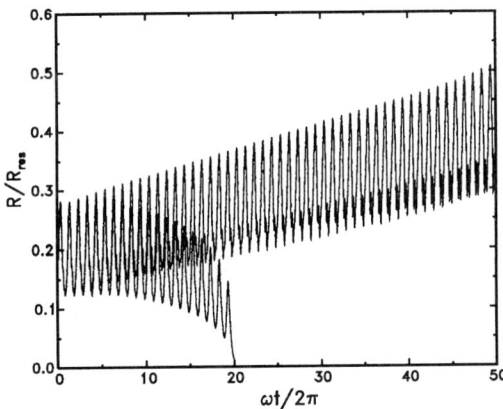

FIGURE 13. Non-dimensional radius versus time of bubbles translating with constant velocity in water at 95 °C and 1 atm static pressure in a 1 kHz, 0.3 atm sound field. The *upper line* is for a velocity of 0.2 m/s, and the *lower line* for 0.3 m/s. The radius is normalized by the resonant radius of a stationary bubble equal to 2.367 mm; the initial radius is 0.5 mm.

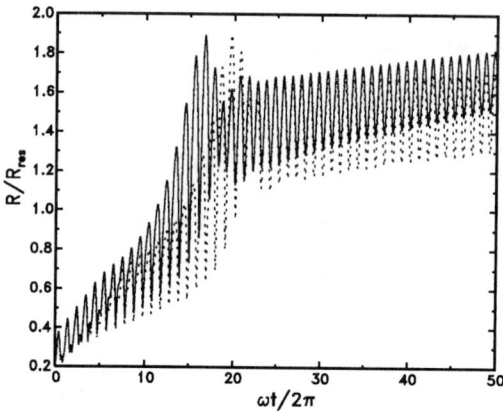

FIGURE 14. As in the previous figure, but with a water temperature of 100°C and for a velocity of 0.3 m/sec (*solid line*) and a stationary bubble.

away from the wall. An example is shown in FIGURE 15: here the bubble is initially placed at a distance of 50 mm from a rigid wall and the wave fronts of the sound are parallel to the wall. For an acoustic pressure amplitude P_A = 30.4 kPa the bubble is attracted by the wall, whereas, for P_A = 50.7 kPa, the primary radiation force wins and pushes the bubble away from the wall. The problem with this outcome (in addition to the large acoustic intensity necessary to bring it about) is that bubbles are usually attached to the wall and, in this case, it is practically impossible to dislodge them

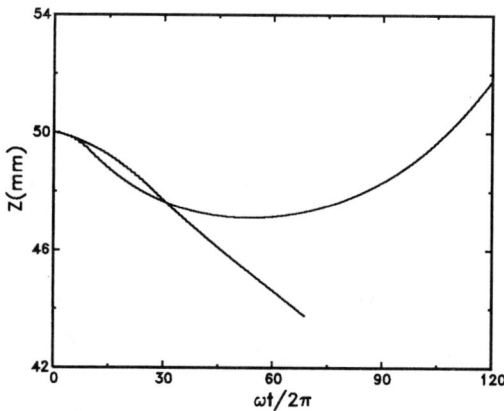

FIGURE 15. Distance between the bubble center and the wall (located at $Z = 0$) as function of time for two pressure amplitudes, P_A = 30.4 kPa (*lower curve*) and P_A = 50.7 kPa; initially the bubble is located at a distance of 50 mm from the wall. The liquid is water in saturated conditions at 1 atm and 100°C, the initial radius 35 μm, and the sound frequency 1 kHz. The acoustic wavefronts are parallel to the wall.

due to the very strong image effect. For this reason, it may be advantageous to operate with wave fronts perpendicular to the wall, which would force the bubble to slide along the wall away from the heat source in a region where, for example, they can be removed by a low-velocity liquid flow.

GAS-VAPOR BUBBLES

The parameter space increases considerably when one considers a bubble containing both vapor and an incondensable gas. Here we limit ourselves to the demonstration of an interesting phenomenon that could have been anticipated from the earlier result of FIGURE 8, which shows that rectified heat diffusion is really due to a non-zero balance between condensing and evaporating fluxes. Indeed, it is well known that even a small amount of incondensable gas can strongly inhibit the condensation of vapor, which leads one to expect that bubble growth by rectified heat diffusion would be substantially faster in the presence of gas.

That this is indeed the case is shown in FIGURE 16, where the growth of a bubble containing CO_2 and water vapor (solid line) is compared with that of a pure vapor bubble. Here $T_\infty = 100°C$, $P_\infty = 101.3\,kPa$, $P_A = 30.4\,kPa$, $R(0) = 1\,mm$, $R_{res} = 7.27\,mm$, and $\omega/2\pi = 400\,Hz$. It is supposed here that the water was saturated with CO_2 at 25°C (with a dissolved gas density of $1.5\,kg/m^3$), and was brought to 100°C with no change in gas content; that is, rapidly enough for the dissolved gas not to equilibrate. The calculation is started with an initial gas mass of $9.85 \times 10^{-9}\,kg$, which corresponds to an initial vapor mass fraction of approximately 25%.

The mechanism underlying the increased growth rate is the accumulation of incondensable gas near the bubble wall, which takes place when vapor condensation

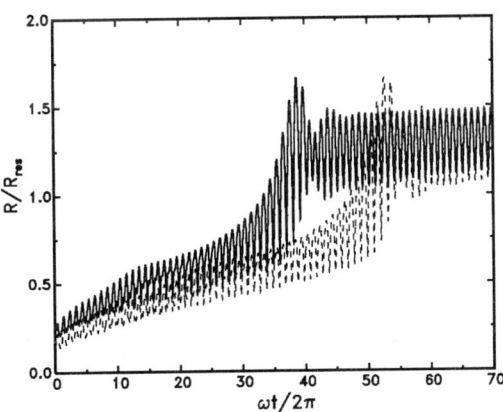

FIGURE 16. Growth of a bubble containing CO_2 and water vapor (*solid line*) compared with that of a pure vapor bubble. Here $T_\infty = 100°C$, $P_\infty = 101.3\,kPa$, $P_A = 30.4\,kPa$, $R(0) = 1\,mm$, $R_{res} = 7.27\,mm$, $\omega/2\pi = 400\,Hz$; the liquid is saturated with dissolved gas at 25°C, and is, therefore, supersaturated by a factor of approximately 15 at the temperature of this calculation.

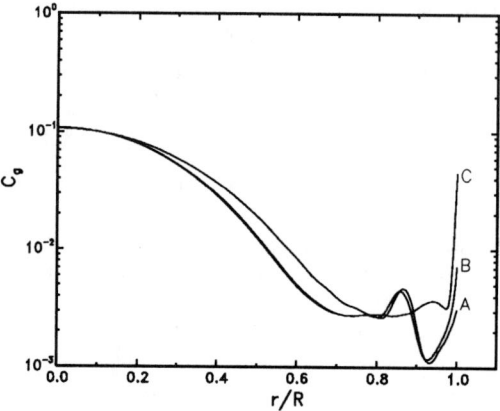

FIGURE 17. Gas mass fraction distribution in the bubble at the three instants of time marked in FIGURE 18.

begins at the start of the compression stage: a concentration boundary layer builds up and offers a barrier to further vapor condensation. The phenomenon is illustrated FIGURE 17 (where the lines are taken at the instants marked by large dots in FIGURE 18), which shows the gas mass fraction in the bubble at different times during the compression. It should be noted that the average gas mass fractions at points A ($\omega t/2\pi = 39.74$) and B ($\omega t/2\pi = 39.84$) are about 0.8%. The phenomenon is, therefore, quite sensitive to even small amounts of incondensables.

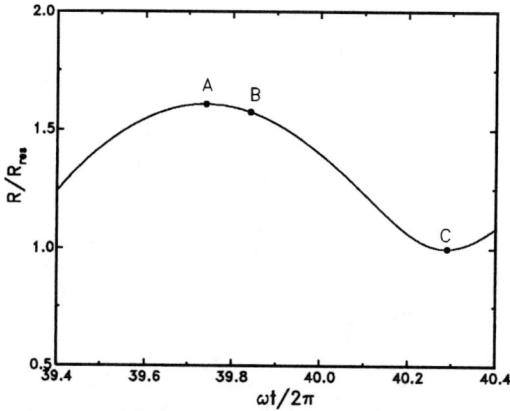

FIGURE 18. Detail of FIGURE 16 with dots marking the times corresponding to the mass fraction distributions shown in FIGURE 17; A, $\omega t/2\pi = 39.74$; B, $\omega t/2\pi = 39.84$; C, $\omega t/2\pi = 40.29$.

ACKNOWLEDGMENT

This work has been supported by NASA under Grants NAG3-1924 and NAG3-2362.

REFERENCES

1. Hao, Y., H.N. Oğuz & A. Prosperetti. 2001. Pressure-radiation forces on pulsating vapor bubbles. Phys. Fluids **13**: 1167–1177.
2. Lamb, H. 1932. Hydrodynamics, 6th edit. Cambridge University Press.
3. Blake, J.R. & P. Cerone. 1982. A note on the impulse due to a vapor bubble near a boundary. J. Aust. Math. Soc. **B23**: 383–393.
4. Oğuz, H.N. & A. Prosperetti. 1990. A generalization of the impulse and virial theorems with an application to bubble oscillations. J. Fluid Mech. **218**: 143–162.
5. Hao, Y. & A. Prosperetti. 1999. The effect of viscosity on the spherical stability of oscillating gas bubbles. Phys. Fluids **11**: 1309–1317.
6. Nigmatulin, R.I. & N.S. Khabeev. 1974. Heat exchange between a gas bubble and a liquid. Fluid Dyn. **9**: 759–764.
7. Nigmatulin, R.I., N.S. Khabeev & F.B. Nagiev. 1981. Dynamics, heat, and mass transfer of vapour–gas bubbles in a liquid. Int. J. Heat Mass Transfer **24**: 1033–1044.
8. Prosperetti, A. 1991. The thermal behaviour of oscillating gas bubbles. J. Fluid Mech. **222**: 587–616.
9. Storey, B.D. & A.J. Szeri. 1999. Mixture segregation within sonoluminescence bubbles. J. Fluid Mech. **396**: 203–221.
10. Moore, D.W. 1965. The velocity of rise of distorted gas bubbles in a liquid of small viscosity. J. Fluid Mech. **23**: 749–766.
11. Hao, Y. & A. Prosperetti. 2000. The collapse of vapor bubbles in a spatially nonuniform flow. Int. J. Heat Mass Transfer **43**: 3539–3550.
12. Hao, Y. & A. Prosperetti. 1999. The dynamics of vapor bubbles in acoustic pressure fields. Phys. Fluids **11**: 2008–2019.
13. Florschuetz, L.W. & B.T. Chao. 1965. On the mechanics of vapor bubble collapse. J. Heat Transfer **87**: 209–220.
14. Wittke, D.D. & B.T. Chao. 1967. Collapse of vapor bubbles with translatory motion. J. Heat Transfer **89**: 17–24.
15. Ruckenstein, E. 1959. On heat transfer between vapour bubbles in motion and the boiling liquid from which they are generated. Chem. Eng. Sci. **10**: 22–30.
16. Magnaudet, J., J. Borée & D. Legendre. 1998. Thermal and dynamic evolution of a spherical bubble moving steadily in a superheated liquid. Phys. Fluids **10**: 1256–1272.
17. Wang, T. 1974. Rectified heat transfer. J. Acoust. Soc. Am. **56**: 1131–1143.
18. Marston, P.L. & D.B. Greene. 1978. Stable microscopic bubbles in helium I and evaporation-condensation resonance. J. Acoust. Soc. Am. **64**: 319–321.
19. Marston, P.L. 1979. Evaporation-condensation resonance frequency of oscillating vapor bubbles. J. Acoust. Soc. Am. **66**: 1516–1521.
20. Patel, G.M., R.E. Nicholas & R.D. Finch. 1985. Rectified heat transfer in vapor bubbles. J. Acoust. Soc. Am. **78**: 2122–2131.
21. Fyrillas, M.M. & A.J. Szeri. 1994. Dissolution or growth of soluble spherical oscillating bubbles. J. Fluid Mech. **277**: 381–407.
22. Gumerov, N.A. 2000. Dynamics of vapor bubbles with non-equilibrium phase transitions in isotropic acoustic fields. Phys. Fluids **12**: 71–88.
23. Alekseev, V.N. 1976. Steady-state behavior of a vapor bubble in an ultrasonic field. Sov. Phys. Acoust. **21**: 311–313.
24. Alekseev, N.V. 1976. Nonsteady behavior of a vapor bubble in an ultrasonic field. Sov. Phys. Acoust. **22**: 104–107.
25. Akulichev, V.A., V.N. Alekseev & V.P. Yushin. 1979. Growth of vapor bubbles in an ultrasonic field. Sov. Phys. Acoust. **25**: 453–457.

Origin and Effect of Thermocapillary Convection in Subcooled Boiling

Observations and Conclusions from Experiments Performed at Microgravity

JOHANNES STRAUB

Technical University Muenchen, Faculty MW,
Thermodynamic Boltzmannstrasse, Garching, Germany

ABSTRACT: During the past two decades we have performed pool boiling experiments under microgravity conditions with saturated and subcooled liquids using various fluids, mostly fluorinated hydrocarbons, and various shapes of heaters. The common observation at subcooled boiling is the development of thermocapillary convection around bubbles. This convection forms jet streams above the head of the bubbles that carry the heat from the bubbles into the ambient liquid. Heat transfer measurements demonstrate that this flow does not directly contribute to the overall heat transfer itself, but that it is an important transport mechanism in subcooled boiling, not only at microgravity, but also under terrestrial conditions. The development of a thermocapillary flow is surprising, because it is well known that for this convection a temperature gradient along the interface of a bubble is necessary. However, it is also well known that the interfacial kinetics of evaporation and condensation is very strong. Thus, small temperature differences generate strong mass flow rates across the interface, coupled with high heat transfer rates that immediately equalize even moderate temperature gradients appearing along the bubble interface. This is confirmed by the observation that, during boiling in saturated liquids under microgravity conditions, thermocapillary convection was never observed, thus indicating uniformity of the temperature along the interface. From this we must conclude that the temperature difference of wall superheat, which generates boiling, cannot be the driving force for the observed thermocapillary flow even in subcooled liquids. Therefore, the question about the origin of this flow arises and is discussed in this paper.

KEYWORDS: pool boiling; heat transfer; mechanism of subcooled boiling; microgravity; bubble dynamics; inert gas; thermocapillary flow; Marangoni convection

NOMENCLATURE:

A	area
a	thermal diffusivity
B	length scale
c_p	isobaric heat capacity
Ma	Marangoni number
m	mass
$\dot{m}$	mass flow

Address for correspondence: Johannes Straub, Technical University Muenchen, Faculty MW, Thermodynamic Boltzmannstrasse 15, 85748 Garching, Germany. Voice/fax: 49 8142 9461. straub@td.mw.tum.de

p		pressure
$\dot{Q}$		heat flux
$\dot{q}$		heat flux density
R		gas constant
T		temperature
t		time
V		volume
x		layer thickness
Greek Symbols		
α		heat transfer coefficient
Δh		phase change enthalpy
η		dynamic viscosity
σ		surface tension
ξ		mass fraction
ρ		density
Subscripts		
B		bubble
G		inert gas
L		liquid
V		vapor
i		interface
ac		accumulation
con		condensation
ev		evaporation
sat		saturation
sub		subcooled
w		wall

INTRODUCTION

The use of microgravity environment proved to be a very important tool for fundamental research in boiling, one especially suited for pool boiling to eliminate the buoyancy effect. Buoyancy is often regarded as the governing mechanism for bubble dynamics and boiling heat transfer. Therefore, initially the question arises: will nucleate pool boiling be maintained in microgravity. Microgravity provides an environment without external forces acting on the bubbles, and offers the possibility for studying the self-dynamics of boiling and its contribution to heat transfer. Until now the availability of microgravity opportunities was insufficient for study of this important field, with its wide impact on technical and industrial applications. Nevertheless, interesting results have been achieved in microgravity research, results that give a much deeper insight into the physical nature of the boiling process than Earth experiments can provide, see Straub.[1] Boiling depends on many interrelated parameters, such as on size, form, geometry, surface condition and material of the heater, thermophysical properties of the fluid and its state, experimental conditions as heat flux, external forces caused by gravity and/or electrical fields, and shear forces appearing at flow boiling. Due to the limited microgravity opportunities until now, we have only been able to investigate only a limited spectrum of these parameters. Among the most obvious results from various experiments are the following:

- The overall heat transfer coefficients at saturated and subcooled nucleate pool boiling up to medium heat fluxes do not diminish with the reduction of the gravity level as predicted by most heat transfer correlations. On the contrary, we generally observed that at lower heat flux densities the heat transfer coefficients are even enhanced up to 15%, but decline with increasing heat flux.

- Likewise, the critical heat flux is higher than predicted by most well known correlations for CHF if the effective gravity value of the microgravity experiment is included in the correlation. However, CHF is influenced by gravity and is lower than under Earth gravity conditions.

- The values of the heat transfer coefficients at saturated and subcooled boiling are both related to the superheated temperature difference $T_w - T_{sat}$ in a nearly identical way, although the noticeable bubble dynamics and transport mechanisms are entirely different.

- For boiling in subcooled fluids, a strong liquid flow around the bubbles is always observed, forming a jet flow above the bubbles. This flow was identified as a quasithermocapillary convection. In this flow a few bubbles sometimes depart and are carried along in the flow.

- Heat transfer measurements indicate that this thermocapillary convection does not significantly contribute to the overall heat transfer. However, this flow is an important transport mechanism in maintaining boiling heat transfer under subcooled conditions in microgravity. Likewise it is also effective under terrestrial conditions, but in this case it is superimposed on gravity convection. The bubbles act as small heat pipes between the bubble base, where the liquid evaporates, and the top, where the vapor condenses in contact with the subcooled liquid. Thermocapillary convection is generated carrying the condensed vapor mass as enthalpy flow into the bulk liquid.

- In contrast, at saturated boiling a thermocapillary flow could never be detected. In this case heat is transported as latent heat, carried along with the bubbles. In absence of buoyancy these bubbles are pushed and transported by the momentum generated by bubble growth and by the momentum and inertia caused by lateral and vertical coalescence processes.

Considering these results, we must conclude that boiling heat transfer under terrestrial, as well as under microgravity, conditions is primarily determined by evaporation mechanisms from the bubble interface in the wedge-shaped microlayer between the bubble and the heater surface, both for saturated and subcooled liquids. This microlayer is formed during bubble growth by static and dynamic forces exerting an effect on the bubble shape. The transport mechanisms for heat from the heater to the ambient liquid are secondary mechanisms, in which the bubbles act as small thermal engines, converting thermal energy into mechanical energy of motion. In microgravity this happens even without the support of gravity. These transport mechanisms are different for saturated and subcooled conditions: in the first case it is, as already mentioned, mainly the momentum and inertia of the liquid during bubble growth and bubble coalescence, in the second case it is the self induced thermocapillary flow and the pumping effect caused by the fast growing and collapsing bubbles.

In the following we describe and discuss observations made during subcooled boiling in microgravity. We then draw conclusions from these observations about the origin and the effect of thermocapillary convection. Marek,[2] and Marek and Straub[3] have already considered this effect for a quasisteady state situation, but in this article we focus our attention mainly on the observation and interpretation of the flow, and why it is generated in subcooled boiling only.

OBSERVATIONS AT SUBCOOLED BOILING

The common characteristics in all experiments of subcooled boiling was the immediate development of a strong flow behind the bubbles. Only few examples are discussed in the following. In FIGURE 1 the onset of boiling and its further development under saturation conditions on the left side and subcooled conditions on the right side, are shown in a sequence of interferometer pictures. The experiment was performed by Zell[4] in a laboratory on Earth with R113 of about 20°C bulk temperature.

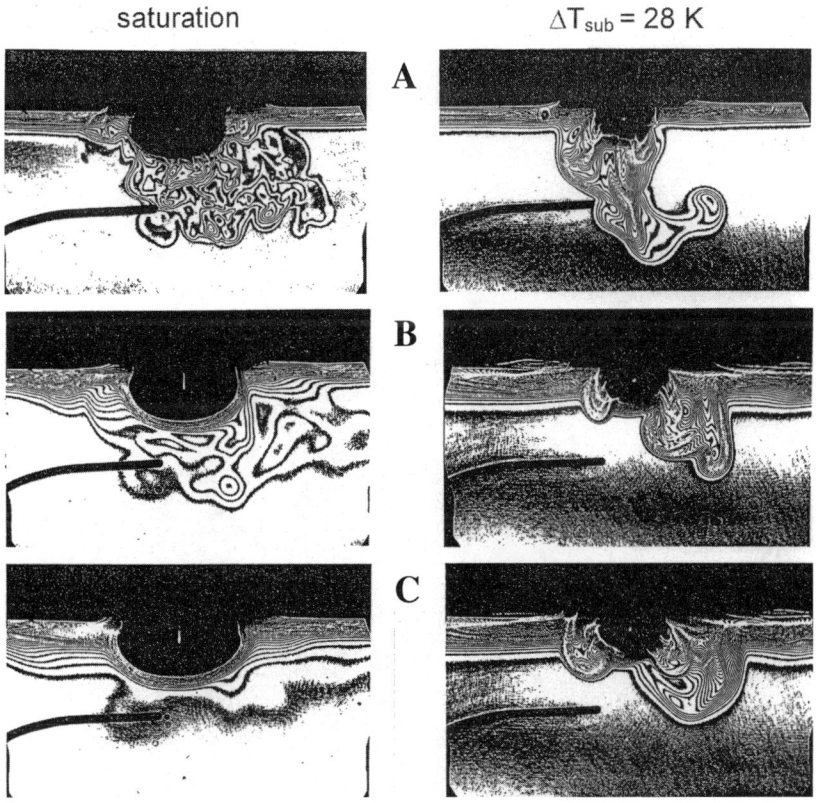

FIGURE 1. Right side, development of thermocapillary flow in subcooled liquid; **left side**, the identical situation in the saturated liquid.[4]

A plane electrical resistance heater with the heating surface facing downwards was prepared with a nucleate site in order to receive a single bubble. At the left side the fluid is under saturated conditions, at the right side the fluid is at the same bulk temperature, but due to a higher pressure the liquid is subcooled at $\Delta T_{sub} = 28\,\text{K}$. Immediately after nucleation the bubble grows and pushes the hot liquid boundary layer away. At saturation, the liquid soon comes to rest and the growth rate slows down. The developed temperature patterns dissolve quickly as seen in pictures Figure 1 B and C, which were taken in a sequence of 0.25-sec time intervals. However, at the subcooled condition, immediately after nucleation, a thermocapillary flow around the bubble sets in and is maintained during bubble growth and subsequently. This flow continuously transports hot liquid at about saturation temperature from the bubble interface into the bulk liquid and carries cold liquid, with a temperature lower than saturation, from the bulk to the interface; due to the vapor in the bubble, it may condense. This thermocapillary flow is strong and is also opposite in direction to buoyancy.

A second example related to a microgravity experiment is shown in FIGURE 2. During the LMS Space Shuttle mission, 1996, a thermocapillary jet flow above bubbles was visualized with the interferometer installation of the BDPU facility (Steinbichler *et al.*[5] and Steinbichler[6]). Boiling was generated on a small circular heater, 1 mm diameter, at a subcooling of 9.9 K in R123. This flow looks like a steady state one, although it did not develop from a single bubble, but probably from a sequence of bubbles that produce thermocapillary flow as they continuously generated and collapsed. The velocity profile of this jet flow was measured by the particle image velocimetry (PIV) method at several distances Δx from the heater surface, with the nearest one at 5.34 mm, as shown in FIGURE 3. Outside of this distance no bubbles are carried with the flow. The extension of the cross-section of the flow from 1 mm heater diameter to 10 mm within the distance of $\Delta x = 5.34\,\text{mm}$ is remarkable.

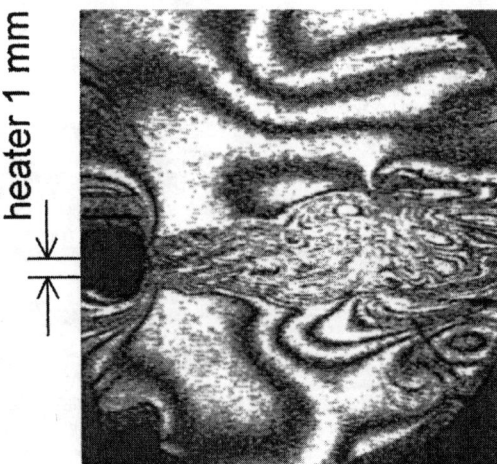

FIGURE 2. Interferometer picture of thermocapillary jet flow generated from a 1 mm-diameter circular heater at subcooled boiling.

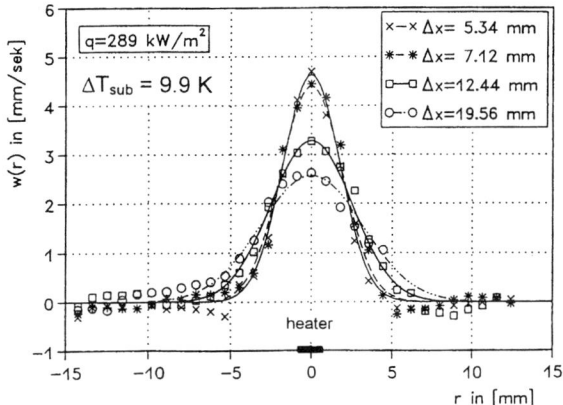

FIGURE 3. Gaussian shaped radial velocity profiles of the jet stream of FIGURE 2. Evaluated with PIW in cross-sections at several axial distances Δx from the heater.

From various measurements with changing heat flux and subcooling we observed that with increasing heat flux the maximum velocity in the center of the stream increases. The mass flow density equivalent with an average velocity increases linearly with the heat flux, FIGURE 4. It is also interesting to note that with increasing degree of subcooling the mass flow density and, likewise, the maximum velocity decreases. Thus, we assume that the maximum and average velocity of the thermocapillary flow depend on the degree of subcooling in the following way. The velocity is zero at saturation, representing zero subcooling, from there it increases rapidly at first, at small subcooling, reaches a maximum, and decreases at higher subcooling to zero (Straub et al.[8,9]). At high subcooling a thermocapillary flow could not be observed, but a hot plume slowly rose from the heater surface, although, it is

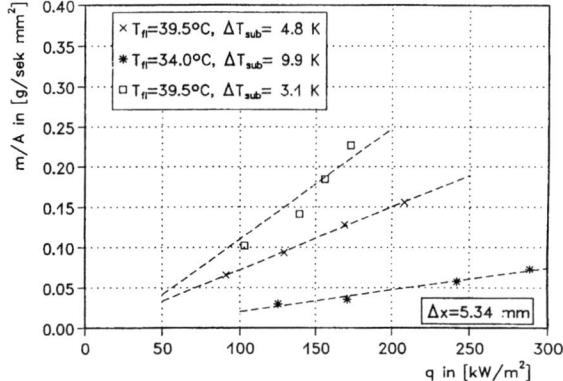

FIGURE 4. Mass flow density, equivalent to the average velocity versus heat flux with subcooling as parameter.

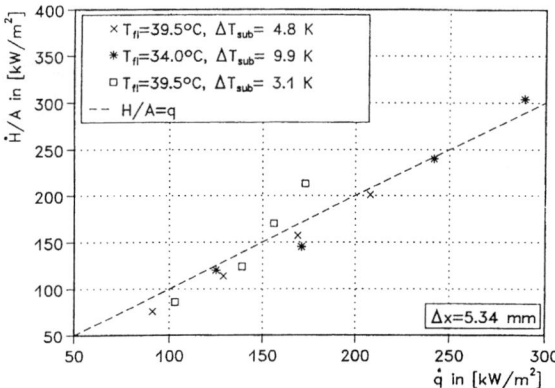

FIGURE 5. Integrated enthalpy flow versus heat flux, demonstrating that all heat is carried by thermocapillary flow, however, with an enlarged flow cross-section.

clear from the heat transfer data that boiling is the mode of heat transfer. We must assume that at high subcooling the process of bubble growth and collapse is so fast that the time necessary to develop noticeable thermocapillary convection is insufficient. In this case the rapid bubble growth and collapse act like a pump, pushing hot liquid of approximately saturation temperature from the heater surface into the bulk liquid, sucking in cold liquid that causes the vapor to condense spontaneously in contact with highly subcooled liquid. The enthalpy flow, integrated over a cross-section in FIGURE 5, demonstrates that the thermocapillary convection is able to transport all the heat $\dot{q}$ generated at the heater, but over an enlarged (nearly 100 times) cross-section into the bulk liquid.

It is interesting to note that in subcooled liquids the thermocapillary flow is even able to maintain film boiling in microgravity on small surfaces, such as wires and spheres. On wires the vapor forms a wavy interface with a wave length determined by the Laplace constant, as can be seen in FIGURE 6 (Micko[7]). A chain of bubbles is formed that are interconnected by a hose of vapor. The liquid interface of the hoses is the nearest position to the heater, where the liquid evaporates. The strong evaporation and the vapor flow in the hoses (Kelvin–Helmholtz instability) prevent rewetting. From the hoses the vapor flows into bubbles with larger radii, in which the pressure is lower due to surface tension. Here, the vapor condenses and generates thermocapillary convection.

Another interesting observation was made at subcooled boiling on a small hemispherical heater, 0.26 mm diameter, using a thermistor.[8,9] For film boiling, a very thin vapor film surrounded the hemispherical heater and at the top of the heater the vapor flowed through a small hose into a large bubble that grow to 10 to 12 mm diameter, roughly speaking 40 times larger than the heater diameter itself. This is illustrated in FIGURE 7. The large bubble provides the cooling area for condensation and the thermocapillary convection developed around it took care of transportation into the bulk liquid, which occurs at the top in the form of a strong jet flow.

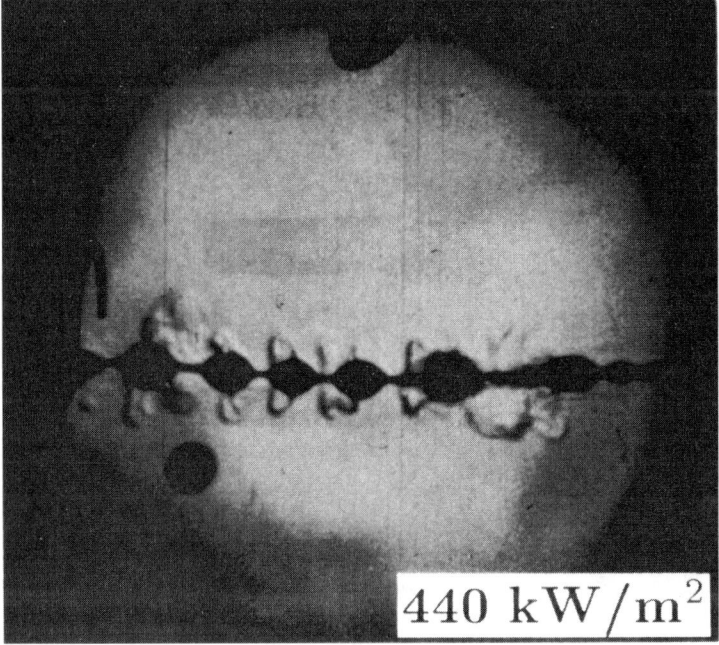

FIGURE 6. Film boiling on a wire in subcooled liquid. Note the thermocapillary convection at the large bubbles.

As already mentioned, the overall heat transfer at moderate and medium subcooling is similar to saturated nucleate boiling and is also almost unaffected by gravity. This means that the observed thermocapillary flow does not directly influence heat transfer from the heater surface. Therefore, thermocapillary convection must be regarded as a secondary transport mechanism that transfers the heat from the bubble interface into the bulk liquid and contributes to the maintenance of subcooled pool boiling. The thermocapillary flow causes reaction forces that push the bubbles to the wall so that they remain there, attached for a longer period. In film boiling heat trans-

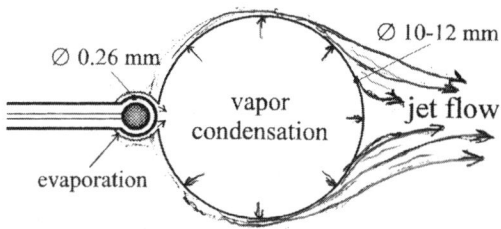

FIGURE 7. Sketch of the thermocapillary jet flow developed at film boiling from a thin film around a 0.26 mm diameter heater.

fer in microgravity is reduced, due to the development of an overall thicker vapor film between the heater and the liquid interface. In a subcooled state film boiling can also be maintained due to thermocapillary convection.

BUBBLE ACTING AS HEAT PIPE: THE MODEL

To generate boiling in subcooled liquids, the liquid boundary layer at the heater must be heated above saturation to the superheat wall temperature, at which nucleation may occur. The bubble first grows in this superheated boundary layer and beyond it. If the top of the bubble penetrates into the subcooled region, the vapor makes contact with the cold liquid, with a temperature below saturation, and begins to condense. Vapor flow inside the bubble is induced from the hot side to the cold side, with evaporation at the base and condensation in the upper part. The bubble can be regarded as a small heat pipe acting between the heater surface and the subcooled liquid. The growth rate of a bubble is balanced by the evaporation rate at the base, $\dot{m}_{ev}$, and condensation rate at the top, $\dot{m}_{com}$, expressed by

$$\frac{dV_B}{dt} = \frac{1}{\rho_V}(\dot{m}_{ev} - \dot{m}_{com}). \tag{1}$$

Bubbles in subcooled liquids control and adjust their size according to the heat and mass balance at the evaporation and the condensation side of the interface. Therefore, their size depends on the heat flux and the subcooled liquid condition. In microgravity the condensation at the top can be maintained only with a self induced thermocapillary convection, otherwise the bubble would grow as far as the interface area; large enough to transfer the heat into the bulk liquid by conduction only. The heat carried with the condensed mass into the bulk by convection can be written

$$\dot{Q}_{con} = \dot{m}_{con}\Delta h = A_{con} \cdot \alpha_{i,L}(T_i - T_L), \tag{2}$$

where $\alpha_{i,L}$ is the heat transfer coefficient between the interface and the ambient liquid and is governed by self-induced convection.

The growth or shrinkage of bubbles is balanced by Equations (1) and (2). If vapor production by evaporation at the base is predominant, the bubble will grow, if condensation at the top is predominant, it will shrink or even collapse. Bubbles even may grow to a quasi steady size, at which the vapor mass flow inside is balanced by $\dot{m}_{ev} = \dot{m}_{con}$, equivalent to $\dot{Q}_{ev} = \dot{Q}_{con}$, but this is an almost unstable situation. Sometimes bubbles collapse immediately after a short term of direct growth at the heater surface. However, bubbles are also observed departing, some of them stand still at a certain distance from the heater and condense there. Other bubbles return to the heater surface, obviously caused by reaction forces on the bubbles due to thermocapillary flow and temperature gradients in the liquid resulting in small temperature differences between the top and the base of the bubbles. At higher subcooling very small bubbles are generated, they grow and collapse directly at the heater wall, as mentioned, without developing thermocapillary flow. Generally speaking the bubble dynamics differ and depend mainly on the heat flux, subcooling, fluid properties, and liquid state conditions.

The first question that arises: What is the origin of this thermocapillary convection and why it is observed only in subcooled boiling?

The intensity of the thermocapillary flow is often expressed using the dimensionless Marangoni number

$$Ma = -\frac{\partial \alpha}{\partial T}\frac{B}{\alpha \eta}[T_{s=0} - T_{s=B}]. \tag{3}$$

This definition is not applicable to subcooled boiling. Without question, one might assume that the characteristic length scale for this problem is the radius or the diameter B of the bubble, and the appropriate temperature difference for $[T_{s=0} - T_{s=B}]$ is the superheat temperature $[T_w - T_{sat}]$. However, as mentioned, at saturated boiling, where this temperature difference is applied to generate boiling, thermocapillary flow could never be observed. Therefore, this temperature difference cannot be the driving force for thermocapillary flow. Indeed, the observation at saturated boiling confirms the fact that the kinetics of evaporation and condensation at the interface is so efficient that a total temperature equalization around the interface can be assumed, with the exception of the microlayer region. The first numerical calculation, due to Stephan and Hammer[10] and Hammer,[11] in that wedged shaped region results in high evaporation rates with extremely high heat transfer rates in the small annular area where the bubble touch the heater. Only near the center of this microwedge are strong temperature gradients along the interface (10^4 K/m) calculated, due to the strongly diverging heat flux. If these gradients are realistic they would certainly provoke thermocapillary flow, as discussed by Straub.[1] However, this convection would appear only in the microregion and it would be extremely difficult to observe with traditional optical methods, because there the liquid layer is very thin and the area microscopically small. However, this topic is outside of our consideration, we discuss here the strong convection that we observed in the upper part of the bubbles. The fact that in saturated boiling no convection was observed confirms that outside the microlayer region the temperature is almost uniform along the bubble interface at saturation temperature. This statement is only valid for pure liquids and cannot be applied to mixtures.

In subcooled boiling one can additionally assume that the appropriate temperature difference in the Marangoni number is the subcooling temperature difference, either $[T_{sat} - T_L]$ or the total difference $[T_w - T_L]$. However, this is in contradiction to the observation that with increasing subcooling the thermocapillary velocities decrease (see FIG, 4). All these considerations lead to the hypothesis that the temperature gradients that generate the observed thermocapillary convection must be induced by the accumulation of inert gases at the interface where the vapor condenses and inert gas may accumulate. This is the only decisive difference from saturated boiling. In any case, the Marangoni number in the conventional definition is not the appropriate dimensionless number to describe thermocapillary flow and, hence, heat transfer in subcooled boiling.

As mentioned, in subcooled boiling vapor flow inside the bubble is induced with evaporation at the base and condensation at the top, see FIGURE 8. Even, if there exists only a small amount of dissolved gas in the liquid, these gas molecules are pulled out by the evaporating molecules at the base of the bubble and are carried with the vapor flow to the cold upper part of the interface. Here, only the vapor molecules will condense, since for the gas molecules the interface is almost a barrier. Certainly, a small part of the gas molecules will dissolve in the liquid again and another small part will diffuse as Stefan flow against the on-streaming vapor, but the larger part of

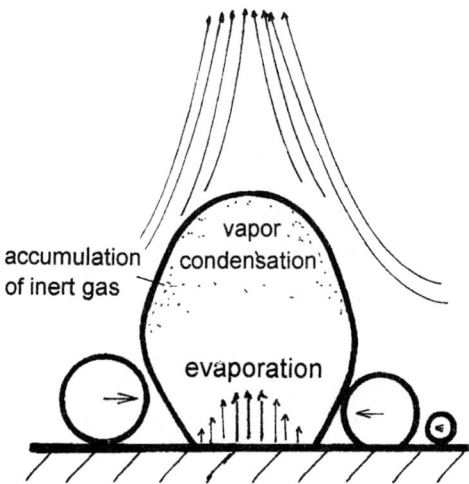

FIGURE 8. Sketch of a bubble acting as microheat pipe in subcooled liquid.

the inert gas will remain and accumulate in front of the interface. Different concentration profiles may be developed according to the local balance of heat and mass transfer at the interface, Equation (**2**). The accumulation of inert gas causes differences in the partial vapor pressure and, with that, also differences in the saturation temperature along the interface, resulting in a positive temperature gradient along the interface towards the bubble base. These temperature gradients generate thermocapillary flow in the opposite direction, from warm to cold, toward the bubble top, which carries the warm condensed liquid away from the interface and transports cold liquid from the bulk to the interface. Consequently, condensation in the upper part of the bubble and the vapor flow through the bubble is maintained.

ACCUMULATION OF INERT GAS

Of course, in all of our space experiments we have used liquids that are as pure as possible. We ordered from the suppliers the highest quality of extra pure liquids paying high prices for them. Before filling the experimental cell with the liquid, we cleaned the cell, purified the liquid by various methods, and analyzed it. The final analysis taken after the mission showed a small increase of inert gas accumulated in the liquid during the mission. For operational reasons, filling the cell and delivery to NASA was necessary about a month before the Shuttle launch. Concerning the high solubility of gases in hydrocarbon liquids, we can assume that during the waiting period a small amount, m_G, of air may have diffused through the sealing material and dissolved in the test liquid.

The dissolved gas is characterized by the mass fraction

$$\xi_G = \frac{m_G}{m}, \tag{4}$$

with $m = m_L + m_G$ as the total mass and m_L the mass of the pure boiling liquid. During the boiling process the inert gas is pulled out with the evaporating vapor mass in the same concentration as it is dissolved. Consequently, in the vapor the concentration of inert gas is the same as in the liquid previously. The total mass flow through the bubble is

$$\dot{m} = \dot{m}_V + \dot{m}_G = \frac{\dot{m}_G}{\xi_G}. \tag{5}$$

The vapor mass flow, $\dot{m}_V$, follows from the energy balance and can be expressed in terms of the heat flux density, $\dot{q}$, and the effective evaporating area A_{ev} at the bubble base. Assuming that all energy from the heater is first transferred to latent heat, and that the solution energy of the inert gas in the liquid can be neglected and does not contribute to the energy balance, we obtain for the heat flux to the bubble

$$\dot{Q}_{ev} = \dot{q} A_{ev} = \dot{m}_V (\Delta h + c_p \Delta T_{sub}), \tag{6}$$

the vapor mass flow into and through the bubble

$$\dot{m}_V = \frac{\dot{q} A_{ev}}{\Delta h + c_p \Delta T_{sub}}, \tag{7}$$

and with Equation (5), the mass flow of inert gas

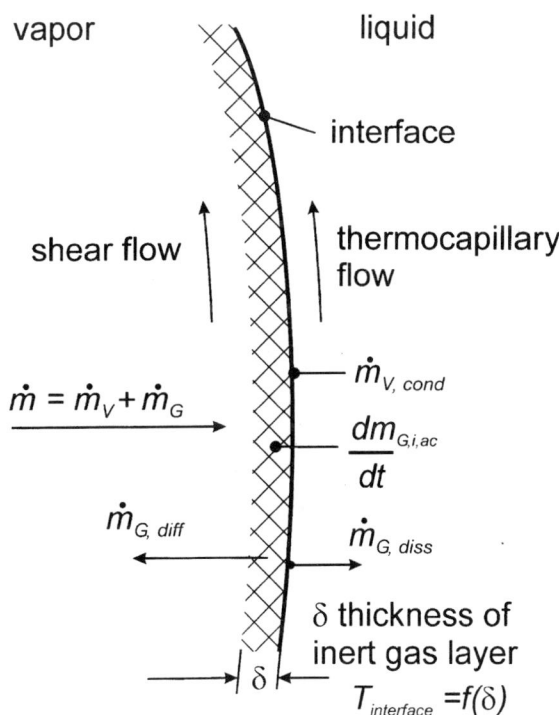

FIGURE 9. Mass balance at the interface of a bubble in subcooled liquid.

$$\dot{m}_G = \frac{\xi_G}{1-\xi_G} \cdot \frac{\dot{q} A_{ev}}{\Delta h + c_p \Delta T_{sub}}. \tag{8}$$

This inert gas flow is carried with the vapor to the interface at the top, where the vapor condenses and the inert gas accumulates on the vapor side of the interface. From the mass balance, FIGURE 9, considering the backward diffusion mass flow, $\dot{m}_{G,\,diff}$, and the mass flow redissolved in the liquid, $\dot{m}_{G,\,diss}$, we obtain the growth of the accumulated gas mass

$$\frac{dm_{G,\,i,\,ac}}{dt} = \frac{d(A_{con}\rho_{G,\,i}\delta)}{dt} = \dot{m}_G - \dot{m}_{G,\,diff} - \dot{m}_{G,\,diss}. \tag{9}$$

The second term describes the growth of the accumulating layer with the thickness δ and the gas density $\rho_{G,i}$ at the interface area A_{con}. This area is the active condensation area, which depends on the bubble size and also on the local condensation resistance. Both $\rho_{G,i}$ and δ depend on the location in the bubble, on the radial distance from the interface, and on the time from the first contact of the bubble with the subcooled liquid. The gas density can be expressed in terms of the partial pressure of the gas, $p_{G,i}$, by using the ideal gas relation

$$\rho_{G,i} = \frac{p_{G,i}}{R_G T}. \tag{10}$$

A rough estimate can be made if we assume that at the first contact of the bubble with the subcooled liquid the condensation area is the same as the evaporation area, and that A_{con}/A_{ev} is constant. Let us further assume that a constant inert gas layer a few molecules thick is sufficient to change the vapor pressure, and let us assume that the inert gas is not redissolved in the liquid, and that backward diffusion can be neglected ($\dot{m}_{G,\,diff} = 0$ and $\dot{m}_{G,\,diss} = 0$), then we may use (8) in the approximate form (9) to obtain

$$\frac{dp_{G,i}}{dt} = \frac{1}{\delta} \frac{A_{ev}}{A_{con}} \frac{R_G \cdot T \cdot \dot{q}}{(\Delta h + c_p \Delta T_{sub})} \frac{\xi_G}{(1-\xi_G)}. \tag{11}$$

The total pressure within the bubble can be assumed to be constant, $p = p_{G,i} + p_V$ is constant, given by surface tension and ambient liquid pressure. With the accumulation of inert gas the partial gas pressure $p_{G,i}$ increases at the interface locally, diminishes the saturation pressure p_V of the liquid, and with it the saturation temperature. For instance, for the liquid R134a in an average temperature range of 40°C the temperature–pressure relation at saturation is $(dT/dp)_{sat} = 3.57$ K/bar. To induce a strong thermocapillary flow only small temperature differences are necessary.[12,13] The thickness δ of the accumulated gas layer, with a layer of several molecules, may be sufficient to change the relation between saturation temperature and vapor pressure. It is arbitrarily but not unrealistic to assume δ to be of the order of 10^{-10} m. With an impurity concentration of $\xi_G \leq 10^{-4}$, only a fraction of a millisecond is necessary for the accumulation of a gas layer to reduce the vapor pressure and induce thermocapillary flow. This estimate does not include the transient behavior during the first stage of bubble growth and the first contact of vapor with the cold interface. In addition, when the first contact of vapor with the cold liquid occurs for a short time, the temperature difference between saturation and subcooled liquid may support the generation of thermocapillary flow.

It is evident that for a realistic case Equation (9) must be numerically integrated over the growth of the bubble, considering the heat and mass balance around the bubble interface, including thermocapillary convection, and gas flow induced in the bubble due to the shear forces from thermocapillary convection in the liquid. Thus, Equation (11) should be regarded as a rough estimate only, but it may confirm the hypothesis of the origin of thermocapillary convection in subcooled boiling, and it may explain the difference from saturated boiling, where no flow was observed.

OSCILLATORY DYNAMICS

In this context we should mention other interesting effects observed[1,4] during TEXUS experiments in 1984 for subcooled boiling; effects that can be explained with the ideas already discussed. At a gold coated electrically heated flat plate the usual nucleate boiling at low heat flux was observed. At a little higher heat flux, a bubble developed that was larger than the uniform size of all other bubbles, see FIGURE 10, and at higher heat flux two large bubbles are generated in some distance from each other. These large bubbles did not depart, they remained at the surface and grew by evaporation from the microlayer and by soaking up all neighboring smaller bubbles. The interface of the large bubbles is in continuous motion, caused on the one hand by continuous coalescence processes with smaller bubbles at the base. On the other hand, in the upper part, interfacial dynamics change the curvature locally from convex to concave shape. This motion is obviously caused by interaction between condensation of the vapor, accumulation of inert gas, development of thermocapillary flow, and gas flow induced by shear stress in the bubble. If the inert gas layer is locally increasing at that position condensation slows down, on the other hand thermocapillary flow is enhanced. The shear forces of thermocapillary flow cause a flow on the gas side of the interface, and move the accumulated gas to the upper part of the bubble. Enhanced liquid flow thins out the accumulated inert gas layer, resulting in diminished local partial gas pressure and, with it, reduced

FIGURE 10. Subcooled boiling on a flat surface. The larger bubble acts as heat pipe, transferring vapor to the interface in the upper part, where thermocapillary convection transports the heat of condensation into the ambient liquid. The small bubbles coalesce with the large bubble.

thermocapillary flow in the liquid. A thinner inert gas layer enforces condensation and renews the formation of a gas layer. The rapid succession of these processes, irregularly distributed over the interface, causes spontaneous condensation and pressure differences that deform the interface. These effects become visible at the interface if the combination of heater and liquid can lead to large bubble generation during subcooled boiling.

The large bubble, in FIGURE 10, stepping out of the others is attached to the heater by reaction forces from the thermocapillary flow. The surrounding smaller bubbles of nearly uniform size seem to be not attached to the heater by a thermocapillary reaction force. We can assume that up to the height of the small bubbles the liquid is nearly at saturation temperature, thus the small bubbles grow up in saturated liquid. These bubbles are freely movable and they migrate toward the large bubbles and coalesce with them. This motion may be caused by the capillary-adhesion flow toward and into the microlayer underneath the large bubble.[1] Only the large bubbles reach into the region of subcooled liquid, where the interface contributes to the condensation and generation of thermocapillary flow. At a little higher heat flux, a second large bubble grows from the smaller bubbles some distance from the first large bubble. If the vapor supply to the large bubbles exceeds the condensation at their tops, the large bubbles grow, touch each other, and join together. Immediately, a very large bubble is formed that penetrates deep into the subcooled region. Strong condensation sets in and the bubble shrink to its former size. A new second large bubble, out of the many small bubbles, forms at the heater surface. This favored bubble grows, mostly by coalescence with others, and once again joins with the other large bubble. It is by such sequences that subcooled boiling is maintained on flat surfaces. However, we call attention to the fact that in the presence of large bubbles, the average heater temperature was higher than the terrestrial values, caused by large dry spots underneath the large bubbles. Immediately after the two large bubbles coalesce the average heater temperature is a little decreased, which indicates that the unified bubble has a smaller dry area than the two bubbles before coalescence. With the growth of the second large bubble the heater temperature slowly increases to its initial value.

CONCLUSION

The complete simulation of such a transient complex system as subcooled boiling with evaporation in the microlayer at the bubble base, vapor–gas flow through the bubble, a moving phase boundary, condensation in the upper part of the bubble, accumulation of inert gas, dissolution of the gas in the liquid, as well as the development of thermocapillary flow around the bubble, shear force induced flow on the gas side in the bubble has not yet been performed due to limited computer capacities and limited knowledge of the basic physical mechanisms. The solution of these problems offers a challenge for the next generation. A simplified simulation model of the effect of inert gas and one that focuses on the essential processes was developed by Marek,[2] and Marek and Straub.[3] In this paper the accumulation of inert gas, the origin of the thermocapillary convection, and basic mechanisms are discussed resulting from our observation of boiling experiments at microgravity. We repeat once again, the use of microgravity for boiling experiments will increasingly support our basic knowledge on the boiling mechanisms that are often hidden on Earth by buoyancy and by strong overlapping effects with thermal convection. Such studies will supply us with important physical input and the necessary boundary conditions to improve the models for numerical simulations.

ACKNOWLEDGMENT

Microgravity research in boiling heat transfer has been financially supported by the Deutsches Zentrum für Luft und Raumfahrt e.V. (DLR) under several contracts during the past years. The author is grateful to DLR for this support and the support for the travel funding. Thanks are also expressed to many individuals at DLR, ESA, and NASA for the disposition of various flight opportunities, and to my former students and coworkers for their engagement in microgravity projects.

REFERENCES

1. STRAUB, J. 2001. Boiling heat transfer and bubble dynamics. *In* Advances in Heat Transfer 35. J.P. Hartnett, T.F. Irvine, Y.I. Cho & G.A. Greene, Eds. Academic Press, New York.
2. MAREK, R. 1996. Einfluß thermokapillarer Konvektion und inerter Gase beim Blasensieden in unterkühlter Flüssigkeit. Dr. Thesis, Technical University Muenchen, Muenchen.
3. MAREK, R. & J. STRAUB. 2001. The origin of thermocapillary convection in subcooled nucleate pool boiling. Int. J. Heat Mass Transfer **44:** 619–632.
4. ZELL, M. 1991. Untersuchung des Siedevorgangs unter reduzierter Schwerkraft. Dr. Thesis, Technical University Muenchen, Muenchen.
5. STEINBICHLER, M., S. MICKO & J. STRAUB. 1998. Nucleate boiling heat transfer on small hemispherical heaters and wires under microgravity. Heat Transfer 1998. Proc. of the 11th IHTC, Kyongju, Korea **2:** 539–544.
6. STEINBICHLER, M. 2000. Experimentelle Untersuchung des gesättigten und unterkühlten Siedens an Miniaturheizflächen unter Mikrogravitation. Dr. Thesis, Technical University Muenchen, Muenchen.
7. MICKO, S. 2000. Sieden am Heizdraht unter reduzierter Schwerkraft. Dr. Thesis, Technical University Muenchen, Muenchen.
8. STRAUB, J., J. WINTER, G. PICKER & M. ZELL. 1995. Boiling on a miniature heater under microgravity—a simulation for cooling of electronic devices. Proc. 30th National Heat Transfer Conf., Portland, OR, 1995. ASME HDT **305:** 61–69.
9. STRAUB, J. 1998. Microscale boiling heat transfer under $0\,g$ and $1\,g$ conditions. Eurotherm Seminar No. 57, Microscale Heat Transfer, Poitiers, France, July 1998. Int. J. Therm. Sci. **39:** 1–8.
10. STEPHAN, P. & J. HAMMER. 1994. A new model for nucleate boiling heat transfer. Heat Mass Transfer **30:** 119–125.
11. HAMMER, J. 1996. Einfluß der Mikrozone auf den Wärmeübergang beim Blasensieden. Forschritts Bericht VDI, Düsseldorf, Ser. 19, No. 96.
12. BETZ, J. 1997. Strömung und Wärmeübergang bei thermokapillarer Konvektion an Gasblasen. Dr. Thesis, Technical University Muenchen, Herbert Utz Verlag Muenchen.
13. BETZ, J. & J. STRAUB. 2001. Numerical and experimental study of the heat transfer and fluid flow by thermocapillary convection around gas bubbles. J. Heat Mass Transfer **37:** 215–227.

High Heat Flux Cooling by Microbubble Emission Boiling

KOICHI SUZUKI,[a] HIROSHI SAITOH,[b] AND KAZUAKI MATSUMOTO[b]

[a]*Department of Mechanical Engineering, Faculty of Science and Technology, Science University of Tokyo, Chiba, Japan*

[b]*Graduate School of Science and Technology, Science University of Tokyo, Chiba, Japan*

ABSTRACT: In subcooled flow boiling of water in a horizontal rectangular channel, microbubble emission boiling occurred at higher subcooling of liquid in transition boiling, and the heat flux increased more than the critical heat flux. The maximum heat flux reached 10 MW/m^2 for a channel with 12 mm × 14 mm cross-section at 40 K liquid subcooling and 0.5 m/sec liquid velocity. For smaller rectangular channels with 14 mm × 5 mm, 14 mm × 3 mm, and 14 mm × 1 mm cross-sections, the maximum heat flux was 7 MW/m^2—more than 20 times the cooling limit of a present day CPU. Microbubble emission boiling is expected to realize high heat flux cooling for electronic devices. In convection boiling with subcooled water jet, the same boiling regime and heat flux were obtained for a downward heating surface and an upward heating surface. In subcooled flow boiling with strong convection, the hydrodynamic force is predominant for vapor–liquid exchange. Accordingly, microbubble emission boiling is expected for high heat flux cooling or high heat flux heat transfer in microgravity.

KEYWORDS: small channel; flow boiling; strong convection; large subcooling; transition boiling; microbubble emission boiling; high heat flux

INTRODUCTION

A previous theoretical analysis indicated that the critical heat flux in pool boiling decreases in proportion to the one-quarter power of gravity.[1,2] Straub *et al.* stated that the critical heat flux of pool boiling obtained in many recent microgravity experiments was 100–300 percent higher than that calculated from theoretical predictions.[3] The same heat flux was obtained by the authors during a parabolic flight experiment on subcooled pool boiling of water.[4] The level of gravity was 0.01–0.04 g. In an observation of bubble behavior, the boiling bubbles stayed on the heating surface at any subcooling of liquid under a reduced gravity condition with high *g*-jitter.

In order to simulate the bubble behavior on the ground, a flat plate was placed over the heating surface with a small clearance so as to hold the boiling bubbles on the surface. The same heating procedure was conducted as was used under reduced

---

Address for correspondence: Koichi Suzuki, Department of Mechanical Engineering, Faculty of Science and Technology, Science University of Tokyo, Noda, Chiba, 278-8510 Japan. Voice: +81-471-24-1501; fax: +81-471-23-9814.
  suzuki@rs.noda.tus.ac.jp

gravity. The same critical heat flux (burnout heat flux) as under reduced gravity was obtained by adjusting the clearance between the plate and the heating surface.[5]

The heating time under reduced gravity was 10–20 seconds until burnout. In the ground experiment, the burnout heat flux was measured by extending the heating time. The heat flux decreased rapidly until about 80 seconds into the heating period and it gradually approached the previous predictions[1,2] after more than 100 seconds.[6] This result suggests that the heat flux in pool boiling is actually lower than the critical heat flux obtained in the recent microgravity experiments for long time heating or steady state heating.

Generally, flow boiling or convection boiling is widely used for industrial applications. It is expected that flow boiling with strong convection gives considerably high heat flux, even under microgravity conditions, because the gravity effect might be very small relative to the hydrodynamic effect of boiling liquid.

In flow boiling with high subcooling of liquid, many microbubbles are emitted from the coalesced bubbles on the heating surface, and the heat flux increases more than the critical heat flux in transition boiling.[7,8] This boiling is called microbubble emission boiling (MEB). In this paper, the authors introduce experimental research on MEB occurring in a horizontal channel, and discuss the availability for microgravity use.

SUBCOOLED FLOW BOILING WITH STRONG CONVECTION FOR UPWARD HEATING SURFACE

In subcooled flow boiling with a parallel liquid jet to the upward heating surface,[9] many fine bubbles were emitted continuously from the coalesced bubbles, and the heat flux increased in transition boiling. The copper heating surface was 14mm in length and 5mm in width. Pure water was used as the boiling liquid. Liquid subcooling was 10–70K and the velocity was 0–1.0m/sec at the nozzle exit. The effect of a subcooled jet on boiling at 20K of liquid subcooling is shown in FIGURE 1 as an example. At above 0.3m/sec of liquid velocity at the nozzle exit, MEB occurred and the heat flux increased more than the critical heat flux in transition boiling, as shown in FIGURE 1. Microbubble emission boiling was investigated by Inada[10] for subcooled pool boiling and by Kumagai[11] for subcooled boiling with weak convection. In subcooled convection boiling with a jet impinging on the heating surface, MEB occurred and the maximum heat flux reached 20MW/m^2 at 40K liquid subcooling and 3m/sec jet velocity.[9]

SUBCOOLED FLOW BOILING WITH STRONG CONVECTION FOR DOWNWARD HEATING SURFACE

Subcooled boiling with a parallel liquid jet to the downward heating surface was investigated for a test section, as shown in FIGURE 2. The heating block is made from copper and it consists of a circular cylinder and a cone part. The heating surface is the bottom surface of the cylindrical part and the diameter is 10mm. The heating

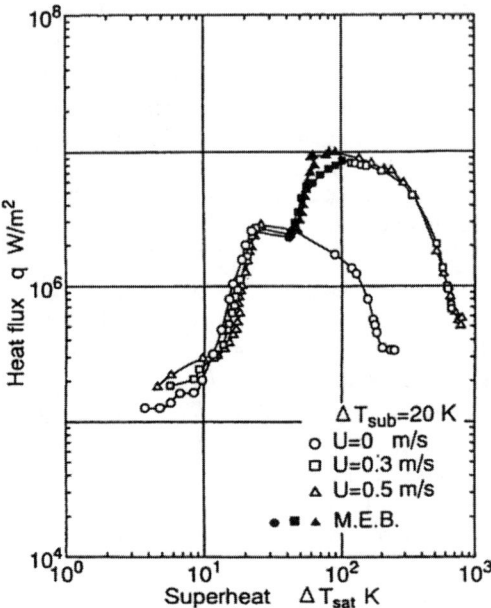

FIGURE 1. Subcooled boiling of water with parallel jet at 20 K liquid subcooling.

block is set vertically to the ground and the side surface is enveloped with insulating materials.

Heat was applied to the upper surface of the block by SiC electric heaters. Three K-type thermocouples were fitted on the center axis of the cylindrical portion with 4 mm separation from the bottom surface. The temperature of the heating surface was determined by extrapolating the temperature distribution to the heating surface,

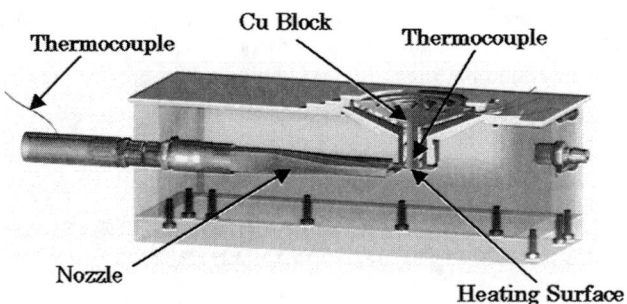

FIGURE 2. Outline of experimental set up for downward heating surface. Heating surface, 10 mm × 10 mm, copper (polished with emery #500); thermocouple, K, sheath, $d = 0.5$ mm; nozzle size, 30 mm × 3 mm.

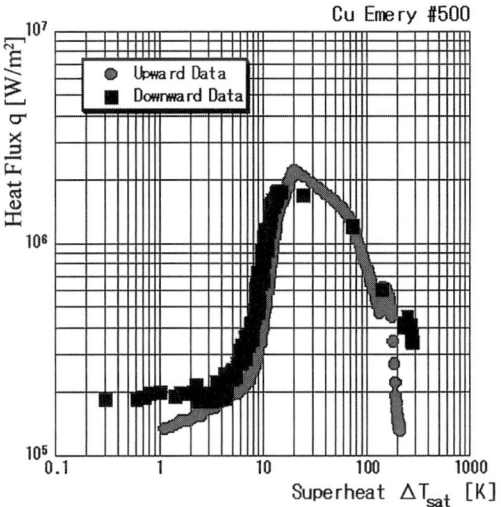

FIGURE 3. Boiling curves for upward heating surface and downward heating surface at 10 K liquid subcooling and 0.5 m/sec liquid velocity.

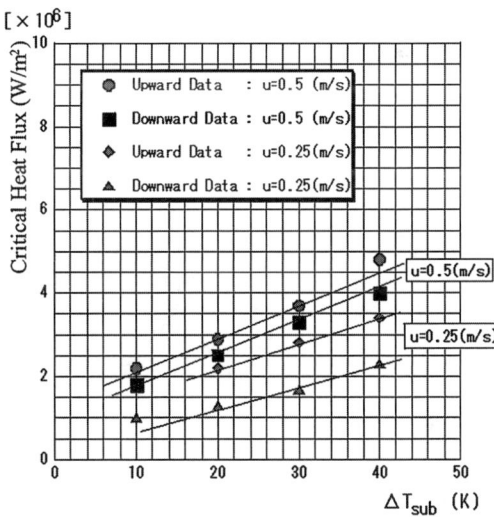

FIGURE 4. Critical heat flux for upward heating surface and downward heating surface with liquid subcooling.

and the heat flux was calculated from the temperature gradient. Subcooled liquid was directed parallel to the heating surface from a flat nozzle.

In subcooled pool boiling, the liquid velocity at nozzle exit was zero, bubbles stayed on the heating surface, and the heating surface dried out at lower heat flux than that of upward heating surface. In the subcooled boiling with a jet, however, similar boiling curves were obtained for the downward heating surface and the upward heating surface at 0.5 m/sec of liquid velocity as shown in FIGURE 3. The liquid jet is considered to accelerate bubble detachment from the heating surface for the downward heating surface.

The critical heat flux for upward and downward heating surfaces at 0.5 m/sec liquid velocity at the nozzle exit, is shown in FIGURE 4. No large differences in the heat flux are indicated at any level of subcooling. According to the experimental results, we consider that the flow boiling with strong convection give good heat transfer even under microgravity conditions.

EXPERIMENT ON SUBCOOLED FLOW BOILING IN A HORIZONTAL CHANNEL

An experimental set up consisting of a fluid loop, a heating section, a test section, and a recording section, as shown in FIGURE 5. Pure water, $1.0 \times 10^7 \Omega cm$, was circulated by a pump in the loop and regulated at predetermined temperature to within $\pm 1 K$ at the entry to the test section. The liquid flow velocity was calculated from the entry area of test section and the volume flow rate.

The rectangular channel and circular channel test sections are shown in FIGURES 6 and 7. Their dimensions are listed in TABLE 1. Both test sections were horizontal during the experiment. The rectangular channels are made from stainless steel and the heating surface, 10 mm × 10 mm, placed on the bottom of the channel. The heating surface is the top of a copper heating block and consists of a rectangular cylinder and a cone. Three sheathed K-type thermocouples, 0.5 mm diameter, are fitted on the center axis of rectangular cylinder, 4 mm from the surface.

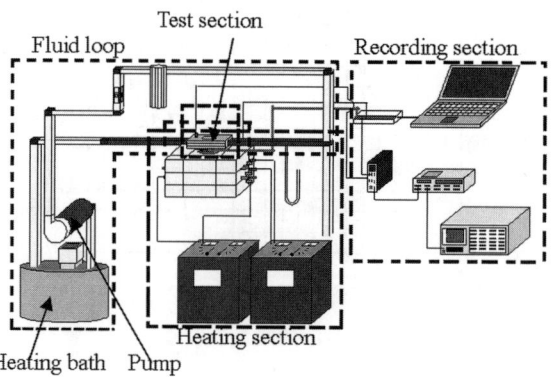

FIGURE 5. Experimental set-up for flow boiling in a horizontal channel.

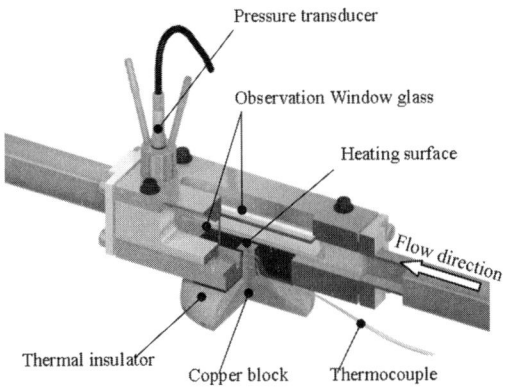

FIGURE 6. Rectangular channel test section.

Heat was applied to the bottom of the heating block from SiC electric heaters. The heating block is covered with insulating materials and the heat loss from the side surface of the block is estimated to within 3% of the calculated heat flux. The temperature of the heating surface was determined by extrapolating the axial temperature distribution, and heat flux was calculated from the temperature gradient. Liquid subcooling was tested for 10–60 K and the liquid velocity was 0.01–1.5 m/sec. Bubble behavior on the heating surface was observed through a transparent glass window assembled on the upper plate and the side part of the test section. Pressure fluctuations were measured with an electronic pressure transducer fitted at about 60 mm downstream of the heating surface.

The circular channel consists of a cylindrical heating block and a circular tube, as shown in FIGURE 7. The heating block is made from copper, and 30 cartridge electric heaters are assembled in the block. Three sheathed K-type thermocouples,

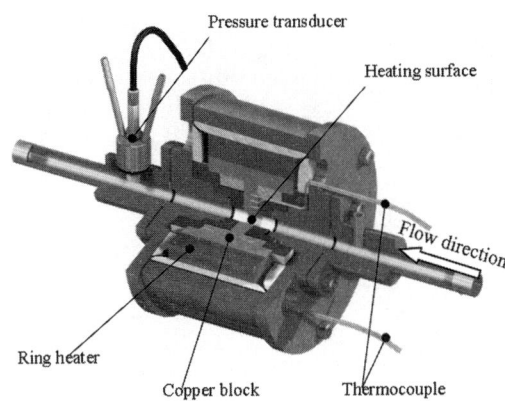

FIGURE 7. Circular channel test section.

TABLE 1. Specifications of test sections

	Height (mm)	Width (mm)	Length (mm)	Diameter (mm)
Rectangular channel	17	12	150	—
	5	14	150	—
	3	14	150	—
	1	14	150	—
Circular channel	—	—	200	10
	—	—	200	16

0.5 mm in diameter, are fitted 4 mm of radial distance from the surface to estimate surface temperature and heat flux. An electronic transducer is fitted at about 80 mm downstream of the heating surface. No observation window is assembled in the circular test section. Subcooling tested was 20 K, 30 K, and 40 K and the liquid velocity was 0.5 m/sec.

RESULTS AND DISCUSSION

Effect of Liquid Subcooling and Liquid Velocity

One effect of liquid subcooling on boiling in a horizontal rectangular channel at 0.5 m/sec liquid velocity is shown in FIGURE 8 for 17 mm × 12 mm channel cross-section to provide a typical example of boiling with MEB. At 10 K liquid subcooling, the heat flux decreases with rise in superheat and the boiling is normal. Above 20 K

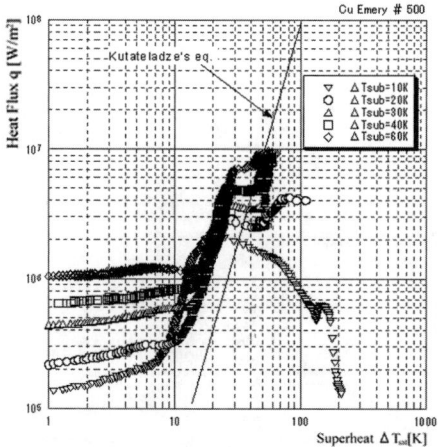

FIGURE 8. Effect of liquid subcooling on boiling in rectangular channel at 0.5 m/sec liquid velocity.

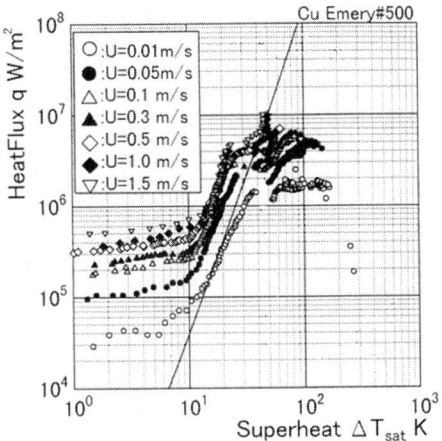

FIGURE 9. Effect of liquid velocity on boiling in rectangular channel at 20 K liquid subcooling.

liquid subcooling, MEB occurs in transition boiling and the heat flux increases more than the critical heat flux. The maximum heat flux reaches 10 MW/m².

FIGURE 9 shows the effect of liquid velocity on boiling at 20 K liquid subcooling for the same rectangular channel. At very low liquid velocity, 0.01 m/sec, MEB is not remarkable and the boiling regime is similar to normal. Above 1.0 m/sec liquid velocity, the maximum heat flux reaches 10 MW/m² for 20 K liquid subcooling. MEB occurred also in the circular channel. An example of boiling with MEB is shown for a channel with 10 mm diameter in FIGURE 10. High heat flux is maintained

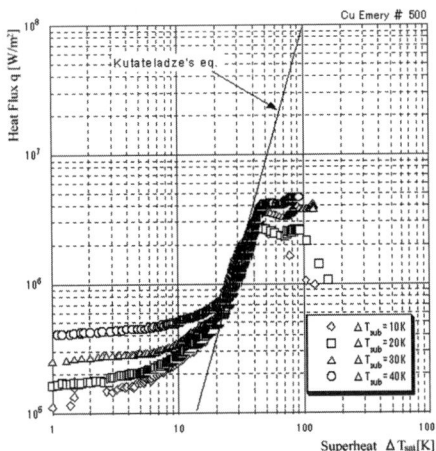

FIGURE 10. Effect of liquid subcooling on boiling in a circular channel 10 mm internal diameter at 0.5 m/sec liquid velocity.

in transition boiling above 20 K liquid subcooling, but the heat flux does not increase as in the rectangular channel. In case of the circular channel, the heating surface is the inner surface of the channel 10 mm in length and then the inner surface of the channel is covered with boiling bubbles. The vapor–liquid exchange rate is considered to be less than that for the rectangular channel.

According to the experimental results, liquid subcooling and liquid velocity are considered to be strong factors for MEB generation in subcooled flow boiling.

Category of MEB

An example of pressure fluctuation in a rectangular channel is shown in FIGURE 11. A periodic pressure fluctuation is indicated in FIGURE 11 B. In this type of MEB, a series of bubble collapses, liquid supply, and bubble growth occur periodically, as shown in FIGURE 12. According to the pressure fluctuation and the bubble behavior on the heating surface, we categorized MEB into three types, as illustrated by the bubble behavior shown in FIGURE 13, and named them MEB I, MEB II, and MEB III. In MEB II microbubbles are emitted continuously without loud sound emission. In MEB III, some of the coalesced bubbles are broken up irregularly. MEB I was observed in a circular channel as a result of pressure fluctuation. However, the causes of MEB-type occurrence are unknown from the present experiment.

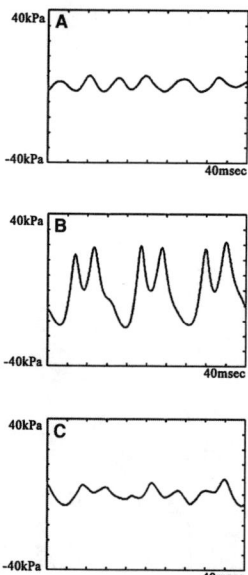

FIGURE 11. Pressure fluctuation in a rectangular channel. **(A)** MEB II, $\Delta T_m = 47.2$ K, $q = 4.03 \times 10^4$ W/m^2. **(B)** MEB I, $\Delta T_m = 56.4$ K, $q = 5.92 \times 10^4$ W/m^2. **(C)** MEB III, $\Delta T_m = 60.8$ K, $q = 7.43 \times 10^4$ W/m^2.

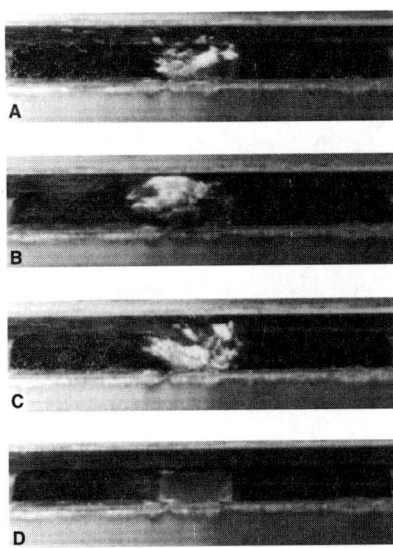

FIGURE 12. Progressive change of boiling bubbles in MEB I. Flow direction is right to left: **(A)** bubble generation, **(B)** bubble growth, **(C)** bubble collapse, and **(D)** liquid supply.

Boiling with MEB in a Small Rectangular Channel

Recently, electronic devices, such as computer CPU and IC power, have been miniaturized and their performance and input power have been increased. Heat generation from the elements has increased with the increase in driving power. It is estimated that the driving power of a CPU will grow to 500 W more than present personal computers within 10 years. It will not be possible to use a single-phase fluid, such as air or liquid, for cooling such high-powered electronic devices.

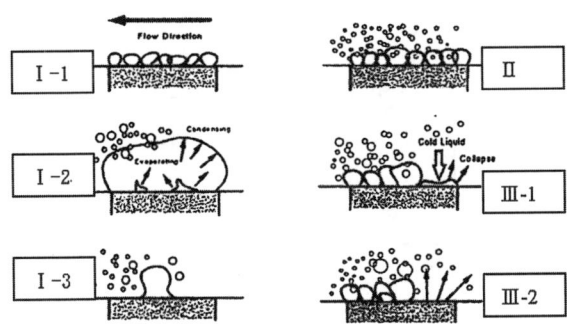

FIGURE 13. Illustration of bubble behavior in MEB. MEB I: I-1, bubble generation; I-2, bubble growth; I-3, bubble collapse. MEB II: continuous microbubble emission. MEB III: III-1, condensed bubbles; III-2, bubble collapse.

Subcooled flow boiling with MEB is considered to be effective as a future cooling method for high-powered electronic devices. In the present study, subcooled flow boiling with MEB was investigated in small rectangular channels for the purpose of obtaining fundamental data for cooling electronic devices.

Three kinds of channels were used for the experiment, as listed in TABLE 1. The heights of the channels were 1 mm, 3 mm, and 5 mm, respectively. The same heating block, with 12 mm × 17 mm cross-section, was fitted into each of the channels.

Boiling curves for the small rectangular channels at 40 K liquid subcooling and 0.5 m/sec liquid velocity are shown in FIGURE 14. The boiling curve for the rectangular channel with 12 mm × 17 mm cross-section is indicated in FIGURE 14. MEB occurred in all the channels and the heat flux increased above CHF in transition boiling. In particular, in the smallest channel (1 mm height) heat flux increased smoothly and reached 7 MW/m^2. This high heat flux is more than 20 times the heat generation of a computer CPU, and it is more than seven times the heat generation for power devices.

Examples of pressure fluctuation in the small channels are shown in FIGURE 15. Only MEBI, the periodic type of MEB, occurred in the channels with height 5 mm and 3 mm, as shown in FIGURE 15. The maximum pressure in MEBI rose for the smaller channel. In the smaller channel, coalesced bubbles expand and almost all the channel space is filled with expanded bubbles.

The relation between heat flux and pressure frequency in MEBI is shown in FIGURE 16. Heat flux increases with increasing pressure frequency. For the rectangular channel with 12 mm × 17 mm cross-section, the heat flux increased linearly with pressure frequency for a liquid subcooling of 30 K, 35 K, and 40 K in FIGURE 16. In

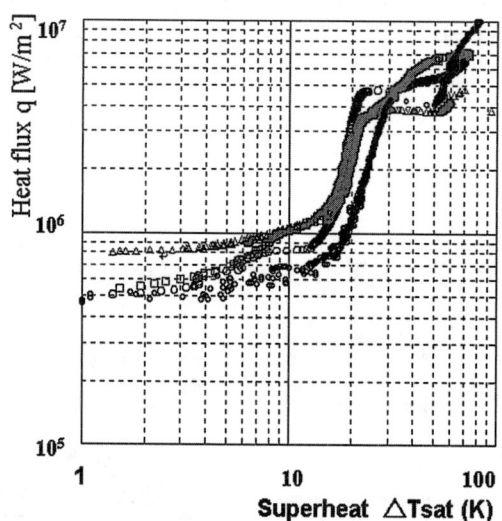

FIGURE 14. Boiling curves for small rectangular channels at 40 K liquid subcooling and 0.5 m/sec liquid velocity. △, 5 mm, $Re = 7,756$; ○, 3 mm, $Re = 5,201$; □, 1 mm, $Re = 1,965$; ●, reference values (12 mm × 17 mm).

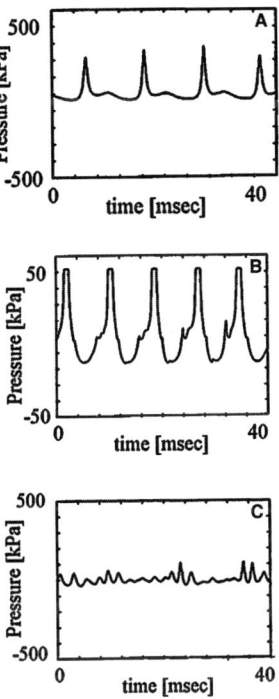

FIGURE 15. Pressure fluctuation in small rectangular channels. (**A**) 5 mm × 14 mm. (**B**) 3 mm × 14 mm. (**C**) 1 mm × 14 mm.

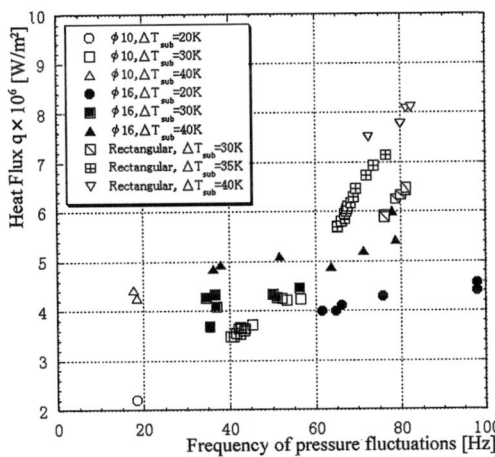

FIGURE 16. Relation between heat flux and pressure frequency in horizontal channels during subcooled flow boiling with MEB I.

MEB I, the pressure frequency is considered to be the frequency of liquid supply and the heat flux q_{MEBI} and is given by the following simple relation.

$$q_{\text{MEBI}} \propto \frac{\rho_l V_v (C_p \Delta T_{\text{sub}} + h v) f}{A} + q_{\text{CHF}}, \quad (1)$$

where A denotes the area of the heating surface (m^2), C_p is the specific heat of the liquid (J/kgK), f is the pressure frequency (Hz), h_v is the latent heat of the liquid (J/kg), ΔT_{sub} is the liquid subcooling (K), V_v is the maximum volume of coalesced bubbles on the heating surface just before collapse (m^3), ρ_l is the density of supplied liquid (kg/m^3), and q_{CHF} is the critical heat flux (W/m^2).

The heat flux and pressure frequency curves are different for the channel configurations and liquid subcooling, as shown in FIGURE 16. We consider that the maximum volume V_v of coalesced bubbles just before collapse depends on the mass of subcooled liquid, channel configurations, and liquid velocity. However, the heat flux and pressure frequency relation may show one of the mechanisms of microbubble emission boiling, as indicated by Equation (1).

According to the experimental results, subcooled flow boiling with microbubble emission boiling gives stably higher heat flux than critical heat flux in transition boiling, and it is also expected to be very effective for high heat flux cooling and high heat flux transfer in microgravity environment.

CONCLUSIONS

In the experimental study on subcooled flow boiling of water in a rectangular horizontal channel and a horizontal circular channel, microbubble emission boiling occurred in transition boiling. In a rectangular channel, the heat flux increased more than the critical heat flux, and the maximum heat flux reached 10 MW/m^2. For small rectangular channels with cross-sections 14 mm × 1 mm, 14 mm × 3 mm, and 14 mm × 1 mm, microbubble emission boiling occurred and the maximum heat flux was 4–7 MW/m^2. This heat flux is more than 20 times that of the heat generated by a computer CPU and more than seven times that of power devices for electric power regulating systems.

Microbubble emission boiling can be categorized in three types depending on observed bubble behavior and pressure fluctuation in the channel. In the periodic type of microbubble emission boiling, heat flux increased with increasing frequency of pressure fluctuation. In the convection boiling with a subcooled water jet, similar boiling curves were observed for upward and downward heating surfaces, and no difference was indicated in the critical heat flux. Flow boiling with strong convection is expected to be effective for microgravity use, and especially high heat flux transfer would be expected in microbubble emission boiling.

ACKNOWLEDGMENTS

Part of the present series of experiments was performed for an international research project founded by New Energy and Industrial Technology Development Organization of Japan (NEDO) in 2000. The authors greatly appreciate the help of

NEDO and the research members, Professors H. Kawamura, H. Ohta, M. Ishizuka, W. Grassi, P. Marco, and J. Straub; Dr. Y. Abe; Mr. H. Iwasaki, K. Kawano, and Y. Horie.

REFERENCES

1. KUTATELADZE, S.S. 1952. Heat transfer in condensation and boiling. USAEC Rep. AEC -tr-3770.
2. ZUBER, N. 1958. On stability of boiling heat transfer. Trans. ASME J. Heat Transfer **80:** 711–720.
3. STRAUB, J., M. ZELL & B. VOGEL. 1991. Boiling under microgravity conditions. Proceedings of the 1st European Symposium on Fluid in Space, Ajacco, France, ESA SP-353. 269–297.
4. SUZUKI, K., H. KAWAMURA, et al. 1997. Burnout in subcooled pool boiling of water under microgravity. Proceedings of Joint Xth European and VIth Russian Symposium on Physical Science in Microgravity, St. Petersburg, Russia. 366–369.
5. SUZUKI, K., H. KAWAMURA, et al. 1999. Experiments on subcooled pool boiling of water in microgravity: observation of bubble behavior and burnout. Proceedings of the 5th ASME and JSME Joint Thermal Engineering Conference. San Diego. USA. AJTE99-6420.
6. SUZUKI, K.,Y. KOYAMA & H. KAWAMURA. 1999. Burnout heat flux in microgravity. Proc. UEF Conference on Microgravity Fluid Physics and Heat Transfer. Hawaii. 144–150.
7. SUZUKI, K. K. TORIKAI, et al. 1999. Boiling heat transfer of subcooled water in a horizontal rectangular channel: observation of MEB and MEB generation. Trans. JSME **65**(637): 161–168.
8. SUZUKI, K., K. SAITO, T. SAWADA & K. TORIKAI. 2000. An experimental study on microbubble emission boiling of water. Proc. 3rd European Thermal Sciences Conference. Heidelberg. Vol. 2. 815–820.
9. TORIKAI, K., K. SUZUKI, A. SUZUKI & T. WATANABE. 1994. Microbubble emission in subcooled transition boiling. Proc. 3rd International Symposium on Multiphase Flow and Heat Transfer, Xi'an, China. 70–77.
10. INADA, S., Y. MIYASAKA, R. IZUMI & Y. OWASE. 1981. Studies on subcooled pool boiling: 1st rep. effect of liquid subcooling on local heat transfer characteristics. Trans. JSME **47**(417): 952–960.
11. KUBO, R. & S. KUMAGAI. 1992. Occurrence and stability of microbubble emission boiling. Trans. JSME **58**(546): 497–502.

Dynamics of Single and Multiple Bubbles and Associated Heat Transfer in Nucleate Boiling Under Low Gravity Conditions

D. QIU,[a] G. SON,[b] V.K. DHIR,[a] D. CHAO,[c] AND K. LOGSDON[c]

[a]*Department of Mechanical and Aerospace Engineering,
University of California at Los Angeles, Los Angeles, California, USA*

[b]*Department of Mechanical Engineering, Sugang University, Seoul, Korea*

[c]*NASA Glenn Research Center, Microgravity Division, Cleveland, Ohio, USA*

ABSTRACT: Experimental studies and numerical simulation of growth and lift-off processes of single bubbles formed on designed nucleation sites have been conducted under low-gravity conditions. Merging of multiple bubbles and lift-off processes during boiling of water in the parabola flights of KC-135 aircraft were also experimentally studied. The heating area of the flat heater surface was discretized and equipped with a number of small heating elements that were separately powered in the temperature-control mode. As such, the wall superheat remained nearly constant during the growth and departure of the bubbles, whereas the local heat flux varied during the boiling process. From numerical calculation it is found that peak of heat flux occurs locally at the contact line of bubble and heater surface. Dry conditions exist inside the bubble base area, which is characterized through a zero heat flux region in the numerical calculation and a lower heat flux period in the experimental results. During the merger of multiple bubbles, dry-out continues. In both the numerical calculations and experimental results, the bubble lift-off is associated with an apparent increase in heat flux. Wall heat flux variation with time and spatial distribution during the growth of a single bubble from numerical simulations are compared with experimental data.

KEYWORDS: dynamics; single bubble; multiple bubbles; heat transfer; nucleate boiling; low gravity

INTRODUCTION

Nucleate boiling is a liquid–vapor phase change process associated with bubble formation and is known to be a highly efficient mode of heat transfer. For space applications boiling is the heat transfer mode of choice since the size of the components can be significantly reduced for a given power rating. Applications of boiling heat transfer in space can be found in such areas as thermal management, fluid handling and control, and power systems. For space power systems based on the Rankine cycle (a representative power cycle), the key issues that need to be addressed are

---

Address for correspondence: V.K. Dhir, Interim Dean, School of Engineering, 48-121 Engineering, University of California at Los Angeles, Los Angeles, CA 90095-1597, USA.
vdhir@seas.ucla.edu

the magnitude of boiling heat transfer coefficient and the critical heat flux under low gravity conditions. The low gravity environment of the KC-135 aircraft provides a less expensive means to experimentally study the liquid–vapor phase change dynamics and the associated heat transfer characteristics and to validate numerical simulations for space applications.

Although, in the past decades, extensive experiments and modeling efforts have been made for boiling processes, a mechanistic model to describe the phenomena and predict the heat transfer coefficient without employing empirical constants has not been available for nucleate boiling on a horizontal surface. As such, the scaling of the effect of gravity on the dynamics and heat transfer during boiling has not been established. Consequently, one cannot extend the results from studies at Earth normal gravity to nucleate boiling under microgravity conditions.

Under microgravity conditions the early data of Keshock and Siegel[1,2] on bubble growth and heat transfer show that the effect of reduced gravity is to reduce the buoyancy and inertia forces acting on a bubble. As a result, under reduced gravity, bubbles grow larger and stay longer on the heater surface. This in turn leads to merger of bubbles on the heater surface and existence of conditions similar to those for fully developed nucleate boiling. Thus, it was concluded that, under microgravity conditions, partial nucleate boiling region may be very short or non-existent.

Merte et al.[3] and Lee and Merte[4] reported results from pool boiling experiments conducted in the space shuttle for a surface of gold film sputtered on a quartz plate with an area $19 \times 38\,\text{mm}^2$. Subcooled boiling during long periods of microgravity was found to be unstable. The surface was found to dry out and rewet at higher surface heat flux. At 11.5°C subcooling, it was observed that after a period of growth of small bubbles and their coalescence, a larger bubble formed that stayed above the surface. The larger bubble acted as a reservoir sucking and removing the smaller bubbles from the surface. The average heat transfer coefficients during the dry-out and rewetting periods were found to be about the same. The nucleate boiling heat fluxes were higher than those obtained on a similar surface at Earth normal gravity, g_e, conditions. It was concluded that subcooling exerts a negligible influence on the steady state microgravity heat transfer coefficient. The effect of Marangoni convection (thermocapillary flow) on the heat transfer was considered to be reduced at microgravity conditions.

Straub, Zell, and Vogel[5] and Straub[6] conducted a series of nucleate boiling experiments using thin platinum wires and a gold-coated flat plate as heaters under low gravity conditions in the flights of ballistic rockets and KC-135. In the experiments for the flat plate heater, with R12 as the test liquid, boiling curves similar to those at g_e were obtained when the liquid was saturated. With R113, rapid bubble growth and large bubbles ($D \geq 2\,\text{cm}$) were observed during the TEXUS ballistic rocket flight at a gravity level of $10^{-4} g_e$. However, no bubbles were seen to lift off from the heater surface, and neither temperature nor heat flux reached a steady state. For subcooled R113 and R12, a reduction in heat transfer coefficient of up to 50% in comparison with that at g_e was obtained. It was observed that larger bubbles occupied the surface and at their edges many smaller bubbles formed, coalesced, and fed the larger ones.

From above description it appears that results from studies conducted to date are inconclusive. Questions remain concerning the stability of nucleate boiling, the roles of liquid microlayer underneath the stationary and translating bubble, the reasons for equivalence of magnitudes of heat transfer coefficients at normal gravity and low gravity conditions, and the overall physics that underlies these phenomena. As such, there is no mechanistic model that describes the observed physical behavior and the dependence of nucleate boiling heat flux on gravity and wall superheat during nucleate boiling. Detailed information on heat transfer characteristics associated with processes of bubble growth and lift-off at a well defined nucleation site is, therefore, needed. However, these characteristics for isolated nucleate boiling is least known.

Rogers and Mesler[7] measured the transient surface temperature in nucleate boiling in the attempt to explore the behavior of the liquid microlayer underneath the bubbles in water at terrestrial gravity using an artificial cavity and thermocouple made on a nichrome strip plated with a thin nickel layer (heater surface). They found that the surface temperature at the nucleation site occasionally dropped 11–16°C in about 2 msec and the temperature recovered in 10–20 msec during single bubble boiling when the heating rate remained constant (5–10 W/cm^2). It was concluded that bubble departure corresponds to temperature recovery and that the sudden temperature drop corresponds to initiation of bubble growth. However, they reported that no significant cooling is apparent as liquid returns to the surface during bubble departure. This fact is opposite to most hypotheses on nucleate boiling.

Recently, Kim and Benton[8] conducted experiments of subcooled nucleate boiling of FC72 on array of miniature serpentine platinum heaters with feedback controls in terrestrial and low gravity conditions. The overall boiling phenomena are observed to be similar to those reported by Merte et al.[3,4] Larger primary bubbles acted as a reservoir, sucking and removing the smaller bubbles from the surface. Large local heat flux fluctuations, for example, from 2 to 20 W/cm^2 for $\Delta T_{sub} = 15°C$ were recorded. A dry spot underneath the primary bubble was observed. The peak in heat flux occurred when the liquid rewetted the surface after the bubble slid over the measuring position. However, the heat flux over the total bubble life was not obtained.

In this paper, efforts to explore the local heat flux associated with bubble growth, merger, and lift-off on the designed nucleation sites are described both by means of numerical simulation and experiments under low gravity conditions.

DESCRIPTION OF NUMERICAL ANALYSIS

The numerical simulation model employed in this work is the same as that developed by Son, Dhir, and Ramanajapu[9] to study bubble dynamics in a saturated liquid at Earth normal gravity. In the model a single bubble formed at a nucleation site is considered and the assumption of axisymmetry is invoked. For hydrodynamic and thermal analysis, the domain of interest is separated into micro- and macroregions, as shown in FIGURE 1. The microregion mainly encompasses the microlayer underneath the bubble. Heat from the solid surface is conducted through the microlayer and is used for evaporation at the interface. Forces acting on the liquid in the microlayer are those due to viscous drag, interfacial tension, long range molecular interactions, and vapor recoil. Neglecting inertia and convection terms in the momentum and energy equations, respectively, the radial thickness of the microlayer during the evolution of the bubble is calculated. For the macroregion, complete conservation equations for both vapor and liquid are used. The level set method is employed to obtain the shape of the evolving interface and the flow and temperature fields in the

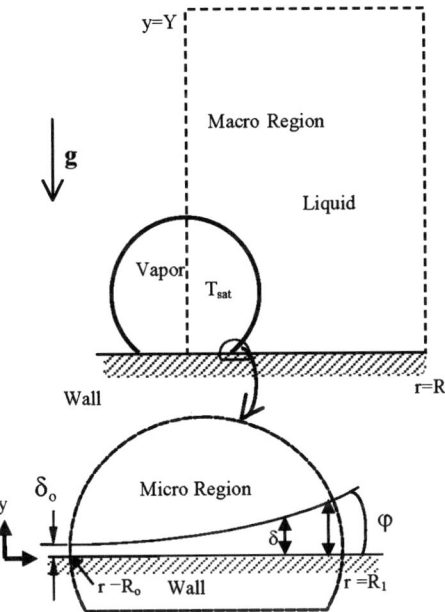

FIGURE 1. Macro- and microregion of the numerical simulation.

vapor and the liquid. The shape of the interface obtained from the solutions for the micro- and macroregions was matched at the outer edge of the microlayer. Also, it was required that tangent to the interface at that location should yield the macroscopic contact angle. The magnitude of the Hamakar constant representing the long range forces was, in turn, related to the macroscopic contact angle. Thus, the effect of wettability of the surface was included in the model through the specification of the value of the Hamakar constant. For the objective of this paper, the calculation was carried out for the variation of surface heat flux during boiling of saturated water at $\Delta T_w = 8.5°C$ and $g_z = 0.01 g_e$.

EXPERIMENTS

The experimental apparatus for the KC-135 flights is shown schematically in FIGURE 2. The system configuration is similar to that used by Merte *et al.*[3] It consists of a test chamber ($D = 15$ cm, $H = 10$ cm), a bellows and a nitrogen (N_2) chamber. Three glass windows are installed on the walls of the test chamber for the visual observation. To control the system pressure transducers are installed in the test chamber and N_2 chamber. The test surface for studying nucleate boiling is installed at the bottom of the test chamber. In the vicinity of the heater surface a rake of six thermocouples is installed in the liquid pool to measure the temperature in the thermal boundary layer, another rake is placed in the upper portion of the chamber to measure the bulk liquid temperature.

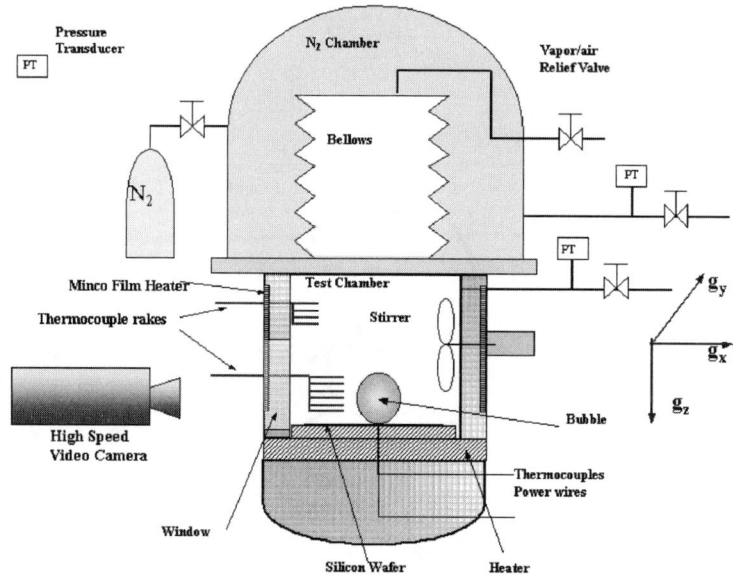

FIGURE 2. Schematic of the experimental set-up.

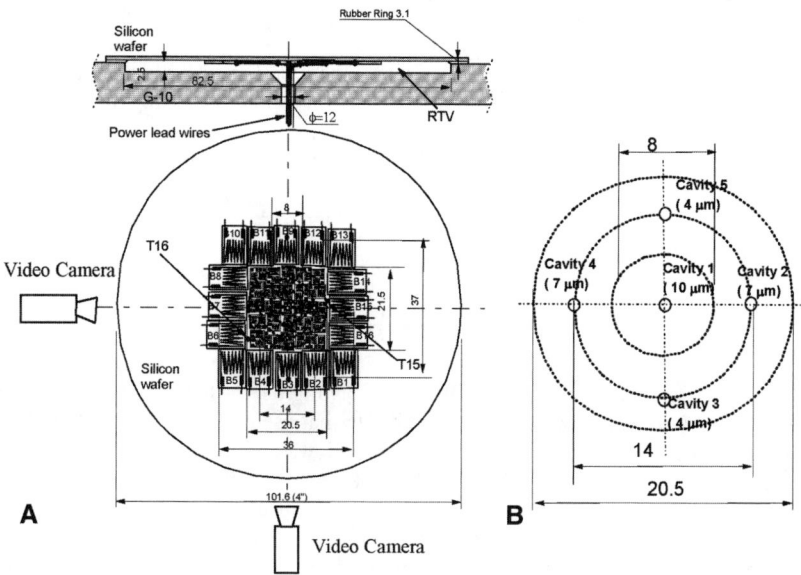

FIGURE 3. Silicon wafer instrumented with strain gage heaters and thermocouples: **(A)** overview, **(B)** cavity positions.

Distilled, filtered, and degassed water was used as the test liquid. Two video cameras operating at 250 frames/second were installed from two directions at an angle of 90° to record the boiling processes. The liquid temperature and pressure in the chamber were controlled according to the set points established by the operator on board. A three-component accelerometer is installed on the frame on which the equipment is mounted.

A polished silicon wafer, 101.6 mm in diameter and 400 μm in thickness, was used as the test surface for the nucleate boiling experiments. From the manufacturer's specification the roughness of the bare polished wafer is less than 5 Å. Five cylindrical cavities with diameters 10 μm (No. 1), 7 μm (Nos. 2 and 4), and 4 μm (Nos. 3 and 5) were etched to a depth of about 100 μm in the wafer center via the deep reactive ion etching technique (DRIE), see FIGURE 3 B.

At the back of the Silicon wafer strain gages were bonded as heating elements. In the central area, miniature elements of size $2 \times 2\,mm^2$ were used and grouped so as to cover small areas of the test surface (FIG. 3 A). In each group, a thermocouple was directly attached to the wafer. The heater surface temperatures in different regions were then separately controlled through a multichannel feedback control system. The surface superheat could be maintained constant during an experimental run and can be automatically changed to the desired set-point. The local heating rate in the individual small area was recorded during the experiments. In the outer area larger elements with an effective heating area of $6.5 \times 6.5\,mm^2$ per element were used. The power lead wires and the thermocouple wires were led out from the hole in the base made from phenolic garolite grade 10 (G-10). The wafer was cast with RTV, silicon rubber, on the G-10 base. The base in turn is mounted in the test chamber.

During the flight experiments, the cavity was activated by energizing the heating elements underneath the cavity before the plane took off. The wall superheat at the cavity area was gradually increased until the inception occurred. After activation, the wall superheat was set at the desired value. The detailed procedure can be found in Qiu et al.[10]

The low gravity condition, $g_z \approx \pm 0.04 g_e$ during the parabolic flights of KC-135 lasted about 20 seconds. Uncertainty in the gravity level measurement is mainly from the synchronization uncertainty between the accelerometer recording and the high speed video recording. The gravity level g_z itself varied within $\pm 0.04 g_e$ due to flight conditions. An examination using a laser pulse with signal recorded in the data files and pulse light displayed in the view of the video recording shows a mismatch in time of less than 0.5 sec, which corresponds to the maximum uncertainty of $\pm 0.002 g_e$ in the level of gravity, g_z, normal to the test surface. The uncertainty in the pressure measurement is $\pm 0.0034\,MPa$. This value yields an uncertainty in evaluating the saturation temperature of $\pm 0.1°C$. The temperature measurements of the heater surface and the liquid have an accuracy of $0.1°C$. The uncertainty in the evaluation of bubble size is $\pm 0.05\,mm$.

RESULTS OF NUMERICAL SIMULATION

FIGURE 4 shows the bubble contours during a typical bubble growth and lift-off cycle. During most time of the bubble life (0–1.39 sec in 1.77 sec) the bubble base is either expanding outward or remaining constant at the maximum diameter. Thereafter, the bubble base begins to shrink rapidly, which indicates that the bubble starts to lift from the surface.

The corresponding heat flux distributions are shown in FIGURE 5 for selected times. A local peak in heat flux is seen to exist at each time before bubble lift-off.

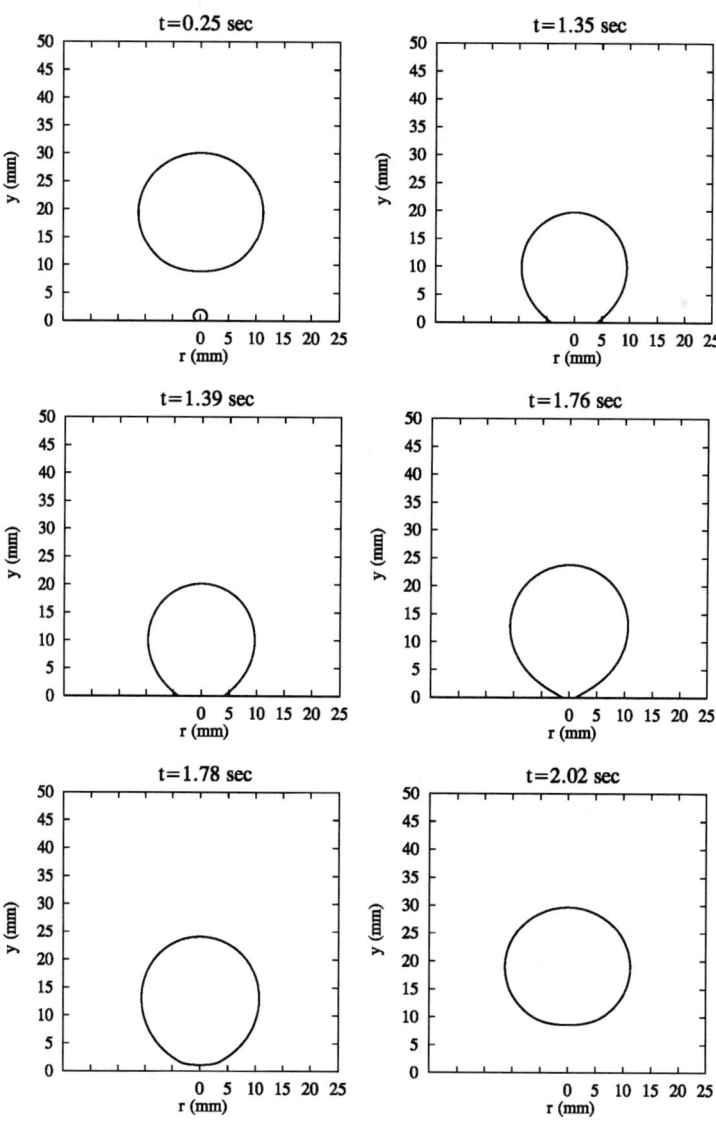

FIGURE 4. Bubble growth pattern at $\Delta T_w = 8.5°C$ and $g_z = 0.01 g_e$.

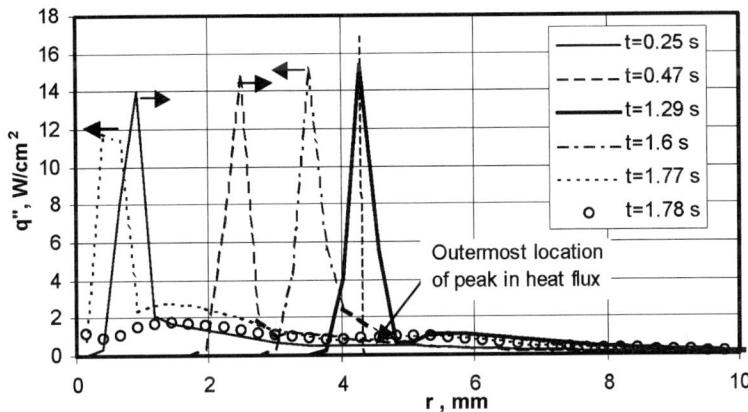

FIGURE 5. Local heat flux from numerical simulation for $\Delta T_w = 8.5°C$ and $g_z = 0.01 g_e$.

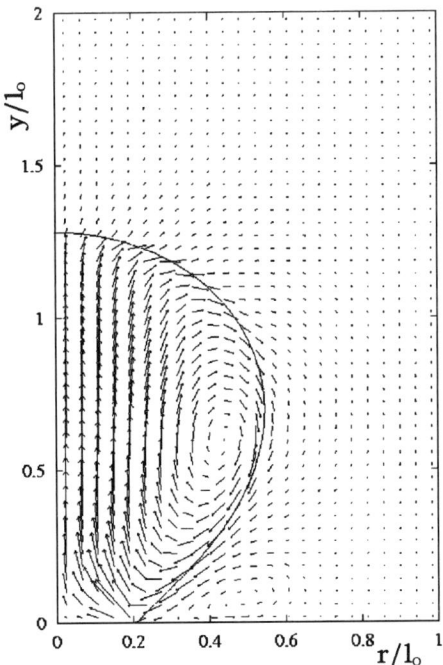

FIGURE 6. Velocity field at $t = 3.4$ sec, $T_w - T_s = 8$ K, $\Delta T_{sub} = 0$ K, $g_z = 0.02 g_e$.[11]

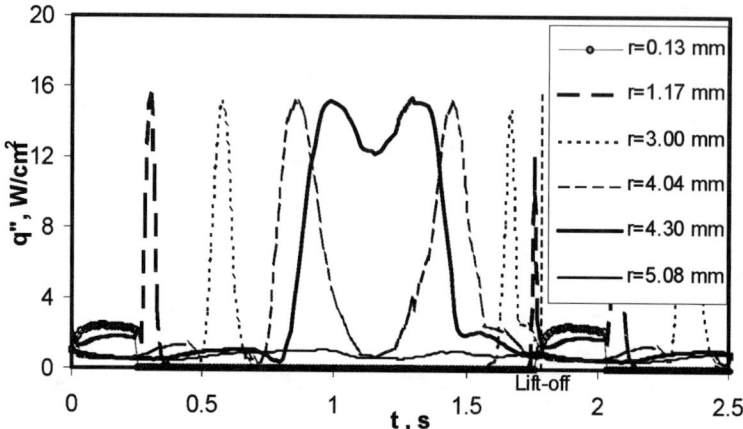

FIGURE 7. Heat flux during a bubble growth/lift-off cycle as a function of time in selected radii.

The peak first moves outward, reaches a position of maximum radius, and then moves back toward the cavity ($r = 0$). The location of the peak approximately corresponds to the position of the contact line between the bubble and the surface. The farthest position of the peak from the center of the cavity occurs when the bubble base diameter acquires its maximum value. The high heat flux at the contact line region is caused by the high rate of evaporation of liquid into the bubble. However, during the lift-off phase of the bubble, as the bubble base shrinks, inflow of cold liquid leads to enhanced heat transfer in the region surrounding the peak. An example of the flow field calculated by Singh and Dhir[11] is given in FIGURE 6. In the area away from the location of peak heat flux, the heat flux decreases with the radius as the thermal boundary layer thickens.

In the region with smaller radius than that of the location of peak in heat flux, the heat flux is very small. A dry region underneath the bubble is assumed to exist in the model. The transition from the maximum in heat flux to the dry region with almost zero heat is seen to occur in a very short distance.

A particular location on the heater surface experiences two peaks in heat flux. One occurs during the bubble base expansion and the other occurs during the lift-off period. This can be seen in FIGURE 7, which shows the heat flux histories at few selected positions. In the present case, the most outward position of the heat flux peak occurs at radius $r \approx 4.3$ mm. Near the center of the calculation domain where the cavity is located, the longest period of low heat flux exists.

EXPERIMENTAL RESULTS

Since the details on single and multiple bubble dynamics at low gravity are reported elsewhere,[10] only the relevant bubble behavior is described here.

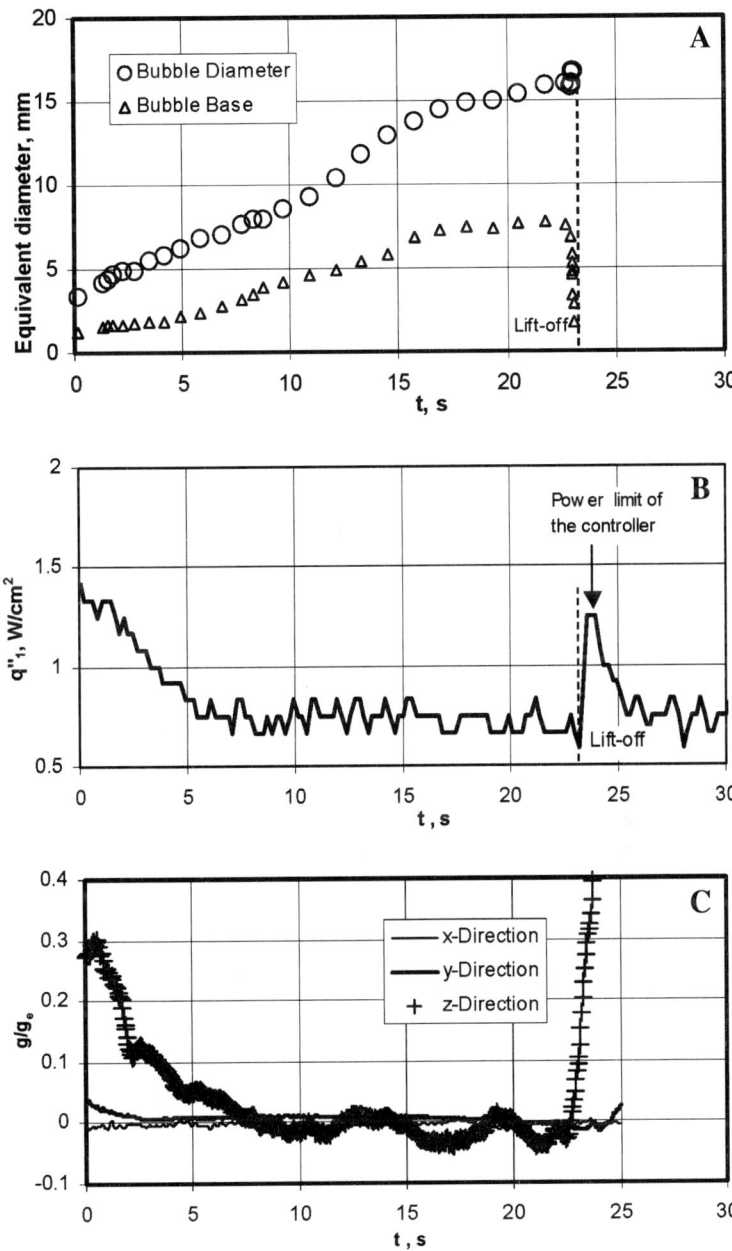

FIGURE 8. Growth rate and heat flux of a single bubble at $\Delta T_w = 5.0°C$, $\Delta T_{sub} = 1.6°C$: **(A)** bubble equivalent diameter, **(B)** heat flux from the elements underneath the cavity, **(C)** gravity level.

Single Bubble Growth and Lift-off

FIGURE 8A shows the bubble equivalent diameter and bubble base diameter during growth and lift-off process for a single bubble on cavity 1. The maximum bubble base diameter is about 7 mm, whereas the bubble lift-off diameter is 16.2 mm. About the same value of bubble lift-off diameter is obtained when scaling is carried out for the gravity level of $g_z = 0.025 g_e$.[10] The corresponding heat flux from the elements underneath cavity 1 is shown in FIGURE 8B. The heat flux value is obtained from the recorded total heat flow to the two heating elements directly under the cavity divided by the area of the elements. It is seen that the heat flux decreases to a nearly constant value during bubble growth. Formation of a dry spot underneath the bubble is believed to be the cause of this behavior. A non-zero value of the power input is a result of heat loss into the backing material and radially outward conduction along the wafer, because the temperature in the surrounding area is set at slightly lower value than the cavity area. The heat flux attains a nearly constant value when the bubble base radius attains its maximum value and remains at that value until just before lift-off of the bubble. A peak in heat flux occurs just after lift-off. The top part of the peak in heat flux is not seen here because this value exceeded the maximum power input to the heater, which was limited to prevent the heating elements from burn-out. It should be noted that, because of thermal inertia of the heater, there is always a delay in the change in power input and in turn in the recorded peak in heat flux. The peaks are believed to be associated with transient conduction into the cooler liquid that flows inward as the bubble lifts off from the heater surface. The segmented heat flux distribution as a function of radius is plotted in FIGURE 9. This shows that the high heat flux mainly appears in the region $r \leq 4$ mm, which corresponds to the maximum base area. There is a shift with time between the occurrence of highest heat fluxes in the regions 0–2 mm and 2–4 mm in radius. This fact reflects the radial outward and inward movement of the local heat flux peak within the region corresponding to maximum bubble base area.

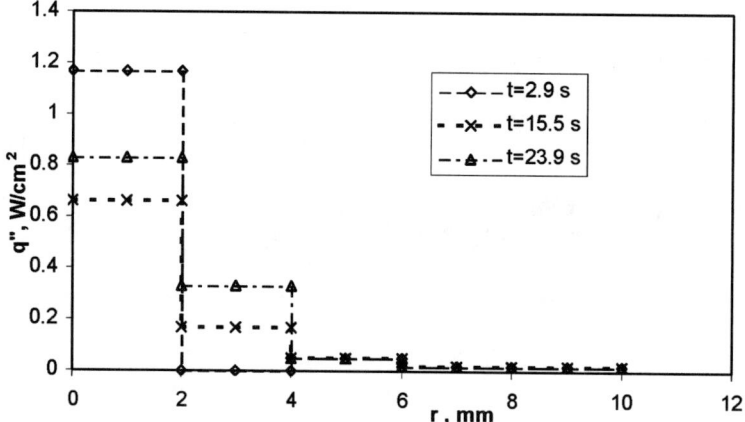

FIGURE 9. Distributions of heat flux at $\Delta T_w = 5.0°C$, $\Delta T_{sub} = 1.6°C$.

FIGURE 10. Pictures during growth and lift-off of three single bubbles at $\Delta T_w = 6.2°C$ and $\Delta T_{sub} = 0.8°C$.

Similar behavior was seen in all cases related to growth and lift-off of single bubbles. FIGURE 10 shows the pictures of three single bubbles during growth and lift-off processes. The dynamic behavior of each of these bubbles was found to be similar to that of a single bubble described above and also previously described by Qiu, et al.[10] The bubbles lifted off when buoyancy exceeded the surface tension force. Because of the phase difference in the growth of the bubbles and time variations in the level of gravity normal to the heater surface, the lift-off diameters are different, as is shown in FIGURES 11 and 12. The corresponding heat flux under cavity 1 is

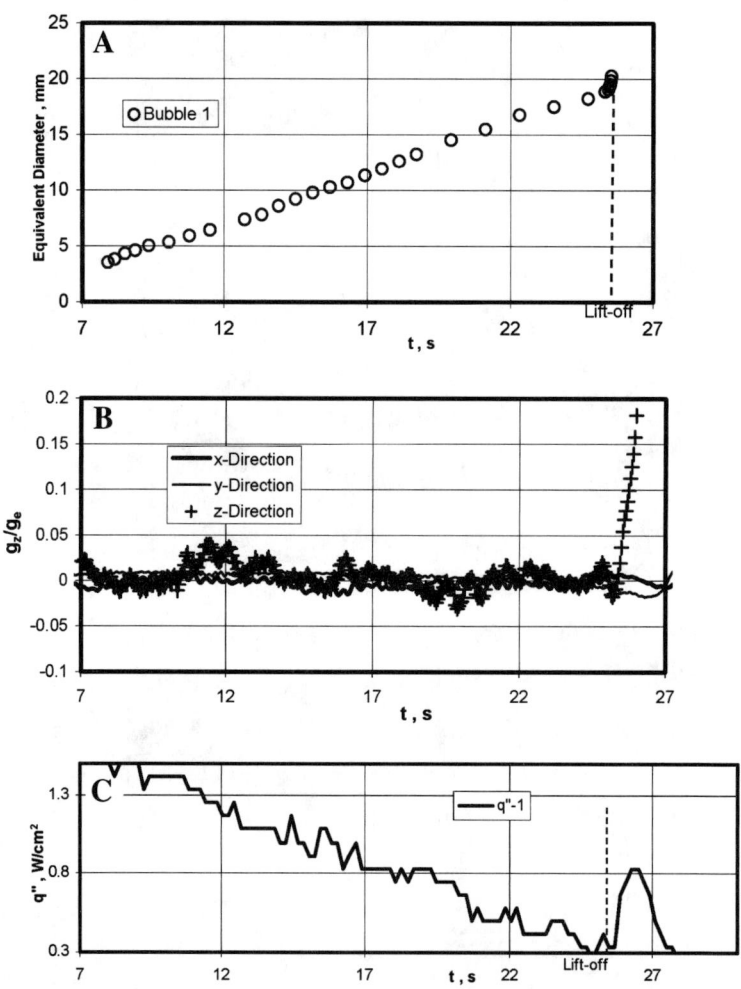

FIGURE 11. Histories of single bubble number 1 at $\Delta T_w = 6.2°C$, $\Delta T_{sub} = 0.8°C$: (**A**) bubble equivalent diameter, (**B**) gravity level, (**C**) heat flux from the heating elements under cavity 1.

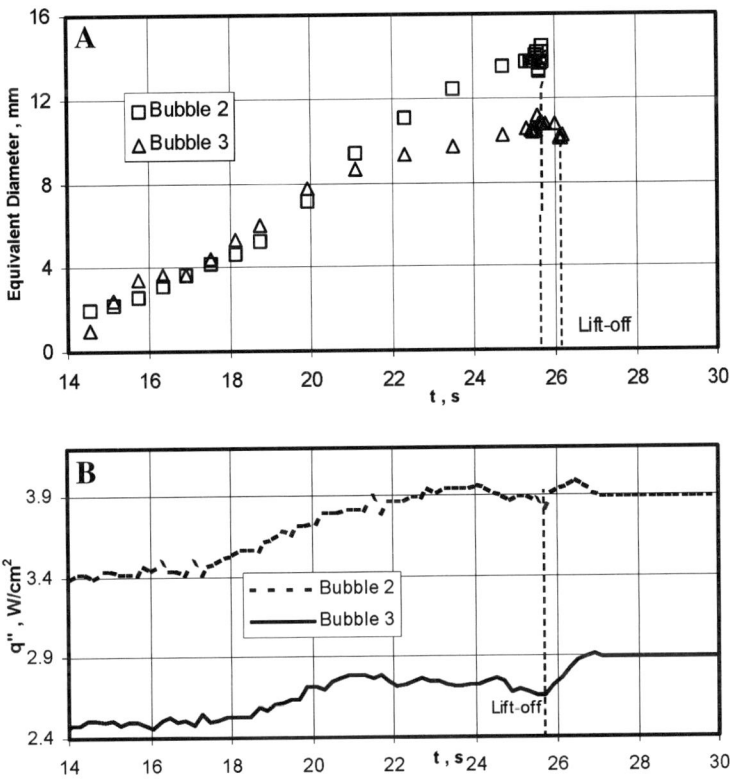

FIGURE 12. Bubble growth history and heat flux for bubbles 2 and 3 at $\Delta T_w = 6.2°C$, $\Delta T_{sub} = 0.8°C$: **(A)** bubble equivalent diameter, **(B)** heat flux for bubbles 2 and 3.

plotted in FIGURE 11C. It can be seen that, during the growth of the bubble, the heat flux decreases and just after lift-off a peak in heat flux occurs. Similar peaks also occur during the lift-off of bubbles 2 and 3, as shown in FIGURE 12B. The higher level of sustained heat flux after the occurrence of the peaks is believed to be due to increased natural convection heat transfer as gravity increases after the plane leaves the parabola.

The continuous decrease of heat flux for bubble 1 during the growth period is a result of the reduced heat transfer in the dry area underneath the bubble as bubble base expands. The non-zero value of heat flux during bubble growth is again due to the radially outward conduction into the silicon wafer and the heat loss into the backing materials. Because the heaters placed under cavities for bubbles 2 and 3 are not symmetric, the conductive heat transfer to the surrounding area more strongly affects the profile of input power to these heaters. Consequently, during the later stages of growth of the bubbles, a small increase in heat flux is observed. This is somewhat contradictory to the observation made for cavity 1 placed in the center of the wafer.

Growth, Merger, and Lift-off of Multiple Bubbles

The heat flux during merger and lift-off processes of multiple bubbles was also measured. In FIGURE 13 the results of two bubbles merger are plotted. The bubble equivalent diameter is plotted in FIGURE 13A and the heat flux underneath cavities 1 and 4 is plotted in FIGURE 13B. It can be seen that the heat flux under cavity 1 decreases and then rapidly increases after the bubble lift-off. The heat flux profile for cavity 4 shows similar behavior, but in a less pronounced way. There is no apparent change in heat flux at the time of merger. This fact indicates that dry areas underneath the merging bubbles continue to persist, even during the merger process. The peak in heat flux after the lift-off for cavity 4 is not as pronounced as that for cavity 1. This may again be the result of asymmetric location of cavity 4 with respect to the center of the wafer. FIGURE 14 shows the area-averaged heat flux in various rings of

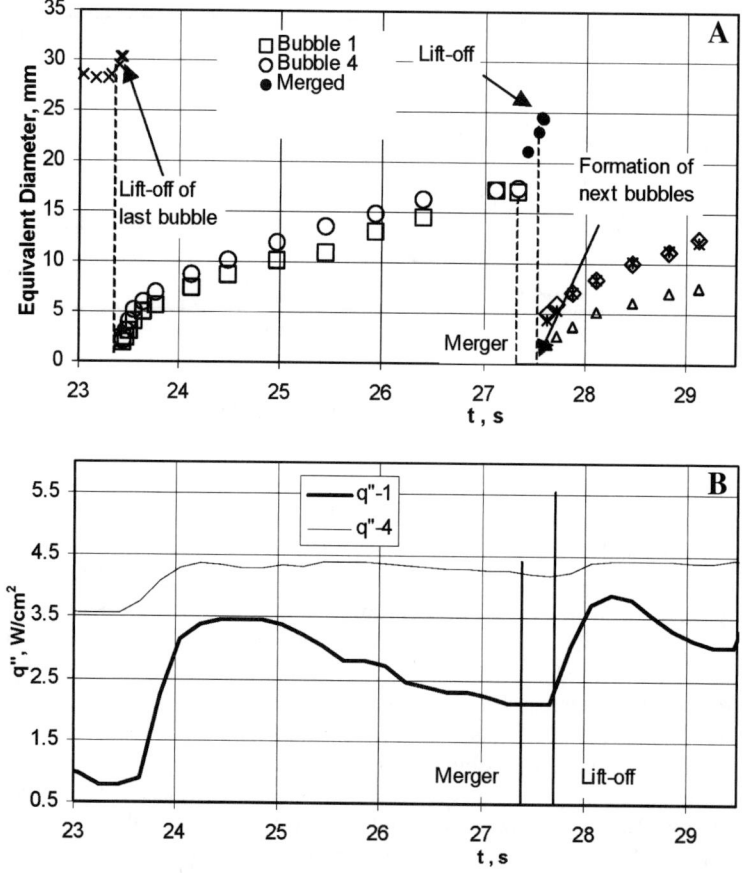

FIGURE 13. Bubble diameters and heat fluxes during two bubble merger and lift-off: (**A**) bubble equivalent diameter, (**B**) heat fluxes from elements under the cavities.

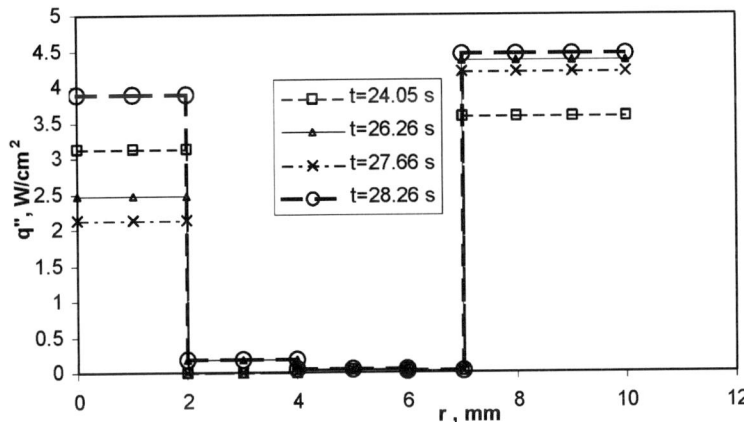

FIGURE 14. Heat flux distribution during a two bubble merger at $\Delta T_w = 6.2°C$, $\Delta T_{sub} = 0.8°C$.

the heaters placed on the back side of the wafer. A reduction in heat transfer in the inner region is seen during the bubble growth. However, heat flux in the outer most region continues to be high.

COMPARISON OF NUMERICAL SIMULATION AND EXPERIMENTAL RESULTS

Because the gravity level during the KC-135 flight experiments varied within a range $\pm 0.04 g_e$ and the test parameters were also somewhat different from that used in the numerical computations, the bubble growth periods differ between the simulations and the experiments. To compare the results, time is normalized using bubble growth period t_g as the characteristic time. Also, the measured heat flux is actually an averaged value over an area supporting a group of heating elements. Thus, the heat flux from the numerical prediction is also averaged over the corresponding radial ring. FIGURE 15 shows heat flux averaged over radial rings as a function of normalized time from the results of numerical simulations as given in FIGURE 7.

In FIGURE 16 a comparison of the heat flux in the central circular area underneath the cavity is made for the case for which results are plotted in FIGURE 11. Qualitative agreement in the variation of the heat flux as a function of time can be seen. The difference in heat flux level between the prediction and the data may seem large, but in reality, when integrated over the area where the heaters are mounted, it is not much beyond the uncertainty in measurements. For three selected times the heat fluxes as a function of radius are also compared in FIGURE 17. The top part of figure is for the period of initial bubble growth and the last is for the time at which bubble lifts off. The middle part of the figure is for the time just before lift-off. As shown in the figure, the trends in prediction and the data are again consistent.

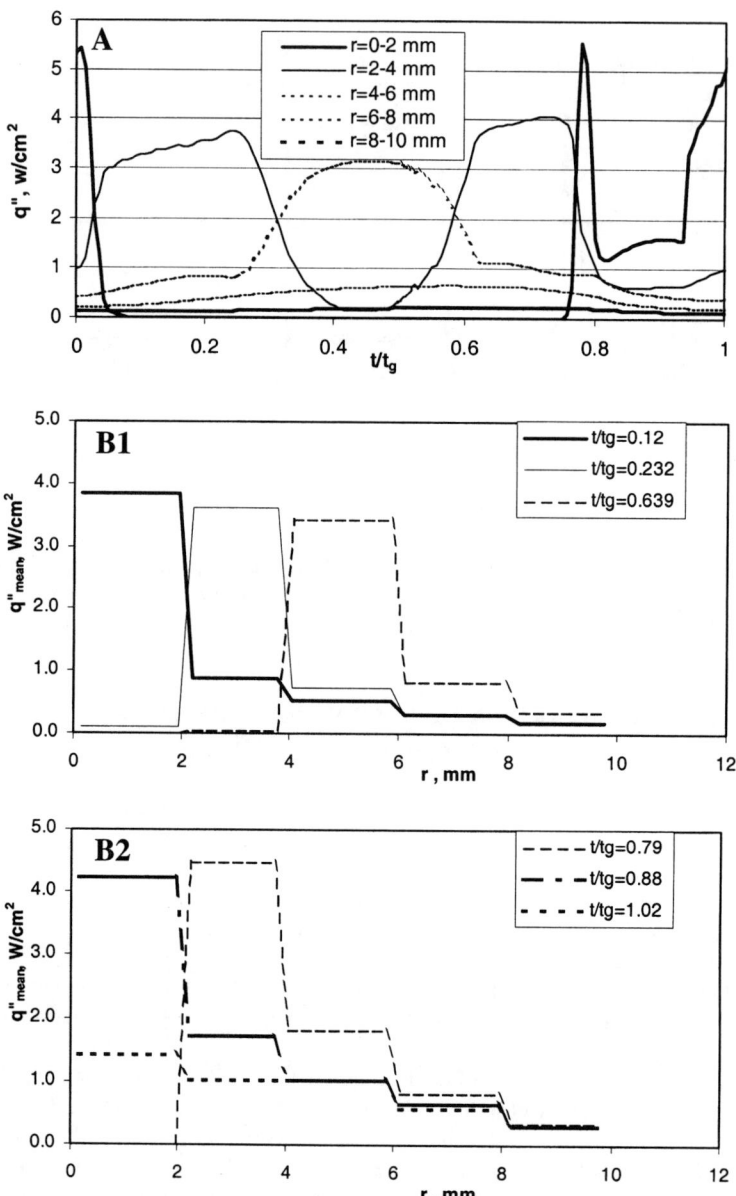

FIGURE 15. Heat flux averaged over radial segments and normalized with time as predicted from numerical simulation for $\Delta T_w = 8.5°C$ and $g_z = 0.01\, g_e$: **(A)** as a function of time, **(B)** as function of radius.

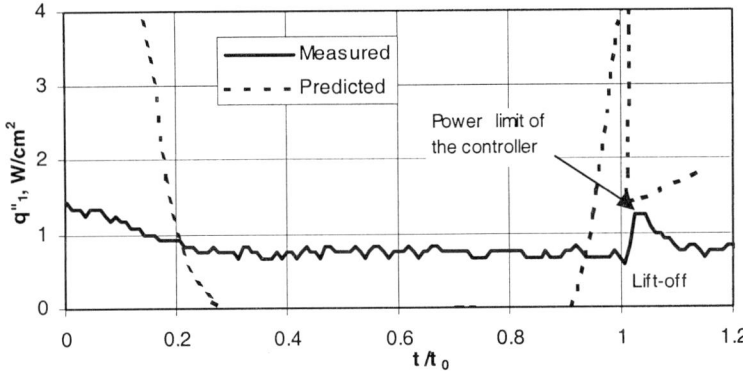

FIGURE 16. Comparison of heat flux for segment $r = 0$–2 mm with data described in FIGURE 11.

CONCLUDING REMARKS

1. Bubble shape and local surface heat flux during growth and lift-off of a single bubble are numerically computed using level set method for the low gravity condition.

2. By making artificial cavities in the polished Silicon wafer, well defined nucleation sites have been activated. Miniature heating elements and thermocouples are used for individual control. This approach has allowed experimental study of dynamics and local heat transfer of single and multiple vapor bubbles during nucleate boiling in low gravity conditions of KC-135.

3. From numerical simulations a peak in local heat flux is predicted to occur at the bubble-surface contact line.

4. A dry region exists in the bubble base area. In numerical simulations this is characterized by a zero heat flux but with a low heat flux in the experimental results.

5. In both the numerical predictions and experimental results the bubble lift-off is associated with an apparent increase in local heat flux.

6. Numerical predictions and experimental data give a consistent trend with respect to time variation of heat flux associated with single bubble growth and lift-off.

7. During the merger of multiple bubbles, a dry region underneath the bubbles continues to exist.

ACKNOWLEDGMENT

This work received support from NASA under the Microgravity Fluid Physics Program.

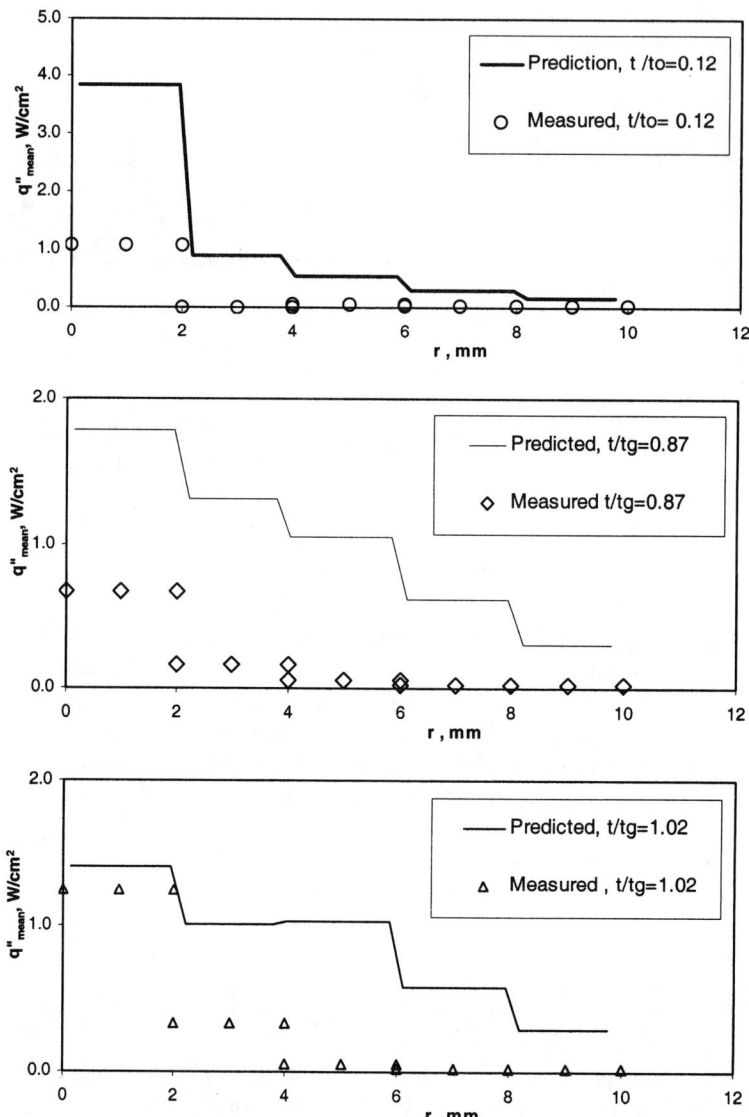

FIGURE 17. Comparison of heat flux for all segments with data described in FIGURE 11.

REFERENCES

1. KESHOCK, E.G. & R. SIEGEL. 1964. Forces acting on bubble in nucleate boiling under normal and reduced gravity conditions. NASA TN-D-2999.
2. SIEGEL, R. & E.G. KESHOCK. 1964. Effect of reduced gravity on nucleate bubble dynamics in water. J. AIChE **10**(4): 509–516.

3. MERTE, H. 1994. Pool and flow boiling in variable and microgravity. Second Microgravity Fluid Physics Conference, Paper No.33, Cleveland, OH, June 21–23.
4. LEE, H.S. & H. MERTE. 1997. Pool boiling curve in microgravity. J. Thermophys. Heat Transfer **11**(2): 216–222.
5. STRAUB, J., M. ZELL & B. VOGEL. 1992. Boiling under microgravity conditions. Proceedings of 1st European Symposium on Fluids in Space, Ajaccio, France, Nov. 18–22.
6. STRAUB, J. 1994. The role of surface tension for two-phase heat and mass transfer in the absence of gravity. Exp. Thermal Fluid Sci. **9**: 253–273.
7. ROGERS, F.T. & R.B. MESLER. 1964. An experimental study of surface cooling by bubbles during nucleate boiling of water. J. AIChE **10**(5): 656–660.
8. KIM, J. & F. BENTON. 2001. Subcooled pool boiling heat transfer in Earth gravity and microgravity. Proceedings of 35th National Heat Transfer Conference, Anaheim, CA, June 10–12.
9. SON, G., V.K. DHIR & N. RAMANUJAPU. 1999. Dynamics and heat transfer associated with a single bubble during nucleate boiling on a horizontal surface. J. Heat Transfer **121**: 623–632.
10. QIU, D.M., V.K. DHIR, M.M. HASAN, et al. 2000. Single bubble dynamics during nucleate boiling under low gravity conditions. Microgravity Fluid Physics and Heat Transfer. V.K. Dhir, Ed.: 62–71. Begell House, New York.
11. SINGH, S. & V.K. DHIR. 2000. Effect of gravity, wall superheat and liquid subcooling on bubble dynamics during nucleate boiling. *In* Microgravity Fluid Physics and Heat Transfer. V.K. Dhir, Ed.: 106–113. Begell House, New York.

Rapidly Expanding Viscous Drop from a Submerged Orifice at Finite Reynolds Numbers

NIVEDITA R. GUPTA,[a] R. BALASUBRAMANIAM,[b] AND KATHLEEN J. STEBE[a]

[a]*Department of Chemical Engineering, The Johns Hopkins University, Baltimore, Maryland, USA*

[b]*National Center for Microgravity Research on Fluids and Combustion, NASA Glenn Research Center, Cleveland, Ohio, USA*

> ABSTRACT: The rapid expansion and subsequent detachment of a drop or bubble are examined in this paper. Establishing the hydrodynamics of the drop growth and detachment process has applications ranging from improving insight into nucleate pool boiling, to improving techniques used to characterize dynamic surface tension. Here, the growth and detachment dynamics of viscous drops are studied numerically to establish an accurate description of the hydrodynamics. The full two-dimensional axisymmetric Navier–Stokes equations are solved using a single fluid volume-of-fluid (VOF) formulation on a fixed Eulerian grid. The interface is represented by marker particles that are convected in a Lagrangian manner on the underlying Eulerian grid. Results for surfactant-free, isothermal droplets from hemispherical caps to near detachment are presented.
>
> KEYWORDS: nonlinear dynamics; viscous drops; submerged orifice; drop expansion; finite reynolds number; VOF; front tracking

INTRODUCTION

Rapidly expanding bubbles occur in several industrial applications, including bubble sparging and pool boiling heat transfer processes. They are also exploited in analytical laboratory techniques, such as the maximum bubble pressure method for characterizing dynamic surface tension. Bubble growth and detachment from a needle have been well studied in the creeping flow limit[1] ($Re \ll 1$), and in the potential flow limit[2] ($Re \gg 1$). Growth and detachment of viscous drops from a capillary tube have been studied numerically in the creeping flow regime[3] and both experimentally and numerically in the finite Reynolds number regimes.[4] The Reynolds numbers for many applications fall into the range between 1 and 1,000. Furthermore, even in the creeping flow limit for bubble growth, the thinning of the neck as the bubble detaches is a rapid process in which inertial effects undoubtedly play a significant role.[1] This motivates the study of a drop/bubble for Reynolds numbers in the intermediate regime, in which both inertial and viscous effects play a role.

Address for correspondence: Kathleen J. Stebe, 221 Maryland Hall, The Johns Hopkins University, 3400 N. Charles Street, Baltimore, MD 21218, USA. Voice: 410-516-7769; fax: 410-516-5510.
  kjs@jhu.edu

Longuet-Higgins et al.[5] developed a quasistatic model by balancing the buoyancy force on a bubble emerging from a submerged orifice with the surface tension force at the line of attachment of the bubble to the orifice. This model, which ignores the complex dynamics that occur during pinch-off, predicts that the detached volume, V_d, is independent of the flow rate, Q, of the gas and is inversely proportional to the Bond number, Bo. Ad hoc inviscid models[6-9] were developed for the radial growth of a bubble in which the bubble detaches when the buoyancy force exceeds the acceleration of the added mass of the liquid. The detached bubble volume predicted by these models depends on the gas flow rate ($V_d \sim Q^{6/5}$). Oguz and Prosperetti[2] solved the inviscid flow equations for a bubble expanding at a needle by carrying out a boundary integral solution. They found that at low volumetric flow rates, the detached bubble volumes are independent of the volumetric flow rate (as predicted by the quasistatic model), and, when the flow rate exceeds a critical flow rate, the detached bubble volumes are proportional to $Q^{6/5}$ (as predicted by the ad hoc inviscid models). The creeping flow limit for bubble evolution was studied numerically, as well as experimentally, by Wong et al.[1] Their predicted detached volumes show a transition from a static case, where V_d is independent of Q, to a Stokes flow regime, where buoyancy balances the Stokes drag ($V_d \sim Q^{3/4}$). Both numerical studies compare well with experimental results in terms of the detached bubble shapes and volumes in the regimes of interest.

In this work, we address the intermediate regime between the Stokes limit ($V_d \sim Q^{3/4}$) and the inertia-dominated limit ($V_d \sim Q^{6/5}$). At elevated flow rates, the details of the neck contraction and break-off are important since small changes in the time scale for snap-off can lead to pronounced changes in the detached volume. Here, we study the dynamics of a viscous Newtonian drop expanding at an orifice submerged in an immiscible Newtonian liquid at finite Reynolds number. The two-dimensional axisymmetric Navier–Stokes equations are solved using the volume-of-fluid (VOF) formulation. The liquid–liquid interface is represented by marker particles that are convected in a Lagrangian manner on the Eulerian grid.

PROBLEM FORMULATION

The two-phase system is shown in FIGURE 1. A drop phase with density ρ_2 and viscosity μ_2 is expanding into an immiscible Newtonian liquid with density and viscosity, ρ_1 and μ_1, respectively. The dimensionless mass and momentum balance equations for a single fluid volume-of-fluid (VOF) formulation for such a system are

$$\nabla \cdot \mathbf{u} = 0 \qquad (1)$$

$$\rho Re\left[\frac{\partial \mathbf{u}}{\partial t} + \mathbf{u} \cdot \nabla \mathbf{u}\right] = -\nabla P + \mu \nabla^2 \mathbf{u} + \frac{1}{Ca}(\kappa - Boz)\mathbf{n}\delta_s, \qquad (2)$$

where $\mathbf{u} = (u, v)$ is the fluid velocity, P represents the modified pressure $(p - \rho \mathbf{g} \cdot \mathbf{x})$, and the relative significance of inertial forces to the viscous forces is incorporated in the Reynolds number, $Re = (\rho_1 R u_{max})/\mu_1$. The surface tension force is written as a concentrated surface force with δ_s representing the surface delta function, κ is the curvature at the interface, $\mathbf{n}$ is the outward pointing unit normal at the interface, and

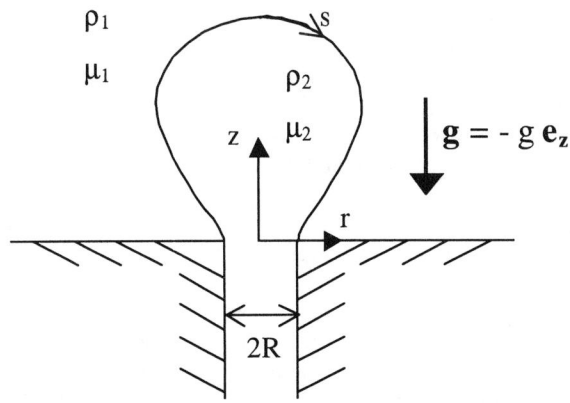

FIGURE 1. Schematic representation of the drop expanding from an orifice.

$Ca = (\mu_1 u_{max})/\sigma_o$ is the capillary number. The dimensionless functions describing the density, ρ, and viscosity, μ, are functions of position and time and are given by

$$\rho(\mathbf{x}, t) = 1 + \phi(\mathbf{x}, t)(\chi - 1)$$
$$\mu(\mathbf{x}, t) = 1 + \phi(\mathbf{x}, t)(\lambda - 1). \quad (3)$$

The volume-of-fluid function, $\phi(\mathbf{x}, t)$, is defined so that it is 1 for the drop phase and 0 for the bulk phase. The ratio of the density of the drop phase to that of the bulk phase is denoted by $\chi = \rho_2/\rho_1$, and the corresponding viscosity ratio is denoted by $\lambda = \mu_2/\mu_1$.

Integration of Equation (2) near the interface gives the interfacial normal stress jump balance as

$$\|-P + \mathbf{n} \cdot \tau \cdot \mathbf{n}\|_s = \frac{1}{Ca}(\kappa - Boz), \quad (4)$$

where $\|\psi\|_s$ represents a jump in the quantity ψ across the interface, and τ is the deviatoric stress tensor. Since the pressure is defined as a modified pressure, gravity appears in the normal stress jump balance as the Bond number, $Bo = ((\rho_1 - \rho_2)gR^2)/\sigma_o$. Continuity of velocity at the interface leads to

$$\|\mathbf{u}\|_s = 0. \quad (5)$$

In this study, we present preliminary results for a viscous drop expanding at an orifice.

NUMERICAL PROCEDURE

The domain, which extends infinitely in the positive r and z direction, is truncated at a distance such that the end effects do not affect the drop evolution or the flow field. The domain is then discretized as a fixed Eulerian grid for the purpose of solving the flow field in the vicinity of the orifice. The governing equations are discretized on the grid with staggered representation of the pressure and velocity fields by using a second-order accurate finite volume approximation. A time splitting method of the following form is used to solve the Navier–Stokes and continuity equations,

$$\frac{\mathbf{u}^* - \mathbf{u}^n}{\Delta t} = -\mathbf{u}^n \cdot \nabla \mathbf{u}^n + \frac{1}{\rho^n Re}\left[\mu^n \nabla^2 \mathbf{u}^n + \frac{1}{Ca}(\kappa - Boz)\mathbf{n}\delta_s\right] \quad (6)$$

$$\frac{\mathbf{u}^{n+1} - \mathbf{u}^*}{\Delta t} = \frac{1}{\rho^n Re}[-\nabla P^{n+1}]. \quad (7)$$

where the superscripts n, $n+1$ signify the time step. In order to ensure incompressibility at the $n+1$ time step ($\nabla \cdot \mathbf{u}^{n+1} = 0$), taking the divergence of Equation (7) gives the pressure Poisson equation,

$$\nabla \cdot \left[\frac{1}{\rho^n}\nabla P^{n+1}\right] = \frac{Re}{\Delta t}\nabla \cdot \mathbf{u}^*. \quad (8)$$

At each time step, the intermediate velocity $\mathbf{u}^*$ is calculated using Equation (6), and is used to obtain the pressure at the $n+1$ time step through an iterative solution of Equation (8). The updated pressure distribution is then used to update the velocity field at the $n+1$ time step using Equation (7).

The interface is represented by marker particles with a parametric description $(r(s), z(s))$ in a manner similar to Popinet and Zaleski.[10] Here, s is the arc length starting from the apex of the drop. The volume fraction field, ϕ, is generated using the interface topology on the underlying flow field grid.

Initially, the drop is described as a spherical cap at the orifice surface. The flow into the orifice is parabolic, $\mathbf{u} = [1 - r_2]\mathbf{e}_z$ independent of time. The average flow rate from the orifice is equal to $\pi/2$. The parametric representation of the interface is used to obtain the VOF field, ϕ. The Navier–Stokes solver is used to solve Equations (6)–(8) for the velocity and pressure fields at the given time step. A simple bilinear interpolation is used to determine the velocity of the marker particles. The marker particles are then advected as material particles giving the new shape of the interface. Since the interface is expanding rapidly, marker particles are added to maintain the same level of discretization of the interface at all time steps. The new parametric representation of the interface is then constructed, the new ϕ distribution calculated, and the procedure is repeated until the drop detaches. In the simulations, the drop is assumed to detach when the radius of the neck that forms beneath the drop is smaller than some prescribed value.

RESULTS AND DISCUSSION

The effects of inertia are evident in the drop shapes fairly early in the simulation, once the drop grows beyond a hemisphere. A comparison of the early shapes realized as a function of Reynolds number and capillary number is given in FIGURES 2 and 3. Even for a low capillary number ($Ca = 0.01$) at small times, the shape deviates from the static shape as seen in FIGURE 2. This is in contrast to the creeping flow results of Wong et al.,[1] who find that for small capillary numbers ($Ca = 0.02$), the bubbles evolve quasistatically up to the onset of pinch-off. As expected, as the capillary number is increased to $Ca = 0.1$, the deviation from the static shape is more pronounced. A comparison of drop shapes with various Reynolds and capillary numbers (all other

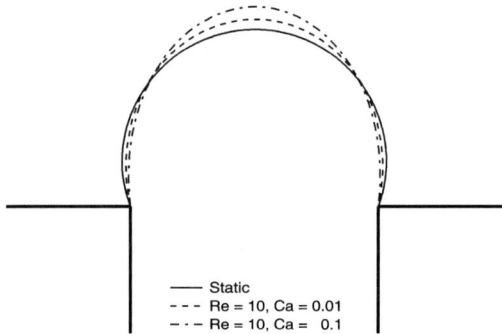

FIGURE 2. Deviation of a drop from the static shape (—) for drops with $\chi = 0.8$, $\lambda = 1$, $Bo = 0.3$, and $Re = 10$ with varying capillary numbers ($Ca = 0.01, 0.1$) at $t = 1$.

parameters constant) for drops far from detachment is shown in FIGURE 3. The early time results suggest that as the Reynolds number (inertia) is increased, the drop shapes are more distended. This trend is clearly observed for drops far from detachment in FIGURE 3 a, b, and d. However, this trend is reversed once the neck forms; in FIGURE 3 c, for $Re = 1$, the drop has begun the process of pinch-off and is more distended than the drop with $Re = 10$.

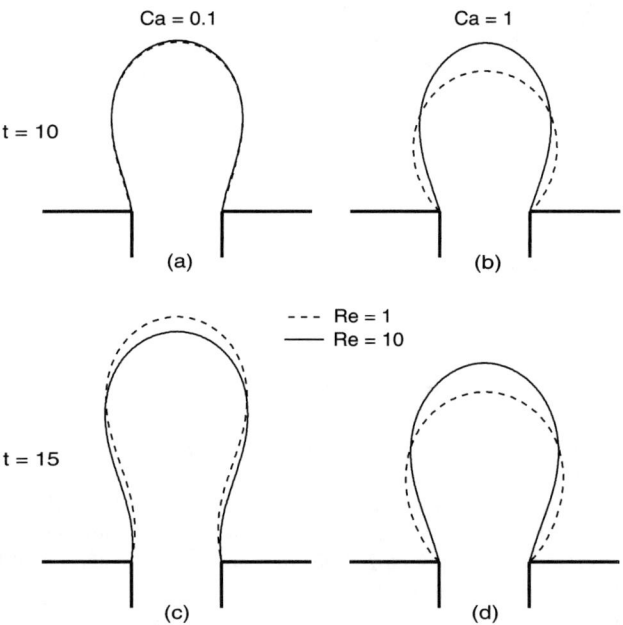

FIGURE 3. Effect of Reynolds number variation on the drop shape.

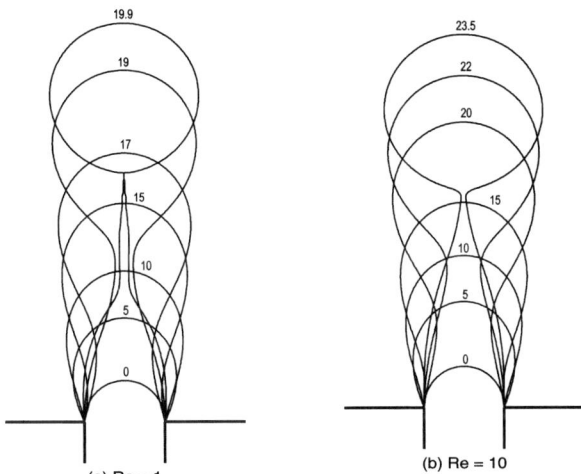

FIGURE 4. Time evolution of drops with $\chi = 0.8$, $\lambda = 1$, $Ca = 0.1$, $Bo = 0.3$, and (**a**) $Re = 1$, (**b**) $Re = 10$.

The time evolution of a viscous drop as a function of the Reynolds number until the point of detachment is seen in FIGURE 4. The simulation was stopped when the neck radius was of the same order as the arc length of discretization, $ds = 0.03$. The volume of the drop increases linearly with time since $dV/dt = \pi/2$. The drop expands almost radially at early times. As the buoyancy force trying to dislodge the drop from the orifice increases, the drop begins to neck. The necking process occurs at a critical arc length. The onset of necking can be seen in the normal velocity of the interface versus arc length plot, FIGURE 5.

The time evolution of the tangential velocity of the interface as a function of the arc length is seen in FIGURE 6. The flow of liquid always takes place from the rim of the drop to the apex. In this case only two stagnation points exist, at the rim and the apex of the drop. As necking proceeds, the tangential velocity of the interface in the necking region increases very rapidly as the liquid in the neck region squeezes out, eventually leading to break up of the drop.

The drop shapes realized in our simulations differ significantly from the bubble shapes observed in previous studies.[1,2] Our droplets are attached to the nozzle by long jets. The droplets neck to detach from these jets at heights several radii above the orifice and form drops that are more spherical. These differences may be attributable to the density and viscosity of the drop phase and the difference in flow geometry (our drop is being injected through a hole in a solid substrate; prior studies considered bubbles injected through needles). Preliminary results indicate that as the viscosity of the drop phase is reduced, the drop deforms more easily, and may require shorter times for drop detachment. The necking takes place closer to the orifice as the viscosity of the drop is reduced; this is illustrated in FIGURE 7. Furthermore, the drop shapes realized in this study show good qualitative agreement with the viscous drop studies,[3,4] reinforcing the suggestion that the viscosity of the drop

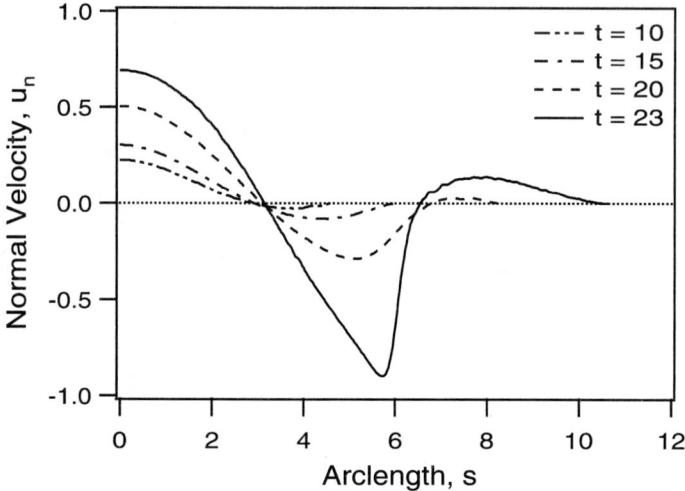

FIGURE 5. Time evolution of the interface normal velocity as a function of the arc length for a drop with $\chi = 0.8$, $\lambda = 1$, $Re = 10$, $Ca = 0.1$, and $Bo = 0.3$.

phase in our simulations is responsible for the long jets. The presence of a wall close to a growing drop may also have a role to play in the development of a long jet before the drop detaches from the orifice. The flow field around an expanding drop for two different times is seen in FIGURE 8. As the drop expands, the continuity of the

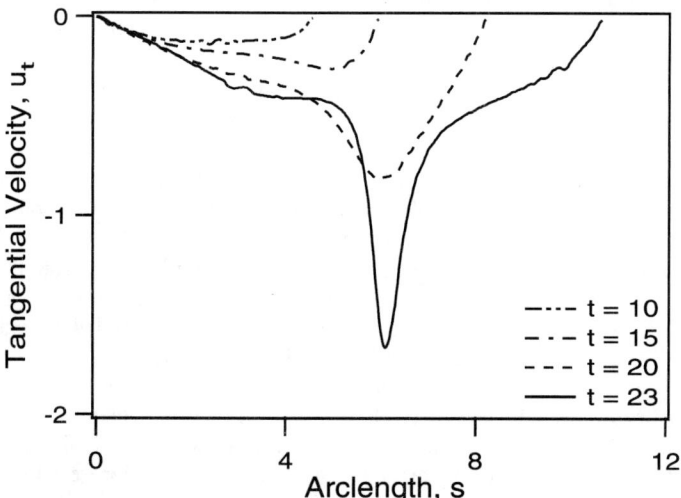

FIGURE 6. Time evolution of the interface tangential velocity as a function of arc length for a drop with $\chi = 0.8$, $\lambda = 1$, $Re = 10$, $Ca = 0.1$, and $Bo = 0.3$.

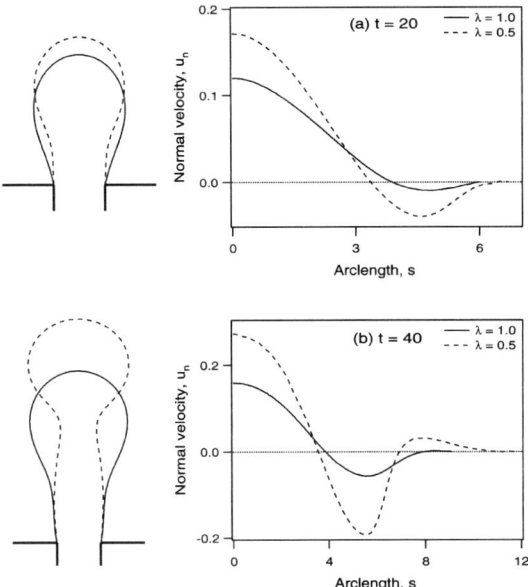

FIGURE 7. Effect of viscosity variation ($\lambda = 0.5, 1$) on the shape and normal interface velocity of drops with $\chi = 0.8$, $Re = 10$, $Ca = 1$, and $Bo = 0.3$.

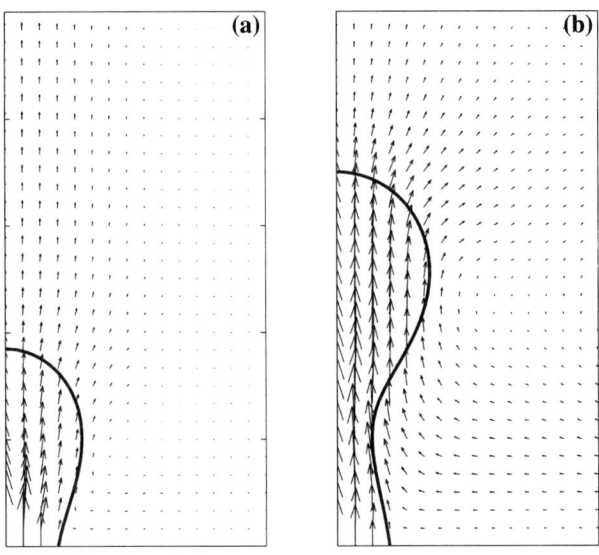

FIGURE 8. Flow field in the drop and bulk phases for a drop with $\chi = 0.8$, $\lambda = 1.0$, $Re = 10$, $Ca = 0.1$, and $Bo = 0.3$ at **(a)** $t = 10$, **(b)** $t = 20$.

velocity at the interface drags the fluid next to it in the upward direction setting up a recirculation vortex next to the drop that rises with the drop. This recirculation vortex may cause the necking region to shift farther away from the orifice.

For both runs presented in FIGURE 4, V_d was calculated and found to be 26.1 for $Re = 1$ and 30.5 for $Re = 10$. If V_d were determined by a static balance of surface tension and buoyancy,

$$V_d = \frac{2\pi}{Bo}\Psi(Bo), \quad (9)$$

where $\Psi(Bo)$ varies weakly with Bo and is always less than 1.[5] This predicts $V_d \leq 21$. Wong et al.[1] reported numerical results for $Ca = 0.1$, for which $V_d(Bo = 0.2) \approx 28$, $V_d(Bo = 0.5) \approx 11$. These values are all of similar magnitude to our results. Thus, V_d is not as sensitive to weak inertia or drop viscosity as were the details of the drop shape during growth and detachment.

To isolate the effects of buoyancy, we vary the Bond number ($Bo = 0.5$ and 1.0) while keeping all other parameters constant. The evolution of drops with $Bo = 0.5$ and $Bo = 1.0$ up to drop detachment is shown in FIGURE 9. A comparison of the drop shapes and normal interface velocities for different time steps is shown in FIGURE 10. As expected, the drop with the lowest Bond number (i.e., weakest buoyancy) is closest to a spherical shape whereas the drop with the largest Bond number is most distended. The necking process begins much earlier for the drop with the highest Bond number and the necking process proceeds much faster. Consequently, the detached drop volumes and the time required for detachment decreases as the Bond number increases.

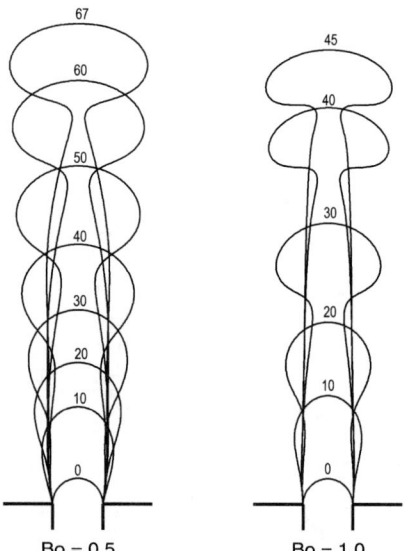

FIGURE 9. Effect of Bond number variation ($Bo = 0.5, 1$) on the time evolution of a drop with $\chi = 0.8$, $\lambda = 1$, $Re = 10$, and $Ca = 1$.

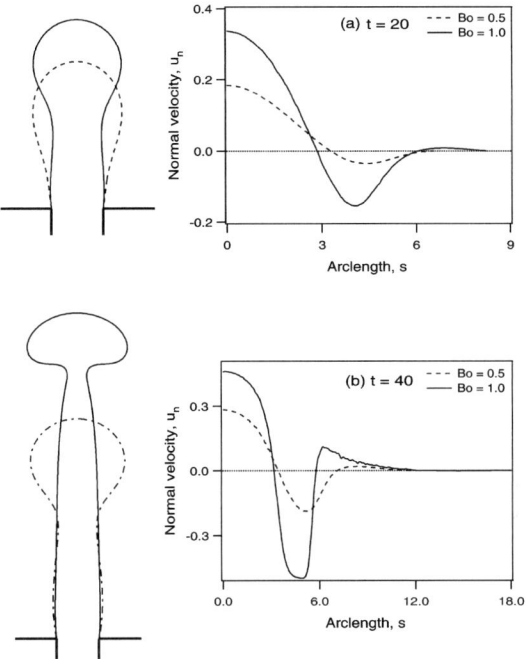

FIGURE 10. Effect of Bond number variation ($Bo = 0.5$ and 1) on the shape and normal interface velocity of drops with $\chi = 0.8$, $\lambda = 1$, $Re = 10$, and $Ca = 1$ at **(a)** $t = 20$ and **(b)** $t = 40$.

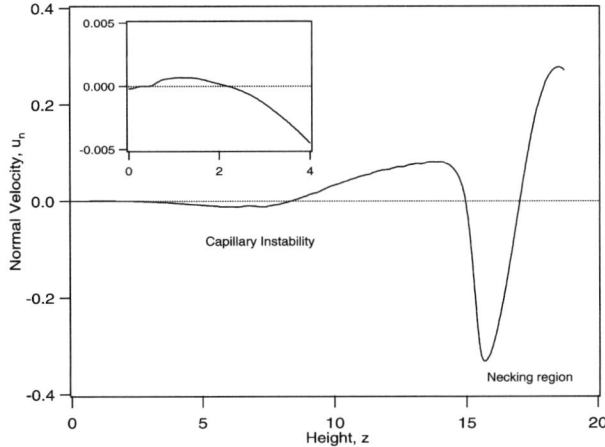

FIGURE 11. Normal velocity as a function of height for a drop with $\chi = 0.8$, $\lambda = 1$, $Re = 10$, $Ca = 1$, and $Bo = 0.5$ showing a capillary instability developing away from the neck.

The formation of long jets before detachment can cause a capillary instability to propagate along the long thread before buoyancy detaches the drop from the jet. This can be clearly seen in the normal interface velocity versus height, z plot for a drop with a long jet (see FIGURE 11). The change of sign in the normal interfacial velocity along the height of the drop corresponds to the neck regions developing along the drop. The first neck region, starting from the apex of the drop, extends between $z \sim 17$ and $z \sim 8.3$, thins very rapidly, and corresponds to the drop pinch-off. A second necking region (seen more clearly in the inset of FIG. 11), which thins slowly, develops in the long jet region and may correspond to a capillary instability.

SUMMARY

We have carried out numerical simulations to study the evolution of a viscous drop expanding at a submerged orifice. In the intermediate regime of Reynolds number, for which inertia and viscosity both play a role, preliminary results show pronounced differences from prior studies in the creeping flow and potential flow limits. Our future work in this problem will focus on understanding how the addition of surfactants changes the interfacial stresses that lead to drop detachment, thereby providing an additional means of controlling this process. An accurate description of the flow field next to an expanding viscous drop provides useful insight into the probable mechanisms of surfactant transport at the interface.

ACKNOWLEDGMENT

The work described here was supported by NASA Physical Science Research Division through Grant NCC3-812 from NASA Glenn Research Center to Johns Hopkins University.

REFERENCES

1. WONG, H., D. RUMSCHITZKI & C. MALDARELLI. 1998. Theory and experiment on the low-Reynolds-number expansion and contraction of a bubble pinned at a submerged tube tip. J. Fluid Mech. **356:** 93–124.
2. OGUZ, H.N. & A. PROSPERETTI. 1993. Dynamics of bubble growth and detachment from a needle. J. Fluid Mech. **257:** 111–145.
3. ZHANG, D.F. & H.A. STONE. 1997. Drop formation in viscous flows at vertical capillary tubes. Phys. Fluids **9:** 2234–2242.
4. ZHANG, X. 1999. Dynamics of drop formation in viscous flows. Chem. Eng. Sci. **54:** 1759–1774.
5. LONGUET-HIGGINS, M.S., B.R. KERMAN & K. LUNDE. 1991. The release of air bubbles from an underwater nozzle. J. Fluid Mech. **230:** 365–390.
6. DAVIDSON, J.F. & B.O.G. SCHULER. 1960. Bubble formation at an orifice in an inviscid liquid. Trans. Instn. Chem. Eng. **38:** 335–342.
7. WALTERS, J.K. & J.F. DAVIDSON. 1963. The initial motion of a gas bubble formed in an inviscid liquid. J. Fluid Mech. **17:** 321–340.
8. LANAUZE, R.D. & I.J. HARRIS. 1972. On a model for the formation of gas bubbles at a single submerged orifice under constant pressure conditions. Chem. Eng. Sci. **27:** 2102–2105.

9. LaNauze, R.D. & I.J. Harris. 1974. Gas bubble formation at elevated system pressures. Trans. Instn. Chem. Eng. **52:** 337–347.
10. Popinet, S. & S. Zaleski. 1999. A front-tracking algorithm for accurate representation of surface tension. Int. J. Num. Meth. Fluids **30:** 775–793.

Review of Existing Research on Microgravity Boiling and Two-Phase Flow

Future Experiments on the International Space Station

HARUHIKO OHTA,[a] ATSUSHI BABA,[b] AND KAMIEL GABRIEL[c]

[a]*Department of Aeronautics and Astronautics, Kyushu University, Higashi-ku, Fukuoka, Japan*

[b]*Office of Research and Development, Thermal Engineering Group, National Space Development Agency of Japan, Tsukuba, Japan*

[c]*College of Engineering, University of Saskatchewan, Saskatoon, Saskatchewan, Canada*

ABSTRACT: This paper describes research objectives on boiling and two-phase flow under microgravity conditions from both scientific and technological points of view. Existing research for various systems in flow boiling are briefly reviewed and problems indicated for research conducted by available facilities with short microgravity duration. A wide range of experimental subjects that become possible by using different types of test sections is clarified and the validity of long term experiments is emphasized. A first step outline is described for a test loop with interchangeable test sections, such as transparent heated tubes, transparent flat heating surfaces, and narrow channels optimized for individual objectives and restricted by specifications of the facility.

KEYWORDS: flow boiling; pool boiling; two-phase flow; microgravity; international space station

NOMENCLATURE:

d	inner tube diameter, mm
E	power input, VA
G	mass velocity, kg/m^{2}sec
g	gravitational acceleration, m/sec^2
g/g_e	ratio of gravity to the Earth value
J	superficial velocity, m/sec
P	pressure, MPa
Q	flow of heat, W
q	heat flux, W/m^2
q_c/q_{co}	ratio of critical heat flux to that in pool on ground
s	gap size, mm
T	temperature, °C or K
ΔT_{sub}	degree of subcooling, K
u	velocity, m/sec
We	Weber number, $u^2 \rho l/\sigma$
x	quality, $\rho_G J_G/(\rho_L J_L + \rho_G J_G)$

Address for correspondence: Haruhiko Ohta, Department of Aeronautics and Astronautics, Kyushu University, 6-10-1 Hakozaki, Higashi-ku, Fukuoka 812-8581, Japan. Voice: +81 92 642 3489; fax: +81 92 642 3752.
ohta@aero.kyushu-u.ac.jp

Greek Symbols

α	heat transfer coefficient, W/m²K
α_{TP}/α_{lo}	ratio of two-phase heat transfer coefficient to that for liquid alone
ρ	density, kg/m³
σ	surface tension, N/m
τ	time, sec

Subscripts

b	bulk
e	Earth
ex	exit
G	gas
H	preheater
in	inlet or input
L	liquid
out	output
P	pump
s	saturated
SG	superficial value for gas
SL	superficial value for liquid
T	test section
w	wall
z	direction perpendicular to a surface

INTRODUCTION

Recent increases in spacecraft size and power requirements for advanced Earth observation satellites and orbiting spacecraft have escalated demands for more effective thermal management and thermal control systems. Thermal systems utilizing boiling and two-phase flow are effective means for the development of high-performance, reliable and safe heat transport systems for future space missions. Boiling heat transfer offers high heat transfer rates associated with nucleate boiling and has the potential to significantly reduce the required surface area in heat exchangers. Transport of the latent heat of vaporization in two-phase flow reduces the flow rate of liquid circulated in the loop for the same amount of heat transport and, in turn, reduces the pump power requirement. Furthermore, two-phase heat transport fluids allow for precise adjustment of the fluid temperature by simply pressurizing the system using an accumulator.

Despite their acknowledged importance, boiling and two-phase flow systems have not yet been fully implemented in new spacecraft, except for small-scale heat pipes and a thermal transport loop planned in the Russian module of International Space Station. This is partially attributed to the lack of a reliable database for the operation of such systems in microgravity. In addition, uncertainty in the critical heat flux conditions discourages space system designers from implementing such systems. Single-phase liquid cooling systems are favored despite the large mass penalty associated with such systems. However, even with single-phase systems, boiling and two-phase flow inevitably occur as a result of, for example, accidental increase in the heat generation rate, or a sudden system depressurization caused by faulty valve operation.

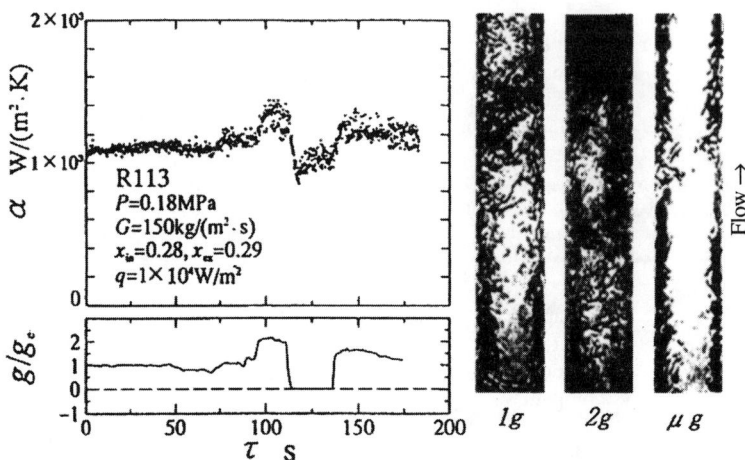

FIGURE 1. Gravity effect observed in the behavior of annular liquid film and in heat transfer coefficient due to two-phase forced convection.[1]

It is safe to say that to date, there is no cohesive database for microgravity boiling and two-phase flow. There is also a prevailing misconception that few differences actually exist between normal and microgravity heat transfer coefficients in flow boiling in the existence of bulk flow. This is not true when bulk flow is not large, as shown in FIGURE 1.[1] In addition to clarifying phenomena in microgravity, the establishment of a coherent database for microgravity flow boiling and two-phase flow would provide fundamental information for the development of large-scale two-phase thermal management systems for possible implementation in future spacecraft and Earth orbiting satellites.

SHORT REVIEW OF EXISTING RESEARCH ON MICROGRAVITY BOILING AND TWO-PHASE FLOW

Historical Review

Boiling experiments in microgravity started late in the 1950s, and now have a history of more than 40 years. In the 1960s, research efforts focused on pool boiling using drop towers of laboratory scale. Merte clarified bubble behavior and heat transfer characteristics in microgravity for a specified system using a very short duration microgravity period, as summarized in his review.[2] The research grew rapidly in the 1980s and the 1990s as a result of the introduction of various platforms that provide longer microgravity duration—aircraft, large-scale drop towers and drop shafts, ballistic rockets, and space shuttles.

Straub et al.[3] conducted various pool boiling experiments on the space shuttle and sounding rockets. Other work on adiabatic two-phase flow of binary components started in the USA, Canada, Europe, and Japan. The following section provides a summary of the main findings of research in these areas.

Flow Boiling in a Tube

Reports on the heat transfer characteristics are very limited, especially those covering wide range of quality. Lui et al.[4] reported an increase in the heat transfer coefficient (up to 25%) when gravity level was reduced in nucleate boiling under subcooled liquid conditions, as shown in FIGURE 2. One of us[1] clarified in a qualitative manner the effects of gravity in a wider range of quality. The results are summarized in TABLE 1. An example of the gravity effect was provided in FIGURE 1, where heat transfer due to two-phase forced convection deteriorates due to reduction of turbulence in the annular liquid film in microgravity. For the critical heat flux there are almost no fundamental data.[5] The existing problems can be summarized as follows: (1) Gravity effect on heat transfer coefficient needs to be clarified in a quantitative manner. (2) There is insufficient data to discuss the mechanisms of heat transfer in various gravity fields. (3) No reliable data on critical heat flux exists. The accuracy of measurements needs to be improved by the experimental opportunity of long-term microgravity duration. Furthermore, effects of gravity on the dryout mechanisms in low and moderate quality regions need to be clarified.

Bubble Dynamics in Liquid Flow

Except for comments related to the structure of bubbles in microgravity, the change in bubble size with gravity reduction had not been investigated well until Qiu and Dhir[6] reported bubble growth rate in microgravity based on observations illustrated in FIGURE 3; and Singh and Dhir[7] investigated the problem using numerical simulation. One of the present authors measured the thickness of microlayer underneath primary bubbles lifting a coalesced bubble. Surface heat flux evaluated from the conduction across the microlayer was compared with that from power input.[8]

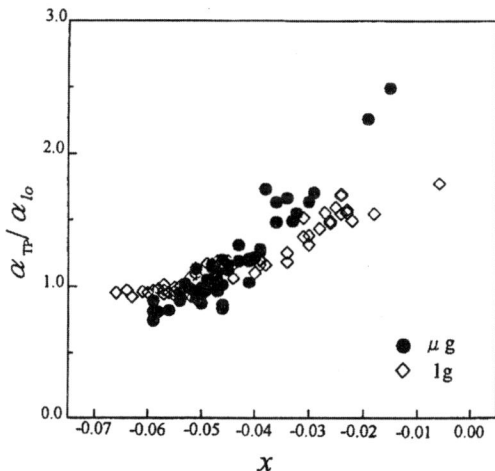

FIGURE 2. Increase of heat transfer coefficient due to subcooled nucleate boiling in microgravity.[4]

TABLE 1. Summary of effects of gravity decrease on liquid–vapor behavior and heat transfer in flow boiling in a tube

			Low quality Bubbly flow regime	Moderate quality Annular flow regime	High quality
Liquid–vapor behavior	Low mass velocity	Low heat flux	increase of detached bubble size in microgravity	decrease of turbulence in annular liquid film of increased thickness in microgravity	no marked gravity effect
		High heat flux		no marked gravity effect	no marked gravity effect
	High mass velocity			no marked gravity effect	
Heat transfer	Low mass velocity	Low heat flux	[nucleate boiling] no marked gravity effect	[two-phase forced convection] heat transfer deteriorates in microgravity	[two-phase forced convection] no marked gravity effect
		High heat flux	[nucleate boiling] no marked gravity effect	[nucleate boiling] no marked gravity effect	[nucleate boiling] no marked gravity effect
	High mass velocity			no marked gravity effect	

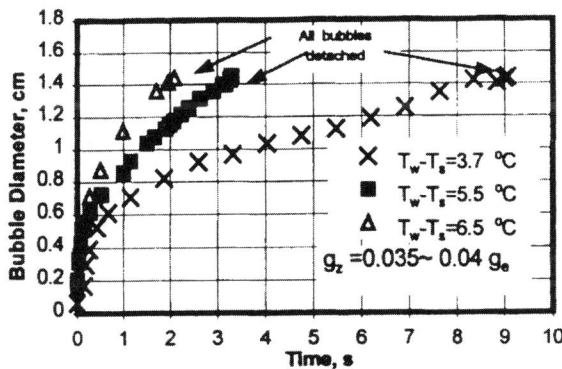

FIGURE 3. Bubble growth at various wall superheats in saturated water in microgravity.[6]

The following topics are yet to be researched and the phenomena need to be quantified. (1) Detailed observation and measurement of the behavior of the microlayer are required for the clarification of heat transfer mechanisms in nucleate boiling in microgravity. (2) Behavior of a single bubble and its microlayer in liquid flow with prescribed velocity needs to be clarified as a more generalized system. (3) According to previous experiments,[1] both effects of heat transfer enhancement and deterioration are possible when gravity is reduced. The criteria may be found through a detailed observation of the bubble attached area and the motion of the microlayer edges under experimental conditions strictly controlled in the long-term microgravity experiments. (4) Under the condition of subcooled boiling, bubble size could be kept constant by balancing the evaporation from the microlayer with the condensation at the bubble surface. Steady-state heat transfer could be possible even in microgravity, as shown in the report by Ohta *et al.*[9] However, in saturated nucleate boiling, no conclusive results have been obtained on whether steady-state heat transfer is possible or not, and the criteria of steady state boiling in microgravity remains to be clarified in a quantitative manner. (5) The criteria are closely related to the bubble detachment, and its mechanism is an important subject for investigation under microgravity conditions. (6) Previous experiments by Lui *et al.*[4] and others have reported small changes in flow boiling heat transfer rates under microgravity compared to normal gravity, despite a significant reduction in the buoyancy force and consequent increase in bubble diameter. The reason for this needs to be clarified, together with an investigation of heat transfer mechanisms.

Two-Phase Flow Dynamics

Research to date has focused on the classification of flow regimes and the development of flow pattern maps for microgravity conditions, as shown in FIGURE 4.[10] Pressure drop data were collected in a small diameter straight tube and the results were compared to those obtained at normal gravity.[11] Heat transfer data were also obtained, as shown in FIGURE 5, for saturated liquid and gas binary mixtures over a wide range of flow rates in both phases.[12] Recently, studies have focused on the transition to annular flow and wave characteristics in fully developed annular flow. Film characteristics were obtained and analyzed using statistical analysis techniques.[13]

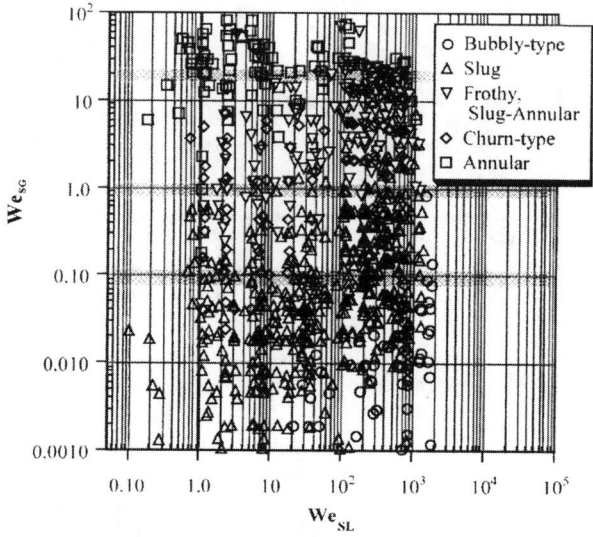

FIGURE 4. Dimensionless flow pattern map and experimental data for microgravity two-phase adiabatic flow.[10]

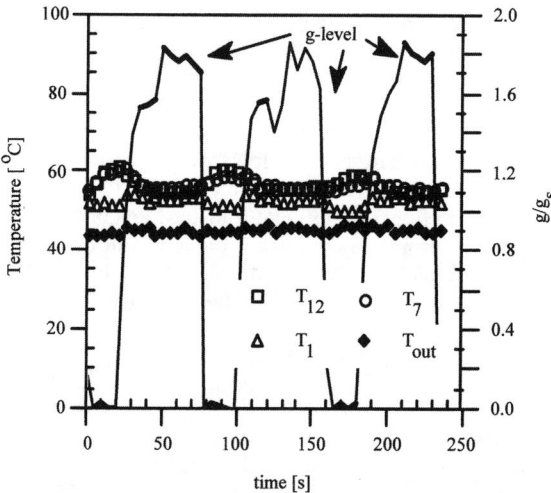

FIGURE 5. Temperature response to gravity change in slug and transition flow regimes.[12]

(1) The existence of annular flow, at low liquid flow rates and a smooth interface, was reported at very low liquid velocities. This needs to be examined further. (2) The effect of gravity on the pressure drop is not yet clear, with conflicting results obtained during short microgravity duration on parabolic flights for vibrations and g-jitter. (3) An increase in the film thickness in annular flow in microgravity was also reported by deJong and Rezkallah.[14] Despite these efforts, there are still no coherent experiments on the behavior of disturbance waves in microgravity. To clarify the effect of gravity on annular liquid film thickness a detailed FFT analysis should be performed.

The ISS provides a longer and more stable microgravity environment allowing for superb conditions for performing the experimental studies mentioned above. The long test time allows for the measurement of film velocity and film flow rates in order to clarify the annular film flow structure.

Flow Boiling in Narrow Channels

There are a few investigations that have focussed on boiling in narrow channels under the variation of gravity level.[15] Matsunaga *et al.*[16] clarified the three different bubble behaviors and corresponding gravity effects. The reduction of gravity deteriorates heat transfer for small gap sizes, where flattened bubbles are generated, whereas almost no gravity effect is observed for large gap sizes without flattened bubbles. To classify the bubble behavior and heat transfer for different

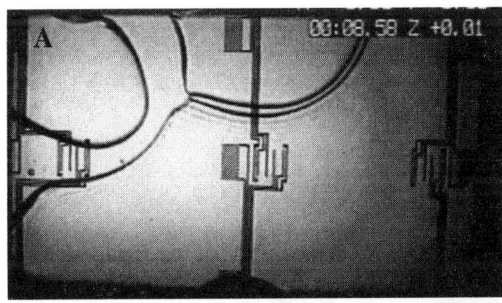

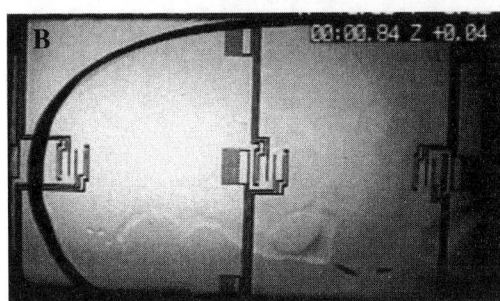

FIGURE 6. Increase of liquid subcooling promotes heat transfer deterioration in microgravity.[16] Water, gap size $s = 2$ mm, $u_{in} = 0.06$ m/sec, $q = 1.4 \times 10^5$ W/m^2. (**A**) $\Delta T_{sub,in} \approx 0$ K, μg. (**B**) $\Delta T_{sub,in} \approx 10$ K, μg.

gravity levels, introduction of a new dimensionless parameter and/or definition of Bond number is required. Confirmation is also required for the observed curious trend indicated in FIGURE 6.[16] The increase in liquid subcooling under microgravity conditions reduces the frequency of bubble detachment and fixes a flattened bubble balancing the evaporation rate with that of condensation at the upstream edge. A large extended dried area underneath the flattened bubble deteriorates the heat transfer from that for saturated inlet conditions. For extremely small gap sizes, where extended dried areas underneath flattened bubbles are quickly wetted by the penetration of bulk fluid as a result of increased growth rate of bubbles, no gravity effect is observed in either bubble behavior or heat transfer. The bubble behavior

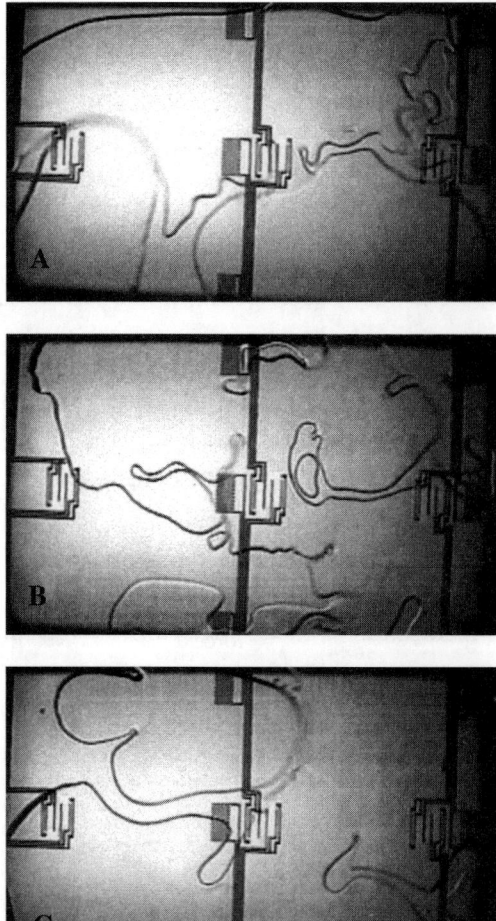

FIGURE 7. Liquid–vapor behavior insensitive to gravity levels in an extremely small gap size.[16] Water, gap size $s = 0.7$ mm, $u_{in} = 0.06$ m/sec, $\Delta T_{\text{sub,in}} = 0$ K, $q = 8 \times 10^4$ W/m^2. **(A)** 1g. **(B)** 2g. **(C)** μg.

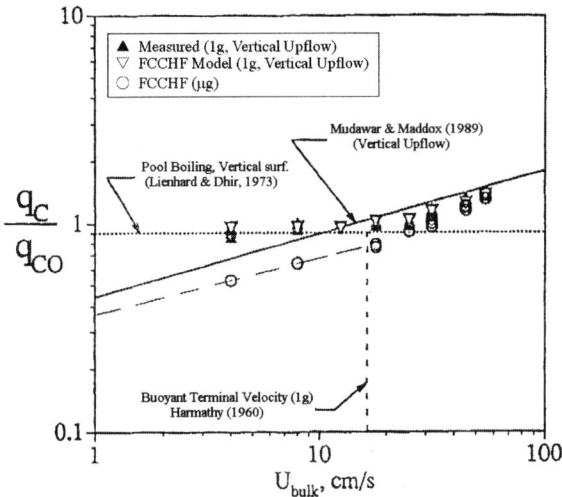

FIGURE 8. Predictions of CHF as a function of the flow velocity in microgravity using FCCHF model compared with CHF for vertical upward flow under $1g$.[17]

is shown in FIGURE 7. This needs further confirmation because the extremely narrow gap size may contribute the reduction of heat exchanger size and weight.

In Earth gravity, the existence of buoyancy provides a well-described lower limit on critical heat flux, leading to the onset of dryout as the imposed bulk fluid velocity is reduced. No such lower limit is known for microgravity conditions, as shown in FIGURE 8.[17] The critical heat flux for narrow ducts or gaps in microgravity remains to be investigated in detail in the presence of bulk flow.

Flow Boiling in Microchannels

There is no reported microgravity experiment for boiling in microchannels. Although in most cases the surface tension force dominates interfacial behaviors and gravitational force plays a minor role, two subjects are worthy of investigation in microgravity for the application to electronic devices integrated in space machines: (1) to determine the boundary between the conditions where the effect of gravitational force disappears, and (2) to determine an effective method for removing generated vapor from the microchannels and ensuring a continuous supply of liquid into microchannels.

EXPERIMENTAL DESIGN METHODS

Experimental Apparatus

Test Loop

A small-scale test loop for boiling experiments is planned within the restriction of size and power supply given to the mission part of fluid physics experiment facility

(FPEF) developed by NASDA. A system of interchangeable test sections is adopted to conduct experiments on the various subjects mentioned above. The test loop is designed under the size limitation 500mm × 300mm × 200mm and maximum electrical power 600W. Test sections are interchanged with the aid of specially designed connectors.

FIGURE 9 shows an example of test loop. The major components are a circulating pump, preheaters, a test section, a condenser, a liquid–vapor separator, and an accumulator. In this example, to reduce the power input, a heat recovery system is introduced by using Pertier elements that transfer the waste heat from the condenser to the preheater 1. The preheater 2 is operated by Joule heating and uses an auxiliary heater at the start of heating the system. The flow of energy in the test loop is shown in FIGURE 10. The power requirements for the loop operation, with and without heat recovery system, are compared in FIGURE 11, where the power requirements without heat recovery system are shown by ten horizontal lines corresponding to the exit quality. Power is reduced from 600W to 440W when saturated liquid is heated to quality 0.2 at the inlet of test section and finally to 0.5 at the exit for the given conditions indicated in the figure.

Actually, two small circulating pump are to be installed in the loop in parallel. Pumps with a self lubrication system are designed with the following specifications: (a) pump type, gear; (b) flow rate, 0.05–0.5L/min (parallel, 0.05–1.0L/min); (c) head, 200kPa; (d) speed, 400–4,000rpm; (e) required power, 1W; (f) weight,

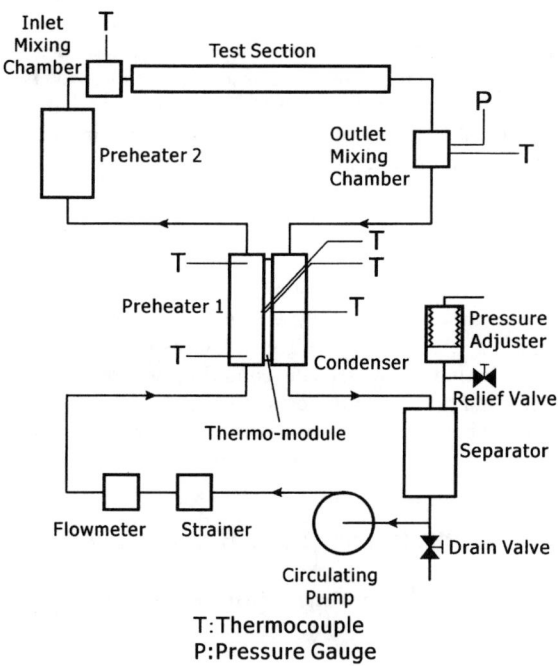

FIGURE 9. Components of a compact two-phase flow loop for preliminary ground tests.

100 g; (g) controller, PID type; and (h) motor type, AC motor. The accumulator operating in thermal control mode adjusts the internal pressure of the loop and works as a damper for pulsating flow. Test sections to be developed include: transparent heated tubes, metal heated tubes, rectangular ducts with a transparent flat heating surface, narrow channels, unheated transparent tubes, microchannels, and porous evaporators applied to a capillary pump loop realized by a bypass tube across the mechanical circulating pump. In a condenser with a heat exchanger, distilled water, and avionics air circulating in Japanese experiment module (JEM) are available as coolants, and Pertier elements are introduced to adjust the capacity of cooling. A serious problem is how to separate liquid from liquid–vapor mixture upstream of the circulating pump. A separator using centrifugal force acting on liquid phase is under development and to be introduced with an auxiliary separator using the surface tension force acting on the interfaces. To prevent cavitation in the circulating pump, subcooled liquid is introduced to the pump under microgravity conditions.

Transparent Heated Tube

The transparent heated tube, as shown in FIGURE 12, is made of a Pyrex tube with a thin gold film, about 0.01 μm thick, uniformly coated on the inner wall. The film is

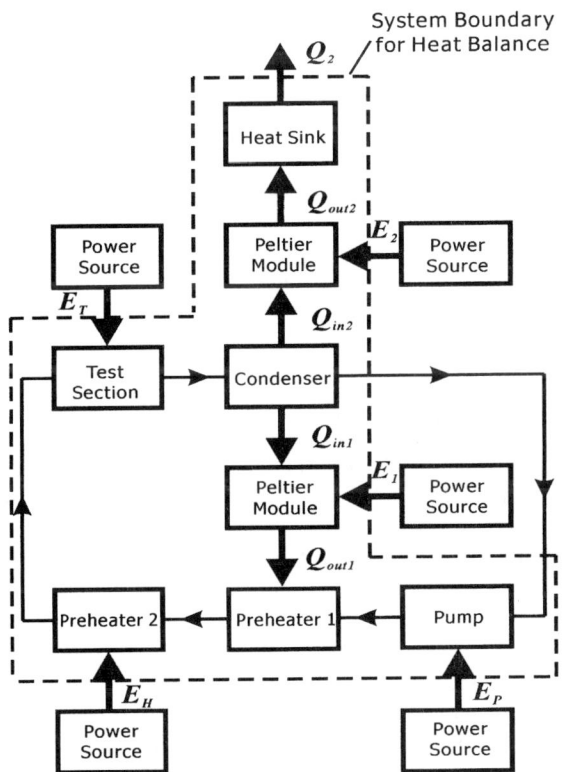

FIGURE 10. Energy flow for compact two-phase flow loop with heat recovery system.

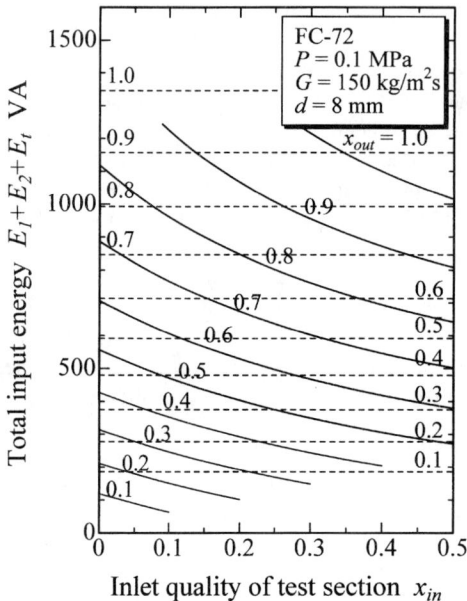

FIGURE 11. Decrease in power input by heat recovery system for various combinations of inlet and exit quality.

used as a heater by passing electrical current directly through it and it is used also as a resistance thermometer to evaluate the inner wall temperature. Thus, this structure makes heating, direct observation through the heated tube wall, and measurement of heat transfer data, simultaneously possible. The minimum inner diameter is 7 mm, but this depends on the coating length. A heating length of 400 mm is possible if an unheated tube is not installed upstream.

Metal Heated Tubes

Stainless steel or copper tubes with inner diameter 5–12 mm are employed for the measurement of CHF values, where the transparent tube cannot be introduced because of its inferior mechanical toughness in both tube substrate and coated film, especially in the case of critical heat flux measurement. In this case, the heated length is restricted to 400 mm under the limited size of FPEF facility. Temperature transition is measured by thermocouples flushed to the inner wall of the tube.

Rectangular Ducts

Ducts with rectangular cross section are employed for the visualization of bubble shapes and for the determination of velocity profiles in the liquid flowing around a bubble by the photochromic dye activation method. The ducts are made from quartz or Pyrex glass. A heating element is attached to the center of transparent flat surface so as to provide a localized nucleation spot and to enable investigation of the departing characteristics of a single nucleating bubble. In particular, the dynamics of the liquid–

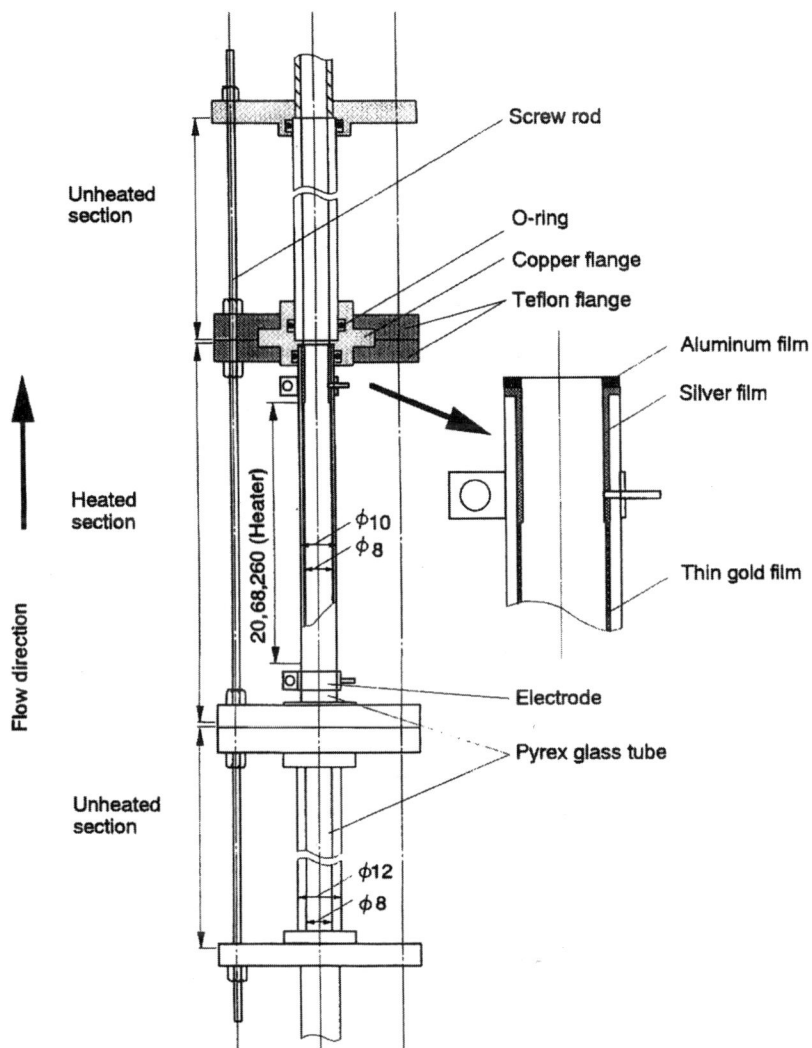

FIGURE 12. Structure of transparent heated tube to make heating, observation, and acquisition of heat transfer data simultaneously possible.

vapor interfaces around the base of the bubble under varying liquid velocities will be studied in detail using the high-magnification video camera of the FPEF facility.

Narrow Channels

A copper block with a built-in heater or a transparent heating plate is installed as the heating element in a narrow gap. FIGURE 13 shows an example of structure for a sapphire plate with an ITO coating heater in the rear side. Temperature distribution is measured by thin film resistance thermometer directly coated on the heat transfer side. The heat flux distribution is evaluated by the solution of transient heat conduction across the substrate plate.

Unheated Transparent Tubes

A quartz tube with inner diameter 5–12 mm is employed for the observation of flow pattern and liquid–vapor distribution, and for the measurement of pressure drop, void fraction, and liquid film thickness. Sensors developed on the ground for the measurement of these parameters are applied to the test section.

Microchannels

Copper microchannels with structures to avoid vapor blanketing and to promote the liquid supply are employed, and the characteristics in the heat transfer coefficients and critical heat flux are examined in detail in comparison with the ground data. A heating surface with 20 mm × 20 mm is planned analogous to a heat spreader planting a small tip.

Porous Evaporators

Aluminum tubes with fine porous structure are employed to produce pressure difference across tube wall by capillary force. Liquid flows inside the porous tube, and heaters around a coaxial outer tube promotes the evaporation from the porous

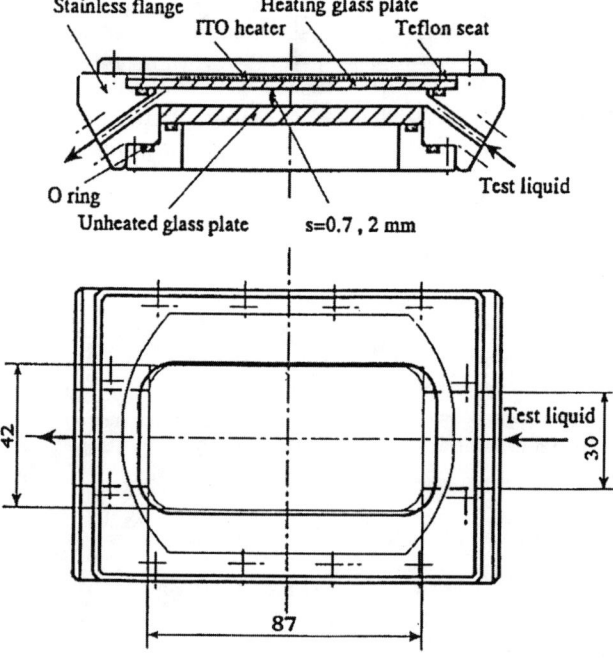

FIGURE 13. Test section of narrow channel with a glass heating surface.

surface partially touching with it. Vapor evaporated from the surface of porous tube passes through the annular cross section and introduced to the condenser. Condensate is circulated in the liquid channel and introduced to the inside of porous tube. Data for performance in heat transport and for detailed operational characteristics are acquired at start-up, steady and unsteady modes.

Test Liquid

The test liquid should be non-toxic and non-flammable because it is kept at the pressurized facility in ISS. Fluorinert-72© (3M Corporation) was selected as test liquid. It has boiling point at 56°C at 0.1 MPa; the temperature value is desirable for safety, ease of handling, and from the demands of cooling capacity. Fluorinert-72 has normal characteristics in both thermal and transport properties as shown in TABLE 2, and the experimental results can be interpreted to apply to other liquid systems. The test liquid can be adopted as a working medium for a large-scale two-phase flow loop for future implementation in space thermal management systems.

CONCLUSIONS

In our review of existing research on boiling and two-phase flow we have only detected some of the phenomena encountered under microgravity conditions. We do not know systematically and quantitatively about the details. The opportunity for experiments under microgravity conditions of long duration is inevitable for the clarification of phenomena and for the establishment of a database. In addition, the opportunity of long-term experiments is significant for the following general reasons: (1) need for a comprehensive and reliable data for qualitative discussion on the effect of gravity on such phenomena; (2) lack of essential data to clarify heat transfer mechanisms in boiling and structures in two-phase flows under microgravity conditions; (3) lack of systematic experiments with sufficient variation in flow parameters due to the limited opportunities in the past using short duration microgravity platforms; and (4) there has been no method in the past to check the stability and reproducibility of the experimental data.

Accumulated knowledge and experience through the existing facilities with short microgravity duration become a powerful means for the development of experimental apparatus that consists of a universal test loop and interchangeable test sections of various types optimized for individual subjects. The feasibility of such experiments with requirements in the apparatus size and in power supply was checked under the assumed application to the FPEF facility developed by NASDA. An outline in the specification of apparatus and experimental conditions was discussed as a preliminary step for the detailed planning of the experiments.

ACKNOWLEDGMENTS

This paper was contributed as a part of project by International Microgravity Two-phase System Working Group (IMT) organized since 1998 for the discussion of ISS experiments. The authors greatly appreciate the collaboration of IMT members;

TABLE 2. Comparison of thermal properties of FC-72 with water and Fron113, $P = 0.1$ MPa

	Boiling point (°C)	Enthalpy of vaporization (kJ/kg)	Density		Viscosity		Isobaric heat capacity		Thermal conductivity		Surface tension (mN/m)
			Liquid (kg/m³)	Vapor (kg/m³)	Liquid (mPa·sec)	Vapor (μPa·sec)	Liquid (J/kgK)	Vapor (J/kgK)	Liquid (mW/mK)	Vapor (mW/mK)	
Fluorinert (FC-72)	55.7	95.7	1,605	12.33	0.44	12.0	1,101	894	54.1	12.9	7.9
Water	100	2,259	960	0.537	0.284	12.1	4,216	2,140	682	24.2	54.4
Fron (R113)	47.57	146	1,507	7.46	0.50	10.0	983.3	674	70.0	7.79	14.7

Dr. Yoshiyuki Abe (AIST, Japan), Professors Gian Piero Celata (ENEA, Italy), Vijay K. Dhir (Univ. California-Los Angels, USA), Walter Grassi (Univ. Pisa, Italy), Masahiro Kawaji (Univ. Toronto, Canada), Herman Merte, Jr. (Univ. Michigan, USA), William W. Schultz (Univ. Michigan, USA), Johannes Straub (Technical University Munich, Germany), and Dr. Martin Zell (Astrium GmbH, Germany).

REFERENCES

1. OHTA, H. 1997. Experiments on microgravity boiling heat transfer by using transparent heaters. Nuclear Eng. Design **175**: 167–180.
2. MERTE, H., JR. 1990. Nucleate pool boiling in variable gravity. Prog. Astronaut. Aeronaut. **130**: 15–69.
3. STRAUB, J., M. ZELL & B. VOGEL. 1990. Pool boiling in a reduced gravity field. Heat Transfer 1990, Proc. 9th Int. Heat Transfer Conf., Kyongju, Korea, **1**: 91–112.
4. LUI, R., M. KAWAJI & T. OGUSHI. 1994. Experimental investigation of flow boiling heat transfer under microgravity. Heat Transfer 1994, Proc. 10th Int. Heat Transfer Conf., Brighton, UK, **7**: 497–502.
5. OHTA, H., K. INOUE & S. YOSIDA. 1997. Experiments on dryout phenomena in microgravity flow boiling. 10th International Symposium on Transport Phenomena in Thermal Science and Process Engineering **2**: 373–378.
6. QIU, D.M., V.K. DHIR, M.N. HASAN, et al. 1999. Single bubble dynamics during nucleate boiling under low gravity conditions. In Microgravity Fluid Physics and Heat Transfer. V.K. Dhir, Ed.: 62–71. Begell House, New York.
7. SINGH, S. & V.K. DHIR. 1999. Effect of gravity, wall superheat, and liquid subcooling on bubble dynamics during nucleate boiling. In Microgravity Fluid Physics and Heat Transfer. V.K. Dhir, Ed.: 106–113. Begell House, New York.
8. OHTA, H., M. KAWAJI, H. AZUMA, et al. 1998. Heat transfer in nucleate pool boiling under microgravity condition. Heat Transfer 1998, Proc. 11th Int. Heat Transfer Conf., Kyongju, Korea, **2**: 401–406.
9. OHTA, H., M. KAWAJI, H. AZUMA, et al. 1997. TR-1A rocket experiment on nucleate pool boiling heat transfer under microgravity. Heat Transfer in Microgravity Systems, Dallas, TX, HDT-354: 249–256.
10. RITE, R. & K.S. REZKALLAH. 1995. New heat transfer data for two-phase, gas-liquid flows under microgravity conditions. In Advances in Multiphase Flow. 521–528. Elsevier Science, New York.
11. ZHAO, L. & K.S. REZKALLAH. 1995. Pressure drop in gas–liquid flow at microgravity conditions. Int. J. Multiphase Flow **21**: 837–849.
12. RITE, R. & K.S. REZKALLAH. 1994. Heat transfer in two-phase flow through a circular tube at reduced gravity. AIAA J. Thermophys. Heat Transfer **8**: 702–708.
13. LOWE, D. & K.S. REZKALLAH. 1999. Flow regime identification in microgravity two-phase flows using void-fraction signals. Int. J. Multiphase Flow **25**: 433–457.
14. DEJONG, P & K.S. REZKALLAH. 1998. A study of annular flow in microgravity conditions. J. Jpn. Soc. Micrograv. Appl. **15**: 226–231.
15. BRUSSTAR, M.J. & H. MERTE, JR. 1997. Effects of heater surface orientation on the critical heat flux—II. a model for pool and forced convection subcooled boiling. Int. J. Heat Mass Transfer **40**(17): 4021–4030.
16. MATSUNAGA, K., H. OHTA, I. HIGAKI, et al. 2000. Effect of gravity on liquid–vapor behavior and heat transfer for flow boiling in narrow gaps. J. Jpn. Soc. Micrograv. Appl. **17**(Suppl.): 136–137.
17. BRUSSTAR, M.J., H. MERTE, JR. & R.B. KELLER. 1995. Relative effect of flow and orientation on the critical heat flux in subcooled forced convection boiling. NASA Rep. UM-MEAM-95-15.

Pool Film Boiling Experiments on a Wire in Low Gravity

Preliminary Results

P. DI MARCO, W. GRASSI, AND F. TRENTAVIZI

LOTHAR, Dipartimento di Energetica, Università di Pisa, Pisa, Italy

ABSTRACT: This paper reports preliminary results for pool film boiling on a wire immersed in almost saturated FC72 recently obtained during an experimental campaign performed in low gravity on the European Space Agency Zero-G airplane, (reduced gravity level 10^{-2}). This is part of a long-term research program on the effect of gravitational and electric forces on boiling. The reported data set refers to experiments performed under the following conditions: (1) Earth gravity without electric field, (2) Earth gravity with electric field, (3) low gravity without electric field, and (4) low gravity with electric field. Although a decrease of gravity causes a heat transfer degradation, the electric field markedly improves heat exchange. This improvement is so effective that, beyond a certain field value, the heat flux is no longer sensitive to gravity. Two main film boiling regimes have been identified, both in normal and in low gravity: one is affected by the electric field and the other is practically insensitive to the field influence.

KEYWORDS: pool boiling; film boiling; boiling enhancement; electric field; microgravity

NOMENCLATURE:

a	vapor layer thickness, m
El^*	electrical influence number
g	gravity acceleration, m/sec^2
I	current intensity, A
k	wave number, m^{-1}
l_L	Laplace length, $\sigma^{1/2}/(\rho_l - \rho_v)^{1/2} g^{1/2}$, m
p	pressure, Pa
q	heat flux, W/m^2
R	radius of the wire, m
R'	dimensionless radius, R/l_L
T	temperature, K
V	applied high voltage, V
Greek Symbols	
α	heat transfer coefficient, W/m^{2}K
ΔT_{sat}	wire superheat, K
ε	relative dielectric permittivity
ε_0	vacuum dielectric permittivity, F/m
λ_d	oscillation wavelength, kg/m^3
ρ	density, kg/m^3
σ	surface tension, N/m

Address for correspondence: W. Grassi, LOTHAR, Dipartimento di Energetica, Università di Pisa, Via Diotisalvi, 2, 56126 Pisa, Italy. Voice: +39.050.569646; fax: +39.050.830116.
w.grassi@ing.unipi.it

Suffixes

corr	corrected
cyl	cylinder
E	in the presence of the electric field
l	liquid
sat	saturated
v	vapor
w	heated wall
*	transition from first to second regime

INTRODUCTION

In saturated film boiling on a horizontal cylinder a stable vapor film entirely surrounds the cylinder separating the heater surface from the surrounding liquid. On ground the vapor layer has a varying thickness, increasing from bottom to top of the cylinder. Contact between liquid and heating wall are prevented and the vapor–liquid interface oscillates. Equally spaced vapor bubbles detach from the interface and rise into the liquid. Conduction, convection, and radiation across the vapor layer are the involved heat exchange mechanisms. Their relative importance depends on the local vapor fluid dynamics (e.g., heater geometry and orientation), thermophysical properties, and wall superheat. In particular, radiation starts playing a significant role at high wall superheat.

The first theoretical model of film boiling on a horizontal cylinder was proposed by Bromley.[1,2] Several further models, all based on refinements of that due to Bromley have been proposed. The most comprehensive correlation for film boiling on wires was developed by Sakurai.[3,4] This correlation fitted a large amount of experimental data, obtained on Earth, both in saturated and in subcooled conditions. In addition, Sakurai *et al.*[5] collected data on interface wavelength, bubble detachment diameter, and frequency in film boiling.

A comparison among the correlations supplied by most of the existing models is provided in Reference 6, with reference to the heater geometry investigated herein. Large discrepancies (up to 350%) among their predictions were noted in the above report.

Apart from this very disappointing situation concerning correlations, there is a general agreement on the fundamental role played by the wavelength of the liquid–vapor interface oscillations and, in particular, by the so called "most dangerous" wavelength. In general this wavelength is recognized to be of the form

$$\lambda_d = 2\pi\sqrt{3}l_L F(R'), \tag{1}$$

where l_L is the capillary Laplace length and $F(R')$ is a function of the dimensionless wire radius R', with

$$R' = \frac{R}{l_L}. \tag{2}$$

That is, the square root of the Bond number, *Bo*. Various relations have been proposed for $F(R')$, among which we recall that due to Lienhard and Wong[7]

$$F_L(R') = \frac{1}{\sqrt{1 + 2/R'}}, \tag{3}$$

and that due to Sakurai *et al.*[5]

$$F_S(R') = \frac{2R'}{2R' + 0.85}. \qquad (4)$$

In a recent paper[8] the present authors obtained a good fit to experimental results by modifying the Sakurai correlation to account for the vapor layer thickness, a. This thickness was estimated simply by assuming a conductive heat transfer mechanism within the vapor film. Thus, in Equation (4), R' was replaced by R'_{corr}, given by

$$R'_{corr} = \frac{R+a}{l_L}, \quad a = R(e^{k_v/\alpha R} - 1). \qquad (5)$$

The brief outline above clearly shows that the gravitational force plays an essential role in this boiling regime, since it exerts a destabilizing action on the liquid–vapor interface. According to some theoretical approaches (Lienhard's modified hydrodynamic theory and Katto's theory accounting for the hydrodynamic instability of vapor stems), this force also exerts a key influence on critical heat flux, CHF. On this basis, a better understanding of film boiling mechanisms might shed light on the physics of CHF.

However, gravity is not the only force affecting the interface behavior. In particular, electric forces heavily influence this behavior, due to the differing electric permittivities of the two phases (liquid and vapor).[8–10] They too tend to destabilize the interface. In general, an increase in both these forces (gravitational and electric) causes a decrease in the oscillation wavelength and a corresponding increase in the detachment frequency of bubbles from the interface.

The effect of an electric field on film boiling has been studied for more than forty years, but only more recently has this research become more systematic and aware of physical aspects often not taken into account in the past (e.g., relaxation time, electric field frequency, and heater size). In addition, the present authors are the first to perform experiments with an electric field in low gravity. Here, we refrain from examining the existing literature in detail. We review briefly previous work that is closely related to the present paper.

The classical paper by Baboi *et al.*,[11] investigating pool boiling of toluene and benzene on a wire heater and applying an electric field between the heater and a parallel wire already referred an exponential increase of the heat transfer coefficient with the applied field. The few measurements of the film wavelength reported show a wavelength decrease with increasing field intensity. The film boiling regime was reported to disappear for a field intensity higher than 50 MV/m, with bubbles also departing from the lower side of the heater.

Jones and Schaeffer[12] and Jones and Hallock[13] performed boiling experiments with R113 in an apparatus quite similar to the present one. They also obtained film boiling data showing a clear heat transfer enhancement in this regime, as well as a decrease in the oscillation wavelength, both due to the electric field. Berghmans[14] improved the previous theoretical models, also accounting for a two-dimensional wave pattern around the wire.

Recently, Verplaetsen and Berghmans[15] modelled the effect of electric field on pool film boiling on a plane heater, developing a correlation showing good agreement with experiments. Carrica *et al.*[16] performed experiments on film boiling of R113 on a platinum wire, 0.2 and 0.3 mm in diameter. For the first time, it was clearly demonstrated that the electrohydrodynamic (EHD) enhanced film boiling regime cannot be sustained indefinitely in this configuration. In fact, a transition takes place at high enough wire superheat, bringing the system back to a film boiling regime that is practically unaffected by the presence of the field. Later, confirmation that the application of an electric field has a weak influence on film boiling performance at high wall superheat was obtained by Di Marco et al.,[17] who used R113 and Vertrel XF (Dupont) as testing fluids. Quite recently Cipriani et al.[8] proposed a simplified model for this transition, also accounting for the vapor film thickness.

The aim of the present paper is to investigate in some detail the effect of electric field on pool film boiling on wires in FC-72, within a broader research program on the effect of force fields on boiling. Electric and gravitation forces are taken into account at the present stage and are dealt with in the following. The choice of wires as heaters is imposed mainly by the requirement of using test samples with low thermal inertia, owing to the short time available (20 seconds) during each microgravity test on the airplane.

EXPERIMENTAL APPARATUS

The experimental apparatus is shown in FIGURE 1. It consists of an aluminium parallelepiped vessel containing the test section. A bellows, connected to this vessel and operated by pressurized nitrogen on the secondary side, is compensated for volume variations due to vapor production and thermal dilatation of the liquid. The

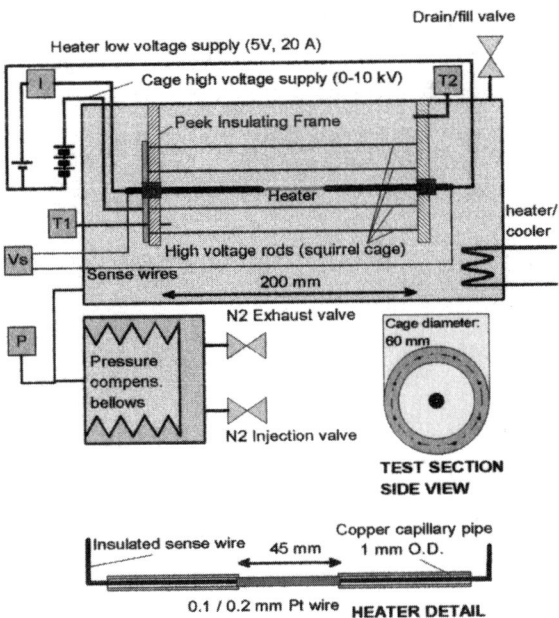

FIGURE 1. Experimental apparatus.

vessel has two windows to allow for visualization of the phenomena by means of a commercial and a high-speed video camera, with appropriate back-illumination. The two cameras can record the occurring phenomena through the same window thanks to the use of a beam splitter. The high speed camera is a Phantom V 4.0 with a memory of 1,024 Mbytes (4,096 frames, equal to four seconds of recording at 1,000 fps, and full resolution of 512×512 pixels) and connected to a dedicated laptop (Sony Pentium III). The camera was operated at 500 fps during most of the experiments. Light is provided by a flat illuminating set (surface size 85.6×108 mm, made by Dolan-Jenner mod. QVABL-48, 150W) An external heating-cooling system, governed by a PID controller, maintains the fluid temperature constant to within 0.5 K. A much more detailed description of the system can be found in Reference 18.

Experiments were carried out using a horizontal platinum wire, 0.2 mm diameter and 45 mm length, heated by Joule effect (dc current) that served as both a resistance heater and a resistance thermometer. The heater (see FIG. 1) was made by brazing the platinum wire (the active heater) coaxially to two copper capillary tubes (1 mm od, 0.2 mm id), designed to work at less than 1 K superheat at the maximum current rate in the experiments. Two further thin insulated wires (0.08 mm diameter) were passed inside the copper tubes and brazed to the copper–platinum junctions for direct voltage sensing and measurement. This design was chosen to eliminate the distortion effects on the electric field that might arise in the presence of external sensing wires at the junctions. The fluid adopted in the tests was FC-72 (C_6F_{14}) a fluoroinert liquid, trademark by 3M, slightly subcooled (1–3 K) at a pressure of 115 kPa.

The electric field was produced applying a high dc voltage (up to 7 kV) between the wire and a coaxial 8-rod cylindrical "squirrel cage", 60 mm diameter and 200 mm length. The electric field was thoroughly characterized via finite element and finite difference analysis,[19] assuming that the fluid was single-phase and homogeneous in the domain investigated. It was shown that the length of the cage is sufficient to avoid side effects along the active part of the heater. For the 0.2 mm wire, the analysis showed that the field obeyed the law $E = CV/r$ (E electric field intensity at a distance r from the wire axis, V the applied high voltage, and $C = 0.1034$, a dimensionless constant) up to 10 mm away from the wire axis. For a solid cylindrical electrode of the same diameter the constant can be calculated analytically and its value is $C = 0.1735$.

Two main types of tests were performed in film boiling during the flight.

- *Constant heat flux tests*—the wall heat flux was kept at the same value during the whole parabola, starting at $g/g_0 = 1$, before the pull-up, and ending at $g/g_0 = 1$, after the pull-out. These tests provided direct results on the influence of acceleration on the involved phenomena and on the related physical quantities.

- *Variable heat flux tests*—stable film boiling was established during level flight ($g/g_0 = 1$). Starting at the beginning of the microgravity phase the heat flux was increased linearly according to a computer aided procedure. This allowed us to obtain various film boiling curves to correspond with the same acceleration value.

During the low gravity phase the mean value of g/g_0 was found to be 0.01 with a standard deviation of 0.02. Signals were acquired by means of a Toshiba Satellite notebook computer equipped a 12-bit PCMCIA data acquisition card (National

TABLE 1. Measurement range and maximum accidental error of the main measured quantities

Quantity	Measurement Range	Maximum Error
Heating current, I	$4 < I < 12\,\text{A}$	0.085% at 4 A
Heater voltage, V_s	$1.5 < V < 5\,\text{V}$	1.25 mV
Liquid temperature, T	$38.4 < T < 70.2\,°\text{C}$	0.3 K
Pressure, p	115 kPa	0.2%
High voltage, V	$0 < V < 7\,\text{kV}$	100 V
Wire temperature, $T_w{}^a$	$200 < T_w < 900\,°\text{C}$	5.7% at 900°C
Wire superheat, $T_w - T_{sat}{}^a$	$150 < T_w - T_{sat} < 850\,\text{K}$	6% at $T_w = 900\,°\text{C}$
Heat flux, q	$140 < q < 900\,\text{kW/m}^2$	5.46%

^{a}Obtained by numerical elaboration.

Instruments DAQ Card AI-16E-4) and conditioned according to a well assessed procedure[8,18] to obtain the data listed in TABLE 1. In particular, the wire temperature was derived from the platinum temperature–resistance curve and it was necessary to correct it analytically to account for the effect of the colder side-ends of the heater. TABLE 1 reports the uncertainty of the main measurements performed. Tests at normal gravity were also performed before and after each flight for comparison.

EXPERIMENTAL RESULTS AND DISCUSSION

A few words are worth saying about the vapor pattern in film boiling on a horizontal cylinder, before going into a detailed discussion of the matter of this article. The classic pattern on Earth, for small cylinders or wires, is a vapor layer surrounding the cylinder, with a varying circumferential thickness: smaller on the lower side and thicker on the upper one. Bubbles detach from the interface on the upper side and are regularly spaced along a coordinate parallel to the cylinder axis. In other words, no waves exist along the cylinder circumference. In this case, we refer to a one-dimensional wave distribution. If the cylinder diameter is large enough it can also accommodate waves along its circumferential periphery. In this case, the wave pattern becomes two-dimensional, similar to what happens on a flat surface. It is very easy to infer that this should at least depend on the cylinder Bond number. More precisely the comparison should be done between the actual interface wavelength and the cylinder diameter. For sake of clarity and brevity, we call the one-dimensional wave pattern Regime A and the two-dimensional one Regime B. At very low gravity level there is no reason why vapor thickness in Regime A should vary, except for the effect of g-jitter. Thus, strictly speaking, we expect a sort of "modified" Regime A, with an increased size of detaching bubbles. The effect exerted by an electric field is the reduction in the wavelength value according to a relation of the form[8]

$$\lambda_{dE} = \lambda_d \frac{\sqrt{3}}{El^* + \sqrt{El^{*2} + 3}}, \qquad (6)$$

FIGURE 2. Regime A: mono-dimensional wave pattern, $p = 115$ kPa, $d = 0.2$ mm, $V = 0$ kV, $q = 400$ kW/m^2, $\Delta T_{sat} = 449$ K.

where the electric influence number El^* is given by

$$El^* = \frac{\varepsilon_{eq} \cdot E_{0v}^2 \cdot l_L}{\sigma} = \frac{\varepsilon_{eq} \cdot E_{0v}^2}{\sqrt{(\rho_l - \rho_v)\sigma g}}. \quad (7)$$

The equivalent electric permittivity for two dielectric materials, assuming that the liquid layer has a thickness far larger than that of the vapor, a, has the form[1]

$$\varepsilon_{eq} = \frac{\varepsilon_0 \varepsilon_l (\varepsilon_l - \varepsilon_v)^2}{\varepsilon_v [\varepsilon_i(ka) + \varepsilon_v]}; \quad (8)$$

and that the electric field is evaluated at the interface, supposed cylindrical with a radius $R + a$,

$$E_l = \frac{V}{R+a}\left(\ln\frac{R+a}{R} + \frac{\varepsilon_v}{\varepsilon_l}\ln\frac{R_2}{R+a}\right)^{-1}, \quad (9)$$

in agreement with what the present authors have always stressed[10] concerning the importance of considering the value of the electric field at the liquid–vapor interface in the study of the instability of this interface.

As already mentioned in the INTRODUCTION, the present authors, on Earth, detected both Regime A and Regime B, depending on the value of the applied electric field. One more regime (Regime C, also in previous papers called "second film boiling regime") was discovered where the process proved to be almost insensitive to the electric field beyond a certain wall superheat, again assuming a one-dimensional wave pattern. The corresponding vapor patterns are shown in FIGURES 2, 3, and 4, at normal gravity. One possible explanation for the transition from Regime B to Regime C was suggested in[8] as follows. The vapor thickness increases with increasing wall superheat. This fact causes a weakening of electric field on the interface, due

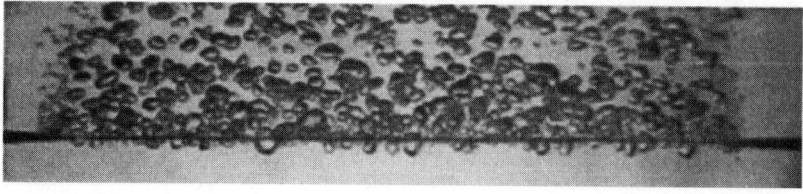

FIGURE 3. Regime B: two-dimensional wave pattern due to the presence of electric field, $p = 115$ kPa, $d = 0.2$ mm, $V = 7$ kV, $q = 400$ kW/m^2, $\Delta T_{sat} = 201$ K.

FIGURE 4. Regime C: independent of applied voltage, one-dimensional oscillatory pattern, $p = 115\,\text{kPa}$, $d = 0.2\,\text{mm}$, $V = 7\,\text{kV}$, $q = 850\,\text{kW/m}^2$, $\Delta T_{sat} = 781\,\text{K}$.

to its cylindrical distribution. Consequently, interface behavior is only poorly affected by the electric forces. A similar approach might be also adopted to explain the very weak influence of an electric field on fully developed nucleate boiling in this same geometry. In TABLE 2, we summarize these observations about the various regimes.

Tests at Various Gravity Levels

In a parabolic flight, various values of acceleration act on the process (see Ref. 19 for a more complete description of the procedure):

- $g/g_0 = 1$ during the levelled (horizontal) flight;
- $g/g_0 = 1.8$ in the pull-up phase;
- $g/g_0 = 0.01$ during the low gravity stage (parabola);
- $g/g_0 = 1.6$ in the pull-out phase.

Thus, data corresponding to these acceleration levels could be obtained.

FIGURE 5 clearly shows the effect of gravity on bubble spacing (wavelength), with the expected increase with reducing gravity. A comparison between the wavelength experimental data set and correlations was performed. The modified correlation by Sakurai et al.[5] has been found to be the most reliable fit to the data, as shown in FIGURE 6 for one of the parabolas. This correlation is (see Eqs. (**3**) and (**4**)) by

$$\lambda_d = 2\pi\sqrt{3}\lambda_L\left(\frac{2R'_{corr}}{2R'_{corr} + 0.85}\right). \tag{10}$$

TABLE 2. Film boiling regimes at normal gravity

Regime	Possible Subregime
First film boiling regime (effect of electric field)	Regime A (1-dimensional)
	Regime B (2-dimensional)
Second film boiling regime (no electric field effect)	Regime C (1-dimensional)

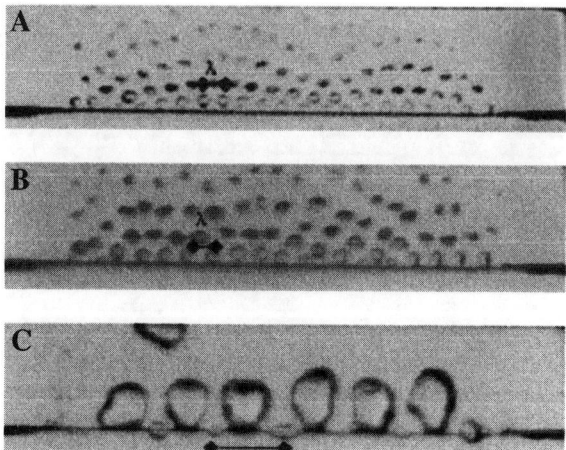

FIGURE 5. Effect of acceleration on wavelength in the absence of electric fields. Wall heat flux $q = 190\,\text{kW/m}^2$. **(A)** $g/g_0 = 1.8$. **(B)** $g/g_0 = 1.0$. **(C)** $g/g_0 = 0.01$.

A correlation between wavelength measurements and the heat transfer coefficient occurring during the various flights was detected. This trend is reported in FIGURE 7. The curve of g/g_0 (values on the right ordinate axis in the graph) clearly shows the different flight stages. The heat transfer coefficient closely follows the acceleration trend. Thus, in this case, the exchange is strictly connected to the wave pattern, and modified by the gravity value, as shown in FIGURE 5.

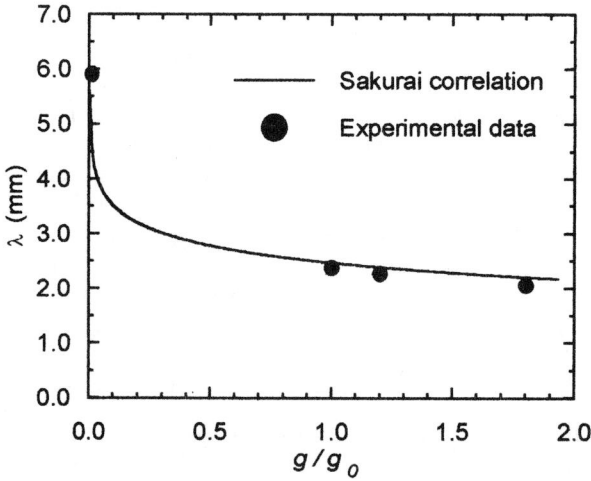

FIGURE 6. Wavelength versus gravity: comparison of the experimental data with the correlation, Equation **(10)**, due to Sakurai *et al.*[5]

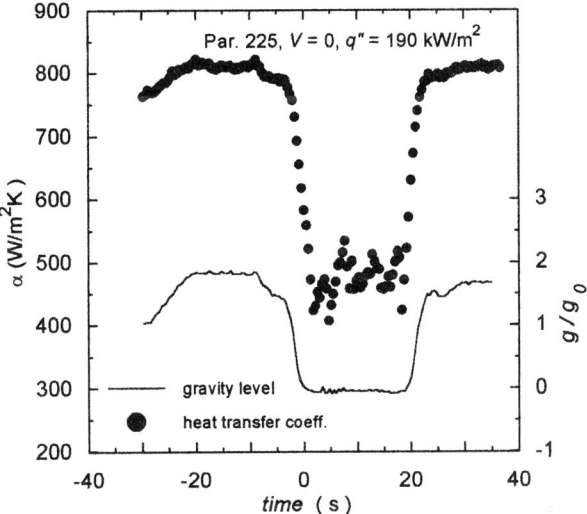

FIGURE 7. Trends of acceleration and corresponding heat transfer coefficient versus time, during parabolic flight. The low gravity phase starts a little before zero and ends around 20 seconds.

Influence of an Imposed Electric Field

The effect of an electric field on the vapor pattern has already been shown in FIGURES 2 and 3. Its effect on the heat transfer coefficient during a flight test is reported in FIGURES 7 and 8, for various values of the high voltage. The corresponding acceleration trend is plotted in FIGURE 7, with the micro-g phase spanning from 0 to about 20 sec.

The graphs of FIGURE 8 compared with that in FIGURE 7, show that the electric field enhances heat exchange throughout the entire range of tested gravity levels, and beyond a high voltage of 2 kV this coefficient does not exhibit appreciable sensitivity to acceleration. This is obviously true for the regime we call "first film boiling regime".

To provide a better evidence of this, film boiling curves (wall heat flux vs. wall superheat) under various conditions were obtained and are shown in FIGURE 9. Each curve is countersigned with the level of gravity (Earth or p.nnn for μ-g, where nnn is the number of the parabola) and the value of the applied high voltage (V).

In the absence of electric field the influence of gravity on the heat exchange can be easily detected by comparing the curve obtained on the ground (Earth, $V = 0$) with that at low gravity (p.309, $V = 0$). As expected a decrease in the acceleration leads to a decrease in the heat exchange. Therefore, a close connection between the wave pattern and the heat exchange is proved to exist, but does not hold for fully developed nucleate boiling in this geometry.[20]

The presence of the electric field improves the heat transfer both on Earth and in microgravity. In particular, it is quite evident from the figure that, above a given field threshold (2 kV in the present case), boiling curves obtained on ground and in low

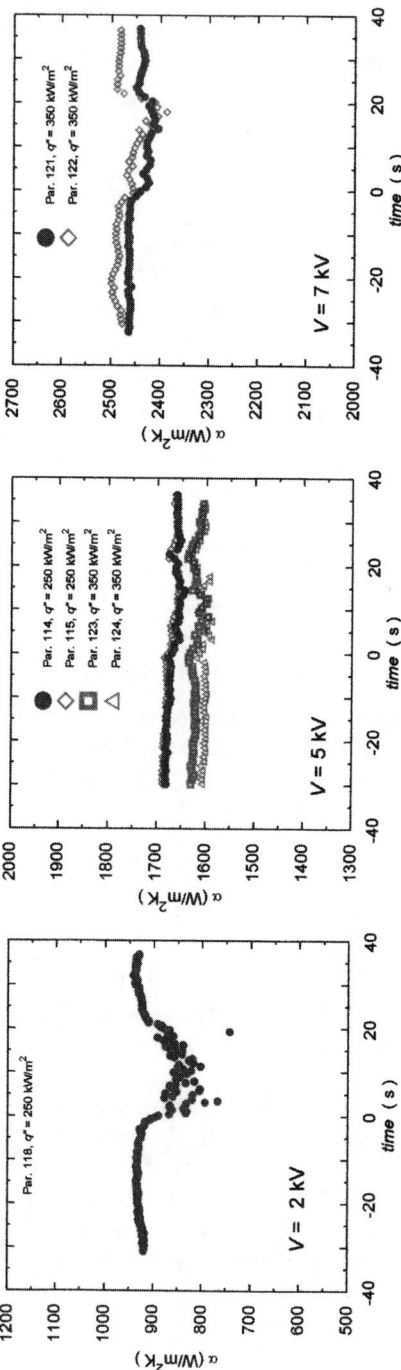

FIGURE 8. Trends of the film boiling heat transfer coefficient versus time during several flight tests. The high voltage value (HV) and the heat flux (q) adopted are indicated in the plots.

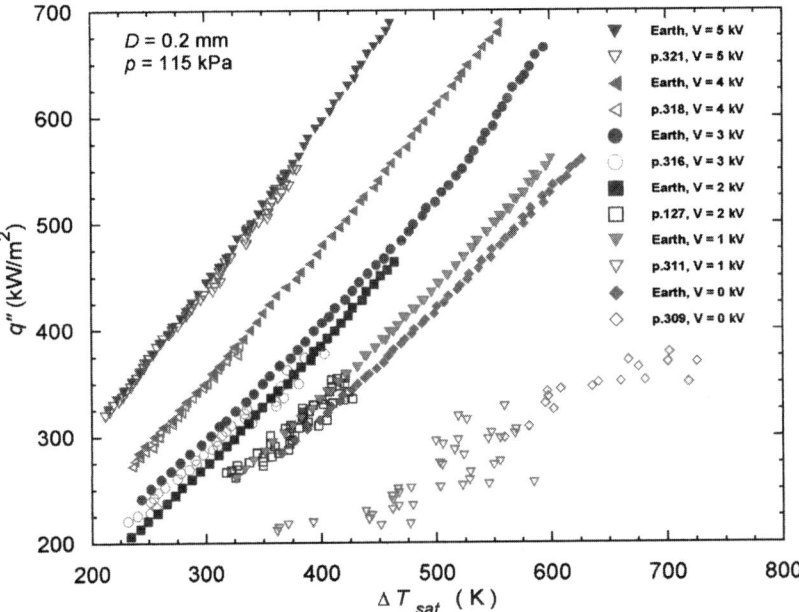

FIGURE 9. Wall heat flux versus wall superheat for the first film boiling regime: effect of gravity and electric field.

gravity, with the same electric field, are indistinguishable. Of course this was already clear from the plots of FIGURE 8. Corresponding to the same high voltage value a change from Regime A to Regime B seems to take place. At the present stage it is possible to infer that above this threshold the vapor–liquid interface wavelength is essentially controlled by the electric field as the electric forces become absolutely prevailing over the gravitational force. This should be considered as a further proof of the fundamental influence of the interface fluid-dynamics on film boiling heat transfer.

It is quite evident how a conclusion like this might also provide an essential contribution about the role played by the interface behavior on CHF, at least for small cylindrical heaters.

Thus far, we have discussed the experimental data obtained with reference to what we named "first film boiling regime". The next paragraph is dedicated to the second regime.

Second Film Boiling Regime (Regime C)

FIGURE 10 shows the film boiling curves that can be obtained on Earth at various values of the applied electric field. Regimes A, B, and C can be clearly identified, with the collapse of the curves into a single curve. We dedicate the following discussion to this transition, comparing the results obtained on Earth with those in low gravity.

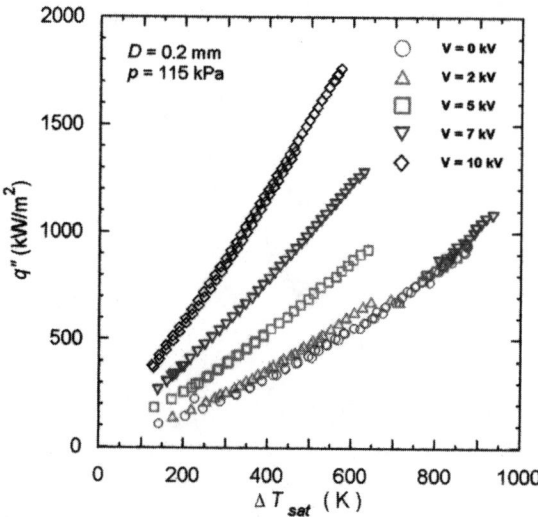

FIGURE 10. Film boiling curves on Earth for FC72.[8] Regime C is located beyond a wall superheat of about 600°C.

The results obtained in three parabolic flights are reported in FIGURE 11. The dotted line, marked AB, represents (in a qualitative manner) the common film boiling curve on which the curves of the first film boiling regime collapse. Even at first glance, it clearly appears that the wall superheat corresponding to the end of the first

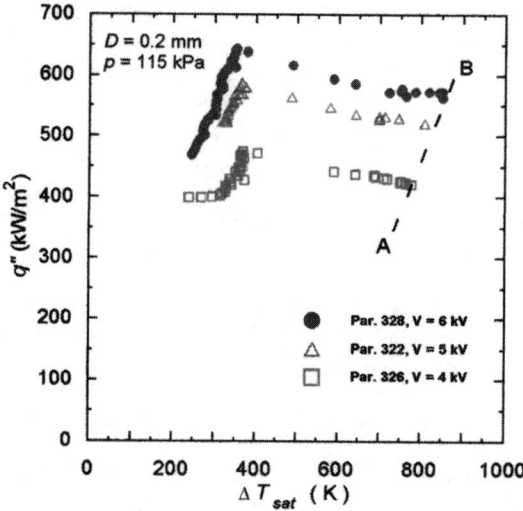

FIGURE 11. Transition from Regime A to Regime C in low gravity. The *dotted line* represents the final curve on which the curves for various electric fields collapse.

boiling regime in microgravity is about 200 K lower than on E. A direct comparison between on-ground and low gravity data, at the same electric field, was, therefore, made to gain a better insight concerning this point. This comparison is shown in FIGURE 12.

Unfortunately, at the present stage, we have too few data (three flights) about this transition, from first to second boiling regime in low gravity, to draw any general conclusion. Therefore, further investigation is needed. Nevertheless the results we obtained, reported in TABLE 3, supply very clear indications about the reduction of the above mentioned wall superheat. This can be explained as follows. In microgravity, the vapor volume surrounding the wire is larger than on ground. Therefore, on an average basis, the liquid–vapor interface is farther from the heater and the corresponding electric field, acting on the interface with the same applied high voltage, is weaker. In addition, TABLE 2 singles out how this transition occurs at the same wall

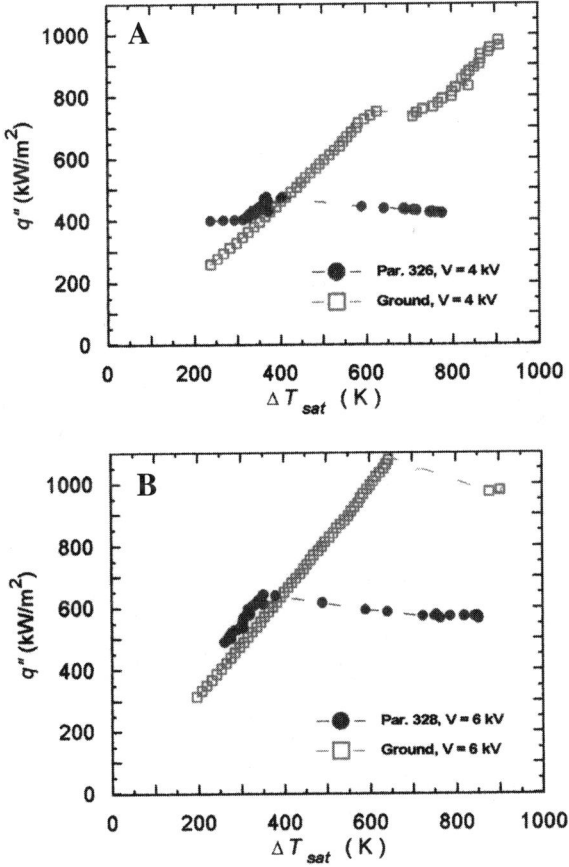

FIGURE 12. Comparison between transitions on earth and in low gravity, with an applied high voltage: **(A)** 4 kV, **(B)** 6 kV.

TABLE 3. Heat flux q^* and wall superheat ΔT^* at the beginning of transition

High voltage (V)	q^* (kW/m^2)	ΔT^* (K)
Microgravity		
4,000	475	368
5,000	587	364
6,000	642	353
On Earth		
4,000	701	558
5,000	874	590
6,000	1,010	580

superheat for each level of gravity. Thus, this superheat (and not the wall heat flux) seems to be the mechanism that controls the phenomenon.

The availability of the high-speed videocamera allowed us to obtain a detailed visual documentation of the transition mechanism in low gravity. A complete set of images is reported in FIGURE 13. The photographs in FIGURE 13 show a much less regular vapor structure than that of FIGURE 5C. This is essentially related to the different value of heat flux (190 kW/m^2 for FIG. 5C and 642 kW/m^2 for FIG. 13) and wall temperature. The random direction of bubbles leaving the heater is attributable to g-jitter. This has no effect on the heat transfer thanks to the cylindrical geometry of the heater that allows bubbles to escape from the test sample in any direction. The same should not be expected for a plane surface, where an upward directed acceleration (boiling on the upper wall of the heater) pushes bubbles onto the surface.

A large level of horizontal bubble coalescence is detectable in the photograph (the most evident are indicated with a circle) with the formation of very large vapor masses. This notwithstanding, a regular structure (regular bubble spacing on the heater) can be observed on the heater. Quite interesting is how the transition takes place: that is, through a vapor front propagation mechanism resembling the mechanism referred to by Lienhard and Dhir[21] for CHF at very low Bond numbers. It is in fact very easy to identify in the set of photographs of FIGURE 13, a movement of the large bubbles front from right to left.

For the sake of comparison, a photograph of the transition at normal gravity is shown in FIGURE 14. In this case too, Regime C starts from a region of the heater, then spreading out over the whole surface. Except for the obvious difference in the vapor pattern that occurs, the transition to this regime seems to take place according to a similar mechanism (vapor front propagation) both in reduced gravity and normal gravity.

CONCLUSIONS

The results obtained about the influence of the gravitational and electrical fields on pool film boiling of FC72 on a heating wire have been presented in this paper. The following experimental conditions were investigated: (1) Earth gravity without

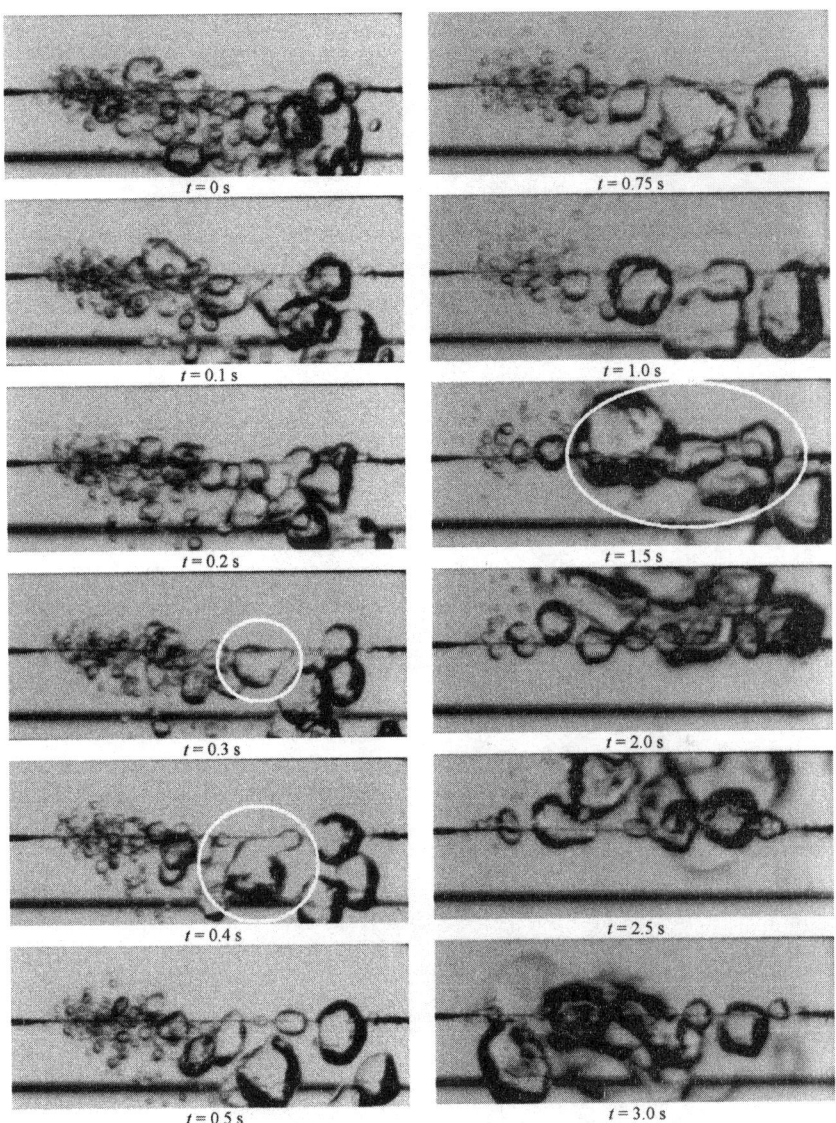

FIGURE 13. Vapor pattern evolution during transition from first to second film boiling regime in low gravity with a high voltage of 6 kV. Time elapsed from the beginning of transition is reported below each photograph.

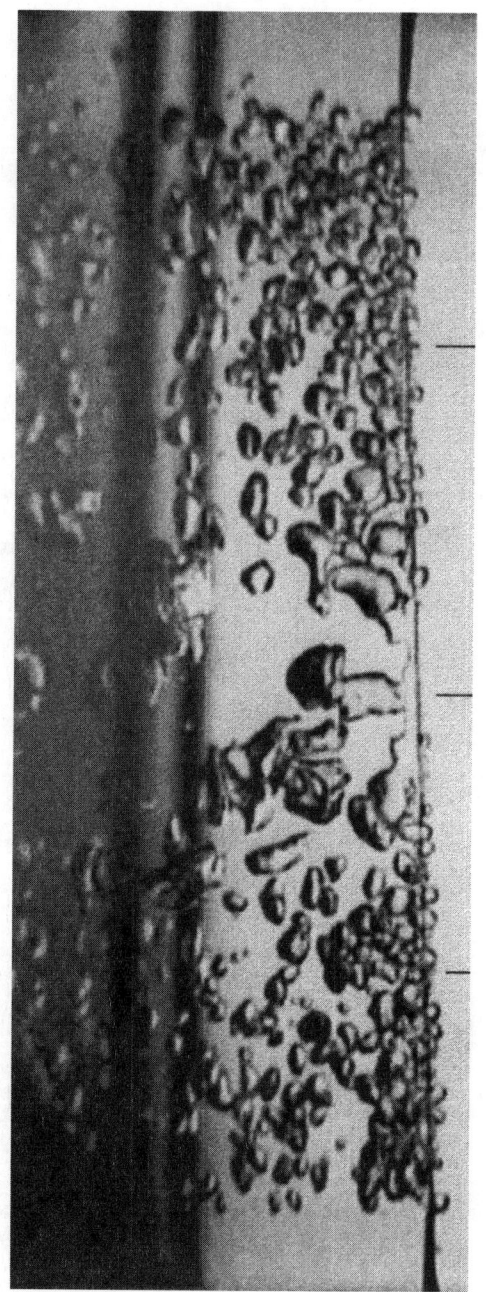

FIGURE 14. Transition from the first to the second film boiling transition (Regime B to Regime C) at normal gravity.

electric field, (2) Earth gravity with electric field, (3) low gravity without electric field, and (4) low gravity with electric field. The main results discussed in the paper can be summarize as follows:

- A very close connection between the vapor morphology on the heating wall and, in particular, the wave structure of the liquid–vapor interface and the heat exchange has been proved to exist. A reduction in gravity causes an increase in wavelength and a decrease in heat transfer. On the other hand, the effect of increasing the value of the applied electric field is a wavelength reduction with a corresponding heat transfer augmentation.

- Heat transfer improvement due to the electric field takes place on Earth and in low gravity, so that the heat transfer coefficient, beyond a certain field threshold, becomes insensitive to gravity changes.

- Film boiling also showed both a one-dimensional and a two dimensional wave pattern under reduced gravity conditions. This confirms previous results found on ground. Thus, two main film boiling regimes (first and second regimes) can be identified. The first is heavily affected by the electric field (independent of the gravity value) and can be separated in two sub-regimes: Regime A, with a one-dimensional wave pattern, and Regime B, with a two-dimensional wave pattern. The transition between the two subregimes is controlled by the electric field. The second film boiling regime (Regime C) implies a one-dimensional wave structure, irrespective of the electric field value. The transition to this regime is controlled by the wall superheat. This superheat value, in turn, depends on the value of gravity.

- The importance of accounting for the vapor film layer thickness has been widely stressed in this paper.

- A mechanism of vapor film propagation has been clearly evidenced to lead from the first to the second film boiling regime.

The reported data show how experimentation in microgravity may shed light on the fundamental physical mechanisms of film boiling and also critical heat flux. This is the reason why the present authors already requested ESA (European Space Agency) to participate to the next parabolic flight campaign to be held in 2002. In particular, various levels of gravity should be available during the above campaign.

ACKNOWLEDGMENTS

The apparatus was designed and built in the laboratories of the University of Pisa, with the assistance of Kayser Italia. Mr. R. Manetti designed and assembled the electronics. Dr. G. Memoli gave his assistance in setting up the optical system. Both Mr. Manetti and Dr. Memoli actively participated to the research campaign in Bordeaux and joined the first two authors on board the A-300 "Zero-G".

This work was funded by ASI (Italian Space Agency) and by ESA (European Space Agency). The authors are indebted to ESA personnel and in particular to the crew on board the aircraft and to the people from Novespace who did a wonderful job. A very particular thanks goes to Dr. Vladimir Pletser, from ESA, a gentle and most effective scientist who did an enormous amount of work to make things go smoothly and made this campaign a real success.

REFERENCES

1. BROMLEY, L.A. 1950. Heat transfer in stable film boiling. Chem. Eng. Prog. **58:** 67–72.
2. BROMLEY A.L. 1952. Effect of capacity of condensate. Ind. Eng. Chem. **44:** 2966.
3. SAKURAI, A., M. SHIOTSU & K. HATA. 1990. A general correlation for film boiling heat transfer from horizontal cylinder to subcooled liquid: Part 1—a theoretical pool film boiling heat transfer model including radiation contribution and its analytical solution. J. Heat Transfer Trans. ASME **112:** 430–440.
4. SAKURAI, A., M. SHIOTSU & K. HATA. 1990. A general correlation for film boiling heat transfer from horizontal cylinder to subcooled liquid: Part 2—experimental data for various liquids and its correlation. J. Heat Transfer Trans ASME **112:** 441–450.
5. SAKURAI, A., M. SHIOTSU & K. HATA. 1984. Effect of system pressure on film boiling heat transfer, minimum heat flux and minimum temperature. Nucl. Sci. Eng. **88:** 321–330.
6. CIPRIANI, M. 1999. Experimental Study on the Influence of the Electric and Gravitational Fields on Pool Film Boiling (in Italian), Degree Thesis, University of Pisa.
7. LIENHARD, J.H. & P.T.Y. WONG. 1964. The dominant unstable wavelength and minimum heat flux during film boiling on an horizontal cylinder. J. Heat Transfer Trans. ASME **86:** 220–226.
8. CIPRIANI, M., P. DI MARCO & W. GRASSI. 2000. Effect of an externally applied electric field on pool film boiling of FC72. Proc. 18th UIT National Conference, Cernobbio, Como (I), ETS, Pisa. 703–714.
9. DI MARCO, P. & W. GRASSI. 1993. Saturated pool boiling enhancement by means of an electric field. J. Enhanced Heat Transfer **1:** 99–114.
10. DI MARCO, P. & W. GRASSI. 1994. Gas–liquid interface stability in presence of an imposed electric field. Proc. 12th UIT National Conference, L'Aquila (I), ETS, Pisa. 299–310.
11. BABOI, N.F., M.K. BOLOGA & A.A. KLYUKANOV. 1968. Some features of ebullition in an electric field. Appl. Electr. Phenom. **2**(20): 57–70.
12. JONES, T.B. & K.R. HALLOCK. 1978. Surface wave model of electrohydrodynamically coupled minimum film boiling. J. Electrostat. **5:** 273–284.
13. JONES, T.B. & R.C. SCHAEFFER. 1976. Electrohydrodynamically coupled minimum film boiling in dielectric liquids. AIAA J. **14:** 1759–1765.
14. BERGHMANS, J. 1978. Prediction of the effect of electric field on the most unstable wavelength during film boiling on small wires. J. Electrostat. **5:** 265–272.
15. VERPLAETSEN, F.M. & J.A. BERGHMANS. 1999. Film boiling of an electrically insulating fluid in the presence of an electric field. Heat Mass Transfer/Waerme-Stoffuebertragung **35:** 235–241.
16. CARRICA, P., P. DI MARCO & W. GRASSI. 1996. Electric field effects on film boiling on a wire. Exp. Heat Transfer **9:** 11–27.
17. DI MARCO, P., W. GRASSI & I. IAKOVLEV. 1997. Single and two-phase EHD enhanced heat transfer—a review of experimental results. 48th International Astronautical Conference, Torino (I), paper IAF-97-J.1.04: 1–12.
18. TRENTAVIZI, F. 2001. Pool Film Boiling Heat Transfer Experiments in Reduced Gravity Conditions (in Italian), Degree Thesis, University of Pisa.
19. DI MARCO, P. & W. GRASSI. 1996. Nucleate pool boiling in the presence of an electric field and in a variable gravity field: results of experiments in parabolic flight. Proc. of Eurotherm Seminar n.48, D. Gorenflo, D. Kenning & C. Marvillet, Eds.: 255–264. ETS, Pisa.
20. DI MARCO, P. & W. GRASSI. 2001. Motivation and results of a long-term research on pool boiling heat transfer in low gravity. Keynote Lecture. Proc. of 5th Conference on Experimental Heat Transfer, Fluid Mechanics and Thermodynamics (ExHFT-5), G.P. Celata, P. Di Marco, A. Goulas & A. Mariani, Eds.: 33–52. ETS, Pisa.
21. LIENHARD, J.H. & V. DHIR. 1973. Extended hydrodynamic theory of the peak and minimum pool boiling heat fluxes. NASA Contractor Report 2270.

Mechanisms of Steady-State Nucleate Pool Boiling in Microgravity

HO SUNG LEE

Department of Mechanical and Aeronautical Engineering, Western Michigan University, Kalamazoo, Michigan, USA

ABSTRACT: Research on nucleate pool boiling in microgravity using R-113 as a working fluid was conducted using a five-second drop tower and five space flights at $a/g \sim 10^{-4}$. A 19×38-mm flat gold film heater was used that allowed cine camera viewing both from the side and the bottom of the heater. It was concluded that for both subcooled and saturated liquids long-term steady-state pool boiling can take place in reduced gravity, but the effectiveness of the boiling heat transfer appears to depend on the heater geometry and on the size and the properties of fluids. Heat transfer is enhanced at lower heat flux levels and the CHF increases as the subcooling increases. It was found that several mechanisms are responsible for the steady-state nucleate pool boiling in the absence of buoyancy. The mechanisms considered here are defined and summarized as bubble removal, bubble coalescence, thermocapillary flow, bubble migration, and latent heat transport.

KEYWORDS: nucleate boiling; pool boiling; microgravity

INTRODUCTION

Nucleate pool boiling in reduced gravity has drawn much attention among researchers and scientists since space exploration began in the late 1950s. In the absence of gravity or buoyancy, questions naturally arose as to whether steady-state nucleate pool boiling could be achieved and, if so, what mechanisms would be acting? What would be the new dynamics of vapor bubble growth and departure? Could existing heat transfer correlations obtained in Earth gravity be extrapolated to predict pool boiling behavior in reduced gravity? A great deal of research has been conducted and reported using facilities such as drop towers, ballistic rockets, parabolic flights, and space shuttles. Unfortunately, although the efforts expended have been considerable, the results appear inconclusive, with various interpretations and, in some cases, contradictory experimental results.

The number of publications dealing with pool boiling in microgravity since the early 1950s has been limited, due mainly to the difficulties in producing a microgravity environment of sufficient duration on Earth. Selected pool boiling experiments conducted in reduced gravity, using various facilities and reported to date, are tabulated in TABLE 1. This table includes descriptive information about the facilities, the

---

Address for correspondence: Dr. Ho Sung Lee, Department of Mechanical and Aeronautical Engineering, Western Michigan University, 1201 Oliver Street, Kalamazoo, MI 49008-5065, USA. Voice: 616-387-6536; fax: 616-387-3558.

hosung.lee@wmich.edu

heaters used, and other details. Previous experiments involving boiling in reduced gravity were conducted and reported by Usiskin and Siegel,[1] Merte and Clark,[2] Siegel and Keshock,[3] Littles and Walls,[4] Straub et al.,[5] Ervin et al.,[6] Lee et al.,[7] Oka et al.,[8,9] Ohta et al.,[10] and Straub.[11] The results to date have been contradictory: Usiskin and Siegel,[1] Merte and Clark,[2] and Straub et al.[11] indicated that the nucleate boiling heat transfer appears to be insensitive to reductions in gravity, whereas the results presented by Littles and Walls,[4] Lee et al.,[7] Ohta et al.,[10] and Straub[12] demonstrate an enhancement of boiling heat transfer in microgravity. In fact, Oka et al.[9,10] demonstrate a deterioration of boiling heat transfer in microgravity. It is believed that the geometry of the heater surface used could be a major factor in the differences observed. In general, experiments conducted with small wires indicate little or no change in the effectiveness of the nucleate boiling heat transfer in reduced gravity or microgravity compared to Earth gravity,[2,5] whereas the use of flat surface heater surfaces tended to show an improvement in microgravity.[4,7,10]

In the light of these inconclusive results, nucleate boiling in reduced gravity indeed appears to be a complex process. Prior to further discussion of the research proposed, it appears essential to first review carefully the previous research conducted in this area. The mechanisms involved will be discussed later. It should be noted that the previous work considered here is intentionally limited to the pure evidence or observations based on the experiments as indicated in TABLE 1.

The early work of Siegel and Keshock,[3] conducted with water using a 22-ft counterweighted drop shaft with 1-sec duration, shows that, in reduced gravity, isolated bubbles become large and that the growth times are long compared with those in Earth gravity. Their observation reveals that a large bubble previously detached absorbs the newly growing bubbles on the heater surface, and thus serves as a temporary vapor sink. They postulated that the absorption could increase turbulence induced near the surface, which would serve to increase the heat transfer coefficient. They also indicated that the bubble departure diameter increased, approximately, as $g^{-1/3}$ for gravity between $a/g = 0.1$ and $a/g = 1$, and as $g^{-1/2}$ for lower gravities.

Merte and Clark[2] conducted pool boiling experiments with liquid nitrogen using a 32-ft drop tower that provided a duration of 1.4-sec, and covered the entire pool boiling curve as a function of gravity, using a 22.2-mm copper sphere as the heater surface. They found that the nucleate boiling regime appears to be insensitive to gravity for saturated liquids. They also stated that, with a saturated liquid, steady-state film boiling is not possible at zero gravity unless the vapor formed can be removed from the vicinity of the heating surface.

Littles and Walls[4] conducted tests with R-113, using a 294-ft drop tower at the Marshall Space Flight Center with an aeroshield to reduce air drag. This provided a gravity level of 10^{-2} a/g with a 4.1-sec duration. They used a relatively wide flat copper heater 50.8×101.6mm in size. Their results indicate that steady-state boiling in reduced gravity can be obtained with saturated liquids. The boiling curve shifted upward when compared to curves obtained in Earth gravity. This corresponds to an enhancement in the process.

Straub et al.[5] studied nucleate boiling with R-113 as a working fluid, using a ballistic rocket (TEXUX Program), with a 0.2-mm diameter wire as a heater and a 20×40-mm gold-coated flat plate. The boiling curve on the wire, including the effect of subcooling at 22K, indicates insensitivity to gravity, whereas the boiling curve on the flat plate shows that the boiling heat transfer in microgravity deteriorates in comparison with those in Earth gravity, regardless of the subcooling up to 48 K. They observed that in subcooled liquids the vapor bubbles remain attached to the surface, postulating that the bubbles contribute to thermocapillary flow or Marangoni convection induced along the liquid–vapor interface while the bubbles are attached to the heater surface. This is attributed to a surface tension variation with temperature at the interface.

Oka et al.[8] conducted pool boiling experiments with water and R-113 using parabolic flights that provided a gravity level of $10^{-2} a/g$ of 25-sec duration. They used a 50×50-mm transparent flat heater (ITO evaporated on a BK-7 glass plate) that allowed video camera viewing through the bottom of the heater. The boiling curves in +1 g and μg show that the boiling heat transfer for water deteriorated throughout the entire nucleate boiling regime, whereas heat transfer for R-113 deteriorated appreciably only at high heat fluxes. From measurements of acceleration during flights, it was determined that the boiling heat transfer is very sensitive to the magnitude of the reduced gravity.

Oka et al.[9] repeated the experiments with water and R-113, using a 710-m drop shaft in Hokkaido, Japan that provided a gravity level of $10^{-4} a/g$ with 10-sec duration. A 30-mm square Pyrex ITO coated plate heater was used that allowed video camera viewing only from the side. It was observed that nucleate boiling was still acting in microgravity. The boiling curves presented indicate that the critical heat flux (CHF) for R-113 appears to be as low as 20–30% of that in Earth gravity, whereas the CHF for water appears to be about half that in Earth gravity

Lee et al.[7] and Lee and Merte[13] conducted pool boiling experiments with R-113 in space flights that provided reduced gravity at $a/g \approx 10^{-4}$ for periods of 280 seconds. A 19×38-mm flat gold film heater was used that allowed cine camera viewing both from the side and through the bottom of the heater. The results demonstrated that not only can a steady pool boiling process exist in microgravity, but that an enhancement in the heat transfer occurs with both saturated and subcooled boiling.

Straub[11] studied pool boiling with R123 using two small plane heaters, 1.4 mm and 3 mm in diameter, during space flights providing reduced gravity at $a/g \approx 10^{-4}$. The results indicate that heat transfer was enhanced with a saturated liquid, but is insensitive to gravity with subcooled liquids.

From this overview, we surmise that, in general, for both subcooled and saturated liquids, long-term steady-state pool boiling can take place in reduced gravity, but the effectiveness of the boiling heat transfer appears to depend on the heater geometry and size, becoming enhanced at lower heat flux levels and with the CHF reduced considerably compared to that in Earth gravity. The CHF increases as the subcooling increases.

Questions now arise as to which mechanisms are responsible for steady-state boiling in the absence of buoyancy for both saturated and subcooled liquids. Observations to date reveal that, in general, vapor bubbles in reduced gravity remain attached to the heater surface and grow to large sizes. When two or more bubbles make contact they coalesce horizontally, or sometimes vertically, into a larger bubble due to capillary action. The larger bubble then detaches from the surface (presumably due to momentum effects) and absorbs the newer growing small bubbles, so that the growth, coalescence, and absorption become periodic. Likewise, the larger

TABLE 1. Nucleate pool boiling experiments in microgravity (continued on opposite page)

Author(s) year	Fluids	Heater type and size (mm)	Approximate heat flux q'' (kW/m²)	Subcool pressure (°C/kPa)	Photo speed side/bottom view	Facility gravity (a/g_e) duration (sec)	Isolated bubble mass boiling	Boil curve comparison to Earth data/results
Siegel and Usiskin (1959)	water	nichrome ribbon 3.2×22.2		saturation slightly subcooled	3,500 fps side	counter-weighted platform 0.7	mass boiling	vapor bubbles remained near heating surface
Siegel and Keshock (1964)	water	flat smooth nickel surface 22.2 dia.	34–56	Saturation /101	3,500 fps side	counter-weighted drop tower 0.014–1/1.0	isolated bubbles	$D_d \approx g^{-1/2}$ $q \approx 35°C$
Merte and Clark (1964)	LN_2	copper sphere 25.4 dia.				drop tower $10^{-2}/1.4$	mass boiling	insensitive to gravity
Littles and Walls (1970)	R113	flat copper 50.8×101.6	22.4–67.8	saturation /101	400 fps	drop tower $10^{-2}/4.1$	mass boiling	enhance
Straub et al. (1990)	R113	wire 0.2 dia.	4–441	0–22 /101	18–100 fps side	ballistic rocket TEXUS/ $< 10^{-4}/360$	mass boiling	insensitive to gravity
Straub et al. (1990)	R113	gold-coated glass 20×40	20–75	0–48 /101	18–100 fps side	ballistic rocket TEXUS/ $< 10^{-4}/360$	mass boiling	deteriorate
Ervin et al. (1993)	R113	flat gold film, SiO_2 19×38	40–80	0.3–11.2 /107–152	400 fps, side/bottom	drop tower/ $10^{-5}/5.1$	mass boiling	transient boiling, dynamic growth

TABLE 1/continued.

Author(s) year	Fluids	Heater type and size (mm)	Approximate heat flux q'' (kW/m²)	Subcool pressure (°C/kPa)	Photo speed side/bottom view	Facility gravity (a/g_e) duration (sec)	Isolated bubble mass boiling	Boil. curve comparison to Earth data/results
Oka et al. (1995)	n-pentane, water, CFC-113	flat ITO BK7 50×50	14–200 5–60	0–21 /81–139	30 fps, side/bottom	parabolic flight/ 10^{-2}/25	mass boiling	deteriorate/ spherical, detached bubbles for water
Oka et al. (1996)	water, CFC-113	flat pyrex plate ITO coating 30×30	15–300 15–60	0–16 & 0–22 /101	30 fps, side	drop shaft Japan/ 10^{-5}/10	mass boiling	deteriorate/ reduction of CHF
Lee et al. (1997)	R113	flat gold film, SiO₂ 19×38	5–80	0.3–22.2 /107–152	10 &100 fps, side/bottom	space shuttle/ 10^{-4}/280	mass boiling	enhance at lower q''/ effect of subcooling
Ohta et al. (1997)	Ethanol	Flat ITO 50 dia.	0.35–80	3–95 /10–480	10 fps, side/bottom	ballistic rocket/ 10^{-4}/360	mass boiling	enhance at lower q''
Qiu and Dhir (1999)	water	flat strain gage-heater, silicon wafer 45.2 dia.	temp. control	0.2–0.5/	250 fps, side	KC-135/ ±0.04	single bubble	N.A./ $D_d \approx g^{-1/2}$
Straub (1999)	R123	flat, small heater 1.4–3 dia.	20–110	0–50 (sat.) /0–0.2 P/P_c		LMS shuttle/ 10^{-4}/	mass boiling	enhance at saturation, but Insensitive to gravity at subcooling

bubble serves as a temporary vapor sink, which is considered to be one of the mechanisms responsible to the steady-state nucleate pool boiling. The experimental study of a single vapor bubble in reduced gravity reveals that the single bubble grows much larger, and remains attached for longer, to the surface and eventually departs from the surface, depending on the gravity level through the relation $D_d \propto 1/\sqrt{g}$.[3,14]

EXPERIMENTS

Pool boiling experiments were conducted on five space flights, designated as PBE-IA, -IB, -IC, -IIB, and -IIA on the STS-47, -57, -60, -72, and -77, respectively, as part of the NASA get away special (GAS) program. Each flight experiment consisting of nine different test runs, conducted during the period 1992 through 1996. Identical test runs were repeated in Earth gravity prior to and/or following each space experiment so that direct comparisons could be obtained between the behavior in microgravity and in Earth gravity.

Measurements of heat transfer and various parameters associated with bubble growth dynamics were obtained with R-113, in body forces on the order of $a/g \approx 10^{-4}$. The five space experiments were virtually identical, differing primarily in heat flux and subcooling and also in the speed and length of the photography. Twenty seven test runs for the first three flights were conducted at three levels of heat flux (2, 4, and 8 W/cm^2) and three levels of subcooling (0, 2.7, and 11.1°C), each lasting up to 120 seconds. The nine test runs of the fourth flight used the same heat flux levels, but with increased subcooling levels (11.1, 16.7, and 22.2°C). The nine test runs of the last flight were conducted with low levels of heat flux (0.5, 1, 2 W/cm^2) and the original three levels of subcooling (0, 2.7, and 11.1°C), each lasting up to 280 seconds. Photographs of the boiling process were obtained simultaneously from the side and from beneath the heater surface at frame rates of 10 and 100 fps with a 16-mm cine camera. The same hardware was used for PBE-IA and IC (STS-47 and -60, respectively); that for PBE-IB, -IIB, and -IIA (STS-57, -72, and -77) was of the same construction. This provided opportunities for observations of repeatability and reproducibility.

Two heater surfaces were placed on a single flat substrate, with one acting as a back-up, installed so as to form one wall of a test vessel with internal dimensions 15.2 cm diameter by 10.2 cm height. Each heater consisted of a 400 Å thick semitransparent gold film sputtered on a highly polished quartz substrate, and served simultaneously as a heater, with an uncertainty of ±2% in heat flux, and a resistance thermometer, with an overall uncertainty of ±1.0°C in mean surface temperature. The heater was rectangular, 19.05 × 38.1 mm (0.75 × 1.5 inch). System subcooling was obtained by increasing the system pressure above the saturation pressure, controlled and measured with an uncertainty of ±0.345 kPa. Degassed commercial grade R-113 was used because of its electrical nonconductivity, compatible for direct contact with the thin gold film heater. Further experimental details may be found in Merte *et al.*[15,16]

MECHANISMS RESPONSIBLE FOR STEADY-STATE POOL BOILING

In the absence of buoyancy, the mechanisms responsible for steady-state pool boiling appear to be quite different from those in Earth gravity. A number of these elements will now be considered.

Bubble Removal Mechanism

Typical nucleate pool boiling associated with bubble removal mechanism is shown in FIGURE 1. Due to surface tension, the small bubbles formed initially agglomerate to become a large bubble that eventually detaches slightly from the heater surface. This is attributed to the momentum and coalescence of the small bubbles growing toward the large bubble.

The large vapor bubble acts as a reservoir for subsequent nucleation and growth of vapor bubbles on the heater surface. It is noted that for subcooled boiling the large bubble hovering just above the heater surface tends to maintain its size due to the balance of the dual processes of condensation at the bubble cap and coalescence with the small bubbles at the base. The latter phenomenon is believed to be responsible for the steady boiling observed in microgravity, even under saturation conditions. Quantitative evidence of the effective steady boiling for FIGURE 1 is illustrated in FIGURES 2 and 3, with the measurements in Earth gravity superimposed. It is noted that an enhancement of about 32% can be seen in the heat transfer coefficient with microgravity in FIGURE 2, whereas some deterioration can be seen in FIGURE 3. The

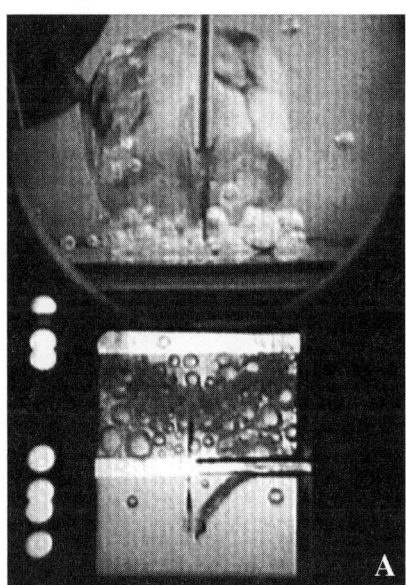 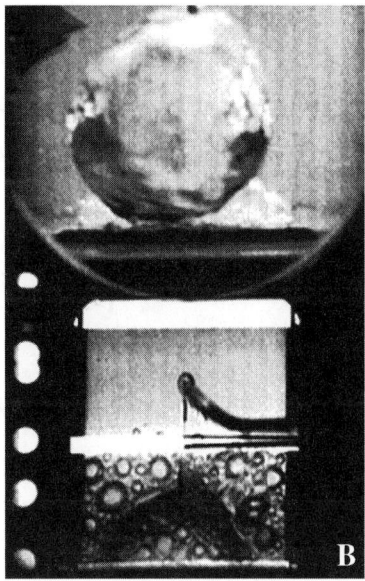

FIGURE 1. Steady pool boiling in microgravity with R113: **(A)** $q'' = 3.6\,\text{W/cm}^2$, $\Delta T_{\text{sub}} = 11.5\,\text{K}$; **(B)** $q'' = 8.0\,\text{W/cm}^2$, $\Delta T_{\text{sub}} = 22.2\,\text{K}$.

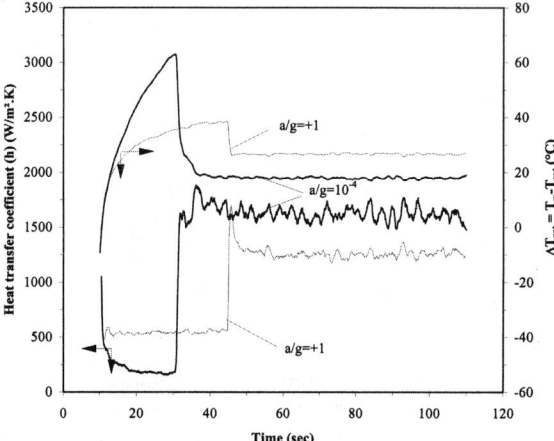

FIGURE 2. Measured mean heater surface temperature and derived heat transfer coefficient for the situation illustrated in FIGURE 1 A.

effectiveness of subcooled pool boiling appears to decrease as the levels of heat flux increase. The oscillations in the heat transfer coefficient seen in FIGURE 2 have a period of approximately 4 sec/cycle. From viewing the sequence of photographs, one cycle in this process corresponds to the agglomeration of the small vapor bubbles on the heater surface to a certain size, causing its mean temperature to rise, followed by the coalescence with the large bubble hovering above the heater surface, which in turn produces turbulence of the liquid–vapor interface, leading to the observed subsequent decrease in the mean heater surface temperature.

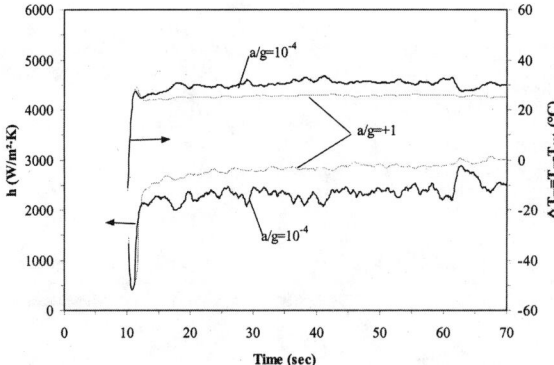

FIGURE 3. Measured mean heater surface temperature and derived heat transfer coefficient for the situation illustrated in FIGURE 1 B.

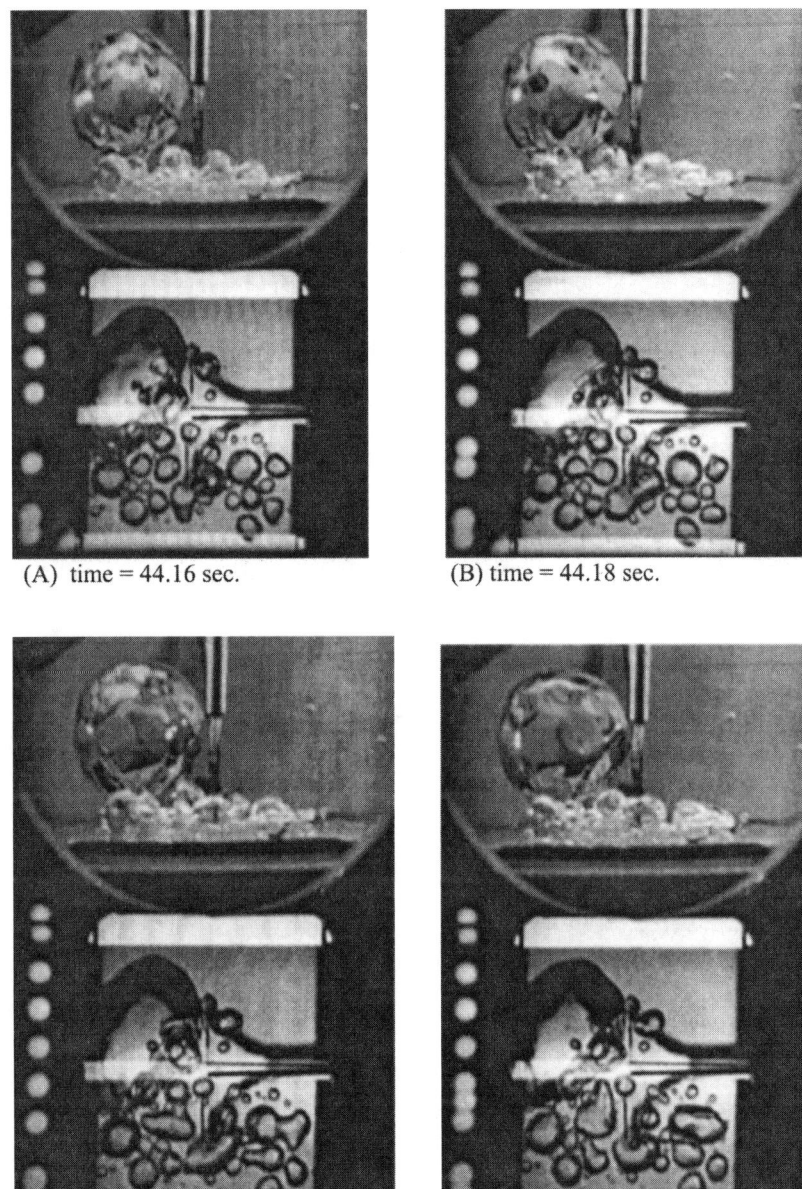

(A) time = 44.16 sec. (B) time = 44.18 sec.

(C) time = 44.20 sec. (D) time = 44.22 sec.

FIGURE 4. Vapor bubble coalescence.

Bubble Coalescence

The coalescence of vapor bubbles on the heater surface in microgravity can be seen more readily at the nominal heat flux level of $2\,\text{W/cm}^2$. The mean heater surface temperature and heat transfer coefficients are shown in FIGURE 5 for the coalescence illustrated visually in FIGURE 4. The photographs show several pairs of vapor bubbles coalescing on contact due to the effects of surface tension, increasing in size, and reducing in number by half during a period of 0.06 seconds. The vapor bubbles attached to the heater surface undergo rapid movements as a result of the coalescence, also inducing corresponding motions in the fluid adjacent to the heater surface. According to the single bubble measurements reported by Qiu and Dhir[14] with water in reduced gravity, the bubble departure diameter can be as large as 20 mm compared to 3 mm in Earth gravity. However, in microgravity the isolated bubble grows to a certain-size depending on the heater size and heat flux applied. The bubbles usually coalesce horizontally before the large predicted size is reached. This can be repeated several times in sequence, and eventually vertical coalescence into a large hovering bubble takes place with the subsequent detachment mentioned previously. Once the coalescence of the bubbles takes place, the heater surface is rewet beneath each bubble by its motion, inducing the microlayer evaporation and the resultant agitation of the hot thermal boundary layer formed on the heater surface. This is one of the mechanisms for the high heat transport taking place from the heater surface to the bulk liquid in microgravity. Oka et al.,[9] Straub,[11] and Lee and Merte[13] observed the coalescence of bubbles during microgravity pool boiling with R-113.

Thermocapillary Flow and Bubble Attachment to Heater

On close examination of motion pictures of pool boiling, the vapor bubbles tend to remain more in the vicinity of the heater surface as the subcooling increases. An

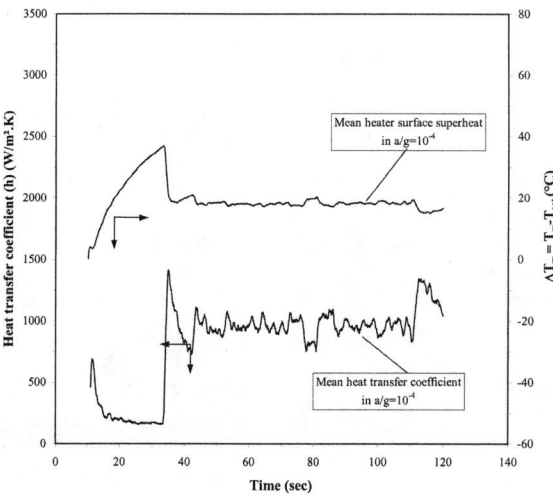

FIGURE 5. Measured mean heater surface temperature and derived heat transfer coefficient for the situation illustrated in FIGURE 4.

induced liquid flow upward from the heater surface was observed to be quite similar to single-phase convection at Earth gravity, shown in FIGURE 6. This is attributed to the combination of condensation at the top of the bubbles and thermocapillary flow. An associated phenomenon is that the larger vapor bubbles formed are often impelled toward the heater surface, where they not only serve as reservoirs attracting small bubbles and decreasing the mean heater surface temperature (as shown in FIGURE 7), but also at times cause partial dryout beneath the large bubble, thereby increasing the mean heater surface temperature. Questions arose as to the mechanism responsible for the bubbles remaining attached to the heater surface and for the induced flow observed. Straub[11] measured the velocity of the flow with R123 using particle image velocimetry (PIV) for a single vapor bubble at a subcooling of 9.9 K. This provided a velocity distribution with a maximum velocity of about 4.7 mm/sec that increases as the heat flux applied is increased. Holographic observations also reveal that the thermocapillary flow pushes the hot thermal boundary layer toward the bulk liquid along the bubble attached to the heater surface.[5,17] Straub[11] postulated that thermocapillary flow, or Marangoni convection, is induced near the bubble due to the temperature gradient along the bubble interface by the heat conduction to the hot thermal boundary layer from the heater surface, and this reacts to press the bubble toward the heater surface.

Indirect evidence for Marangoni convection was observed by Wozniak and Balasubramanian,[18] who conducted experiments in microgravity on thermocapillary

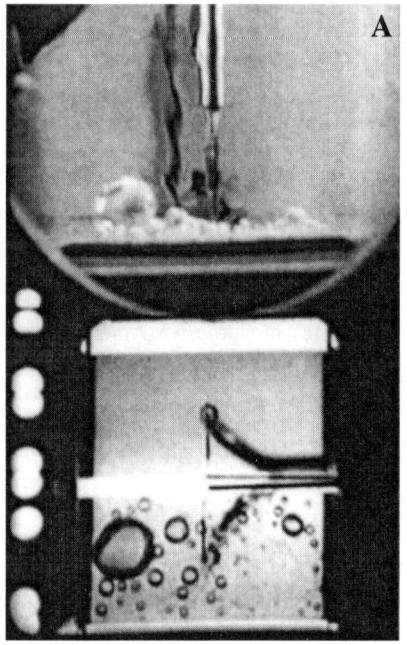

FIGURE 6. Subcooled boiling with R-113: **(A)** $q'' = 2\,\text{W/cm}^2$, $\Delta T_{sub} = 22.2\,\text{K}$; **(B)** $q'' = 4\,\text{W/cm}^2$, $\Delta T_{sub} = 22.2\,\text{K}$.

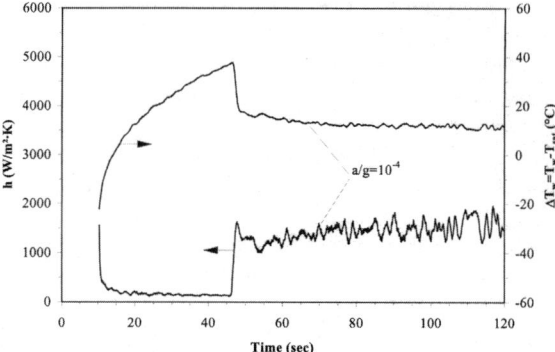

FIGURE 7. Measured mean heater surface temperature and derived heat transfer coefficient for the situation illustrated in FIGURE 6A.

motion with a single air bubble in a silicon liquid. They confirmed the air bubble motion toward the hotter liquid, along with the temperature field prevailing around the air bubble, using a Schlieren interferometer and obtaining a measured mean velocity of 2.7 mm/sec for a the air bubble at a temperature gradient of 0.33 K/mm. This appears to be of the same order of the maximum velocity for thermocapillary flow of 4.7 mm/sec measured by Straub.[11] The difference between the air bubble[18] and the vapor bubble[11] is that for the vapor bubble a phase change is taking place at the liquid–vapor interface, whereas no phase change occurs in the air bubble. However, if phase change (whether evaporation or condensation) indeed occurs at the saturation temperature, which corresponds to the vapor pressure in the bubble, usually assumed to be uniform, no temperature gradient will exist along the interface in the vapor bubble. In this case, the concept as to the source of the thermocapillary flow around the vapor bubble lacks justification. Nevertheless, it can be speculated that, for organic fluids with subcooled boiling, evaporation near the bottom of the vapor bubble takes place at an elevated saturation temperature because of the local vapor recoil taking place, whereas condensation near the top of the bubble occurs at a reduced saturation temperature due to the local pressure drop associated with the condensation. This is somewhat similar to the operation of a heat pipe.[11] Thus, a considerable upward vapor flow, particularly during the early stages of growth, is induced in the vapor bubble. This flow will slow down in the later stages due to the dryout of the liquid microlayer. although still speculation, a possible explanation for the strong thermocapillary flow observed is that it is due to the significant temperature gradient along the vapor bubble, as reported in the interferograms of Abe and Iwasaki[17] and Straub.[11] It is anticipated that the thermocapillary flow around the bubble on the heater surface plays an important role in the heat transport from the heater surface to the bulk liquid, due to the local transportation of the hot thermal boundary layer, in addition to the latent heat due to evaporation of the liquid microlayer and macrolayer.

For subcooled boiling with water in microgravity there was no indication of thermocapillary flow.[9,12] This somewhat surprising result should be studied further in detail with water. It is possible that, for the lower or intermediate levels of heat flux, the strong surface tension of water reduces any local pressure changes along the interface of the vapor bubble due to evaporation and condensation, so that an effectively uniform saturation temperature prevails along the interface of the vapor bubble. Consequently, with a lack of thermocapillary flow it would not be unexpected that the microgravity boiling curve for water would show a deterioration over the entire range of heat flux, compared to those in Earth gravity.[9]

Bubble Migration

A most unusual phenomenon was observed under conditions of high subcooling (16.7–22.2°C) at low heat fluxes (2–4 W/cm^2). Numerous very small bubbles growing on the heater surface appear unstable, moving, actually sliding on the heater surface, toward the large bubble with a measured velocity of approximately 2.5 cm/sec, as shown in FIGURE 6A. The sliding bubbles on the heater surface appear to keep growing due to microlayer evaporation. This phenomenon is responsible for the effective steady-state pool boiling and enhanced heat transfer, up to 40%, which is defined as bubble migration. This phenomenon would not have been disclosed without both the long-term space experiments and a gravity level on the order of $a/g \approx 10^{-4}$.

Microlayer Evaporation (Latent Heat Transport)

In Earth gravity, it is postulated by Rohsenow[14] that the high heat transfer coefficient in nucleate boiling could be attributed mainly to both natural convection and local agitation resulting from liquid flowing in behind the wake of a departing bubble, also called microconvection. In microgravity, the vapor bubbles tend to grow larger in size while attached to the surface, until other bubbles interfere with the growth by coalescence or because of the geometric constraints of the heater. This indicates that during the process of nucleate boiling in microgravity the vapor bubbles appear much greater in size than those in Earth gravity. It can be postulated that the microlayer formed beneath the growing bubbles rapidly evaporates, so that latent heat plays an important role in heat transport from the heater surface during this evaporation period as long as the coalescence and bubble removal mechanism take place effectively to avoid dryout beneath the bubbles.

A single vapor bubble growth was observed during microgravity experiments, shown in FIGURE 8. This growth occurred under a thick thermal boundary layer near the surface at a wall superheat of 26 K. It was found that the bubble in microgravity grows to a larger bubble, more than 1.5 cm in diameter, compared to less than 1 mm under Earth gravity with R-113, although the larger bubble is not illustrated in FIGURE 8. The bright spot beneath the bubble probably indicates the contact line of the bubble with the heater surface. It is known that microlayer evaporation occurs during the early stages and dryout during the late stages. However, at present, it is difficult to determine at what size the bubble ceases to result in effective microlayer evaporation and commence dryout at the bottom of the bubble. Some previous data showed the enhanced heat transfer compared to those in Earth gravity, which indicate that a number of growing bubbles coalesce with each other before reaching dryout-size bubbles.

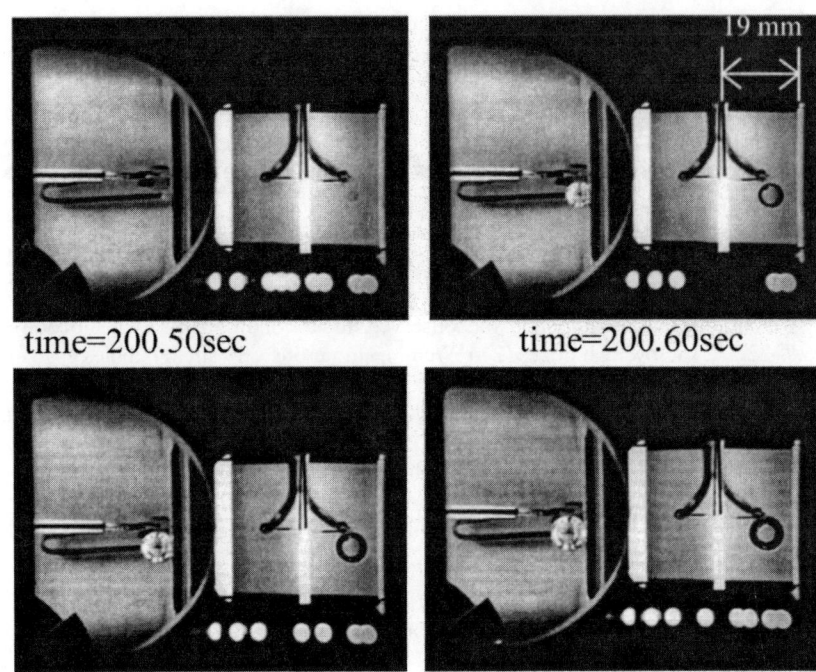

FIGURE 8. A series of photographs showing vapor bubble growth in microgravity.

BOILING CURVES IN MICROGRAVITY AND EARTH GRAVITY

Pool boiling experiments in microgravity with R113 were conducted in space flights, providing more than 280 seconds of reduced gravity at $a/g \approx 10^{-4}$.[7] Complete pool boiling curves in both microgravity and Earth gravity as a function of subcooling were obtained and are illustrated in FIGURE 9, in which all available steady-state data for R113 are plotted. The dark symbols indicate data in microgravity; the open symbols indicate data in Earth gravity. The effect of subcooling of the boiling curves in both microgravity and Earth gravity appear distinctive. In general, wall superheat decreases with increasing subcooling. Heat transfer in the lower levels of heat fluxes is enhanced in the nucleate boiling regime relative to that in Earth gravity. It is noted that effective steady-state boiling in microgravity could be reached up to a heat flux of 8 W/cm^2 with subcooling of 22.2 K, although the heat transfer coefficient in microgravity appears slightly deteriorated compared to that in Earth gravity. It is interesting that effective nucleate boiling in microgravity takes place in the regime of the natural convection in Earth gravity. At an almost saturation condition, it is also possible to achieve steady-state nucleate boiling as long as a large bubble hovering near heater absorbs small growing bubbles. Photographic observation indicates that film boiling as known in Earth gravity does not occur in microgravity, rather a large bubble covers the entire heater, resulting in dryout.

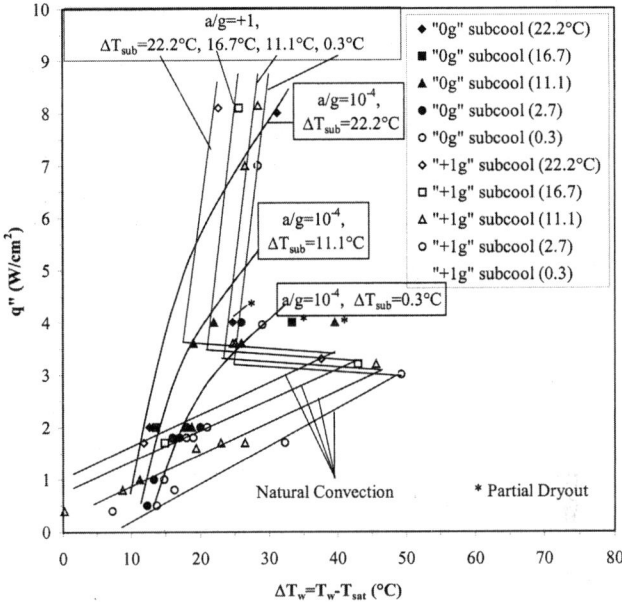

FIGURE 9. Nucleate pool boiling curves in microgravity and in Earth gravity with R-113 as a function of subcooling.

In general, nucleate pool boiling in microgravity (μg) differs considerably from that in Earth gravity ($1g$): Bubble diameters in μg are much larger than that in $1g$. The effects of heater geometry and liquid subcooling are much more significant. Further experimental and analytical studies are anticipated, particularly with water.

CONCLUSIONS

It appears that long-term steady-state nucleate boiling can take place on a flat heater surface in microgravity. Photographic observations reveal that several mechanisms are responsible for steady-state pool boiling in microgravity. These mechanisms are bubble removal, bubble coalescence, thermocapillary flow, bubble migration, and latent heat transport.

ACKNOWLEDGMENT

Support of this work under NASA Grant NAG-1684 is sincerely appreciated.

REFERENCES

1. USISKIN, C.M. & R. SIEGEL. 1961. An experimental study of boiling in reduced and zero gravity fields. J. Heat Transfer **83**(3): 243–253.

2. MERTE, JR., H. & J.A. CLARK. 1964. Boiling heat transfer with cryogenic fluid at standard, fractional, and near-zero gravity. J. Heat Transfer **86C:** 315–319.
3. SIEGEL, R. & E.G. KESHOCK. 1964. Effects of reduced gravity on nucleate boiling bubble dynamics in saturated water. AIChE J. **10**(4): 509–517.
4. LITTLES, J.W. & H.A. WALLS. 1970. Nucleate pool boiling of freon 113 at reduced gravity levels. Proc. Fluid. Eng., Heat Transfer and Lubr. Conference, Detroit, MI., May 24–27, 1970. ASME, 70-HT-17.
5. STRAUB, J., M. ZELL & B. VOGEL. 1990. Pool boiling in a reduced gravity field. Proceedings of the Ninth International Heat Transfer Conference, Jerusalem. **1:** 91–112.
6. ERVIN, J.S., H. MERTE, JR., R.B. KELLER & K. KIRK. 1992. Transient pool boiling in microgravity. Int. J. Heat Mass Transfer **35**(3): 659–674.
7. LEE, H.S., H. MERTE, JR., & F. CHIARAMONTE. 1997. Pool boiling curve in microgravity. J. Thermophys. Heat Transfer **11**(2): 216–222.
8. OKA, T., Y. ABE, Y.H. MORI & A. NAGASHIMA. 1995. Pool boiling of n-pentane, CFC-113, and water under reduced gravity: parabolic flight experiments with a transparent heater. J. Heat Transfer **117:** 408–417.
9. OKA, T., Y. ABE, Y.H. MORI & A. NAGASHIMA. 1996. Pool boiling heat transfer in microgravity. JSME Intl. J. Series B **39**(4): 798–807.
10. OHTA, H., M. KAWAJI, H. AZUMA, et al. 1997. 1A rocket experiment on nucleate pool boiling heat transfer under microgravity. DSC-Vol.62/HTD **354:** Microelectromechanical Systems [MEMS] ASME.
11. STRAUB, J. 1999. Pool boiling in microgravity. Proceedings of the International Conference on Microgravity Fluid Physics and Heat Transfer held at Oahu, Hawaii, on September 19–24. 114–125.
12. STRAUB, J. 1994. The role of surface tension for two-phase heat and mass transfer in the absence of gravity. Exp. Thermal Fluid Sci. **9:** 253–273.
13. LEE, H.S. & H. MERTE, JR. 1999. Pool boiling mechanisms in microgravity. Proceedings of the International Conference on Microgravity Fluid Physics and Heat Transfer held at Oahu, Hawaii, on September 19–24. 126–135.
14. QIU, D.M. & V.K. DHIR. 1999. Single bubble dynamics during nucleate boiling under low gravity conditions. Proceedings of the International Conference on Microgravity Fluid Physics and Heat Transfer held at Oahu, Hawaii, on September 19–24. 62–71.
15. MERTE, JR. H., H.S. LEE & R.B. KELLER. 1996. Report on pool boiling experiment flown on STS-47 [PBE-IA], STS-57 [PBE-IB], and STS-60 [PBE-IC]. NASA Contractor Report CR-198465, Contract NAS 3-25812. NASA, Cleveland, OH.
16. MERTE, JR., H., H.S. LEE & R.B. KELLER. 1998. Dryout and rewetting in the pool boiling experiment flown on STS-72, and STS-77. NASA Grant NAG-1684, Report No. UM-MEAM-98-01.
17. ABE, Y. & A. IWASAKI. 1999. Single and dual vapor bubble experiments in microgravity. Proceedings of the International Conference on Microgravity Fluid Physics and Heat Transfer held at Oahu, Hawaii, on September 19–24. 55–61.
18. WOZNIAK, G., R. BALASUBRAMANIAN, P.H. HADLAND & R.S. SUBRAMANIAN. 1999. Temperature fields in a liquid due to the thermocapillary motion of bubbles and drops. Proceedings of the International Conference on Microgravity Fluid Physics and Heat Transfer held at Oahu, Hawaii, on September 19–24. 160–166.

Heat Transfer Mechanisms in Microgravity Flow Boiling

HARUHIKO OHTA

Department of Aeronautics and Astronautics, Kyushu University, Higashi-ku, Fukuoka, Japan

ABSTRACT: The objective of this paper is to clarify the mechanisms of heat transfer and dryout phenomena in flow boiling under microgravity conditions. Liquid–vapor behavior in annular flow, encountered in the moderate quality region, has extreme significance for practical application in space. To clarify the gravity effect on the heat transfer observed for an upward flow in a tube, the research described here started from the measurement of pressure drop for binary gas–liquid mixture under various gravity conditions. The shear stress acting on the surface of the annular liquid film was correlated by an empirical method. Gravity effects on the heat transfer due to two-phase forced convection were investigated by the analysis of velocity and temperature profiles in the film. The results reproduce well the trends of heat transfer coefficients varying with the gravity level, quality, and mass velocity. Dryout phenomena in the moderate quality region were observed in detail by the introduction of a transparent heated tube. At heat fluxes just lower and higher than CHF value, a transition of the heat transfer coefficient was calculated from oscillating wall temperature, where a series of opposing heat transfer trends—the enhancement due to the quenching of dried areas or evaporation from thin liquid films and the deterioration due to the extension of dry patches —were observed between the passage of disturbance waves. The CHF condition that resulted from the insufficient decrease of wall temperature in the period of enhanced heat transfer was overcome by a temperature increase in the deterioration period. No clear effect of gravity on the mechanisms of dryout was observed within the range of experiments.

KEYWORDS: flow boiling; two-phase flow; microgravity; heat transfer mechanism; critical heat flux; pressure drop

NOMENCLATURE:
a	thermal diffusivity, m^2/sec
b	constant
C	constant
c_p	specific heat, J/kgK
D	inner tube diameter, m
Fr	Froude number, $(\rho_L J_L + \rho_G J_G)^2 / \rho_L^2 gD$
f_i	interfacial friction factor
f_{i0}	interfacial friction factor at assumed zero-gravity
G	mass velocity, kg/m^{2}sec
g	gravitational acceleration, m/sec^2
g/g_e	ratio of gravity to the earth value

Address for correspondence: Haruhiko Ohta, Department of Aeronautics and Astronautics, Kyushu University, 6-10-1 Hakozaki, Higashi-ku, Fukuoka 812-8581, Japan. Voice: +81 92 642 3489; fax +81 92 642 3752.
ohta@aero.kyushu-u.ac.jp

h	heat transfer coefficient, W/m²K
J_L	liquid superficial velocity, m/sec
J_G	gas superficial velocity, m/sec
n	number
N	total number
P	pressure, MPa
Q	volumetric flow rate, m³/sec
q	heat flux, W/m²
q_w	heat flux to fluid, W/m²
q_0	nominal heat flux supplied to heated metal film, W/m²
r	radius, m
T	temperature, °C
u	velocity, m/sec
z	length along flow direction, m
X	Martinelli parameter, $[(-dP/dz)_L/(-dP/dz)_G]^{0.5}$
x	quality, $\rho_G J_G/(\rho_L J_L + \rho_G J_G)$
y	distance from wall $(r_w - r)$, m

Greek Symbols

α	void fraction, or heat transfer coefficient, W/m²K
δ	thickness of annular liquid film, m
ϕ_L	ratio of two-phase pressure drop to that for liquid flow alone $(-dP/dz)_{TP}/(-dP/dz)_L$
e	eddy viscosity or thermal diffusivity, m²/sec
m	viscosity, Pa·sec
ρ	density, kg/m³
t	shear stress, N/m², or time, sec

Subscripts

G	gas
ex	exit of heated tube
i	interface
L	liquid
D	disturbance wave
F	film
TP	two phase
w	wall

INTRODUCTION

Boiling phenomena are utilized for the development of high-performance heat exchangers in space. A small size for heat exchangers reduces the volume occupied in a spacecraft and minimizes the launch weight. From the viewpoint of heat transport, the latent heat involved in a two-phase mixture can reduce the mass flow rate from that for single-phase systems, which in turn reduces the pump power required. Furthermore, two-phase systems are much easier to control by a change of heat load on simply adjusting the system pressure (i.e., corresponding saturation temperature) using an accumulator.

Existing studies on isothermal two-phase flow in microgravity are focused so excessively on the flow pattern classification that it is hard to find systematic data on pressure drop. Chen et al.[1] employed the single-component system of R114. A larger pressure drop in microgravity was reported when stratified flow is changed to annular flow in microgravity. The difference in pressure drop, however, is still large even at high quality where annular flow was observed to be independent of gravity level. Zhao and Rezkallah[2] tested air–water mixture in the regimes from bubble to annular, and concluded that frictional pressure drop in microgravity is comparable with that in normal gravity under the same flow conditions. The present author investigated the effect of gravity on the heat transfer over the wide range of quality,[3] where the heat transfer due to two-phase forced convection at moderate quality and for annular flow varies with gravity and deteriorates in microgravity.

It is of great importance for the safe operation of space thermal systems to know the dryout conditions as a limitation on heat removal. However, fundamental research for CHF in microgravity is quite limited.[4] Because of limited opportunities for microgravity experiments, the mechanism of CHF first needs to be clarified. On the ground, Ueda and Kim[5] used a vertical annulus with a heated core rod in the bubbly flow regime and concluded that CHF conditions are established when the increase of wall temperature due to the partial disruption of the liquid film becomes greater than the temperature decrease due to quenching during the following period of liquid passage.

This paper reviews recent work by the author. The effects of gravity in heat transfer due to two-phase forced convection are analyzed, starting from the correlation of frictional pressure drop in a vertical tube. The mechanisms of dryout are investigated, based on the observation of liquid–vapor behavior through a transparent heated tube.[3]

PRESSURE DROP DATA AND A CORRELATION METHOD FOR THE ANNULAR FLOW REGIME

Experimental Apparatus

A test section was made from acrylic tube with I.D. 8 mm and length 723 mm. The tube was oriented vertically under gravity fields. The pressure drop was measured by two transducers for air–water mixtures under the conditions: pressure, $P = 0.093\,\mathrm{MPa}$; liquid and gas superficial velocity, $J_L = 0.0497$–$0.199\,\mathrm{m/s}$, and $J_G = 1.99$–$15.9\,\mathrm{m/s}$, respectively. Experiments were conducted on board DAS MU-300 where, along a parabolic trajectory, 15-sec periods of hypergravity ($2g$) and 20-sec periods of reduced gravity, 0.01–$0.03\,g$ (denoted by μg in this paper), are succeeded by periods of normal gravity ($1g$). In most of experimental conditions, annular flow was realized independently of gravity level. This is confirmed by the flow regime map developed by Dukler et al.[6] for μg.

Interfacial Behavior in Annular Flow

FIGURE 1 shows the liquid–gas behavior. For the combination of low flow rates $J_L = 0.0663\,\mathrm{m/sec}$ and $J_G = 1.99\,\mathrm{m/sec}$ (FIG. 1A), an annular liquid film contains small bubbles at $1g$. At $2g$ both the disturbance of interface and number of bubbles entrapped by the annular film are increased. At μg the annular film becomes smooth with almost no bubbles.

When gas velocity is increased to $J_G = 15.9\,\mathrm{m/sec}$ keeping J_L constant (FIG. 1B), the surface disturbance of annular film becomes larger, in order, at $2g$, $1g$, and μg,

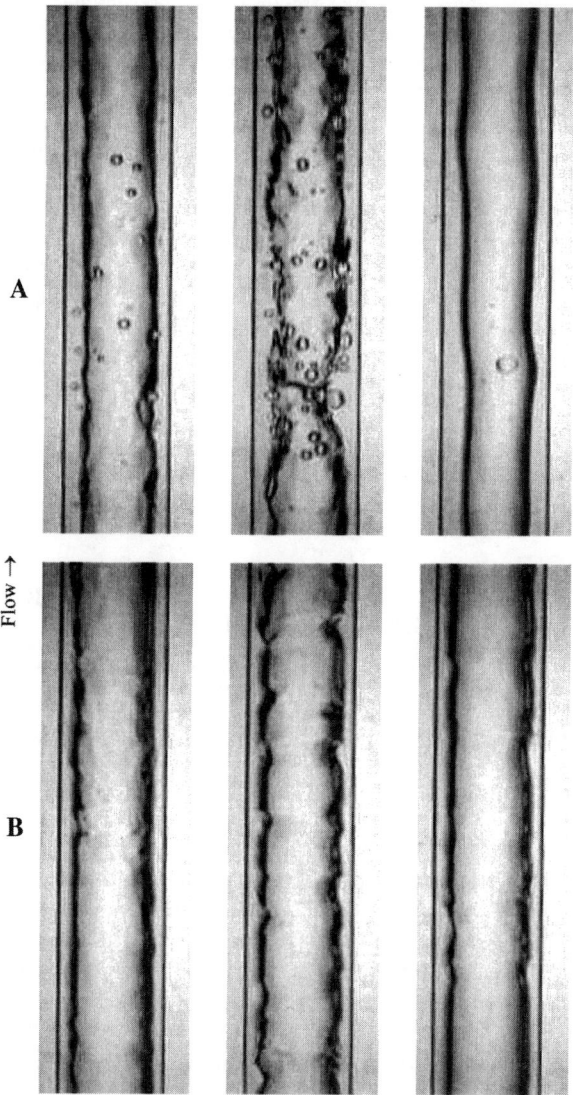

FIGURE 1. Behavior of liquid–gas interface: **(A)** $J_G = 1.99\,\text{m/sec}$, $J_L = 0.0663\,\text{m/sec}$; **(B)** $J_G = 15.9\,\text{m/sec}$, $J_L = 0.0663\,\text{m/sec}$.

but the difference is smaller than that observed at low J_G. The passing frequency of the disturbance wave is markedly increased at $2g$. The thickness of annular liquid film is decreased at larger J_G.

At large liquid velocity $J_L = 0.199$ m/sec, the liquid–gas behavior is similar to that for $J_L = 0.0663$ m/sec in the present experimental range for J_G.

Pressure Drop Data

Values for time-averaged frictional pressure drop are plotted in FIGURE 2. In the annular flow regime, the pressure drop increases in $2g$ and decreases in μg. Gravity effect becomes small at higher J_L or J_G. In μg the pressure drop increased markedly with an increase of J_L or J_G. At low J_L, the value of pressure drop for $1g$ and $2g$ is not sensitive to J_G.

Data for the two-phase pressure drop $-dp/dz$ are correlated using the method due to Lockhart and Martinelli.[7] In FIGURE 3, experimental data are plotted for the relation between the ratio of two-phase pressure drop to that for liquid flowing alone and

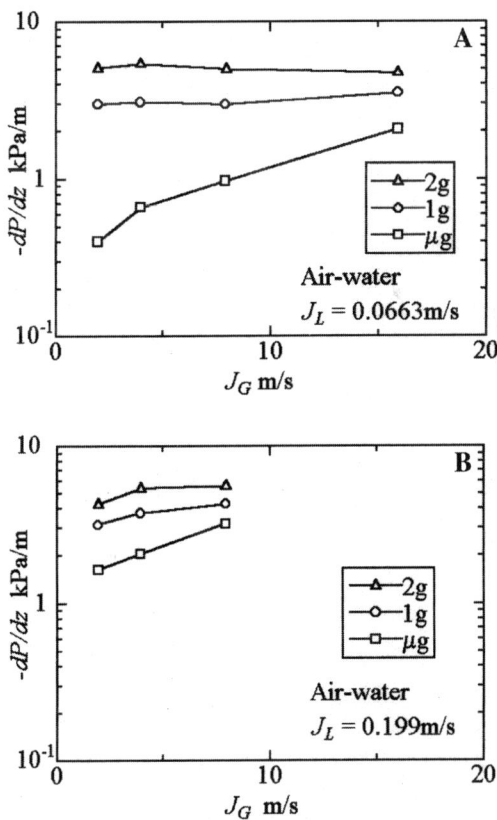

FIGURE 2. Effect of gravity on pressure drop: (**A**) $J_L = 0.0663$ m/sec, (**B**) $J_L = 0.199$ m/sec.

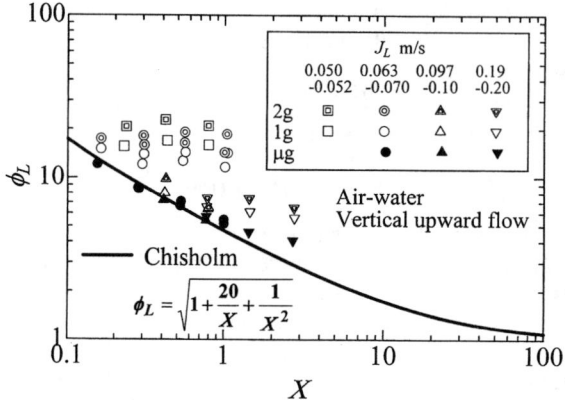

FIGURE 3. Relation between measured pressure drop and values prediction using the Chisholm correlation.[8]

Martinelli parameter X. The solid line is calculated from the correlation due to Chisholm,[8]

$$\phi_L^2 = 1 + \frac{C}{X} + \frac{1}{X^2}, \quad (1)$$

where $C = 20$ is given for turbulent flow in both phases.

The correlation is proposed inherently for stratified flow in a horizontal tube on ground, whereas a vertical tube in annular flow is investigated here. The pressure drop data at μg agree well with the prediction from Equation (1), which offers no physical interpretation, but there is no pressure gradient due to gravity along flow direction in either case. Experimental data for lower J_L or lower J_G (i.e., higher X) deviate from (1) because of a larger gravity effect, as shown in FIGURE 2. At high J_G (i.e., low X) the experimental data for all gravity levels agree well with the correlation as a result of decreased gravity effect in this condition.

Correlating Method

The effect of pressure drop observed here is correlated under the assumption that surface roughness of the annular liquid film changes for the gas core flow. By introducing an interfacial friction factor f_i, the force balance becomes

$$\tau_i = f_i \rho_G \frac{(u_G - u_i)^2}{2} \approx f_i \rho_G \frac{u_G^2}{2} = f_i \rho_G \frac{(J_G/\alpha)^2}{2} \quad (2)$$

$$-\frac{dP}{dz} = \frac{4\tau_i}{D - 2\delta} = \frac{4\tau_i \alpha^{0.5}}{D}, \quad (3)$$

where τ_i is the interfacial shear stress, δ the thickness of annular liquid film, α the void fraction, and u_G and u_i are the mean gas core velocity and velocity at the surface of annular liquid film, respectively. Substitution of (2) into (3) gives a relation between pressure gradient $-dp/dz$ and f_i for given J_G and α. The data are well reproduced by the following correlation, giving a ratio of interfacial friction factors for an arbitrary gravity level f_i to that for the assumed zero-gravity f_{i0}.

$$\frac{f_i}{f_{i0}} = 1 + 0.035 Fr^{-1}\left(\frac{1-x}{x}\right)^{0.65}, \tag{4}$$

$$Fr \equiv \frac{(\rho_G J_G + \rho_L J_L)^2}{\rho_L^2 g D}, \quad 10^{-2} g_e \leq g \leq 2 g_e, \tag{5}$$

where Fr is the Froude number, g the gravitational acceleration, and x is the quality defined by

$$x \equiv \frac{G_G}{G_G + G_L} = \frac{\rho_G J_G}{\rho_G J_G + \rho_L J_L}. \tag{6}$$

Values of pressure gradient for $2g$ or μg are calculated as ratios to the values for $1g$ assuming no serious difference in α for both gravity levels and that the values for $1g$ are calculated by using Equation (1).

PREDICTION OF GRAVITY EFFECT ON HEAT TRANSFER IN MODERATE AND HIGH QUALITY REGIONS

Heat Transfer Model

To clarify the mechanisms for the gravity effect, a simple model was developed. The outline of the model is illustrated in FIGURE 4. The thickness of an annular liquid film and profiles of velocity and temperature in the film are determined by the solution of momentum and energy equations, respectively.

The following assumptions are made. (1) Annular flow is composed of an annular liquid film (base film), disturbance waves, and vapor core flow. (2) The annular liquid film has uniform thickness in both longitudinal and circumferential directions and the thickness of the annular liquid film does not change with time. (3) The velocity and temperature profiles are symmetric with respect to the tube axis, and are not changed by the passage of disturbance waves. (4) The influence of disturbance waves is reflected only through the change in the rate of liquid flow in the annular liquid film. (5) The flow in the annular liquid film is turbulent.

Analysis

The volumetric flow rate of vapor Q_G and volumetric flow rate of liquid Q_L are represented by mass velocity G and quality x as follows:

$$Q_G = \frac{G x \pi r_w^2}{\rho_G}, \tag{7}$$

$$Q_L = \frac{G(1-x) \pi r_w^2}{\rho_L}, \tag{8}$$

where r_w is the inner tube radius, ρ_L is the liquid density, and ρ_G is the vapor density. Q_L consists of Q_{LF}, due to annular liquid film, and Q_{LD}, due to disturbance wave. The flow rate by entrained liquid droplets Q_{LE} in the vapor core is included in Q_{LD} in this model.

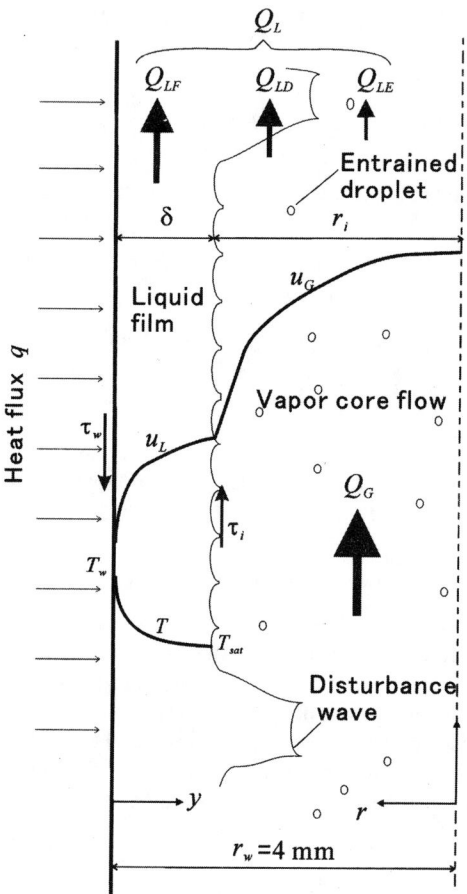

FIGURE 4. Simplified model for simulating two-phase forced convective heat transfer.

$$Q_L = Q_{LF} + Q_{LD}. \tag{9}$$

Equations of momentum and energy for the fully developed annular liquid film are

$$\frac{1}{r}\frac{\partial}{\partial r}(r\tau_r) + \left(\frac{dP}{dz} + \rho_L g\right) = 0, \tag{10}$$

$$\frac{1}{r}\frac{\partial}{\partial r}(rq_r) = 0, \tag{11}$$

where r is the radius, τ_r the shear stress, and q_r is the heat flux. Integrating of Equation (10) across the thickness of annular liquid film gives shear stress at the tube wall τ_w,

$$\tau_w = \frac{r_i}{r_w}\tau_i - \frac{r_w^2 - r_i^2}{2r_w}\left(\frac{dP}{dz} + \rho_L g\right), \tag{12}$$

where r_i is the radius at the interface (i.e., film surface) and τ_i is the interfacial shear stress. For the vapor core flow, Equation (10), using ρ_G instead of ρ_L, can be integrated to give the pressure gradient

$$\frac{dP}{dz} = -\frac{2\tau_i}{r_i} - \rho_G g. \tag{13}$$

For the turbulent liquid film,

$$\tau_r = -\rho_L(\nu + \varepsilon_{mr})\frac{\partial u}{\partial r}, \tag{14}$$

where ν is the kinematic viscosity, and the eddy viscosity ε_{mr} is given by the relation due to Sleicher[9] in the range $y^+ < 30\text{–}50$ for single-phase flow:

$$\frac{\varepsilon_{mr}}{\nu} = b^2(y^+)^2, \quad b = 0.091, \tag{15}$$

where y^+ is defined by friction velocity $\sqrt{\tau_w/\rho_L}$ and the distance from the inner tube wall $y = r_w - r$ as follows:

$$y^+ = \frac{y}{\nu_L}\sqrt{\frac{\tau_w}{\rho_L}}. \tag{16}$$

The velocity profile in the annular liquid film is obtained by integrating Equation (10) under the following boundary conditions:

$$r = r_w : u = 0, \quad r = r_i : \tau = \tau_i. \tag{17}$$

The volumetric flow rate of annular liquid film, Q_{LF}, is obtained by integrating the velocity profile

$$Q_{LF} = 2\pi \int_{r_i}^{r_w} u r \, dr. \tag{18}$$

In the present analysis, there are six unknown parameters, r_i, τ_w, τ_i, dP/dz, Q_{LD}, and Q_{LF} for the given Equations (9), (12), (13), and (18). A solution is obtained if values for Q_{LD} and τ_i, for example, are specified.

The temperature profile is obtained by substituting

$$q_r = \rho_L c_p (a + \varepsilon_{hr})\frac{\partial T}{\partial r} \tag{19}$$

into (11). The eddy thermal diffusivity, ε_{hr}, is equated to ε_{mr} and integrated under the following boundary conditions:

$$r = r_w : q_r = q_w, \quad r = r_i : T = T_{sat}, \tag{20}$$

where q_w is the heat flux at the tube wall and T_{sat} ($= T_i$) is the saturation temperature (interfacial temperature). The heat transfer coefficient α is defined by

$$\alpha = \frac{q_w}{T_w - T_{sat}}. \tag{21}$$

Predicted Gravity Effects

In FIGURE 5A, predicted values of the heat transfer coefficient α and the thickness of annular liquid film δ ($= r_w - r_i$) at mass velocity $G = 150\,\text{kg/m}^2\text{sec}$ for three different gravity levels are presented as a function of quality x. The liquid film becomes thicker under μg. The effect of gravity on α becomes small with increasing x. In FIGURE 5B, for higher mass velocity, $G = 300\,\text{kg/m}^2\text{sec}$, the effect of gravity on both α and δ decreases over the entire quality range. Both trends reproduce well the experimental results for Fron 113.[3] In the figure, the heat transfer coefficients from the correlation by Dengler and Addoms[10] are represented by lines that give the same level of values as the prediction except for quality greater than 0.8. For Fron 113, the heat transfer coefficient increases by 25% for hypergravity. It decreases by 7% for reduced gravity for $G = 150\,\text{kg/m}^2\text{sec}$, inlet quality $x_{in} = 0.28$, and heat flux $q = 1 \times 10^4\,\text{W/m}^2$. According to the present analysis, the gravity effect is smaller for water than Fron113 under the same conditions of pressure, mass velocity, and quality. This will be confirmed experimentally by further investigation.

In the above analysis, volumetric flow rate due to disturbance wave Q_{LD} is neglected. The value Q_{LD} sometimes occupies a large part of total liquid flow rate and it varies with gravity level. When the effect of a disturbance wave is simply

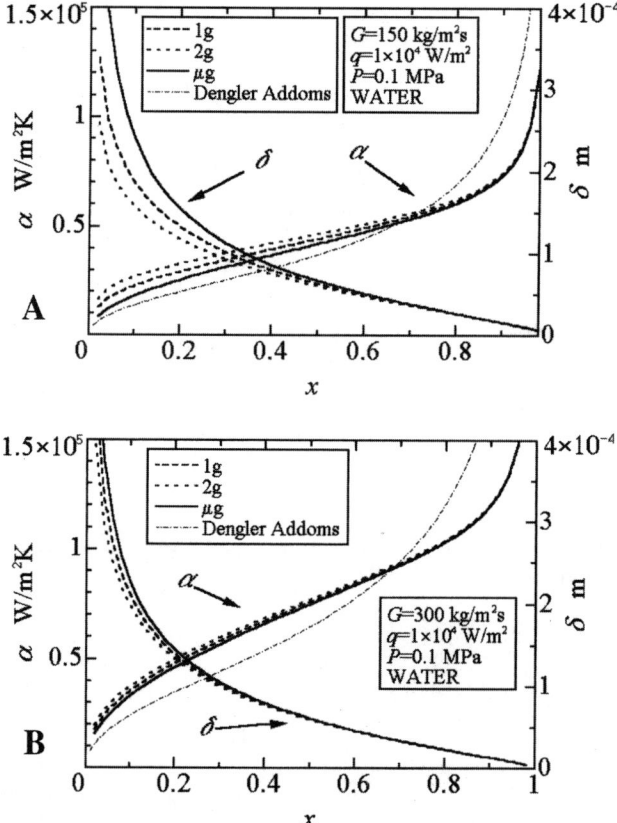

FIGURE 5. Predicted gravity effects on film thickness and heat transfer coefficient in the annular flow regime: (**A**) $G = 150\,\text{kg/m}^2\text{sec}$, (**B**) $G = 300\,\text{kg/m}^2\text{sec}$.

treated as the reduction of film flow rate, the effect was known to be small. The effect of developing a thermal boundary layer on the heat transfer coefficient is also confirmed to be quite small because of the small film thickness.[11]

DRYOUT MECHANISMS IN FLOW BOILING

Experimental Apparatus and Procedure

A transparent heated tube was made from a Pyrex glass tube with I.D. 8 mm and thickness 1 mm. A thin gold film was coated uniformly on the inner wall of the tube to operate as a heater and a temperature thermometer.[3] The distribution of wall temperature along the flow direction was evaluated by separating the film into segments by additional thin electrodes.[12] Since confirmation of temperature–electric resistance calibration curves after each dryout test run was quite difficult for the limited time interval between parabolic flights by DAS MU-300 aircraft, a short tube with 17 mm heated length was employed, and the fluctuation of wall temperature directly related to the liquid–vapor behavior in the section.

The experiments were conducted by using Fron 113 under $P = 0.093\,\text{MPa}$, $G = 150\,\text{kg/m}^2\text{sec}$, $x_{in} = 0.2$ and 0.5, and $q_o \leq 3.0\times 10^5\,\text{W/m}^2$. The experiments were carried out in μg only. Heat flux to fluid q_w was evaluated from the supplied value q_o and inner wall temperature $T_w(\tau)$ by the numerical solution of transient heat conduction across the glass tube wall. A high speed video camera, 240 fps, was introduced for the observation.

Observation of Dryout Phenomena

FIGURE 6 A and B show the liquid–vapor behavior just before the passage of disturbance wave at $q_o = 1.6\times 10^5$ and $2.2\times 10^5\,\text{W/m}^2$, that is, lower and higher than the CHF value, respectively. Inlet quality is $x_{in} = 0.2$ and exit qualities are $x_{ex} = 0.26$ and 0.29 for given heat flux levels. Disturbance waves play a role in supplying liquid to the tube wall. Nucleate boiling occurs in the annular liquid film and even at heat flux lower than CHF dry patches extended locally before the rewetting by a disturbance wave (FIG. 6A). Under CHF conditions, a large dried area is observed just before the passage of disturbance wave (FIG. 6B), and the trend is emphasized at higher wall temperature after the elapse of time.

FIGURE 6 C and D are for higher inlet quality $x_{in} = 0.5$ at heat fluxes $q_o = 2.2\times 10^5$ and $3.0\times 10^5\,\text{W/m}^2$, that is, lower and higher than CHF, respectively. Corresponding exit qualities are $x_{ex} = 0.59$ and 0.62. Since the thickness of annular liquid film becomes smaller at higher quality, the film easily evaporates and disrupts to form rivulets. Smaller bubbles are observed, but the existence of dry patches at lower heat flux than CHF is also true for this case (FIG. 6C). At heat flux higher than CHF, larger extents of dry patches can be observed just before the passage of a disturbance wave followed. However, small pieces of liquid rivulet still remain on the dried area (FIG. 6D), which is quite different from the behavior at lower quality.

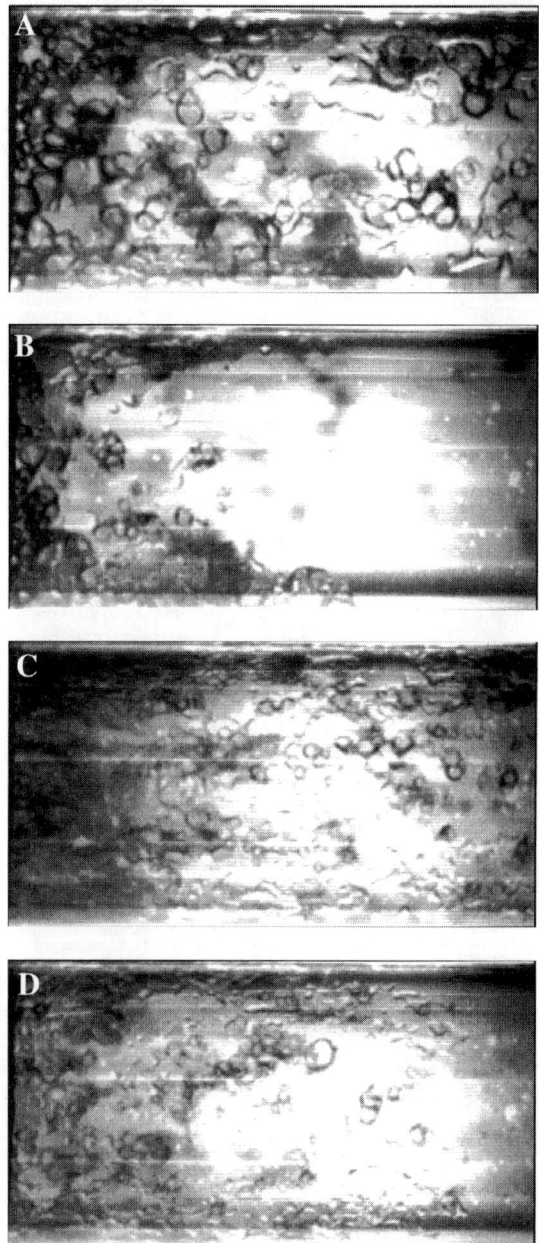

FIGURE 6. Behavior of annular liquid film in microgravity: (**A**) $x_{in} = 0.2$, $q_o = 1.6 \times 10^5 \text{W/m}^2$; (**B**) $x_{in} = 0.2$, $q_o = 2.2 \times 10^5 \text{W/m}^2$; (**C**) $x_{in} = 0.5$, $q_o = 2.2 \times 10^5 \text{W/m}^2$; (**D**) $x_{in} = 0.5$, $q_o = 3.0 \times 10^5 \text{W/m}^2$.

Heat Transfer Data

FIGURE 7 shows the transition of inner wall temperature T_w, heat flux to fluid q_w, and heat transfer coefficient α for $x_{in} = 0.2$ at $q_o = 1.6 \times 10^5$ and $2.2 \times 10^5 \,\mathrm{W/m^2}$. Gray bands indicate the passing period of disturbance waves recognized from video pictures.

The top parts of FIGURE 8 A and B show the magnified transition of q_w and α for $x_{in} = 0.2$ at nominal heat fluxes q_o lower and higher than CHF. The fluctuation in q_w and α values is completely synchronized with the passage of disturbance waves.

At heat flux lower than CHF (FIG. 8A), the value α has the minimum and the maximum when the intervals between consecutive disturbance waves are long. The first decrease of α is due to the increase in the thickness of annular liquid film just after the passage of a disturbance wave and also due to the deactivation of nucleation sites. The increase of α followed by the first decrease is resulted from the reduction of film

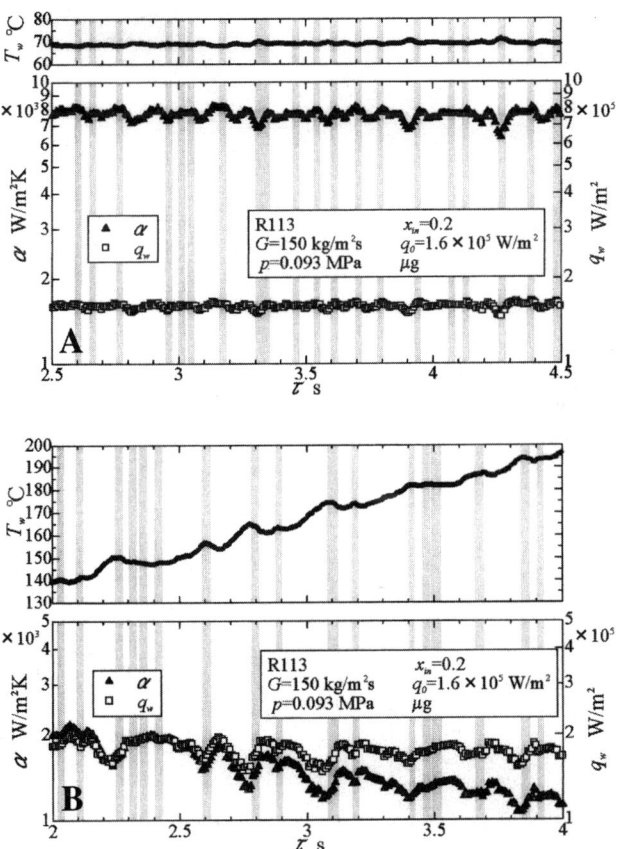

FIGURE 7. Transition of inner wall temperature, heat flux to fluid and heat transfer coefficient in microgravity: (**A**) μg, $x_{in} = 0.2$, $q_o = 1.6 \times 10^5 \,\mathrm{W/m^2}$ (< q_{CHF}); (**B**) μg, $x_{in} = 0.2$, $q_o = 2.2 \times 10^5 \,\mathrm{W/m^2}$ (> q_{CHF}).

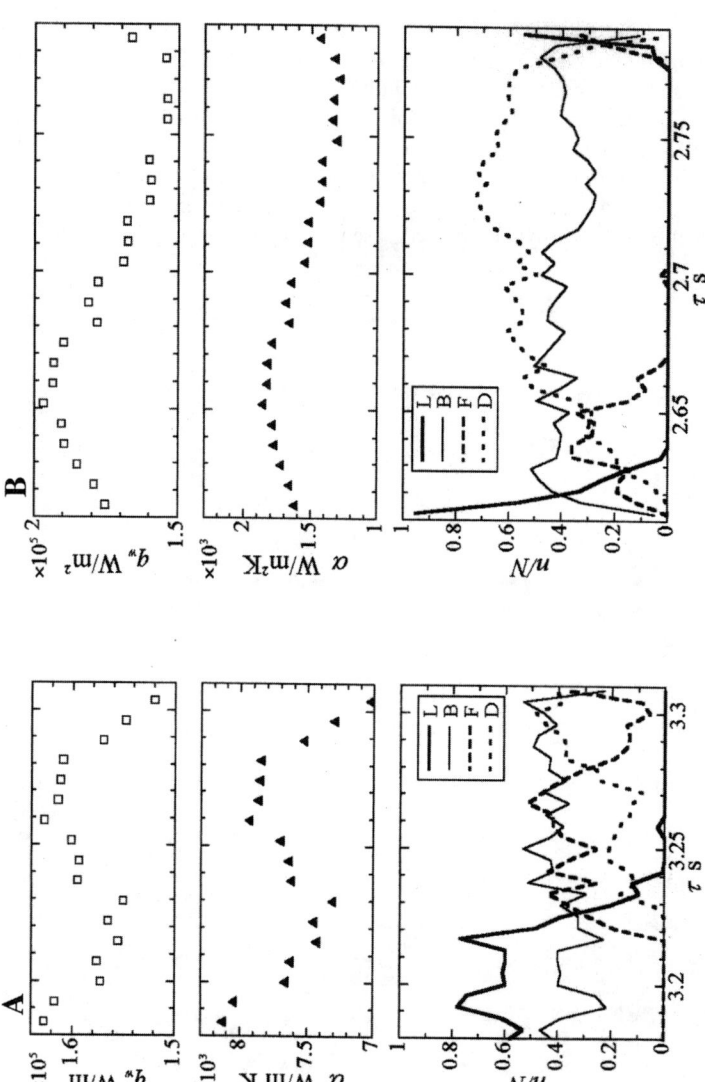

FIGURE 8. Area ratio of patterns for different liquid–vapor behavior in microgravity: L, thick liquid film; B, nucleate boiling; D, dried area. (**A**) μg, $x_{in} = 0.2$, $q_o = 1.6 \times 10^5 \text{W/m}^2$ ($< q_{CHF}$); (**B**) μg, $x_{in} = 0.2$, $q_o = 2.2 \times 10^5 \text{W/m}^2$ ($> q_{CHF}$).

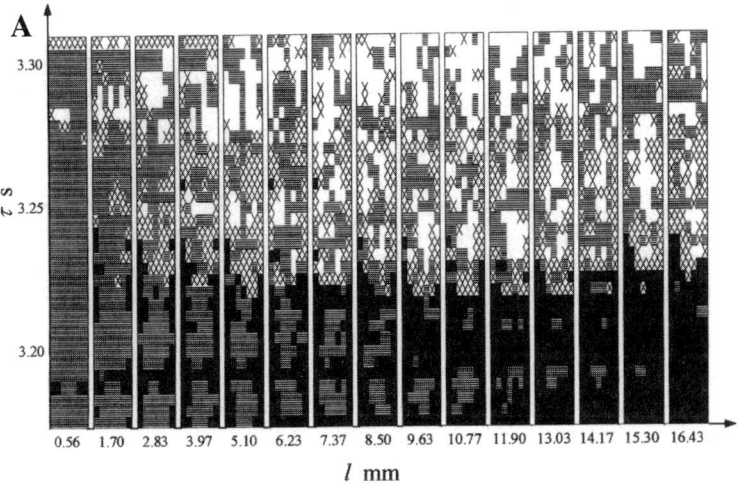

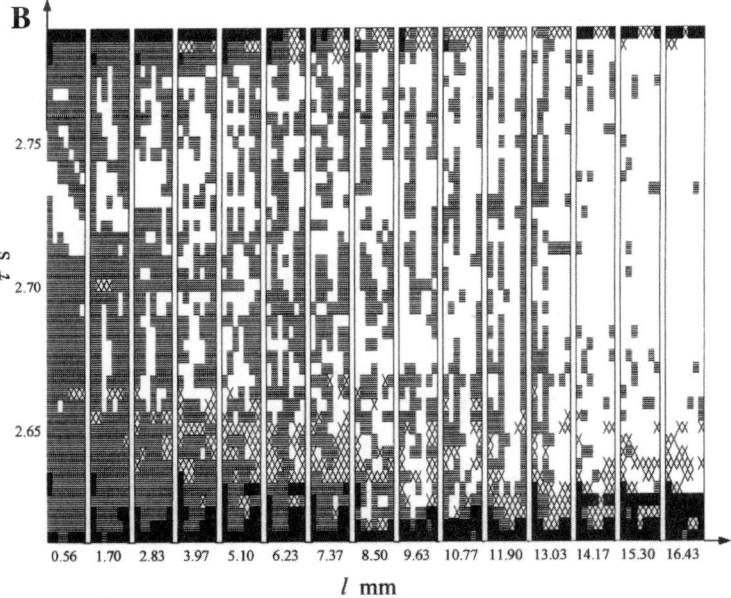

FIGURE 9. Classification of liquid–vapor behavior on the segment areas of tube wall and its transition. **(A)** μg, $x_{in} = 0.2$, $q_o = 1.6\times10^5$W/m^2 ($< q_{CHF}$); **(B)** μg, $x_{in} = 0.2$, $q_o = 2.2\times10^5$W/m^2 ($> q_{CHF}$). ■, disturbance wave (L); ⊠, thin liquid film (F); ☰, nucleate boiling (B); ☐, dried area (D).

thickness by the evaporation and also from reactivation of nucleation sites under the thin liquid film of increased superheat. The second decrease of α is caused by the increase of total area occupied by dry patches.

At heat flux larger than CHF (FIG. 8B), the first decrease of α is not observed because of rapid reduction of liquid film thickness just after the passage of a disturbance wave. The value of heat flux q_w to liquid is smaller than the nominal value, q_o, based on the power input by the amount consumed for the heating of the tube substrate.

For $x_{in} = 0.5$, the fluctuation of q_w and α at heat flux lower than CHF is quite small because of increased passing frequency of disturbance waves. Even at heat flux higher than CHF, small liquid films or rivulets still remain in the extended dry areas. The difference between q_o and q_w is smaller and thus the rate of increasing T_w is smaller than that observed for $x_{in} = 0.2$.

Discussion

The passing frequency of disturbance waves increases markedly at higher quality, but the effect of heat flux on the frequency is very small. Then the situations of liquid supply to the dry patches are almost the same for heat fluxes lower and higher than CHF. The liquid–vapor behavior at 7×15 segments on the front half of the tube can be classified into four patterns: thick liquid film or disturbance wave (L), nucleate boiling (B), thin liquid film (F), and dried area (D). The results are shown in FIGURE 9 A and B, where the ordinate is time elapsed and the abscissa is the location along the tube axis. The variation of phenomena along seven different circumferential positions on the tube is taken into account at each time. The ratios of area occupied by each pattern to the total n/N are plotted against the time at the bottom of FIGURE 8 A and B.

The ratio of the area for nucleate boiling does not change markedly in one interval. The ratio for thin liquid film directly affects the transition of q_w and α. Since there is almost no qualitative difference in either the liquid–vapor or the heat transfer behavior at heat fluxes just lower and higher than CHF value, CHF conditions are established when increase of wall temperature due to local dryout in the latter half of the interval exceeds the temperature decrease by the evaporation of thin liquid film in the first half of the next interval.

CONCLUSIONS

To investigate the mechanisms of the gravity effect on the heat transfer due to two-phase forced convection, the interfacial shear stress acting on the surface of annular liquid film was correlated on the basis of pressure drop data obtained under various gravity conditions. (1) The gravity effect on the pressure drop is well reproduced by a correlation for interfacial friction factors represented by the Froude number and vapor quality. (2) The analysis reproduces well the gravity effect on heat transfer due to two-phase forced convection, where heat transfer is deteriorated by the reduction of gravity as a result of reduced turbulence in the annular liquid film. (3) The analysis shows a decrease in the gravity effect on the heat transfer at higher quality or higher mass velocity, which is consistent with the trends previously confirmed experimentally.[3]

To clarify dryout mechanisms in the moderate quality region, a transparent heated tube was employed to relate the transient heat transfer to the observed liquid–vapor behavior. (4) Even at heat flux below CHF, dry patches extend in the annular liquid film just before the passage of a disturbance wave playing a role in rewetting the dried surface. (5) Even at heat flux above CHF, heat transfer enhancement due to evaporation of thin liquid film is observed in the first half of an interval between succeeding disturbance waves. (6) The fluctuation of heat transfer rate observed at heat fluxes lower and higher than CHF is related exactly to the liquid–vapor behavior observed through the transparent tube. (7) Under the present experimental conditions, the temperature excursion at CHF conditions is interpreted as an accumulated increment of wall temperature when increase of wall temperature due to local dryout exceeds the decrease in temperature due to evaporation of thin liquid film in a interval between disturbance waves. Under the present experimental conditions, no distinct gravity effect was observed in either transition of wall temperature or behavior of both phases. Experiments in the same quality ranges but with a longer heated tube and also investigation of gravity effects on DNB at lower quality are planned for further study.

When long-term microgravity experiments could be realized, the accumulated knowledge of pool boiling research on, for example, thermocapillary flow[13] and bubble behavior in subcooled liquid[14] should be reflected in the flow boiling studies in addition to the results from isothermal two-phase flow in microgravity.

ACKNOWLEDGMENT

This research was conducted under the Funds for Basic Experiments Oriented to Space Station Utilization by ISAS (Institute of Space and Astronautical Science). The authors appreciate the support by ISAS and the collaboration by DAS (Diamond Air Service).

REFERENCES

1. CHEN, I., R. DOWING, E.G. KESHOCK & M. AI-SHARIF. 1991. Measurements and correlation of two-phase pressure drop under microgravity conditions. J. Thermophys. **5**(4): 514–523.
2. ZHAO, L. & K.S. REZKALLAH. 1995. Pressure drop in gas–liquid flow at microgravity conditions. Int. J. Multiphase Flow **21**(5): 837–849.
3. OHTA, H., K. INOUE, Y. YAMADA, et al. 1995. Microgravity flow boiling in a transparent tube. Proc. 4th ASME-JSME Thermal Engineering Joint Conf., Maui, Hawaii. **4:** 547–554.
4. OHTA, H., K. INOUE & S. YOSHIDA. 1997. Experiments on dryout phenomena in microgravity flow boiling. Proc. 10th International Symposium on Transport Phenomena in Thermal Science and Process Engineering, Kyoto, Japan. **2:** 373–378.
5. UEDA, T. & K. KIM. 1986. Heat transfer characteristics during the critical heat flux condition in a subcooled flow boiling system. Proc. 8th Int. Heat Transfer Conf., San Francisco. **5:** 2203–2208.

6. DUKLER, A.E., J.A. FABRE, J.B. MCQUILLEN & R. VERNON. 1988. Gas–liquid flow at microgravity conditions: flow patterns and their transitions. Int. J. Multiphase Flow **14**(4): 389–400.
7. LOCKHART, R.W. & R.C. MARTINELLI. 1949. Proposed correlations of data for isothermal two-phase two-component flow in pipes. Chem. Eng. Prog. **45:** 39–48.
8. CHISHOLM, D. 1967. A theoretical basis for the Lockhart–Martinelli correlation for two-phase flow. Int. J. Heat Mass Transfer **10:** 1767–1778.
9. SLEICHER, C.A., JR. 1958. Experimental velocity and temperature profiles for air in turbulent pipe flow. Trans. ASME **80:** 693–704.
10. DENGLER, C.E. & J.N. ADDOMS. 1956. Heat Transfer mechanism for vaporization of water in a vertical tube. Chem. Eng. Prog. Symp. Ser. **18**(52): 95–103.
11. OHTA, H., K. INOUE & H. FUJIYAMA. 1998. Analysis of gravity effect on two-phase forced convective heat transfer in annular flow regime. Proc. 3rd Int. Conference on Multiphase Flow, Lyon, France, 389 CD Rom.
12. OHTA, H. & H. FUJIYAMA. 1999. Local heat transfer measurement and visualization of liquid–vapor behavior for microgravity flow boiling experiments. Two-Phase Flow Modeling Exp. **1:** 255–262.
13. STRAUB, J. 2000. Pool boiling in microgravity. *In* Microgravity Fluid Physics and Heat Transfer. V.K. Dhir, Ed.: 114–125. Begell House, New York.
14. LEE, H.S. & H. MERTE, JR. 2000. Pool boiling mechanisms in microgravity. *In* Microgravity Fluid Physics and Heat Transfer. V.K. Dhir, Ed.: 126–135. Begell House, New York.

Criteria for Approximating Certain Microgravity Flow Boiling Characteristics in Earth Gravity

HERMAN MERTE, JR., JAESEOK PARK, WILLIAM W. SHULTZ, AND ROBERT B. KELLER

University of Michigan, Ann Arbor, Michigan, USA

ABSTRACT: The forces governing flow boiling, aside from system pressure, are buoyancy, liquid momentum, interfacial surface tensions, and liquid viscosity. Guidance for approximating certain aspects of the flow boiling process in microgravity can be obtained in Earth gravity research by the imposition of a liquid velocity parallel to a flat heater surface in the inverted position, horizontal, or nearly horizontal, by having buoyancy hold the heated liquid and vapor formed close to the heater surface. Bounds on the velocities of interest are obtained from several dimensionless numbers: a two-phase Richardson number, a two-phase Weber number, and a Bond number. For the fluid used in the experimental work here, liquid velocities in the range $U = 5\text{--}10 \, \text{cm/sec}$ are judged to be critical for changes in behavior of the flow boiling process. Experimental results are presented for flow boiling heat transfer, concentrating on orientations that provide the largest reductions in buoyancy parallel to the heater surface, varying ± 5 degrees from facing horizontal downward. Results are presented for velocity, orientation, and subcooling effects on nucleation, dryout, and heat transfer. Two different heater surfaces were used: a thin gold film on a polished quartz substrate, acting as a heater and resistance thermometer, and a gold-plated copper heater. Both transient and steady measurements of surface heat flux and superheat were made with the quartz heater; only steady measurements were possible with the copper heater. R-113 was the fluid used; the velocity varied over the interval 4–16 cm/sec; bulk liquid subcooling varied over 2–20°C; heat flux varied over 4–8 W/cm².

KEYWORDS: flow boiling; microgravity; Earth gravity

NOMENCLATURE:
a	Earth gravity acceleration component parallel to heater surface
Bo	Bond number, Equation **(8)**
C_b	buoyancy coefficient
C_s	surface tension coefficient
F	force
g, g_e	gravity and Earth gravity acceleration
Nd	dimensionless number, Equation **(6)**
R	bubble radius
Re	Reynolds number
$Ri_{2\varphi}$	two-phase Richardson number, Equation **(5)**
T	temperature
U	velocity
$We_{2\varphi}$	two-phase Weber number, Equation **(7)**

Address for correspondence: Herman Merte, Jr., Mechanical Engineering Dept., 2026 G.G. Brown Building, University of Michigan, Ann Arbor, MI 48109-2125, USA. Voice: 734-764-5240; fax: 734-647-3170.
merte@umich.edu

Greek Symbols

θ	heater surface orientation angle, FIGURE 4
Δρ	density difference between liquid and vapor phase
σ	surface tension

Subscripts

b	buoyant
d	departure or drag
l	liquid phase
rel	relative
ST	surface tension
sat	saturation
sub	subcooled

INTRODUCTION

The elements constituting the flow boiling process can be delineated as follows:

1. Nucleation—or the onset of boiling. These are not necessarily identical. The latter occurs once in a transient heat up, whereas nucleation may take place repetitively at a given location.

2. Vapor bubble growth and collapse. These may occur at the nucleation site or as the bubble is transported downstream by the bulk liquid.

3. Structure of the two-phase flow—for example, bubbly, slug, annular, and the corresponding pressure drop.

4. The local heater surface heat flux associated with the two-phase flow structures and with the various types of liquid–vapor phase changes possible—for example, nucleate boiling, film boiling (or dryout-CHF), and droplet–liquid film evaporation. This is to be expressed in terms of the existing significant local parameters, including heater surface temperature, void fraction, fluid properties, and heater surface geometry and microgeometry.

It is readily obvious, as is well documented by Hewitt[1] and others, that the phenomenon under consideration is quite complex: The advancement of its truly fundamental understanding requires considerable simplification in experimentation before synthesis of its description can take place. The present work hopefully contributes in this direction.

The opportunity to study flow boiling in the absence of buoyancy is intrinsically important, not only because of the potential space applications, but for Earth bound applications as well. The removal of gravity simplifies the boiling process in that certain features, otherwise disturbed by buoyancy, can now be disclosed or examined. However, the elevated cost of long-term space flights, which provide the high quality stable microgravity environment necessary, dictate that the maximum prior use be made of ground testing, including approximations to microgravity behavior. Observations of pool boiling in long-term space flights[2,3] with relatively large flat heater surfaces indicate that the growing vapor bubbles tend to remain in the vicinity of the heater surfaces. The imposition of a liquid velocity parallel to such a flat heater surface in the inverted position, horizontal or nearly horizontal relative to Earth gravity, might then somewhat approximate certain aspects of the corresponding flow boiling process in microgravity in that buoyancy now holds the heated liquid and vapor formed close to the heater surface.

The use of a flat heater surface simplifies the relationships between the bulk liquid flow velocity, heater surface orientation, and vapor motion. The process is further simplified in the work here in that only the onset of flow boiling is considered with initially zero void fraction, and only a pure single component fluid is used whose properties are well defined and not close to the thermodynamic critical state.

It is understood that for a sufficiently large liquid velocity the presence or absence of buoyancy becomes irrelevant: estimates of upper and lower velocity (when buoyancy is irrelevant) and lower velocity (when velocity is negligible) limits for nucleate flow boiling with buoyancy were presented by Kirk and Merte[4] and for the flow boiling critical heat flux with and without buoyancy by Brusstar et al.[5] Certain aspects of these are presented below.

BASIS FOR APPROXIMATING MICROGRAVITY IN EARTH GRAVITY

Flow boiling behavior is governed by the dynamic behavior of the vapor generated, which itself is a consequence of the forces acting at the interface. These can be more precisely expressed in terms of local stresses. For present purposes, however, only the global nature of the forces will be considered, consisting of:

1. Buoyancy arising from a body force acting on the liquid surrounding the vapor bubble.

2. Momentum associated with the externally imposed liquid flow around the vapor bubble and acting parallel to the flow.

3. Interfacial (surface) tensions, considered here as liquid–vapor and liquid–solid–vapor. The former primarily controls the liquid–vapor surface geometry, whereas the latter controls the adhesion or departure of the vapor bubble from the solid heater surface.

4. Viscous shear arising principally from the liquid viscosity.

5. Lift arising from the externally imposed liquid flow around the vapor bubble and acting normal to the flow.

6. Internal vapor pressure generated by the mean liquid–vapor interface temperature provides the driving force for vapor bubble growth and collapse dynamics.

For the purpose of assessing the degree to which flow boiling in Earth gravity may approximate that in microgravity, advantage will be taken of ratios of certain of the above forces, in particular, those associated with the removal of the vapor bubble from its nucleation site at the heater in the inverted horizontal or nearly horizontal position relative to Earth gravity. Since the bulk liquid flow is parallel to the heater surface, only those components of the forces parallel to the heater are considered, including buoyancy.

The normal component of buoyancy as a vapor bubble removal mechanism is modified by inversion of the heater. However, the influence of lift, which is also normal to the heater surface, must be given consideration, and would be of special significance in microgravity. For spherical particles moving in an inviscid fluid parallel to a plane wall, Li et al.[6] demonstrate that the lift moves the sphere toward the wall. With a viscous fluid, on the other hand, the resulting velocity gradient in duct flow causes the sphere to migrate away from the stationary wall.[7] With Earth gravity buoyancy acting to move the vapor bubble toward the wall for the case considered here, calculations by Ervin[8] for parameters approximating those present show that such lift can be neglected relative to the buoyancy. In addition, the internal vapor pressure giving rise to the bubble dynamics is neglected for present purposes. This leaves the forces related to buoyancy, momentum (or inertial drag), surface tension, and viscous shear.

Several dimensionless numbers, defined in terms of ratios of these forces acting on the vapor bubbles, are now examined to estimate the conditions under which certain aspects of the flow boiling process in microgravity might be approximated in Earth gravity. These are:

$$\text{Richardson number} \equiv \frac{\text{buoyancy}}{\text{inertial drag}} \equiv (\text{Froude number})^{-2}.$$

$$\text{Weber number} \equiv \frac{\text{inertial drag}}{\text{surface tension}}.$$

$$\text{Bond number} \equiv \frac{\text{buoyancy}}{\text{surface tension}}.$$

$$\text{capillary number} \equiv \frac{\text{viscous drag}}{\text{surface tension}}.$$

The capillary number can be expressed as the ratio of the Weber number to the Reynolds number: if the viscous drag in the capillary number were to be replaced by the inertial drag, which would correspond to a Reynolds number of unity, the capillary number would become equivalent to the Weber number. As was demonstrated by the computations of Ervin,[8] the capillary number affects the lift force acting on a bubble in determining its deformation due to the shear flow. Since lift is being neglected, the capillary number is given no further consideration here.

An expression for a two-phase Richardson number was developed by Kirk and Merte,[9] in which the buoyant and drag forces acting on a vapor bubble are defined in terms of the bubble radius, the relative velocity between the bubble and the bulk liquid, U_{rel}, and coefficients, C_b and C_d, given by

$$F_b = C_b \frac{4}{3}\pi R^3 g_e \Delta\rho, \tag{1}$$

$$F_d = C_d \pi R^2 \rho_l \times \frac{U_{rel}^2}{2}. \tag{2}$$

The buoyancy coefficient, C_b, accounts for the difference between a spherical bubble of radius, R, and that of a non-spherical bubble of equal mean radius, and Cd is the drag coefficient, determined empirically. The drag force can be considered to be a manifestation of the net bulk flow momentum effects acting on the vapor bubble. The characteristic bubble radius, R, in Equations (1) and (2) is taken to be the departure bubble radius R_d. Once the vapor bubble has departed its local nucleation site it can be considered to no longer directly influence the local heat transfer processes. An expression for the departure radius was obtained from a force balance on a vapor bubble growing from its nucleation site, using Equations (1) and (2), together with a surface tension force, F_s, at the solid–liquid–vapor contact line,

$$F_s = C_s \sigma R. \tag{3}$$

The surface tension coefficient, C_s, was obtained empirically from the balance of buoyancy and surface tension forces for the case of pool boiling by equating (3) and (1), with $C_b = 1$. Using the measured departure radius 0.50 mm for R-113, the surface tension coefficient was found to be approximately unity.

An estimate of the drag coefficient, C_d, was determined empirically from a balance of the buoyant and drag forces, Equations (1) and (2), parallel to a heating surface at which an immediately adjacent spherical bubble was moving. Measurements of the bubble rise velocities, sizes, and liquid velocities were made for various orientations θ of the flat heating surface relative to the horizontal face-up position (see FIG. 4), such that g_e in (1) is replaced by

$$g = g_e \sin\theta. \tag{4}$$

Over a range of bubble Reynolds numbers from 24 to 920 the drag coefficient, C_d, is roughly constant and equal to 0.56. This compares well with the turbulent drag coefficient of 0.50 for a solid sphere. Further details are available from Kirk and Merte.[4]

Taking the ratio of (1) and (2) as the two-phase Richardson number, together with the coefficients given above,

$$Ri_{2\varphi} = \frac{1}{2}(\sqrt{1 + 21.7 N_d} - 1), \tag{5}$$

where

$$N_d = \frac{g \Delta\rho \sigma}{\rho_l^2 U^4}. \tag{6}$$

The liquid velocity U is taken as U_{rel} in (2), and g is given by (4).

Equation (5) is plotted in FIGURE 1 for R-113 at atmospheric pressure, for three orientation angles, and the range of velocities used here.

A two-phase Weber number is obtained by taking the ratio of (2) to (3), resulting in

$$We_{2\varphi} = \frac{0.1}{N_d}(\sqrt{1 + 21.7 N_d} - 1) = 0.2x \frac{Ri_{2\varphi}}{N_d}. \tag{7}$$

Equation (7) is shown in FIGURE 2 for the similar conditions as those in FIGURE 1.

In a similar manner the Bond number is obtained by taking the ratio of (1) to (3), and can be represented as

$$Bo = 0.42\left(21.7 - 4\frac{Ri_{2\varphi}}{N_d}\right), \tag{8}$$

or, equivalently, as

$$Bo = 9.1(1 - We_{2\varphi}). \tag{9}$$

Equation (8) is plotted in FIGURE 3 for similar conditions as those in FIGURES 1 and 2.

For θ = 180°, where buoyancy is acting normal to the flat heater surface, these three dimensionless number take on singular values, given by

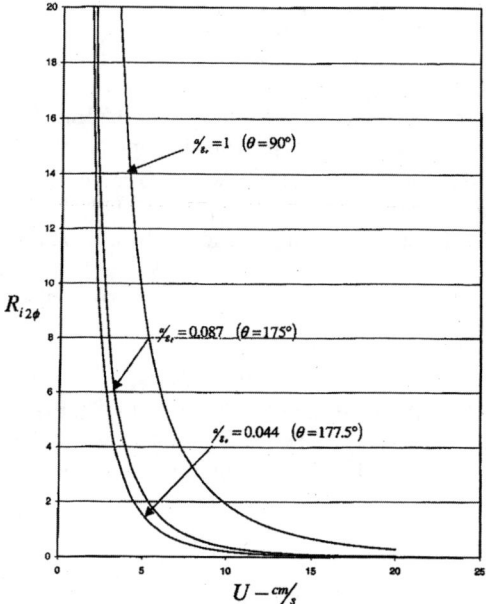

FIGURE 1. Two-phase Richardson number for R-113 at $P = 1$ atmosphere.

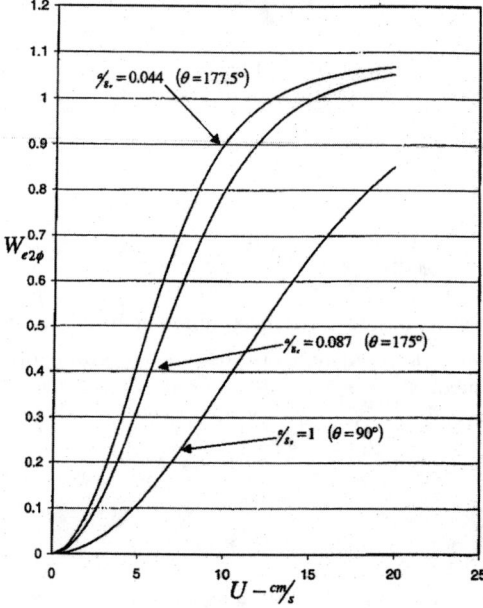

FIGURE 2. Two-phase Weber number for R-113 at $P = 1$ atmosphere.

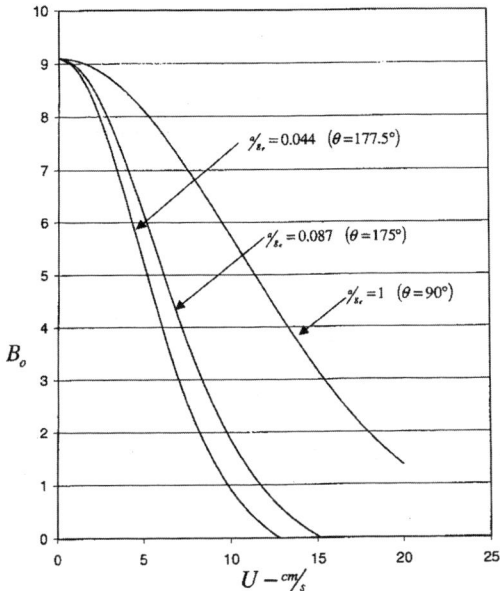

FIGURE 3. Bond number for R-113 at $P = 1$ atmosphere.

$$\left.\begin{array}{l} Ri_{2\varphi} \to 0 \\ We_{2\varphi} \to 1 \\ Bo \to 0 \end{array}\right\} \quad (10)$$

Of course, the assumptions under which the vapor bubble departure sizes are computed are no longer valid here.

Making use of FIGURES 1–3, the conditions for possibly simulating microgravity flow boiling, or certain elements of it, in Earth gravity are now examined:

Two-Phase Richardson Number. FIGURE 1

$$Ri_{2\varphi} = \frac{F_{buoy}}{F_{drag}}, \text{ desire small } Ri_{2\varphi}, \text{ but low } U.$$

At $\theta = 175°$ and $177.5°$ with $U \approx 5\text{–}10\,\text{cm/sec}$; $Ri_{2\varphi} < 2$.

Two-phase Weber Number. FIGURE 2

$$We_{2\varphi} = \frac{F_{drag}}{F_{ST}}, \text{ desire small } We_{2\varphi}, U < 5\,\text{cm/sec}.$$

At $\theta = 175°$ and $177.5°$ with $U \approx 5\text{–}10\,\text{cm/sec}$; $We_{2\varphi} \approx 0.3\text{–}0.9$.

Bond Number. FIGURE 3

$$Bo_{2\varphi} = \frac{F_{buoy}}{F_{ST}}, \text{ desire small } Bo, \text{ but not too low } U.$$

At $\theta = 175°$ and $177.5°$ with $U \approx 5\text{–}10\,\text{cm/sec}$; $Bo_{2\varphi} \approx 1\text{–}6$.

It appears that liquid velocities in the range of $U = 5–10$ cm/sec might be judged as critical or illuminating for changes in behavior of the flow boiling process. The next phase is to examine the behavior of the various facets of flow boiling over this range of liquid velocities with orientations in the vicinity of horizontal facing downward.

DESCRIPTION OF EXPERIMENTAL HARDWARE

A schematic of the forced convection boiling loop is given in FIGURE 4. The loop is mounted on an axis near its center of gravity such that it can be rotated a full 360 degrees while in operation in order to obtain the various heater surface orientations. The loop consists of five basic components: the test section, the condenser-cooler, the flow system, the pressure control system, and the preheater system. Degassed R-113 was employed as the test fluid. The sufficiency of the degassing process was demonstrated by comparing the measured vapor pressure curve with published saturation properties over the domain of interest.

The test section provides a rectangular flow cross section 10.80 cm wide by 1.27 cm high by 35.56 cm long ($4.25 \times 0.50 \times 14.0$ in), capable of accommodating three heater surfaces on each side. Viewing windows made from optical grade quartz permit a transverse view of the boiling process on the flat heaters. The condenser-cooler system condenses R-113 vapor leaving the test section, and subcools the

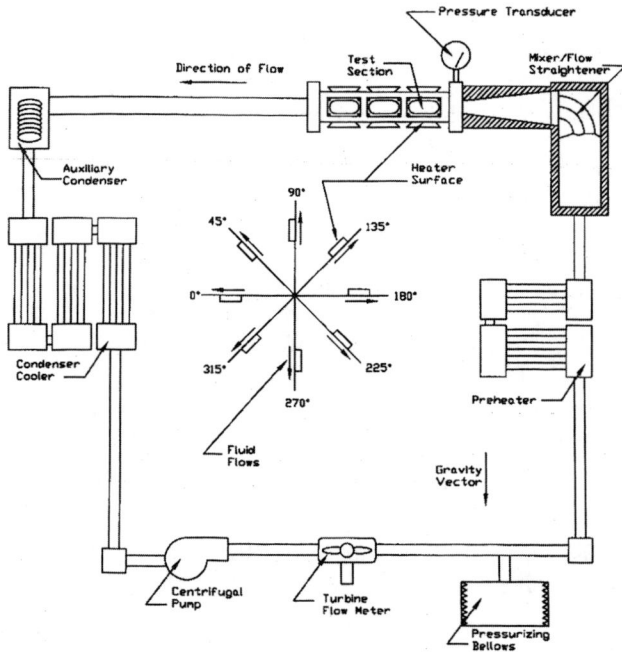

FIGURE 4. Forced convection flow loop.

liquid sufficiently to avoid cavitation at the pump inlet. The pressure control system actively maintains the pressure in the flow loop to within ±0.34 kPa (0.05 psi), and consists of four main components: a bellows, valves, electronic control circuits, and a pressure transducer. The 17.78 cm (7.0 in) diameter stainless steel bellows separates the test fluid from the air in the chamber surrounding the bellows. Air is continuously admitted to the chamber through a manually adjusted needle valve, and is vented from the chamber through an electronically controlled proportional valve. The preheater is used to raise the temperature of the R-113 to the desired level at the inlet to the test section.

The heater surfaces installed in the test section were of dimensions larger than the bubbles growing on them in order that they be more representative of boiling surfaces encountered in engineering applications. Two types of relatively large test surfaces, with dimensions 1.905×3.81 cm (0.75×1.50 in.), were employed in this study: a thin gold film heater deposited on a quartz substrate, and a gold-coated copper substrate heater, referred to as the "metal" surface.

The semitransparent thin gold film surface functions simultaneously as a heater and as a resistance thermometer, and consists of nominally 400 Å thick gold sputtered on a highly polished quartz substrate, which was presputtered with 20 Å of tantalum for improved gold adhesion. Measurement of the resistance of the thin film heater provides a spatially averaged mean surface temperature, with an uncertainty of ±1.0°C.

The metal heater consists of a copper block with a film heater on the underside and embedded in an electrical encapsulating potting compound. A smooth copper foil 0.0254 mm (0.001 in) thick was soldered to the copper block and to the stainless steel holder to eliminate artificial boiling, which otherwise would result at the heater edge. The copper foil was sputtered with gold, approximately 1,000 Å thick, to provide the identical surface–liquid energy combination as for the gold deposited on quartz. A chromel–constantan thermocouple with a bead of 0.64 mm (0.025 in) diameter was inserted into a 0.76 mm (0.030 in) diameter hole in the copper heater in order to estimate the temperature of the copper foil. The hole for the thermocouple was then filled with a thermally conducting copper compound. The tip of the thermocouple was about 0.76 mm (0.030 in) from the gold-sputtered copper foil. The metal heater surface temperature had an error of ±0.1°C, while that of assuming that the temperature of the thermocouple was equal to the surface temperature was estimated to be less than 0.002°C.

Because of their different constructions, the thin gold film heater provides a uniform heat flux, with a spatial temperature variation in the flow direction; the metal copper heater provides a uniform surface temperature with a spatial variation in the heat flux in the flow direction. Both transient and steady measurements of heat flux and heater surface superheats were made with the thin film heater; only steady measurements were possible with the metal copper heater. Additional details of the flow loop, heaters, and experimental procedures followed are available elsewhere.[4,10]

RESULTS

The results presented here were obtained with the two flat heater surfaces described above facing horizontally downward θ = 180°, with variations in the orientation angles within ±5 producing very small changes in the buoyancy components parallel to the heat surfaces. The bulk liquid velocity at the inlet to the test section was varied over the range of 4–16 cm/sec; the bulk liquid subcooling was varied over the range of 2.5–20.2°C by changing the system pressure appropriately; the input heat flux was varied over 4–8 w/cm^2. A total of 314 experiments were conducted over the range of parameters given above, the complete set of results is included in Merte et al.[11] However, only selected results are included below that demonstrate the interrelationships between buoyancy parallel to the flow direction, both aiding and in opposition to the flow, and velocity, subcooling, and heat flux. Prior work covering wider ranges of each of the above parameters (except for subcooling) are summarized in the APPENDIX.

The transient heating experiments with the gold film heater on a quartz substrate are presented first, followed by the steady state experiments with the copper substrate heater. The transient heating entails a step input in heater power and naturally reverts to the steady state over a sufficiently long period of time. Some steady results with the thin gold film heater are superimposed on the steady state results with the copper substrate heater for purposes of comparison.

TRANSIENT HEATING WITH THE THIN GOLD FILM ON QUARTZ

Since the thermal conductivity (and the thermal diffusivity) of the quartz substrate is on the order of 20 times that of the R-113, the initial transient heat flux to the R-113 is less than the steady state level, with the rate of increase in the mean (and local) heater surface temperature depending on the fluid velocity. This becomes obvious from FIGURES 5–7, in which the mean heater surface superheats are plotted as a function of time, all for fixed values of subcooling (11.1°C) and input heat flux (4 w/cm^2), with four successive increases in velocity (4–16 cm/sec) for each of the three orientations. In interpreting the behavior observed, it must be kept in mind that for orientation angles θ < 180°, the component of buoyancy, whether with a heated liquid or with a vapor present, is acting in the same direction as the flow velocity, whereas for θ > 180° it is acting counter to the flow velocity. For θ = 180° buoyancy acts normal to the flow velocity. Also included in each of FIGURES 5–7, for reference purposes, are the heater surface superheats computed assuming one-dimensional heat conduction for the case of an infinitely large heater with a stagnant liquid. This approximates the transient heating process for pool boiling in microgravity. When corrections were made for three-dimensional conduction effects in the quartz substrate, the computed mean surface temperatures agree quite well with measurements, as reported in Merte et al.[2] Attention is to be drawn to several items in FIGURES 5–7:

1. As the velocity increases with each orientation, not only does the rate of heater surface temperature rise decrease because of the increase in heat transfer to the liquid, accompanied by an increase in the time required for the onset of boiling,

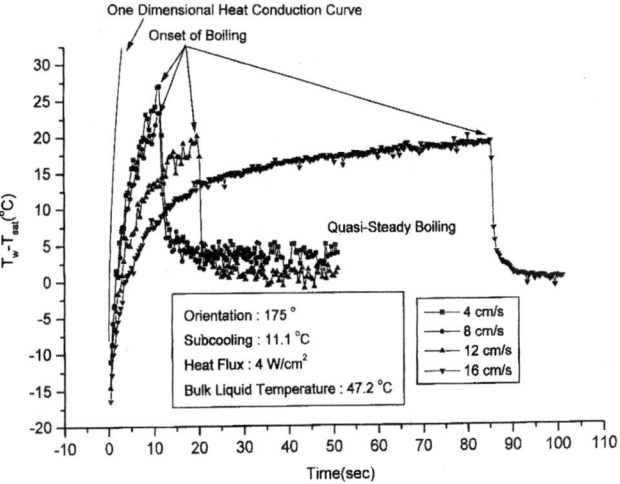

FIGURE 5. Transient flow boiling—quartz substrate. Effect of velocity for $\theta = 175°$, $\Delta T_{sub} = 11.1°$, $q''_{in} = 4\,\text{W/cm}^2$.

as indicated on the figures, but the heater surface superheat at which nucleation takes place is in general reduced. This is interpreted as supporting the concept that the temperature gradient in the liquid in the vicinity of the heater surface has a significant effect on the nucleation process, as discussed in Merte and Lee.[12]

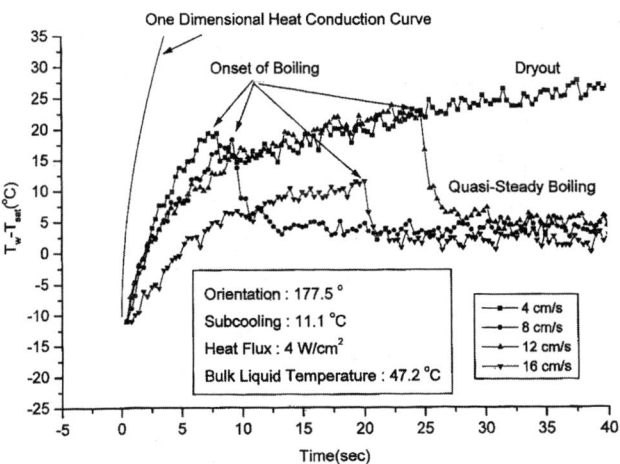

FIGURE 6. Transient flow boiling—quartz substrate. Effect of velocity for $\theta = 177.5°$, $\Delta T_{sub} = 11.1°$, $q''_{in} = 4\,\text{W/cm}^2$.

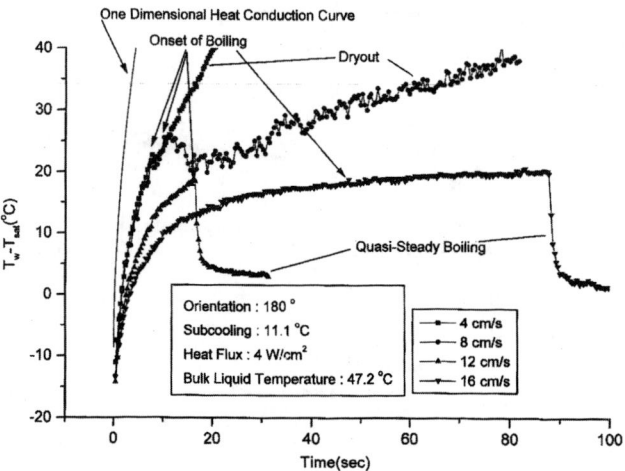

FIGURE 7. Transient flow boiling—quartz substrate. Effect of velocity for $\theta = 180°$, $\Delta T_{sub} = 11.1°$, $q''_{in} = 4\,\text{W/cm}^2$.

2. No dryout (DO) occurs at any velocity in FIGURE 5 for $\theta = 175°$, where the buoyancy in the flow direction is the strongest. With a reduction in this buoyancy for $\theta = 177.5°$ in FIGURE 6, dryout takes place at the lowest velocity $U = 4\,\text{cm/sec}$. With a further reduction of buoyancy to zero with $\theta = 180°$ in FIGURE 7, dryout occurs at the two lower velocities of $U = 4$ and $8\,\text{cm/sec}$. For this same orientation and the velocity of $U = 4\,\text{cm/sec}$, dryout took place at all levels of liquid subcooling used here, up to $\Delta T_{sub} = 20.2°\text{C}$. As is observed in FIGURE 8 below, for this same orientation and a velocity of $U = 8\,\text{cm/sec}$, dryout occurs at all but the highest level of liquid subcooling. The data are not shown here, but for this same orientation of $\theta = 180°$, heat flux $q''_{in} = 4\,\text{W/cm}^2$ and the lowest subcooling $\Delta T_{sub} = 2.5°\text{C}$, dryout took place following nucleation for all velocities $U = 4\text{--}16\,\text{cm/sec}$.

3. For each orientation, the indicated quasisteady boiling processes, so called since the length of time used was not sufficient to assure true steady conditions, become more effective with increasing velocity, as indicated by a decrease in the heater surface superheat. For $U = 8\,\text{cm/sec}$, the quasisteady boiling effectiveness is virtually unchanged between the buoyancy changes in FIGURES 5 and 6, and the same is true for $U = 12$ and $16\,\text{cm/sec}$ between FIGURES 5–7.

FIGURES 8 and 9 present the transient heater surface superheat for the fixed values of orientation with no buoyancy component in the flow direction ($\theta = 180°$), and input heat flux ($4\,\text{w/cm}^2$), with four successive increases in subcooling (2.5–20.2°C) for each of the velocities $U = 8$ and $12\,\text{cm/sec}$.

Several observations can be made about FIGURES 8 and 9:

1. These velocities cross the upper limit of the range $U = 5\text{--}10\,\text{cm/sec}$, postulated earlier as critical for changes in behavior of the flow boiling process.

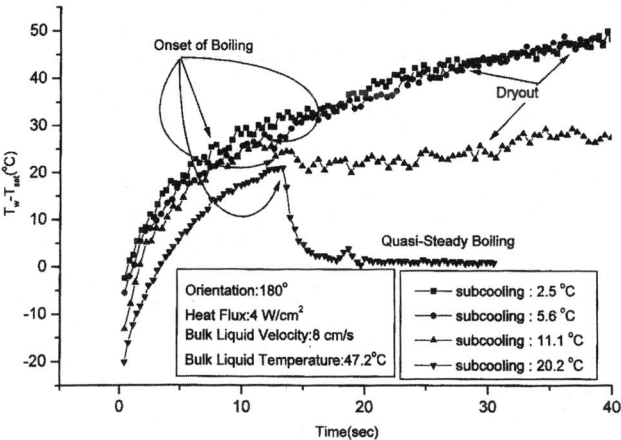

FIGURE 8. Transient flow boiling—quartz substrate. Effect of subcooling for $U = 8\,\text{cm/sec}$, $\theta = 180°$, $q''_{in} = 4\,\text{W/cm}^2$.

2. Whereas FIGURE 7 presented the transient flow boiling behavior for $\theta = 175°$ for various flow velocities, FIGURES 8 and 9 show the influence of subcooling for the same orientation of $\theta = 180°$ for $U = 8$ and $12\,\text{cm/sec}$, respectively.

3. At the velocity $U = 8\,\text{cm/sec}$ in FIGURE 8, except for the two lowest subcoolings, where no change is detected, increasing the subcooling results in modest decreases in the mean heater surface superheat at nucleation. At the higher velocity, $U = 12\,\text{cm/sec}$, in FIGURE 9, no changes again are detected between the two lowest

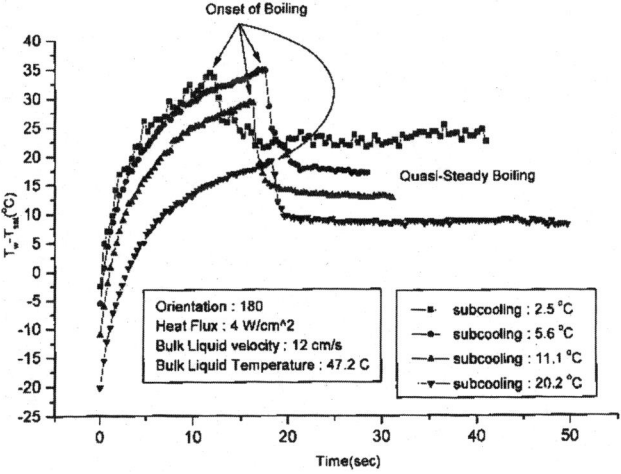

FIGURE 9. Transient flow boiling—quartz substrate. Effect of subcooling for $U = 12\,\text{cm/sec}$, $\theta = 180°$, $q''_{in} = 4\,\text{W/cm}^2$.

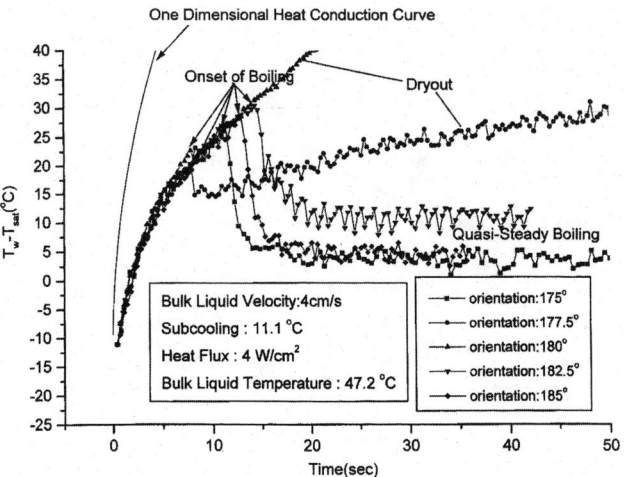

FIGURE 10. Transient flow boiling—quartz substrate. Effect of orientation for $U = 4$ cm/sec, $\Delta T_{sub} = 11.1°$, $q''_{in} = 4$ W/cm^2.

subcoolings, although the heater surface superheats at nucleation here are about 8°C higher than with $U = 8$ cm/sec, whereas a dramatic decrease in the heater surface superheat at nucleation occurs at the highest subcooling, indeed decreasing to below that for $U = 8$ cm/sec in FIGURE 8. This may again be a consequence of the influences that the velocity and subcooling level exert on the liquid temperature gradient adjacent to the heater surface.

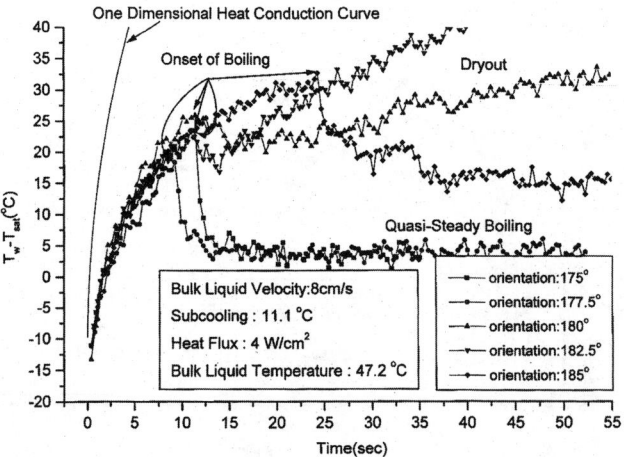

FIGURE 11. Transient flow boiling—quartz substrate. Effect of orientation for $U = 8$ cm/sec, $\Delta T_{sub} = 11.1°$, $q''_{in} = 4$ W/cm^2.

4. It is noted that, following the onset of boiling at $U = 8\,\text{cm/sec}$ in FIGURE 8, dryout occurs at all except the highest subcooling, although no dryout takes place when the velocity is increased to $U = 12\,\text{cm/sec}$ in FIGURE 9.

5. For the quasisteady nucleate boiling taking place in both FIGURES 8 and 9, with the highest subcooling of $\Delta T_{sub} = 20.2°\text{C}$, the lower velocity of $U = 8\,\text{cm/sec}$ in FIGURE 8 exhibits a considerably more effective heat transfer process, with a mean heater surface superheat of about 1°C, compared to about 6°C in FIGURE 9.

The effects of buoyancy relative to liquid velocity on the transient mean heater surface superheat are illustrated in FIGURES 10–12 for fixed levels of subcooling of $\Delta T_{sub} = 11.1°\text{C}$ and input heat flux $q''_{in} = 4\,\text{W/cm}^2$. The velocities are increased to $U = 4, 8, 12\,\text{cm/sec}$ for each of the three figures, respectively, whereas the orientation angles are incremented within each figure from $\theta = 175°$ to $\theta = 185°$ in steps of $\Delta\theta = 2.5°$. Attention is drawn to the following features:

1. As the orientation angles within each of FIGURES 10–12 increase, beginning with $\theta = 175°$, the buoyancy component acting in the flow direction parallel to the heater surface decreases successively, to zero at $\theta = 180°$, then increases successively in opposition to the flow direction, ending at $\theta = 185°$.

2. The heater surface superheats at nucleation are given in TABLE 1 for the experiments in FIGURES 10–12. For each of the orientations except $\theta = 177.5°$, where relatively small changes are observed, the superheats decrease as the liquid velocity increases, with the most dramatic decreases occurring at $\theta = 182.5°$ and $\theta = 185°$ for velocity increase from $U = 8\,\text{cm/sec}$ to $U = 12\,\text{cm/sec}$. This is where the liquid velocity is acting counter to buoyancy, and is believed to reflect the combined effects of velocity and buoyancy on the liquid temperature gradient at the heater surface on nucleation. This is supported by the relatively small changes in super-

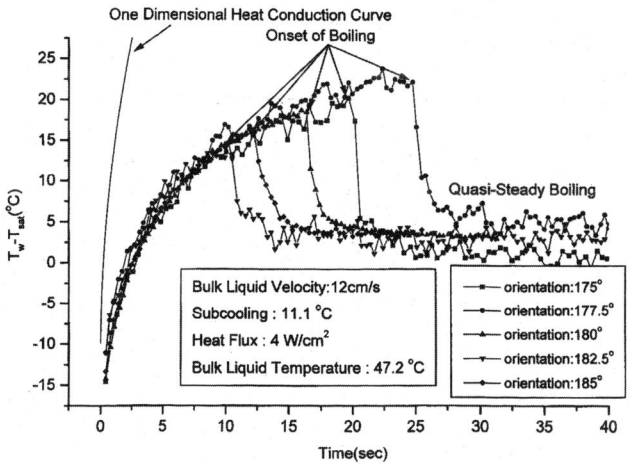

FIGURE 12. Transient flow boiling—quartz substrate. Effect of orientation for $U = 12\,\text{cm/sec}$, $\Delta T_{sub} = 11.1°$, $q''_{in} = 4\,\text{W/cm}^2$.

TABLE 1. Heater surface superheats at nucleation for FIGURES 10–12

	Surface Superheat (°C)		
θ (°C)	$U = 4$ cm/sec	$U = 8$ cm/sec	$U = 12$ cm/sec
175	26	22.5	19.5
177.5	18	18	22
180	22.5	25	18
182.5	30	22.5	14
185	27.5	32.5	15.5

heat at θ = 175° and θ = 177.5°, where buoyancy is acting in the flow direction, and changes in flow velocities have less effect on the liquid temperature gradient.

3. For the lowest velocity, $U = 4$ cm/sec in FIGURE 10, dryout takes place with θ = 177.5° and θ = 180°. With an increase in velocity to $U = 8$ cm/sec in FIGURE 11, dryout now takes place with θ = 180° and θ = 182.5°, whereas as the velocity increases further to $U = 12$ cm/sec in FIGURE 12, no dryout takes place.

4. Where quasisteady nucleate boiling is taking place, the large increases in heat transfer effectiveness take place as the velocity is increased to $U = 12$ cm/sec for the orientations of θ = 182.5° and θ = 185°, where buoyancy is acting counter to the flow direction.

STEADY STATE WITH COPPER SUBSTRATE HEATER

FIGURE 13 shows the typical hysteresis taking place with the copper substrate heater. The established nucleate boiling with increasing and decreasing heat flux are identical. The point from the onset of boiling to the next higher heat flux level have virtually the same input heat flux. The significant increase in heat flux taking place on nucleation is due to the reduction in the calibrated heat loss associated with the reduced heater surface temperature. Calibration of the heat loss is described in detail in Kirk and Merte.[4] The onset of boiling in FIGURE 13 takes place with a heater surface superheat of 32°C and mean heat flux of 1.2 W/cm². For the same bulk subcooling and liquid velocity, when the orientation is changed to θ = 175°, the corresponding levels of superheat and heat flux are 38°C and 1.2 W/cm². For comparison of the mean heater surface superheat on nucleation with the quartz substrate, FIGURE 9 shows a value of 19°C, compared with 32°C for the identical conditions with the copper substrate.

FIGURES 14–17 present the steady results with the copper substrate in three orientations, θ = 175°, 180°, and 185°, within each figure, with successively increasing levels of the liquid velocity $U = 4, 8, 12$, and 16 cm/sec, all for a moderate liquid subcooling level of $\Delta T_{sub} = 5.6$°C. Also superimposed on FIGURES 14–17, for comparison purposes are the quasisteady results with the quartz substrate heater surface, for the same operating parameters of subcooling, velocity, and orientations. It is noted that the steady nucleate boiling results for the two surfaces are not too dissimilar at

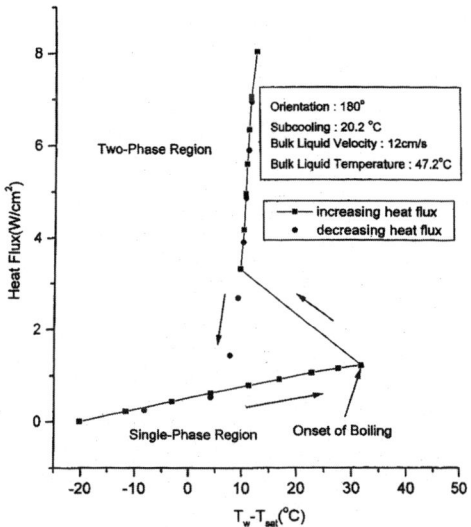

FIGURE 13. Steady flow boiling—copper substrate. Demonstration of hysteresis for θ = 180°, U = 12 cm/sec, ΔT_{sub} = 20.2°.

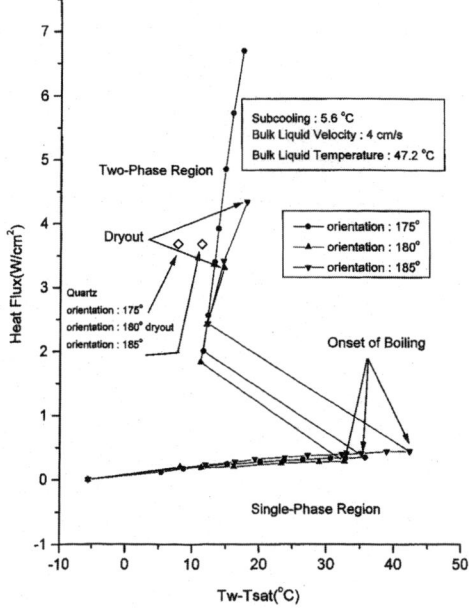

FIGURE 14. Steady flow boiling—copper substrate. Effects of orientation for U = 4 cm/sec, ΔT_{sub} = 5.6°.

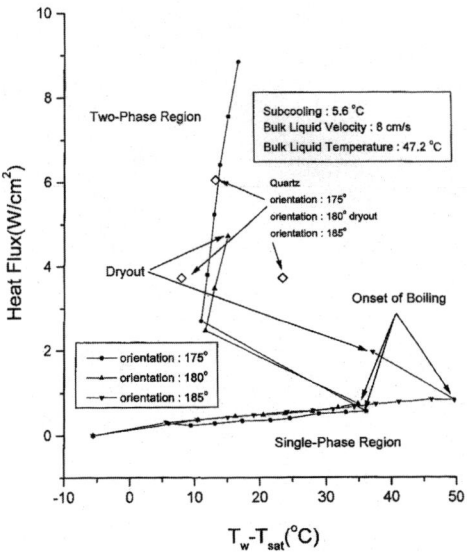

FIGURE 15. Steady flow boiling—copper substrate. Effects of orientation for $U = 8$ cm/sec, $\Delta T_{sub} = 5.6°$.

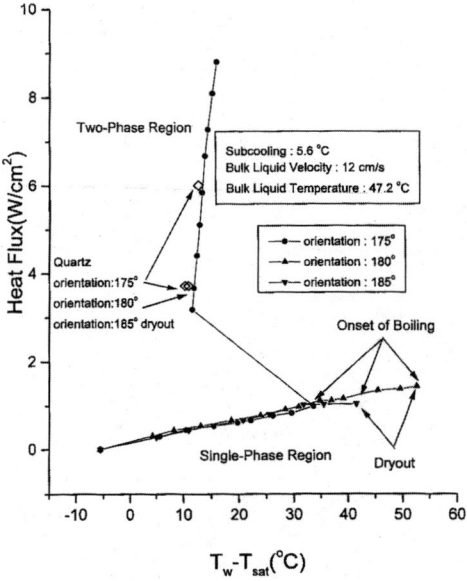

FIGURE 16. Steady flow boiling—copper substrate. Effects of orientation for $U = 12$ cm/sec, $\Delta T_{sub} = 5.6°$.

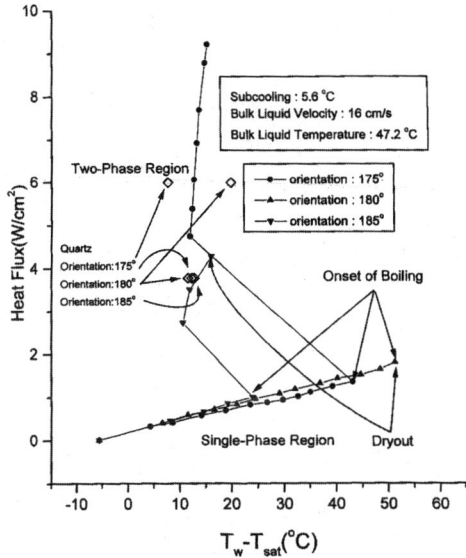

FIGURE 17. Steady flow boiling—copper substrate. Effects of orientation for $U = 16\,\text{cm/sec}$, $\Delta T_{sub} = 5.6°$.

$U = 4\,\text{cm/sec}$ and $U = 12\,\text{cm/sec}$ in FIGURES 14 and 16, respectively. Dryout takes place at a relatively low heat flux level with the copper substrate heater surface in FIGURE 15, associated with the adverse buoyancy effect, but does not occur with the quartz substrate. In FIGURE 16, with $U = 12\,\text{cm/sec}$, dryout occurs on nucleation with the copper substrate for orientations of $\theta = 180°$ and $185°$, whereas it only takes place at $\theta = 180°$ with the quartz substrate. The difference is attributed to the larger heater surface superheat existing at nucleation with the copper substrate. As the liquid subcooling is increased to $20.2°C$ with the copper substrate heater for this same velocity, in FIGURE 13, the heater surface superheat at nucleation is reduced to $32°C$, compared with $53°C$ in FIGURE 16, and dryout does not take place. However, dryout on the copper substrate heater occurs even with the larger velocity $U = 16\,\text{cm/sec}$ in FIGURE 17. This has the lower liquid subcooling level of $5.6°C$ compared to $20.2°C$ in FIGURE 13, which contributes to a larger amount of vapor generation.

The steady results for $\theta = 180°$ with the copper substrate heater in FIGURES 14–17 are combined in FIGURE 18 to clarify the effects of changing the liquid velocity alone. Increasing the velocity increases the single-phase heat transfer coefficient, as demonstrated by the increasing slope. The two highest velocities result in dryout immediately following nucleation, somewhat contrary to expectations. As stated earlier, the higher superheat at nucleation produces a large amount of vapor, inhibiting the rewetting process. This dryout does not take place with the quartz substrate heater, as can be seen in FIGURE 9, for $U = 12\,\text{cm/sec}$, nor at $U = 16\,\text{cm/sec}$, not included here. The role of liquid subcooling for the copper substrate, which was not shown in FIGURES 14–17, is presented in FIGURE 19 for $\theta = 180°$ and the velocity

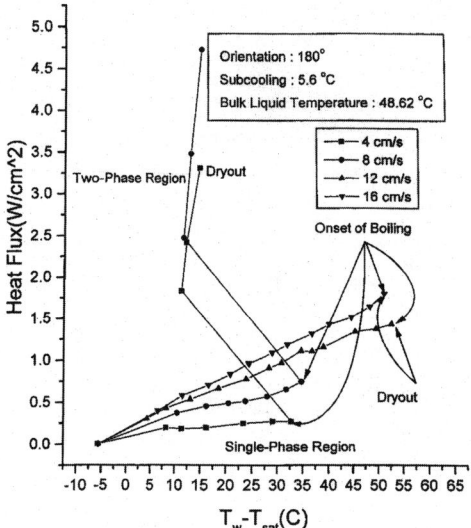

FIGURE 18. Steady flow boiling—copper substrate. Effects of velocity for $\theta = 180°$, $\Delta T_{sub} = 5.6°$.

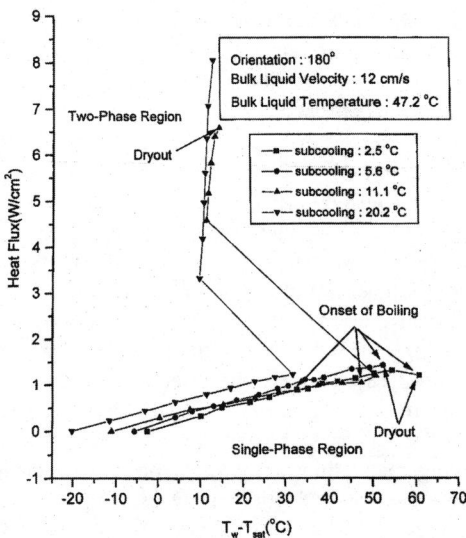

FIGURE 19. Steady flow boiling—copper substrate. Effects of subcooling for, $\theta = 180°$, $U = 12$ cm/sec.

$U = 12$ cm/sec. It is to be noted that dryout now only takes place at the lower two levels of subcooling, $\Delta T_{sub} = 2.5°C$ and $5.6°C$.

CONCLUSIONS

It has been demonstrated that distinct changes in behavior of nucleation, nucleate boiling, and dryout take place in flow boiling in earth gravity in the inverted orientation, corresponding to simulated microgravity, in velocities between 8–12 cm/sec for the fluid used here. This is in the same range determined from the three dimensionless numbers: a two-phase Richarson number, a two-phase Weber number, and a Bond number.

Whether similar behavior occurs in the long-term microgravity of space remains to be demonstrated. However, based on observations with pool boiling in long-term microgravity, it is believed that long-term microgravity flow boiling experiments conceived and carried out with care, covering the full range of parameters expected to be encountered in applications of interest, will generate surprising results. It also becomes clear from the work carried out here that extrapolating transient experiments to steady-state conditions must be carried out with care. Aircraft parabolic flights involve relatively short experimental times, and certain flow boiling behaviors are quite sensitive to the variations of $\pm 0.02\,g$ inherently present.

ACKNOWLEDGMENTS

The research presented here was conducted with NASA support under Grant NAG3-1900. The encouragement of Dr. Fran Chiaramonte of the NASA Glenn Research Center is gratefully recognized.

REFERENCES

1. HEWITT, G.F. 1996. Phenomenological issues in forced convective boiling. *In* Convective Flow Boiling, John C. Chen, Ed.: 3–14. Taylor and Francis, Washington, D.C.
2. MERTE, H. JR., H.S. LEE & R.B. KELLER. 1995. Report on pool boiling experiment flown on STS-47 (PBE-IA), STS-57 (PBE-IB), STS-60 (PBE-IC). NASA Contract NAS 3-25812, Report No. UM-MEAM-95-01. Department of Mechanical Engineering and Applied Mechanics, University of Michigan, Ann Arbor, Michigan. February 1995.
3. MERTE, H., JR., H.S. LEE & R.B. KELLER. 1998. Dryout and rewetting in the pool boiling experiment flow on STS-72 (PBE-IIB), STS-77 (PBE-IIA). Final Report, NASA Grant NAG-1684, Report NO. UM-MEAM-98-01. Department of Mechanical Engineering and Applied Mechanics, University of Michigan, Ann Arbor, Michigan. January 1998.
4. KIRK, K.M. & H. MERTE, JR. 1992. A study of the relative effects of buoyancy and liquid momentum in forced convection nucleate boiling. Final Report, NASA Grant NAG3-1310, Report No. UM-MEAM-92-06. Department of Mechanical Engineering and Applied Mechanics, University of Michigan, Ann Arbor, Michigan. November 1992.

5. BRUSSTAR, M.J., H. MERTE, JR. & R.B. KELLER. 1995. Relative effects of flow and orientation on the critical heat flux in subcooled forced convection boiling. Final Report, NASA Grant NGT-50928, Report No. UM-MEAM-95-15. Department of Mechanical Engineering and Applied Mechanics, University of Michigan, Ann Arbor, Michigan. July 1995.
6. LI, L., W.W. SCHULTZ & H. MERTE, JR. 1993. The velocity potential and the interacting force for two spheres moving perpendicularly to the line joining their centers. J. Eng. Math. **27**: 147–160.
7. CHERUKAT, P. & J.B. MCLAUGHLIN. 1993. Wall-induced lift on a sphere. Intl. J. Multiphase Flow **16**(5): 899–907.
8. ERVIN, E.A. 1993. Computations of Bubbles and Drops in Shear Flow. Ph.D. Thesis, The University of Michigan, Ann Arbor, Michigan.
9. KIRK, K.M. & H. MERTE, JR. 1994. A mixed natural/ forced convection nucleate boiling heat transfer criteria. *In* Heat Transfer 1994. **7**: 479–484.
10. KIRK, K.M., H. MERTE, JR. & R.B. KELLER. 1995. Low-velocity subcooled nucleate flow boiling at various orientations. ASME J. Heat Transfer **117**: 380–386.
11. MERTE, H., JR., J. PARK, W.W. SCHULTZ & R.B. KELLER. 2001. Simulation of low velocity subcooled microgravity boiling in earth gravity. Final Report, NASA Grant NAG3-1900, Report No. UM-ME-01-02. Department of Mechanical Engineering, University of Michigan, Ann Arbor, Michigan. April 2001.
12. MERTE, H., JR. & H.S. LEE. 1997. Quasi-homogeneous nucleation in microgravity at low heat flux: experiments and theory. ASME J. Heat Transfer **119**: 305–312.
13. NESTEL, S.U. & H. MERTE, JR. 1995. The effects of heater surface length in forced convection nucleate boiling. NASA Grant NAG3-1310, Report No. UM-MEAM-95-01. Department of Mechanical Engineering and Applied Mechanics, University of Michigan, Ann Arbor, Michigan. July 1995.
14. TORTORA, P.R., H. MERTE, JR., H.S. LEE & R.B. KELLER. 1999. Buoyant and inertial effects in subcooled forced convection nucleate boiling on flat inverted surfaces. NASA Grant NAG3-1900, Report No. UM-MEAM-99-03. Department of Mechanical Engineering and Applied Mechanics, University of Michigan, Ann Arbor, Michigan. April 1999.
15. LIU, Q., H. MERTE, JR., H.S. LEE & R.B. KELLER. 1999. Boiling incipience in subcooled forced convection with a highly-wetting fluid and variable orientation. NASA Grant NAG3-1900, Report No. UM-MEAM-99-09. Department of Mechanical Engineering and Applied Mechanics, University of Michigan, Ann Arbor, Michigan. July, 1999.

APPENDIX

Prior Flow Boiling Results with Various Orientations

Kirk and Merte:[4] nucleate boiling + some dryout.
  Heaters: gold film on quartz + copper (slim). Steady.
  q'', 0–11 W/cm^2
  ΔT_{sub}, 4.4–1.1°C
  U, 4–32 cm/sec
  θ, 0–360°, 45° and 90° increments.

Brusstar *et al.*:[5] critical heat flux.
  Heaters: copper (heavy) + some gold film. Steady.
  q'', CHF 58.7 W/cm^2 max
  ΔT_{sub}, 5.6–22.2°C
  U, 4–55 cm/sec
  θ, 0–360°, 15° and 45° increments.

Nestel and Merte:[13] effect of heater length.
  Heaters: gold film on quartz + copper (slim). Steady.
  q'', 0–10 W/cm^2
  ΔT_{sub}, 2.8°C, 5.6°C, and 11.1°C
  U, 12, 18, and 32 cm/sec
  θ, 0°, 90°, 135°–225°, and 270°.

Tortora et al.:[14] nucleate boiling, hysteresis.
  Heaters: gold film on quartz + copper (heavy). Steady.
  q'', 0–12 W/cm^2 (quartz). Some CHF
    0–35 W/cm^2 (copper) to CHF
  ΔT_{sub}, 5.5°C, 11°C
  U, 8, 18, and 32 cm/sec
  θ, 0, 90°–270°, in 15° increments.

Liu et al.:[15] incipient boiling.
  Heaters: gold film on quartz + copper (heavy). Steady.
  q'', 0–7 W/cm^2 (quartz)
    0–10 W/cm^2 (copper)
  ΔT_{sub}, 5–17°C
  U, 4–28 cm/sec
  θ, 0–360°, 45° increments.

Merte et al.:[11] transient flow boiling near inverted position.
  Heaters: gold film (transient + steady) + copper (slim). Steady.
  q'', 4, 6, and 8 W/cm^2 (gold film)
    0–9 W/cm^2 (copper) some D.O.
  ΔT_{sub}, 2.5–20.2°C
  U, 4, 8, 12, and 16 cm/sec
  θ, 175°, 180°, and 185°.

Microgravity Studies of Cells and Tissues

GORDANA VUNJAK-NOVAKOVIC,[a] NANCY SEARBY,[b]
JAVIER DE LUIS,[c] AND LISA E. FREED[a]

[a]*Massachusetts Institute of Technology, Cambridge, Massachusetts, USA*

[b]*NASA–Ames Research Center, Moffett Field, California, USA*

[c]*Payload Systems Inc., Cambridge, Massachusetts, USA*

> ABSTRACT: Controlled *in vitro* studies of cells and tissues under the conditions of microgravity (simulated on Earth, or actual in space) can improve our understanding of gravity sensing, transduction, and responses in living cells and tissues. This paper discusses the scientific results and practical implications of three NASA-related biotechnology projects: ground and space studies of microgravity tissue engineering (JSC-Houston), and the development of the cell culture unit for use aboard the International Space Station (ARC-Ames).
>
> KEYWORDS: microgravity; cell culture; tissue engineering; bioreactor; ground studies; flight studies

INTRODUCTION

Are single cells capable of sensing and responding to alterations in gravity? If so, what are the underlying mechanisms? Are the effects of gravity on cells, tissues, and whole organisms direct (via a *gravity sensor*), indirect (via related changes in *mass transfer and shear rates* in the cell microenvironment), or both? These questions provide a philosophical and scientific underpinning for why NASA has invested in gravity-related studies of cells and tissues. It is well established that a stay in space can affect biological systems at a variety of levels[1,2] including loss of bone mass, muscle strength, and cardiovascular fitness in astronauts even when they exercise regularly,[1,3] changes in plant cell growth and metabolism,[4,5] and in the swimming behavior of aquatic organisms.[6,7]

Flight studies offer a unique opportunity to assess the effects of gravity on cell and tissue function. However, only controlled biological experimentation can improve our understanding of the measured effects and underlying mechanisms. *In vitro* studies of cells and tissues can be designed to distinguish (1) effects of reduced gravity from the complex milieu of factors involved in space flight, and (2) direct effects of gravity from the effects of the associated changes in mass transfer and shear rates. This, in turn, requires the use of at least three different environments for each experiment: the actual microgravity of space (in an experiment rack), the unit-gravity environment in space (artificial gravity in a space centrifuge), and the 1 *g* environment on Earth.

Address for correspondence: Gordana Vunjak-Novakovic, Massachusetts Institute of Technology, 77 Massachusetts Ave. E25-330 Cambridge, MA 02139, USA. Voice: 617-452-2593; fax: 617-258-8827.

gordana@mit.edu

Many flight experiments carried out so far have had only a ground control and the observed effects were attributed to the "microgravity environment of space" (including launch, landing, microgravity, cosmic radiation, and possibly other factors) as compared to the "unit-gravity environment of Earth", rather than to any particular factor from either environment. Some experiments had access to a small on-board centrifuge, providing a unit-gravity control in space that in several cases yielded results different from those obtained in unit-gravity ground controls.[1] Previous space studies have resulted in valuable scientific information, and they also provided much needed feasibility data that now form a basis for the design and development of the new generation hardware.

This paper reviews results and experiences from microgravity studies of cells and tissues: (1) ground-based tissue engineering studies carried out in rotating bioreactors, (2) a long-term space study of cartilage tissue engineering carried out aboard Mir, and (3) the ongoing development and testing of a new cell culture unit (CCU) for use aboard the International Space Station.

MICROGRAVITY CULTURE VESSELS

TABLE 1 gives an overview of the operating characteristics of the culture systems used for microgravity studies: rotating bioreactor, rotating perfused bioreactor, and the cell specimen chamber that is a component of the cell culture unit.

The rotating bioreactor[8] developed at NASA's Johnson Space Center (JSC) is configured as an annular space holding 110 mL of culture medium and up to 12 tissue constructs. The vessel is operated by solid body rotation in a horizontal plane. Viscous coupling induces a rotational flow field that can suspend tissue constructs without external fixation.[9] Vessel rotation speed can be adjusted so that the contents (i.e., inoculated cells or tissue constructs) either rotate synchronously with the device, or are maintained suspended by a dynamic equilibrium between the acting forces. The rotating bioreactor randomizes the effects of unit Earth gravity on freely suspended biological specimens and exposes them to dynamic fluctuations in fluid velocity, pressure, and shear.

The rotating wall perfused vessel[10] is a part of the biotechnology system (BTS), a cell culture flight hardware system developed by NASA-JSC. The BTS contains a bioreactor connected to a recirculation loop consisting of a silicone membrane gas exchanger and a peristaltic pump. The unit is maintained at 37°C, has access ports for infusing fresh media and gas (10% CO_2 in air), sampling, and removing media (see FIGURE 1). The bioreactor is configured as a 125 mL volume annular space between two cylinders. A disc attached at one end of the inner cylinder serves as a viscous pump. In microgravity, the inner and outer cylinders are differentially rotated (e.g., at rates of 10 and 1 rpm, respectively) whereas on Earth, the vessel is rotated as a solid body (e.g., at 10–40 rpm). The BTS has approximate dimensions of 17×20×25 inches (two half-middeck Shuttle single middeck lockers) and weighs 137 lbs.

A perfused chamber is a part of the cell culture unit (CCU), a cell culture flight hardware system currently being developed by Payload Systems Inc., Massachusetts Institute of Technology (M.I.T.) and NASA Ames Research Center (ARC) for use

TABLE 1. Overview of microgravity culture vessels

Cultivation vessel	Rotating vessel	Rotating wall perfused vessel	Perfused chamber
Volume (cm^3)	110	125	3, 10, or 30
Biological specimens	Freely settling tissue constructs 5 mm dia × 2 mm thick discs $n \leq 12$ per vessel	Freely settling tissue constructs 5 mm dia × 2 mm thick discs $n \leq 10$ per vessel	Cell monolayers, suspensions or tissue constructs
Medium exchange	Batch-wise (3 cm^3 per tissue per day)	Medium recirculation (80 cm^3 every 6h) and exchange (5–10 cm^3 per tissue per day)	Medium recirculation (0.1–10 cm^3/min) and exchange (10–50% per day)
Gas exchange	Continuous, via an internal membrane	Continuous, via an external membrane	Continuous, via an external gas exchanger
Mixing mechanism	Construct settling in rotational flow	µg: differential rotation, viscous pumping 1g: construct settling	Recirculating flow; mechanical stirring
Fluid flow	Laminar	Laminar	Laminar or turbulent
Mass transfer	Convection (due to construct settling)	Convection (due to differential rotation, viscous pumping or construct settling)	Convection (due to recirculation and stirring)
References	9, 12, 14, 15, 17, 24	10	11, 23, 25

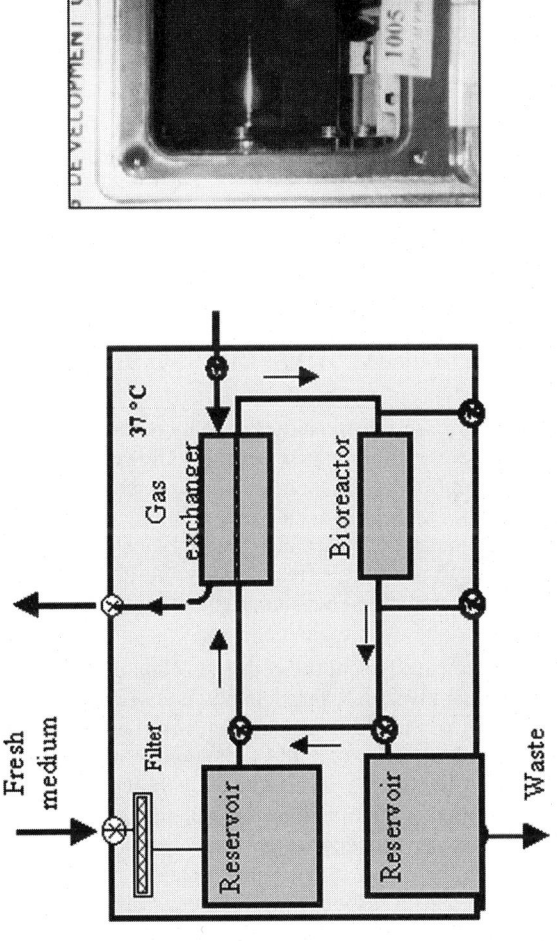

FIGURE 1. NASA–JSC developed biotechnology system (BTS) containing a rotating perfused bioreactor.[10]

aboard the ISS.[11,23,25] The CCU is designed to accommodate diverse specimens (animal, microbial, and plant cells; suspension and attachment cultures; tissues; and non-feeding aquatic specimens), and to meet the requirements of long duration and sequential experiments, both in reduced and in unit gravity. The CCU contains individual cell specimen chambers (CSCs) (8, 12, or 24 chambers with respective volumes of 30, 10, or 3 mL), each connected to a silicone tubing gas exchanger by a recirculation loop consisting of a peristaltic pump and reservoirs for fresh media, additives, fixatives, and waste. The CSC contains a porous membrane that restrains biological specimens within its central compartment. Medium enters the central compartment and exits from the annular cell-free space. Stir bar rotation speeds and perfusion flow are adjusted to provide the required mass transfer rates and to maintain spatially uniform cell distribution (for cultures of cell suspensions). The CCU also provides the capabilities for on-line video microscopy, automated collection and storage of up to 60 samples of cells or cell-free media, transfer of cultures from one CSC to another, additions of reagents and medium supplements and lighting (e.g., for studies of photosynthetic plant cells). The CCU has dimensions of 18.5×21×21 inches (two Shuttle middeck lockers), and weighs 147 lbs. One CCU will be operated within a habitat holding rack; a second CCU will be operated within the ISS centrifuge to serve as an on-board 1 g control; a third CCU will be operated as a ground control.

ROTATING BIOREACTORS: GROUND STUDIES

In non-perfused rotating vessels, flow studies done by particle tracking or by correlation of sequential images of illuminated particles allowed quantitation of the velocity vector at every location in the plane of the light sheet.[12] Maximal shear stresses at the construct surface were calculated to be on the order of 0.8dyne/cm^2 for a vessel containing one model construct and operated such that the inner and outer cylinders were rotated at 13 and 37 rpm, respectively. Residence time distribution studies demonstrated that mixing efficiency due to gravitational construct settling in rotating vessels was comparable to that due to stirring in spinner flasks.[13]

Rotating vessels have been used to culture cells (i.e., chondrocytes, mesenchymal progenitor cells, and cardiac myocytes, respectively, isolated from cartilage, bone marrow, and cardiac tissue) on three-dimensional synthetic, biodegradable polymer scaffolds (i.e., meshes of polyglycolic acid, PGA, or sponges of polylactic coglycolic acid, PLGA) to engineer functional tissues. The chondrocytes formed engineered cartilage[14] that responded to compressive loads in a characteristic manner.[15] Bone marrow derived progenitors formed engineered cartilage or bone.[16] The cardiac myocytes formed cardiac muscle that expressed cardiac-specific proteins and contracted both spontaneously and in response to electrical stimulation.[17–19] Rotating bioreactors were also used to culture a variety of normal and transformed cells.[20,24]

Cartilaginous constructs cultured for six weeks in rotating bioreactors were continuously cartilaginous over their entire cross-sectional areas (6.7 mm diameter × 5 mm thick)[14] and contained an interconnected network of collagen fibers that resembled that of articular collagen with respect to overall organization, fiber diameter[21] and molecular structure (type II collagen represented more than 90% of the total collagen[14]). In contrast, constructs grown in static flasks accumulated matrix components mainly in a 1-mm thick peripheral region, whereas constructs grown in mixed flasks formed peripheral capsules consisting of multiple layers of elongated cells[15] (see TABLE 2A). As compared to native articular cartilage, constructs cultured for six weeks in rotating vessels had comparable cellularities, 68% as much GAG and 33% as much collagen type II per unit wet weight.[14]

Constructs cultured in rotating vessels progressively assembled cartilaginous tissue matrix (TABLE 2B). Over seven months of *in vitro* cultivation, the fraction of glycosaminoglycan and equilibrium modulus increased to values comparable to those measured for adult native cartilage. However, other properties of seven month constructs remained subnormal (e.g., the fractions of collagen and pyridinium crosslinks and dynamic stiffness were respectively 34%, 30%, and 46% as high as in native cartilage[10,21]). Mechanical properties (i.e., equilibrium modulus, hydraulic permeability, dynamic stiffness, and streaming potential) correlated with the tissue fractions of matrix components.[15]

Contractile cardiac-like tissues were also grown in rotating vessels and shown to contain the gap-junctional protein connexin-43, creatine kinase isoform MM, sarcomeric myosin heavy chain protein,[22] cardiac troponin-T, and sarcomeric tropomyosin.[19] The structure and function of cardiac constructs cultured in rotating bioreactors were markedly better than those cultured in flasks.[19] In particular, after one week of cultivation in rotating vessels, cardiac myocytes were electrically

TABLE 2A. Effects of bioreactor vessel (six weeks of culture) on structural and functional properties of engineered cartilage

Culture vessel	Static flask	Mixed flask	Rotating vessel	Cartilage explants
GAG (% wet weight)	2.73 ± 0.2	2.19 ± 0.17	4.71 ± 0.41	6.06 ± 0.29
Collagen (% wet weight)	1.41 ± 0.08	2.94 ± 0.16	3.79 ± 0.05	8.84 ± 1.34
Equilibrium modulus (kPa)	53.3 ± 11.8	50.8 ± 4.6	172.4 ± 35.5	928.4 ± 30.2
Permeability ($\times 10^{-15}$ m^4/Ns)	47.82 ± 21.6	21.28 ± 17.4	10.43 ± 5.0	2.14 ± 0.5
Dynamic stiffness (Mpa, 1 Hz)	0.88 ± 0.33	1.09 ± 0.22	2.91 ± 0.97	16.8 ± 1.1

TABLE 2B. Effects of cultivation time (rotating vessel) on structural and functional properties of engineered cartilage

Cultivation time	3 days	6 weeks	7 months	Cartilage explants
GAG (% wet weight)	0.71 ± 0.03	4.71 ± 0.41	8.83 ± 0.93	6.06 ± 0.29
Collagen (% wet weight)	0.48 ± 0.08	3.79 ± 0.05	3.68 ± 0.27	8.84 ± 1.34
Equilibrium modulus (kPa)	—	172.4 ± 35.5	932.0 ± 49	928.4 ± 30.2
Permeability ($\times 10^{-15}$ m^4/Ns)	—	10.43 ± 5.0	3.72 ± 0.17	2.14 ± 0.5
Dynamic stiffness (Mpa, 1 Hz)	—	2.91 ± 0.97	7.75 ± 0.30	16.8 ± 1.1

connected through gap junctions and they sustained macroscopically continuous impulse propagation.[22]

ROTATING BIOREACTORS: SPACE STUDY

Engineered constructs based on bovine calf chondrocytes and PGA scaffolds were cultured for three months in rotating vessels on Earth and then transferred into the BTS for an additional four months of cultivation either in space on the Mir Space Station or on Earth. One BTS containing ten constructs was transferred to the Mir Priroda module via Shuttle STS-79 (9/16/96 launch) and brought back via STS-81 (1/22/97 landing). A second BTS with ten constructs served as an otherwise identical study conducted on Earth at NASA-JSC. Concentration gradients within the bioreactors were minimized by differential rotation of the inner and outer cylinders at 10 and 1 rpm, respectively (in the microgravity of space), and by the convection associated with gravitational construct settling during solid body rotation of the bioreactor at 28 rpm (in unit gravity).[10] Medium was recirculated between the bioreactor and the gas exchanger at 4 mL/min for 20 min four times per day, and 50 to 100 mL of fresh medium was infused into the system approximately once per day. Under these conditions, metabolic parameters were maintained within previously established target ranges.

Constructs from both groups were cartilaginous and contained live, metabolically active cells. Constructs grown on Mir (which floated freely in the bioreactor) tended to become more spherical whereas those grown on Earth settled, tumbled, and collided with the vessel walls and maintained their discoid shape. Therefore, Mir-grown constructs were exposed to uniform shear and mass transfer, whereas the settling of Earth-grown constructs tended to align their flat circular areas perpendicular to the direction of motion, increasing shear and mass transfer circumferentially. On Earth, construct weights increased 1.7-fold between three and seven months of culture, which could be attributed to increasing amounts of cartilage—specific tissue components; on Mir, construct weights increased only 1.3-fold. In the flight constructs, the fraction of type II collagen decreased, but not significantly between the time of launch and landing, demonstrating relatively good maintenance of the chondrocyte phenotype.[10] The equilibrium modulus of Earth-grown constructs was comparable to that of native calf cartilage and was three-fold higher than for Mir-grown constructs, as assessed in radially confined compression.

The observed differences in structural and functional properties of constructs grown on Earth and aboard Mir could be attributed only to the respective differences between the two culture environments. However, we could not distinguish between the relative contributions from microgravity, launch, landing, and local environmental factors (e.g., cosmic radiation), due to the lack of unit-gravity control aboard Mir. This experiment demonstrated the feasibility of long-term cell culture in space, and yielded results that were consistent with effects observed for the whole organisms exposed to comparable microgravity environments, but also emphasized the need for control studies carried out at unit gravity in space.

CELL CULTURE UNIT: DESIGN AND TESTING

The development of the cell culture unit (CCU) has involved the design, fabrication, and functional testing of the prototypes of the subsystems (e.g., single loops containing one cell specimen chamber, CSC, apiece) (see FIGURE 2) and the system as a whole (science evaluation unit, SEU) (see FIGURES 3 and 4). The development philosophy was to perform functional testing in single loops in order to define cultivation parameters and establish operating protocols, and then verify the system performance in subsequent SEU experiments.

Six reference cell types have been selected for the developmental testing of single loops and the SEU. These include: three types of animal cells (C2C12 muscle monolayer, primary bone cells, and avian muscle organoids), two types of plant cells (nonfeeding aquatic organism Euglena, and BY2 tobacco cell suspension) and one microorganism (*Saccharomyces cerevisiæ*). Each of these cell types challenges the functionality of the CCU hardware in a specific way; together, the selected reference cell types are representative of the entire range of experiments that may be carried out aboard the ISS. In particular:

1. C2C12 cells are used as a model to study mammalian myogenesis. The main requirements include cell attachment and growth in monolayers, cell differentiation on a molecular level (e.g., expression of muscle-specific proteins) and cell fusion into contractile myotubes.

2. Bone cells are studied as a model of primary mammalian cells relevant for the development and maintenance of skeleton, under normal and pathological conditions. The main requirements include cell attachment, proliferation, differentiation (e.g., expression of bone-specific markers) and the formation of mineralized matrix.

3. Primary avian pectoral myoblasts cultured under static tension to form a three-dimensional muscle structure are studied as a model of muscle development and atrophy. The main requirements include the formation of functional muscle organoids and their maintenance without myofiber atrophy.

4. Euglena gracilis, a motile unicellular organism, is studied as a model of unicellular aquatic species whose behavior is affected by microgravity (gravitaxis) and by changes in light intensity (phototaxis) and chemical environment (chemitaxis). The main requirements include cell growth, motility, and metabolism without cell escape, membrane blockage, or cell segregation, during serial sub-cultivations.

5. Tobacco cell suspensions are studied as a model for plant cell growth, biosynthesis of secondary metabolites and genetic manipulations during serial sub-cultivations. The main requirements include cell growth without aggregation and settling, normal cell metabolism, and morphology.

6. Yeast cell suspension is studied as a model of fast-growing microorganism releasing carbon dioxide, and a well established model for studies of molecular mechanisms of cell growth and function and genetic manipulations. The main requirements include normal cell growth in suspension without cell settling, cell escape or membrane blockage, prevention of bubble formation due to the release of carbon dioxide, and the maintenance of cell phenotype.

The CSC within a perfusion loop (FIG. 2A) is the heart of the CCU. The CSC provides a controlled environment for cell cultivation (FIG. 2B) whereas the perfusion

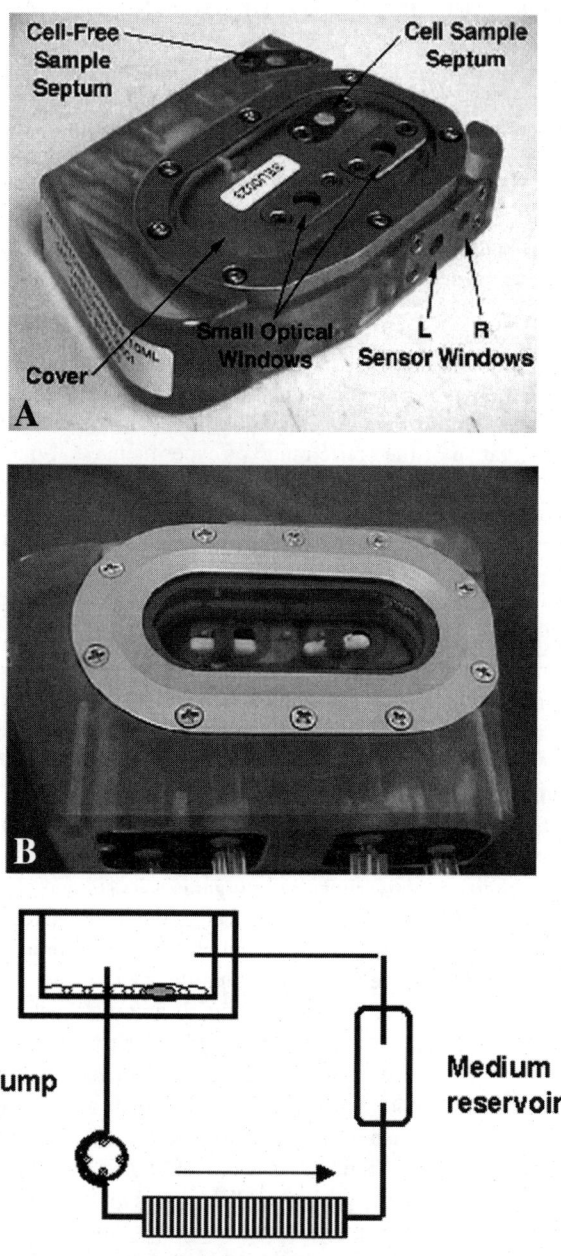

FIGURE 2. Cell specimen chamber (CSC) and the recirculation loops: 10 ml CSC (**A**), 10 ml CSC with stir bars (**B**), recirculation loop (**C**).

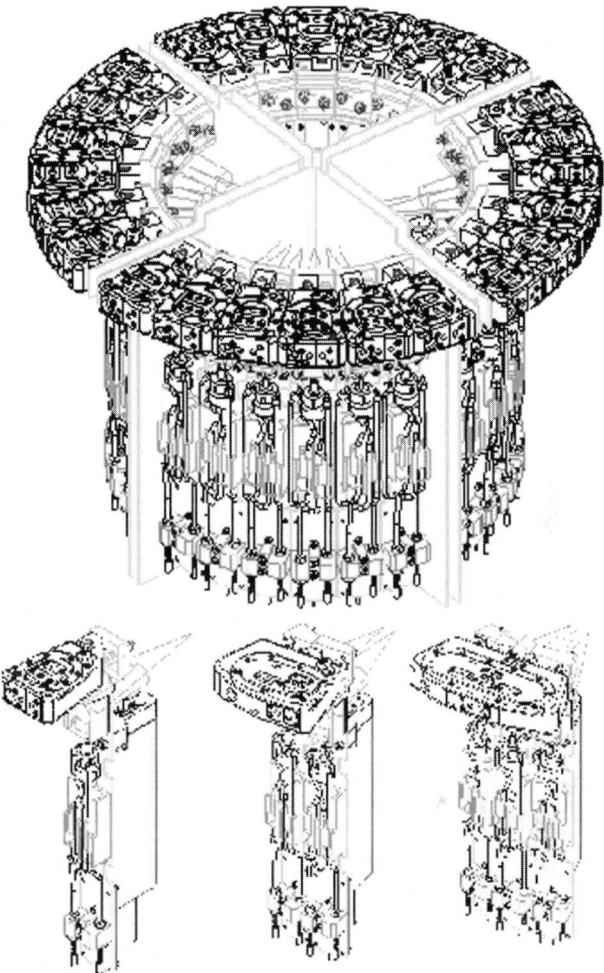

FIGURE 3. CSC support module **(top)** of the CCU with details of recirculation loops for 3 ml **(bottom left)** 10 ml **(bottom middle)**, and 30 ml CSCs **(bottom right)**.

loop provides medium and gas exchange (FIG. 2C). Each CSC contains a membrane that restrains the biological specimens within its central compartment. A removable optical window on the CSC bottom serves as a growing platform for cell monolayers. To promote and control cell attachment, the optical window can be coated (e.g., using proteins, such as laminin or collagen, or mixtures of proteins and growth factors, such as Matrigel). The CSC top has a condenser window for video microscopy that contains septa for sterile addition and removal of cell-free culture medium and cell suspensions, sampling, inoculation, and cell transfer from one CSC to another during subculture.

The CSC is completely filled with culture medium at all times and there is no air space within the CSC or recirculation loop. Conditioned media (media that has been passed through the gas exchanger, with or without addition of fresh components, and equilibrated with respect to pH and oxygen) enters the central cell culture space; media drawn from the annular cell-free space is returned into the gas exchanger. Depending on the CSC size, there are one, two, or three sets of inlets and outlets (for 3, 10, and 30 ml CSC, respectively). The perfusion loop consists of flexible tubing for medium recirculation (with low gas permeability), a small peristaltic pump, connections for fresh and used medium, and silicone-coil gas exchanger (with high permeability for oxygen and carbon dioxide).

Specific regimes of medium recirculation and gas exchange can be established to meet the requirements of specific experiments. Examples of established flow algorithms include:

FIGURE 4. Science evaluation unit (SEU), a working prototype of the CCU (photograph).

1. Low-rate, intermittent flow for mammalian cell monolayers, designed to maintain the levels of oxygen and pH by medium recirculation through the gas exchanger, while minimizing the effects of shear on attached cells.

2. Forward–reverse flow for cell suspensions (e.g., Euglena) in order to maintain cell suspension by using a shaker-like mechanism commonly utilized in standard laboratory cultures, and to prevent membrane blockage by the cells.

3. Forward–reverse flow in conjunction with mechanical stirring for all situations where gas exchange and the maintenance of cell suspension have to be provided by separate mechanisms. One representative situation of this kind is the cultivation of yeast, where gas exchange can be precisely controlled by medium flow to remove carbon dioxide and avoid bubble formation, while stir bars are used to maintain cells in suspension. The other is the cultivation of tobacco, which, due to high oxygen demand and extreme shear sensitivity of the cells, has a requirement to provide high rate of oxygen supply to culture medium along with gentle mixing patterns.

The CSC support module shown in FIGURES 3 and 4 provides a mounting mechanism and fluid connections for multiple CSCs within the SEU. Note that the SEU (FIG. 4) contains only one quadrant and represents a reduced version of the future CCU (FIG. 3) containing four quadrants. In either case, the individual CSCs, each connected to its recirculation loop, are mounted in a circle, with the gas exchangers, pumps, valves, and other related hardware mounted behind and below each CSC. All CSCs and their recirculation loops are independent from each other. The CSC support module is divided into four quadrants, designed to provide a subdivision of the CSCs into four groups (e.g., each corresponding to one subculture, or one particular thermal environment). The CSCs are mounted on a turntable that can rotate ±180° in order to allow positioning of any individual CSC for on-line video microscopy or for sampling by the automated sampling module (ASM).

The video microscopy subsystem provides either bright field or differential interference contrast imaging of cells within the CSCs at one selected magnification, within a range 40×–200×. A specific imaging configuration is selected prior to launch. The ASM provides sample acquisition and storage, on command or as directed by an operator.[25] Each sample is taken into a separate container (a total of 60 per experiment) that are stored as required. The sample containers can be preloaded with fixatives or additives. The ASM can also provide cell transfer from one CSC to another (i.e., subculture), and the addition of small amounts of medium supplements (e.g., growth factors) into the individual CSCs. In each case, the specific ASM configuration prevents contamination and spillage.

At the present time, we have successfully cultured C2C12 cells, avian muscle organoids, Euglena, tobacco, and yeast cells in single loops. In all cases, cell growth, differentiation, morphology, and the maintenance of cell phenotype were comparable to standard laboratory controls. Formal SEU testing has been done for yeast cell culture (two tests at Payload Systems Inc. during spring 2001; one test at NASA–Ames during August 2001). Further testing has been performed at NASA–Ames using an additional yeast strain, in a SEU that is now permanently installed at NASA–Ames. Before the critical design review (CDR) we plan to carry out formal SEU tests of C2C12 and tobacco cell suspensions.

SUMMARY

Controlled microgravity studies can improve our understanding of the fundamental role of gravity in the development and function of cells, tissues, and whole organisms. In particular, controlled *in vitro* studies (e.g., in the actual microgravity of space, at $1g$ in a space centrifuge, and on Earth) can be designed to identify the individual effects of gravity; for example, direct (due to *gravity sensing*) and indirect (due to the related changes in *mass transport and shear rates*). Experiments of this kind may have implications both for the space flight (the development of countermeasures for long duration missions) and basic science.

In space, a rotating bioreactor containing engineered cartilage was studied aboard Mir for a period of four months and an identical bioreactor was operated as a ground control. Constructs from the two groups had markedly different structure and function, and the observed differences were consistent with previous reports that musculoskeletal tissues remodel in response to physical forces and are adversely affected by space flight. One major limitation of this study was the lack of a $1g$ control in space that would allow the separation of the effect of microgravity from other factors present during a space flight.

In ground studies, rotating bioreactors have been used in conjunction with biodegradable scaffolds to engineer functional tissue constructs. Chondrogenic cells isolated from cartilage or bone marrow formed large (approximately 1 cm in diameter, 6 mm thick) mechanically functional cartilaginous constructs. Likewise, heart cells formed cardiac constructs that contracted synchronously and contained electromechanically coupled cells expressing cardiac-specific tissue markers.

We are now involved in the development of the CCU, a fully automated system for cell and tissue culture in microgravity (in the habitat holding rack), at $1g$ in space (in the space centrifuge), and on Earth (at $1g$ and in hypergravity). CCU can accommodate diverse biological specimens in up to 24 individual culture chambers, each operated within a recirculation loop containing a gas exchanger, supply of medium and additives, a set of sensors, on-line sampling, and video microscopy. Ground testing in single-loop and SEU prototype hardware have shown that diverse cell types can be cultured in the CCU in a wide range of experimental conditions. The CCU is thus expected to enable rigorous biological experimentation, in space and on Earth.

ACKNOWLEDGMENTS

The studies reviewed in this paper were carried out by research and development teams at the Massachusetts Institute of Technology, Payload Systems Inc., NASA JSC, and NASA ARC. Funding was provided by NASA (Grants NCC8-170 and CCU-97001).

REFERENCES

1. CHURCHILL, S.E. 1997. Fundamentals of Space Life Sciences. Krieger Publishing, Malabar.
2. NICOGOSSIAN, A.E., C.L. HUNTOON & S.L. POOL. 1994. Space Physiology and Medicine. Lea and Febiger, Philadelphia.
3. HUGHES-FULFORD, M. & M.L. LEWIS. 1996. Effects of microgravity on osteoblast growth activation. Exp. Cell Res. **224:** 103–109.

4. TRIPATHY, B.C., C.S. BROWN, H.G. LEVINE & A.D. KRIKORIAN. 1996. Growth and photosynthetic responses of wheat plants grown in space. Plant Physiol. **110:** 801–806.
5. BONTING, S.L., Ed. 1992. Effects of spaceflight on growth and cell division in higher plants. *In* Adv. Space Biol. Med., Vol. 2. JAI Press, Greenwich.
6. DE JONG, H.A., E.N. SONDAG, A. KUIPERS & W.J. OOSTERVELD. 1996. Swimming behavior of fish during short periods of weightlessness. Aviat. Space Environ. Med. **67:** 463–466.
7. WIEDERHOLD, M.L., W.Y. GAO, J.L. HARRISON & R. HEJL. 1997. Development of gravity-sensing organs in altered gravity. Gravitational Space and Biology Bulletin **10:** 91–96.
8. SCHWARZ, R.P., T.J. GOODWIN & D.A. WOLF. 1992. Cell culture for three-dimensional modeling in rotating-wall vessels: an application of simulated microgravity. J. Tiss. Cult. Meth. **14:** 51–58.
9. FREED, L.E. & G. VUNJAK-NOVAKOVIC. 1995. Cultivation of cell-polymer constructs in simulated microgravity. Biotechnol. Bioeng. **46:** 306–313.
10. FREED, L.E., R. LANGER, I. MARTIN, *et al.* 1997. Tissue engineering of cartilage in space. Proc. Natl. Acad. Sci. USA **94:** 13885–13890.
11. SEARBY, N.D., J. DE LUIS & G. VUNJAK-NOVAKOVIC. 1998. Design and development of a space station cell culture unit. J. Aerospace **107:** 445–457.
12. NEITZEL, G.P., R.M. NEREM, A. SAMBANIS, *et al.* 1998. Cell function and tissue growth in bioreactors: fluid mechanical and chemical environments. J. Jpn. Soc. Microgr. Appl. **15:** 602–607.
13. FREED, L.E. & G. VUNJAK-NOVAKOVIC. 1997. Tissue culture bioreactors: chondrogenesis as a model system. In Principles of Tissue Engineering. R. Langer & W. Chick, Eds.: 150–158. R. G. Landes. Austin.
14. FREED, L.E., A.P. HOLLANDER, I. MARTIN, *et al.* 1998. Chondrogenesis in a cell–polymer–bioreactor system. Exp. Cell Res. **240:** 58–65.
15. VUNJAK-NOVAKOVIC, G., I. MARTIN, B. OBRADOVIC, *et al.* 1999. Bioreactor cultivation conditions modulate the composition and mechanical properties of tissue engineered cartilage. J. Orthop. Res. **17:** 130–138.
16. MARTIN, I., V.P. SHASTRI, R.F. PADERA, *et al.* 2001. Selective differentiation of mammalian bone marrow stromal cells cultured on three-dimensional polymer foams. J. Biomed. Mater. Res. **55:** 229–235.
17. FREED, L.E. & G. VUNJAK-NOVAKOVIC. 2000. Tissue engineering bioreactors. In Principles of Tissue Engineering. R.P. Lanza, R. Langer & J. Vacanti, Eds.: 143–156. Academic Press, San Diego.
18. BURSAC, N., M. PAPADAKI, R.J. COHEN, *et al.* 1999. Cardiac muscle tissue engineering: toward an in vitro model for electrophysiological studies. Am. J. Physiol. **277:** H433–H444.
19. CARRIER, R.L., M. PAPADAKI, M. RUPNICK, *et al.* 1999. Cardiac tissue engineering: cell seeding, cultivation parameters and tissue construct characterization. Biotechnol. Bioeng. **64:** 580–589.
20. UNSWORTH, B.R. & P.I. LELKES. 1998. Growing tissues in microgravity. Nat. Med. **4:** 901–907.
21. RIESLE, J., A.P. HOLLANDER, R. LANGER, *et al.* 1998. Collagen in tissue-engineered cartilage: Types, structure and crosslinks. J. Cell. Biochem. **71:** 313–327.
22. PAPADAKI, M., N. BURSAC, R. LANGER, *et al.* 2001. Tissue engineering of functional cardiac muscle: molecular, structural and electrophysiological studies. Am. J. Physiol. Heart Circ. Physiol. **280:** H168–H178.
23. FREED, L.E. & G. VUNJAK-NOVAKOVIC. 2002. Space bioreactors: past experiences and ongoing work. *In* Advances in Space Biology and Medicine, Vol. 8: Cell Biology and Biotechnology. A. Cogoli, Ed. Elsevier Science. In press.
24. MADRY, H., R. PADERA, J. SEIDEL, *et al.* 2002. Gene transfer of a human insulin-like growth factor I cDNA enhances tissue engineering of cartilage. Hum. Gene Ther. **13**(13): 1621–1630.
25. DE LUIS, J., G. VUNJAK-NOVAKOVIC & N. SEARBY. 2002. Design and testing of the ISS cell culture unit. Proc. 51st Congress of the International Astronautical Federation, Rio de Janeiro, October 2–6, 2000.

Flow Field Measurements in the Cell Culture Unit

STEPHEN WALKER, MIKE WILDER, ARSENIO DIMANLIG, JUSTIN JAGGER, AND NANCY SEARBY

NASA Ames Research Center, Moffett Field, California, USA

ABSTRACT: The cell culture unit (CCU) is being designed to support cell growth for long-duration life science experiments on the International Space Station (ISS). The CCU is a perfused loop system that provides a fluid environment for controlled cell growth experiments within cell specimen chambers (CSCs), and is intended to accommodate diverse cell specimen types. Many of the functional requirements depend on the fluid flow field within the CSC (e.g., feeding and gas management). A design goal of the CCU is to match, within experimental limits, all environmental conditions, other than the effects of gravity on the cells, whether the hardware is in microgravity (μg), normal Earth gravity, or up to $2g$ on the ISS centrifuge. In order to achieve this goal, two steps are being taken. The first step is to characterize the environmental conditions of current $1g$ cell biology experiments being performed in laboratories using ground-based hardware. The second step is to ensure that the design of the CCU allows the fluid flow conditions found in $1g$ to be replicated from microgravity up to $2g$. The techniques that are being used to take these steps include flow visualization, particle image velocimetry (PIV), and computational fluid dynamics (CFD). Flow visualization using the injection of dye has been used to gain a global perspective of the characteristics of the CSC flow field. To characterize laboratory cell culture conditions, PIV is being used to determine the flow field parameters of cell suspension cultures grown in Erlenmeyer flasks on orbital shakers. These measured parameters will be compared to PIV measurements in the CSCs to ensure that the flow field that cells encounter in CSCs is within the bounds determined for typical laboratory experiments. Using CFD, a detailed simulation is being developed to predict the flow field within the CSC for a wide variety of flow conditions, including microgravity environments. Results from all these measurements and analyses of the CSC flow environment are presented and discussed. The final configuration of the CSC employs magnetic stir bars with angled paddles to achieve the necessary flow requirements within the CSC.

KEYWORDS: international space station; life science; cell growth; PIV; cell specimen chamber; cell culture unit; CFD; flow visualization

NOMENCLATURE:

C	concentration
C_0	initial concentration
A	absorbance
T	transmittance

Address for correspondence: Stephen Walker, NASA Ames Research Center, MS/260-1 Moffett Field, CA 94035, USA. Voice: 650-604-6720; fax: 650-604-4511.
swalker@mail.arc.nasa.gov

INTRODUCTION

The growth of cells in an Earth-based laboratory enables investigations into how cells function. These investigations explore the relationship between cell growth, function, and response to various environmental conditions. A number of environmental variables must be controlled to promote successful cell growth. Cells receive the nutrients and gases required for life from the fluid environment. This fluid environment is called the cell culture medium. Cell culture medium formulations are specific for each cell type, but the same four basic components are required to support all types of cell growth. These four basic components are inorganic salts, a simple sugar, organic supplements (e.g., vitamins and amino acids), and trace elements.[1-3] As a result of the unique formulations for the cell culture medium, the fluid properties of the medium are also unique but similar to water. Therefore, initial testing will be conducted primarily in water to simplify the initial analyses and tests.

Each cell type has specific requirements of the fluid environment. The important parameters to be investigated include mixing additives into the cell culture medium, flushing additives from the medium, maintaining cell suspension for cell types that require suspension, and minimizing shear forces and fluid-induced accelerations within experimental limitations.

Mixing of the medium around the cells is necessary for successful cell biology experiments. Mixing is the mechanism that provides for uniform distribution of gases, concentrated additives, fixatives, and nutrients to cells. Mixing is also responsible for waste removal from cells. Furthermore, proper mixing is necessary to provide a uniform environment for all cells within the cell specimen chamber (CSC). Uniformity is important for successful cell growth and, in addition, it is important for sampling and removing subcultures of cells. Proper mixing results in the uniform distribution of all the materials within the cell culture medium throughout the CSC volume.

Flushing is the mechanism by which materials, such as fixatives, are removed from the CSC volume. The requirements for mixing and flushing conflict. For mixing, it is desired to promote uniformity within the cell culture medium, but for flushing, the requirement is to quickly exchange the cell culture medium to a new configuration of constituent materials. For example, to terminate some experiments, the CSCs must be flushed of medium and filled with buffered saline, then fixative, and finally buffered saline. The goal in flushing is to minimize the amount of fluid and time required to flush the volume within the CSC.

Suspension cell types must remain suspended within the CSC. The force to maintain cells in suspension must come from the fluid motion of the medium. Therefore, the fluid environment must provide for gentle motion that enables suspension of the cells throughout the volume of the CSC without undue force.

Shear forces are a concern for all cell types, but specific cell types are more susceptible to the effects of shear. For example, adherent cells require a substrate for cell attachment, and under extreme amounts of shear, these cell types may be separated from the substrate. In addition, fluid-induced shear forces have been shown to alter cellular function, and this effect may mask the effects of changing gravity. Similarly, fluid-induced accelerations may also alter cellular function and mask gravitational effects.

Within the CSC, mixing, flushing, and suspension require fluid motion, and that fluid motion necessarily imparts shear forces and fluid-induced accelerations to the cells. Therefore, in order to perform accurate experiments it is important that these effects be quantified in standard $1\,g$ laboratory conditions and then compared to the CSC conditions to ensure that the fluid dynamics are within known experimental limits. Many $1\,g$ cell biology experiments using suspension cells take place in Erlenmeyer flasks on orbital shaker tables. The motion of the orbital shaker table provides for mixing and suspension. This testing configuration can serve as a baseline for comparison of fluid-induced acceleration and shear.

To ensure that the cell culture unit (CCU) perfusion loop and CSC provide adequate mixing, flushing, and suspension without inducing substantial shear force and acceleration, several flow testing and analysis techniques are being used, including flow visualization, particle image velocimetry (PIV), and computational fluid dynamics (CFD). These techniques are used to evaluate and quantify the flow field environment within the CSC as well as the current $1\,g$, Earth-based laboratory environment on orbital shaker tables.

Flow visualization studies are applied in order to determine the effectiveness of the CSC for mixing and flushing. Flow visualization is a test technique that can rapidly give an indication of the effectiveness of a particular CSC operation configuration. Flow visualization studies using dye within the CSCs were performed at $1\,g$ and at $2\,g$.

PIV is a measurement technique that determines three-dimensional velocities in a plane. PIV is used to characterize the flow fields for current $1\,g$ Earth-based testing in order that these same flow conditions can be replicated within the CSC. In particular, the velocity field has been measured using PIV, and the PIV data is being used to calculate the hydrodynamic forces that the cells encounter. PIV is being used to characterize the flow field on orbital shaker tables, as well as in the CSC.

CFD has also been performed on a CSC model and velocity fields have been calculated. CFD results allow analysis of various changes in the flow field caused by different flow input conditions. CFD can calculate the velocity and pressure fields within the CSC. In addition, CFD permits determination of the flow field at microgravity and up to $2\,g$.

CELL CULTURE UNIT AND CELL SPECIMEN CHAMBER

The same cell life support conditions that are normally provided in $1\,g$ cell biology experiments will be provided in the environment of space by the CCU that is being developed for use on the international space station (ISS). A significant obstacle that CCU operations must address is the absence of natural convection in microgravity conditions. Convection and diffusion are the processes by which nutrients and gases are transported to cells, and also the processes by which waste products are transported away from cells. Convection operates due to small density gradients, which under the influence of gravity results in the transport of waste products and nutrients. In the absence of gravity, the transport of materials to and from cells becomes diffusion-limited in a stationary system, and diffusion is not sufficient for long-term cell growth. The CCU solves this problem by using a perfusion-based system that generates forced convection via the motion of the cell culture medium. In the final configuration of the CSC, magnetic stir bars are used for additional control of the fluid flow field. The medium is recirculated through a gas exchanger that helps to maintain the concentration of nutrients, gases, or waste products at acceptable levels within the medium. Periodically, new cell culture medium can be added and used medium removed. The forced convection requirement is one of the primary design requirements and becomes the basis of the CCU and CSC design.

FIGURE 1 shows a perfusion-based flow loop designed to support a single CSC within the CCU. In operation, the nutrient medium recirculates through the CSC to feed the cell culture and remove waste products. The medium passes through a gas exchanger to replenish gases. There is a medium bag for periodically replenishing fresh medium, additive bags to provide researcher-selected additives, and two waste bags for removing spent medium from the perfusion loop. The design permits additions and removal of medium without contaminating the perfusion loop or other medium containers within the perfusion loop. Using ground-based testing, researchers can set the medium replenishment and additive addition flow rates, durations, and delivery schedule. Researchers may also designate the rotation rate and operation schedule of the magnetic stir bars. The schedule for replenishment of gases within the gas exchanger can also be set by the researcher.

In addition to the perfusion of the cells, the CCU provides specimen lighting for those cell types that require lighting. The specimen temperature is controlled at a set point in the range of 4 to 45°C. The CCU and CSC also provide for automated sampling, fixation, and reagent additions via an automated sampling module through sample ports. Video-microscopy imaging of the cells is also provided for researchers. The CCU is designed to slide into a habitat holding rack or the centrifuge rotor, both of which will be located inside the centrifuge accommodation module of the ISS. Additional details of the flight hardware design can be found in Searby,[4] de Luis,[5] and Searby et al.[6]

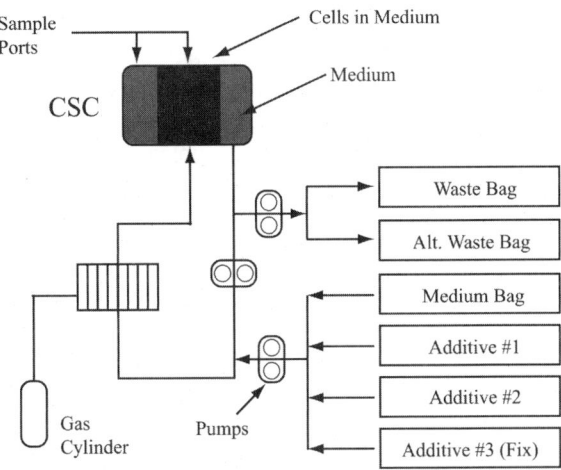

FIGURE 1. Schematic of CCU perfusion loop for one CSC.

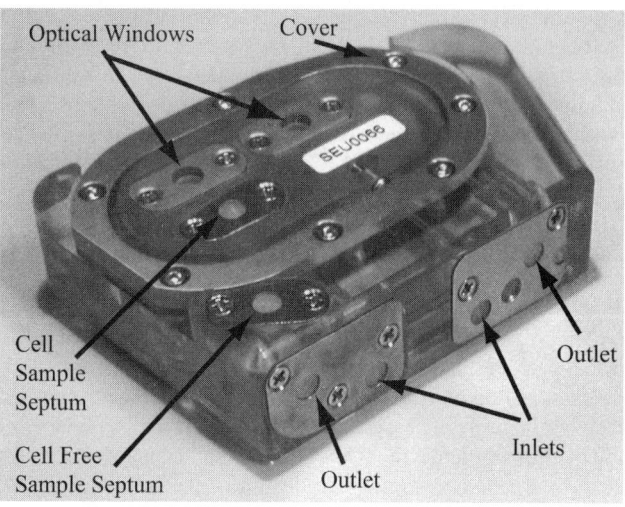

FIGURE 2. Ten-milliliter cell specimen chamber.

The CSCs are designed to culture either cells in suspension or cells attached to the lower surface of the chamber. The lower surface may also act as an optical window. There are three sizes of CSC. The size differences are designated by the different volumes within the CSC. The three volumes are 3, 10, and 30 ml. FIGURE 2 shows a photograph of the 10 ml CSC. The 10 ml CSC is configured with two outlets and two inlets. The 3 ml CSC has one inlet and one outlet and the 30 ml CSC has three inlets and three outlets. Each inlet and outlet can be attached to a perfusion loop as illustrated in FIGURE 1. Additional ports are provided in the chamber to remove samples of cells and cell-free medium. Optical windows assist in viewing the cells using the CCU video-microscopy system. The lower surface of the CSC is also an optical window and can be removed study adherent cells attached to this lower surface.

A porous membrane encompasses the inner chamber to prevent cells in the medium from escaping the chamber. FIGURE 3A illustrates the inner shape of the 10 ml CSC and represents the grid model used for CFD calculations. The CFD model of the membrane is shown in FIGURE 3B. The porous membrane has the same oval shape as the inner shape of the 10 ml CSC. The membrane sits on the shelf indicated in FIGURE 3A and is sealed with gaskets to the shelf and to the CSC lid.

FLOW ANALYSIS AND TEST RESULTS

The three flow field diagnostic techniques applied to characterize the CSC flow environment are flow visualization, PIV, and CFD. Each technique is applied to analyze a particular element of the CSC, but there will be sufficient overlap of the data from each of the techniques to permit verification of the results in the future.

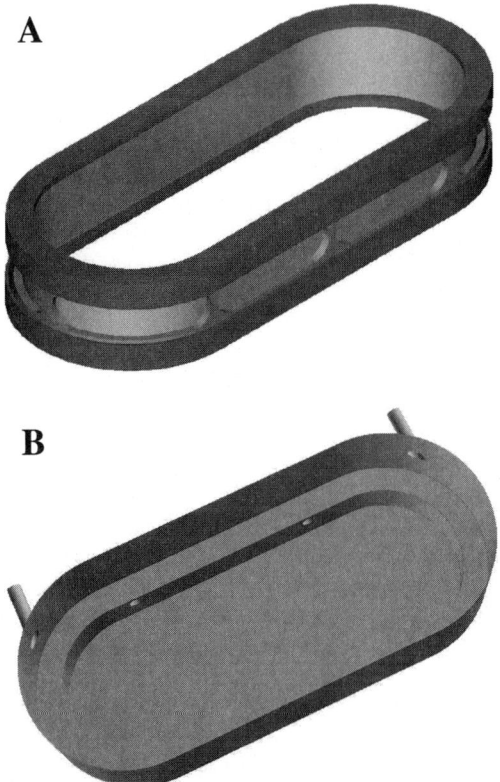

FIGURE 3. (**A**) Computated surface representation of inner 10-ml cell culture volume and membrane. (**B**) Computated representation of the 10-ml membrane.

Flow Visualization

Flow visualization has been used to characterize the general flow environment within the CSC. The ability to mix in additives or new medium and to flush out reagents, such as fixatives, has also been evaluated using flow visualization. For all cases investigated, the criteria for good mixing was that a dye mixed quickly throughout the chamber and indicated no flow stagnation regions where dye would either accumulate or not occupy them at all. However, providing good mixing of the medium is not the only design challenge. When the CSC needs to be flushed, before or after fixation for example, good mixing is undesirable because it requires long times and prohibitively large fluid volumes to change the fluid in the chamber. Flushing was evaluated by determining the amount of fluid needed to remove the visibly apparent dye present in the CSC chamber. In addition to the visual methods employed to evaluate mixing and flushing, the effectiveness of mixing or flushing were also evaluated by performing species concentration studies that quantitatively indicated the concentration of dye present in a sample.

Basic Flow Field

The first flow visualization investigated the mixing behavior of the basic flow field inside the 3, 10, and 30 ml CSC. The basic flow field did not use magnetic stir bars within the CSC. The basic flow field studies used steady flow rates through the inlets and outlets of the CSCs. An example of flow visualization data of the basic flow in the 10 ml CSC is shown in FIGURE 4. The photograph in FIGURE 4 was acquired through the large plate optical window on the lower surface of the CSC. Note that the small optical windows and cell sample septum located on the top cover (see FIG. 2) are seen out of focus in this image since these objects are on the opposite side of the CSC from the large viewing window. The flow rate was 4 ml/min at each inlet. The inlet on the right half of the photograph had approximately 0.1 ml of dye added to the inlet flow, and the left inlet had only water entering the chamber. The image shown was acquired at 0.5 min after the introduction of the dye. FIGURE 4 illustrates that the right inlet established the flow field within the right half of the CSC, and little or no mixing occurred across the center line of the chamber.

FIGURE 5 shows flow-visualization images of the basic 10 ml CSC when the flow rate was reduced to 2 ml/min at both inlets. Approximately 0.1 ml of red dye was input through the left inlet starting at time zero and mixed through three-quarters of the CSC within one minute. Similar to the flow pattern at 4 ml/min (see FIG. 4), little of the dye entered the right quarter of the chamber due to the strong recirculation regions established by the stagnation of the inlet jets on the opposite wall of the CSC. Approximately 0.1 ml of blue dye was input through the right inlet starting at five minutes. Again, the dye mixed through all but the opposite quarter of the CSC within about one minute. After 20 minutes of flushing with water (80 ml of water) there was still a visible concentration of dye in the right and left quarters of the CSC. The poor mixing into these regions was due to the strong discontinuity created by the inlet jets themselves.

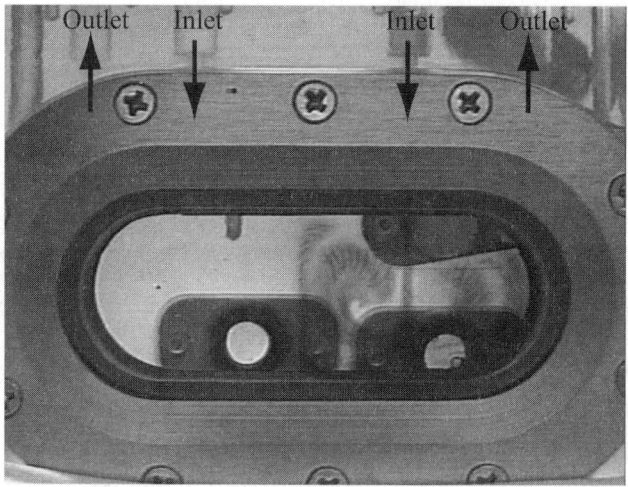

FIGURE 4. Ten-milliliter CSC. Dye injection during flow visualization.

Flow visualization was used to perform analysis of flushing times for the basic flow field also. Two methods were used to determine the time to flush the 3 ml CSC. In the first, dye (blue food coloring mixed with distilled water) was pumped into the chamber initially filled with distilled water. In the second, distilled water was pumped into the dye-filled chamber. For both methods the concentration of dye at the outlet of the CSC was determined from transmittance measurements made with a Spectronic 20 spectrometer set at 630 nm, which is the adsorption peak for this dye. The sample volumes were 0.5 ml, collected at various intervals in time. Each sample was diluted with 4.0 ml of distilled water to provide adequate volume for the Spectronic 20. Measurements were made of percent transmittance (%T), with the instrument adjusted, such that distilled water gave 100%T. The absorbance (A), proportional to the concentration (C), was determined from the %T measurements, $A = 2 - \log 10\ (\%T)$. The concentration of dye at the outlet was normalized by the concentration of the dye at the inlet at the start of the dye-in experiment (C_0), such that $C/C_0 = A/A_0$.

Measurements were made for the steady-flow conditions, 0.5 ml/min, 1.0 ml/min, 1.5 ml/min, and 2.0 ml/min in the 3-ml CSC, for both the dye-in and the water-in methods. (FIGURE 6 shows the 3-ml CSC with the top cover removed.) The 0.5 ml/min dye-in experiment was repeated once. The results are shown in FIGURES 7 and 8 for the dye-in and water-in methods, respectively. The collapse of the curves onto a single line shows that the volume of fluid pumped through the chamber is the appropriate scaling parameter. The same volume of fluid is required to flush the chamber, regardless of the flow rate. For FIGURE 7 the dye-in condition, the outlet concentration reached about 99% of the inlet concentration after about 14 ml of fluid was pumped. For FIGURE 8 the water-in condition about 30 ml of fluid was required (20 ml for the 0.5 ml/min case) for the outlet concentration to reach approximately 99% of the inlet concentration. This difference between the dye-in and water-in cases was due to dye becoming trapped in various locations within the CSC after the CSC was completely filled with dye. A small amount of water similarly trapped during the dye-in experiments had a negligible effect on the transmittance measurements.

In conclusion, the basic configuration resulted in poor mixing, and high flushing times and volumes. The next design for the CSC sought to reduce the flushing volume and to achieve better mixing.

Split-Membrane Stir Bar Flow Field

The criterion for good mixing was rapid dispersion of dye introduced into the CSC. The basic configuration is described in the BASIC FLOW FIELD section. It was determined that the basic configuration resulted in poor mixing throughout the CSC chamber. This result is illustrated in FIGURE 5. As a result two new modifications for controlling the flow field within the CSC were devised. The first modification utilized a new design for the membrane, which is called the split-membrane, and the second modification introduced magnetic stir bars within the cell volume.

The split-membrane was developed to enhance mixing and cell suspension. Furthermore, the split-membrane is useful in clearing the membrane due to membrane fouling from cell structures. The design of the split-membrane is sketched in FIGURE 9. For the split-membrane CSC, the membrane is split by a barrier that separates the cell-free volume into two parts, with each half interacting directly with

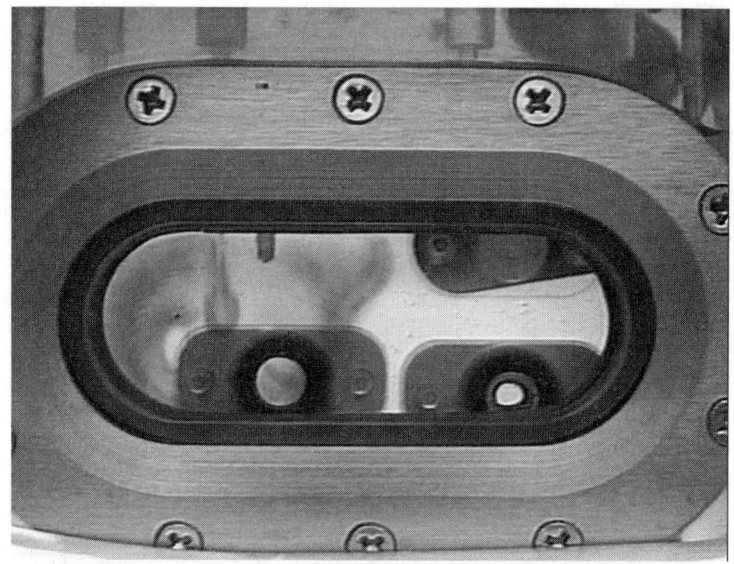

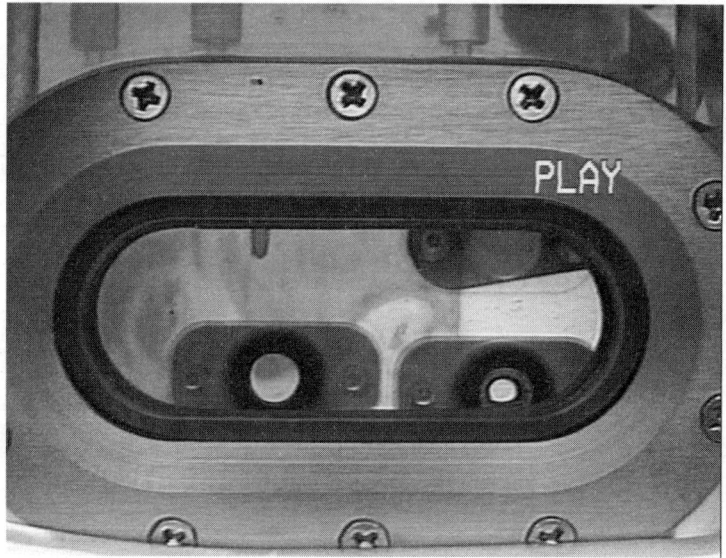

FIGURE 5. (**A**) $t = 0.5$ min; red dye injected at $t = 0$. (**B**) $t = 1$ min. (**C**) $t = 6$ min; blue dye injected at $t = 5$ min. (**D**) $t = 20$ min. (Figure continues on opposite page.)

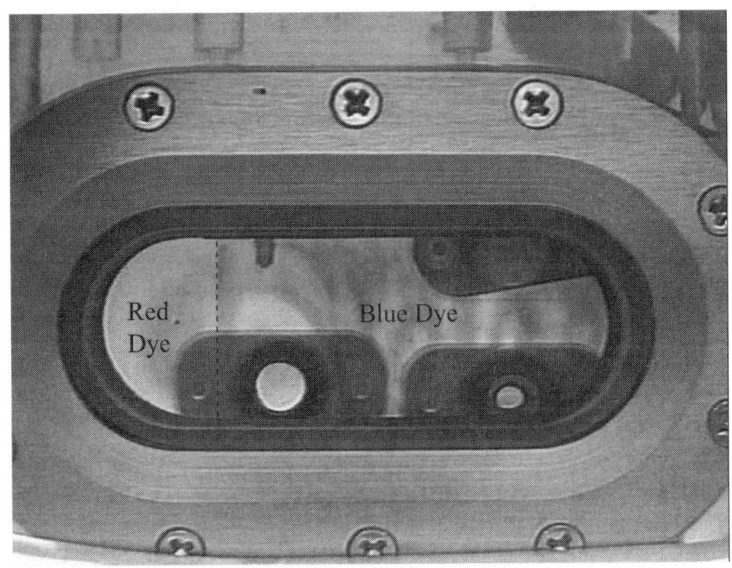

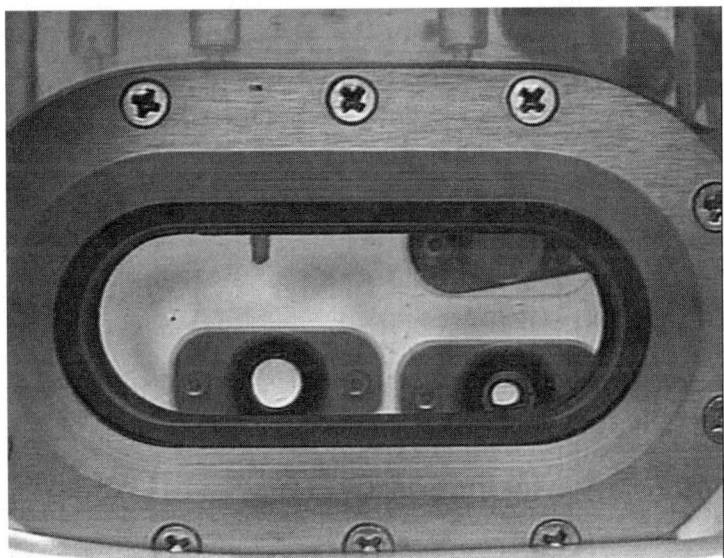

FIGURE 5/continued.

FIGURE 6. Three-milliliter cell specimen chamber with top cover removed.

only one outlet. The cell-free volume is a small volume on the outside of the membrane, as shown in FIGURE 9. The membrane prevents cells from entering this small volume. A set of valves allows switching of the flow direction through the membrane such that one half of the split-membrane functions as an outlet while the other half functions as an inlet. This concept was developed in particular for suspension-cell cultures to clear the membrane of cells that are drawn to the membrane along with the medium exiting the chamber.

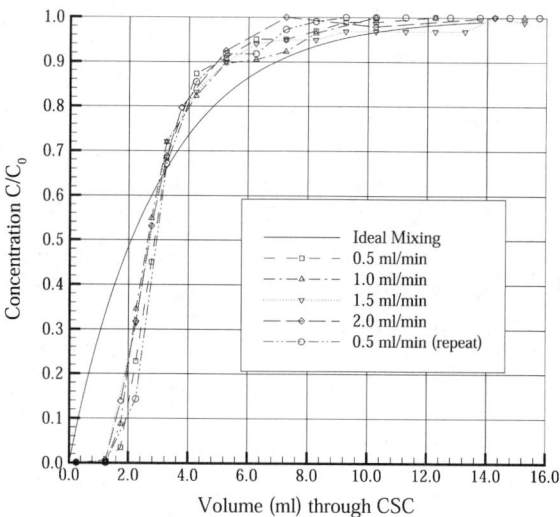

FIGURE 7. Volume to flush 3-ml CSC. Dye pumped into water.

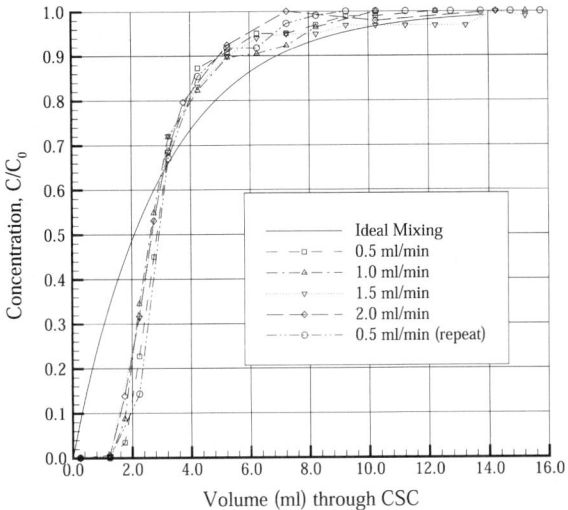

FIGURE 8. Volume to flush 3-ml CSC. Water pumped into dye.

In addition, a new degree of freedom for CSC operation was introduced through the use of the magnetic stir bars. The magnetic stir bars permit the CSC to be operated in various flow configurations as needed. Since mixing and flushing require two different flow field regimes, it was conceived that when good mixing was required the CSC may be operated in one configuration, and when flushing was required a different flow field configuration could be used. The stir bars could be operated continuously when good mixing was required, but when flushing the CSC, the stir bars would not be operated. The stir bars are shown in FIGURE 10. The stir bars rate and periodicity can be varied to optimize mixing and suspension independently of the net flow rate of medium through the chamber, allowing for greater flexibility and increased probability of success in maintaining multiple cell cultures under variable conditions.

Once stir bars were identified as a mechanism for mixing and suspension, the process of adapting them for use in a CSC began. Each representative cell type posed a unique set of requirements on the hardware being developed and stir bars were no exception. In general, a high stir rate provided adequate suspension, but damaged the cells. A slow stir rate was gentle enough for the cells, but failed to provide adequate suspension. Several design iterations were employed in an effort to increase the effectiveness of stir bars at slower stir rates. The first successful design modification incorporated onto the stir bars was the inclusion of paddles on the stir bars. The paddles increased the effective surface area of the stir bar so that more momentum was delivered to the fluid in the CSC per revolution of the stir bars. By angling the paddles, the momentum was given direction and the momentum was more easily sustained. For example, these design changes resulted in the reduction of the stirring rate for tobacco cells from 200 rpm using stir bars without paddles to 60 rpm when using stir bars with angled paddles, which in turn resulted in maintaining a healthy tobacco cell growth study.

530   **ANNALS NEW YORK ACADEMY OF SCIENCES**

FIGURE 11A shows a photograph of the stir bars in the 10-ml CSC with a split-membrane during a flow visualization test. The paddles on these stir bars were at an angle of 45°. In the configuration shown in FIGURE 11A, the 10-ml CSC was being tested in the HyFaCC centrifuge at NASA Ames Research Center. During the test, the simulated gravitational field was $2g$, and the flow rate into the CSC was 3 ml/min. FIGURE 11A shows the dye front initially after adding 0.10 ml of dye. After three seconds, the dye was nearly mixed (FIG. 11B). Full mixing was achieved in less than 15 seconds (FIG. 11C). Preliminary testing at the same input flow rate indicates that flushing can be achieved in approximately 2.5 minutes, which corresponds to approx-

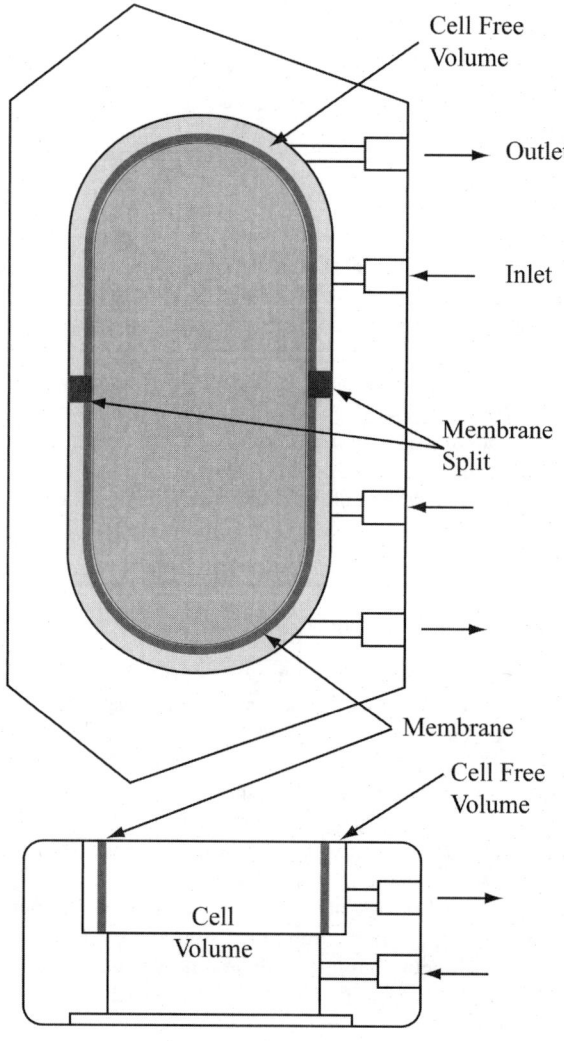

FIGURE 9. Illustration of 10-ml split-membrane CSC.

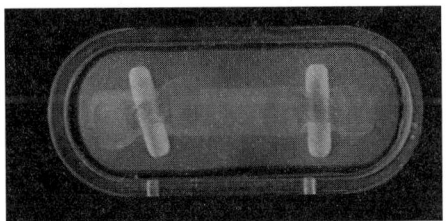

FIGURE 10. Illustration of stir bars with angled paddles in the 10-ml CSC.

imately 7.5 ml of fluid required to flush the CSC. Under the basic flow field conditions, approximately 20 ml of fluid was required to flush the CSC. From these results, it can be seen that the stir bars may be used to achieve excellent mixing and flushing.

Flow Visualization Conclusions

The optimization of net flow rate and direction through the chamber combined with the flexibility of stir bars and paddles allow the CSC to be tailored to suit the

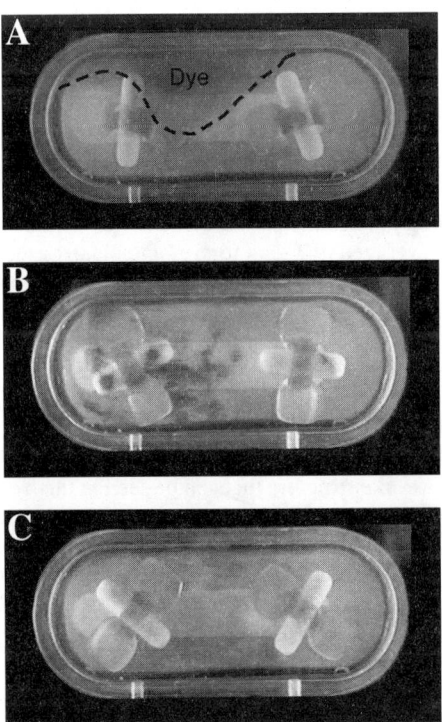

FIGURE 11. Stir bars with angled paddles and split-membrane within the 10-ml CSC. (**A**) $t = 0$ sec; (**B**) $t = 3$ sec, dye nearly mixed; (**C**) $t = 15$ sec, dye fully mixed.

needs of many cell types. Nutrient delivery and gas management are controlled by net flow rate and stir bar mixing, membrane clearing is achieved by varying flow direction, and uniformity of suspension is provided by stirring rate, direction, and periodicity. With the tools for maintaining cell cultures available, determining the cell specific flow and stirring parameters would be the task of the researcher to ensure success in the CCU with his/her specific cell type.

PARTICLE IMAGE VELOCIMETRY

The magnitude of the hydrodynamic forces that the cells would be exposed to in a standard laboratory Erlenmeyer flask on a shaker table and in the CSC flow environment is being quantified using particle imaging velocimetry (PIV). This involves making PIV measurements of the three-dimensional velocity field. In order to make these velocity measurements, PIV uses small tracer particles (10–20µm) within the fluid. Tracer particle positions are illuminated by using a plane of high-powered laser light. Digital images are captured of the tracer particle positions. By capturing two images at successive instants in time, one may calculate the velocities within a plane. Two cameras are used to get a stereoscopic view of the fluid, and by utilizing the stereoscopic views all three components of velocity may be determined.[7] The shear forces and accelerations that the cells encounter can be calculated from measured velocities over time and space, using standard numerical methods.

At present, the flow field for the Erlenmeyer shaker flask is being quantified. FIGURE 12 shows the PIV setup to study the Erlenmeyer flask. The flask used for this study is a 250 ml Erlenmeyer flask. In FIGURE 12, the two cameras are visible on the lower deck of the orbital shaker table. These two cameras provide stereoscopic viewing of the plane illuminated by the laser sheet. The laser beam propagates through the laser sheet forming optics, and passes through the flask as a horizontal sheet.

FIGURE 13 shows averaged PIV data for a single plane of measurement within the Erlenmeyer flask. For the image shown in FIGURE 13, the orbital shaker was operating at a rate of 120 rpm, and the flask contained 100 ml of fluid. The PIV images were captured with a time interval of 350 µs between successive image capture. FIGURE 13 illustrates the three-dimensional velocity field at a height of 2 mm from the bottom of the flask. The vectors indicate the swirling in-plane velocities, and the shaded contour plot indicates the out-of-plane velocity (designated by W) in units of m/sec. The predominant positive and negative out-of-plane velocities correspond to positions before and after a wave-like structure that can be seen in the free surface between the liquid and air interface inside of the flask. FIGURE 14 illustrates the free surface of the liquid-air interface within the flask at the time of data collection for the PIV data in FIGURE 13. FIGURE 14 was reconstructed from images of the free surface. The coordinate system indicated in FIGURE 14 corresponds to the same coordinate system shown in FIGURE 13.

Data such as those shown in FIGURE 13 were used to determine the acceleration field in the Erlenmeyer flask. A sample of the calculated acceleration field from two averaged three-dimensional velocity fields is given in FIGURE 15, which illustrates the magnitude of the acceleration (designated by MAGdvdt) in units of m/sec^2 for the same data plane as was shown in the velocity field of FIGURE 13. The majority of the accelerations occur in the narrow region on the right side of the image. The maximum accelerations measured in this plane are approximately twice the magnitude of the gravitational field.

FIGURE 16 shows a sample of the calculated velocity gradients (designated by MAGTAU). The data of FIGURE 16 represent the spatial velocity gradients, whereas the accelerations are temporal velocity gradients. FIGURE 16 represents the magnitude of all the velocity gradients in units of 1/sec. The velocity gradients can be used to determine shear stress by multiplying by the appropriate dynamic viscosity for the cell culture medium. By using the viscosity for water, it can be seen that the measured shear stress within the Erlenmeyer flask may be approximately $0.015\,\text{N/m}^2$ for the data shown in FIGURE 16.

Based on the results obtained thus far, it appears that suspension cells experience a range of hydrodynamic forces depending on their position in the Erlenmeyer flask, and the accelerations induced can be twice that due to gravity. The entire volume within the Erlenmeyer flask will be measured to determine the maximum hydrodynamic forces that cells may encounter within the $1\,g$ cell biology experiments. The

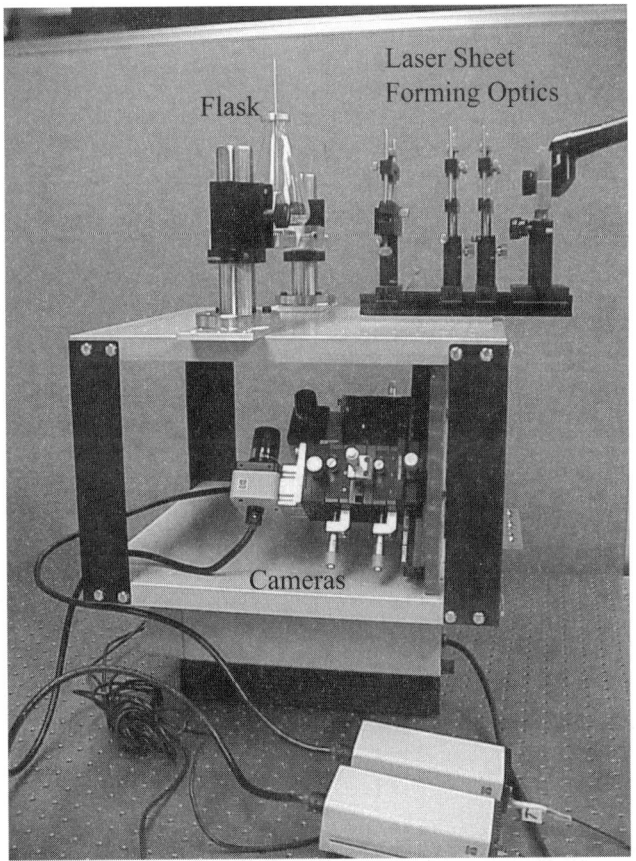

FIGURE 12. PIV setup of Erlenmeyer flask on orbital shaker table.

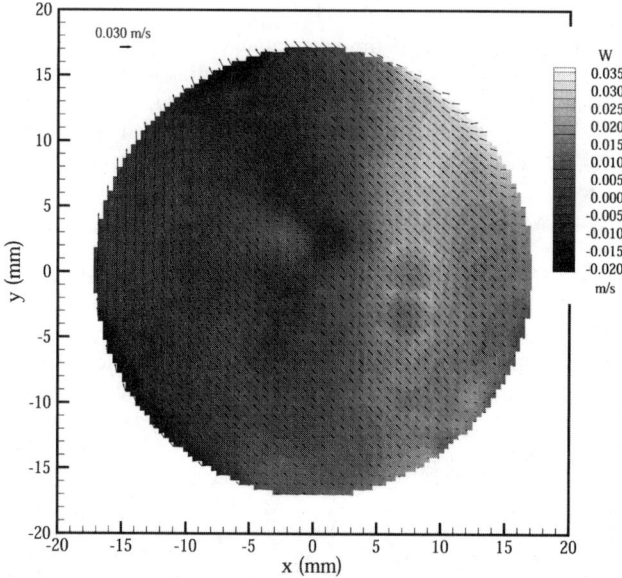

FIGURE 13. Averaged PIV velocity data in Erlenmeyer flask.

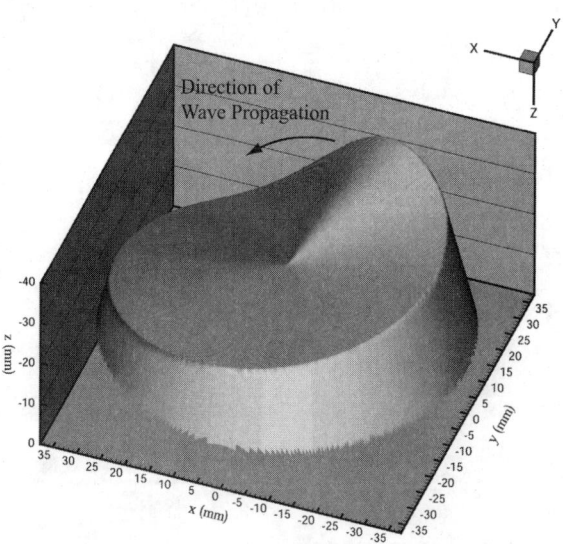

FIGURE 14. Illustration of the free surface of the liquid air interface within the Erlenmeyer flask.

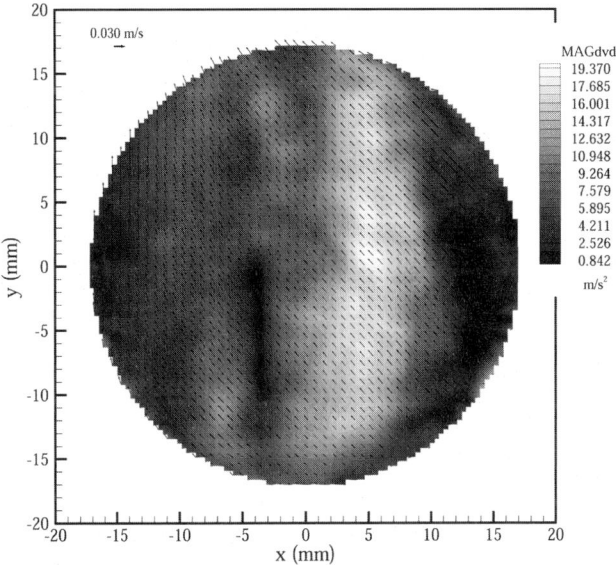

FIGURE 15. Acceleration (temporal velocity gradient) magnitudes calculated using PIV data.

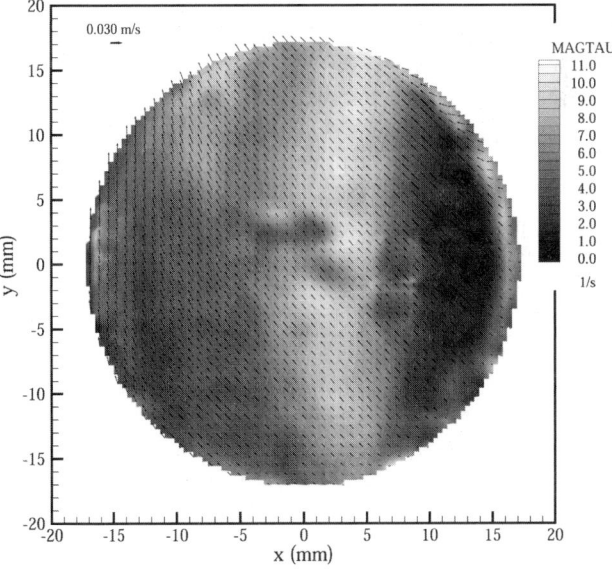

FIGURE 16. Spatial velocity gradients calculated using PIV data.

data of FIGURES 13, 15, and 16 are preliminary results. Once the entire volume has been measured, the shear forces will also be calculated, as well as the accelerations. The second phase of the PIV study will be to quantify the flow field within the CSC. Eventually, the forces generated by the flow field that cells encounter within the CSCs will be compared to the results from the shaker flask testing to ensure that they are within the bounds of typical laboratory experiments.

COMPUTATIONAL FLUID DYNAMICS

Computational fluid dynamics (CFD), a technique for modeling fluid flow fields using numerical simulations, is being used to model the flow field within the CSC for the gravitational fields of $1g$, $2g$, and microgravity. A detailed CFD model of the 10ml CSC, which includes a computational grid, has been developed to predict the flow field within the CSC. The computational grid consists of 29 zones that contain 2.3 million points, and the fluid properties are calculated at each of the 2.3 million grid points within the 10ml volume. The model was shown in FIGURE 3A. The process by which the flow field is solved uses a Navier–Stokes solver, which accommodates the non-linear nature of the governing equations. The CFD code used to solve the Navier–Stokes equations is OVERFLOW. CFD was used to solve for steady and time varying solutions for the flow field. The gravitational field was implemented via a body force source term. The flow equations are non-linear, and as a result the equations are solved by an iterative process. For each of the cases shown, the solutions were obtained through 15,000 iterations. These calculations were performed on a computer with 16 parallel processors and the solutions required to approximately 21 hours to achieve the results shown.

FIGURES 17 and 18 show CFD results for a calculated flow field, where the data are shown with particle paths instead of velocity vectors. The particle paths yield a more global view of the full flow field. For completeness, the velocity vector fields are shown in FIGURES 19 and 20 for the same cases as in FIGURES 17 and 18. FIGURES 17 and 19 are the result of CFD calculations performed without the cell containment membrane present. FIGURES 18 and 20 are the result of calculations with the membrane present. In both cases, a steady flow rate of 10ml/min was injected through both inlets. The membrane was modeled using a pressure drop that is scaled based upon the inlet velocity and the density of the fluid. It can be seen from this CFD data that the membrane greatly affects the velocity field and results in increased circulation of fluid before the fluid is extracted through the outlet. Thus, the membrane is effective in increasing mixing. Presently, a model for the stir bars is being created for inclusion in the iterative calculations. In FIGURES 17 through 20 the volume flow rate was 10ml/min.

The accuracy of the CFD model will be evaluated by comparing the model to flow visualization results for experiments performed at $1g$ and $2g$. The CFD results will also be compared to PIV results from future CSC measurements.

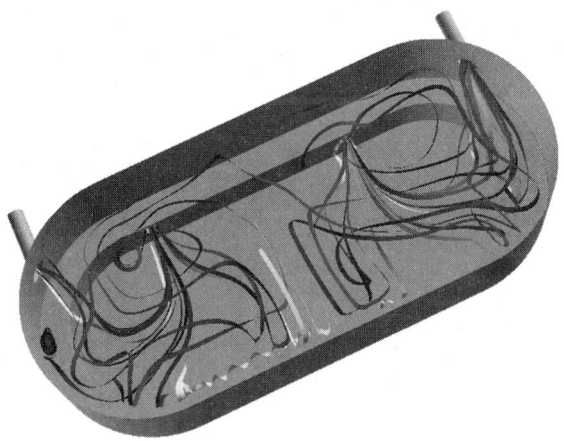

FIGURE 17. CFD calculated particle paths without membrane present.

DISCUSSION AND CONCLUSIONS

The CCU and CSCs are designed to meet the diverse demands necessary for cell culture studies at various gravitational conditions from μg to $2g$. The CCU design permits researchers the ability to control a variety of environmental conditions important for cell growth. CCU and CSCs provide an environment that can be adapted for successful cell growth studies for a large variety of animal and plant cells with very different requirements. The CCU and CSC designs have been successful in cell growth studies for a variety of different cell types at $1g$. These cell growth

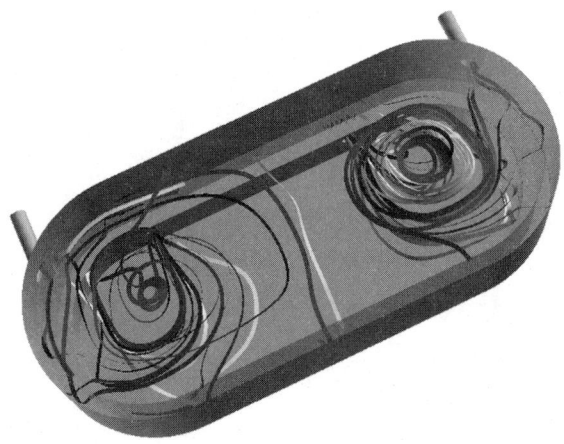

FIGURE 18. CFD calculated particle paths with membrane present.

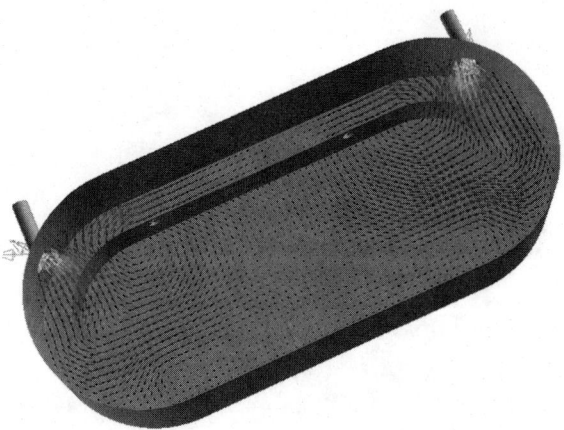

FIGURE 19. CFD calculated velocity field without membrane present.

experiments have served to test the CCU and CSC hardware designs and to make improvements to the final CCU flight design for the ISS.

To achieve the goals of maintaining known flow field environmental conditions, flow visualization, PIV, and CFD have been applied in tandem with the development of the CCU and CSC in order to determine the optimal flow field for cell growth studies. Furthermore, the test techniques that have been applied will help to quantify the flow field and improve researchers understanding of the factors that influence cell growth.

The flow visualization study was performed to investigate flow field characteristics associated with various inlet and outlet flow configurations. The primary results

FIGURE 20. CFD calculated velocity field with membrane present.

obtained from the flow visualization were used to analyze mixing and flushing characteristics. Many approaches for providing good mixing and flushing have been tested during development of the CCU, including mechanical devices, such as nozzles, sprinklers, paddles and stir bars, and various flow algorithms that vary the flow rate or provide pulsed flow, periodic flow, or alternating forward and reverse flow. From this multitude of configurations, the split-membrane stir bar configuration using angled paddles offers the greatest flexibility for satisfying all the requirements for mixing, flushing, and suspension. Flow visualization was the testing method that was used to quickly determine effective configurations within the CSC that resulted in faster mixing or flushing times as compared to the baseline flow field.

PIV measurements are being used to characterize the hydrodynamic forces that the cells encounter within the CSC. In future, the PIV measurements within the CSC will be used to ensure that these hydrodynamic forces are similar to the hydrodynamic forces that cell cultures experience with current $1\,g$ cell biology experiments. To achieve this goal, the first step for PIV testing was to measure the hydrodynamic forces within an Erlenmeyer flask mounted on an orbital shaker. Testing that configuration will yield the baseline shear forces and accelerations that the cells encounter.

CFD results have indicated the quantitative three-dimensional flow field within the 10-ml CSC. Initial CFD analyses have demonstrated that the membrane does contribute significantly to the flow pattern. The inclusion of the CFD stir bar model into the current CSC computational model will permit a full analysis of the flow field within the CSC, and future efforts will identify the effects of changing the gravitational field from $1\,g$ to microgravity and to $2\,g$.

Researchers are continuing to test the CCU hardware. Based on continued optimization of the cell environment from flow testing and analyses as well as cell growth experiments, the CCU and CSCs have been successful in growing yeast, C2C12 murite miyoblast, Euglena gracilis, muscle organoids, and BY2 tobacco cells. Efforts with PIV and CFD continue to provide a better understanding of flow field and of the effects of changing gravity from microgravity to $2\,g$.

ACKNOWLEDGMENTS

The authors would like to acknowledge the following contributors: The CCU team at NASA Ames Research Center and Lockheed Martin, the CCU contractor, Payload Systems Inc. and the Massachusetts Institute of Technology, and the consultant hardware testing teams. The authors would like to credit the following individuals: Isaac Berzin, Ljiljana Kundakovic, Carmen Preda, Liping Sun, and Gordana Vunjak for their work with the biological use of the CSC that directed the flow parameters. Special thanks to Gerard Heyenga for his development of the split-membrane design.

REFERENCES

1. GEORGE, E.F. 1987. Plant Culture Medium Vol. I: Formulations and Uses. Exegetics Limited, Westbury, Wiltshire, England.
2. POLLARD, J.W. & J.M. WALKER. 1990. Methods in Molecular Biology, Vol: VI: Plant Cell and Tissue Culture. Humana Press, Clifton.

3. FRESHNEY, R.I. 1994. Culture of Animal Cells: A Manual of Basic Technique. Wiley-Liss, New York.
4. SEARBY, N.D. *et al.* 2001. Life support from the cellular perspective. ICES Paper 01ICES-331. Presented at the 31st International Conference on Environmental Systems. Orlando, FL. July 9–12.
5. DE LUIS, J., G. VUNJAK-NOVAKOVIC, N.D. SEARBY. 2000. Design and testing of the ISS cell culture unit. IAF Paper IAF-00-G.4.0.6. International Astronautical Federation Conference, Rio de Janeiro, Brazil. October.
6. RAFFEL, M., *et al.* 1998. Particle Image Velocimetry. Spring Verlag, New York.
7. SEARBY, N.D, J. DE LUIS, G. VUNJAK-NOVAKOVIC. 1998. Design and development of a space station cell culture unit. SAE Technical Paper Series 981604. Presented at the 28th International Conference on Environmental Systems. Danvers, MA. July 13–16.

Compact Optical Sensor for Real-Time Monitoring of Bacterial Growth for Space Applications

R.C. VAN BENTHEM,[a] D. VAN DEN ASSEM,[a] AND J. KROONEMAN[b]

[a]*National Aerospace Laboratory, NLR, Amsterdam, The Netherlands*
[b]*Bioclear Environmental Biotechnology, Groningen, The Netherlands*

ABSTRACT: During long-duration manned space missions, complex chemical and biological processes need to be managed accurately for recycling human wastes and to produce human consumables. As a result, there is increasing interest in how the characteristics of microbes are influenced by microgravity. Compact optical instrumentation allows for real-time and non-invasive measurement of bacterial growth parameters during flight experiments. In close collaboration, the National Aerospace Laboratory of the Netherlands (NLR) and Bioclear Environmental Biotechnology developed and tested an on-line optical biomass sensor successfully. The sensor concept is based on a turbidity measurement technique operating in the VIS-blue part of the light spectrum with use of blue LED sources. A diagnostic tool has been developed using compact spectrometers and optical fibers to characterize bacterial cultures. As a result a few sensor applications operating at different colors and sensor layouts are discussed in the paper.

KEYWORDS: turbidity; turbidity sensor; bacteria growth

NOMENCLATURE:
Acronyms

3T	3T B.V. Measurement, Control & Data-registration, Enschede
ATS	autonomous turbidity sensor
BAF	biological air filter
Bioclear	Bioclear Environmental Biotechnology, Groningen
DESC	Dutch Experiment Support Centre
FOTON	Russian unmanned satellite
ITS	in-line turbidity sensor
LED	light emitting diode
Melissa	micro-ecological life support alternative
OD	optical density
RPM	random positioning machine
RTS	turbidity sensor on the RPM
WTS	wireless turbidity sensor
XA GJ10	bacterium *Xanthobacter Autotrophicus* GJ10

Symbols

α	absorptivity function
β	scattering function
$\Delta\lambda$	part of wavelength spectrum

Address for correspondence: R.C. van Benthem, National Aerospace Laboratory, NLR, Antony Fokkerweg 2, P.O. Box 90502, 1006 BM, Amsterdam, The Netherlands. Voice: +31 527 24 8236; fax: +31 527 24 8210.
vdassem@nlr.nl

λ	wavelength of the transmitted light
$[c]$	concentration of scatters (e.g., cells)
d	transmission light path
D	detection function
I	intensity function
R	ratio of scattering to transmission signal
R^*	geometrical mean of ratio R
S	detected signal
Subscripts	
λ	wavelength
λ_1	lower wavelength value
λ_2	upper wavelength value
det	detector
medium	medium without cells
cells	cells in the sample

INTRODUCTION

Recycling human waste and production of human consumables are almost mandatory necessities for the success of long-term manned space missions. For this purpose complex biological and chemical processes are incorporated in the life support systems on board the spacecraft. These processes need to be managed accurately. Applied biotechnology recycling technologies, for example, biological air filters (BAF), make use of microbial activities.[1] Consequently, there is increasing interest in the influence of microgravity on bacterial growth.

Recently several microbiological growth experiments have been performed in a microgravity environment on board a space laboratory. Until now, the bacterial growth parameters obtained with these experiments were analyzed after flight in the usual laboratory environment on ground. This type of research will benefit considerably from in-flight real-time monitoring and online data recording.

This paper covers a two-year period of research that was started with laboratory investigations on a compact optical turbidity sensor, because of its interesting features for microgravity experiments.[2] The turbidity sensor concept has been developed into a sensor for real-time bacteria growth measurement. Recently the sensor was used for the determination of growth characteristics of *Xanthobacter Autotrophicus GJ10*.[3] In the sensor concept, two techniques are effectively combined—simultaneous measurement of transmission (optical density) and scattering measurement.

This paper provides a short discussion of the sensor concept. In order to optimize the operation of the sensor for new applications, a diagnostic tool was developed. Finally, a few applications and sensor configurations are discussed such as a wireless turbidity sensor (WTS) prototype, a turbidity sensor on a random positioning machine (RTS), and an in-line (ITS) version for continuous biomass production monitoring for an experimental bioreactor. Applications of the sensor for milking machines and hydraulic oil systems are currently under investigation.

TRANSMISSION MEASUREMENTS

Optical laboratory techniques for bacteria growth measurement have for a long time been based on the observation that bacterial cultures become visibly turbid when they multiply. As such, the bacteria density, or biomass, is related to a measure for the turbidity. Usually the biomass is obtained by measurement of the optical density (OD) at a predefined wavelength; for example, with use of a photospectrometer. The OD value is calibrated to the bacteria concentration by counting the bacteria number. For these measurements, as for instance for the growth of the bacteria *Xantobacter Autrophicus GJ10*, blue light is used at $\lambda = 450\,\text{nm}$. The OD is defined as the logarithm of the ratio of the undisturbed light intensity and the intensity of the light transmitted through the absorbing material. According to Beer's law, the detected signal obtained after transmission through the absorbing material can be expressed as an exponential function as follows:

$$S_{\text{det}} = I_\lambda \cdot e^{-\alpha_\lambda d} \cdot D_\lambda, \tag{1}$$

where, at specified wavelength λ, S_{det} is the detected signal, I_λ the input intensity, α_λ the absorptivity, d the transmission light path, λ the wavelength of the transmitted light, and D_λ the detector function.

For a culture medium with, and without growing bacteria, the detected signals according to Equation (1) are, respectively,

$$S_{\text{det, sample}} = I_\lambda \cdot e^{-(\alpha_{\lambda,\text{cells}}\{c\} + \alpha_{\lambda,\text{medium}})d} \cdot D_\lambda \tag{2}$$

and

$$S_{\text{det, medium}} = I_\lambda \cdot e^{-\alpha_{\lambda,\text{medium}} d} \cdot D_\lambda, \tag{3}$$

where, at the specified wavelength λ, $S_{\lambda,\text{det,sample}}$ is the detected sample signal, $S_{\lambda,\text{det,medium}}$ is the detected medium signal, $\alpha_{\lambda,\text{cells}}$ is the absorptivity of the cells per unit concentration, $\alpha_{\lambda,\text{medium}}$ is the absorptivity of the medium, and $[c]$ is the bacterial concentration.

By definition the OD is calculated from the ratio of (2) and (3), in order to compensate for random errors and characteristic instrument component parameters, such as the output spectrum of the light source, the spectral sensitivity of the detector, and absorptivity of the medium

$$\text{OD}_\lambda = -\log_{10}(S_{\text{det, sample}}/S_{\text{det, medium}}) \approx \alpha_{\lambda,\text{cells}}[c]d. \tag{4}$$

Thus, the measured OD_λ can be approximated by a linear function of the cell concentration for small values of $\alpha_{\lambda,\text{cells}}[c]d$.

In a photometer (see FIGURE 1) a light source, for instance a halogen lamp, emits light that is transmitted through a sample, for which the OD is measured. A detector collects the transmitted light. An optical wavelength bandfilter in front of the light source transmitted part of the wavelength spectrum. Within the transmitted spectral band the transmission is assumed to depend only on the sample concentration, provided that the other parameters have been kept constant.

The linear measurement range is limited to densities OD < 1, due to increasing concentration and the influence of forward scattering. Since bacteria growth can extent to a period of a few days and may reach high optical density values, measuring bacteria growth using a photospectrometer has its limitations. It requires frequent

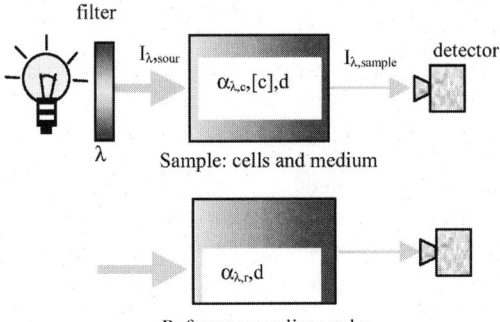

FIGURE 1. Optical density (OD) transmission measurements with a photospectrometer using a selected spectrum from source and detector. The OD is calculated from a reference (medium only) and a sample measurement.

manual extraction of measurement samples, dilution of samples, in case the OD < 1 and the taking of reference samples. In addition it is sensitive to pollution of the optical transmission surfaces.

TURBIDITY MEASUREMENTS

In industry and laboratories, particle concentrations of turbid suspensions are usually measured with use of light scattering. A white light source illuminates a suspension that contains the scattering particles (see FIGURE 2). The scattered light is collected on a detector. The sensor signal is an integral function, proportional to light source intensity, detector response, scattering properties of the particles, and the particle concentration.

$$S = \int_{\lambda_1}^{\lambda_2} I_\lambda \beta_\lambda [c] D_\lambda d\lambda, \qquad (5)$$

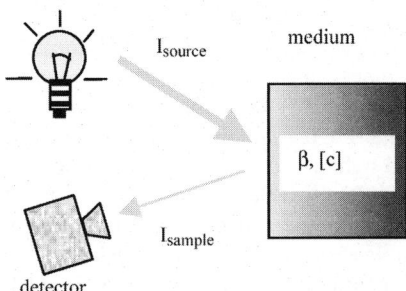

FIGURE 2. Scattering (turbidity) measurements using a (white) light source and detector. Instrumentation errors are not compensated for.

where, at the specified wavelength λ, S is the detected signal, I_λ is the input intensity function, β_λ is the scatter function, $[c]$ is the concentration of scatters, D_λ is the detector function, λ_1 is the lower wavelength value, and λ_2 is the upper wavelength value.

In order to be able to measure the concentration, the output of the light source, the detector sensitivity, and particle scattering properties must be constant. Note that for long duration measurements this is usually not the case. Frequent maintenance is a disadvantage of online turbidity monitoring. In addition, the signal response is poor for low concentration values, since fewer scatterers are present in the medium. Source and/or detector degradation errors are not compensated for. It requires regular calibration using a turbidity standard solution, such as Formazine.

IMPROVED TURBIDITY SENSOR

An improved turbidity sensor concept is based on the dual ratio turbidity measurement. It applies simultaneous transmitted and scattered light measurement. Based on this concept, the Dutch firm 3T BV (Enschede, The Netherlands) originally started with the development of a real-time turbidity sensor for a high turbidity applications. Accordingly, the sensor has been adapted for bacterial growth measurements under micro-gravity conditions onboard a space laboratory.[2]

From an instrumental point of view the sensor has several advantages[a,b] with respect to single OD or scattering measurements. It is well suited for real-time monitoring and data analysis. It compensates for instabilities in the light source and detector circuitry and it is insensitive to pollution of optical surfaces. It can be build solidly with compact dimensions and inexpensive components.[3]

The turbidity sensor (see FIGURE 3) consists of two light emitting diodes (LEDs) and two photodiodes that are located around a medium under investigation. The LEDs are operated one by one, producing a transmission and a less than 90° scatter signal collected by the photodiodes.

The light intensity $I_{\Delta\lambda}$, the detector sensitivity $D_{\Delta\lambda}$, specific transmission $\alpha_{\Delta\lambda}$, and scattering properties $\beta_{\Delta\lambda}$ are assumed constant over the spectral bandwidth $\Delta\lambda$ of the LED, which is normally on the order of 60 nm to 80 nm. The cell concentration is $[c]$ and the dimensions of the sensor (d). Amplification factors are neglected since they only add linear terms to the equations. Each measurement cycle produces four signals, two transmission and two scattering signals.

When LED1 is on,

$$S_{1,\Delta\lambda} = I_{1,\Delta\lambda} e^{-\alpha[c]d} D_{1,\Delta\lambda}, \tag{5a}$$

$$S_{2,\Delta\lambda} = I_{1,\Delta\lambda} \beta_{\Delta\lambda} [c] D_{2,\Delta\lambda}. \tag{5b}$$

When LED 2 is on,

---

[a]Note that the scattered light intensity is usually a few orders of magnitude weaker than the transmitted light intensity. The dual ratio method is only valid when the scatter intensity is above detection limit.

[b]R^* also compensates for (inhomogeneous) window fouling. Pollution adds additional linear terms in (5)–(9). The ratios compensate, therefore, until the windows are completely blocked.

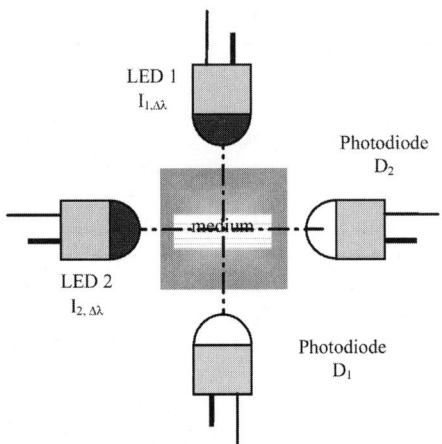

FIGURE 3. Turbidity sensor concept using two LEDs and two photodiodes placed around a medium. The LEDs are operated one after each other to measure both ratios of the transmitted and scattered light using the photodiodes. The dual ratio R^* is calculated from the ratios, thus compensating for instrumentation errors.

$$S_{3,\Delta\lambda} = I_{2,\Delta\lambda} e^{-\alpha[c]d} D_{2,\Delta\lambda}, \tag{6a}$$

$$S_{4,\Delta\lambda} = I_{2,\Delta\lambda} \beta_{\Delta\lambda}[c] D_{1,\Delta\lambda}. \tag{6b}$$

The ratios R_1 and R_2 (obtained from **(5a)**, **(5b)**, **(6a)**, and **(6b)**) of the scattered and transmitted light signals collected by the two photodiodes produce two signals that compensate for instabilities of the LED.

For LED 1,

$$R_{1,\Delta\lambda} = \frac{S_{2,\Delta\lambda}}{S_{1,\Delta\lambda}} = \beta_{\Delta\lambda} e^{\alpha[c]d} \frac{D_{2,\Delta\lambda}}{D_{1,\Delta\lambda}}. \tag{7}$$

For LED 2:

$$R_{2,\Delta\lambda} = \frac{S_{4,\Delta\lambda}}{S_{3,\Delta\lambda}} = \beta_{\Delta\lambda} e^{\alpha[c]d} \frac{D_{1,\Delta\lambda}}{D_{2,\Delta\lambda}}. \tag{8}$$

Calculating the geometrical mean value of the two ratios compensates for both detector sensitivity variations. This yields

$$\begin{aligned} R^*_{\Delta\lambda} &= \sqrt{R_{1,\Delta\lambda} \cdot R_{2,\Delta\lambda}} = \beta_{\Delta\lambda}[c] e^{\alpha[c]d} \\ &\approx \beta_{\Delta\lambda}[c] + \alpha_{\Delta\lambda}\beta_{\Delta\lambda}[c]^2 d + \alpha_{\Delta\lambda}^2 \beta_{\Delta\lambda}[c]^3 d^2 + \dots \\ &= a_1 + a_2[c] + a_3[c]^2 + \dots \, . \end{aligned} \tag{9}$$

The dual ratio, $R^*_{\Delta\lambda}$ is expressed here as an exponential expansion series, compensates for the instrumentation errors. $R^*_{\Delta\lambda}$ is a polynomial function of the cell concentration $[c]$. The coefficients a_i are sensitive to the specific transmission and scattering properties of the cells, considered to be constant. Note that $R^*_{\Delta\lambda}$ depends on the selected wavelength (LED color) and cell geometry. This means that the

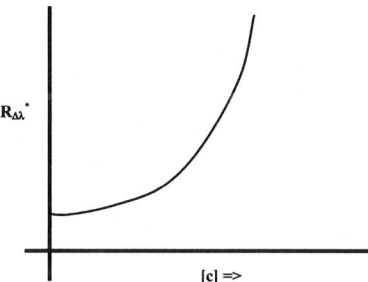

FIGURE 4. The turbidity sensor is calibrated using known cell concentrations $[c]$ in the operational range, measuring the response $R^*_{\Delta\lambda}$ of the sensor.

sensor has to be calibrated for mutually different types of cells for which the growth must be measured. Therefor, the sensor signal R^* must be measured for various known cell concentrations. FIGURE 4 presents an example.

General Approach for LED Selection

For each new application of the improved turbidity sensor, a spectral characterization test must be performed to select the LED color. For this reason a new diagnostic tool was developed, as is described in the next section. After selection of the LED sensor calibration is performed to relate the sensor signal to known sample concentrations (see FIGURE 5).

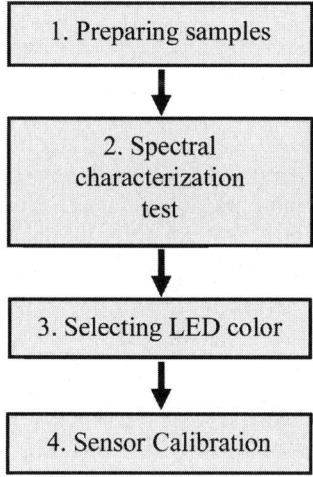

FIGURE 5. General approach for selecting the LED color for a new application of the improved turbidity sensor.

TABLE 1. Comparing optical techniques for real-time bacteria growth measurements

Method	Advantages	Disadvantages
Transmission (optical density)	Compensates for instrumentation errors. Linear with the cell concentration for OD < 1. Accurate and spectrally selective.	Requires reference measurements. Limited range: non- linear above OD > 1. Dilution of samples required. Off-line: manual extraction of samples required. Laboratory environment. Sensitive to pollution.
Scattering (turbidity)	On-line measurements possible. Coarse: less spectrally sensitive. Industrial environments: robust and cost effective. Large turbidity value applications.	Does not compensate for instrumentation errors. Needs regular calibration with standard solutions. Limited range: weak signal at low turbidities. Sensitive to pollution.
Improved turbidity sensor (Dual ratio of transmission and scattering)	On-line measurement and control applications. Robust and in-sensitive to pollution and instrumentation errors. Small dimensions and cost effective. Large range: small and large turbidity applications. Spectrally selective, possibly allowing for multiparameter analysis using more than one sensor. Spectrally applicable: new diagnostic tool	Non-linear response: calibration required. Spectral characterization test needed

The advantage of applying the improved turbidity sensor for real-time measurements with respect to the more traditional sample extraction methods is that it is insensitive to instrumentation errors, exploiting the dual ratio. It is compact and cost effective, has a large dynamic range and is spectrally selective, making multiparameter analysis a possibility. The pro and cons of the discussed techniques are summarized in TABLE 1.

DIAGNOSTIC INSTRUMENTATION

From Equation (9) it is clear that R^* is spectrally selective and sensitive for the scattering properties of cells. Therefore, for a new application the medium needs to be spectrally characterized with diagnostic instrumentation for the selection of the LED spectrum to be used as the sensor (see FIGURE 6).

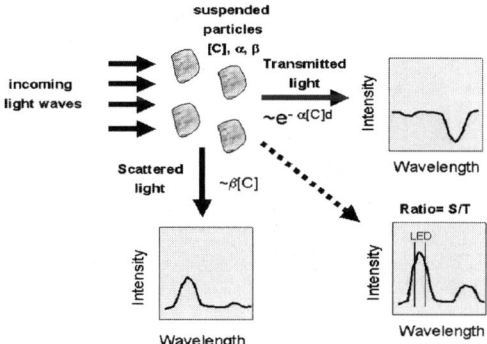

FIGURE 6. A light source illuminates a medium with suspended particles; it leads to spectrally sensitive transmission and scattering. The maximum ratio gives the LED color to be selected for the improved turbidity sensor.

For this reason a diagnostic instrument was developed that explores the dual ratio technique by calculating $R^*_{\Delta\lambda}$, to compensate for instrumentation errors (see FIGURES 7 and 8). It encompasses a halogen light source, an optical switch, a sample holder, and two photo-spectrometers ($\lambda = 530$ nm to $\lambda = 1,180$ nm) mutually connected via optical fibers. A cuvette containing the culture sample is placed in the sample

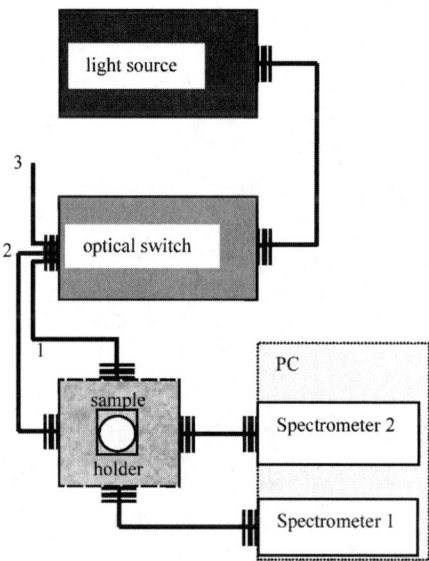

FIGURE 7. Diagnostic instrument for new applications of the turbidity sensor using a light source, an optical switch and two photospectrometers ($\lambda = 530$ nm to $\lambda = 1,180$ nm) exploring the dual ratio technique to compensate for instrumentation errors.

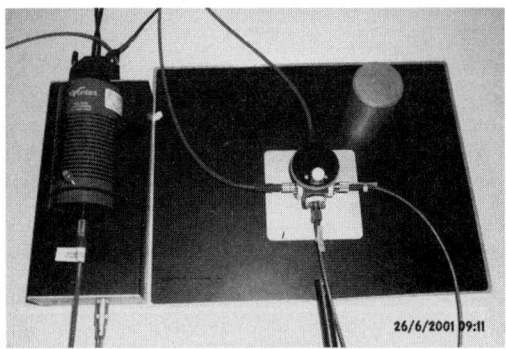

FIGURE 8. Diagnostic instrument with, from *left* to *right*, a light source, an optical switch, a sample holder with cuvette and protection cover that are mutually connected via optical fibers.

holder. The optical switch adjusts the light of port 1 or 2. Port 3 is used for dark signal measurement. The transmitted and the 90° scattered light intensities are transferred via fibers to the two photo-spectrometers. The integration time for the scattering and transmitted light is automatically adjusted in the software to obtain an optimum signal response. The dark current and integration time are compensated for. A high value for $R^*_{\Delta\lambda}$ indicates high scattering and low transmission—a suitable choice for the LED spectrum to be used for the turbidity sensor. With the diagnostic

FIGURE 9. Set-up used for the on-line test of the *XA GJ10* bacterium at Bioclear premises.

instrumentation a successful characterization test was carried out for the selection of the spectrum for the *Rhodospirillum Rubrum* measurement (see APPLICATION IV)

APPLICATIONS

Application I: The On-Line Biomass Turbidity Sensor

A prototype dual ratio turbidity sensor was used to measure the growth curve of the bacterium *Xanthobacter autotrophicus GJ10*. The experiment was carried out at Bioclear B.V, Environmental Technology premises.[3] FIGURE 9 shows a photograph of the set-up. FIGURE 10 presents the test results. A square cuvette (with air reservoir) was placed in a sample holder. The measurements were performed with blue LEDs (λ = 450nm). A dark cover protected the set-up from environmental light.

The measurements show good agreement with reference measurements. An interesting feature is the abrupt signal drop at the end of the exponential growth (FIG. 10d) when the bacteria run out of nutrient. Apparently its optical properties change.

Application II: Turbidity Sensor (RTS) for the Random Positioning Machine (RPM)

An experiment was developed in close cooperation with the Dutch Experiment Support Centre (DESC, at the Free University of Amsterdam, The Netherlands) and Fokker Space (Leiden, The Netherlands). The three-dimensional random rotations of the RPM around the Earth's gravity vector "simulate" a microgravity environment in the center of the machine.

A sample holder, equipped with a double turbidity sensor was mounted on the platform of the Desktop RPM (see FIGURES 11 and 12).

Disposable laboratory tubes can be placed in the sample holder. The sensor is optimized for monitoring two specimens (e.g., E-colie and yeast cells). With the set-up two-parameter analysis is possible, or scattering can be measured simultaneously in various directions with respect to the incident light beam.

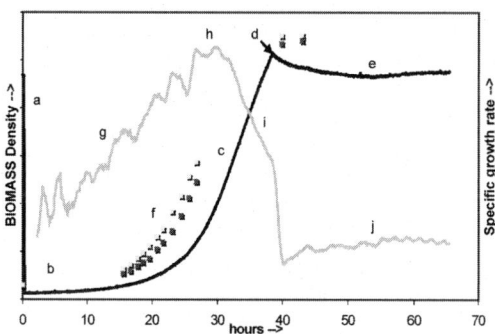

FIGURE 10. Test results with the improved turbidity sensor showing exponential growth of the *XA GJ10* bacterium (c) and specific growth rate (g). Off-line reference measurements (f) show a comparable biomass density and specific growth rate.

FIGURE 11. The desktop random positioning machine (RMP) of Fokker Space.

Application III: Wireless/Autonomous Turbidity Sensor (WTS/ATS) for a FOTON Flight

In cooperation with IDENTO Electronics (Ens, The Netherlands), a wireless implementation of the sensor is being developed (see FIGURE 13) with use of a RF technique. The sensor itself is enclosed in an RF transparent container. An external

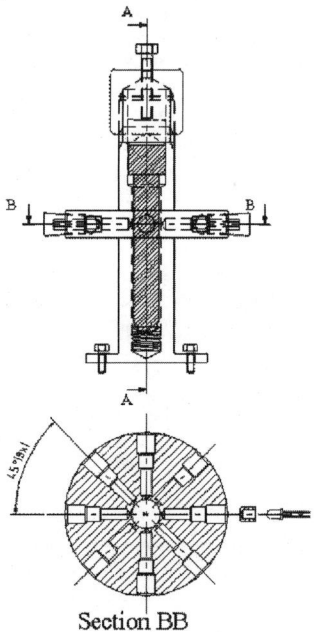

FIGURE 12. RTS sample holder with double sensor design.

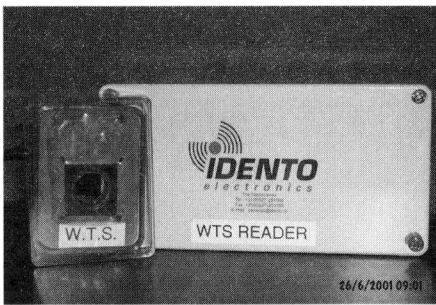

FIGURE 13. Prototype wireless turbidity sensor using RF field for power and data transfer and remote operation electronics as developed in cooperation with IDENTO Electronics.

RF field supplies the required energy and transmits the data to a reader. The distances between the RF field supply, data transfer antenna, and the sensor are limited to a few centimeters.

At present, an autonomous version of the turbidity sensor (ATS) is being developed. A battery supplies the energy and a memory is used for data storage.

In the near future, the ATS will be used in a self-supporting bacterial growth experiment, accommodated in an autonomous experiment box on a FOTON flight at the end of 2002. The experiment is being prepared and set up in cooperation with Bioclear (Bioclear B.V, Environmental Technology) and the European Space Agency (ESA).

Application IV: An In-Line Turbidity Sensor (ITS) for a Melissa Bioreactor

A latest development, in cooperation with ESA, is the application of the turbidity sensor to control the biomass production of the bacterium *Rhodospirilum rubrum* in an experimental bioreactor of Melissa.[4] FIGURE 14 shows a photograph of the set-up.

FIGURE 14. Experimental bioreactor setup of Melissa.

Melissa is a project for the development a biological life support system for long-duration manned space missions. The real-time sensor is placed in an overflow appendage of the bioreactor. Primarily it has been verified that red LEDs are well suited for the *Rhodospirillum Rubrum* population measurements (see FIGURE 15). The sensor output signal will be correlated to the biomass dry-weight.

CONCLUSIONS

To image biological processes there is a demand for on-line and compact optical diagnostic sensors for Earth and space applications. Traditional techniques showed less suited for continuous, long-term monitoring (e.g., for control purposes) of bacteria growth. The dual ratio turbidity sensor, discussed in the present paper, copes with these problems. A determination of the growth rate of *Xanthobacter Autotrophicus GJ10* showed its performance, although interpretation of some of the results obtained with the new technique is still open. The need to perform spectral analysis for selection of the LED color induced the development of a new diagnostic instrument. A first successful diagnostic test, using the instrument was carried out on *Rhodospirillum rubrum*. New applications of the sensor, such as for the RPM and the Melissa experimental bioreactors are in progress. The sensor is cost effective, robust, accurate, and reliable, and capable of performing real-time measurements. New applications of the sensor concept are foreseen in future based on in-depth interpretation of the results. Therefor, more theoretical and practical research with respect to optical (scattering) properties of cells and particles is required.

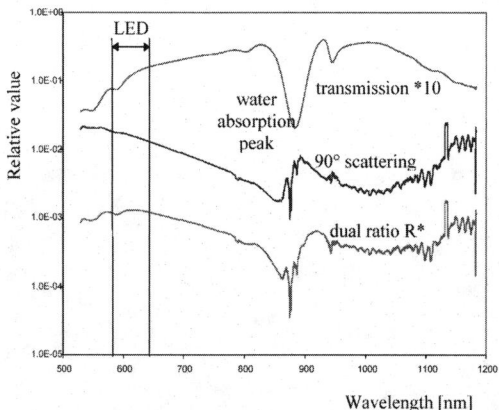

FIGURE 15. Spectral characterization test on the *Rhodospirillum rubrum* using the diagnostic instrumentation. The transmission value is multiplied by a factor 10 to shift it upward in the figure. Note that the y-scale is logarithmic. The maximum dual ratio $R^*_{\Delta\lambda}$ value is found at about $\lambda = 620$ nm. In this case red LEDs must be used for the improved turbidity sensor.

ACKNOWLEDGMENTS

We owe much gratitude to 3T for providing their original turbidity sensor and breadboard electronics and Janneke Krooneman, Jaap van der Waarde en Hilma Hamminga (Bioclear) for their support during the experiments on the *Xanthobacter Autotrophicus* performed at their site. We highly appreciate cooperation with Ron Huijser, Pepijn Esman (Fokker Space), and Jack van Loon (DESC) during the preparation of the turbidity sensor for the RPM. Thanks to Jack Cöp and Ted Camu (IDENTO Electronics) for their work on the prototype WTS and sensor electronics. Christel Paille and Christophe Lasseur (ESA) we thank for their support and interest during the development of the in-line sensor for Melissa.

REFERENCES

1. WAARDE, J.J., S. VAN DER KEUNING, P.G. PAUL & R.G. BINOT. 1997. Determination of space influence on the kinetics for biodegradation of organic volatile contaminants. Proceedings of the 6th European symposium on Space Environmental Control Systems, Noordwijk, 20–22 May.
2. VAN DEN ASSEM, D. 1999. Development of a biomass sensor. Dual ratio versus transmission trade-off, NLR-TR-99444.
3. VAN DEN ASSEM, D., R.C. VAN BENTHEM & A.A. CASTELEIJN. 2000. Self compensating real-time biomass sensor. Proceedings of the First International Symposium on Microgravity Research and Applications in Physical Sciences and Biotechnology, Sorrento, Italy, 10–15 September 2000, ESA SP-454, January 2001.
4. LASSEUR, CH. & I. FEDELE, Eds. 1999. Melissa Final Report for 1999, Activity ESA/EWP-2092.

Bioactive, Degradable Composite Microspheres

Effect of Filler Material on Surface Reactivity

QING-QING QIU,[a] PAUL DUCHEYNE,[a,b] AND PORTONOVO S. AYYASWAMY[c]

[a]*Department of Bioengineering, Center for Bioactive Materials and Tissue Engineering, University of Pennsylvania, Philadelphia, Pennsylvania, USA*

[b]*Department of Orthopedic Surgery, Center for Bioactive Materials and Tissue Engineering University of Pennsylvania, Philadelphia, Pennsylvania, USA*

[c]*Department of Mechanical Engineering and Applied Mechanics, Center for Bioactive Materials and Tissue Engineering, University of Pennsylvania, Philadelphia, Pennsylvania, USA*

> ABSTRACT: Composite microspheres with two different fillers were developed using a solid-in-oil-in-water (s/o/w) emulsion solvent removal method. Two types of bioactive ceramic powders, specifically calcium hydroxyapatite (HA) and modified bioactive glass (MBG), were incorporated into degradable poly(lactic acid) (PLA) polymer matrix to form composite microspheres. For each filler material, microspheres with three different weight ratios of filler material to polymer, namely, 1:1, 1:3, and 1:9, were synthesized. *In vitro* immersion using simulated physiological fluid (SPF) was employed to evaluate the surface reactivity of the microspheres. SEM analysis revealed that after a 14-day immersion the surface of the microspheres containing 50% MBG was fully transformed into a bone-like apatite. In contrast, a limited number of mineral nodules were present on the surface of microspheres containing HA. The solution chemical analyses performed to determine changes of Ca, P, and Si concentrations as a function of the immersion time showed that the ion concentration profiles were similar for all microspheres, except the [Si] profile. A higher Si concentration was detected in the SPF immersed with MBG-containing microspheres. These data show that the MBG filler significantly enhances the surface reactivity of the composite microspheres. This observation enables us to conclude that the composite MBG-containing microspheres are the preferable microspheres for three-dimensional bone tissue engineering.
>
> KEYWORDS: calcium phosphate ceramic; bioactive glass; degradable microspheres; drug delivery; three-dimensional bone tissue engineering

INTRODUCTION

Bone tissue engineering *in vitro* is gaining widespread interest as a means for alleviating profound limitations of both autograft and allograft bone tissues. Our previous

---

Address for correspondence: P. Ayyaswamy, Department of Mechanical Engineering and Applied Mechanics, School of Engineering and Applied Science, University of Pennsylvania, Philadelphia, PA 19104-6315, USA. Voice: 215 898 8362; fax: 215 573 6334.
 ayya@seas.upenn.edu

studies on three-dimensional bone-like tissue formation in bioreactors revealed that a resorbable, bioactive, and light microsphere is desired.[1-9]

A variety of microspheres made from various polymers, including dextran, polystyrene, and polyacrylamide, have been developed and used in tissue cultures as well as in sustained delivery of biologically active compounds.[10-15] Among these polymers, poly(lactic acid) (PLA) and its copolymers with glycolic acid (PLGA) have received considerable attention for their biocompatibility and resorbability.[16,17] There is recently growing interest in the development and use of ceramic-biodegradable polymer composite for bone repair, tissue engineering, and drug delivery.[18-20] In order to overcome the mechanical and biological limitations of the polymers, bioceramics, such as calcium hydroxyapatite, have been incorporated into polymers to form composites useful in orthopedic applications.[21,22]

In our recent study, modified bioactive glass (MBG) powders were incorporated into poly(lactic acid) (PLA) matrix using a solid-in-oil-in-water (s/o/w) emulsion solvent removal method to form composite microspheres for three-dimensional bone tissue engineering.[7] It was observed that the MBG filler enhanced the reactivity of the polymeric surface of the composite microspheres. This finding led us to ask if other bone bioactive ceramics, such as calcium hydroxyapatite, would have a similar effect in that they could lead to a bone-like apatite layer formation at the same rate.

To investigate the effects of filler materials on surface reactivity of composite microspheres, in this study we synthesized composite microspheres with two different fillers—specifically, calcium hydroxyapatite (HA) and modified bioactive glass (MBG) powders. For each filler material, microspheres with different filler-to-polymer ratios were synthesized. *In vitro* immersion was employed to evaluate and compare the ability of the microspheres to induce the formation of bioactive bone-like apatite on their surfaces. The surface morphology and microstructure of the microspheres after various periods of immersion were evaluated using scanning electron microscopy (SEM) and Fourier transform infrared spectroscopy (FTIR).

MATERIALS AND METHODS

Synthesis of Composite Microspheres

Calcium Phosphate Powders

Calcium hydroxyapatite colloid was synthesized according to procedures described elsewhere.[6] The fine calcium hydroxyapatite powders were collected using a sequence of centrifugation, filtration, and air drying at room temperature. Subsequently, the powders were stored in a desiccator.

Modified Bioactive Glass Particles

Bioactive glass powders with a size less than 20 µm (as measured by sieving) and a nominal composition in weight percent of 45% SiO_2, 24.5% CaO, 24.5% Na_2O, and 6% P_2O_5 were modified as described elsewhere.[7] Briefly, the bioactive glass powders were immersed in simulated physiological fluid (SPF) containing ions to represent the ionic concentrations of plasma (see Radin *et al.*[23]). The bioactive glass powders were incubated in SPF prewarmed to 37°C at a particle-to-solution ratio equal to 2.5 mg/ml for three days. Subsequent to immersion, the particles were collected using centrifugation, filtered, dried at room temperature, and stored in a dessicator. The structure and composition of the surface reaction layer were analyzed using Fourier transform infrared spectroscopy (FTIR: 5DXC, Nicolet, Madison, WI). The FTIR analysis of modified glass powders showed that crystalline calcium phosphate was formed, indicated by the presence of divided bending vibration band of the PO_4 groups (P–O bend) at $570 cm^{-1}$. The C–O bands present at 1,410 and $874 cm^{-1}$ further indicated the formation of carbonated calcium hydroxyapatite.[7]

Composite Microspheres

A solid-in-oil-in-water (s/o/w) emulsion solvent removal method was conceived to form composite microspheres (for details see Qiu *et al.*[7]). Two kinds of bioactive ceramic powders, specifically calcium hydroxyapatite (HA), and modified bioactive glass (MBG), were incorporated into poly(lactic acid) (PLA, MW12845-80928, Alkermes, OH) polymer matrix, at a filler-to-polymer (weight) ratio 1:1, 1:3, 1:9, respectively. The fabrication procedures are illustrated in FIGURE 1. Briefly, 600 mg of poly(lactic acid) (PLA) was dissolved in 5 ml of methylene chloride. Next, 65, 200, or 600 mg of bioactive ceramic powders was mixed with the PLA solution via sonication for 15 min. The mixture was added drop by drop into 200 ml of 0.5% (w/v) polyvinyl alcohol (PVA, 87–89% hydrolyzed, MW = 85,000–146,000, Aldrich, WI) solution. The mixture was vigorously stirred in a 500-ml beaker for four hours at room temperature. Then the microspheres were collected by centrifugation, filtered, washed, dried, and stored in a desiccator. Subsequently, the microspheres were immersed in SPF prewarmed to 37°C at a particle-to-solution ratio equal to 0.5 mg/ml for up to two weeks.

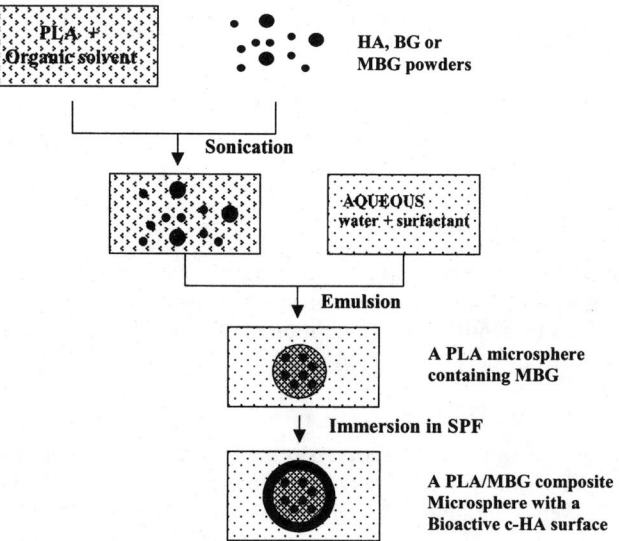

FIGURE 1. Schematic of fabrication procedures of composite microspheres.

Characterization of Microspheres

To evaluate the surface reactivity of the composite microspheres, the microspheres were immersed in SPF at a particle-to-solution ratio equal to 0.5 mg/ml at 37°C for one hour to two weeks. The surface morphology, microstructure, and chemical composition of the microspheres before and after immersion were analyzed. The surface morphology was analyzed using scanning electron microscopy (SEM, JOEL 6300). The microspheres were dried, mounted on metal stubs, and coated with carbon prior to examination. Chemical composition and microstructure of the microspheres was analyzed using energy dispersive x-ray (EDX, Kevex, San Carlos, CA) and Fourier transform infrared spectroscopy. FTIR was performed on powder–KBr mixtures in the diffuse reflectance mode.

The postimmersion and control solutions were analyzed by wet chemical methods. The Si and Ca concentrations were measured using flame atomic absorption spectrophotometry (Perkin-Elmer, model 5100, Norwalk, CT), and the PO4 concentration was measured using a UV-visible light spectrophotometer (molybdenum yellow, 400 nm) (Ultrospec Plus, Pharmacia LKB).

Statistical Analysis

Three samples per composite microspheres were immersed in SPF. All ion concentration measurements were performed in triplicate and expressed as mean ± standard errors.

RESULTS AND DISCUSSION

Two different composite microspheres with filler-to-polymer ratio equal to 1:1 (w:w) were immersed in SPF for one to two weeks to evaluate and compare their surface reactivity. After seven days, a small number of mineral nodules appeared on the surfaces of the microspheres containing HA (see FIGURE 2A). In contrast, a large number of mineral precipitates were present on the surface of MBG-containing microspheres, and the mineral layer covered most of the surface (FIG. 2B). After 14

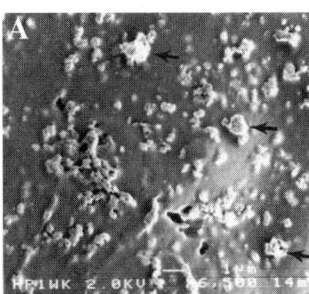 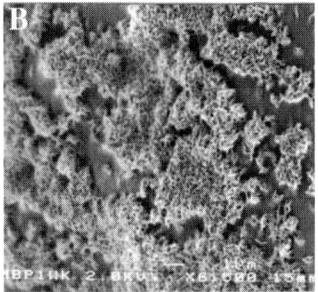

FIGURE 2. SEM micrographs of surfaces of composite microspheres immersed in simulated physiological fluid for seven days. (**A**) Microsphere containing HA. Note the small number of mineral nodules formed on its surface (*arrows*). (**B**) Microspheres containing MBG. Note the surface is mostly covered by mineral deposition.

days, the surfaces of the MBG/PLA microspheres were totally covered by the mineral layer, whereas the PLA microspheres exhibited a limited number of mineral nodules on their surfaces.

Previous studies have shown that calcium phosphate (Ca-P) mineral nodules were formed on the surface of bioactive glass fiber reinforced-polysulfone composites.[24] Both the surface of glass fiber and the surface of the polymer in the regions surrounding the glass fibers exhibited Ca-P deposition after immersion in a simulated body fluid for 10 days. Kokubo et al.[25] demonstrated the formation of an apatite layer on the surface of polymers using a two-step *in vitro* immersion method. The polymers were immersed in a solution containing bioactive SiO_2–CaO glasses and subsequently transferred to a simulated body solution with 1.5 times the ion concentration of SPF, resulting in the formation of an apatite layer on the surfaces of the polymers. In our study, an analysis of the surface morphology using scanning electron microscopy showed that the composite microspheres containing modified bioactive glass powders exhibited more extensive mineral deposition on their surfaces than those containing HA. These results demonstrate that modified glass powders, when used as filler material, can enhance surface reactivity of the composite microsphere. In our study, we have also noticed that composite microspheres containing unmodified bioactive glass powders as fillers develop cracks during synthesis (data not shown). Furthermore, we have noticed that unmodifed bioactive glass significantly increases the pH of SPF solution, whereas modified bioactive glass showed negligible effect on pH.[7] It is also known that the decomposition rate of PLA polymer increases in an acidic or a basic environment. The modification process that leaches out Na^+ ions from the glass can avoid a significant change of local pH on the inside and the outside of the microsphere and, as a result, prevent extensive cracking and fast decomposition of the polymer matrix. Therefore, we conclude that the modification of bioactive glass powders not only enhances the bioactivity, but also avoids the accelerated decomposition of the microsphere.

In the immersion studies, microspheres with various fillers, HA, and MBG, at a filler-to-polymer ratio of 1:1, were immersed in SPF for time periods ranging from one hour to 14 days. The solution chemical analyses showed that all microspheres had similar Ca- and P-uptake profiles (see FIGURE 3). Microspheres containing MBG exhibited cumulative Si release into the solution (see FIGURE 4).

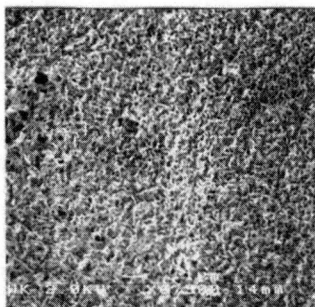

FIGURE 3. SEM micrograph of surface of a composite microsphere containing MBG immersed in simulated physiological fluid for 14 days. Note the extensive mineral deposition on the surface.

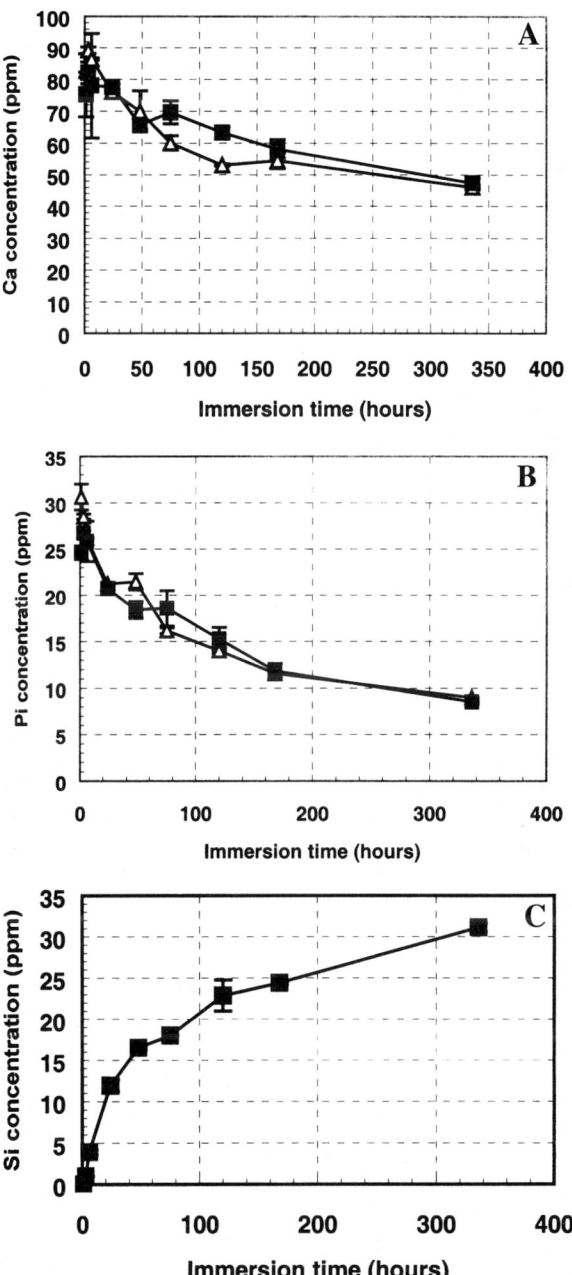

FIGURE 4. Change in (**A**) calcium, (**B**) phosphorus, and (**C**) silicon concentrations of simulated physiological fluid immersed with composite microspheres containing 50% (w:w) calcium hydroxyapatite (△) and MBG (■) powders.

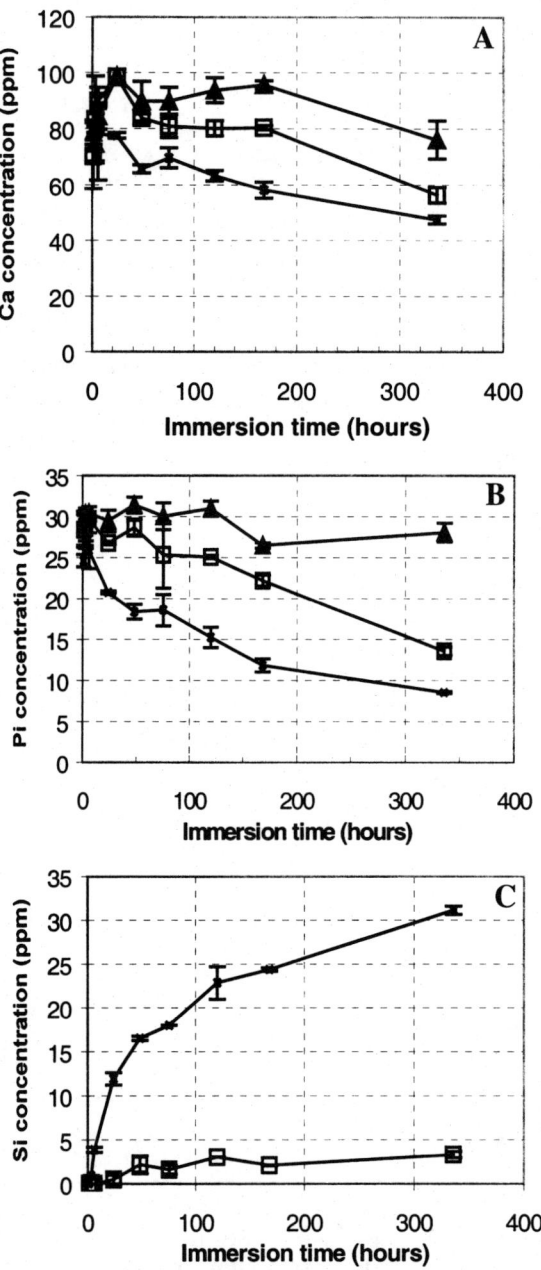

FIGURE 5. Change in (**A**) calcium, (**B**) phosphorus, and (**C**) silicon concentrations of simulated physiological fluid immersed with composite microspheres containing 10% (▲), 25% (□), and 50% (■) MBG powder.

In order to investigate the effects of weight ratios of the MBG filler material on the surface reactivity of the microspheres, composite microspheres with MBG-to-PLA ratios, specifically 1:9, 1:3, and 1:1 (w:w), were immersed in SPF for various time periods ranging from one hour to 14 days. The chemical analyses of the solutions revealed that the greater percentage of MBG the higher Ca and P uptake (see FIGURE 5A and B). Furthermore, the higher percentage of MBG results in a greater amount of Si release into the solution (FIG. 5C).

The similarity in Ca- and P-uptake profiles with various fillers indicates that the amount of Ca-P deposited onto the microspheres is similar. However, the SEM analysis showed that more CaP mineral deposited on the surface of a microsphere containing MBG. This feature can be explained by postulating that the CaP primarily forms on the surfaces of HA powders encapsulated inside the microspheres, whereas for the microspheres containing MBG more deposition occurs on their polymeric surface. A higher Si concentration in the solution with MBG-containing microspheres suggests that the formation of CaP on the polymer surface may have been promoted in this case. This observation is compatible with previous findings by Kokubo et al.[25] These results support the idea that the silicon released from glass particles in solution is critical in enabling the polymers to form a calcium phosphate layer in the subsequent immersion.

Previous studies have shown that the presence of a layer of biologically active apatite, namely, carbonate-containing calcium hydroxyapatite (c-HA), at the implant-bone interface is a common characteristic of bone-bonding bioactive materials.[1] The formation of a c-HA surface layer during immersion in a simulated body fluid (SBF) has been used to predict the bone-bonding ability of ceramics and polymers.[26,27] In this study, incorporation of modified bioactive glass powders was shown to significantly enhance the surface reactivity of composite microspheres. The MBG-containing microspheres show greater bioactivity compared to microspheres containing other filler materials, such as HA and unmodified bioactive glass.

ACKNOWLEDGMENTS

This work was supported by NASA-NRA Grants NAG 8-1483 and NAG 9-817.

REFERENCES

1. DUCHEYNE, P. & J.M. CUCKLER. 1992. Bioactive ceramic prosthetic coatings. [Review]. Clin. Orthop. Rel. Res. **276:** 102–114.
2. DUCHEYNE, P. & Q. QIU. 1999. Bioactive materials: the effects of surface reactivity on bone formation and bone cell function. [Review]. Biomaterials **20**(23/24): 2287–2303.
3. GAO, H., P.S. AYYASWAMY & P. DUCHEYNE. 1997. The dynamics of a microcarrier particle in a rotating wall vessel. Micrograv. Sci. Technol. **X/3:** 154–165.
4. GAO, H., P.S. AYYASWAMY, P. DUCHEYNE & S. RADIN. 2001. Surface transformation of bioactive glass in bioreactors simulating microgravity conditions: Part II. Numeric simulation. Biotechnol. Bioeng. **75**(3): 379–385.
5. QIU, Q., P. DUCHEYNE, H. GAO & P.S. AYYASWAMY. 1998. Formation and differentiation of three-dimensional rat marrow stromal cell culture on microcarriers in a rotating wall vessel. Tissue Eng. **4**(1): 19–35.

6. QIU, Q., P. DUCHEYNE & P. AYYASWAMY. 1999. Fabrication, characterization and evaluation of bioceramic hollow microspheres used as microcarriers for 3-D bone tissue formation in rotating bioreactors. Biomaterials **20**: 989–1001.
7. QIU, Q., P. DUCHEYNE & P. AYYASWAMY. 2000. New bioactive, degradable composite microspheres as tissue engineering substrates. J. Biomed. Mater. Res. **52**(1): 66–76.
8. QIU, Q., P. DUCHEYNE & P.S. AYYASWAMY. 2001. Three-dimensional bone tissue engineering with bioactive microspheres in simulated microgravity. *In Vitro* Cell. Devel. Biol.—Animal **37**: 157–165.
9. RADIN, S., P. DUCHEYNE, P.S. AYYASWAMY & H. GAO. 2001. Surface transformation of bioactive glass in bioreactors simulating microgravity conditions: Part I. Experimental investigation. Biotechnol. Bioeng. **75**(3): 369–378.
10. CROTTS, G. & T.G. PARK. 1998. Protein delivery from poly(lactic-co-glycolic acid) biodegradable microspheres: release kinetics and stability issues. J. Microencaps. **15**: 699–713.
11. JOHANSON, A. & V. NIELSON. 1980. Biosil a new microcarrier. Dev. Biol. Stand. **46**: 125–129.
12. LANGER, R. New methods of drug delivery. 1990. Science **249**: 1527–1533.
13. LEVINE, D.W., D.I.C. WANG & W.G. THILLI. 1979. Optimization of growth surface parameters in microcarrier cell culture. Biotech. Bioeng. **21**: 821–845.
14. MATHIOWITZ, E., W.M. SALTZMAN, A. DOMB, et al. 1988. Polyanhydride microspheres as drug carriers. J. App. Polym. Sci. **35**: 755–774.
15. VAN WEZEL, A.L. 1967. Growth of cell-strains and primary cells on microcarriers in homogeneous culture. Nature **216**: 64–65.
16. HOLLINGER, J.O. 1983. Preliminary report on the osteogenic potential of biodegradable copolymer of polylactide (PLA) and polyglycolide (PGA). J. Biomed. Mater. Res. **17**: 71–82.
17. MILLER, R.A., J.M. BRADY & D.E. CUTRIGHT. 1997. Degradation rates of oral resorbable implants (polylactates and polyglycolates): Rate modification with changes in PLA/PGA copolymer ratios. J. Biomed. Mater. Res. **11**: 711–716.
18. DEVIN, J.E., M.A. ATTAWIA & C.T. LAURENCIN. 1996. Three-dimensional degradable porous polymer-ceramic matrices for use in bone repair. J. Biomater. Sci. Polym. Ed. **7**: 661–699.
19. HIGASHI, S., T. YAMAMURO, T. NAKAMURA, et al. 1986. Polymer-hydroxyapatite composites for biodegradable bone fillers. Biomaterials **7**: 183–187.
20. THOMSON, R.C., et al. 1998. Hydroxyapatite fiber reinforced poly α-hydroxy ester foams for bone regeneration. Biomaterials **19**: 1935–1943.
21. KIKUCHI, M., Y. SUETSUGU, J. TANAKA & M. AKAO. 1997. Preparation and mechanical properties of calcium phosphate/copoly-L-lactide composites. J. Mater. Sci. Mater. Med. **8**: 361–364.
22. VERHEYEN, C.C., J.R. DE WIJIN, C.A. VAN BLITTERSWIJK & K. DE GROOT. 1992. Evaluation of hydroxyapapte/poly (L-lactide) composites: mechanical behavior. J. Biomed. Mater. Res. **26**: 1277–1296.
23. RADIN, S.R. & P. DUCHEYNE. 1993. The effect of calcium phosphate ceramic composition and structure on *in vitro* behavior. II. Precipitation. J. Biomed. Mater. Res. **27**: 35–45.
24. MARCOLONGO, M., P. DUCHEYNE & W.C. LACOURSE. 1997. Surface reaction layer formation *in vitro* on a bioactive glass fiber/polymeric composite. J. Biomed. Mater. Res. **37**: 440–448.
25. KOKUBO, T., M. TANAHASHI, T. YAO, et al. 1993. Apatite-polymer composites prepared by biomimetic process: Improvement of adhesion of apatite to polymer by glow-discharge treatment. *In* Bioceramics, Vol. 6. P. Ducheyne & D. Christiansen, Eds.: 327–332. Butterworth-Heinemann Ltd., Oxford.
26. LI, P., D. BAKKER & C.A. VAN BLITTERSWIJK. 1997. A bone-bonding polymer (Polyactive(R) 80/20) induces hydroxycarbonate apatite formation *in vitro*. J. Biomed. Mater. Res. **34**: 79–86.
27. OHTSUKI, C., T. KOKUBO & T. YAMAMURO. 1992. Mechanism of apatite formation on $CaO-SiO_2-P_2O_5$ glasses in a simulated body-fluid. J. Non-Cryst. Solids **143**: 84–92.

Vapor Transport Growth of Organic Solids in Microgravity and Unit Gravity

Some Comparisons and Results to Date

MARIA ITTU ZUGRAV,[a] WILLIAM E. CARSWELL,[b] GLEN B. HAULENBEEK,[b] MOHAN SANGHADASA,[c] SUE K. O'BRIEN,[d] BOGDAN C. GHITA,[e] AND WILLIAM E. GATHINGS[f]

[a]*Center for Microgravity and Materials Research, University of Alabama in Huntsville, Huntsville, Alabama, USA*

[b]*Alliance for Microgravity Materials Science and Applications, University of Alabama in Huntsville, Huntsville, Alabama, USA*

[c]*Department of Physics, University of Alabama in Huntsville, Huntsville, Alabama, USA*

[d]*Center for Automation and Robotics, University of Alabama in Huntsville, Huntsville, Alabama, USA*

[e]*Department of Mechanical and Aerospace Engineering, University of Alabama in Huntsville, Huntsville, Alabama, USA*

[f]*Consortium for Materials Development in Space, University of Alabama in Huntsville, Huntsville, Alabama, USA*

ABSTRACT: Thin films of an organic nonlinear optical (NLO) material, N,N-dimethyl-*p*-(2,2-dicyanovinyl) aniline (DCVA), have been grown in space and on the ground by physical vapor transport in an effusive ampoule arrangement. The thin film growth technique developed on the ground is a direct result of information gleaned from experiments in microgravity. This paper covers the results of our experimental investigations for establishing "ideal" terrestrial conditions for deposition of a DCVA film. The active control during the deposition process was exercised by three deposition variables: the material source temperature, the background pressure external to the growth ampoule and the substrate temperature. Successful growth occurred when the difference in temperature between the source material and the copper substrate was 14°C and the background nitrogen pressure was such that the transport was either diffusive or convective. A qualitative diffusion limited boundary was estimated to occur at a pressure of approximately 20 torr. We have probed the DCVA thin films with visible-near infrared reflection absorption spectroscopy, polarized Fourier transform infrared spectrometry, differential interference contrast optical microscopy, and stylus profilometry.

KEYWORDS: vapor transport; organic solids; microgravity; nonlinear optical; thin films

Address for correspondence: Dr. Maria Ittu Zugrav, 7246 Marlow Place, University Park, FL 34201, USA.
zugravm@email.uah.edu

INTRODUCTION

Vapor transport process is at the heart of ongoing efforts to develop the nanoelectronics, including computer memory, and to create functional films containing single or multilayer structures of different materials. These efforts are closely related to those devoted to understanding and demonstration of all-optical devices, where the optical-electrical-optical conversion is eliminated. The carrier is all optical from the beginning to the end and nonlinear optical processors carry out the control functions. As Assanto et al. point out,[1] progress in materials with a phase-matchable second-order susceptibility has stimulated a novel approach to all-optical integrated processors based on quadratic nonlinearities. Progress with the new approach is strongly dependent on the availability of materials. It is well known that the inorganic crystals are limited by their low nonlinearities. The use of organic crystals with their great potential for higher nonlinearities may provide the key to progress.[2-4] The organic material used in our experiments is N, N-dimethyl-p-(2,2-dicyanovinyl) aniline (DCVA), a donor/acceptor-substituted aromatic compound, with a highly polarizable dissymmetric π-electron cloud induced by the donor–acceptor interactions across the conjugated system. Thin films of DCVA have been grown in space and on the ground and the results pertaining to vapor transport growth may be found in References 5–8. In this paper the comparison of the microgravity and unit gravity grown thin films is extended to include characterization by visible-near infrared reflection absorption spectroscopy, polarized Fourier transform infrared (FTIR) spectrometry, differential interference contrast (DIC) optical microscopy, and stylus profilometry.

EXPERIMENTAL DETAILS

A versatile flight-qualified moderate temperature facility (MTF) has been developed for materials processing in space.[9] The MTF is capable of growing crystals and thin films on Earth and in space on a wide variety of carriers, including the International Space Station. Four successful flights on the US Space Shuttles have demonstrated the MTF materials processing capability. Two growth cells on the consortium complex autonomous payload (CONCAP) IV-02 on STS-59 and eight growth cells on CONCAP IV-03, STS-69, were devoted to the vapor growth of DCVA in space. The growth parameters of the two DCVA cells of CONCAP IV-02 were set to reproduce laboratory experiments that yielded small, bulk crystals. The results, however, were surprising. The space experiment yielded DCVA thin films rather than bulk crystals. Repeated laboratory tests continued to grow bulk crystals. The next space shuttle experiment, CONCAP IV-03, was therefore tailored to focus specifically on DCVA in an attempt to understand the dilemma. Several growth parameters were changed and each cell, in total eight, had a different set of growth parameters in an attempt to study their influences on the thin film growth process. All eight yielded thin films in microgravity and, in summary, the film growth process was found to be very robust.

An evaluation of the flight and ground experiments data revealed a slower heat-up rate than was typical of the laboratory experiments. No background pressure measurements were made in space, but the temperature profile is an indication of a much higher background nitrogen pressure than the millitorr range used in the laboratory. This is due to the presence of a 60-micron effluence filter and a small quantity of molecular sieve material placed in our vacuum line, as required by the National Aeronautics and Space Administration (NASA) for safety. This changes in an uncontrolled way the background nitrogen pressure, the actual internal growth cell gas pressure, and also affects the temperature profile.

In previous papers[5–8] more introductory information is presented on the MTF experiments and the conditions of the microgravity and unit gravity experiments. The experiments have used effusive ampoule physical vapor transport (EAPVT) as a process to deposit organic thin films of DCVA on a variety of substrates. In this technique, extensively described in References 10 and 11, incongruent or impurity vapor components are continuously removed from the vicinity of the growing crystal through a calibrated leak to vacuum. Once the diffusion transport barriers are removed or sufficiently reduced, even for small temperature difference, ΔT, between source and crystal, the transport rate may be too high for interfacial attachment to occur in a manner that assures single crystal growth with low defect concentrations. This can be counteracted by either lowering the temperatures of both source and crystal (thus, lowering the vapor pressure and transport flux), and/or by a flow-restriction between the source and crystal.

GROUND (UNIT GRAVITY) CONTROL EXPERIMENTS

Ideally, the EAPVT parameters of the ground control experiments should exactly duplicate those of the flight experiments. In practice, this is a difficult task. Exactly reproducing all none-gravity EAPVT parameters will be very difficult because some of them are completely different (e.g., temperature profile) and others could not be measured (for instance the background pressure external to the growth cell). The active control during the deposition process was exercised by three deposition variables: the source material temperature, the background pressure external to the growth ampoule and the substrate temperature. Successful growth occurred when the difference in temperature between the source material and the copper substrate was 14°C.

The substrate temperature was determined to have a significant effect on film quality. It plays a major role, along with supersaturation, in determining the surface texture of the grown films. The surface kinetics, which are governed by surface temperature, govern the relative roles played by island and layer growth, and can even transition the system to a polycrystalline growth process. Specifically, our results show that at substrate temperatures around 90°C the growth process no longer produces single crystal films, but results instead in polycrystalline deposition.

Laboratory tests with ground control hardware showed efficient, if somewhat moderately slower, evacuation with the 60-μm filter and molecular sieve material in place. Therefore, initial laboratory experiments were run with low nitrogen pressure conditions, on the order of millitorrs. It is under these conditions that bulk crystals form. Eventually, however, it was realized that an unique condition existed on orbit that was not reproduced by ground control experiments. During the space shuttle launches, vibrations caused some of the molecular sieve material to powderize. This did not happen in the laboratory experiments. Furthermore, in the weightless environment the powder, small in quantity though it was, did not settle to the "bottom" of the evacuation tube but was swept into the 60-μm filter by the vacating gasses, effectively clogging the filter. This would not have happened in the laboratory even if the powder had been formed. Thus, the evacuation rate in the space experiment was much, much less than the evacuation rate in the ground control experiment. Once these conditions became apparent it was decided to examine the opposite extreme to vacuum conditions and laboratory growth runs were carried out at higher pressures including full atmospheric nitrogen pressure. Experiments at varying pressures and ΔT values appear to indicate that there is a transition region at which growth changes from thin films to bulk crystals. One particular experiment dramatically made this point at the pressure of 2×10^{-1} torr.

CONSIDERATION OF TRANSPORT REGIMES

Preliminary calculations of the Knudsen number in the growth cell (i.e., the ratio of the mean free path to a characteristic growth cell dimension) estimate that at pressure levels below about 10^{-1} torr the flow is molecular and the molecule–wall collisions dominate. Vaporization is intense and the DCVA molecules pass through the vacuum in straight lines until they strike a surface; that is, the walls of the growth cell or the substrate on which the film is to be deposited. It is remarkable to notice that, in one of our ground experiments, the pressure level of 2×10^{-1} torr was the primary contributing factor in changing the growth form from thin film to bulk crystal.

For experiments carried out at atmospheric nitrogen pressure, the Prandtl number (Pr), (i.e., the ratio of the kinematic viscosity, ν, to the thermal diffusivity, α), which is independent of geometry, is 0.71.[12] One finds that $Pr \approx 0.7$ is rather representative for gases over a wide range of temperatures. Because the values of ν and α in gases are quite close to each other, the thermal and momentum boundary layers are of comparable width. However, both boundaries extend quite far into the fluid and the buoyancy driven effects extend over a longer distance from the wall.

The Rayleigh number (Ra) describes the interplay between buoyancy generating motion and thermal and viscous diffusive effects dissipating it. For the geometry used in our system one can define $Ra = \beta g \Delta T L^3 / \nu \alpha$ where $\beta = 1/T$, in which β is the thermal expansion coefficient, g is the gravitational acceleration, ΔT is the temperature difference between source and substrate, L is the length (1.27 cm) of the transport tube, ν is the kinematic viscosity, and $\alpha = \nu/Pr$ is the thermal diffusivity. The thermophysical values for the evaluation of Ra were: $\beta = 3.4 \times 10^{-3} K^{-1}$, $g = 980 \text{cm/sec}^{-2}$, $\Delta T = 14.0 K$, $\nu = 0.156 \text{cm}^2 \cdot \text{sec}^{-1}$ (from Ref. 12), $\alpha = 0.218 \text{cm}^2 \cdot \text{sec}^{-1}$.

The balance occurs when Ra is equal to unity. Below unity, the diffusive effects are dominant. Our calculations for nitrogen, at atmospheric pressure, give the value $Ra \approx 2,820$. Knowing that ν and α are functions of density and, implicitly Ra is a function of the density, we can calculate the nitrogen density required to reduce Ra to unity. Using the relation $Ra \rho_0^2 = Ra_0 \rho^2$, where ρ is the density and subscript 0 indicates the values at the atmospheric pressure, we obtain $\rho = 0.019 \rho_0$. If we substitute this value in the equation of state for an ideal gas we obtain, for an average $T = 412 K$, the value of about 20 torr for the nitrogen pressure. At approximately this pressure and below, the diffusive effects start to dominate. Consequently, the partial pressure gradient of the DCVA between the source and the growing interface is approximately linear; the rate-limiting step is the mass diffusion and the transport rate is less than in the molecular flow regime. The partial pressure of the DCVA at the source is determined by the source temperature whereas at the growing interface by mass transport and substrate temperature.

During the growth experiments the nitrogen background pressure is kept under control at a fixed value by an electromagnetically operated mini valve (Pfeiffer Vacuum GmbG, Germany). The control electronics for the valve allows two modes of control, automated and manual. The automated mode utilizes a relay circuit in the pressure measurement system. A pressure sensor is located between each valve and its corresponding oven (each oven consists of two independently controlled growth cells). When the pressure is too high, the relay is closed sending a signal to the TattleTale 5F controller, which controls the growth cell. Every minute, the TattleTale 5F reports experimental data to a computer via an RS-232 connection. At that time, if the signal from the relay is detected, the valve is powered to open using a transistor circuit. The length of time that the valve is open each minute is preset in the computer program and is typically five seconds. The manual mode uses a switch to open and close the valve. We found advantageous to place the thin film growth system in the diffusive or convective regime, at higher nitrogen pressure, to repress and control the nucleation. The transport due to convection may annul the DCVA partial pressure gradient across the transport path except at the source and growth interface.

SUBSTRATE AND SOURCE MATERIAL VARIATION

In addition to our ability to grow DCVA thin films on copper substrates (see FIGURE 1), we have defined the parameters for depositing DCVA on glass substrates (see FIGURE 2). In order to achieve high reflectivity, device quality glass substrates were coated with a sandwich of 500 Å indium tin oxide (ITO) and 25 Å top layer of

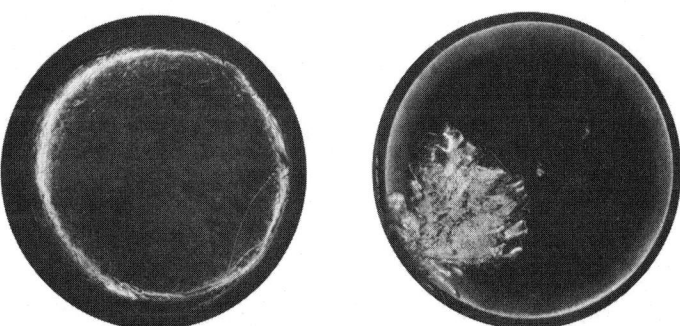

FIGURE 1. DCVA thin film deposited on copper substrates. **Left panel** shows the 1.4 μm thick film grown in the laboratory. **Right panel** shows the 3.7 μm thick film grown in space. The thinner film allows the surface relief of the copper substrate to be seen. The space film has a polycrystalline clump (lower left quadrant) grown on top of it.

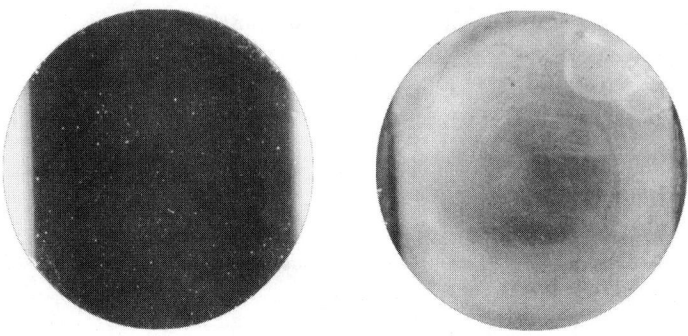

FIGURE 2. DCVA deposited on a device quality glass substrate. **Left panel** shows the substrate prior to thin film growth. The **right panel** is the same substrate after DCVA deposition.

copper. The capability has been demonstrated of depositing films on glass coated with ITO, an industrially relevant substrate, and copper-coated glass.

This is remarkable evidence in support of the "Van der Waals" nature[14–16] of the DCVA thin film epitaxial growth process. In such a process, the Van der Waals bonded DCVA compound can be deposited on a variety of substrates in crystalline thin film form even though lattice matching is not achieved. The binding to the substrate provides a preferred orientational alignment between the thin film and the substrate lattices. The Van der Waals forces are weak and short-range in their effect. Therefore, the shear stress at the interface with the substrate is not sufficient to induce a high density of defects in the first lattice mismatched monolayer. Hence, subsequent layers are grown in their relaxed, bulk structural configuration.

The EAPVT technique was recently successfully applied to the growth of thin films of organometallic compounds such as triethylphosphine sulfide (TPS) and 8-hydroxyquinoline aluminum salt (AlQ3). TPS is a potential NLO candidate. The crystal structure is hexagonal and polar with P=S bonds lined up along the unique axis, perfect for optimum tensor summation to give the maximum electrooptic effect. AlQ3 is currently used for making organic light emitting diodes (OLED) on glass substrates coated with ITO. This improved knowledge of the applicability of the EAPVT technique to a new class of compounds, such as organometallic, will open the way for new molecular devices for the electronics industry. The new challenge in the field of organometallic compound is the development of routes to structures that combine organometallic donor and acceptor groups, as in the DCVA, to make use of the great power of transition metal complexes to stabilize positive charges. These organometallic dipoles have nonlinear optical properties.

FILM CHARACTERIZATION TECHNIQUES

We have probed the DCVA thin films with visible-near infrared reflection absorption spectroscopy, polarized Fourier transform infrared spectrometry, differential interference contrast optical microscopy, and stylus profilometry. Detailed comparisons have been made between the microgravity grown and unit gravity (laboratory) grown samples. DCVA thin films have been successfully deposited on both copper and glass substrates on the ground, however, the thin films obtained from experiments on the Space Shuttle were on copper substrates only. Therefore, Earth grown films on copper substrates only were used for comparative testing.

REFLECTION ABSORPTION SPECTROSCOPY

The total and diffuse reflectance measurements were made using a Perkin Elmer Lambda 900 spectrophotometer. Measurements were made from ultraviolet (UV) to near infrared (IR) at the wavelength range from 330 nm to 2,300 nm. The sample was placed at the near normal incidence with an angle of incidence of 8° so that the data were collected with or without specular reflection. Both ground and space grown DCVA samples show a gradual decrease in total and diffuse reflectance toward the shorter wavelengths in the visible region (see FIGURES 3 and 4). However, the reflectance has become almost constant in the IR region.

FIGURES 5 and 6 illustrate the comparative total and diffuse reflectance spectra of the substrate alone and the two DCVA samples. In the infrared region, the total and diffuse reflectance of both the ground and space grown DCVA thin films are greater than the reflectance observed for the copper substrate. The comparison of diffuse and total reflection data indicates that, in the infrared region, the specular reflection of the ground deposited DCVA thin films is stronger than that of the space produced sample. However, the spectral behavior of the two samples is not significantly different in the visible region. The level of specular reflection is directly proportional to the overall quality of organic thin films. The stronger specular reflection observed for the ground grown thin films implies enhanced smoothness and better homogeneity. In summary, these results indicate that ground produced DCVA thin films may be better than space produced thin films. This statement has full validity only when the compared films have the same thickness.

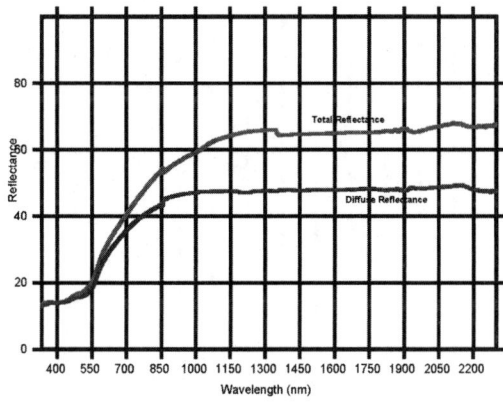

FIGURE 3. Total and diffuse reflectance spectra of the ground grown DCVA thin film.

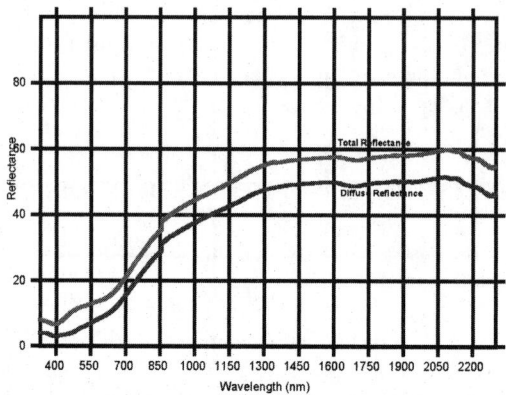

FIGURE 4. Total and diffuse reflectance spectra of the space grown DCVA thin film.

POLARIZED FOURIER TRANSFORM INFRARED (FTIR) REFLECTION ABSORPTION SPECTROMETRY

In an FTIR spectrum an absorption peak is most intense when the plane of oscillation of the infrared radiation is parallel to the direction of the corresponding chemical bond or the motion of the corresponding functional group that may exist in the molecules. Therefore, the relative comparison of the absorption peaks obtained using linearly polarized infrared light provides some insight into the relative orientation of bonds and groups in a molecule.

The FTIR spectrometer used in this work was BioRad Model FTS-60A with a ceramic source and a fast TRS wideband MCT mid IR detector (cooled with liquid nitrogen). Before the measurements were made, the instrument was calibrated using the built-in calibration unit. Samples were mounted on a three-dimensional translation stage so that their position could be adjusted precisely.

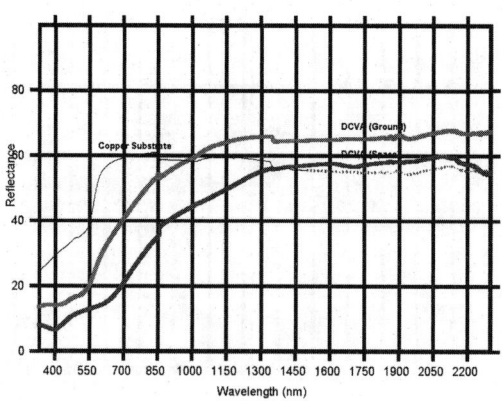

FIGURE 5. Total reflectance spectra of copper, DCVA ground, and space thin films.

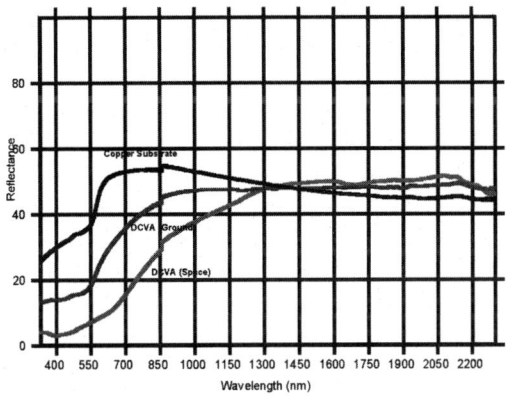

FIGURE 6. Diffuse reflectance spectra of copper, DCVA ground, and space thin films.

The reference spectrum was obtained with the copper substrate. A mid infrared range grid polarizer (Molectron) was used to polarize the incident infrared light either parallel to or perpendicular to the plane of incidence. This could be easily achieved just by rotating the polarizer without disturbing the experimental setup. However, when the polarizer was rotated, the intensity of both input and output infrared radiation were affected by the depolarization or the preference to a specific state of polarization by the reflective and refractive optical elements in the spectrometer.

FIGURES 7 and 8 show the FTIR reflection absorption spectra obtained for the space grown sample using horizontally and vertically polarized light, respectively. Among the many peaks in the spectra, the pronounced peak around 2,215–2,219 cm^{-1} is identified to correspond to the –CN groups. The small peak at 2,809.7 cm^{-1} may correspond to the N-(CH$_3$)$_2$ group. The numerous peaks below

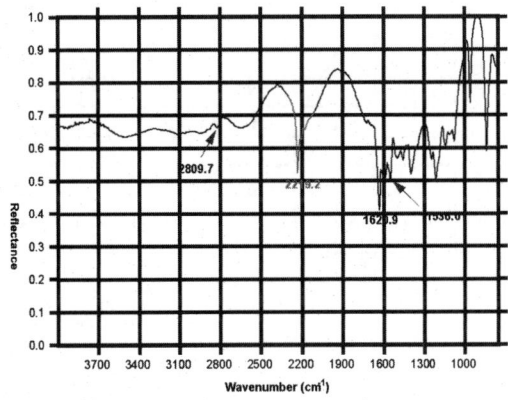

FIGURE 7. FTIR reflectance spectrum of space samples using horizontally polarized light.

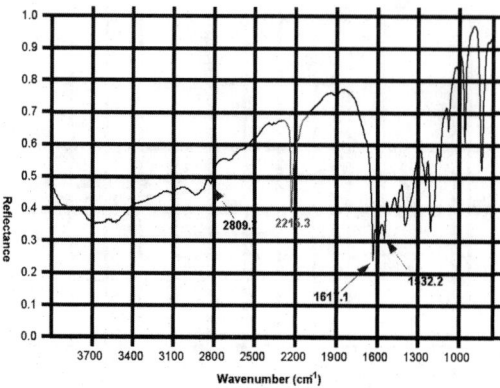

FIGURE 8. FTIR reflectance spectrum of space samples using vertically polarized light.

$1,000\,cm^{-1}$ seem to correspond to the benzene ring. The pronounced peak around $1,617$–$1,621\,cm^{-1}$ may correspond to the C=C bond in the molecule.

FIGURES 9 and 10 show the FTIR reflection absorption spectra obtained for the ground grown sample using horizontally and vertically polarized light, respectively. The peaks corresponding to the various bonds and groups of DCVA molecules in the sample are not as strong as those observed with the space grown sample. The reason for the weak peaks is due to the smaller thickness of the ground grown sample. However, some of the peaks could be identified. The peak corresponding to the –CN groups is located around $2,207$–$2,212\,cm^{-1}$.

In order to accurately and unequivocally identify the molecular species present after thin film deposition, the spectrum of DCVA was obtained in bulk form by supporting a small amount of the material in a KBr matrix. The resulting spectrum

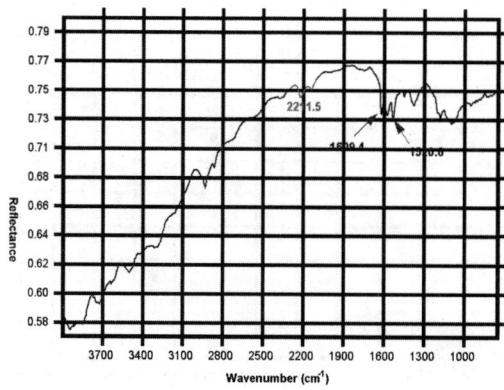

FIGURE 9. FTIR reflectance spectrum of ground samples using horizontally polarized light.

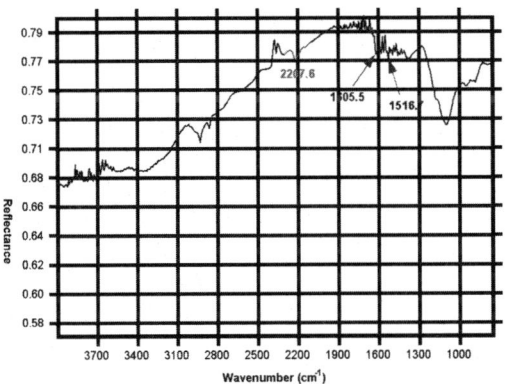

FIGURE 10. FTIR reflectance spectrum of the ground sample using vertically polarized light.

is shown in FIGURE 11 and represents a randomized sample, in contrast to FIGURES 7–10. The stretching of the C≡N (cyano) bond causes the strong peak at 2,210cm^{-1}. For a saturated nitrile, the C≡N stretch appears from 2,260 to 2,240cm^{-1}.[13] For an aromatic nitrile, as in the DCVA, the C≡N stretch appears from 2,240 to 2,210cm^{-1}. It is lower than in saturated nitriles because of conjugation.

The characteristic vibration bands due to the cyano groups are clearly observed in FIGURES 7–10, in good agreement with the transmission spectrum (FIG. 11) of the same material in a KBr matrix. The small shifts in the wavenumber observed from one spectrum to another and the different intensities may be due to the Fermi resonance.[13] In Fermi resonance, two vibrations of similar wavenumber and proper symmetry repel each other, appearing at wavenumbers above and bellow where they are normally expected. In addition, the band positions of the two vibrations are altered, and the intensity of the bond stretch is less than would be expected.

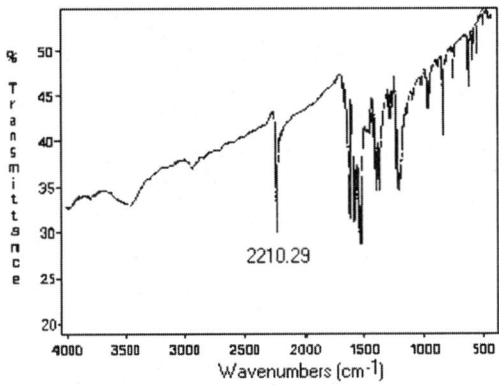

FIGURE 11. FTIR transmission spectrum of bulk DCVA prepared in a KBr matrix.

OTHER POSSIBLE EXPLANATIONS FOR THE CHANGES IN PEAK INTENSITIES IN INFRARED SPECTRA

An additional factor that determines the peak intensities (heights) in IR spectra is the concentration of the molecules in the sample. The equation that relates concentration to absorbance is Beer's law, $A = \varepsilon l c$, where A is the absorbance, ε is the absorptivity, l is the path length, and c is the concentration. The absorbance is measured as a peak height in the IR spectrum. The path length is typically on the order of microns for solids. The absorptivity, ε, is the proportionality constant between concentration and absorbance. For a given molecule and wavenumber, the absorptivity is a fundamental physical property of the molecule. It appears, from the recorded IR spectra, that the concentration is at its maximum in the bulk DCVA material (FIG. 11) and at its minimum in the ground grown DCVA thin film (FIGS. 9 and 10). The difference in the peak intensities between space and ground grown thin films is consistent with the difference of their thickness.

A comparison of each pair of figures corresponding to the two different states of polarization shows many different variations in the spectra, indicating strong polarization dependence. The difference is larger in the spectra obtained using the ground grown thin sample. Therefore, investigation of the alignment of –CN group in the molecules relative to the direction of the electric field is complicated. If the sample is placed in the yz plane, the electric field of the light polarized perpendicular to the plane of incidence (xz plane) is parallel to the surface of the sample (see FIGURE 12A). However, when the light is polarized parallel to the plane of incidence, one component of the electric field is parallel to the sample and the other component is perpendicular to the sample (FIG. 12B). The component perpendicular to the sample becomes stronger when the angle of incidence increases. However, in the experiment, the spectrometer allowed only for a small angle of incidence to be used. Therefore, the electric field perpendicular to the sample was either zero or had a very small value.

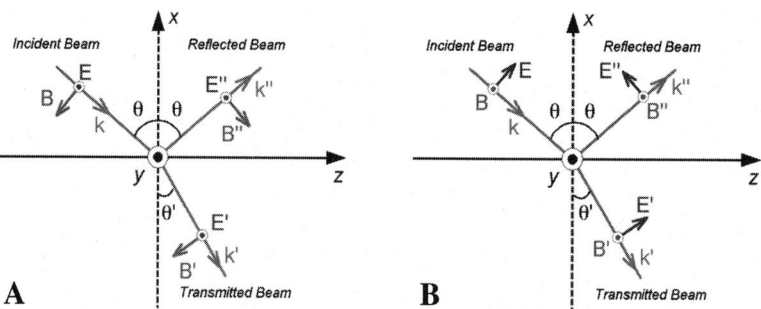

FIGURE 12. The reflection and transmission at the interface (yz plane) with light polarized (**A**) perpendicular to and (**B**) parallel to the plane of incidence (xz plane). B is the electromagnetic field vector, E is the electric field vector, K is the wave vector, θ is the angle of incidence, and θ^1 is the angle of refraction.

DIFFERENTIAL INTERFERENCE CONTRAST (DIC) MICROSCOPY

The surface relief of the following samples were observed using an optical microscope (Olympus Model BX60 F5) in both bright field (BF) and DIC configurations: (1) space grown DCVA thin film on copper substrate, (2) ground grown DCVA thin film on copper substrate, (3) ground grown DCVA thin film on ITO coated glass substrate, and (4) bare copper substrate. Images were optically magnified using objective lenses with various magnifications between ×5 and ×100. They were further magnified digitally by a color CCD camera (Sony Model CCD 1R1S). The camera was connected to a computer to observe, capture, and save the images in files with a special graphics format. Because of the small working distance in the microscope and the large thickness (cylinder with 1.2 cm diameter and 2.3 cm height) of the copper substrate (used as a temperature controlled heater), samples (1), (2), and (4) could not be directly mounted on the x–y translational stage assembly attached to the microscope. Therefore, the samples were mounted on a custom made holder and attached to the movable parts of the translational stage. At higher optical magnifications, the objective lens was located very close to the sample surface, which had a slight concave shape in sample (2). Therefore, the maximum optical magnification was limited to ×50 in many cases in order to protect the surface of the sample from any possible damages. However, in the case of sample (3), no special holder was necessary and images were observed even at the maximum optical magnification of ×100.

In the BF mode, the intensity of the illumination was adjusted to achieve the best contrast of the images. In the DIC mode, the contrast of the sample surface located at the various levels could also be adjusted by tuning the interference pattern observed. Clear images of sample (2) could not be observed because of its very small thickness. The surface characteristics of copper were dominant in the images observed. Illumination of the sample at large angles of incidence could have allowed the light to be reflected from the surface of the film and the characteristic of the surface could have been visible. However, due to the closeness of the objective lens and the sample, such an illumination was not possible.

The smooth and flat, mirror-like surface of the space grown thin film is pictured in FIGURE 13. The color is dark red and it becomes lighter near the edges allowing

FIGURE 13. Surface relief of a DCVA film of 3.7 μm thickness and 78 μm × 58 μm area on copper substrate.

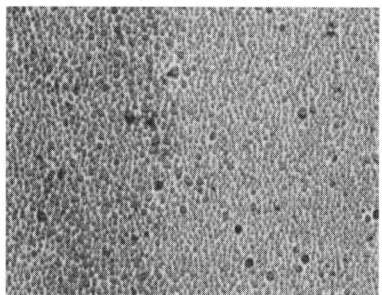 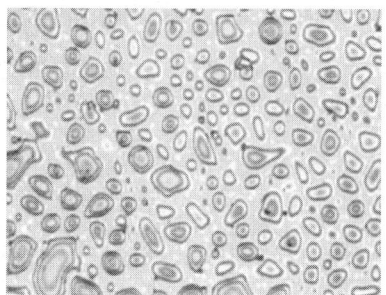

FIGURE 14. A set of DIC images showing the small and large islands on the glass substrate. **Left image** area is 0.5 mm × 0.3 mm. **Right image** area is 78 μm × 58 μm.

the eyes to see the pinkish hue of the underlying copper substrate. The $78 \times 58 \mu m^2$ area imaged in FIGURE 13 is featureless. In contrast, crater like features and porous structure were found in the film deposited on the edge of the substrate. The DIC micrograph images of FIGURE 14 appear to be the island growth of the thin film on an ITO coated glass substrate. As more DCVA is deposited, the islands eventually coalesce and form a continuous, hole-free film. The liquid-like behavior of the growing film is obvious. Pashley[17,18] and Vook[19–21] gave a semi-quantitative explanation for this phenomenon in terms of the surface diffusion of molecules. The driving force for the mass transfer is the reduction in surface energy. The corners of initially well defined, crystallographically shaped islands are the most effective sources of fresh mobile molecules, so they round off rapidly. What we see in FIGURE 14 is a reduction in area and an increase in height of the compound (the largest) islands, to lower their total surface energy. The secondary islands grow until they touch a neighbor and, if this happens to be a much larger island, they coalesce and become completely incorporated in the large island. In order to minimize the elastic strain at each interface, the islands have joined so that a void remains in the region between them. The event is characterized by a decrease in total area of the islands grown on top of the initial film layers, as in the Stranski–Krastanov[17–19] mode of thin film growth, or on the substrate, as in Volmer–Weber[17–19] mode of thin film growth.

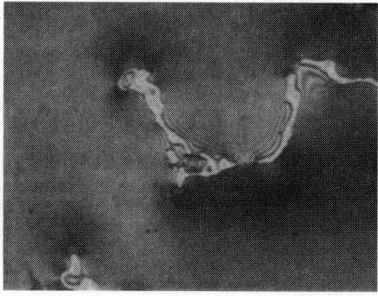

FIGURE 15. Layer by layer growth of an island. Image area is 10 μm × 7.5 μm.

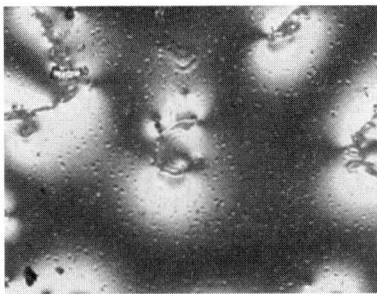

FIGURE 16. Grain boundaries observed at the edges of large double positioned crystallites. Image area is 10 μm × 7.5 μm.

The DCVA films grow by the coalescence of the three dimensional islands and each island grows, layer by layer, as shown in FIGURE 15. When the islands coalesce to form a continuous film, their structures are either in phase, and consequently large crystallites are formed, or they are out of phase and defect structures are formed. The most common defects observed are grain boundaries. As shown in FIGURE 16, the large crystallites are rotated by 180° with respect to each other (double positioning). Consequently, they are out of phase and the grain boundaries are formed around the edges.

STYLUS PROFILOMETRY

Stylus profilometry was used to measure the thickness of the film deposited on the substrate. This technique has the unique ability to measure a range of dimensional changes from 1 mm to 1 nm. A Taylor–Hobson Talystep instrument was used for profilometer measurements. The deposited thin film could easily be scratched down to the substrate interface with a fine tungsten wire.

CONCLUSIONS

The work presented in this paper is a prime example of a novel thin film technology developed on the ground as a direct result of information gleaned from experiments in microgravity. The key to success was analyzing and understanding the processes occurring in the EAPVT system: vapor transport and thin film growth. The reason for choosing a higher nitrogen background pressure is to repress the nucleation by placing the system in the diffusive or convective regime. We have idealized the system in our calculations. Therefore, the 20 torr diffusion–convection boundary should be regarded as qualitative rather than absolute. We have recently used similar conditions to grow thin films of triethylphosphine sulfide, and hydroxyquinoline aluminum salt.

ACKNOWLEDGMENTS

Support of this work by the National Aeronautics and Space Administration under Grants NAGW-812 and NAG8-1456 is gratefully acknowledged.

REFERENCES

1. ASSANTO, G., G.I. STEGEMAN & R. SCHIEK. 1998. Thin Solid Films **331**: 291–297.
2. NICOUD, J.F. & R.J. TWIEG. 1987. Nonlinear Optical Properties of Organic Molecules and Crystals. D.S. Chemla & J. Zyss, Eds.: Academic Press, Orlando.
3. MEREDITH, G.R. 1986. Nonlinear Optics: Materials and Devices. C. Flytzanis & J.L. Oudar, Eds.: Springer-Verlag, Berlin.
4. KONDO, T. & R. ITO. 1994. Molecular Nonlinear Optics, Materials, Physics, and Devices. J. Zyss, Ed.: Academic Press, Boston.
5. ITTU ZUGRAV, M., F.C. WESSLING, W.E. CARSWELL, et al. 1997. J. Crystal Growth **174**: 130–138.
6. ITTU ZUGRAV, M., W.E. CARSWELL, C.A. LUNDQUIST, et al. 1997. Materials Research in Low Gravity, SPIE **3123**: 110–125.
7. CARSWELL, W.E., M. ITTU ZUGRAV, G. HAULENBEEK, et al. 2000. J. Crystal Growth **211**: 428–433.
8. ITTU ZUGRAV, M., W.E. CARSWELL & F.C. WESSLING. 2000. Ground and space processing of nonlinear optical organic materials. In Recent Research Developments in Crystal Growth. K. Benz, D. Schwabe & I. Higashi, Eds.: 123–140. Transworld Research Network, Trivandrum-8, India.
9. WESSLING, F.C. & W.E. CARSWELL. 1995. Microgravity Sci. Technol. VIII/3: 196–201.
10. ITTU ZUGRAV, M. & F. ROSENBERGER. 1997. Materials Research in Low Gravity, SPIE **3123**: 135–143.
11. ITTU ZUGRAV, M. & F. ROSENBERGER. 1996. Space Processing of Materials, SPIE **2809**: 271–276.
12. ECKERT, E.R.G. & R.M. DRAKE. 1987. Analysis of Heat and Mass Transfer. McGraw-Hill, New York.
13. SMITH, B.C. 1998. Infrared Spectral Interpretation—A Systematic Approach. CRC Press, Washington, D.C.
14. FORREST, S.R. & Y. ZHANG. 1994. Phys. Rev. **B49**: 11297–11308.
15. FORREST, S.R., P.E. BURROWS, E.I. HASKAL, et al. 1994. Physical Review **B49**: 11309–11321.
16. FENTER, P., P.E. BURROWS, P. EISENBERGER, et al. 1995. J. Crystal Growth **152**: 65–72.
17. PASHLEY, D.W., M.J. STOWELL, M.H. JACOBS, et al. 1964. Philos. Mag. **10**: 127–158.
18. PASHLEY, D.W. 1965. Adv. Phys. **14**: 327–416.
19. VOOK, R.W. 1982. Intl. Metals Rev. **27**: 209–245.
20. VOOK, R.W. 1984. Optical Engin. **23**: 343–348.
21. VOOK, R.W. 1988. Mat. Res. Symp. Proc. **103**: 3–11.

Sliding-Cavity Fluid Contactors in Low-Gravity Fluids, Materials, and Biotechnology Research

PAUL TODD,[a] JOHN C. VELLINGER,[a] SHRAMIK SENGUPTA,[b] MICHAEL G. SPORTIELLO,[c] ALAN R. GREENBERG,[c] AND WILLIAM B. KRANTZ[d]

[a]*Space Hardware Optimization Technology, Inc., Greenville, Indiana, USA*

[b]*University of Minnesota, Minneapolis, Minnesota, USA*

[c]*University of Colorado, Boulder, Colorado, USA*

[d]*University of Cincinnati, Cincinnati, Indiana, USA*

ABSTRACT: The well-known method of sliding-cavity fluid contactors used by Gosting for diffusion measurements and by Tiselius in electrophoresis has found considerable use in low-gravity research. To date, sliding-cavity contactors have been used in liquid diffusion experiments, interfacial transport experiments, biomolecular crystal growth, biphasic extraction, multistage extraction, microencapsulation, seed germination, invertebrate development, and thin-film casting. Sliding-cavity technology has several advantages for spaceflight: it is simple, it accommodates small samples, samples can be fully enclosed, phases can be combined, multiple samples can be processed at high sample density, real-time observations can be made, and mixed and diffused samples can be compared. An analysis of the transport phenomena that govern the sliding-cavity method is offered. During sliding of one liquid over another flow rates between 0.001 and 0.1 m/sec are developed, giving Reynolds numbers in the range 0.1–100. Assuming no slip at liquid–solid boundaries shear rates are of the order $1 \sec^{-1}$. The measured consequence is the transfer of 2–5% of the content of a cavity to the opposite cavity. In the absence of gravity, buoyancy-driven transport is assumed absent. Transport processes are limited to (1) molecular diffusion, in which reactants diffuse toward one another at rates that depend on their diffusion coefficient and concentration gradient (Fick's second law), (2) solutocapillary (Marangoni) flow driven by surface-tension gradients, (3) capillary flow (drop spreading) at liquid–solid three-phase lines leading to immiscible phase demixing, and (4) vapor–phase diffusive mass transfer in evaporative processes. Quantitative treatment of these phenomena has been accomplished over the past few years in low-gravity research in space and on aircraft.

KEYWORDS: biotechnology research; shear cell; liquid contactors; sliding cavity; sliding block; crystal growth; film casting; phase separation; diffusion cell

NOMENCLATURE:
d diameter of cavity, cm
g acceleration due to gravity, cm/sec^2
h height of cavity, cm
Re Reynolds number

Address for correspondence: Paul Todd, Space Hardware Optimization Technology, Inc., 7200 Highway 150, Greenville, IN 47124, USA. Voice: 812-923-9591; fax: 812-923-9598.
ptodd@shot.com

v_x	velocity in the direction of translation, cm/sec
x	distance in the direction of sliding, cm
z	distance in the axial direction of a cylindrical cavity
γ	interfacial tension, dyn/cm
μ	dynamic viscosity, g/cm·sec
θ	liquid–solid contact angle
ρ	density

INTRODUCTION

The well known method of sliding-cavity fluid contactors utilized by Gosting[1] for diffusion measurements and by Tiselius[2] in electrophoresis has found considerable use in low-gravity biotechnology research, especially in studies of bioprocessing methods. To date, sliding-cavity contactors have been used in single-phase liquid diffusion experiments, interfacial transport experiments, biomolecular crystal growth, biphasic extraction, multistage extraction, microencapsulation, seed germination or imbibition, invertebrate development, and thin-film casting. Sliding-cavity technology has several advantages for spaceflight applications: it is simple, it accommodates small samples, samples can be fully enclosed, multiple phases can be combined, multiple samples can be processed at high sample density, real-time observations or measurements can be made, and mixed and diffused samples can be compared. An analysis of the transport phenomena that govern the sliding-cavity method is offered.

ANALYSIS OF FLUID MOTION

Simplified dimensionless and Newtonian-fluid analyses lead to the following generalizations and approximations.

Reynolds Numbers

During sliding of one liquid over another flow rates between 0.001 and 0.1 m/sec are developed, giving Reynolds numbers in the broad range 0.1–100. A generalized diagram of the steps involved in exposing one liquid to another using sliding cavities is given in FIGURE 1. Cavity diameters, d, range from 0.3 to 1.0 cm. They are typically filled with fluids similar to water or aqueous solutions with viscosity $\mu = 0.01$–0.02 g/cm·sec and density $\rho = 1.0$–1.2 g/cm^3.

$$Re = \frac{\rho d v_x}{\mu}, \quad (1)$$

where v_x is the cavity relative velocity in the direction of translation (x-direction) and can vary from 1 cm/sec to 1 cm/min.

For smaller dimensions, slow translation, and higher viscosity,

$$Re_{\min} = (1.0\,\text{g/cm}^3)(0.3\,\text{cm})(1\,\text{cm/60sec})/(0.02\,\text{g/cm·sec}) = 0.25;$$

and for larger dimensions, rapid translation, and lower viscosity,

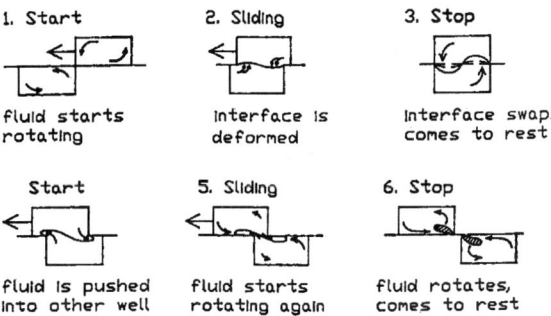

FIGURE 1. Generalized diagram of the steps involved in exposing one liquid to another using sliding cavities. The shaded zones at step 6 represent εV, the volume transferred due to fluid motion.[6]

$$Re_{max} = (1.2 \text{g/cm}^3)(1.0 \text{cm})(1 \text{cm/sec})/(0.01 \text{g/cm·sec}) = 120.$$

In some experiments, such as protein crystallization, polymeric membrane casting and biphasic aqueous systems the viscosity can be greater than 1.0 g·cm/sec,[3] in which case, for a 1-cm diameter cavity moving at 1 cm/min,

$$Re = (1.1 \text{g/cm}^3)(1.0 \text{cm})(1.0 \text{cm/60 sec})/(1.0 \text{g·cm/sec}) = 0.018.$$

In one polymeric membrane casting application cavity movement was spring driven,[4,5] and v_x exceeded 100 cm/sec.

$$Re = (1.1 \text{g/cm}^3)(1.0 \text{cm})(100 \text{cm/sec})/(1.0 \text{g·cm/sec}) = 110.$$

Thus, in the slower, higher-viscosity cases creeping (Stokes) flow prevails, but laminar (Poiseuille) flow occurs in larger, faster sliding cavities. Using these assumptions, specific Re values are listed in TABLE 1 for specific low-gravity sliding-cavity instruments.

TABLE 1. Approximate operating values for versions of three sliding-cavity devices designed for space-flight experiments and approximate calculated values of parameters of fluid motion in each

EXAMPLE	d (cm)	h (cm)	v_x (cm/sec)	Re ($\mu = 0.01$)	Re ($\mu = 1.0$)	ε (%)	$2v_x/h$ (sec^{-1})
ADSEP	1.0	0.5	0.03	3.3	0.03	1	0.12
AGBA	0.63	0.63	2.0	140	1.4		6.3
MCA	1.0	0.015	2.0		2.2		267
MCA	1.0	0.015	100		110		
MDA	0.63	1.6	0.2	14	0.14		0.24
MDA	0.32	1.6	0.2	7	0.07	0.37	0.24

NOTE: Fluid density in all cases was assumed to be 1.1. Re values are given at $\mu = 0.01$ g/cm·sec (water) and $\mu = 1.0$ g/cm·sec (polymer solutions). Only measured values of epsilon (ε) are listed.

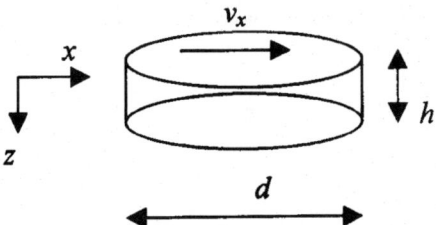

FIGURE 2. System of cavity coordinates and dimensions for analysis of fluid motion. Dimensions of specific sliding-cavity instruments are given in TABLE 1.

Shear Rates

Assuming no slip at liquid–solid boundaries shear rates are of the order $1\,\text{sec}^{-1}$. The extreme cases are calculated by assuming a Newtonian fluid and approximating the cavity cross-section as a rectangle, using the coordinates defined in FIGURE 2,

$$\frac{dv_x}{dz} \approx \frac{2v_x}{h}. \qquad (2)$$

This can be shown to be the *average* shear rate in the cavity. In laminar flow cases ($Re > 1$) the maximum shear rate, which occurs near the interface, is higher. Shear stresses are approximated by multiplying shear rate by the dynamic viscosity in each specific case.

MASS TRANSFER

Mass Transfer During Cavity Motion

The measured consequence of the combination of inertial motion and shear is the immediate transfer of 2–5% of a cavity volume to the opposite cavity during transit. Obviously (viewing FIG. 1) the percentage of fluid transferred will depend on the depth of the cavity, since the total volume transferred is the quantity dictated by inertial motion and shear. The degree of fluid transfer that occurs is critical to the overall value of the experiment being performed and its interpretation. First, second, and third transfer fractions have been defined as parameters that measure the fractions transferred when the cavities slide into alignment (ε_1), during contact (ε_2), and as the cavities slide away from each other (ε_3). The sum is the observed transfer at the end of the experiment,

$$\varepsilon = \varepsilon_1 + \varepsilon_2 + \varepsilon_3. \qquad (3)$$

This fraction was measured in several experiments using ADSEP plates (shown in FIGURE 3). In each experiment a solution (0.2 g/L) of methylene blue was placed in six upper cavities and six lower cavities. The dye-containing cavities were then contacted with cavities containing pure water for 15, 30, 45, or 60 sec, and the amount of dye transferred was measured, after uniformly mixing the contents of each cavity, by spectrophotometer. When dye was initially in the lower cavity, a maximum of

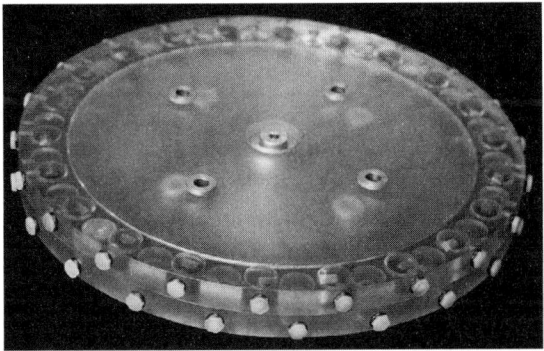

FIGURE 3. A pair of polycarbonate ADSEP plates (see TABLES 1 and 2) in the launch (half-stepped) position. Each plate has 22 cavities, 0.5 mL each, machined into it. The cavities are loaded with fluids from the side using syringes. Upper and lower cavities are brought into alignment by rotating one plate with respect to the other using a motorized drive unit that is coupled via the four lug holes visible in the photograph.

2% was transferred; when dye was initially in the upper cavity an average of 6.6% was transferred indicating that Rayleigh convection was significant even with $\Delta\rho/\rho < 0.0002$. No dependence of the transferred fraction on the contact interval was apparent; therefore, at least in the case of bottom-to-top transfer ε_2 may be considered negligible. When the cavities are once again separated the fraction ε_3 depends on the active process that occurred during contact. In the case of electrophoresis in the presence of gravity, for example,[7] ε_3 from 6% up to 9% was estimated.[8]

A similar set of experiments performed with the MDA (see TABLES 1 and 2) is reported elsewhere.[9,10] To summarize, the fraction transferred on the ground was $0.37 \pm 0.16\%$ and in low gravity $0.25 \pm 0.12\%$. Viewing these values in TABLE 1 shows that the fraction transferred scales, as expected, with the diameter of the cavity and inversely with cavity depth.

There are two additional mechanisms of unplanned fluid volume transfer immediately following the contacting of cavities. Capillary intrusion occurs along the walls of one of the chambers in the presence of gravity according to the force-balance relationship

$$F_{cap} = F_{grav},$$

$$\pi d \gamma \cos\theta = \frac{\rho g \pi d^2 h}{4}, \qquad (4)$$

$$h = \frac{4\gamma \cos\theta}{\rho g d},$$

where γ is interfacial tension (73 dyn/cm at the air-water interface), and θ is the contact angle at the 3-phase line. Water in air, for example, would rise nearly 0.1 cm in a cavity on Earth. It is apparent from Equation (4) that low gravity can lead to very large intrusion distances. Indeed, to model flow through porous media in low gravity significant departures from the traditional modeling approach are required.[12]

TABLE 2. Definitions of four sliding-cavity devices and references in which they are described

Example	Name	Reference
ADSEP	*ad*vanced *sep*arations	Deuser et al.[11]
AGBA	*a*utomatic *g*eneric *b*ioprocessing *a*pparatus	Pohlmann[6]
MCA	*m*embrane *c*asting *a*pparatus	Konagurthu et al.[4]
MDA	*m*aterials *d*ispersion *a*pparatus	Cassanto et al.[9]

Preliminary attempts to measure this intrusion volume in low gravity were made using MDAs (see TABLES 1 and 2) during a seven-minute sounding rocket flight. However, even in near-zero gravity, very little evidence for large intrusion distance is found, possibly due to the low value of γ between the two immiscible aqueous phases chosen for study, the similarity between their contact angles with the polymeric surface of the cavities and the short duration of low gravity.

The second immediate cause of volume transfer is the inertial motion of fluid upon stopping cavity movement. This has been discussed for the special case of polymer thin-film casting in MCAs (see TABLES 1 and 2).[5] The spring-loaded version of the MCA (see FIGURE 4) causes cavities to travel at speeds exceeding 1 m/sec. The abrupt deceleration of the lower cavity containing polymer solution causes the solution to rise at the distal end of the lower cavity, and this has been observed in the form of non-uniform membrane thickness after casting. This phenomenon has been termed "sloshing" and is depicted in FIGURE 5.

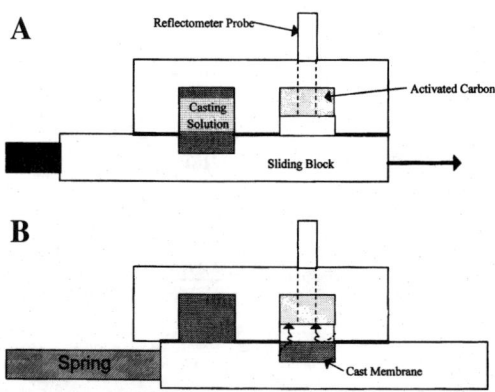

FIGURE 4. Membrane casting apparatus (MCA). The sliding of the lower cavities is driven by a spring (rapid) or by hand (slow): **(A)** launch position; **(B)** activated position. Upon activation solvent evaporates and is absorbed by granular activated carbon in the receiving upper cavity or chimney.

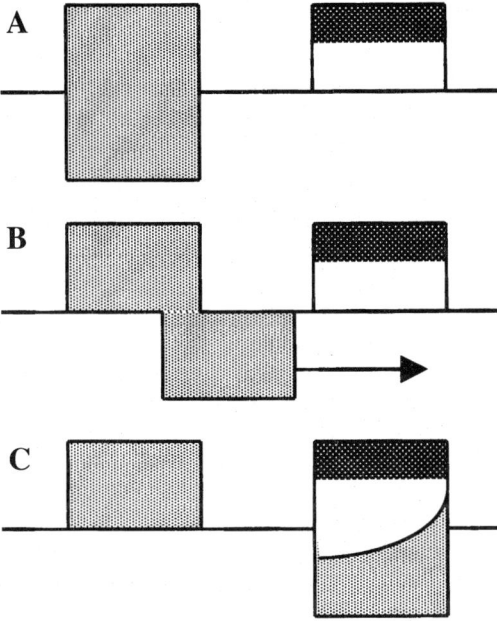

FIGURE 5. Illustration of fluid motion at the end of spring-driven motion of cavities in the MCA shown in FIGURE 4: **(A)** launch position; **(B)** during sliding; **(C)** just after sliding. The rise of fluid at the leading edge during deceleration has been termed "sloshing".

Mass Transfer During Contact Interval

Mass transfer during the contact interval falls into two categories: fluid-volume transfer during quiescence (just discussed, ε_2) and solute or dispersed-phase transfer due to an experimentally-applied driving force. The former is generally not desired, and the latter is usually associated with a designed variable of the experiment.

In the absence of inertial acceleration (gravity) buoyancy-driven transport is assumed absent. Subsequent transport processes are limited to (1) molecular diffusion, in which reactants (such as a solute and its precipitant) diffuse toward one another at rates that depend on their diffusion coefficient and their concentration gradient (Fick's law), (2) solutocapillary flow driven by surface-tension gradients (Marangoni flow), (3) capillary flow (drop spreading) at liquid–solid three-phase lines leading to immiscible phase demixing, (4) magnetically or electrically driven flow, (5) electrokinetic transport, and (6) vapor-phase diffusive mass transfer in evaporative processes. Some examples are described in the next section.

APPLICATIONS

The liquid–liquid diffusion method of crystal growth from solution[13] is readily automated in sliding-cavity technology.[9,14–18] Indeed, with two sliding movements, the nucleation and growth of protein crystals can be treated as two separate events in sequence.[19]

During phase demixing in the casting of polymer thin films from polymer solutions, large voids form (macrovoids) apparently due to solutocapillary motion, which is opposed by buoyancy. Sliding cavities permit the automated formation of thin films in thin (0.015 cm) cavities as shown in FIGURE 4.[4,5,20,21]

Surface forces have been exploited in low-gravity experiments on aqueous two-phase systems. To meet the challenge of phase separation in low gravity, wall materials were chosen to be wetted by specific phases, so that, after thorough demixing and mass transfer drops adhering to hydrophobic walls spread and attract the hydrophobic liquid phase while hydrophilic drops adhere to hydrophilic wall material and nucleate the hydrophilic phase.[11,22] Successful phase separation using this approach in space was demonstrated on shuttle flight STS-95.[18,23]

By placing magnetic microspheres in one cavity and a magnet at the end of its opposite counterpart, magnetic particles are driven toward the magnet and into the freshly contacted cavity. This motion can be exploited in the mixing of a viscous precipitating solution with a protein solution to be crystallized[24] or in the sorting of magnetically labeled cells or magnetic particles. Likewise, an electric field can be applied between cavities to collect particles or solutes on the basis of electrophoretic mobility.[7,8]

Two well-worked examples of vapor-phase transport have been implemented in sliding cavities. By placing a membrane between a solution in one cavity and a vapor space in the opposing cavity it is possible to transfer vapor through the membrane. The opposing cavity must provide a driving force for solvent vapor, and this usually takes the form of a concentrated solution. Such an arrangement facilitates the simultaneous concentration of solute and precipitant and drives the nucleation and growth of crystals from solution.[17] Evaporation of solvent also drives the phase change required for the formation of polymer thin films. In this case the driving force of choice is a dry absorber of the solvent, granular activated carbon, as shown in FIGURE 4, the MCA for polymer film casting in low gravity.[4,5,21]

DISCUSSION

All of the values in TABLE 1 are approximations. These realistic values are presented to indicate to potential users of sliding-cavity methods the historical range of fluid motion parameters used. Low shear rates, low Re and low transfer fractions are clearly desirable. These are obviously achieved by using cavities that are slowly translated, deep, and narrow. The selection of viscosity is not normally under control of the investigator but is dictated by the experimental solution required for the experiment. Relatively little rigor has been applied to obtain the above results, and there remains room for the application of more sophisticated approaches, such as computational fluid dynamics (CFD).

CONCLUSIONS

*This is the lesson concerning low gravity
Reactions can start in a small sliding cavity.
The hardware is lacking in bells and in whistles
But truly excels in the growing of crystals
For casting of films in a chamber robotic
Or the formation of capsules exotic
Even extractions in aqueous phases
Bring to the sliding cavities praises.
The MDAs, ADSEPs and then MCAs
Show they are used in multiple ways
But watch what you're doing, the experienced say
For pitfalls await in a most subtle way
They do not work well at all on the ground
And even in space troubles are found.
Make Reynolds and Rayleigh numbers lie low
And physics won't deal you a terrible blow.
These statements are not the end to be
More work should be done using CFD.*

ACKNOWLEDGMENTS

This paper is dedicated to the late Dr. Z. Richard Korszun of Brookhaven National Laboratory, who, while at the University of Pennsylvania in 1985, was coinventor of the MDA and later performed the first rocking curve on a space-grown crystal.

We acknowledge support from the U.S. National Aeronautics and Space Administration for a portion of this work done at the University of Colorado through Grants NAG8-1165, NAGW-1197, NAG8-1475, NCC8-131. We thank Instrumentation Technology Associates, Inc. Space Hardware Optimization Technology, Inc. was the recipient of NASA SBIR Grants NAS8-39340, NAS8-39914, and NAS9-98088.

The participation of Robin M. Stewart, Dr. John J. Pellegrino, Dr. Dennis R. Morrison, Derek Gregory, Ryan P. Cooper, and Nathan Thomas in various aspects of this research is gratefully acknowledged.

REFERENCES

1. GOSTING, L.J. 1956. *In* Advances in Protein Chemistry, Vol. 11. M.L. Anson, K. Bailey & J.T. Edsall, Eds.: 429.
2. TISELIUS, A. & P. FLODIN. 1953. Adv. Protein Chem. **8:** 461. Also LONGSWORTH, L.G. 1942. Chem. Rev. **30:** 323.
3. KONAGURTHU, S. 1998. Macrovoids in Dry-cast Polymeric Membranes: Growth Mechanisms, Non-invasive Detection, and Effects on Performance. Ph.D. Dissertation, University of Colorado, Boulder.
4. KONAGURTHU, S., W.B. KRANTZ & P. TODD. 1996. Use of low g to test alternative hypotheses for macrovoid growth in polymeric membranes. Proc. of Euromembrane Meeting, Bath, UK, 1995. Vol. 1, 256–261.
5. TODD, P., M.R. PEKNY, J. ZARTMAN, *et al.* 2001. Instrumentation for studying polymer film formation in low gravity. *In* Polymer Research in Microgravity. Polymerization and Processing. J. Pojman & J.P. Downey, Eds.: 126–137. ACS Symposium Series 793.

6. POHLMANN, C. 1993. M.S. Thesis, University of Colorado, Boulder.
7. TODD, P., K.S.M.S. RAGHAVARAO, S. SENGUPTA, J.F. Doyle, *et al.* 2000. Multistage electrophoresis system for the separation of cells, particles and solutes. Electrophoresis **21**: 318–324.
8. SENGUPTA, S. 2000. Multistage Electrophoresis. M.S. Thesis, University of Colorado, Boulder, Colorado.
9. CASSANTO, J.M., W. HOLEMANS, T. MOLLER, *et al.* 1990. Low-cost low-volume carrier (minilab) for biotechnology and fluids experiments in low gravity. Prog. Astronaut. Aeronaut. **127**: Space Commercialization: Platforms and Processing. F. Shahrokhi, G. Hazelrigg & R. Bayuzick, Eds.: 199–213. Am. Inst. Aeronaut. Astronaut., Washington, D.C.
10. TODD, P. 1990. Separation physics. *In* Progress in Low-Gravity Fluid Dynamics and Transport Phenomena. J.N. Koster and R.L. Sani, Eds.: 539-602. Progress in Astronautics and Aeronautics, Vol. 130. Am. Inst. Aeronaut. Astronaut., Washington, D.C.
11. DEUSER, M.S., J.C. VELLINGER, R.J. NAUMANN, *et al.* 1995. Apparatus for aqueous two-phase partitioning for terrestrial and space applications. *In* Terrestrial Applications of Biomedical Research in Space. Am. Inst. Aeronaut. Astronaut., Washington, D.C.
12. SCOVAZZO,P., T.H. ILLANGASEKARE, A. HOEHN & P. TODD. 2001. Modeling of two-phase flow in membranes and porous media in low gravity as applied to plant irrigation in space. Water Resour. Res. **37**: 1231–1243.
13. SALEMME, F.R. 1971. Enzyme purification and related techniques. *In* Methods in Enzymology, Vol. 22. W.B. Jakoby, Ed. Academic Press, New York.
14. HOLEMANS, J., J.M. CASSANTO, T.W. MOLLER, *et al.* 1991. The BIMDA shuttle flight mission: a low cost microgravity payload. Microgravity Q. **1**: 235–247.
15. TRAKHANOV, S.D., A.I. GREBENKO, V.A. SHIROKOV, *et al.* 1991. Crystallization of protein and ribosomal particles in microgravity. J. Cryst. Growth **110**: 317–321.
16. STODDARD, B.L., R.K STRONG & G.R. FARBER. 1991. Design of apparatus and experiments to determine the effect of microgravity on the crystallization of biological macromolecules using the Mir Space Station. J. Crystal Growth **110**: 312–316.
17. TODD, P., M.G. SPORTIELLO, D. GREGORY, *et al.* 1993. Transport phenomena in the crystallization of lysoyme by osmotoc dewaterng and liquid–liquid diffusion in low gravity. Am. Inst. Aeronaut. Astronaut. Preprint Series AIAA-93-0720, Washington, DC.
18. VELLINGER, J., M.S. DEUSER, P. TODD, *et al.* 2001. Biotechnology experiments in the "ADSEP" payload on space shuttle flight STS-95 (1998). Spacebound 2000 Proceedings, Canadian Space Agency, St-Hubert, PQ, Canada.
19. SYGUSCH, J., R. COULOMBE, J.M. CASSANTO, *et al.* 1996. Protein crystallization in low gravity by step gradient diffusion method. J. Cryst. Growth **162**: 167–172.
20. ZARTMAN, J., V. KHARE, M.R. PEKNY, *et al.* 2001. Solutocapillary-convection-driven macrovoid defect formation in polymeric membranes. AIAA Paper 2001-0461, Am. Inst. Aeronaut. Astronaut., Washington, DC.
21. PEKNY, M.R., J. ZARTMAN, A.R. GREENBERG, *et al.* 2000. Influence of buoyancy and surfactant solutes on macrovoid defect formation in dry-cast cellulose acetate membranes. Polymer Preprints **41**(1): 1060–1061.
22. BAMBERGER, S., J.M. VAN ALSTINE, D.E. BROOKS, *et al.* 1990. Phase partitioning in low gravity. *In* Progress in Low-Gravity Fluid Dynamics and Transport Phenomena, vol. 130. J.N. Koster & R.L. Sani, Eds.: 603–617. Progress in Astronautics and Aeronautics. Am. Inst. Aeronaut. Astronaut., Washington, D.C.
23. BROOKS, D.E., A. TAKACS-COX & R. NORRIS-JONES. 2001. Surface energy-driven localization of phase-separated aqueous polymer mixtures in microgravity and the partition of latex particles. Spacebound 2000 Proceedings, Canadian Space Agency, St-Hubert, PQ, Canada.
24. TODD, P., J.F. DOYLE, P. CARTER, *et al.* 1999. Magnetic space shuttle experiments. J. Magnet. Magnetic Mat. **194**: 96–101.

Microgravity Protein Crystallization

Are We Reaping the Full Benefit of Outer Space?

NAOMI E. CHAYEN[a] AND JOHN R. HELLIWELL[b]

[a]*Biological Structure and Function Section, Division of Biomedical Sciences, Faculty of Medicine, Imperial College, London, UK*

[b]*Department of Chemistry, The University of Manchester, Manchester, UK*

ABSTRACT: Protein crystallography, which has a growing role in the human genome project, is one of the most powerful techniques in modern biology. However, it can only be applied providing suitable crystals can be obtained. The ability to produce suitable crystals is currently the major bottleneck to structure determination. Microgravity, which is used as a tool for improving crystal growth, is a high-profile enterprise, especially with the upcoming prospects for utilizing the International Space Station, yet the issue of crystallization in microgravity is an extremely controversial one. In spite of numerous experiments conducted in microgravity during 16 years, the reported success rate, measured in terms of various improvements witnessed in the perfection of protein crystals, has been a mere 20%. In this paper we present experimental evidence, supporting previous fluid physics calculations to show that in many cases the potential benefits of the microgravity environment have not been fully exploited. These findings offer an explanation for the low rate of success and open up the possibility for enhancing the efficiency of experimentation in microgravity. Furthermore, these findings extend into a number of other disciplines involving fluid physics.

KEYWORDS: protein crystallization; nucleation; microgravity

INTRODUCTION

Microgravity eliminates sedimentation and convective mixing, thus offering a more homogenous medium for crystal growth compared to growth on Earth. It is, therefore, likely to improve the degree of perfection of the crystals.

Justification for using microgravity as a platform for growing crystals of biological macromolecules has been the subject of vehement debate and disagreement among scientists ever since the first protein crystals were grown in Space.[1] Analysis of 16 NASA Shuttle missions flown between 1985 and 1994 show that only 20% of the proteins examined were found to exhibit better morphologies or better quality diffraction data than their Earth-grown counterparts.[2] There is no evidence to date that this low rate of success has changed, yet the continuous global interest concerning this issue is evident.[3–5] In this article we demonstrate that, although many trials

Address for correspondence: N.E. Chayen, Biological Structure and Function Section, Division of Biomedical Sciences, Faculty of Medicine, Imperial College, London SW7 2BZ, UK. Voice: +44-207-594-3240; fax: +44-207-594-3169.
n.chayen@ic.ac.uk

have been conducted in microgravity, it appears that the benefits of the microgravity environment have not been fully exploited.

CONSIDERATIONS FOR CRYSTALLIZATION IN MICROGRAVITY

The reasons for the low success rate are only now beginning to emerge. With new apparatus constructed and made available by Dornier and the European Space Agency, which includes CCD video and interferometry, it is possible to monitor in detail the protein crystal growth process in microgravity. Experiments can now be conducted in a far more systematic way than previous trials, which were heavily reliant on luck. As a result, researchers are currently beginning to realize the crucial factors on which to focus when conducting experiments in microgravity.

Two major issues need to be addressed: (1) Are the optimal conditions for crystallization (i.e., concentration of ingredients, temperature, and pH) in microgravity being applied? (2) Does the choice of crystallization method have any bearing on the outcome of the results?

(1) Conditions for Crystallization

It has been shown that the optimal conditions for crystallization in microgravity are not necessarily identical to those on Earth.[6,7] Hence, the optimal conditions for crystallization in microgravity need to be determined in order to ensure that the full benefit of the microgravity environment is attained. Finding the optimal conditions is a matter of trial and error, which requires numerous trials. However, this cannot be achieved on a typical flight, where each experimenter is allocated a fairly small number (three to ten) of experimental reactors.[8] Failure to optimize the conditions for crystallization in microgravity has even resulted in reactors returning from missions without any crystals at all. Moreover, when success is achieved, the experimenter is unable to determine with certainty, whether the result would be reproducible or whether it was just a "one off" coincidence.[9] An additional problem with most of the reactors available is that typically 50µl or more of protein solution is required per trial; the minimum quantity that can be usefully applied is 15µl (on the ground, quantities of 1µl, and even less, can be used). This severely restricts the number of proteins that can be tested in microgravity.

(2) The Method of Crystallization

There are four methods for crystallizing proteins: batch, vapor diffusion, dialysis, and free interface diffusion. Vapor diffusion involves an aqueous protein drop that equilibrates via vapor pressure with a reservoir solution. For the other methods, two similar solutions are in contact with each other without a liquid–vapor phase boundary.

Vapor diffusion is the most widely used technique for crystallization on the ground, hence, it naturally became the dominant method for crystallization in microgravity.[2] It is only recently that experiments that facilitate monitoring of the crystal growth process by CCD and interferometry have become available. This enabled us to compare the crystallization results using different methods of crystallization. We have demonstrated that the method of crystallization in microgravity may be as crucial to the results as that of determining the optimal conditions. Our attention was drawn to this issue when analyzing, jointly with Drs. Edward Snell and Titus Boggon, the results of several different missions by correlation of the X-ray analysis results with CCD observation of the crystal growth at various time intervals throughout the flights.

Crystals of α-crustacyanin, lysozyme, and apocrustacyanin C_1 were grown in microgravity by three different crystallization methods using the protein crystallization facility (PCF) on the Eureca satellite and the advanced protein crystallization facility (APCF) on the SpaceHab, IML–2, and USML–2 shuttle missions. Crystals grown in microgravity were analyzed and compared to respective crystals grown in ground controls. Overall, the best crystals of these proteins were microgravity grown.[10–14] However, even when improvement in diffraction quality was found in the case of apocrustacyanin C_1, which was grown by the vapor diffusion method, it was not as pronounced as the improvement seen in microgravity grown lysozyme crystals, which were grown by the dialysis method. In the case of lysozyme *all* the microgravity-grown crystals tested were superior to their Earth-grown counterparts.[10] In the case of apocrustacyanin C_1, only *some* of the crystals were superior.[11,14]

The partial improvement in quality of the microgravity-grown apocrustacyanin C_1 crystals compared to the Earth-grown ones may be explained by correlation of the X-ray analysis results with CCD observations of apocrustacyanin C_1 crystals at time intervals throughout the IML–2 flight. The crystals displayed a cyclic motion within the hanging drop (see FIGURE 1),[15] which can be attributed to Marangoni convection that serves to reduce concentration gradients along the interface between the protein solution and the vapor.[16]

In contrast, for lysozyme, grown by dialysis (without a liquid–vapor phase boundary), crystal motion was not visible[17] and there was a very clear increase in crystal quality. Another experiment involved CCD video monitoring of the crystallization of α–crustacyanin by free interface diffusion on the Eureca mission. The Space crystals were stationary throughout the long mission,[18] and they were far larger than their Earth-grown counterparts[12] in which continuous turbulence throughout the growth was observed.

Cyclic movement of the crystals may well be a limiting factor in the ultimate perfection of the protein crystals that can be obtained. The crystals that grow near the surface of the hanging drop would be most affected by Marangoni convection and would, therefore, not get the full benefit of the microgravity environment. Consequently, these crystals are not likely to be better ordered than Earth-grown crystals, as theoretically predicted.[19] The crystals grown deeper inside the drops (which showed far less movement than the crystals at the surface; FIG. 1) are likely to be the crystals that show significant improvement over their Earth-grown counterparts.[14]

Previous experiments[20] reported that canavalin crystals grown in microgravity in the free interface reactors of the APCF exhibited a significant increase in diffraction resolution limit. This was not the case when the canavalin crystals were grown by the vapor diffusion method using the VDA apparatus. Other experiments, conducted on the LMS mission,[21] compared the quality of crystals of alcohol dehydrogenase grown in microgravity in APCF dialysis chambers with those grown in the APCF

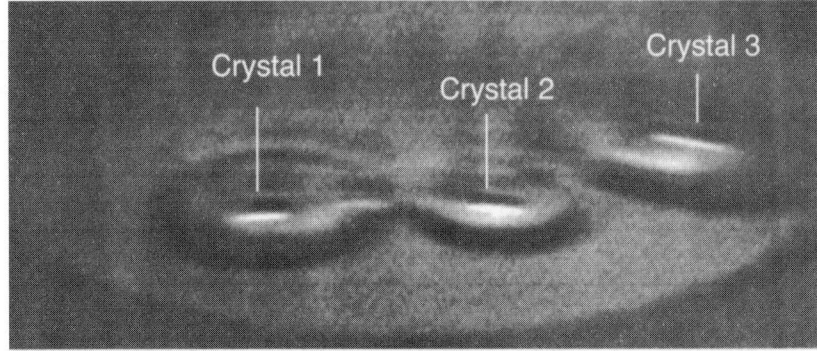

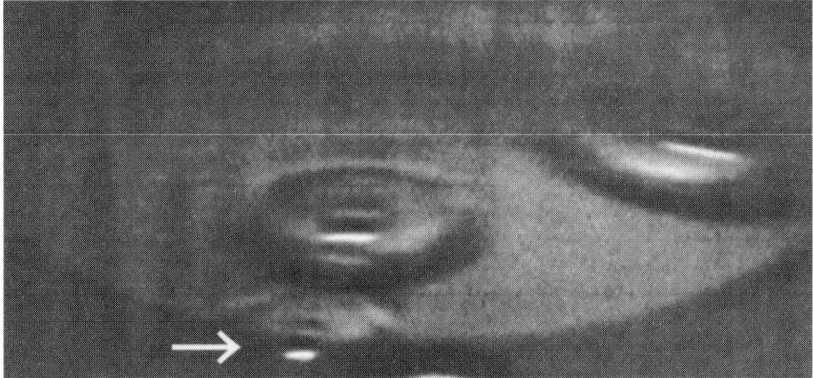

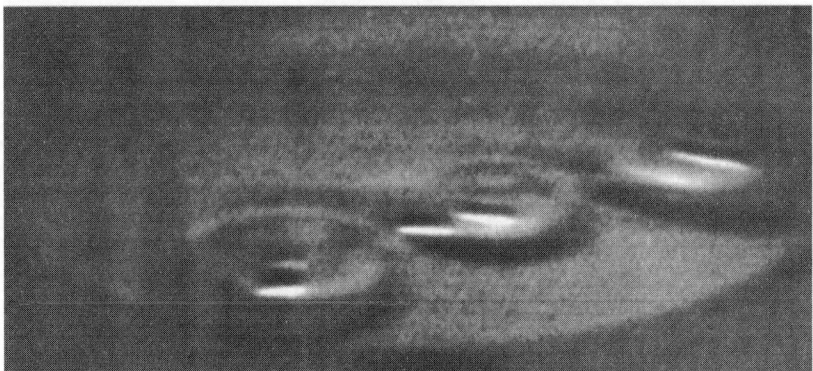

FIGURE 1. CCD video images of apocrustacyanin C_1 crystals taken at various times during the IML-2 flight. A pattern of movement is shown that then repeats itself throughout the flight. The *arrow* highlights the crystal that has changed its position.

using vapor diffusion. It was shown that crystals grown by the dialysis method were better ordered than their Earth-grown counterparts, whereas the vapor diffusion crystals were no different to the Earth-grown crystals.

Other causes of crystal motion during growth, besides Marangoni convection, have been witnessed by CCD video. Such movements were not cyclic and they were more gentle, thereby not affecting the quality of the resulting crystals.[22,23] However, once the crystals have grown, even sudden considerable movement does not appear to harm them. For example, a very large and sudden incident occurred on SpaceHab 01, late into the mission, after protein crystal growth had essentially stopped. This motion was attributed to retrieval of the Eureca satellite. The perfection of the lysozyme crystals (grown by the dialysis method) examined by ultra-long distance synchrotron Laue diffraction, showed the most perfect protein crystals ever grown.[10] It appears that sudden movements in flight, and also re-entry g-forces through the Earth's atmosphere, are damaging to crystals that are in the process of growing,[24] but they are unlikely to affect already grown protein crystals. There may be exceptions; that is, cases where crystals are especially fragile. In another experiment on IML2, involving lysozyme, correlation of CCD video with accelerometer (g–jitter) data showed spurts and lulls of crystal growth coincident with g–jitter bursts caused by astronaut exercise periods.[17] Such spurts and lulls during crystal growth may explain the residual mosaic character of protein crystals seen in X–ray topographs.[11]

CONCLUDING REMARKS

The fact that there is some success in microgravity crystallization, albeit 20%, indicates that microgravity is potentially a useful tool for improving protein crystal order. It is apparent that the problem is that we may have not (yet) grasped how best to use the microgravity environment. It may also be that only a subset of protein crystals would benefit from more stable fluid conditions. Perhaps those protein crystals displaying strong disorder diffuse X–ray scattering[25] would most benefit from microgravity growth, as seen in the case of a small molecule microgravity versus Earth crystal growth.[26] The correlation of the X–ray analyses of the resulting crystals with the CCD video observations of the protein crystal growth demonstrate that the method of crystallization has an important bearing on the results. The findings reported here have shed light as to why this is so. A new generation of diagnostic apparatus that will carry larger numbers of samples using smaller quantities of protein material, is currently being built by ESA, NASA, and NASDA. This will enable deeper insight into protein crystal growth in microgravity. We strongly believe that using such equipment, combined with the application of the optimal conditions for crystallization, and the use of the right methodology will significantly increase the success rate of protein crystallization in microgravity.

ACKNOWLEDGMENTS

The authors are very grateful to ESA for providing the APCF apparatus and flight opportunities on NASA flights that enabled this research.

REFERENCES

1. LITTKE, W. & C. JOHN. 1984. Protein single crystal growth under microgravity. Science **255**: 203–204.
2. DELUCAS, L.J., et al. 1994. Recent results and new hardware developments for protein crystal growth in microgravity. J. Cryst. Growth **135**: 183–195.
3. NORMILE, D. 1995. Search for better crystals explores inner, outer space. Science **270**: 1921–1922.
4. MCPHERSON, A. & L. DELUCAS. 1999. Crystal-growing in space. Science **283**: 1459.
5. CHAYEN, N.E. & J.R. HELLIWELL. 1999. Space-grown crystals may prove their worth. Nature **398**: 20.
6. STRONG, R.K., et al. 1992. Long duration growth of protein crystals in microgravity aboard the MIR space station. J. Cryst. Growth **119**: 200–204.
7. CHAYEN, N.E. 1995. Microgravity protein crystallization aboard the Photon satellite. J. Cryst. Growth **153**: 175–179.
8. MCPHERSON, A. 1997. Recent advances in the microgravity crystallization of biological macromolecules. Trends Biotechnol. **15**: 197–200.
9. ASANO, K., et al. 1992. Crystal growth of ribonuclease S under microgravity. J. Cryst. Growth **122**: 323–329.
10. SNELL, E.H., et al. 1995. Improvements in lysozyme protein crystal perfection through microgravity growth. Acta Crystallogr. **D51**: 1099–1102.
11. CHAYEN, N.E., et al. 1996. Trends and challanges in experimental macromolecular crystallography. Q. Rev. Biophys. **29**: 227–278.
12. ZAGALSKY, P.F., C.E. WRIGHT & M. PARSONS. 1995. Crystallization of α-crustacyanin, the lobster carapace astaxanthin-protein: results from EURECA. Adv. Space Res. **16**: 91–94.
13. CHAYEN, N.E., E.J. GORDON & P.F. ZAGALSKY. 1996b. Crystallization of apocrustacyanin C_1 on the International Microgravity Laboratory (IML-2) mission. Acta Crystallogr. **D52**: 156–159.
14. SNELL, E.H., et al. 1997. Partial improvement of crystal quality for microgravity-grown apocrustacyanin C_1. Acta Crystallogr. **D53**: 231–239.
15. CHAYEN, N.E., et al. 1997. CCD video observation of microgravity crystallization: apocrustacyanin C1. J. Cryst. Growth **171**: 219–225.
16. MOLENKAMP, T., P.B.M. JANSSEN & J. DRENTH. 1994. Protein crystallisation and Marangoni convection. Final reports of sounding rocket experiments in fluid science and materials science. ESA Report SP-1132, 22–43.
17. SNELL, E.H., et al. 1997. CCD video observation of microgravity crystallization of lysozyme and correlation with accelerometer data. Acta Crystallogr. **D53**: 747–755.
18. BOGGON, T.J., et al. 1998. Protein crystal movements and fluid flows during growth. Phil. Trans. R. Soc. Lond. A. **356**: 1–17.
19. SAVINO, R. & R. MONTI. 1996. Buoyancy and surface-tension-driven convection in hanging drop protein crystallizer. J. Cryst. Growth **165**: 308–318.
20. KOSZELAK, S., et al. 1995. Protein and virus crystal growth on international microgravity laboratory-2. Biophys. J. **69**: 13–19.
21. EPOSITO, L., et al. 1998. Protein crystal growth in the advanced protein crystallization facility on the LMS mission: a comparison of *Sulfolobus solfataricus* alcohol dehysdrogenase crystals grown on the ground and in microgravity. Acta Crystallogr. **D54**: 386–390.
22. LORBER, B., et al. 2000. Growth kinetics and motions of thaumatin crystals during UMSL-2 and LMS microgravity missions and comparison with earth controls. J. Cryst. Growth **208**: 665–673.
23. CAROTENUTO, L., et al. 2001. Video-observation of protein crystal growth in the advanced protein crystallization facility aboard the space shuttle mission STS-95 J. Cryst. Growth **232**: 481–488.
24. OTALORA, F., et al. 2001. Experimental evidence for the stability of the depletion zone around a growing protein crystal under microgravity. Acta Crystallogr. **D57**: 412–417.

25. GLOVER, I.D., *et al.* 1991. The variety of X-ray diffuse-scattering from macromolecular crystals and its respective components. Acta Crystallogr. **D47:** 960–968.
26. AHARI, H., *et al.* 1997. Effect of microgravity on the crystallization of a self-assembling layered material. Nature **388:** 857–860.

Space-Grown Protein Crystals Are More Useful for Structure Determination

JOSEPH D. NG

Laboratory for Structural Biology and the Department of Biological Sciences, University of Alabama in Huntsville, Huntsville, Alabama, USA

ABSTRACT: The usefulness of X-ray data derived from space-grown protein crystals for calculating a more accurate structure is reviewed here for three model proteins. These include the plant sweetening protein, thaumatin, from *Thaumatococcus daniellii*; the aspartyl-*t*RNA synthetase from *Thermus thermophilus*; and pea lectin from *Pisum sativum*. In all three cases, X-ray diffraction data collected from protein crystals obtained under reduced gravity lead to better defined initial electron density maps, facilitating model building and improved crystallographic statistics. With thaumatin, the phasing power of the anomalous scattering atom, sulfur, is used to determine protein crystal quality in terms of its usefulness for *ab initio* structure determination. Thaumatin crystals grown under microgravity provided improved phasing statistics compared to those of Earth-grown crystals. Consequently, generating a *de novo* protein model of higher quality was facilitated using X-ray diffraction data from space-grown crystals. This lends evidence to the possibility that a microgravity environment can favor protein crystal growth and, subsequently, be more useful for structure determination.

KEYWORDS: aspartyl-*t*RNA synthetase; crystal quality; crystallization microgravity; mosaicity; pea lectin; protein; thaumatin

INTRODUCTION

Macromolecular crystallization experiments in microgravity have been conducted for about two decades aboard the Space Shuttle[1] and now in the International Space Station.[2] The propelling interest to study crystal growth in space is based on the concept that the microgravity environment influences macromolecular crystal growth by minimizing buoyant convection flows, limiting nucleation sites, and eliminating crystal sedimentation.[3,4] Although the effectiveness of obtaining improved protein crystals in space has been greatly criticized,[5] there is much evidence that space grown protein crystals diffract X-rays to higher resolution with an improved diffraction intensity.[6,7] In the case of small proteins, the positive effects of microgravity are particularly observed.[8–14] Crystallization of large multisubunit proteins also benefits through microgravity, but their examples are fewer.[8,15] The most well studied case is lysozyme[16–19] and insulin,[20] where improved crystal quality has been observed compared to Earth controls.

Address for correspondence: Joseph D. Ng, Laboratory for Structural Biology and the Department of Biological Sciences, University of Alabama in Huntsville, Huntsville, AL 35899, USA. Voice: 256-824-3715; fax: 256-824-3204.
  NgJ@email.uah.edu

Three general methods have been used to assess protein crystal quality. The first examines a crystal with an optical microscope by direct visualization to document size and presence of any defects in the crystalline habit. This process is the most straightforward and is commonly used for initial discrimination against imperfect crystals. However, there are many instances where a protein crystal, of large volume (greater than $0.5\,\text{mm}^3$) without any observable defects, does not diffract X-rays beyond $3\,\text{Å}$ atomic resolution. On the other hand, there are cases where a crystal contains physical imperfections and is small in size, yet it diffracts X-rays to atomic resolution. The popular phrase "looks are not everything" indeed applies to crystal quality assessment using visual means. Even though better looking crystals were considered to be of better quality during the earlier evaluation procedures in assessing space-grown crystals,[1] it is no longer a sufficient parameter.

The second method to determine crystal quality is by measuring the scattered X-ray intensity, I, of reflections over their background noise, $s(I)$, within a specific range of resolution.[8,9,21] This method of evaluation is the most widely used to assess the quality of space-grown crystals compared to that of Earth-grown crystals. A complete X-ray data set is collected from both crystals at a conventional laboratory or at a synchrotron source. However, crystals can have excellent $I/s(I)$ versus resolution profiles and yet may not be suitable for structure determination as a result of twinning or reflections that can not be processed for other reasons.

The third means of crystal quality assessment is the shape of the scattered X-ray reflections, because this reflects the intrinsic order of the crystalline lattice. Mosaicity is a refined parameter of the X-ray data that quantifies the shape of the scattered X-ray reflection. Thus, the mosaicity of the reflections has been utilized as a measurement of crystal quality.[9,15,20,22–23] The value of this parameter is probably the most accurate assessment of crystal quality, because the internal arrangement of molecules affect the spatial resolution, reflection spot overlap, as well as the signal to noise ratio observed in the diffraction image. Specific reflections of space-grown crystals have been reported to have lower mosaic values than equivalent reflections of their Earth-grown controls.[9,15,24,25] The difficult aspect of this procedure is that it requires a highly parallel, monochromatic synchrotron radiation source with very fine oscillation angle changes ($\Delta 0.001°$). In addition, the measured mosaicity of one particular reflection does not necessarily represent the entire crystal. The mosaicity of reflections throughout a single crystal may vary as much as $0.010°$. The quality of the crystals may not be uniform and the value of the mosaicity measurements may vary in the crystal volume. More recent studies, using X-ray topography[26] and average mosaicity measurements of hundreds of reflections from a single crystal,[20] have provides the most accurate assessment to date for evaluating crystal quality. These examples are few and technically very demanding.

Since 1984, more than 20 percent of the biomacromolecules crystallized under microgravity were claimed to be superior to those of Earth-grown crystals as determined by one or all of the three previously described criteria for quality assessment.[7] However, no satisfactory evidence shows that improved crystal quality can result in a more accurate structure and, thus, be more useful for structure determination than that obtained on Earth.

To truly support the idea that microgravity is advantageous for protein crystal growth and, in turn, more beneficial for structure determination, crystals must be analyzed for their effectiveness and ease of producing an accurate crystallographic model. Here I review some examples from our own flight experiences, where protein crystal quality is evaluated not only by X-ray diffraction data, but in terms of "usefulness" toward calculating an accurate crystallographic protein model.

RESULTS AND DISCUSSION

Protein Crystal Quality Determined by Crystallographic Structure Assessment

Thaumatin

Crystallization of the sweetening protein from the arils of the African shrub *Thaumatococcus danielli*[27,28] has been studied on numerous space flights.[9,13] In all published cases, the crystals were shown to be superior to the Earth-grown equivalents by virtue of their crystal size and diffraction quality. Our best example was on a two-week flight aboard the space shuttle using the advance protein crystallization facility (APCF)[9,29] in which the protein crystals obtained under microgravity environment achieve a larger volume and fewer crystals per experimental sample than Earth-grown controls prepared in parallel. The diffraction properties of space-grown crystals were improved compared to their controls since they diffracted X-rays to a higher resolution and with a significantly lower mosaicity. Our most recent crystallization experiment in microgravity with thaumatin on the International Space Station using the enhanced gaseous nitrogen Dewar system (EGN Dewer) during a 45-day flight (transported aboard the space shuttle, Atlantis [STS-106]) showed similar results. As in previous studies, the crystals obtained from space, compared to their Earth equivalents, diffracted to higher resolution and showed improved $I/s(I)$ over resolution profiles (see FIGURE 1).

To examine if space-grown crystals produce a more accurate crystallographic model compared to Earth-grown crystals, complete X-ray data sets were used for initial electron density map calculations. Only reflections that were in common between the two data sets collected from Earth and space crystals were used in the 12–1.7 Å resolution range, so that no bias as a result of different contributions from reflections would occur.

The positive effects of microgravity were evident from the initial electron density map calculations. The quality of the Fourier $2F_o-F_c$ maps (2 Å) from Earth and space data were compared after one cycle of rigid body minimization and simulated annealing refinement against the original Earth model. FIGURE 2 displays a typical example, where the electron density tracing for an amino acid is not clear in the map calculated from the Earth data whereas the side chain fits easily in the map calculated from the space data.

The accuracy of the initial models was evaluated by analyzing the statistical distribution of atomic distances and conventional R-factors. The preliminary refinement statistics derived from Earth control and space thaumatin crystals are outlined in TABLE 1. The initial refinement statistics show a favored model from the space-grown crystals.

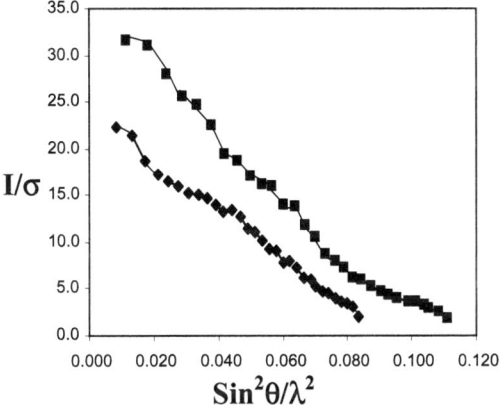

FIGURE 1. Profile of diffraction intensities represented as the intensity over estimated error ratio (I/σ) versus resolution for space- (■) and Earth-grown (♦) thaumatin crystals. The space-grown crystals showed as much as 40% more intensity and diffracted to higher resolution relative to Earth grown crystals in the range of resolution analyzed. Diffraction data was collected from capillary-mounted crystals at 22°C using a MSC R-Axis IV image plate detector with a crystal-to-detector distance of 250 mm. The X-rays were generated by a Rigaku rotating-anode generator operated at 50 kV and 100 mA and focused with MSC OSMIC confocal mirrors. Complete data sets were collected at a 1.0 degree oscillation angle with an exposure time of 15 min. Data were indexed using DENZO and reduced using SCALEPACK from the program package HKL2000.[41]

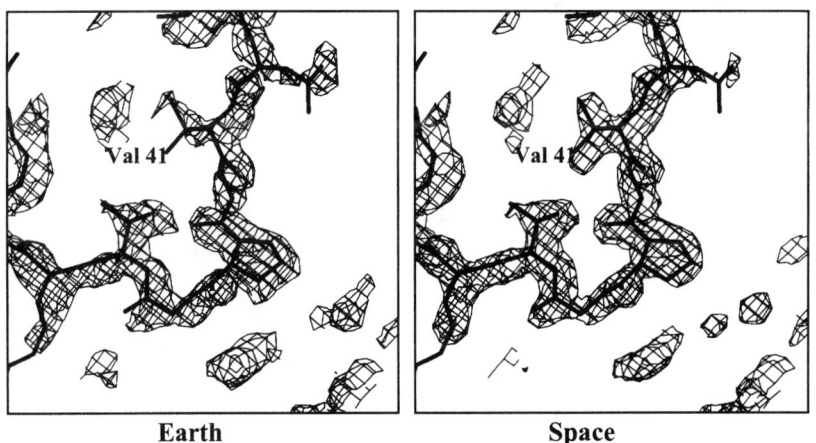

FIGURE 2. Comparative initial $2F_o$–F_c Fourier difference electron-density maps (contoured at 2σ) calculated from Earth- (**left panel**) and space-grown (**right panel**) data of thaumatin. The side chain, Val41, exemplifies some differences observed in the clarity of the map that are found throughout the molecule. The electron density maps were calculated using CNS 1.0[42] and visualized using O.[43]

TABLE 1. Preliminary refinement statistics derived from Earth- and space-grown thaumatin crystals

	Earth	Space
Refinement		
Resolution range (Å)	12–1.7	12–1.7
R_{free} factor (%)	26.5	24.7
R factor (%)	23.5	22.3
Number of reflections used	45,053	45,053
Number of reflections in the test set (%)	5	5
R.M.S. deviation from ideality		
Bond lengths (Å)	0.025	0.022
Bond angles (°)	3.0	2.8
Mean B-factors (Å^2)	18.0	15.0

NOTE: The program CNS[42] was used for the crystallographic structure refinement.

Aspartyl-tRNA Synthetase

The aspartyl-tRNA synthetase (AspRS) was selected to represent a multisubunit protein typical of many functional biomacromolecules. The protein is a homodimer of 122 kD with each subunit composed of 580 amino acids[30] and is involved in the catalytic reactions involved in translating genetic information. The protein was cloned from *Thermus thermophilus*, expressed in *Escherichia coli*, crystallized and structure determined to 2.5 Å.[31,32] The crystallization of AspRS has been studied in the laboratory of R. Giegé (IBMC, Strasbourg).[15,33] On a US Space Shuttle mission designated life microgravity space laboratory (LMS) mission STS-78, AspRS was crystallized in microgravity using the dialysis reactors of the APCF provided by Dornier Deutche Aerospace.[29] The experiment was conducted at 20°C during a 16-day mission. Crystals were grown in space and on Earth from a single protein preparation and set-up in the same crystallization reactors.

Space AspRS protein crystals grown in microgravity were observed to diffract X-rays more intensely, have lower mosaicity, and provided a more accurate crystallographic structure. The detailed results of this study are reported elsewhere.[15] The principal evidence lending support that space grown crystals serve to be more useful for structural determination is revealed in the initial electron density maps.

As described previously for thaumatin, only reflections that were in common between the two AspRS data sets collected from Earth and space crystals were used in the resolution range of 12–2.0 Å. Initial electron density map calculations were performed and the quality of the Fourier $2F_o-F_c$ maps are compared in FIGURE 3. The electron density differences between the Earth and space maps are particularly pronounced in the side chains of aromatic amino acids and their tracings are stronger and more prominent in the map calculated from the space crystals. A better-defined electron density map facilitates the model tracing and improved placement of the amino acid side chains. The overall local structures obtained from microgravity

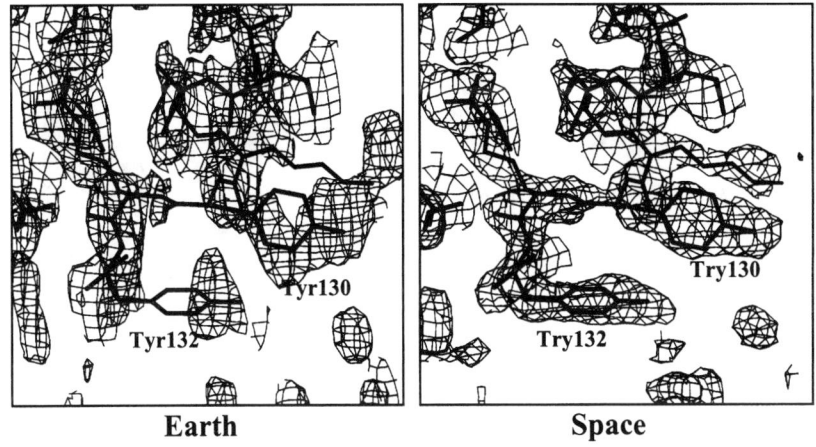

FIGURE 3. Comparative initial $2F_o-F_c$ Fourier difference electron-density maps (contoured at 2σ) calculated from Earth- (**left panel**) and space-grown (**right panel**) data of AspRS crystals. The electron density maps were calculated and visualized as described in FIGURE 2. Residues Tyr130 and Try132 are indicated to show a higher accuracy of model fitting from the map derived from space data compared to that obtained from Earth. The maps were calculated against the lower resolution AspRS structure determined by Delarue et al.[44]

crystals showed improvements in geometry, including the statistical distribution of atomic distances, and deviations from standard atomic volumes. In addition, increased solvent density, corresponding to weak interactions between the AspRS domains, was observed from the electron density maps derived from space-grown molecules (data not shown).

Pea Lectin

The mitogenic protein, pea lectin, was crystallized in the orthorhombic space group on the International Space Station in EGN. The protein crystallized during an effective 45 day flight on the International Space Station as described previously for thaumatin. The space-grown pea lectin crystals are examples of visually imperfect crystals that diffract to much higher resolution and with much better diffraction intensity than their Earth equivalents. FIGURE 4 shows a pea lectin crystal obtained from microgravity compared to that of the best Earth grown in the laboratory. Control pea lectin protein crystals prepared in the EGN, and employed under the same crystallization conditions did not produce any suitable crystals for X-ray diffraction. However, the best pea lectin crystals grown in our laboratory were larger and possessed very good crystal habits with no visual defects. These crystals do not diffract beyond 3 Å. The Earth-grown pea lectin X-ray data had higher I/σ values as a function of resolution with higher resolution limits compared to the best equivalent crystals grown on Earth. The most striking improvement was in resolution, with the space grown diffracting to 1.2 Å, which is 0.5 Å better than the highest resolved terrestrial crystal. The initial electron density map calculations were performed using the X-ray data collected from the space grown crystals to produce a Fourier $2F_o-F_c$

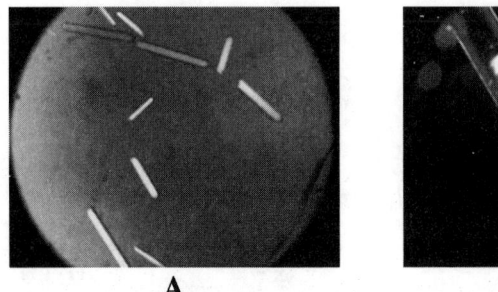 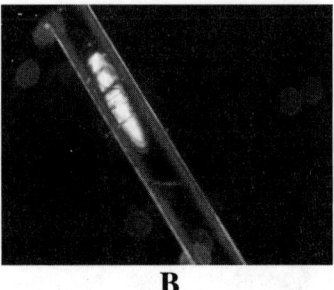

FIGURE 4. Earth- and space-grown crystals of pea lectin. **Panel A** shows the best pea lectin crystals obtained in the laboratory by the hanging drop vapor diffusion method.[45] The Earth-grown crystals were of triclinic habit and the biggest crystals observed measured to be 0.3 mm in longest dimension. These crystals do not diffract to more than 3 Å resolution. **Panel B** displays a space-grown pea lectin crystal obtained in a tygon tube installed in EGN. Crystals typically grown from space do not have well-formed morphology, but the diffract to high resolution. Crystallization conditions are described by Prasthofer *et al.*[46]

map (see FIGURE 5). The best terrestrial crystal was obtained from recombinant protein, while the higher resolution space grown crystal of pea lectin was obtained directly from crude plant material.

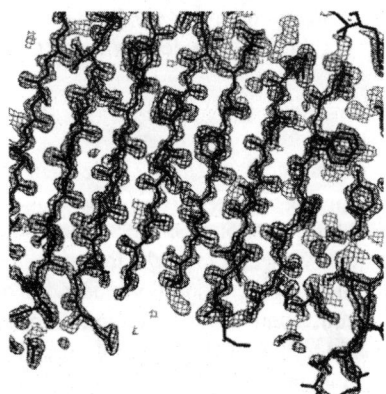

FIGURE 5. Initial $2F_o-F_c$ difference electron-density map (contoured at 2σ) calculated from a 1.2 Å room-temperature space data set against the published pea lectin model.[46] The pea lectin model was easily fitted after rigid-body refinement while preserving proper stereochemistry. The data from these space-grown crystals were the best ever obtained, giving rise to the highest resolution structure solved for pea lectin. The complete model refinement is in progress. The data collection was performed at the Synchrotron Radiation Laboratory at 1.0 degree oscillation angles with an exposure time of 30 seconds. Data were indexed DENZO and reduced using SCALEPACK from the program package HKL2000.[41] The electron density map was calculated using CNS 1.0[42] and visualized using O.[43]

Additional Views on Crystal Quality with Thaumatin

Protein crystals can be evaluated by their ability to diffract X-rays sufficiently for *ab initio* structure determination without any *a priori* structural information. The electron density, ρ, in a crystal is calculated from an X-ray data set by Fourier summation,

$$\rho(x, y, z) = \frac{1}{V}\sum_{hkl} |F(h, k, l)| e^{-2\pi i(hx + ky + lz) + i\alpha(h, k, l)}, \quad (1)$$

where V and x, y, z are the volume and coordinates in the unit cell, respectively. $F(h, k, l)$ is the structure factor amplitude of reflection (hkl) and $\alpha(h, k, l)$ is the phase angle. The molecular trace of an initial model is built from the calculated electron density. Although most values in this calculation can be obtained from the diffraction data, there is no known information on the phase angles. This is known as the phase problem in crystallography. The techniques available to solve the phase problem include isomorphic replacement methods,[34] multiple or single wavelength anomalous diffraction,[35] and molecular replacement.[36] When there is no *a priori* structural information available for the targeted protein, the former two techniques must be used. They rely on the presence of an atom that strongly scatters X-rays relative to C, N, and O that predominate in protein molecules (whether it is intrinsic to the protein or incorporated). Consequently, the quality of a protein crystal corresponds to how *useful* its diffraction data can be for the accurate calculation of phase angles. The quality of the protein crystal, therefore, influences the accuracy of the initial protein model.

If we can quantify the potential for a protein crystal to provide accurate phase information for *de novo* structure determination, a measure of its usefulness for crystallographic structural determination can be obtained. Many proteins containing intrinsic atoms that scatter X-rays anomalously are valuable for multiple or single wavelength anomalous diffraction analysis. Because most proteins contain covalently bound sulfur atoms, the order of the sulfur substructure in the crystalline lattice can represent the overall order of the protein and its potential for phase determination.[37,38] Therefore, the quality of the sulfur atom contribution can be expressed as the phasing power, defined as follows:

$$\left[\frac{\langle |F_{s'} - F_s|^2 \rangle}{\int_\phi P(\phi)(||F_s|e^{i\phi} + \Delta F_{s''}| - |F_{s'}|)^2 d\phi} \right]^{1/2}, \quad (2)$$

where $P(\phi)$ is the experimental phase probability distribution, F_s corresponds to the structure factors of the sulfur atoms, $F_{s'}$ corresponds to that of its Friedel mate, and $F_{s''}$ is twice the anomalous components of the sulfur structure factors. In addition to general assessment methods, the quality of protein crystals grown from space can be compared to those grown on Earth in terms of the phasing statistics for sulfur and, in turn, their usefulness in structure determination.

Measurement of phasing power from anomalous scattering atoms is not limited to sulfur. Many proteins have naturally incorporated metal ions, such as iron or cobalt, associated with enzyme cofactors that can also be used for measurement of phasing power. We have begun assessing crystal quality of space-grown protein crystals compared to that of Earth-grown by measuring the phasing power of sulfur and its potential for *ab initio* phase determination for thaumatin.

TABLE 2. Phasing statistics of thaumatin Earth- and space-grown crystals

	Earth		Space	
	Crystal 1	Crystal 2	Crystal 1	Crystal 2
Phasing power	1.3	1.5	2.1	2.2
Generalized R_{cullis}	0.60	0.59	0.55	0.52

NOTE: Calculations were performed using the CNS software suite for macromolecular structure determination.[42] The phasing power and generalized R_{cullis} are defined in Equation (2) and in the text.

Complete data sets with more than 15-fold redundancy were recorded for each single thaumatin crystal from space and on Earth with our home laboratory X-ray source. The phasing power was calculated as defined above and the corresponding generalized R_{cullis},

$$R_{cullis} = \frac{\langle LOC^2 \rangle^{1/2}}{\langle |\Delta F_{s''}|^2 \rangle^{1/2}},$$

is also included in the phasing statistics (see TABLE 2) where LOC is the *lack of closure error*.[35]

Both thaumatin crystals grown under microgravity showed improved phasing statistics over those of Earth-grown crystals. The phasing potential for *ab initio* structure determination was at least 1.4 times higher for crystals obtained from space compared to the those grown on Earth. The sulfur anomalous signals allowed a more accurate determination of the sulfur positions from the space-grown than from the Earth-grown crystals. When the signal of the sulfur can be measured accurately, solvent flattening procedures are more effective leading to improved estimates for protein phases.[39] Consequently, initial electron-density Fourier difference maps, suitable for atomic fitting, can be obtained.[40]

CONCLUSION

The crystallization of thaumatin, aspartyl-*t*RNA synthetase, and pea lectin under microgravity resulted in higher quality crystals as determined by their usefulness for crystallographic determination. Comparative experiments revealed that space-grown compared to that of Earth-grown crystals diffracted more intensely, had lower mosaicity, and provided an easier calculation in an initial crystallographic model.

Using thaumatin, the effectiveness of a crystal for *ab initio* structure determination is shown to be reflected on the quality of the protein crystal. The phase determination depends on the accuracy of the sulfur anomalous signal that reflects the crystalline order. The phasing power of a protein crystal thus becomes a valuable measure on how useful it can be for structure determination and, thus, is a more reliable measure of quality. In the case of thaumatin, microgravity resulted in crystals with improved phasing statistics compared to those obtained on Earth. The ability to grow superior crystals in space that are more useful for X-ray crystallography further supports the benefits of microgravity protein crystal growth.

ACKNOWLEDGMENTS

Protein preparation, crystallization, data collection, and X-ray analysis for thaumatin and pea lectin were performed by Anna M. Holmes, José A. Gavira-Gallardo, Eddie Snell, and Mark van der Woerd. Great appreciation is extended to Gloria Borgstahl and Henry Bellamy for providing synchrotron beam-time on line 7-1 at the Stanford Synchrotron Radiation Laboratory. Participants involved in the crystal growth and crystallographic analysis of AspRS included Claude Sauter, Bernard Lorber, John Arnez, and Richard Giegé. The home source X-ray facility was managed by Liqing Chen.

Special thanks go to Zhi-Jie (James) Liu, B.C. Wang, and Edward Meehan for their invigorating discussion and scientific input and Joyce Looger for her invaluable technical support. The author also thanks Alex McPherson and Greg Jenkins for their collaboration using EGN. Much appreciation goes to Marilyn Yoder for reviewing this manuscript.

This work was funded by the National Aeronautics and Space Administration (NASA), the Centre National d'Etudes Spatiales (CNES), the European Space Agency (ESA), and the Centre National de la Recherche Scientifique (CNRS).

REFERENCES

1. LITTKE, W. & C. JOHN. 1984. Protein single crystal growth under microgravity. Science **225**: 203–204.
2. DELUCAS, L.J., K.M. MOORE & M.M. LONG. 1999. Protein crystal growth and the International Space Station. Gravit. Space Biol. Bull. **12**: 39–45.
3. MOORE, K.M., M.M. LONG & L.J. DELUCAS. 1999. Space Technology and Applications Int. Forum. M.S. EI-Genk, Ed. **458**: 217–224. American Institute of Physics.
4. MCPHERSON, A. 1999. Crystallization of Biological Macromolecules. Cold Spring Harbor Press, Cold Spring Harbor, New York.
5. COUZIN, J. 1998. Negative review galls space crystallographers. Science **281**: 497–498.
6. MCPHERSON, A. 1997. Recent advances in the microgravity crystallization of biological macromolecules. Trends Biotech. **15**: 197–200.
7. KUNDROT, C.E., R.A. JUDGE, M.L. PUSEY & E. SNELL. 2000. Microgravity and macromolecular crystallography. Cryst. Growth Design **1**: 87–99.
8. MCPHERSON, A. 1996. Macromolecular crystal growth in microgravity. Crystallogr. Rev. **6**: 157–305.
9. NG, J.D., B. LORBER, R. GIEGÉ, et al. 1997. Comparative analysis of thaumatin crystals grown on earth and in microgravity. Acta Crystallogr. **D53**: 724–733.
10. CARTER, D.C., K. LIM, J.X. HO, et al. 1999. Lower dimer impurity incorporation may result in higher perfection of HEWL crystals grown in microgravity. A case study. J. Cryst. Growth **196**: 623–637.
11. CHAYEN, N.E. & J.R. HELLIWELL. 1999. Space-grown crystals may prove their worth. Nature **398**: 20.
12. DECLERCQ, J.-P., et al. 1999. A crystal of a typical EF-hand protein grown under microgravity diffracts X-rays beyond 0.9 Å resolution. J. Cryst. Growth **196**: 595–601.
13. LORBER, B., C. SAUTER, M.-C. ROBERT, et al. 1999. Crystallization within agarose gel in microgravity improves the quality of thaumatin crystals. Acta Crystallogr. **D55**: 1491–1494.
14. DELUCAS, L.J., et al. 2000. Space Technology and Application International Forum. CP504. 488.

15. NG, J.D., C. SAUTER, B. LORBER, et al. 2002. Comparative analysis of space- and Earth-grown crystals of an aminoacyl-tRNA synthetase: space-grown crystals are more useful for structure determination. Acta Crystallogr. **D58:** 645–652.
16. VANEY, M.C., M. MAIGNAN, M. RIÈS-KAUTT & A. DUCRUIX. 1996. High-resolution structure (1.33 Å) of a HEW lysozyme tetragonal crystal grown in the APCF apparatus. Data and structural comparison with a crystal grown under microgravity from Spacehab-01 mission. Acta Crystallogr. **D52:** 505–517.
17. DONG, J., T.J. BOGGON, N.E. CHAYEN, et al. 1999. Bound-solvent structures for microgravity-, ground control-, gel-, and microbatch-grown hen egg-white lysozyme crystals at 1.8 Å resolution. Acta Crystallogr. **D55:** 745–752.
18. OTÁLORA, F., M.L. NOVELLA, J.A. GAVIRA, et al. 2001. Experimental evidence for the stability of the depletion zone around a growing protein crystal under microgravity. Acta Cryst. **D57:** 412–417.
19. SAUTER, C., F. OTÁLORA, J.A. GAVIRA, et al. 2001. Structure of tetragonal hen egg-white lysozyme at 0.94 Å from crystals grown by the counter-diffusion method. Acta Crystallogr. **D57:** 1119–1126.
20. BORGSTAHL, G.E.O., A. VAHEDI-FARIDI, H. BELLAMY & E.H. SNELL. 2001. A test of macromolecular crystallization in microgravity: large, well ordered insulin crystals. Acta Crystallogr. **D57:** 1204–1207.
21. DELUCAS, L.J., C.D. SMITH, H.W. SMITH, et al. 1989. Protein crystal growth in microgravity. Science **246:** 651–654.
22. HELLIWELL, J.R. 1988. Protein crystal perfection and the nature of radiation damage. J. Cryst. Growth **90:** 259–272.
23. WEISGERBER, S. & J.R. HELLIWELL. 1993. Chicken egg-white lysozyme crystal grown in microgravity and on Earth for a comparison on their respective perfection. Joint CCP4 & ESF-EACBM Newsletter on Protein Crystallography **29:** 10–13.
24. SNELL, E., H.S. WEISGERBER, J.R. HELLIWELL, et al. 1995. Improvements in lysozyme protein crystal perfection through microgravity growth. Acta Crystallogr. **D51:** 1099–1102.
25. FERRER, J.L., J. HIRSCHLER, M. ROTH & J. FONTECILLA-CAMPS. 1996. ESRF Newsletter 27.
26. LORBER, B., C. SAUTER, J.D. NG, et al. 1999. Characterization of protein and virus crystals by quasi-planar wave X-ray topography: a comparison between crystals grown in solution and in agarose gel. J. Cryst. Growth **204:** 357–368.
27. VAN DER WEL, H., T.C. VAN SOEST & E.C. ROYERS. 1975. Crystallization and crystal data of thaumatin I, a sweet-tasting protein from *Thaumatococcus daniellii benth.* FEBS Lett. **56:** 316–317.
28. MCPHERSON, A. & J. WEICKMANN. 1990. X-ray analysis of new crystal forms of the sweet protein thaumatin. J. Biomol. Struct. Dyn. **5:** 1053–1060.
29. BOSCH, R., P. LAUTENSCHLAGER, L. POTTHAST & J. STAPELMANN. 1992. Experimental equipment for protein crystallization in microgravity facilities. J. Cryst. Growth **122:** 310–316.
30. BECKER, H.D., J. REINBOLT, R. KREUTZER, et al. 1997. Existence of two distinct aspartyl-tRNA synthetases in *Thermus thermophilus*. Structural and biochemical properties of the two enzymes. Biochemistry **36:** 8785–8797.
31. POTERSZMAN, A., P. PLATEAU, D. MORAS, et al. 1993. Sequence, overproduction and crystallization of aspartyl-tRNA synthetase from *Thermus thermophilus*. Implications for the structure of prokaryotic aspartyl-tRNA synthetases. FEBS Lett. **325:** 183–186.
32. DELARUE, M., A. POTERSZMAN, S. NIKONOV, et al. 1994. Crystal structure of a prokaryotic aspartyl-tRNA synthetase. EMBO J. **13:** 3219–3229.
33. NG, J.D., B. LORBER, J. WITZ, et al. 1996. Crystal growth of macromolecules from precipitation: evidence for Ostwald ripening. J. Cryst. Growth **168:** 50–62.
34. PHILLIPS, D.C. 1966. Advances in Protein Crystallography, Advances in Structure Research by Diffraction Methods. 75–140. F. Wiweg, Brauschweig.
35. DRENTH, J. 1995. Principles of Protein X-ray Crystallography. 153–182. Springer-Verlag, New York.
36. ROSSMAN, M.G. 1972. The Molecular Replacement Method. Gordon and Breach, London.

37. DAUTER, Z., M. DAUTER, E. DE LA FORTELLE, *et al.* 1999. Can anomalous signal of sulfur become a tool for solving protein crystal structures? J. Mol. Biol. **289:** 83–92.
38. LIU, Z.-J., E.S. VYSOTSKI, C.-J. CHEN, *et al.* 2000. Structure of the Ca^{2+}-regulated photoprotein obelin at 1.7 Å resolution determined directly form its sulfur substructure. Protein Sci. **9:** 2085–2093.
39. WANG, B.C. 1985. Resolution of phase ambiguity in macromolecular crystallography. Meth. Enzymol. **115:** 90–112.
40. GAVIRA, J.A., D. TOH, J. LOPÉZ-JARAMILLO, *et al.* 2002. *Ab initio* crystallographic structure determination of insulin from protein to electron density without crystal handling. Acta Crystallogr. **D58:** 1147–1154.
41. OTWINOWSKI, Z. & W. MINOR. 1997. Processing of X-ray diffraction data collected in oscillation mode. Meth. Enzymol. **276:** 307–326.
42. BRÜNGER, A., A.T. ADAMS, P. D. CLORE, *et al.* 1998. Crystallography and NMR system: a new software suite for macromolecular structure determination. Acta Crystallogr. **D54:** 905–921.
43. JONES, T.A., J.-Y. ZOU, S.W. COWAN & M. KJELDGAARD. 1991. Improved methods for building protein models in electron density maps and the location of errors in these models. Acta Crystallogr. **A47:** 110–119.
44. DELARUE, M., A. POTERSZMAN, S. NIKONOV, *et al.* 1994. Crystal structure of a prokaryotic aspartyl-*t*RNA synthetase. EMBO J. **13:** 3219–3229.
45. MCPHERSON, A. 1982. Preparation and Analysis of Protien Cystals. John Wiley & Sons, New York.
46. PRASTHOFER, T., S.R. PHILLIPS, F.L. SUDDATH & J.A. ENGLER. 1989. Design, expression and crystallization of recombinant lectin from the garden pea (*Pisum sativum*). J. Biol. Chem. **264:** 6793–6796.

Mathematical Model for Diffusion of a Protein and a Precipitant about a Growing Protein Crystal in Microgravity

ONOFRIO ANNUNZIATA[a,b] AND JOHN G. ALBRIGHT[a]

[a]*Department of Chemistry, Texas Christian University, Fort Worth, Texas, USA*

[b]*Department of Physics, Massachusetts Institute of Technology, Cambridge, Massachusetts, USA*

> ABSTRACT: Equations are presented that model diffusion of a protein to the surface of a growing crystal in a convection-free environment. The equations apply to crystal growth solutions that contain both a protein and a protein precipitant. The solutions are assumed ternary and the equations include all four diffusion coefficients necessary for the full description of the diffusion process. The four diffusion coefficients are assumed constant. Effects of crystal/solution moving boundary and the effect of a protein adsorption barrier at the crystal interface are included. The equations were applied to the system lysozyme chloride + NaCl + H_2O, which has served as the primary model system for the study of crystal growth of proteins and for which there are now published ternary diffusion coefficients. Calculated results with and without the inclusion of cross-term diffusion coefficients are compared.
>
> KEYWORDS: multicomponent; diffusion; crystal growth; microgravity; lysozyme; NaCl

INTRODUCTION

Obtaining protein crystals of good structural quality is often the main issue for three dimensional, atomic resolution structure studies of biological macromolecules. Crystallization is an intrinsically non-equilibrium process, and concentration gradients occur around the crystal. The protein crystallizes, reducing its concentration at the moving face of the growing crystal. This creates a protein gradient between the bulk solution and the crystal. This gradient, in turn, causes multicomponent diffusive transport of protein and precipitant. The salt may move toward or away from the crystal face due to the concentration gradient of the protein adjacent to the crystal face and depending on the sign of D_{21}.[1–3] The gradients produced are related to several mechanisms: both the intrinsic kinetic rates of crystal growth and transport to the crystal surface determine the path the system takes toward equilibrium. Any variable, whose change modifies those processes, changes the properties of the resultant final crystal.

Address for correspondence: John G. Albright, Department of Chemistry, Box 298860, Texas Christian University, Fort Worth, TX 76129, USA.
j.Albright@tcu.edu

Crystallization experiments conducted under microgravity conditions have yielded protein crystals that provide diffraction data of significantly higher resolution than the best crystals of these proteins grown under normal gravity conditions.[3] Since a clear difference between microgravity and normal gravity based experiments is the magnitude of the buoyancy forces in the solution, the role of convection in protein crystal growth is of great interest.

Lin et al.[2] have analyzed, by numerically modeling lysozyme crystal growth, the difference between the concentration fields that occur at $0\,g$ and $1\,g$ around a crystal. With convection, the lysozyme concentration in the bulk solution is more uniform. At $0\,g$, a concentration depletion zone of considerable width develops about the growing crystal. Such boundary layers, in which the probability for nucleation of parasitic crystals is strongly reduced, have often been cited as a reason for obtaining better protein crystal in microgravity. Lin et al.[2] also raised the issue of coupled transport between protein and precipitant. However, the lack of information on the protein-salt transport interaction has made it difficult to develop accurate and meaningful models.

In this paper, a mathematical model is developed for describing the concentration profiles around a growing spherical crystal in a microgravity environment, in which convection is not considered.

GEOMETRY OF THE MODEL

The geometry of the model is based on a spherical crystal surrounded by an aqueous fluid containing protein and precipitant. This configuration can be prepared experimentally by adding a crystal to a uniform supersaturated solution in a microgravity environment. However, it is important to note that, due to vibrations on the spacecraft, the crystal may move within the solution. We do not consider this effect in the model. The dimensions of the cell are considered large with respect to the size of the crystal contained inside. Model symmetry, together with other reasonable assumptions, furnishes analytic solutions of the concentration profiles around the crystal. In a real experiment, the dimensions of the crystal may not be negligible relative to the size of the cell, and the crystal shape will not be spherical.[4] Nevertheless, the aim of the model is to highlight the role of the coupled diffusion and relate the effects to only a few relevant variables.

Fick's second law is the cardinal equation for the model:

$$\frac{\partial c_1}{\partial t} = D_{11}\nabla^2 c_1 + D_{12}\nabla^2 c_2,$$
$$\frac{\partial c_2}{\partial t} = D_{21}\nabla^2 c_1 + D_{22}\nabla^2 c_2. \quad (1)$$

Subscript "1" denotes the protein and "2" the precipitant, c_i is the molar concentration of species i, D_{ij} denotes a diffusion coefficient (the volume frame reference is assumed), and t is the time. For a spherical crystal, it is convenient to express the Laplacian in spherical coordinates. In this representation, in fact, the radial distance from the center of the crystal r is the only position variable. We can transform the concentration Laplacian into[5] and by setting $\gamma_i = c_i r$, from (1) we obtain

$$\nabla^2 c_i = \frac{\partial^2 c_i}{\partial r^2} + \frac{2}{r}\frac{\partial c_i}{\partial r}, \tag{2}$$

$$\frac{\partial \gamma_1}{\partial t} = D_{11}\frac{\partial^2 \gamma_1}{\partial r^2} + D_{12}\frac{\partial^2 \gamma_2}{\partial r^2},$$

$$\frac{\partial \gamma_2}{\partial t} = D_{21}\frac{\partial^2 \gamma_1}{\partial r^2} + D_{22}\frac{\partial^2 \gamma_2}{\partial r^2}. \tag{3}$$

INITIAL CONDITIONS

As an initial condition (time $t = 0$), we assume the solution to have uniform concentrations of all components throughout; that is,

$$c_i = c_i^0, \quad \gamma_i = c_i^0 r = \gamma_i^0 \quad (i = 1, 2). \tag{4}$$

Several partial differential equations can be reduced to ordinary differential equations by performing the Laplace transformation $L\{\cdot\}$ that incorporates the initial conditions. The Laplace transformation of a function $f(t)$ is defined in the s domain by the following operation:

$$\bar{f}(s) \equiv L\{f(t)\} = \int_0^\infty f(t) e^{-st} dt. \tag{5}$$

Its application to the concentration quantities gives

$$L\{\gamma_i\} = \bar{\gamma}_i$$

$$L\left\{\frac{\partial \gamma_i}{\partial t}\right\} = s\bar{\gamma}_i - \gamma_i^0, \tag{6}$$

so that (1) becomes

$$s\bar{\gamma}_1 - \bar{\gamma}_1^0 = D_{11}\frac{\partial^2 \bar{\gamma}_1}{\partial r^2} + D_{12}\frac{\partial^2 \bar{\gamma}_2}{\partial r^2}$$

$$s\bar{\gamma}_2 - \bar{\gamma}_2^0 = D_{21}\frac{\partial^2 \bar{\gamma}_1}{\partial r^2} + D_{22}\frac{\partial^2 \bar{\gamma}_2}{\partial r^2}. \tag{7}$$

GENERAL SOLUTIONS

Define the quantities μ_i from the following equations:

$$\bar{\gamma}_1 = H\mu_1 - F\mu_2$$

$$\bar{\gamma}_2 = -G\mu_1 + E\mu_2, \tag{8}$$

where E, F, G, and H (following Fujita and Gosting symbolism[6]) are

$$E = \frac{D_{11}}{|D|}, \quad F = \frac{D_{12}}{|D|}, \quad G = \frac{D_{21}}{|D|}, \quad H = \frac{D_{22}}{|D|}, \qquad (9)$$

with the determinant

$$|D| = \begin{vmatrix} D_{11} & D_{12} \\ D_{21} & D_{22} \end{vmatrix}.$$

Differentiating (8) with respect to r, we obtain

$$\frac{\partial^2 \mu_1}{\partial r^2} = D_{11}\frac{\partial^2 \bar{\gamma}_1}{\partial r^2} + D_{12}\frac{\partial^2 \bar{\gamma}_2}{\partial r^2}$$

$$\frac{\partial^2 \mu_2}{\partial r^2} = D_{21}\frac{\partial^2 \bar{\gamma}_1}{\partial r^2} + D_{22}\frac{\partial^2 \bar{\gamma}_2}{\partial r^2}. \qquad (10)$$

Inserting (8) and (10) into (7), we obtain

$$\frac{\partial^2 \mu_1}{\partial r^2} = -\gamma_1^0 + s\bar{\gamma}_1 = -\gamma_1^0 + s(H\mu_1 - F\mu_2)$$

$$\frac{\partial^2 \mu_2}{\partial r^2} = -\gamma_2^0 + s\bar{\gamma}_2 = -\gamma_2^0 + s(E\mu_2 - G\mu_1). \qquad (11)$$

Differentiating (11) with respect to r twice, we obtain

$$\frac{\partial^4 \mu_1}{\partial r^4} = s\left(H\frac{\partial^2 \mu_1}{\partial r^2} - F\frac{\partial^2 \mu_2}{\partial r^2}\right)$$

$$\frac{\partial^4 \mu_2}{\partial r^4} = s\left(E\frac{\partial^2 \mu_2}{\partial r^2} - G\frac{\partial^2 \mu_1}{\partial r^2}\right). \qquad (12)$$

It now remains to uncouple the two differential equations (12) so that μ_1 and μ_2 do not depend on each other. From Equations (11) we can write:

$$\frac{\partial^2 \mu_1}{\partial r^2} = -\gamma_1^0 - sF\mu_2 + \frac{H}{G}\left(-\gamma_2^0 + sE\mu_2 - \frac{\partial^2 \mu_2}{\partial r^2}\right)$$

$$\frac{\partial^2 \mu_2}{\partial r^2} = -\gamma_2^0 - sG\mu_1 + \frac{E}{F}\left(-\gamma_1^0 + sH\mu_1 - \frac{\partial^2 \mu_1}{\partial r^2}\right), \qquad (13)$$

and inserting (13) into (12), we have

$$\frac{\partial^2 \mu_1}{\partial r^4} - s(H+E)\frac{\partial^2 \mu_1}{\partial r^2} + s^2(EH - FG)\mu_1 = s(E\gamma_1^0 + F\gamma_2^0)$$

$$\frac{\partial^2 \mu_2}{\partial r^4} - s(H+E)\frac{\partial^2 \mu_2}{\partial r^2} + s^2(EH - FG)\mu_2 = s(G\gamma_1^0 + H\gamma_2^0). \qquad (14)$$

These are two uncoupled non-homogeneous differential equations. The characteristic equation related to the corresponding homogeneous differential equation, is

$$\lambda^4 - s(H+E)\lambda^2 + s^2(EH - FG) = 0, \tag{15}$$

from which we obtain two real negative physical solutions

$$\lambda_1 = -\left\{\frac{s}{2}[H + E + \sqrt{(E+H)^2 + 4FG}]\right\}^{1/2} = -\sqrt{\sigma_1}\sqrt{s}$$

$$\lambda_2 = -\left\{\frac{s}{2}[H + E + \sqrt{(E+H)^2 + 4FG}]\right\}^{1/2} = -\sqrt{\sigma_2}\sqrt{s}. \tag{16}$$

From (14) the form of the particular integrals is

$$\mu_i' = [\text{constant}]r. \tag{17}$$

Thus, imposing

$$\frac{\partial^4}{\partial r^4}\mu_i' = \frac{\partial^2}{\partial r^2}\mu_i' = 0$$

from (17), we obtain

$$\mu_1' = \frac{1}{s}\frac{E\gamma_1^0 + F\gamma_2^0}{EH - FG}$$

$$\mu_2' = \frac{1}{s}\frac{G\gamma_1^0 + H\gamma_2^0}{EH - FG}. \tag{18}$$

Defining

$$\alpha_i(r) = e^{-\sqrt{s\sigma_i}r}, \tag{19}$$

the solutions have the following form:

$$\mu_1 = \theta_{11}\alpha_1(r) + \theta_{12}\alpha_2(r) + \frac{1}{s}\frac{E\gamma_1^0 + F\gamma_2^0}{EH - FG}$$

$$\mu_2 = \theta_{21}\alpha_1(r) + \theta_{22}\alpha_2(r) + \frac{1}{s}\frac{G\gamma_1^0 + H\gamma_2^0}{EH - FG}, \tag{20}$$

where the θ_{ij} values are determined by the boundary conditions. Differentiating (20) with respect to r twice,

$$\frac{\partial^2 \mu_1}{\partial r^2} = s[\theta_{11}\sigma_1\alpha_1(r) + \theta_{12}\sigma_2\alpha_2(r)]$$

$$\frac{\partial^2 \mu_2}{\partial r^2} = s[\theta_{21}\sigma_1\alpha_1(r) + \theta_{22}\sigma_2\alpha_2(r)]. \tag{21}$$

Substituting (21) into (11),

$$\theta_{11}s\sigma_1\alpha_1(r) + \theta_{12}s\sigma_2\alpha_2(r)$$
$$= -\gamma_1^0 + sH[\theta_{11}\alpha_1(r) + \theta_{12}\alpha_2(r) + \mu_1'] - sG[\theta_{21}\alpha_1(r) + \theta_{22}\alpha_2(r) + \mu_2']$$

$$\theta_{21}s\sigma_1\alpha_1(r) + \theta_{22}s\sigma_2\alpha_2(r)$$
$$= -\gamma_2^0 + sE[\theta_{21}\alpha_1(r) + \theta_{22}\alpha_2(r) + \mu_2'] - sG[\theta_{11}\alpha_1(r) + \theta_{12}\alpha_2(r) + \mu_1'].$$
(22)

The identity between corresponding coefficients produces

$$\theta_{21} = \frac{H - \sigma_1}{F}\theta_{11} = \frac{G}{E - \sigma_1}\theta_{11}$$

$$\theta_{12} = \frac{F}{H - \sigma_2}\theta_{22} = \frac{E - \sigma_2}{G}\theta_{22}.$$
(23)

The reduction to two coefficients, allows us to set

$$K_1 = \theta_{11},$$
$$K_2 = \theta_{22},$$
(24)

so that (20) becomes

$$\mu_1 = K_1\alpha_1(r) + \frac{F}{H - \sigma_2}K_2\alpha_2(r) + \mu_1'$$

$$\mu_2 = \frac{G}{E - \sigma_1}K_1\alpha_1(r) + K_2\alpha_2(r) + \mu_2',$$
(25)

and from (8), we obtain:

$$\bar{\gamma}_1 = \sigma_1 K_1\alpha_1(r) + \frac{F\sigma_2}{H - \sigma_2}K_2\alpha_2(r) + \frac{\bar{\gamma}_1^0}{s}$$

$$\bar{\gamma}_2 = \frac{G\sigma_1}{E - \sigma_1}K_1\alpha_1(r) + \sigma_2 K_2\alpha_2(r) + \frac{\bar{\gamma}_2^0}{s}.$$
(26)

BOUNDARY CONDITIONS

Further specialization of integral solutions (26) needs the determination of K_i values, which are obtained by specifying the boundary conditions. Since we have assumed the cell is large with respect to the crystal dimensions, the restrictions imposed by the cell wall are not considered (free diffusion). The only boundary conditions that need to be considered are related to the crystal-solution interface.

The crystal growth rate, da/dt, where a is the crystal radius, is in general a complicated function of the supersaturation $\sigma = (c_1^i - c_1^e)/c_1^e$ at the crystal interface, where c_1^i is the concentration of the protein at the interface and c_1^e is the concentration in equilibrium with the crystal for a given concentration c_2^i of the precipitant. However, in several cases, the growth rate can be reasonably and conveniently expressed by the following linear function of the supersaturation:[2]

$$\frac{da}{dt} = \beta(\sigma - \sigma_0),$$
(27)

where β is a kinetic constant and σ_0 indicates the minimum supersaturation required for the crystal to grow.

The solutes fluxes, J_1^i and J_2^i, at the interface $r = a$, are defined by

$$-J_1^i = D_{11}\frac{\partial c_1}{\partial r}\bigg|_{r=a} + D_{12}\frac{\partial c_2}{\partial r}\bigg|_{r=a}$$
$$-J_2^i = D_{21}\frac{\partial c_1}{\partial r}\bigg|_{r=a} + D_{22}\frac{\partial c_2}{\partial r}\bigg|_{r=a}.$$
(28)

The mass balance at the interface is described by the following equations:[2]

$$-J_1^i = \left(c_1^s - c_1^i\frac{\rho_s}{\rho_l}\right)\frac{da}{dt}$$
$$-J_2^i = \left(c_2^s - c_2^i\frac{\rho_s}{\rho_l}\right)\frac{da}{dt},$$
(29)

where c_i^s denotes the concentration of solute i in the solid phase, ρ_s is the solid phase density, and ρ_l is the liquid phase density. Balance equations **(29)** regard the net fluxes as the difference between the rate of solute insertion into the crystal and the rate of solutes rejection due to the replacement of the fluid solution with the new crystalline phase. Note that since the crystal is growing in the positive r direction, the protein flux towards the crystal is negative whereas its interface gradient is positive.

Rigorously speaking, the growing crystal gives rise to a moving boundary problem that will cause changes in a. If the crystal geometry were rectangular, both setting the frame at the crystal interface and using the free diffusion condition would not give rise to any loss of rigor in the treatment. However, the spherical geometry associated with the growing crystal imposes a curvature change that should be taken into account. If the radius is large and the curvature changes are small relative to the crystal growth rate, then the approximation of letting a be constant becomes reasonable.[2]

The analytic expressions for the inverse Laplace transformation are available[7] if the boundary equations can be expressed in the following way:

$$D_{11}\frac{\partial c_1}{\partial r}\bigg|_{r=a} + D_{12}\frac{\partial c_2}{\partial r}\bigg|_{r=a} = h_1(c_1^i - c_1^*)$$
$$D_{21}\frac{\partial c_1}{\partial r}\bigg|_{r=a} + D_{22}\frac{\partial c_2}{\partial r}\bigg|_{r=a} = h_2(c_1^i - c_1^*),$$
(30)

where the h_i and the c_1^* are constants. The expression for these parameters can be obtaining inserting **(27)** and **(29)** into **(28)**,

$$-J_1^i = \left(c_1^s - c_1^i\frac{\rho_s}{\rho_l}\right)\frac{\beta}{c_1^e}[c_1^i - c_1^e(1+\sigma_0)] = h_1(c_1^i - c_1^*)$$
$$-J_2^i = \left(c_2^s - c_2^i\frac{\rho_s}{\rho_l}\right)\frac{\beta}{c_1^e}[c_1^i - c_1^e(1+\sigma_0)] = h_2(c_1^i - c_1^*),$$
(31)

where

$$c_1^* = c_1^e(1 + \sigma_0)$$

$$h_1 = \left(c_1^s - c_1^i \frac{\rho_s}{\rho_l}\right) \frac{\beta}{c_1^e}$$

$$h_2 = \left(c_2^s - c_2^i \frac{\rho_s}{\rho_l}\right) \frac{\beta}{c_1^e}.$$

(32)

We can see that c_1^* is constant if and only if c_1^e is constant. From solubility theory we expect that the protein equilibrium concentration is given approximately by the following equation:[8,9]

$$c_1^e = c e^{-k c_2^i}, \quad (33)$$

where c and k are constants. Since we expect that c_2^i is not so different from c_2^0, we can reasonably assume that $c_1^e \approx c_1^{e0} \equiv c e^{-k c_2^e}$ is roughly constant. We can then write

$$c_1^* \approx c_1^{e0}(1 + \sigma_0). \quad (34)$$

The concentration c_1^i is small with respect to c_1^s, whereas the crystal density is only slightly larger than the solution density, thus[2] $c_1^s \gg c_1^i \rho_s/\rho_l$. We can now reasonably assume that h_1 is constant and equal to

$$h_1 \approx \beta \frac{c_1^*}{c_1^{e0}}. \quad (35)$$

The concentration of salt in the solid can be estimated by the repartition constant, α, defined by the following equation:[10]

$$\alpha = \frac{c_2^s/c_1^s}{c_2^i/c_1^i} \cong \frac{c_2^s/c_1^s}{c_2^0/c_1^0}. \quad (36)$$

It is reasonable to assume that h_2 is constant and given by

$$h_2 \approx \beta \frac{c_2^0}{c_1^{e0}}\left(\alpha \frac{c_1^s}{c_1^0} - \frac{\rho_s}{\rho_l}\right). \quad (37)$$

Note that h_1 is always positive because it describes the protein mass transfer. On the other hand, h_2 is negative because the precipitant rejection caused by the crystal growth is larger than the contribution due to the salt inclusion into the crystal. Hence, the protein flux is negative, whereas the precipitant flux is positive.

CONCENTRATION PROFILES

From the definition of γ_i, we can write

$$\frac{\partial c_i}{\partial r} = \frac{\partial}{\partial r}(\gamma_i/r) = \frac{1}{r}\frac{\partial \gamma_i}{\partial r} - \frac{1}{r^2}\gamma_i, \quad (38)$$

and by defining $\bar{\gamma}_i^* \equiv c_1^* a/s$, (**30**) becomes:

$$D_{11}\left(\frac{\partial\bar{\gamma}_1}{\partial r}\bigg|_{r=a} - \frac{\bar{\gamma}_1^i}{a}\right) + D_{12}\left(\frac{\partial\bar{\gamma}_2}{\partial r}\bigg|_{r=a} - \frac{\bar{\gamma}_2^i}{a}\right) = h_1(\bar{\gamma}_1^i - \bar{\gamma}_1^*)$$
$$D_{21}\left(\frac{\partial\bar{\gamma}_1}{\partial r}\bigg|_{r=a} - \frac{\bar{\gamma}_1^i}{a}\right) + D_{22}\left(\frac{\partial\bar{\gamma}_2}{\partial r}\bigg|_{r=a} - \frac{\bar{\gamma}_2^i}{a}\right) = h_2(\bar{\gamma}_1^i - \bar{\gamma}_1^*).$$
(39)

To obtain the K_i, we need to insert the general solutions into (39). First note that from (39),

$$-\frac{\partial\bar{\gamma}_1}{\partial r}\bigg|_{r=a} = \sqrt{s}\left[\sigma_1^{3/2}K_1\alpha_1(r) + \frac{F\sigma_2^{3/2}}{H-\sigma_2}K_2\alpha_2(r) + \frac{c_1^0}{s}\right]$$
$$-\frac{\partial\bar{\gamma}_2}{\partial r}\bigg|_{r=a} = \sqrt{s}\left[\frac{G\sigma_1^{3/2}}{E-\sigma_1}K_1\alpha_1(r) + \sigma_2^{3/2}K_2\alpha_2(r) + \frac{c_2^0}{s}\right]$$
(40)

and

$$-\left(\frac{\partial\bar{\gamma}_1}{\partial r}\bigg|_{r=a} - \frac{\bar{\gamma}_1^i}{a}\right) = \sigma_1^{3/2}K_1\alpha_1(r)\left(\frac{1}{a} + \sqrt{s}\sqrt{\sigma_1}\right) + \frac{F\sigma_2^{3/2}}{H-\sigma_2}K_2\alpha_2(r)\left(\frac{1}{a} + \sqrt{s}\sqrt{\sigma_2}\right)$$
$$-\left(\frac{\partial\bar{\gamma}_2}{\partial r}\bigg|_{r=a} - \frac{\bar{\gamma}_2^i}{a}\right) = \frac{G\sigma_1^{3/2}}{E-\sigma_1}K_1\alpha_1(r)\left(\frac{1}{a} + \sqrt{s}\sqrt{\sigma_1}\right) + \sigma_2^{3/2}K_2\alpha_2(r)\left(\frac{1}{a} + \sqrt{s}\sqrt{\sigma_2}\right),$$
(41)

then, inserting (41) into (39)

$$(P_{11} + \sqrt{s}Q_{11})\alpha_1(r)K_1 + (P_{12} + \sqrt{s}Q_{12})\alpha_2(r)K_2 = \frac{B_1}{s}$$
$$(P_{21} + \sqrt{s}Q_{21})\alpha_1(r)K_1 + (P_{22} + \sqrt{s}Q_{22})\alpha_2(r)K_2 = \frac{B_2}{s},$$
(42)

where

$$P_{11} = \left(D_{11}\sigma_1 + D_{12}\frac{G\sigma_1}{E-\sigma_1}\right)\frac{1}{a} + h_1\sigma_1$$
$$P_{12} = \left(D_{11}\frac{F\sigma_2}{H-\sigma_2} + D_{12}\sigma_2\right)\frac{1}{a} + h_1\frac{F\sigma_2}{H-\sigma_2}$$
$$P_{21} = \left(D_{21}\sigma_1 + D_{22}\frac{G\sigma_1}{E-\sigma_1}\right)\frac{1}{a} + h_2\sigma_1$$
$$P_{22} = \left(D_{21}\frac{F\sigma_2}{H-\sigma_2} + D_{22}\sigma_2\right)\frac{1}{a} + h_2\frac{F\sigma_2}{H-\sigma_2}$$
(43)

$$Q_{11} = \left(D_{11}\sigma_1 + D_{12}\frac{G\sigma_1}{E-\sigma_1}\right)\sqrt{\sigma_1}$$
$$Q_{12} = \left(D_{11}\frac{F\sigma_2}{H-\sigma_2} + D_{12}\sigma_2\right)\sqrt{\sigma_2}$$
$$Q_{21} = \left(D_{21}\sigma_1 + D_{22}\frac{G\sigma_1}{E-\sigma_1}\right)\sqrt{\sigma_1}$$
$$Q_{22} = \left(D_{21}\frac{F\sigma_2}{H-\sigma_2} + D_{22}\sigma_2\right)\sqrt{\sigma_2}$$
(44)

$$B_1 = -h_1 a(c_1^0 - c_1^*)$$
$$B_2 = -h_2 a(c_1^0 - c_1^*). \tag{45}$$

By setting

$$R_1 = B_1 P_{22} - B_2 P_{12}$$
$$R_2 = B_2 P_{11} - B_1 P_{21}$$
$$S_1 = B_1 Q_{22} - B_2 Q_{12}$$
$$S_2 = B_2 Q_{11} - B_1 Q_{21} \tag{46}$$
$$T = P_{11} P_{22} - P_{12} P_{21}$$
$$U = Q_{11} P_{22} + P_{11} Q_{22} - Q_{12} P_{21} - P_{12} Q_{21}$$
$$V = Q_{11} Q_{22} - Q_{12} Q_{21}$$

we obtain

$$K_1 = \frac{1}{s}\alpha_1(-a)\frac{R_1 + S_1\sqrt{s}}{T + U\sqrt{s} + Vs} = \frac{1}{s}\alpha_1(-a)\left(\frac{M_{11}}{\sqrt{s} - s_1} + \frac{M_{12}}{\sqrt{s} - s_2}\right)$$
$$K_2 = \frac{1}{s}\alpha_2(-a)\frac{R_2 + S_2\sqrt{s}}{T + U\sqrt{s} + Vs} = \frac{1}{s}\alpha_2(-a)\left(\frac{M_{12}}{\sqrt{s} - s_1} + \frac{M_{22}}{\sqrt{s} - s_2}\right), \tag{47}$$

where

$$s_1 = \frac{1}{2}\left[+\sqrt{\left(\frac{U}{V}\right)^2 - 4\frac{T}{V}}\right]$$
$$s_2 = \frac{1}{2}\left[-\sqrt{\left(\frac{U}{V}\right)^2 - 4\frac{T}{V}}\right] \tag{48}$$

$$M_{11} = \frac{S_1 s_1 + R_1/S_1}{V} \frac{1}{s_1 - s_2}$$
$$M_{12} = -\frac{S_1 s_2 + R_1/S_1}{V} \frac{1}{s_1 - s_2}$$
$$M_{21} = \frac{S_2 s_1 + R_2/S_2}{V} \frac{1}{s_1 - s_2} \tag{49}$$
$$M_{22} = -\frac{S_2 s_2 + R_2/S_2}{V} \frac{1}{s_1 - s_2}.$$

This defines the K_i as functions of s. We can now apply the inverse Laplace transformation $L^{-1}\{\cdot\}$ to $\bar{\gamma}_i$. Since,[7]

$$L^{-1}\left\{\frac{1}{s}\frac{\alpha_i(r-a)}{\sqrt{s} - s_j}\right\} = -\frac{1}{\sqrt{s_j}}\chi_{ij}(r-a;t), \tag{50}$$

where

$$\chi_{ij}(r-a;t) = \mathrm{erfc}\left[\frac{\sqrt{\sigma_i}(r-a)}{2\sqrt{t}}\right] - e^{-s_j\sqrt{\sigma_i}(r-a)+s_j^2 t}\,\mathrm{erfc}\left[\frac{\sqrt{\sigma_i}(r-a)}{2\sqrt{t}}-s_j t\right] \quad (51)$$

and $\mathrm{erfc}(z) = 1 - \mathrm{erf}(z)$. We finally obtain the analytic expressions for the concentration profiles

$$\begin{aligned} c_1 &= c_1^0 + \frac{1}{r}\left[\sigma_1 \Phi_1(r-a;t) + \frac{F\sigma_2}{H-\sigma_2}\Phi_2(r-a;t)\right] \\ c_2 &= c_2^0 + \frac{1}{r}\left[\frac{G\sigma_1}{E-\sigma_1}\Phi_1(r-a;t) + \sigma_2\Phi_2(r-a;t)\right], \end{aligned} \quad (52)$$

where:

$$\begin{aligned} \Phi_1(r-a;t) &= \frac{M_{11}}{s_1}\chi_{11}(r-a;t) + \frac{M_{12}}{s_2}\chi_{12}(r-a;t) \\ \Phi_2(r-a;t) &= \frac{M_{21}}{s_1}\chi_{21}(r-a;t) + \frac{M_{22}}{s_2}\chi_{22}(r-a;t). \end{aligned} \quad (53)$$

RESULTS

For a typical lysozyme crystal growth process in aqueous sodium chloride, we consider the following experimental conditions and system physicochemical properties:[11–15]

$c_1^0 = 0.0035\,\mathrm{mol\cdot dm^{-3}}$ $a = 0.1\,\mathrm{cm}$ $D_{11} = 0.1\times 10^{-5}\,\mathrm{cm^2\,sec^{-1}}$

$c_2^0 = 0.5\,\mathrm{mol\cdot dm^{-3}}$ $\beta = 1\times 10^{-7}\,\mathrm{cm/sec}$ $D_{12} = 0.0001\times 10^{-5}\,\mathrm{cm^2\,sec^{-1}}$

$\sigma = 15$ $\sigma_0 = 3$ $D_{21} = 15\times 10^{-5}\,\mathrm{cm^2\,sec^{-1}}$

    $\alpha = 0.01$ $D_{22} = 1.5\times 10^{-5}\,\mathrm{cm^2\,sec^{-1}}$

$c_1^s = 0.06\,\mathrm{mol\cdot dm^{-3}}$

$\rho_s = 1.2\,\mathrm{g\cdot cm^{-3}}$

$\rho_l = 1.0\,\mathrm{g\cdot cm^{-3}}$

FIGURE 1 gives the *protein* concentration profiles around the crystal for a given set of time values (1′, 10′, 30′, 2h, 6h, 1 day, and 5 days). The presence of a depletion zone around the crystal is visible, and its width is expanding with time. However, the absence of large salt gradients and the consequent small contribution in the protein flux gives rise to marginal coupling effects, so that the protein concentration profiles are well approximated by the uncoupled transport process. Thus, the coefficient D_{12} is not a relevant parameter for the crystal growth scenario; however, this conclusion does not preclude the importance of the coefficient during the nucleation stage not contained in the model.

FIGURE 2 gives the *precipitant* concentration profiles around the crystal for the same set of times and for both a coupled and an uncoupled scenario. The salt concentration profiles seem to be significantly affected by the interaction of the fluxes, and at the interfacial concentration, the results seem to be larger for the coupled case.

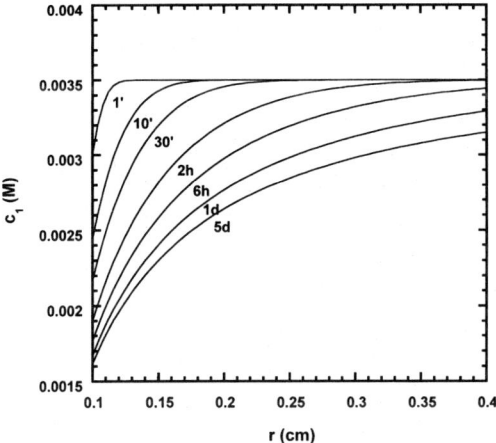

FIGURE 1. Protein concentration profiles around the crystal for time values 1', 10', 30', 2h, 6h, 1 day, and 5 days.

A parameter that may quantify the effect of the coupled transport on the precipitant interfacial concentration is the stationary salt concentration excess defined by

$$\chi = \lim_{t \to \infty} \left(\frac{c_2^i}{c_2^0} \right). \quad (54)$$

From FIGURE 2, we see that slight precipitant concentration excesses are present for the uncoupled case, which are also being caused by the crystal growth wall effect. A

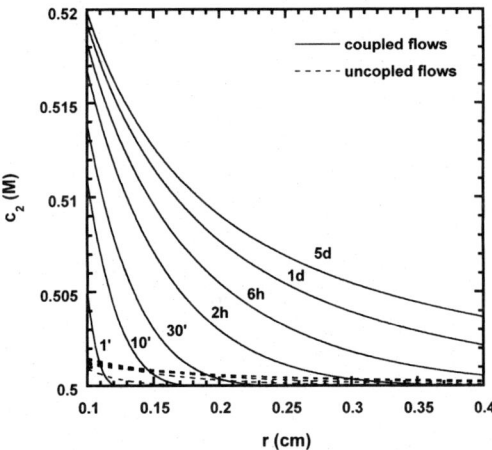

FIGURE 2. Precipitant concentration profiles around the crystal for time values 1', 10', 30', 2h, 6h, 1 day, and 5 days. The *solid lines* refer to the coupled scenario; the *dashed lines* refer to the uncoupled scenario.

numerical analysis was performed to examine how the relevant system variables may influence χ. The results were as follows:

1. changing D_{21}/D_{11} from 0 to 400 changes χ from 1.004 to 1.128;
2. changing D_{21}/D_{22} from 10 to 30 changes χ from 1.052 to 1.152; and
3. changing σ from 5 to 100 changes χ from 1.030 to 1.072.

Another important variable for the overall crystallization process is the kinetic Peclet number[16] defined by $Pe_k = \beta\delta/D_{11}$, where δ is the width of the diffusion layer and can be estimated from

$$\delta \approx \frac{c_1^0 - c_1^i}{\left.\dfrac{\partial c_1}{\partial r}\right|_{r=a}}.$$

In our case, the diffusion layer width was found to be roughly 0.1 cm. The kinetic Peclet number is the controlling parameter that gives the relative importance of the transport mechanism with respect to the incorporation kinetics. For instance, values of $Pe_k \ll 0.1$ or $Pe_k \gg 0.1$ indicate, respectively, purely kinetic or transport controlled growth. A numerical analysis shows that changing the kinetic Peclet number from 0.1 to 10, χ changes from 1.014 to 1.070.

Lin et al.[2] reported an incorrect interpretation of the coupling transport between salt and protein. The coupled transport was attributed to the association between the macromolecules and the ions giving rise to the depletion of the salt concentration at the crystal interface. In contrast, this simulation yields an interfacial salt enrichment that is caused by electrostatic coupling between both ions and excluded volume effects. Since the partition constant α defined by (**36**) was used to describe the precipitant concentration inside the crystal, the model cannot predict non-uniform precipitant distribution within the solid phase. However, due to the time dependence of the interfacial precipitant concentration (and of the precipitant chemical potential), non-uniform precipitant incorporation can be presumed. This may cause strain in the crystal and ultimately compromise crystal size and quality.

CONCLUSIONS

For the system lysozyme chloride + NaCl + H_2O at relatively low protein concentrations the effect of the inclusion of cross-term diffusion coefficients does effect the calculation of precipitant concentration distribution about the growing crystal and should be included when modeling crystal growth. This will be possible if general schemes for estimating the diffusion coefficients in multicomponent systems containing a protein become available. However for the model calculations considered here the effect causes only small changes of crystal face precipitant concentrations, but this effect could nevertheless be important. Finally, it should be noted that the diffusion coefficients depend on concentration and that if the diffusion coefficient changes are sufficiently large within a system, then numerical analysis methods are necessary.

ACKNOWLEDGMENTS

J.G.A. notes that this paper is taken directly from chapter 19 of the Ph.D. dissertation[17] (ISBN 0-493-12797-6) of the first author who deserves full credit for this work. The support of the NASA Microgravity Biotechnology Program through Grant NAG8-1356 is gratefully acknowledged. Support from Texas Christian University Grant RCAF-11950 is also gratefully acknowledged.

REFERENCES

1. ROSENBERGER, F. 1986. Inorganic and protein crystal growth—similarities and differences. J. Cryst. Growth **76**: 618–636.
2. LIN, H., et al. 1995. Convective-diffusive transport in protein crystal growth. J. Cryst. Growth **151**: 153–162.
3. MCPHERSON, A., et al. 1999 The effects of microgravity on protein crystallization: evidence for concentration gradients around growing crystals. J. Cryst. Growth **196**: 572–586.
4. MCPHERSON, A. 1999. Crystallization of Biological Macromolecules. CSHL Press, New York.
5. CRANK, J. 1975. The Mathematics of Diffusion. Clarendon Press, Oxford.
6. FUJITA, H. & L.J. GOSTING. 1960. A new procedure for calculating the four diffusion coefficients of three-component systems from Gouy diffusiometer data. J. Phys. Chem. **64**: 1256–1263.
7. CHURCHILL, R.V. 1958. Operational Mathematics. McGraw-Hill, New York.
8. ARAKAWA, T. & S.N. TIMASHEFF. 1985. Theory of protein solubility. In Methods in Enzymology, Vol. 114. H.W. Wyckoff, C.H.W. Hirs & S.N. Timasheff, Eds.: 49–77. Academic Press, New York.
9. GREEN, A.A. 1932. Studies in the physical chemistry of proteins. X. The solubility of hemoglobin in solutions of chlorides and sulfates of varying concentrations. J. Biol. Chem. **95**: 47–66.
10. VEKILOV, P.G., et al. 1996. Repartitioning of NaCl and protein impurities in lysozyme crystallization. Acta Crystallogr. Sect. D **52**: 785–798.
11. ALBRIGHT, J.G., et al. 1999. Precision measurements of binary and multicomponent diffusion coefficients in protein soutions relevant to crystal growth: lysozyme chloride in water and aqueous NaCl at pH 4.5 and 25°C. J. Am. Chem. Soc. **121**: 3256–3266.
12. ANNUNZIATA, O., et al. 2000. Extraction of thermodynamic data from ternary diffusion coefficients. Use of precision diffusion measurements for aqueous lysozyme chloride-NaCl at 25°C to determine the charge of lysozyme chloride chemical potential with increasing NaCl concentration well into the supersaturated region. J. Am. Chem. Soc. **122**: 5916–5928.
13. GOSTING, L.J. 1956. Measurement and interpretation of diffusion coefficients of proteins. In Advances in Protein Chemistry, Vol. 11. M.L. Anson, K. Bailey & J.T. Edsall, Eds.: 429–555. Academic Press, New York.
14. STEINRAUF, L.K. 1959. Preliminary X-ray data for some new crystalline forms of β-lactoglobulin and hen egg-white lysozyme. Acta Cryst. **12**: 77–78.
15. MONACO, L.A. & F. ROSENBERGER. 1993. Growth and etching kinetics of tetragonal lysozyme. J. Cryst. Growth **129**: 465–484.
16. VEKILOV, P.G., et al. 1996. Nonlinear response of layer growth dynamics in mixed kinetics-bulk transport regime. Phys. Rev. E **54**: 6650–6660.
17. ANNUNZIATA, O. 2001. Analysis of the Protein-Salt Coupled Transport. Ph.D. Dissertation, Texas Christian University, Fort Worth.

Index of Contributors

Adamson, J., 57–67
Albert, L.B.J., 146–156
Albright, J.G., 610–623
Alexander, J.I.D., 146–156
Allegro, L., 288–305
Amara, K., 10–28
Andrews, J.B., 102–109
Annunziata, O., 610–623
Asano, H., 316–327
Ayyaswamy, P.S., 556–564

Baba, A., 410–427
Bakhtiyarov, S.I., 132–145
Balasubramaniam, R., 193–200, 398–409
Banish, R.M., 146–156
Bankoff, S.G., 1–9
Betz, J., 220–245

Carswell, W.E., 565–580
Chao, D., 378–397
Chayen, N.E., 591–597
Choi, B., 316–327
Costanza, G., 68–78

de Luis, J., 504–517
Dhir, V.K., xi, 378–397
Dimanlig, A., 518–540
Di Marco, P., 428–446
Dreyer, M.E., 246–260
Ducheyne, P., 556–564

Freed, L.E., 504–517
Friesen, T.J., 288–305
Fujii, T., 316–327

Gabriel, K.S., 306–315, 410–427
Gathings, W.E., 565–580

Gauzzi, F., 68–78
Ghita, B.C., 565–580
Grassi, W., 428–446
Greenberg, A.R., 581–590
Griffing, E.M., 1–9
Gupta, N.R., 398–409

Hao, Y., 328–347
Haulenbeek, G.B., 565–580
Hayes, L.J., 102–109
Hegseth, J., 10–28
Helliwell, J.R., 591–597

Jagger, J., 518–540

Kamada, K., 79–86
Kawaji, M., 288–305
Keller, R.B., 481–503
Kennedy, A.P., 87–101
Knobloch, E., 201–219
Krantz, W.B., 581–590
Krooneman, J., 541–555

Lee, H.S., 447–462
Leyse, R.H., 261–273
Logsdon, K., 378–397

MacGillivray, R.M., 306–315
Martel, C., 201–219
Matsumoto, K., 364–377
Merte, H., Jr., 481–503
Michaelis, M., 246–260
Minagawa, H., 79–86
Misener, D.L., 157–163
Misener, L., 57–67
Montanari, R., 68–78

Nagai, H., 79–86
Nakata, Y., 79–86
Narayanan, R., 42–56
Naumann, R.J., 29–41
Ng, J.D., 598–609

O'Brien, S.K., 565–580
Ohsaka, K., 124–131
Ohta, H., xi, 410–427, 463–480
Okubo, T., 164–175
Okutani, T., 79–86
Oram, G.R.J., 157–163
Overfelt, R.A., 132–145

Park, J., 481–503
Plawsky, J.L., 274–287
Pourpoint, T.L., 146–156
Prosperetti, A., 328–347

Qiu, D., 378–397
Qiu, Q.-Q., 556–564

Rashidnia, N., 193–200
Rath, H.J., 246–260
Rednikov, A., 124–131

Sadayappan, M., 110–123
Sadhal, S.S., xi, 124–131
Saho, M., 110–123
Saitoh, H., 364–377
Sanghadasa, M., 565–580
Schluter, R.A., 1–9
Searby, N., 504–517, 518–540
Sekerka, R.F., 146–156
Sengupta, S., 581–590
Shultz, W.W., 481–503

Smith, R.W., xi, 57–67, 110–123, 157–163, 176–192
Smith, T.J.N., 57–67, 157–163
Son, G., 378–397
Sportiello, M.G., 581–590
Stebe, K.J., 398–409
Straub, J., xi, 220–245, 348–363
Sugimoto, K., 316–327
Suzuki, K., 364–377

Tadesse, S., 87–101
Takahira, H., 288–305
Thomas, A.M., 42–56
Todd, P., 581–590
Trentavizi, F., 428–446
Tsuchida, A., 164–175
Tunnicliffe, M.C., 57–67
Turnbull, W.N.O., 157–163

Van Benthem, R.C., 541–555
van den Assem, D., 541–555
Vega, J.M., 201–219
Vellinger, J.C., 581–590
Vunjak-Novakovic, G., 504–517

Walker, S., 518–540
Wang, Y., 274–287
Wayner, P.C., Jr., 274–287
Wilder, M., 518–540

Yang, B.J., 110–123
Yasuda, Y., 288–305

Zheng, L., 274–287
Zhu, X., 57–67
Zugrav, M.I., 565–580